BIOLOGY TODAY

An Issues Approach

Third Edition

BIOLOGY TODAY

An Issues Approach

Third Edition

ELI C. MINKOFF · PAMELA J. BAKER

Vice President: Denise Schanck
Text Editor: Emma Hunt
Managing Production Editor: Emma Hunt
Editorial Assistants: Emma Catherall, Frances Morgan
Production Assistant: Nicola Tidman
Copy Editor: Bruce Goatly
Illustrator: Nigel Orme
Cover Designer: Joan Greenfield
Indexer: Jill Halliday

THE COVER
Soya bean, Holt Studios International Ltd.; Woodpecker (carpenter) finch, Auscape International; Unmyelinated nerves, © Lester V. Bergman/CORBIS; Double helix, © Digital Art/CORBIS; Cardiovascular scan, © Howard Sochurek, Inc./CORBIS; Danube delta, © AFP/CORBIS.

Visit the *Biology Today* Web site:
http://www.garlandscience.com

Library of Congress Cataloging-in-Publication Data
Minkoff, Eli C.
    Biology today: an issues approach/Eli C. Minkoff, Pamela J. Baker.—3rd ed.
        p. cm.
    Includes bibliographical references (p.).
    0-8153-4157-1 (pbk. : alk. paper)
    1. Biology. I. Baker, Pamela J., 1947–II. Title.

QH315.M63 2004
570—dc21   2003011485

Published by Garland Publishing, a member of the Taylor & Francis Group
29 West 35th Street, New York, NY 10001-2299

Printed in the United States of America
15 14 13 12 11 10 9 8 7 6 5 4 3 2 1

Preface

Our book takes an issues-oriented approach to the teaching of biology, one that covers all of the major biological concepts. Our approach aims to educate citizens, biologists and nonbiologists alike, with an understanding that will enable them to evaluate scientific arguments and make informed decisions affecting their own lives and the well-being of society. We are committed to teaching science as a human activity that impinges upon other aspects of society and gives rise to social issues that require discussion. Individuals are increasingly called upon to deal with science-based issues throughout their lives. Each of us makes food choices daily and medical decisions nearly as often. DNA evidence is used more and more in solving crimes and in predicting susceptibility to disease. Stem cell technologies confront us with real possibilities only imagined a few years ago. Our waste disposal habits affect the environment in which we live, and our transportation and manufacturing choices affect the very air we breathe. Citizens, legislators, juries, and corporate managers need to make important decisions, affecting many lives, based in part on the findings of science. Everyone needs to be aware of science, the way that scientists work, and how science can be used and misused.

The issues themselves are not our focus, however; instead, they form a context in which to teach basic biology. This is very different from teaching the biology specific to a particular issue, which often leaves students thinking that the biological concept applies only to that case. Some other texts are now using current issues, but our approach is unique because in our text the issues are central to the pedagogy, not "add-ons" presented as case studies or in side-bars, boxes or separate pamphlets. Consequently, we have selected issues that are not only of current importance, but that also lend themselves as vehicles for teaching the major concepts of biology. For example, we use the chapter on the population explosion as a vehicle to teach the biology of reproduction; also osmosis and photosynthesis are covered in a chapter called Plants to Feed the World. Wherever possible, we have presented a concept in more than one part of the book, developing the concept in more than one context. Procaryotic biology and asexual reproduction, for example, are introduced in a chapter called Classifying Nature and then are further developed in a new chapter called New Infectious Threats.

Years ago, a group of educators called the Biological Sciences Curriculum Study developed a list of key concepts in biology. Each of these concepts is covered in our book. In the years since the Biological Sciences Curriculum Study list was developed, biology itself has evolved, so we have added new concepts as they have risen in importance (see our Web site under Resources: Key biological concepts, for both the list of concepts and where they are covered). In this third edition, we have brought the text up-to-date by expanding certain topics and adding new concepts. Three new chapters have been added. Chapter 4, Genomics and Genetic Engineering, covers concepts flowing from the Human Genome Project, including genomics, proteomics and bioinformatics. Chapter 17, New Infectious Threats, discusses the microbiology and ecology of emerging infections and the issue of bioterrorism. Chapter 19, Protecting the Biosphere, covers the continuing evolution of our atmosphere and the effects of acid rain, ozone depletion and global warming.

Other chapters have been reorganized to incorporate new concepts. Evolution and Classification are now covered in separate chapters and we have expanded the discussion of each. Intelligent design is now included in Chapter 5, and Chapter 6 now includes cladistics, comparative genomics, and other taxonomic methods, as well as the three-domain classification system. We have expanded the section on reproductive anatomy and physiology as well as reproductive technologies in Chapter 9, and, obesity as well as cellular respiration is covered in Chapter 10. Stem cells and cloning are added to Chapter 12 as are the ethical considerations surrounding these controversial topics. A greater emphasis on the diversity of sensory systems in various animals is found in Chapter 13 and the human effects on biomes are detailed in Chapter 18.

The issues have also been adjusted to stay current. Students (especially those not majoring in biology) are more likely to be interested in and develop an understanding of material if it is meaningfully related to issues of concern to them, and these change over time. Infectious diseases are much in the news, and these form the basis for the new Chapter 17. More on obesity and on newly recognized micronutrients have been added to Chapter 10. For issues of continuing interest, such as AIDS or cancer or drugs, we have incorporated the latest statistics as well as the latest biological advances. We recognize that the average undergraduate student is now older than twenty-two; therefore, we have picked issues with a wide age-appeal. We have also attempted to use geographically diverse examples. We know our readers, regardless of where they attend college, are from around the world. By the combination of text and Thought Questions, we encourage students to think beyond themselves, both globally and locally.

In addition to its real-life appeal, the issues approach allows for a more comprehensive view of biology. As a discipline biology has become fragmented to the extent that different perspectives on the same problem, for example, molecular perspectives and environmental perspectives, are often taught in separate courses with no reference to each other. The current understanding of each issue is covered from different perspectives, which often include cellular and molecular perspectives, organismal or individual perspectives, and global or population perspectives, combined as appropriate. Our approach accordingly helps students to experience the interdisciplinary nature of today's biology. Each chapter ends with a section called "Connections to Other Chapters" that further emphasizes this point.

Our approach also examines the intimate connections between biological and social issues. We have chosen to teach 'facts' in a context that emphasizes how they are produced, organized, and used to solve problems. Other books often expose students to the results of biology without gaining understanding of biology as a process of discovery. Instead, we hope to instill in students an understanding and appreciation of this process. To help students, we have presented multiple interpretations or points of view as much as possible. Societal and ethical issues are mentioned wherever relevant, and part of the initial chapter is devoted to an examination of ethical principles. We encourage teachers to set aside time for class discussions to further stimulate student thought, or for students to set up such discussions among themselves informally. With *Biology Today* we aim to stimulate critical thinking and questioning rather than memorization. Thought Questions, suitable for class discussions are provided at the end of each section, and, in this new edition, new Thought Questions have been added.

As *Biology Today* goes into its third edition, we have continued to benefit from the input of many people in producing a book that remains scientifically accurate and is optimally organized for student understanding. We would like to take this opportunity to thank the many people who reviewed portions of our text and provided us with helpful suggestions. In alphabetical order; they were: Lee Abrahamsen, Bates College; Paul Biersuck, Nassau Community College; Robert Benda, Prince William Sound Community College; Virginia L. Bliss, Framingham State College; Neil Buckley, SUNY Plattsburgh; Craig Coleman, Brigham Young University; Patricia Flower, University of San Diego; Wendy Jean Garrison, University of Mississippi; Denis Goulet, University of Mississippi; Tamar L. Goulet, University of Mississippi; Eric G. Haenni, Hendrix College; Heidi Hawkins, College of Southern Idaho; Joseph Hawkins, College of Southern Idaho; Bernard Hauser, University of Florida; Pat Hauslein, St. Cloud State University; Melissa Ishler, Mansfield University; Eric Jellen, Brigham Young University; Walter Johnson, Merritt College; Charles D. Kay, Wofford College; Nancy Kleckner, Bates College; Michael E. Kovach, Baldwin-Wallace College; Glen Lawson, Bates College; Mary Lehman, Longwood College; Marilyn Mathis, Howard

Payne University; Debra Mayers, Southwest Missouri State University; Nancy Minkoff; Neil Minkoff, Partners Community HealthCare; Cheryl McCormick, Bates College; Robert Moss, Wofford College; Valeri Olness, Augustana College; Brian K. Paulson, California University of Pennsylvania; Joseph G. Pelliccia, Bates College; Karen Rasmussen, Maine Cancer Research and Education Foundation; Julio G. Soto, San Jose State University; Carol St. Angelo, Hofstra University; Shawn Stover, Davis & Elkins College; Joyce Tamashiro, University of Puget Sound; Morgan Wilson, University of Mississippi.

Special thanks to Denise Schanck, who encouraged our work on all three editions. We would also like to thank the staff at Garland Science Publishing, who were most helpful throughout the editing and production of this edition. They include: Emma Catherall, Antonella Collaro, Jackie Harbor, Emma Hunt, Frances Morgan, and Nicola Tidman. Emma Hunt read through the entire book and provided many helpful suggestions for this edition. Nigel Orme drew most of the illustrations. We could never have brought this project to fruition without their help.

Eli C. Minkoff
Pamela J. Baker

Biology Today Web Site

Biology Today offers two complementary Web sites that serve as a complete teaching and learning resource—an essential supplement to an issues-oriented biology course.

Where the Web site icon is featured in the book, students are encouraged to go to the *Biology Today* Web site to find additional content. We have indicated the names of these resources in the text itself and each resource is organized by chapter on the Web site. Also organized by chapter are notes and outlines, and a selection of ethical discussion topics. These are completely new to this edition, and designed to invite in-class discussions. In addition, you can find a bibliography, the glossary, a table of biological concepts, a list of useful Web links and sample term papers. Please visit this Web site at:

http://www.garlandscience.com/biologytoday

The Garland Science Classwire Web site offers extensive instructional resources. In addition to containing all materials on the first site, it provides a testbank, and curriculum advice and assistance for those teaching an issues-oriented biology course for the first time. It also contains a sample syllabus, advice about laboratory manuals, and all the answers to the practice questions from the textbook. All the images from the textbook are also available in a downloadable, Web-ready, as well as Power Point-ready, format. Instructors can choose whether they wish to make these resources available to students.

Garland Science Classwire also does much more than offer supplementary teaching resources. It is a flexible and easy to use course management tool that allows instructors to build Web sites for their classes. It offers features such as a syllabus builder, a course calendar, a message center, a course planner, virtual office hours and a resource manager. No programming or technical skills are needed. Garland Science Classwire is offered free of charge to all instructors who adopt Biology Today for their course. Resources for all other Garland textbooks are also available.

Please visit Garland Science Classwire at:

http://www.classwire.com/garlandscience

About the Book

This third edition of *Biology Today* has undergone considerable revision and reorganization. The basic biology content has been expanded on several topics while current issues and recent science have been both expanded and updated.

Here is an overview of the study aids and features included in this edition. Each chapter opens with a page containing three lists: Issues, Biological Concepts and a Chapter Outline.

Issues: Every chapter raises a number of critical issues, and asks the student to consider a number of important, pivotal questions. These are presented as a bulleted list. Most will be of personal interest to the students regardless of their age. They encourage students to think beyond themselves and to consider their place in the world. Most are issues that students will have encountered in the news. By raising questions, we prompt students to think beyond what they have seen or heard in the news, to question how and by whom "facts" are produced, and to recognize when information is incomplete. These underlying questions are linked with other critical thinking sections within the chapter, thought questions, and the linkage is indicated by the "perspectives" icon.

Biological Concepts: Our main purpose is to cover key biological concepts. This list summarizes the biology that is covered within the chapter. The list is nested so that it can serve as a concept map to guide student's analysis. For example in Chapter 8, one of the entries on the list is: Population ecology (populations, regulating population size). The Biological Concepts headings are taken primarily from the lists of key biological concepts developed by the Biological Sciences Curriculum Study. A complete table of the biological concepts recommended by this group as being key to a basic understanding of biology can be found on our Web site (under Resources: Key biological concepts). This table is fully cross-referenced to indicate where the material is covered in the book. As can be seen by reference to this table, all of the concepts are covered somewhere in the book, and many are covered in more than one context.

Chapter Outline: Chapters begin with an outline listing the first and second level section headings. The first level section headings are sentences stating the main theme of that section. The second level headings are short summary phrases of the main content topic of that section. Thus, these section headings can serve as a study guide for the student to the biological and issue-related content covered within the chapter.

Opening Narrative: The opening narrative scenario immediately introduces the science covered in the chapter by placing it within an interesting and relevant societal context.

Thought Questions: Sets of thought questions appear at the end of each chapter section. They are linked to the chapter opening issues list by the perspectives icon. Although a few thought questions have factual 'right' answers, most do not. Some questions require students to do further reading and many encourage students to think about the limitations of available data or the applications and implications of science. We encourage students with differing viewpoints to discuss these questions among themselves and to ask, "What further information would help us resolve our differences or reach decisions?" These questions can form the basis for discussion either in class or in informal study groups.

Illustrations: Since many of the concepts of biology can be understood and remembered visually, the book is well illustrated with photographs and drawings. Many of the illustrations are new since the second edition and many are visually simplified. The captions to these illustrations often provide another important avenue to understanding. All figures from the book are available on *Art of Biology Today* CD-ROM in two convenient formats: PowerPoint and JPEG. In addition, they are available on our Classwire Web site.

Tables: Throughout the book are numerous tables; these summarize important sections of text or provide specific examples. Several new summary tables are included.

Boxes: Interspersed throughout the text are boxes some of which supplement the text while others review key principles.

Important Vocabulary: Important vocabulary is highlighted in bold type in the text. Each of these terms is defined in the glossary at the end of the book and on our Web site. The terms that are most important conceptually are used in sentences at the end of each chapter in the chapter summary.

Chapter Summary: A bulleted list of the most important ideas from the chapter, with the key vocabulary in bold. This format is useful as a study guide.

Connections to Other Chapters: *Biology Today* aims for an integrated view of biology. The connections section at the end of each chapter helps reinforce this integrated view, with related material presented in different contexts at other places in the text. In addition, "connections" icons placed throughout the book indicate cross-references to different contexts for the same content. Pedagogically this serves to discourage students from compartmentalizing their new knowledge, developing the critical thinking level of generalizing concepts from a variety of contexts.

Practice Questions: Learning biological concepts still requires some memorization of the facts. End-of-chapter practice questions are review questions for students to answer.

Glossary: Each of the bolded terms is defined in an alphabetically arranged glossary at the end of the book and on the *Biology Today* Web site.

Contents

List of Headings

Biology: Science and Ethics

Issues

- How do we know what we know?
- How do scientists make discoveries and advance our knowledge?
- What constitutes a 'discovery' in science?
- How is science creative?
- Does science contain absolute truths?
- How do ethics and morals fit into science?
- How do scientists make ethical decisions in a social context?
- How are decisions made on social issues, and to what extent can science help in these decisions?
- What rights do animals have? How do we safeguard those rights?
- How do we safeguard the rights of experimental subjects?

Biological Concepts

- Properties of living organisms (organization, metabolism, selective response, homeostasis, growth and biosynthesis, genetic material, reproduction, population)
- Hypotheses and theories
- Experimental science versus naturalistic science
- Normal science and paradigm shifts
- Science and society
- Biological ethics

Chapter Outline

Science Develops Theories by Testing Hypotheses

Hypotheses

Hypothesis testing in science

Theories

A theory describing the properties of living systems

Scientists Work in Paradigms, Which Can Help Define Scientific Revolutions

Paradigms and scientific revolutions

Molecular genetics as a paradigm in biology

The scientific community

Scientists Often Consider Ethical Issues

Ethics

Resolving moral conflicts

Deontological and utilitarian ethics

Ethical decision-making

Ethical Questions Arise in Decisions About the Use of Experimental Subjects

Uses of animals

The animal rights movement

Humans as experimental subjects

1 Biology: Science and Ethics

Biology is the scientific study of living systems. Our gardens, our pets, our trees, and our fellow humans are all examples of living systems. We can look at them, admire them, write poems about them, and enjoy their company. The Nuer, a pastoral people of Africa, care for their cattle and attach great emotional value to each of them. They write poetry about—and occasionally to—their cattle, they name themselves after their favorite cows or bulls, and they move from place to place according to the needs of their cattle for new pastures. They come to know individual cattle very well, almost as members of the family. The Nuer have also acquired a vast store of useful knowledge about the many animal and plant species in their region. Many other people who live close to the land have a similar familiarity with their environment and the many species living in it. Scientific understanding of the world around us grew out of this kind of familiarity with nature, supplemented by a tradition of systematic testing. In this chapter we examine the methods of science in general and the application of those methods to the study of living systems.

Because living systems are complex and continually changing, an understanding of these systems often requires special methods of investigation or ways of formulating thoughts. This chapter describes the special methods that have come to be called science. Many people think that science is defined by its subject matter, but this is not correct. Science is defined by its methods.

The scientific method does not answer questions about values and, therefore, cannot by itself answer questions such as whether certain types of research should be done, or to what uses scientific results should be put. Such decisions often involve a branch of philosophy called ethics. Many issues confronting societies today have a scientific dimension. Policy decisions on such issues involve both science and ethics.

Science Develops Theories by Testing Hypotheses

The essence of science is the formulation and testing of certain kinds of statements called hypotheses. At the moment of its inception, a hypothesis is a tentative explanation of events or of how something works. What makes science distinctive is that hypotheses are subjected to rigorous testing. Many hypotheses are falsified (rejected as false) by such testing. Eliminating one hypothesis often helps us to frame the next hypothesis. If a hypothesis is repeatedly tested and not falsified, it may be put together with related hypotheses that have also withstood repeated testing. Such a group of related hypotheses may become recognized as a theory.

Hypotheses

Hypotheses must be statements about the observable universe, formulated in such a way that they can be tested. Observations of the material world that we make with our senses (aided in some cases by scientific instruments) are called empirical observations. To be a hypothesis, a statement must be capable of being tested by comparison with such empirical observations. Observations gathered for testing hypotheses are called **data**. Karl Popper, a philosopher of science, said that scientific hypotheses must always be tested in such a way that they can be rejected if they turn out to be inconsistent with observations—in his words, they must always be **falsifiable**.

Certain types of statements cannot be used as scientific hypotheses because they are not subject to testing and falsification. This includes esthetic judgments about what is valuable, beautiful, or likable, e.g., "I like my kind of music, whatever you may say about it." Moral judgments and religious concepts are also not scientific hypotheses because observational data are not sufficient to test their truth. This means that a devout person's belief in God cannot be shaken by any demonstration of an empirical fact or observation. To a devout believer, no such demonstration is even possible. The same is true of strongly held beliefs about the goodness of human equality or the wrongfulness of inflicting death.

Hypotheses, therefore, must be: (1) testable and (2) falsifiable. A third characteristic is that of simplicity—a problem should be stated in its basic and simplest terms. When several hypotheses fit the facts of a problem, scientists usually choose the simplest hypothesis. An example is the appearance of crop circles: the simplest hypothesis—that human activity created the patterns—is more likely than the more complex hypothesis— alien activity—because in the latter case we would also need to explain who the aliens are, where they came from, how they came, and so on.

Specific versus general hypotheses. Hypotheses that are easy to verify generally tell us very little. For example, the hypothesis "this frog will jump if I touch it" can be tested by touching the frog and observing what happens. If the frog does jump, then our hypothesis is verified or confirmed; if the frog does not jump, then our hypothesis is falsified or disconfirmed. However, the confirmation of this hypothesis about a particular frog is far from an important scientific discovery. It is relatively unimportant because it is too specific, which is exactly what makes it verifiable.

Suppose, now, that we examine the much bolder hypothesis "all animals will react when stimulated." We can test this second hypothesis in the same way that we tested the first, by touching or otherwise stimulating some animals, and we could also declare that the hypothesis would be falsified if one animal failed to respond. (Even then, we could never be sure, for it might respond in a way that we do not immediately notice, e.g., by remembering the event and responding at a later time.) But what if the animals tested do all respond? Does this verify that *all* animals will respond? Suppose we test 5 animals in a row, or 5000? No finite number of successes would be sufficient to verify the hypothesis for all animals and all circumstances. This is the kind of hypothesis that science usually examines: hypotheses that could potentially be falsified each time that we test them, but cannot be absolutely verified for all possible occurrences.

Falsified hypotheses are rejected, and new hypotheses (which may in some cases be modifications of the original hypotheses) are suggested in their place. To extend the previous example, the hypothesis, "all frogs will jump if touched" would be falsified if 2 out of 2000 frogs did not jump.

We could, however, modify the original hypothesis to one that is consistent with our data, e.g., "frogs will usually jump if touched." In practice, we might also want to be specific about the nature of the stimulus (how sharp the object, how firm the touch), the response (how frequent, how strong), and other particulars (the species or size of frog used, the temperature, and so on).

If testing a hypothesis does not reject it, we may want to generalize the hypothesis. For example, if a hypothesis tested using rats has not been falsified, we might want to apply the hypothesis to people as well, or to all animals. However, we can never know how far we can extrapolate (generalize) results unless we continue to try to falsify our premise under different conditions. In this way, the testing of hypotheses allows us to draw conclusions about the observable world, but only to the extent that we have tested many possible circumstances and conditions.

Ways of devising hypotheses. One form of reasoning is called **deduction**, defined as reasoning that guarantees the truth of a conclusion if we accept the truth of the premises. Deduction is frequently used to set up testable hypotheses: "*If* organisms of type X require oxygen to live, *then* this individual of type X will die if I put it in an atmosphere without oxygen." If the organism lives, then one of the premises must be rejected: either organisms of type X do not always require oxygen, or else this individual was not of type X. Another type of reasoning, called **induction**, may be defined as reasoning that does not guarantee the truth of any conclusions drawn, except in terms of probabilities. Science often uses induction to generalize from specific hypotheses, such as when we reason from five frogs to all frogs or from frogs and rats to all animals. Induction is also commonly used in everyday life: "I have liked pizza in each of the restaurants where I have ever eaten it; therefore I will like pizza in any other restaurant." Induction allows us to reason beyond what we know with certainty, but in statistical terms only: I have always liked pizza, and I will probably like the next one that I try. However, because induction never guarantees the truth of any result, generalizations made by induction always need to be tested further. If I actually tried pizza in 450 restaurants that I had never tried before, I might discover that I only like the pizza in 442 of them, and that there are 8 places that serve pizza that I don't like. The probability that I will like the pizza in a randomly chosen restaurant is thus 442 out of 450, or about 98.2%.

One common way of reasoning in science is to start with data already gathered, use induction to generalize and to formulate a hypothesis, use deduction to set up a test situation, and test the hypothesis through further observation. In our example, I can start with the five frogs that respond to touch, create the hypothesis that "all animals will respond to touch," reason by deduction that, if all animals will respond to touch, then this rat and that starfish will respond, and finally set up an experiment to test the hypothesis and see if it works as expected. This process is often called the **scientific method**. In reality, few scientists adhere rigidly to this prescription.

Deduction and induction are only two of the many ways in which scientists go about the business of formulating hypotheses. Other ways include (1) intuition or imagination, (2) esthetic preferences, (3) religious and philosophical ideas, (4) comparison and analogy with other processes, and (5) serendipity, or the discovery of one thing while looking for something else. Moreover, these ways may be mixed or combined. For example, Albert Einstein declared that he arrived at his hypotheses about the physics of the universe by considering esthetic qualities such as beauty or simplicity and by asking, "if I were God, how would I have made the world?". Einstein also said that "imagination is more important than knowledge," a remark that is particularly true for the formulation of hypotheses (Figure 1.1). Nobel Prize-winning physicist Niels Bohr said that his hypothesis of atomic structure (the heavy nucleus in the center, with the electrons circling rapidly around it, "like a miniature solar system") first occurred to him by analogy with our solar system. Alexander Fleming found the first antibiotic as the result of a laboratory accident: on dishes of bacteria that should have been thrown away earlier, he observed clear areas where fungi had overgrown the bacteria. His hypothesis, that a product of the fungi had killed the bacteria, was validated by tests, and that fungal product is what we now know as penicillin. As these several examples show, *hypotheses are formed by all kinds of logical and extralogical processes*, which is one more reason why they must be subjected to rigorous testing afterward.

Biology: hypothesis testing in living systems. Animals, plants, and bacteria are complex and variable. So are other living systems, large and small, from ecosystems to individual cells. No individual animal or plant is exactly like any other animal or plant. At any moment, living systems that are otherwise similar may differ in external conditions,

Figure 1.1

Imaginative hypotheses may originate from various logical or extralogical processes, especially from young scientists. Does the idea shown here qualify as a scientific hypothesis? Why, or why not? Is it testable?

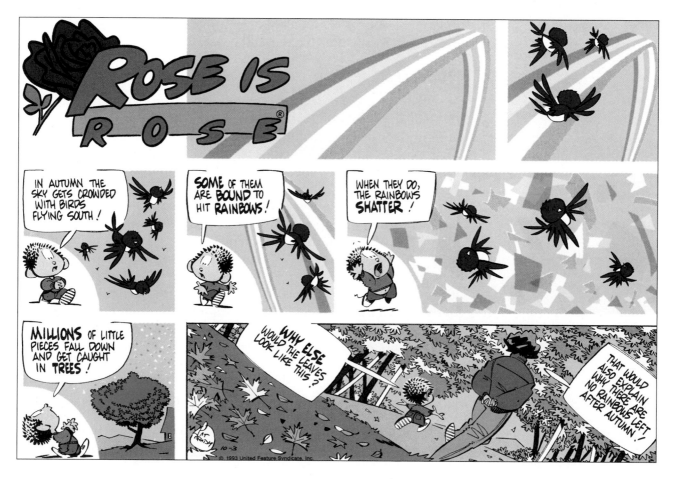

internal conditions, or in the way in which these conditions interact. Further, the same individual is not exactly the same from one day to the next. Because living systems vary, tests must be repeated. If the hypothesis is tested in one animal, or one cell, and a particular response occurs, the result is far less reliable as a means of prediction than if 10 animals, or 100 cells, all responded in the same way. What often happens, however, is that 9 out of 10 animals, or 94 of 100 cells, respond in one way and the remainder in another way. Interpretation of the results from tests on variable systems usually requires statistical treatment to ascertain whether the observed differences are 'real' or can be explained by random variation. The differing responses may come from a source of variation that has not yet been identified, and scientists who study the anomalous cases sometimes discover new, previously overlooked phenomena.

A definition of science. Science may now be defined as a method of investigation based on the testing of hypotheses by organized comparisons with empirical evidence. Notice that this makes scientific statements tentative, or provisional, and subject to possible falsification and rejection. Repeated exposure of our hypotheses to possible refutation increases our confidence in these hypotheses when test results agree with predictions, but no amount of testing can guarantee absolute truth.

Any hypothesis that is tested again and again, always successfully, is considered well supported and comes to be generally accepted. It may be used as the basis for formulating further hypotheses, so there is soon a cluster of related hypotheses, supported by the results of many tests, which is then called a theory. (Some people say that a widely accepted theory becomes a 'law,' but most philosophers of science do not recognize a 'law' as anything but a useful generalization.)

Hypothesis testing in science

Scientists test hypotheses by comparing them with the real world through empirical observations. Scientists differ from one another, however, in the ways in which hypotheses are tested. Some scientists do all their work in laboratories with specially designed equipment; other scientists gather data and specimens in the field for analysis and interpretation (Figure 1.2).

Some scientists test hypotheses by conducting **experiments**—artificially contrived situations set up for the express purpose of testing some hypothesis. Most **experimental sciences** seek to answer questions of the form "How does X work?". The scientist designs an experiment such that, if the hypothesis is true, a certain outcome is expected (or not expected). Then, the results of the experiment are determined 'objectively,' which means, in this context, *without bias either for or against the hypothesis being tested*.

In many experiments an experimental situation or group is compared with a control situation or with a **control group**. Ideally, the control group exactly matches the experimental group in all variables except the one being tested. For example, animals given a new drug are compared with animals in a similar group—the control group—that are not given the drug. The control group is given a substance similar to whatever is given to the experimental group, but lacking the one ingredient being tested. The two groups are selected and handled so as to be equivalent in every other way: similar animals, similar cages, similar temperatures, similar diets, and so on.

As an example of the experimental approach, consider the following experiment in bacterial genetics that was conducted by Joshua and Esther Lederberg. (This experiment was part of the basis for Joshua's subsequent Nobel Prize.) Most bacteria are killed by streptomycin, but the Lederbergs exposed the common intestinal bacterium *Escherichia coli* to this drug and were able to isolate a number of streptomycin-resistant bacteria. They allowed these bacteria to reproduce and were able to show that resistance to streptomycin was inherited by their offspring. In other words, the change to streptomycin resistance was a permanent genetic change; such changes are called mutations (see Chapter 3, pp. 67–69). The discovery of resistance gave the Lederbergs two hypotheses to test. The first (the induced-mutation hypothesis) was that the mutation had been induced, or caused, by exposure to the streptomycin. The second (the prior-mutation hypothesis) was that the bacteria had mutated before exposure to the streptomycin, in which case the mutation would be inde-

Figure 1.2

Scientists at work.

pendent of the exposure. To distinguish between these hypotheses, the Lederbergs devised the experiment shown in Figure 1.3. In this experiment, a copy, or replica, of the original plate of bacteria was made. Only the replica, not the original, was exposed to streptomycin, and the position of each bacterial colony was noted. The induced-mutation hypothesis predicted that bacteria exposed to streptomycin would mutate, but that unexposed bacteria would not. In fact, most of the bacteria died, but a few survived and were thus identified as being streptomycin resistant. To test the prior-mutation hypothesis, the Lederbergs went back and tested the colonies from the original plate. They discovered that the same colonies that were streptomycin resistant on the replica plate were also streptomycin resistant on the original plate. These findings support the prior-mutation hypothesis for this particular sample of bacteria.

The prior-mutation hypothesis for drug resistance had been tested and not falsified in the case of one mutation in one species of

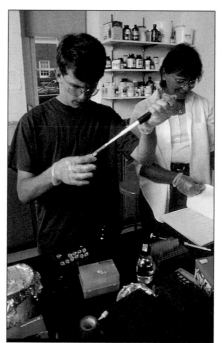

Transferring and examining solutions in a biochemistry laboratory.

Examining cells with an electron microscope.

Preparing cultures in a bacteriology laboratory.

Surveying diversity in each square meter along a transect line perpendicular to a rocky shore.

CONNECTIONS CHAPTER 3

Figure 1.3

The replica-plating experiment of Lederberg and Lederberg.

bacteria. How far could the finding be generalized? From this one experiment alone, one cannot tell. However, other investigators repeated the experiment for other mutations and other species of microorganisms. So far, the hypothesis of prior mutation has not been falsified. It is difficult to test the hypothesis in large or long-lived organisms, but most scientists

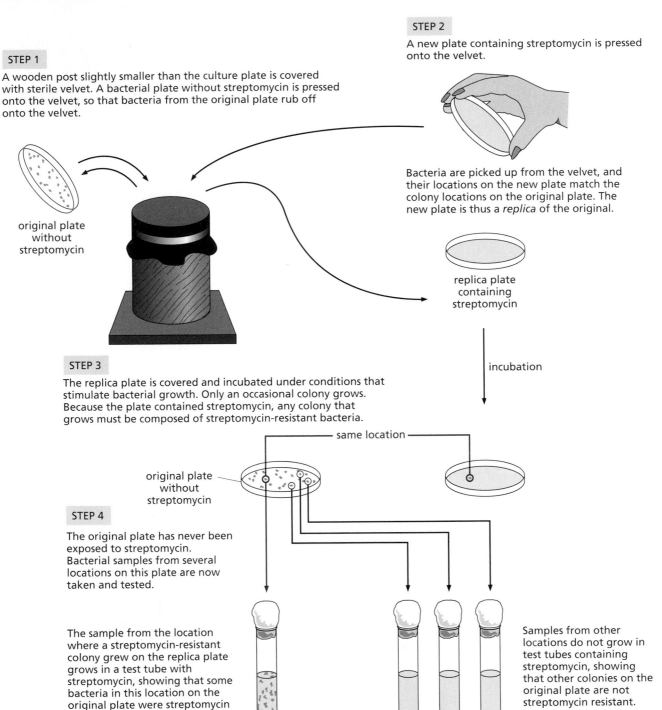

STEP 1

A wooden post slightly smaller than the culture plate is covered with sterile velvet. A bacterial plate without streptomycin is pressed onto the velvet, so that bacteria from the original plate rub off onto the velvet.

original plate without streptomycin

STEP 2

A new plate containing streptomycin is pressed onto the velvet.

Bacteria are picked up from the velvet, and their locations on the new plate match the colony locations on the original plate. The new plate is thus a *replica* of the original.

replica plate containing streptomycin

incubation

STEP 3

The replica plate is covered and incubated under conditions that stimulate bacterial growth. Only an occasional colony grows. Because the plate contained streptomycin, any colony that grows must be composed of streptomycin-resistant bacteria.

— same location —

original plate without streptomycin

STEP 4

The original plate has never been exposed to streptomycin. Bacterial samples from several locations on this plate are now taken and tested.

The sample from the location where a streptomycin-resistant colony grew on the replica plate grows in a test tube with streptomycin, showing that some bacteria in this location on the original plate were streptomycin resistant before they were exposed to the streptomycin in the experiment.

Samples from other locations do not grow in test tubes containing streptomycin, showing that other colonies on the original plate are not streptomycin resistant.

CONCLUSION

The result falsifies the hypothesis of induced mutations, but is consistent with the hypothesis that the bacteria mutated before they were exposed to streptomycin.

are willing to assume the truth of the hypothesis for *all* organisms. There are many species (and thousands of mutations for each species) that have never been tested in this way, which leaves opportunities for the hypothesis to be falsified in the future.

An introduction to naturalistic science. Another type of hypothesis testing is one in which direct experimental manipulation is either impossible or undesirable. For example, if an animal behaviorist wishes to study mating behavior *under natural conditions*, then any experimental manipulation that alters these natural conditions must be avoided. Thus we see ornithologists hiding in blinds to study birds, while other naturalists photograph their subjects by using telephoto lenses. The extinct species studied by paleontologists cannot be recreated in the laboratory to permit an experiment. We can refer to these sciences as **naturalistic sciences** because their method is based primarily on naturalistic observation rather than experiment. Naturalists do test hypotheses, but they do so by patient observation and record keeping. Naturalists often use control groups by comparing observations made when certain conditions are present with similar observations made when one of the conditions is not. The major difference between the experimental and the naturalistic sciences is that experimentalists set up and control their experiments, while naturalists can only observe and record those 'experiments' that occur in nature. This often means that naturalists must search the world over for the right circumstances, or must wait patiently for the right circumstances to occur.

The naturalistic sciences, moreover, are historical sciences. Any scientist seeking to understand why mammals differ from reptiles, or why the U.S. economy differs from the Japanese economy, will soon realize that the histories of animals, or of economies, form an important part of the explanation.

There are many types of questions in the naturalistic or historical sciences. The most characteristic type of question in these sciences is, "How did X get to be that way?". For example, a scientific team led by Rebecca Cann examined the DNA inside the mitochondria (the major energy-producing cell parts) of a large number of human populations. Mitochondrial DNA is always inherited from the mother, never from the father. Cann and her co-workers found that the chemical structure of the mitochondrial DNA in certain populations was very similar to its structure in other populations, allowing groups of related populations to be recognized. These scientists hypothesized that populations with similar mitochondrial DNA sequences share a close common descent through female lineages. This hypothesis explains the patterns of similarity among mitochondrial DNA sequences by a series of progressively 'smaller' hypotheses about the past histories of a given set of populations: that the populations of the Americas all share a common descent, that the populations of New Guinea all share a common descent, and so on. Some of these smaller hypotheses are falsified by the data, and must be replaced by modified hypotheses: New Guinea, for example, forms two clusters, and we can set up the hypothesis that it was colonized twice, with each line of descent forming a separate cluster. As modified in this manner, these smaller hypotheses of geographic dispersal are now interpreted as part of a common pattern of descent (see Figure 7.7, p. 224),

CONNECTIONS
CHAPTER 7

with an area of origin in Africa. The data are consistent with a hypothesis that all human populations are descended from an ancestral African population, or from a single ancestral female, nicknamed 'Eve' in the popular press. Like most other explanations in the natural sciences, the 'Eve hypothesis' explains present conditions on the basis of their past history, an evolutionary or historical mode of explanation.

Theories

A theory is a coherent set of related and well-tested hypotheses that explain a broad set of observations and that guide scientific research. Theories are usually developed by the testing of hypotheses, aided in many cases by mathematical and logical analysis of the resulting data. Theories then inspire future hypotheses and future tests. Most theories contain explanatory language that helps us understand some observed phenomena. One of the most important features of a good theory is that it may suggest new and different hypotheses. A theory of this kind is a stimulus to further research and is sometimes called a productive theory. A theory may be productive for a while and then no longer stimulate new research. The theories that last are the ones that remain productive the longest, while the less productive ones are often abandoned. Sometimes they are abandoned without ever being fully disproved. In other cases, it is the falsification of one of its hypotheses (or the failure of a crucial test) that causes a theory to be rejected. Remember that the hypotheses that make up a theory are always subject to possible refutation. Even a long-cherished theory may be abandoned (or greatly modified) if it no longer holds predictive or explanatory power.

Theoretical language and models. A theory usually contains language that helps communicate its subject matter. Many theories also use a simplified mathematical or visual form, called a **model**. Such a model, while not a formal part of the theory, can nevertheless be an important teaching tool in helping communicate it to other people. For example, Bohr's conceptualization of the atom in terms of electrons circling around the nucleus "like a miniature solar system" was the model of atomic structure for generations of students. However, models are analogies. Like other analogies, models are comparable to the phenomena they describe only so far, and no further. Attempts to determine *how far* an analogy holds often suggest new hypotheses to test or new ways to test old hypotheses. The planetary model of atomic structure is a case in point. With the development of quantum physics, it became clear that the solar system model was inadequate to explain the behavior of subatomic particles. Scientific theories are tentative. Even the best-cherished theoretical models can be supplanted by other models—either because an important hypothesis is falsified or because a more satisfactory explanation or model is proposed. John A. Moore, an embryologist and science historian, stated that "great art is eternal, but great science tends to be replaced by even greater science."

Most theoretical models describe observable events in terms of underlying causes described in the special language of the theory. Atoms, for example, are described by a theory that explains much of the observable behavior of matter, and heredity was explained in terms of genes long before anyone really knew what genes were. Theoretical concepts are typically studied in terms of their observable effects: the properties of matter are described in terms of atoms, and heredity is described in terms of inherited genes. A scientist who describes heredity in terms of genes may never observe those particular genes, but the theory allows her to make further predictions regarding animals or plants and their inherited characteristics. In science, calling something "theoretical" simply means that, even if we cannot directly observe it, we can use the theory to make predictions and study the observable effects.

A theory describing the properties of living systems

Animals, plants, and bacteria are examples of living systems that share many properties distinguishing them from nonliving systems. Each of these properties was initially a hypothesis about how living and nonliving systems differ. Each has been repeatedly tested and verified by observation across a wide variety of organisms, compared with various nonliving systems.

Properties shared by living systems, and the testable hypotheses about those properties, may be summarized as in Table 1.1 and listed as follows:

- *Organization.* The fundamental unit of life is called a cell. All living systems are composed of one or more cells.

- *Metabolism.* Living systems take energy-rich materials from their environment and release other materials that, on average, have a lower energy content. Some of the energy fuels life processes, but some accumulates and is released only upon death.

- *Selective response.* Living systems can respond selectively to certain external stimuli and not to others. Many organisms respond to offensive stimuli by withdrawing. Living organisms can distinguish needed nutrients from other chemicals and use only certain chemicals from among those available in their surroundings.

- *Homeostasis.* Living systems have at least some capacity to change potentially harmful or threatening conditions into conditions more favorable to their continuing existence, e.g., by converting certain toxic chemicals into less harmful ones.

- *Growth and biosynthesis.* Living systems go through phases during which they make more of their own material at the expense of some of the materials around them.

- *Genetic material.* Living systems contain hereditary information derived from previously living systems. This genetic material is a nucleic acid (either DNA or RNA) in all known cases.

- *Reproduction.* Living systems can produce new living systems similar to themselves by transmitting at least some of their genetic material.

- *Population structure.* Living organisms form populations. Populations can be defined retrospectively as groups of individual organisms related by common descent. Among organisms capable of sexual processes, a population is all those organisms that can interbreed with one another.

Table 1.1

Properties of living systems.

Cellular organization
Metabolism
Selective response
Homeostasis
Growth and biosynthesis
Genetic material
Reproduction
Population structure

Implicit in this listing of properties is also the testable assertion that anything that is alive by one criterion usually meets the other criteria as well.

As new organisms were discovered, each of the above hypotheses has been tested further and sometimes modified. For example, "breathing" was once considered an essential property of life, but this was modified several times to include other forms of gas exchange and metabolism. The invention of the microscope, in about 1700, led to the discovery of bacteria, which caused us to expand our concepts of selective response and led to the cell theory in the 1830s. The discovery of viruses early in the twentieth century strained this theory even more: seven of the eight hypotheses apply, yet viruses are not cellular and cannot reproduce on their own—they must use the cellular machinery of other organisms to reproduce themselves.

Together, these hypotheses form a working theory about the characteristics of living things that continues to be productive and to suggest new hypotheses to test. They are not rigid definitions that must be met. If something lacks one of the properties of life, we would not be forced to define it as nonliving; we could instead modify the hypotheses to define the limits of life more precisely. Viruses, for example, fulfill many of the properties of life but cannot reproduce without the help of another organism. We may therefore need to modify the criterion of reproduction to say that living things can bring about or can direct the production of new living things similar to themselves.

The above hypotheses have been tested in a wide variety of living systems. We can therefore gain confidence that the properties summarized in Table 1.1 form a coherent theory about how living systems differ from the nonliving.

THOUGHT QUESTIONS

1 In a group, discuss the hypothesis shown in Figure 1.1. Is it testable? If you believe so, then explain what sorts of observations you would make to test it.

2 Scientists seek to provide evidence to 'support' hypotheses. Why don't scientists say that their evidence 'proves' hypotheses?

3 Which of the following are experimental tests, and which are naturalistic observations?
a. Measurements made on the bones of an extinct species are compared with similar measurements made on the bones of a related living species.
b. The activity of white blood cells in a blood sample taken from stressed rats is compared with the activity of white blood cells taken from unstressed rats.

c. A group of animals is fed a certain chemical to see whether they will get cancer as a result.
d. A list of the species found in a particular square meter near the coast is compared with another list of species found 20 meters farther inland.

4 Viruses strain any definition of living systems: they contain genetic material, yet they replicate only inside the cells of and with the help of some other organism. Should we think of viruses as alive and devise a theory (or definition) of life that includes them, or should we think of them as lifeless and devise a theory (or definition) that excludes them? Would one of these theories be right and the other wrong, or are we free to choose either option?

Scientists Work in Paradigms, Which Can Help Define Scientific Revolutions

In a book that was itself considered revolutionary when it was first published in 1962, Thomas Kuhn, a philosopher and science historian, proposed a new method of looking at the ways in which science accommodates to new discoveries. Kuhn's observations were based on his studies of historical revolutions in science.

Paradigms and scientific revolutions

According to Kuhn, everything that we have described thus far is part of **normal science**, science that proceeds by the piecemeal discovery and gradual accumulation of new but small findings. Normal science in Kuhn's theory is always channeled by what he calls a **paradigm** (pronounced 'para-dime'). A paradigm is much more than a theory; it includes a strong belief in the truth of one or more theories and shared opinions as to what problems are important, what problems are unimportant or uninteresting, what techniques and research methods are useful, and so on. The research methodology and sometimes the instrumentation are important parts of the paradigm. Normal science proceeds cumulatively, in small steps, within the context of an existing paradigm. Paradigms, according to Kuhn, are best represented by science textbooks, which are written for the purpose of training new scientists within the paradigm. Students trained by these textbooks are taught not just facts, they are taught attitudes, approaches, values, and a vocabulary that teach them to think in certain ways.

Once in a great while, says Kuhn, science proceeds in a very different way: a **scientific revolution** occurs and is marked by the emergence of a new paradigm, which often requires replacement of an older one. Few scientists educated in the old paradigm support the new one at first. Most support for the new paradigm comes from new scientists just beginning their careers, and the founder of the revolution is usually either young or a new entrant into that particular scientific field. Once a scientific revolution occurs, its new paradigm opens up a new field of investigation or rejuvenates an old one. Such an infrequent event is called a **paradigm shift**. Scientists will not be attracted to a new paradigm unless they feel that it is somehow superior to the old paradigm, usually because it explains a wider variety of phenomena or because it explains certain new findings better than the old paradigm did. Often the vocabulary terms of the old paradigm are redefined or newer terms are adopted; terms that are no longer useful to the new paradigm may be abandoned. A paradigm shift is not just a triumph of logic or of experimental evidence. It is decided, at least in some measure, by a political-style process in which allegiances and influences shift. New paradigms succeed when scientists find them to be fruitful or productive of new approaches to research. Some examples of paradigm shifts are given in Table 1.2.

Paradigms are sometimes so powerful as to allow anomalies—observations that 'do not fit'—to be ignored. To scientists working within a

Table 1.2

Some examples of paradigm shifts in biology.

OLD PARADIGM	NEW PARADIGM
Natural Theology, Lamarckism, and several other competing paradigms	Darwinism (since 1859)
Blending inheritance and various folk ideas	Classical Mendelian genetics (since 1865 or 1900)
Various beliefs: bad humors, bad air or water, evil spirits, and many others	Germ theory of disease (Pasteur, Koch, since 1880)
Competing paradigms, including Darwinism, mutationism, population genetics, neo-Lamarckism	Modern evolutionary theory (since 1940)
Classical Mendelian genetics	Molecular genetics (since 1950s)
Various theories of territorial behavior, sexual behavior, etc.; also psychological theories (gestalt, behaviorism, ethology)	Sociobiology (since 1975)
Descartes' mechanistic theories and dualism	Mind–body connections (since 1980s)
Classic germ theory: pathogenicity as a characteristic of pathogen only	Pathogenicity as an interaction of pathogen and host (since 1990s)

paradigm, anomalies are small problems that they agree to ignore, believing that the integrity and success of the paradigm are more important than trying to accommodate the unexplained anomaly. Scientists who become interested in the anomaly may become founders of their own new paradigms, and the increased attention that they draw to the anomaly may precipitate a scientific revolution. Paradigms become successful in large measure by the students that they attract. Paradigms that no longer attract students die out.

Molecular genetics as a paradigm in biology

As an example of a scientific paradigm in biology, we describe here the field of molecular genetics as it has existed since about 1950. Other examples of scientific paradigms are described in subsequent chapters, including Darwinian evolution in Chapter 5, sociobiology in Chapter 8, and the connection between the mind and the body in Chapter 15.

CONNECTIONS CHAPTERS 2, 5, 8, 15

The paradigm of molecular genetics (or molecular biology) emerged in the decades following the determination of the structure of DNA by James Watson and Francis Crick in 1953. The structure of DNA was itself simply a hypothesis that gained rapid acceptance because it explained many known facts and allowed many new predictions to be made. The 'central dogma' of molecular biology (so named by the molecular biologists themselves) was that DNA was used to make RNA and RNA was used to make protein. (Further details of this process are described in Chapter 2.) Both DNA and RNA were said to contain information, and the making of one molecule from another was said to be a form of information transfer. As in other paradigms, the central dogma was more than just a theory because it also suggested a new vocabulary and drove a new research program. The language used within a paradigm often reveals much about how the paradigm is understood by the scientists working within it. How did DNA make copies of itself? This was called *replication* long before any of its details became known. How was information from DNA transferred to RNA? This was called *transcription*. How was information from RNA transferred to protein? This was called *translation*. How replication, transcription, and translation occurred were among the major problems to be solved. The terminology in molecular biology, like that in many fields, was part of an elaborate analogy that drew its inspiration from a comparison with linguistics and included such new vocabulary words as *code* (the language itself) and *codons* (items in the code). Also, words such as *transcription* (rewriting within the same language) and *translation* (changing from one language to another) were deliberately chosen for literal meanings that matched the biological theory. Textbook descriptions were replete with verbs like *read*, *copy*, and *translate*. There were also a number of laboratory methods, inherited from the field of biochemistry, plus a few extra technical advances, such as the use of high-speed centrifuges. Together, this all formed an orderly paradigm that outlined not only what was known, but also what remained to be discovered, what was thought to be important, and how the details were to be investigated and described. DNA was championed as the most important 'master molecule,' RNA was almost as important, and protein was important only until its synthesis was completed. Protein that was completely synthesized was no longer deemed interesting, except for a few enzymes that helped in the working of DNA or RNA. The paradigm thus defined the boundaries of the field.

The paradigm of molecular genetics guided research on DNA, RNA, and protein synthesis throughout the 1950s and 1960s; much of the work begun in those decades continues today. For its workers, the paradigm defined a set of shared beliefs (including the central dogma), a vocabulary, a set of research techniques, and, most of all, a set of problems to be solved. These problems included the mechanisms by which replication, transcription, and translation took place, as well as how to crack the genetic code. Once this last problem had been solved, the 'coding dictionary' (i.e., the list of correspondences between RNA sequences and protein sequences) was given a prominent place in every genetics book and most general biology texts.

As the molecular genetics paradigm matured, some of its early tenets were modified. Information flow, once thought to be unidirectional, is now thought to be more complex, sometimes flowing in both directions. Also, the idea of 'master' molecules that 'make' or 'control' other molecules is slowly being replaced by a vocabulary that speaks in terms of cells 'communicating' with other cells (sending and receiving signals) or 'influencing' other cells (in both directions). Likewise, attention has shifted to new questions, such as how the environment of a cell influences that cell to transcribe certain portions of its DNA at certain times. Still more recently, technological advances have permitted the rapid determination of many DNA sequences, including the sequencing of whole genomes, which has allowed some new questions to be raised. The molecular biology paradigm, like other paradigms before it, has gradually changed over time, although its core beliefs remain unshaken.

The scientific community

Is science something that only scientists can do? On the contrary, many people use scientific methods in their everyday lives. For example, if my car fails to start, I might formulate one hypothesis after another as to the possible cause. To test the hypothesis that the car is out of gas, I would examine the gas gauge. Additionally, I could add some gasoline to the tank and then try to start the car. If the car starts, I conclude that it was out of gas. The Swiss child psychologist Jean Piaget has written that children often behave as little scientists, formulating possibilities (hypotheses) in their minds and then testing them. "I can take the toy away from my little brother" can be tested by trying to take it away; the hypothesis would be falsified if brother successfully resisted or if an adult intervened.

It is unusual for a single person to formulate a hypothesis, test it, and then critically evaluate the results. For this reason, it is important for scientists to communicate with one another so that all these steps can be performed. Early written examples of hypothesis testing are found in the writings of the Greek historian Herodotus (fifth century B.C.). Early scientists in China, India, and elsewhere also wrote down their ideas by hand, but the spread of printing using movable type greatly speeded up the spread of scientific ideas after about 1500 A.D. Early European scientists (Copernicus, Galileo, Harvey, Descartes, Newton, and others) wrote their ideas in the form of books, pamphlets, and private letters. A major advance, however, occurred in seventeenth-century England, with the founding of the Royal Society in about 1660. This marked the first time in history that a permanent, *organized community* of scientists had communicated with one another and shared their results in a scientific journal (*Philosophical Transactions*). Now there was a written and permanent record of experiments performed and conclusions reached—a shared record that encouraged scientists to check one another's work in a systematic way. Because of this written record, scientists of the past continue to be part of the scientific community when their ideas are tested, even generations later. The scientists shown in Figure 1.4, whose accomplishments are each described elsewhere in this book, are still part of this scientific community even though they published over a time-span of about 150 years.

Many of the ways in which today's scientists behave toward one another may be viewed as efforts to maintain their ability to do the kind of systematic checking described above, including the ability to test hypotheses. Every test must be conducted in such a way as to make it possible for the hypothesis to be falsified, if indeed it is false, and the testing of hypotheses should be described as publicly as possible so as to permit the test to be repeated by other scientists. As David Hull points out, "Scientists rarely refute their own pet hypotheses, especially after they have appeared in print, but that is all right. Their fellow scientists will be happy to expose these hypotheses to severe testing." (D. Hull. *Science as a Process.* Chicago: University of Chicago Press, 1988, p. 4.) Skeptics who doubt a particular result unless and until they have seen it themselves can best be won over by a tradition that allows them to hear about repetitions of the test or to repeat the test themselves and to make their own observations. For example, Galileo, the astronomer and early scientist, invited critics who doubted his observations to look for themselves through his telescope.

So, the process of science is conducted in the public forum as well as in the laboratory. The publishing and dissemination of results (both in print and increasingly on the Internet) and the repetition of observations and experiments by others are thus valued among scientists. This is why the human genome sequence and an increasing number of scientific journals are available through the Internet. Scientists are expected not to work in isolation, but to discuss their results with other interested scientists, allowing them to build upon the results of previous scientists. They can repeat experiments and confirm the results, but they do not need to start from scratch and repeat *all* earlier work in their field. Science is a cumulative process in which it pays for individual scientists to begin with some of the groundwork laid by others, rather than to start always from scratch. As Isaac Newton once said, "If I have seen further than others who have gone before me, it is because I have stood on the shoulders of giants."

Figure 1.4

A few notable scientists of the nineteenth and twentieth centuries.

Charles Darwin (1809–1882), evolutionary biologist. Darwin's theories are described in Chapter 5.

Gregor Mendel (1822–1884), botanist and geneticist. Mendel's experiments in genetics are described in Chapter 2.

Ernest Everett Just (1883–1941), marine biologist and embryologist. Just's contributions to understanding the processes of sexual reproduction are described in Chapter 2.

Barbara McClintock (1902–1992), agricultural geneticist and Nobel Prize winner. Some of her contributions to genetics are described in Chapter 2.

Luis W. Alvarez (1911–1988), Nobel Prize-winning physicist, and Walter Alvarez (1940–   ), geologist. Walter's right hand rests on a layer of clay 65 million years old, at the boundary between the Age of Reptiles and the Age of Mammals. The Alvarezes' hypothesis to account for the extinction of dinosaurs and many other species across this boundary is described in Chapter 18.

Sarah B. Hrdy (1946–   ), sociobiologist, helping care for a friend's baby, a form of behavior whose importance she emphasizes among primates in general and humans in particular. Some of Hrdy's work in sociobiology is described in Chapter 8.

THOUGHT QUESTIONS

1 Why is it so difficult for a scientist to work outside the prevailing paradigm? Give at least three reasons.

2 In what ways is the paradigm of molecular genetics more than just a scientific theory?

3 Many companies conduct what they call research and development, yet many of these companies zealously guard their results and do not publish them. Are they doing science?

4 Researchers have deciphered the complete genetic blueprints of over 800 viruses and many disease-causing bacteria. These sequences are available on the Internet. How might free access to the genetic make-up of disease-causing pathogens be used? How could it be abused?

Scientists Often Consider Ethical Issues

Science itself can never tell us whether certain research should be done or how the results should be used by society; for those answers we turn to the branch of philosophy called ethics. Many topics in this book have an ethical dimension. Are some applications of specific biological research morally right and other applications morally wrong? Should society place legal restrictions on scientific research? Should biologists concern themselves with the ethics, applications, and implications of their work? This section describes some of the ways in which individuals and societies make ethical decisions.

We each use beliefs concerning what is right or wrong, proper or improper, to guide our own behavior. It is right to come to class at the scheduled time and in general to keep appointments that one has agreed to. It is wrong to steal, to lie, to murder, or to park in the NO PARKING zone. It is proper to wait for the traffic light to turn green and to wait for one's turn in line. All these moral rules, or **morals**, are products of societies. Anthropologists who have compared societies from around the world tell us that moral rules differ from one society to the next; they also change over time as society changes.

Any personal decision about whether to follow a moral rule may be called a moral decision: for example, should I park in the NO PARKING zone? Moral decisions are often made with the knowledge that society will attempt to enforce the rules with penalties or sanctions. Formal sanctions include fines and jail time; informal sanctions, which operate more often, include being criticized or avoided by others and ending up with fewer friends.

Ethics

Ethics is a discipline dealing with the analysis of moral rules and the ways in which moral judgments are made and justified. Descriptive ethics, the study of how these judgments are actually made, is a social science that investigates human behavior using scientific methods. In contrast, normative ethics, a branch of philosophy, deals with the logical analysis of how ethical judgments *should* be made, an analysis for which observational data are insufficient—no data can either confirm or refute a moral law (such as "Thou shalt not kill"). In its simplest form, normative ethics is an attempt to reduce moral codes to a minimum set of basic rules (maxims). For example, I should come to class on time because my signing up for a course is like making an appointment. Appointments should be kept because they are promises or contracts. An ethic of keeping appointments is part of a larger ethic of keeping promises.

Some rules of conduct are simply inventions of a society for the convenience of its members, such as waiting for the green light, driving on the right side of the street (in North America), and the observing of NO PARKING zones (Figure 1.5). We cannot all drive through the intersection at the same time, and traffic lights are a convenient (if arbitrary) contractual way of arranging whose turn is next. The contractual nature of such agreements is obvious because there are usually publicly controlled processes (like city council meetings) to decide where to put NO PARKING zones. We promise to observe traffic laws when we apply for a driver's license, so following these laws may be viewed as another form of promise keeping.

Waiting one's turn in a line or waiting for a green traffic light are both ways of introducing order and fairness into a situation that would otherwise be chaotic and conducive to unnecessary disputes. A major difference, however, is that waiting one's turn in line is not enforced by law or traffic code. It is enforced informally by the tacit agreement of those who are present.

A simple moral code might therefore instruct us to keep our promises, not to interfere with the rights of others, and to observe the common social conventions. This could easily be expanded into a more general code of benevolence, cooperation, and mutual aid.

Resolving moral conflicts

There are occasions when conflicts arise within sets of moral rules. I know I should obey the traffic laws, but what if I am taking an injured person to the hospital and the person's life is in danger? Does the duty to save a life justify driving above the speed limit, driving through a red light, or parking in a loading zone? Can I justify disobeying traffic laws to keep an appointment? Does it matter how important I think the appointment is? Resolving conflicts of this kind is one of the major goals of ethics.

In most cases, the resolution of such moral conflicts is made by determining that one rule or goal is more important than another: saving a life is more important than obeying traffic rules, for example.

Figure 1.5

Would you park in this space? Give reasons to explain your decision.

Thus, there are *exceptions* to most moral rules: obey traffic rules and other useful conventions *except* when obeying them causes greater harm or violates a more important rule. Notice that this ranks certain rules as more important than other rules, allowing us to justify an exception to one rule by invoking a 'higher' rule.

Although ethics is a branch of philosophy, ethical arguments arise in everyday life and also in science. For example, a scientific researcher may have financial interest in the success of a particular company, which might create a bias in scientific research regarding that company's products.

More and more scientific endeavors are raising ethical issues that are of practical interest to people in all walks of life. Virtually all institutions that conduct research now have policies and procedures for managing conflicts of interest. The U.S. government has sponsored research, meetings, and publications in the field of biological ethics. Many government programs, notably the Human Genome Project (see Chapter 4), have set aside portions of their budgets for the examination of the ethical implications of science. Our grounding in ethics in this chapter will support our examination of many issues with far-reaching ethical implications in the chapters that follow.

Deontological and utilitarian ethics

An ethical system is a set of rules for resolving ethical questions or for judging moral rules. Of the many possible ethical systems, we will describe the two major types that have received the most attention and attracted the most followers. Other ethical systems are described on our Web site, under Resources: Other ethical systems.

Deontological systems. An ethical system is called **deontological** if each person has a duty to perform certain acts and to avoid others, but the rightness or wrongness of an act depends on the act itself and not on its consequences. To a deontologist, the wrongness of murder is in the act itself, not stemming from its results or its effects on society. Similarly, a deontologist who believes in keeping promises does so apart from any consequences.

Historically, most deontologists have developed moral codes based on religious traditions. The Bible, the Koran, and the sacred texts of other religions have been the source of many moral codes. People who share the same religious tradition can often reach agreement quickly under such a system. But a deontological system based on a particular religion may have less influence on people not belonging to that religion.

The German philosopher Immanuel Kant (1724–1804) devised a deontological system without a religious basis. Kant based all ethical statements on a single precept: act only according to rules that you could want everyone to adopt as general legislation. He called this rule the **categorical imperative**. His test of the morality of an act is whether the act can be universalized, that is, applied to all people at all times. Thus, killing is (always) wrong because I could not possibly want people always to kill one another—I would be willing my own death and the death of my loved ones. Keeping promises can be universalized, and promise-keeping is therefore (always) moral. Respect for all human beings is an important part of Kant's system, and fundamental **rights** are based on respect for the dignity and autonomy of all persons. If you respect the dignity of all human beings, then you cannot ever will the death of any person, nor can you deny them their fundamental rights, nor can you use them as objects for your own personal gratification in any way. If you respect their selfhood, then you cannot morally abridge their freedom.

There has been considerable disagreement among philosophers, and even more variations in historical practice, over the types of beings to which various rights apply. At various times in the past, certain groups of persons (including women, children, slaves, the lower classes of stratified societies, impoverished people, foreigners, members of various races, mental patients, and persons unable to speak for themselves) were denied the rights that were afforded to other members of society (Figure 1.6). Many people now invoke dignity and autonomy criteria in discussions of whether certain rights should also now be extended to unborn fetuses or to animals.

An argument against rights-based deontology is that there are many circumstances in which one right conflicts with another, resulting in a moral dilemma. Unless there is a clear way of deciding between conflicting rights, moral dilemmas are inevitable. An obvious way out is to declare one particular right (such as the right to life or the right to freedom of action) supreme over all others. Aside from the problem that different deontologists would choose different rights to take precedence over the others, there is the more serious objection that insistence on a single right leads to the dangers of absolutism. Historically, many atrocities have been perpetrated by the followers of systems that put absolute adherence to a single principle above all others.

Utilitarian systems. A **utilitarian** system of ethics is one in which acts are judged right or wrong according to their consequences: rightful acts are those whose consequences are beneficial, whereas wrongful acts are those with harmful consequences. To a utilitarian, murder is wrong because the death of the victim is an undesirable outcome under most circumstances. Also, on a larger scale, murder is additionally wrong because it produces a society in which people live in fear.

A challenge to all utilitarian systems is to find a way of measuring the goodness or badness of consequences. Over the years, utilitarian philosophers have come up with different criteria by which to judge consequences: the greatest happiness for the greatest number of individuals, the greatest excess of pleasure over pain, and so on. All utilitarian systems require that value judgments be made between outcomes that are difficult to measure and quantify (Figure 1.7).

Utilitarianism can be summarized by the rule "always act so as to maximize the amount of good in the universe." The first major utilitarian philosopher was Jeremy Bentham (1748–1832), who said that we should always strive to bring about "the greatest good for the greatest number." To decide which actions produce the greatest good, Bentham suggested a type of *cost–benefit analysis* that he called "a

Figure 1.6

Do you have a deontological reason for agreeing or disagreeing with the premise that all persons share the same basic rights? This woman was protesting the fact that U.S. President Woodrow Wilson supported the rights of poor Germans in World War I, while women in the United States were denied the right to vote.

Figure 1.7

What benefits could come to society from this nuclear power plant? What costs or risks are present? How would a utilitarian argue in favor of this power plant? How might another utilitarian argue against it?

calculus of pleasures and pains." Other notable utilitarian philosophers include John Stuart Mill (1806–1873) and G.E. Moore (1873–1958).

One of the major criticisms of utilitarianism is that its cost–benefit approach reduces the status and dignity of human beings, and in some cases violates their rights. Can the killing of one person be justified if it results in saving the lives of other people? Deontologists argue that certain individual rights must be protected regardless of whether society as a whole benefits. To do any less, they argue, deprives human beings of their fundamental dignity as individuals and makes them nothing more than a cluster of costs and benefits. Even if a larger benefit can be demonstrated in a given instance, say these critics, it is still unethical to violate an individual's fundamental right, because "the ends do not justify the means."

Ethical decision-making

Individuals who are faced with moral choices are 'moral agents,' meaning that they are held responsible for their decisions and are praised for good decisions and criticized or punished for bad decisions, at least in a just world. Scientists are moral agents because they encounter moral decisions in their work. Science has also presented society at large with moral choices concerning the ways in which the findings of science may be used. Some moral decisions are left up to individual choice, while other decisions are made by institutions or by society as a whole. Many moral decisions involve conflicts between the rights and interests of individuals and the broader interests of large institutions or of society at large. Decisions that affect many people at once are usually made collectively, at least in a democracy, and many people may seek to influence such collective decisions.

Individuals making moral choices are often guided by principles of duty or moral imperatives, such as those listed in Table 1.3.

In facing a moral decision, a person using deontological reasoning asks, "what are my duties?". If several duties are in conflict, the question then becomes, "which is the more important or higher duty?". In a utilitarian approach, the costs (or risks) and benefits of each possible choice must be compared, including the example that may be set for others. The question then becomes, "which set of costs and benefits is preferable?" meaning "which set achieves a maximum of benefits at a minimum of costs?"

In the case of a moral dilemma—where no option is entirely good—we often seek further guidance in reaching a decision. The following procedural steps can help us in reaching moral choices: (1) gather all the facts and check them for correctness; (2) identify the ethical problem; (3) identify all the parties (stakeholders) that would be affected; (4) analyze the problem (e.g., with a deontological or utilitarian approach); (5) present the alternatives in priority order; (6) make and implement the decision; (7) if possible, reevaluate the decision as its consequences

unfold. On our Web site we have included ethical case studies for you to consider (under Resources: Ethical discussion topics). You may wish to refer to these steps to help you complete the assignments, or to help make moral choices in the real world.

Collective ethical decisions. Moral choices are in many cases made by groups of people rather than by individuals. These groups can be legislative bodies, professional groups, or private corporations. Many of these choices can affect large numbers of people or society as a whole. Individuals have an important role in such group decisions; however, the decisions made by a particular group can sometimes conflict with one's own self-interest or one's own moral judgment. How, then, should ethical choices be made by collective groups?

In culturally homogeneous societies in which people share common values and religious beliefs, it may be possible to reach consensus about which acts are wrong and which are right. However, most societies today are *pluralistic* in the sense that their populations include people with differing cultural and religious backgrounds, who are likely to be diverse in their opinions, values, and ethical approaches. Reaching collective decisions on ethical issues is more difficult in pluralistic societies, most of which are democracies or are becoming more democratic. We therefore examine how people in modern pluralistic democracies reach agreement on issues with moral dimensions.

In a pluralistic society, some people come to the public forum with utilitarian assumptions, others come with deontological assumptions, and others may have different positions. Even within these ethical traditions, people differ in their outlook: utilitarians have differing evaluations of costs and benefits, and deontologists have differing rankings of rights. One way of reconciling these different views is to have a public debate (so that all views are heard) and then vote. The voting procedure should be structured in a way that all parties recognize as fair. **Fairness** is the principle that states that all people should be treated impartially and equally; it is a way of ensuring that rights are not violated. In most cases, a system that works in this way results in the least displeasure with the decision. Of course, total agreement on a decision is hard to achieve in any large pluralistic society.

Social policy includes all those laws, rules, and customs that people follow in making individual decisions in a society. The making or changing of a new social policy is called a **policy decision**. Most social policies and policy decisions involve ethical considerations. An increasing number of policy decisions also involve some aspect of science, and it is often convenient to divide these into three phases: scientific issues, science policy issues, and policy issues.

- *Scientific issues.* What possible explanations (hypotheses) exist that might explain the available data? Can these hypotheses be tested? Do the tests support or falsify each hypothesis? Are alternative explanations available? What additional data are needed to evaluate these hypotheses? These are often characterized as 'purely' scientific issues on which scientists of different political or ethical persuasions may be expected to agree if sufficient data are available.

Table 1.3

Some moral imperatives.

Nonmaleficence ("above all, do no harm")

Beneficence (doing good to others)

Veracity (telling the truth)

Fidelity (being faithful, honoring commitments, and keeping promises)

Autonomy (respecting the rights and individual integrity of others)

Justice (treating all people with equal respect)

Preserving important social institutions

Preserving life

Setting a good example for others

(This list is not exhaustive, and should not be taken as a priority ordering. Modified from Ruth Purtilo, Ethical Dimensions in the Health Professions, 3rd edn. Philadelphia: W.B. Saunders Company, 1999.)

- *Science policy issues*. What would be the consequences of this or that particular legislation or policy change? What would be lost or gained from each proposed plan of action? Can the probability of uncertain consequences be estimated? The probability of any outcome, especially one not desired, is called a **risk**. Can we calculate or estimate the risks? In a cost–benefit analysis, what are the costs and what are the benefits? How certain are we of the estimated values? These are still scientific issues in the sense that they are evaluated from data, but disagreements on the data or their significance are expected between experts with different political or ethical viewpoints. If the experts disagree, how shall we evaluate their respective positions?

- *Policy issues*. Once we have evaluated the possible consequences of various possible policies, which one should we choose? These are ethical decisions in which values have a prominent role (e.g., is it worth risking the disruption of the economy to stop global warming?). In general, is some predicted but uncertain benefit worth the calculated risks? If a proposed change can be made only by incurring a certain cost to society, is this social cost worth the intended benefit? In all these cases, the cost–benefit analysis is used not to make the final decision, but rather to provide the necessary data on which a policy decision can be intelligently based.

Who makes the decisions? In most societies, scientific issues are frequently decided by scientists with little input from interested citizens. Science policy issues are often decided in the court of public opinion by an interplay of scientists and other 'experts' under the scrutiny of policy advocates for one side or the other. Public discussion and disclosure of evidence is useful in exposing and eliminating faulty information. Although evidence is used, it is more like courtroom evidence, obtained and evaluated by cross-examining witnesses, than like the evidence of science, obtained by the formulation and testing of alternative hypotheses.

As for the final policy decisions, who makes them: the scientists, the public, the media, or the government? In most democracies, the decisions are made either by the public or by government agencies acting (in theory at least) in the public interest. Decision makers are often influenced by scientists, the media, particular interest groups, and public pressures of various forms—marketplace decisions (decisions by individuals about where to spend their money), letters and e-mailings, organized demonstrations (Figure 1.8), enthusiasm at public gatherings, public opinion surveys, and direct votes on referendum questions. All of these influences certainly have a role in what is essentially a political process.

Figure 1.8

Why do people have a right to express their opinions through public demonstrations like this one? What role do such public demonstrations have in shaping social policy?

THOUGHT QUESTIONS

1 Together with other students, make a list of five to ten laws or rules that are generally followed on your campus or in your community. For each, try to discover:
a. Why such a rule is considered important (or why it *was* considered important when the rule was adopted),
b. Whether there is a more general moral concept of which this rule is just a special case, and
c. Whether society is better off with rules of this general kind than without them, and, if so, why?

2 Try to justify the wrongness of the following acts under the two ethical systems discussed in this section:
• murder
• rape
• bank robbery
• failure to repay a debt
• racial segregation

• driving over the speed limit
• parking in the handicapped space shown in Figure 1.5 if you are not handicapped
Which ethical system makes it easy to explain the wrongness of these acts? Under which ethical system are these explanations difficult?

3 A scientist is paid by a drug company to test a new drug for safety and effectiveness. What are her duties to the drug company? What ethical conflicts might arise? What safeguards are needed? Does it make a difference if she owns stock in the drug company?

4 What scientific issues are raised by the nuclear power plant pictured in Figure 1.7? What science policy issues? What data would you seek on which to base a policy decision one way or the other?

Ethical Questions Arise in Decisions About the Use of Experimental Subjects

So far we have discussed ethical issues in very general terms. We now turn, for illustrative purposes, to a specific issue, the use of experimental subjects in biological research. This issue involves moral choices at all the levels we have discussed: individual, institutional, and societal. First, we consider animal rights and the uses of animals in scientific experiments. Second, we contrast animal experimentation with experiments on humans. Of the many ethical issues surrounding biology today, few are as divisive as those touched upon here.

Uses of animals

Human societies have kept animals at least since the origin of agriculture. There are few societies in which animals are not kept as food, as pets, or as workmates. Most societies that practice agriculture use animals for all three purposes. Love of animals and use of animals can go hand in hand.

By far the largest number of animals used by most societies are raised for consumption as food for humans. Animal products are used for clothing. Animals are also the targets of recreational hunting, fishing, and trapping, even in many industrial societies. Many people keep pets or 'companion animals.' Work animals pull and carry loads, help in police work, and help handicapped people. Finally, animals are often used in research, although the number is only a tiny fraction of the numbers consumed as food.

Nearly all new drugs, cosmetics, food additives, and new forms of therapy and surgery are tested on animals before they are tested on humans. Many people regard animal research as critical to continued progress in human health. Over 40 Nobel Prizes in medicine and physiology have been awarded for research that used experimental animals. Organ transplants, open-heart surgery, and various other surgical techniques were performed and perfected on animals before they were performed on humans. All vaccines were tested on animals before they were used on human patients. In most cases, animals are used in research as stand-ins for humans. If we did not use animals for these purposes, humans would be the experimental subjects tested. Most animal testing is limited to the initial development of a drug or surgical procedure, but the human benefit continues for many generations or longer.

Few people object to a use of animals that saves human lives. Most people also agree that animals should not suffer unnecessarily, whether they are pets, work animals, or research subjects. Scientists need to use healthy, well-treated animals in research, and the U.S. Guide for the Care and Use of Laboratory Animals reflects this concern. There are standards such as cage sizes that must be followed by scientists using animals. All research using live animals must, by law, be scrutinized and approved by supervisory committees, and the committees require the investigators to minimize both the number of animals used and the amount of pain that those animals experience, and to substitute other types of tests where possible. According to statistics from the U.S. Department of Agriculture, 62% of animals used in research experienced no pain, and another 32% were given anesthesia, pain killers, or both, to alleviate pain. Only 6% of animals suffered pain without benefit of anesthesia. Federal law in the United States requires the use of anesthesia or pain killers in animal research wherever possible. Exceptions are allowed only when the use of anesthesia would compromise the experimental design and when no alternative method is available for conducting the test.

The animal rights movement

Those concerned with animal rights vary from traditional humane societies such as the Society for the Prevention of Cruelty to Animals (S.P.C.A.) and various national, state, and local humane societies, through groups such as People for the Ethical Treatment of Animals (PETA, founded in 1980), to groups such as the Animal Liberation Front (ALF, founded in 1972). Some animal rights advocates use a utilitarian ethic to advocate their position; others are deontologists.

Do animals have rights? Bernard Rollin, an American philosopher who supports animal rights, argues that there is no good reason for drawing an ethical distinction between mentally competent adult humans, other human beings (including children, comatose patients, mentally ill or brain-damaged persons), and animals. Animals are therefore, in his view, worthy of any moral consideration that would normally be given to babies, comatose patients, and other people unable to speak for themselves or to articulate their own viewpoints.

Historically, animals have been treated legally as property. Animal owners have property rights such as the right to sue for damages if their animals are killed or injured, but the animals themselves have no legal rights. Many philosophers have attempted to justify this approach by asserting that animals cannot be held morally responsible for their actions and are therefore not moral agents. Only moral agents are considered to be capable of entering into contractual agreements or of having any rights.

The question of animal rights is actually part of a broader question: how far do we extend the scope of any rights that we recognize? Many societies have historically denied even the most basic of rights to certain classes of persons on the basis of economics, gender, race, ethnicity, or religious beliefs. The extension of certain basic rights to all humans, including children and convicted criminals, is now considered so fundamental to the ethical sensibilities of most people that we refer to these as *human* rights. International agreements, such as the Geneva Convention on the treatment of prisoners of war and the International Convention on Human Rights (the Helsinki Accord), attest to the importance given to these human rights in world affairs. But should we stop there? Animal rights advocates say that we should extend these same rights to all beings capable of sensing pleasure and pain. A few people go even further, asserting that even trees have such rights as the right to go on living or not to have their air and soil poisoned. To go still further, a few people assert that habitats themselves, including mountains and forests, have rights not to be despoiled.

This book's Web site (see Resources: Animal rights) contains further discussions about animal rights and the use of animals in experiments.

Humans as experimental subjects

Many animal experiments are undertaken to determine the effects of some new drug or other therapy. The real question is usually about what the effects would be in humans, and the animals are merely used as stand-ins. Is it safe to extrapolate to humans results obtained from non-human species? In most cases in which data are adequate to answer this question, human physiological reactions have turned out to be comparable to those of experimental animals. Even when differences between humans and other species are known, they are often known in sufficient detail that the different responses to testing can help us understand the human system better, which still makes the animal tests valuable.

A few instances are known in which humans and certain commonly used experimental animal species respond differently. Saccharin, for example, causes cancer in rats, but has never been shown to cause harm to humans.

Direct experimentation on humans avoids the question of comparability between species, meaning that the results can be used more directly than results obtained from other species. Also, results obtained from psychologists, epidemiologists, and others who study humans with naturalistic methods can be applied even more directly. For example, one could not ethically force-feed cholesterol to an experimental group of human subjects, but one could observe the diets that different people choose on their own and study how people with high-cholesterol diets differ from people with low-cholesterol diets. The diets in such a study are more directly comparable to the diets of other humans than are those of experimental animals fed with different amounts of cholesterol in their food. As explained in many later chapters, more than one type of method must be applied to answer many scientific questions. We should not view animal studies versus naturalistic studies of humans as either/or choices; in most cases, both approaches are needed.

Among possible experimental subjects, humans have a special status. On the one hand, inflicting pain on human subjects, or exposing them to experimental risks, raises more ethical objections than does the similar treatment of nonhuman animals. On the other hand, human subjects can tell us how they feel, or when and where they experience discomfort or pain. Certain types of drug side-effects, like headaches or impairment of problem-solving ability, are difficult to assess without using human subjects.

Voluntary informed consent. A purely ethical consideration is that human subjects can voluntarily consent to serve as experimental subjects, which is something that nonhuman animals cannot do. Humans are considered to be autonomous beings who have the right to consent to putting themselves at risk, whether in a space capsule, a bungee jump, or an experiment. An important consideration, however, is that the person serving as a subject must give consent voluntarily. This is a legal as well as a moral issue, because persons who did not consent voluntarily can sue for damages if any harm comes to them. Consent is usually obtained in writing on a form that informs the person of the possible benefits and risks of the experimental procedure and is therefore called **informed consent**. If an experiment may bring direct benefit to a subject (as when a disease or its symptoms are being treated), potential subjects may be more willing to undergo certain risks than they would otherwise.

Special questions arise in the case of persons who may not have the full capacity to understand all the possible risks and benefits, including mentally deficient persons, unconscious persons, or children. Most people would now consider it unethical to use such a person as an experimental subject unless there was obvious great promise of direct benefit and only minimal risk of harm. In most jurisdictions, parents are considered to have the legal right to make such decisions on behalf of their children. In past decades, prison populations were often used as sources for experimental subjects, but this practice is now frowned upon because the consent of a prisoner may not be truly voluntary if she or he thinks—rightly or wrongly—that cooperating might result in a sentence reduction.

Guidelines for human experimentation. As a safeguard against possible abuses, research on human subjects is now usually reviewed by institutional committees set up for that purpose. As is true in reviews of animal experimentation, the review process ensures that someone other than the researchers evaluates the ethics of the proposed experiment. Federally sponsored research in the United States and in many other countries requires that such committees authorize all experiments in which humans are used as subjects. In addition to ensuring that proper voluntary consent has been obtained, such committees also have the obligation to suggest ways in which risks can be reduced or benefits increased without impairing the validity of the experiment. Many scientists work within the ethical tradition in which exposing humans to experimental risks is more objectionable than exposing animals of other species to those same risks. Guidelines have been developed that specify testing to be carried out on animals first, then on small numbers of carefully chosen and carefully monitored human subjects, and only last on large and diverse human populations. In the United States, federally sponsored research and research on new drugs seeking federal approval are required to follow this procedure.

Avoiding gender bias. It is now considered poor practice to extrapolate experimental results to both sexes if the tests were done on men only. However, such studies were once fairly common, and many women therefore received drug doses that had been based on studies of men only. Guidelines for human testing were revised only after the ethical issue of gender bias was raised by Dr Bernadette Healy. This is further discussed on our Web site, under Resources: Gender bias.

THOUGHT QUESTIONS

1 For each of the following acts, try to give:
   a. A deontological argument against the act
   b. A deontological argument justifying the act
   c. A utilitarian argument against the act
   d. A utilitarian argument justifying the act
   • beating your horse
   • taking a canary into a coal mine so that, if it dies from toxic gases, miners would be warned to evacuate
   • raising broiler chickens or beef cattle for human consumption
   • testing a drug on rats (or cats) before giving it to humans
   • testing a drug on human prison convicts

2 How widely can experimental results be extrapolated? If a drug is tested on inbred male rats, is it certain that the results are applicable to humans? Is it likely? Is the drug likely to have similar effects on both sexes? What issues of methodology or of ethics are raised by experiments that used only inbred male rats?

3 Is it ethical to infect a few people with a deadly disease to study its effects in the hopes of saving many more lives in the future? How do you justify your answer? Give both deontological and utilitarian reasons.

4 One hundred patients are to be enrolled in a study of a new drug. Half of the enrolled patients will be given the new drug while the other half will be used as a control and will therefore not receive the new drug. How many of the patients need to give informed consent? Why?

Concluding Remarks

In science, we know what we know through a process often called the scientific method. Scientists formulate tentative ideas (hypotheses) about living systems and about the world in general, and submit these ideas to extensive testing by comparing their ideas with observations made in the material world around them. Scientific knowledge is forever tentative and is never 'proved' because it is always subject to change if new observations do not fit our existing theories. The language of science is often metaphorical. Scientists often use words with specialized meanings. Scientific paradigms give science its vocabulary, imagery, attitudes, and value judgments. Science is conducted in a social context that includes a community of scientists sharing their ideas and testing each other's hypotheses.

Ethical decisions can be made either by judging actions themselves (deontological ethics) or by judging actions on the basis of their consequences (utilitarian ethics). Science may sometimes confront individuals with moral choices. Many scientific issues have implications for large numbers of people or for society as a whole. All citizens, not just scientists, should help make these decisions collectively. However, scientists should bear some responsibility for educating others about the science issues and science policy issues that can inform these decisions. Scientists should educate themselves as to the ethical dimensions of their work, including both the treatment of experimental subjects and the possible uses and misuses of scientific findings.

Chapter Summary

- **Science** is based on the testing of **hypotheses** under conditions in which it is always possible to observe results inconsistent with the hypothesis.
- Hypotheses originate by many types of reasoning and inspiration, including both **induction** and **deduction**.
- A group of well-tested hypotheses forms a **theory**. A theory may be communicated by special vocabulary or by a descriptive analogy (a model).
- **Biology** is a science because biologists use hypothesis testing to study living systems.
- Living systems exhibit cellular organization, growth, metabolism, homeostasis, and selective response. They contain genetic material, and they belong to populations, most of whose members are capable of reproducing.
- Biologists test hypotheses either by studying natural conditions as they occur (**naturalistic science**) or by conducting **experiments** under conditions that the scientists help to create (**experimental science**). After data are gathered, the interpretation of results usually involves comparison with **control groups**, often with the help of statistical methods.
- Normal science proceeds by testing hypotheses one at a time, under the guidance of a **paradigm**. A new paradigm may replace an older one if it can explain things better than the old one did; such a shift in paradigms brings about a scientific revolution.
- **Scientific methods** are used in everyday life, but scientists use these methods more often and more systematically.
- Certain values are held by members of the scientific community that ensure them the continuing ability to test, falsify and change each other's hypotheses.
- **Morals** are rules that guide our conduct. **Ethics** is the discipline that examines moral rules and attempts to explain or justify them.
- Two major types of ethical systems are **deontological** and **utilitarian**. Deontologists judge the rightness or wrongness of an act by characteristics

of the act itself, apart from its consequences. Utilitarians judge the rightness or wrongness of an act on the basis of its consequences. Utilitarian analysis often includes a comparison of the undesirable effects (costs) of an act with its desirable effects (benefits).

- Science can confront individuals and societies with moral choices.

- Individual moral choices can be guided by principles such as nonmaleficence, beneficence, autonomy, and **fairness**.

- In facing collective ethical issues involving scientific questions, it is often useful to distinguish between scientific issues, science policy issues, and policy issues.

- Animals are used in our society for food, for labor, for companionship, and for laboratory experimentation. Biological experiments often use living organisms as subjects. In many cases, laboratory animals are used as stand-ins for humans, and their use is often justified on a utilitarian basis (the cost–benefit ratio is lower if animals are tested before humans) or on a deontological basis (humans have rights and animals do not, or human rights supersede animal rights).

- Before any experimentation on animals or humans can take place, the proposed experiments must pass an ethics review. If humans are used, their voluntary **informed consent** must be obtained.

CONNECTIONS TO OTHER CHAPTERS

This chapter connects to the remainder of the book because the methods of discovery outlined in this chapter were used to explore all of the topics described in subsequent chapters. The characteristics of life listed at the beginning of this chapter are referred to throughout the book. Also, many applications of science have ethical dimensions, including the following: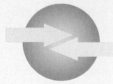

Chapter 3   The Human Genome Project and human genetic testing raise ethical questions.

Chapter 7   Ethical objections have been raised against the ways in which biology has supported racism.

Chapter 8   Evolution has resulted in various behaviors, including both moral and immoral acts among humans.

Chapter 9   The need for population control conflicts with the ethic of allowing reproductive freedom.

Chapter 10   Patterns of food consumption and distribution raise ethical issues.

Chapter 11   Some people object to genetically engineered foods on ethical grounds.

Chapter 12   Cancer research often involves the use of animal experimentation.

Chapter 13   Brain research involves animal experimentation and also the use of fetal tissue.

Chapter 14   Drugs are usually tested on animals before giving them to people.

Chapter 16   Many ethical issues surround the transmission of AIDS, testing for AIDS, and the prevention of AIDS.

Chapter 18   Many ethical issues are raised by habitat destruction and species extinction.

Chapter 19   Global patterns of pollution raise ethical issues.

PRACTICE QUESTIONS

1. Which property of life is exhibited by each of the following?

   a. The frog jumps around when I touch it.

   b. The bread rises because the yeast has given off carbon dioxide bubbles.

   c. Blood samples from healthy humans always have about the same pH and salt concentration.

   d. Wherever I find one mosquito, I usually find many.

   e. Only a few rabbits were brought to Australia, but now there are millions.

   f. Puppies usually resemble their parents.

   g. Baby animals get bigger and become adults.

   h. A bright light at night always attracts moths.

2. Which of the following are testable hypotheses? For each statement that you think is testable, explain what sort of test you might conduct and what possible outcome would falsify the hypothesis.

   a. 'N SYNC is a better musical group than the Rolling Stones.

   b. In a maze that they have never seen before, rats will turn right just about as often as they will turn left.

   c. If these two plants are crossed, approximately half of the offspring will resemble one parent and half will resemble the other.

   d. It is wrong to inflict pain on a cat.

   e. Restaurant A is better than restaurant B.

   f. The average science major at this school gets better grades than the average humanities major.

3. Which of the following examples of reasoning use induction? Which use deduction?

   a. If all adult female birds lay eggs, then this female chick will lay eggs if raised to maturity.

   b. If all known species of birds are egg-laying, then the next bird species to be discovered will be egg-laying too.

   c. If all known enzymes are made of protein, and I discover a new enzyme, then it, too, will be made of protein.

   d. If the amounts of protein X are increased under stress, then I should be able to increase the amount of protein X in these frogs by subjecting them to stressful conditions.

   e. If this species mates in April, then I should be able to observe more mating on April 10 than on June 10.

4. Which of the following reflect the community nature of science?

   a. A scientist presenting a talk at a scientific meeting.

   b. Another scientist asking a question at that same meeting.

   c. A field naturalist tracking a rare species.

   d. The same field naturalist publishing her findings.

   e. A scientist feeding a new chemical to mice to study its effects.

   f. A bacteriologist using techniques developed by Louis Pasteur and Robert Koch for growing bacteria in laboratory cultures.

   g. A scientist displaying his experimental results over the Internet.

5. For each of the following, identify (a) whether the argument is based on utilitarian or deontological ethics, and (b) whether any assumptions are made about whether animals have no rights, some rights, or rights equal to those of human beings.

   a. Hunting is wrong because the victim is part of nature and it is wrong to interfere with nature.

   b. Hunting is justified because the death of one animal makes such a small difference to most hunted species.

   c. Hunting is wrong because it makes the hunter more prone to future violence.

   d. Hunting is justified if the animal is used as food but not as a trophy.

   e. Raising beef cattle for human consumption is justified because people need to eat.

   f. Raising beef cattle for human consumption is wrong because cows are sacred.

   g. Raising beef cattle for human consumption is wrong because people would be healthier if they ate more plant foods instead.

   h. Raising beef cattle for human consumption is wrong because it causes pain and suffering to the animals.

   i. Be kind to your pet because you will be rewarded with loving companionship.

Issues

- How have our concepts of genes developed?
- What are the limitations of Mendelian genetics?
- Does Mendelian genetics explain inheritance in all species?
- What do we not understand about genes, chromosomes and DNA?

Biological Concepts

- The gene
- Patterns of inheritance (trait, phenotype and genotype; sex determination; sex-linked traits)
- Mitosis and meiosis
- DNA (the genetic material, DNA structure, DNA replication)
- Human genetic conditions

Chapter Outline

Mendel Observed Phenotypes and Formed Hypotheses

Traits of pea plants

Genotype and phenotype

The Chromosomal Basis of Inheritance Explains Mendel's Hypotheses

Mitosis

Meiosis and sexual life cycles

Gene linkage

Confirmation of the chromosomal theory

Genes Carried on Sex Chromosomes Determine Sex and Sex-linked Traits

Sex determination

Sex-linked traits

Chromosomal variation

Social and ethical issues regarding sex determination

The Molecular Basis of Inheritance Further Explains Mendel's Hypotheses

DNA and genetic transformation

The structure of DNA

DNA replication

2 Genes, Chromosomes, and DNA

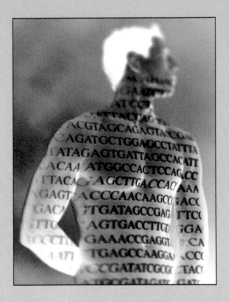

2 Genes, Chromosomes, and DNA

How do offspring come to resemble their parents physically? This is the major question posed by the field of biology called genetics, the study of inherited traits. Genetics begins with the unifying assumption that biological inheritance is carried by structures called genes. The discovery of what genes are and how they work has been the subject of many years of research. Genes are carried on chromosomes and much has been learned about genetics from the study of chromosomes. Among the earliest findings was the fact that the same basic patterns of inheritance apply to most organisms. The inheritance of some human traits, such as albinism, can be explained by hypotheses first formulated from the study of pea plants, whereas the inheritance of other human traits, such as sex determination, is a much more complicated affair.

Mendel Observed Phenotypes and Formed Hypotheses

No two individual organisms are exactly alike. Folk wisdom going back to ancient times taught people that a child or an animal resembles both its mother and its father by showing a mixture of traits derived from the two sides of the family. This suggested a concept that came to be called 'blending inheritance,' in which heredity was compared to a mixing of fluids, often identified as 'blood.' The research that we are about to describe caused this theory to be abandoned.

Traits of pea plants

During the nineteenth century, Gregor Mendel, a Czech scientist living under Austrian rule, worked out the principles of inheritance for simple traits that he described in 'either/or' terms. Mendel was a priest who grew pea plants (*Pisum sativum*, kingdom Plantae, phylum Anthophyta, also called Angiospermae) in the garden of his monastery. Mendel was curious about differences that he observed among different varieties of pea plants, so he decided to breed them and keep careful records.

Why were the peas a good species for Mendel's experiments? Pea plants have many distinctive traits (Figure 2.1), and other practical advantages: Mendel could easily grow them in the monastery garden, and many varieties were locally available, including some with yellow peas and others with green peas, and some with round and others with wrinkled peas. Mendel knew that peas reproduce sexually—that is, a new individual forms by the union of an egg from a female with a sperm from a male, an event called **fertilization**. Unlike most animals, an individual plant may have both female and male reproductive organs. This is true of many plant species, including peas (Figure 2.2). Within the pea flower are male structures called anthers that produce pollen grains, each of which contains a sperm. The pea flower also has a female structure called a stigma that receives pollen grains (a step called 'pollination') and permits the sperm to travel to the ovary to reach the eggs. Self-pollination occurs when pollen from a plant is deposited on a stigma of the same plant. In some of his experiments, Mendel sewed together the margins of one large petal to enclose the anthers and stigma together and ensure self-pollination. At other times, he cross-pollinated his peas by dusting pollen from a flower of one plant onto the stigma of a flower of another plant, after first removing the anthers from the recipient plant.

Mendel organized his work so as to answer specific questions, a procedure that we recognize today as good experimental design. Unlike most of his predecessors who failed to discover how plant offspring inherit their parents' traits, Mendel followed certain careful procedures:

1. First, for each of the traits he studied, Mendel used peas belonging to pure lines. A 'pure line' is one that breeds true from generation to generation, always producing offspring that express the same form of the trait as the parents. For example, tall parents from one pure line always produce tall offspring and short parents from another pure line always produce short offspring.

2. He chose which plants would mate by either cross-pollinating (crossing) the flowers or closing up the flower parts to ensure self-pollination (see Figure 2.2).

3. He first studied only one trait at a time, until he understood its pattern of inheritance. Later, he studied two and three traits at a time. His predecessors, on the other hand, often began by examining several or many traits at once.

4. He counted the offspring of each cross, and was thus able to recognize ratios between them. (Those few of his predecessors who

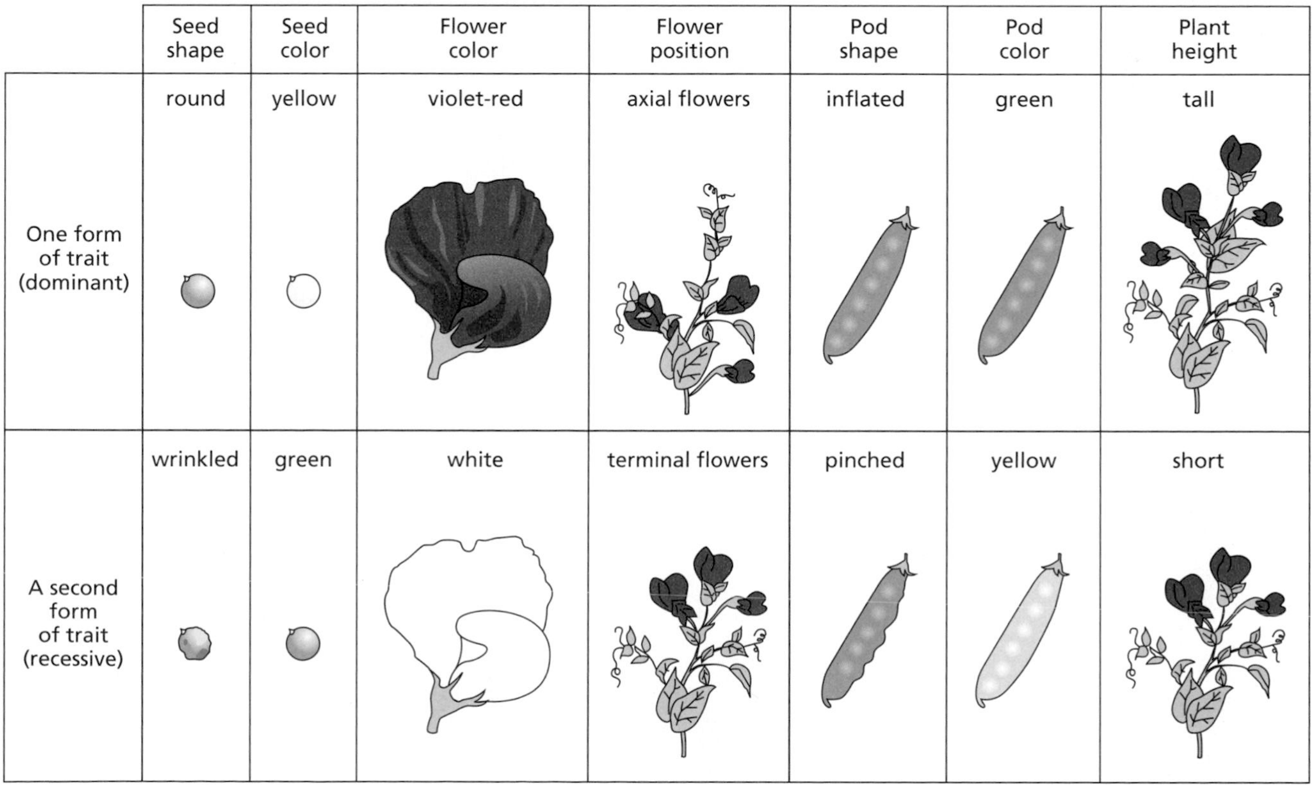

Figure 2.1

The seven traits studied by Mendel in peas.

	Seed shape	Seed color	Flower color	Flower position	Pod shape	Pod color	Plant height
One form of trait (dominant)	round	yellow	violet-red	axial flowers	inflated	green	tall
A second form of trait (recessive)	wrinkled	green	white	terminal flowers	pinched	yellow	short

Figure 2.2

The structure of a pea flower. A more complete view of the sexual reproductive structures of flowering plants is shown in Figure 6.9 (p. 183).

pea flower with petals enclosing sexual parts inside

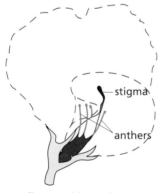

pea flower with petals removed, to show male and female flower parts

looked at single traits never counted the offspring of each type and thus failed to find ratios.)

5. He continued each experiment through several pea generations.

Mendel studied one trait at a time. For example, he crossed plants having white flowers with plants having violet-red flowers, but which were alike in all other traits. He found that all the first generation of offspring plants (symbolized as F_1) had violet-red flowers, but none of the plants had white flowers. When he crossed tall and short plants, all the offspring were tall. Mendel introduced the term **dominant** for the form of the trait that appeared in the first-generation offspring of his initial cross; the trait that did not show up he called **recessive**. Thus, violet-red flowers are dominant to white, and tall is dominant to short. Mendel found for each of the seven pairs of either/or traits that one form of the trait was dominant and one was recessive. The forms of the traits shown in the upper row of Figure 2.1 are dominant; those in the lower row are recessive. The traits did not blend. Violet-red-flowered plants crossed with white-flowered plants produced violet-red-flowered offspring, not pink-flowered offspring. Tall plants crossed with short plants produced offspring of the height of the tall parents, not halfway between.

Genotype and phenotype

The tall F_1 plants in Mendel's experiments were just as tall as their tall parents. They looked the same as their tall parents; that is, their **phenotype** for height was tall. But Mendel realized that the hereditary makeup of the F_1 plants was different from that of their parents: the tall parents had come from a pure line, so all their hereditary make-up had been tall, but each F_1 plant had both tall and short parents. Could this difference in hereditary make-up, or **genotype**, be made visible? Would it show up in future generations? Mendel found out by mating the plants of the F_1 generation with themselves (self-pollinating them) and raising a second generation, symbolized as F_2. He obtained both tall plants and short plants in the F_2, but no plants of intermediate height. When he counted them, he found that the tall plants were approximately three times as numerous as the short ones. Mendel conducted similar experiments with other traits, such as flower color, and in each experiment he produced an F_2 generation. In each case he discovered the same thing: in the F_2 generation, the dominant phenotype outnumbered the recessive phenotype in the approximate ratio of 3:1, so that three-quarters of the F_2 plants had the dominant phenotype and one-quarter had the recessive one.

Mendel's explanation for inheritance of single traits. Mendel proposed a multipart hypothesis to explain his results. Using modern terminology and some modern understanding, we can list the points covered by his hypothesis.

1. The inheritance of traits is controlled by hereditary factors; today these factors are called genes.

2. Each individual has two copies of the gene for each trait. If each gene is represented by a letter, then the genetic makeup (genotype) of an individual for a trait can be represented with two letters.

3. Each gene exists in different forms; these variant forms of the same gene are called **alleles**. For example, the gene for flower

color in peas has an allele that produces violet-red flowers and another allele that produces white flowers. Alleles that are dominant produce only the dominant phenotype even when the recessive allele is present. Mendel designated the dominant and recessive alleles controlling the same trait by a single letter of the alphabet, using a capital letter for the dominant allele and the lowercase of the same letter for the recessive allele. For example, the allele V for violet-red flowers is dominant to the allele v for white flowers. (Today, many genes are designated by two- and three-letter combinations, and different alleles of the same gene by superscripts.)

4. An individual whose genotype contains two identical alleles, such as VV or vv, is said to be **homozygous** for the trait. An individual whose genotype combines dissimilar alleles, such as Vv, is said to be **heterozygous**.

5. Dominant alleles always show up in the phenotype, but recessive alleles are masked by their dominant counterparts. When a dominant and a recessive allele for the same trait are present in a heterozygous individual, the dominant allele produces the phenotype. Recessive alleles produce the phenotype only when they are homozygous—only, in other words, when the corresponding dominant allele is absent.

6. The genes behave as particles that remain separate instead of blending. Recessive genes are masked during the F_1 generation, which shows the dominant phenotype, but are not changed by being in a heterozygous F_1 individual.

7. When the F_1 individuals produce eggs and sperm, the dominant and recessive alleles separate from one another, or 'segregate.' The eggs and sperm are called **gametes**, and each gamete receives only one allele of each gene after they segregate. The principle of a **segregation** of alleles is often called Mendel's first law. (The term 'law' was used more frequently in the past; in this case, it simply means a concept devised to explain the data.)

8. During sexual reproduction, the gametes combine so that each F_2 individual has two alleles, one contributed by the egg and one by the sperm. This explains both why the recessive phenotype disappears in the F_1 and why it reappears in about one-quarter of the F_2 individuals.

Figure 2.3 shows Mendel's multipart hypothesis applied to one of the traits he studied. A pure-line plant with violet-red flowers and the genotype VV, which produces V gametes, is crossed with a pure-line, white-flowered vv plant, which produces v gametes. All F_1 plants, having received V from one parent and v from the other, have violet-red flowers and the heterozygous genotype Vv. Half of the eggs produced by the F_1 carry the dominant allele V and half carry the recessive allele v, and the same is true for the sperm, as indicated in the margins of the square in Figure 2.3. When the gametes from the self-pollinated F_1 combine at random to form the F_2 generation, one out of four new individuals is homozygous for the dominant allele VV and has violet-red flowers, two out of four are heterozygous Vv with violet-red flowers, and one out of four is homozygous for the recessive allele vv and has white flowers. The ratio of violet-red to white phenotypes is thus 3:1. A square like this, showing how gametes combine to produce genotypes and phenotypes, is called a Punnett square.

Independent assortment. After Mendel had investigated the inheritance of single traits, he proceeded to study the inheritance of two traits together. In one of his experiments, the parents differed in both seed shape and seed color. The parents in one group had yellow, round seeds, were homozygous for both traits (*YYRR*), and thus produced all *YR* gametes. The other parents had green, wrinkled seeds, were homozygous for both traits (*yyrr*), and thus produced all *yr* gametes. The first-generation offspring (F$_1$) were all *YyRr*, heterozygous for both traits. All these F$_1$ plants produced seeds that were yellow (because yellow is dominant to green) and round (because round is dominant to wrinkled). No new principles were involved so far.

Mendel expected half of the gametes produced by the F$_1$ to contain the dominant allele *Y* and the remainder the recessive allele *y*. He also expected half of the gametes to contain allele *R* and the other half the allele *r*. But would a gamete's receiving *Y* rather than *y* influence whether it received *R* or *r*? To find out, Mendel raised an F$_2$ generation by self-pollinating some F$_1$ plants. He obtained the 9:3:3:1 ratio of phenotypes shown in Figure 2.4, with 9/16 of the F$_2$ individuals showing both dominant traits (they were both yellow and round); while 3/16 were round but green; 3/16 were yellow but wrinkled; and only 1/16 showed both recessive traits (green and wrinkled). Mendel then reasoned that this ratio could be explained if he assumed that all four possible types of gametes (*YR*, *Yr*, *yR*, and *yr*) were produced in equal proportions, as shown in the margins of the square in Figure 2.4. This means that the inheritance of the trait of seed color has no influence on the inheritance of the trait of seed shape. This principle is called **independent assortment** or Mendel's second law.

Figure 2.3

One of Mendel's experiments with peas differing in a single trait.

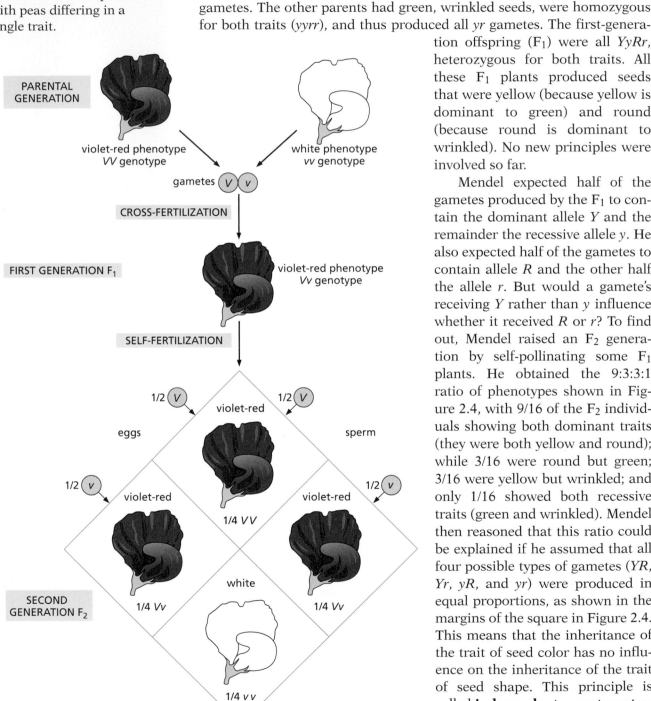

All of the traits that Mendel worked with assorted independently in this way, but, as we will soon see, there are exceptions to Mendel's second law.

Mendel's results, published in 1865, were ignored by most scientists. The reasons for the lack of impact of his theories are many, but one contributing factor was that he presented his theories in the language of mathematics, to which most botanists of his generation were not accustomed. Another factor was that Mendel's findings were published in an obscure journal that was not widely read. Later, in 1900, each of three other scientists conducted experiments similar to Mendel's, reached similar conclusions, and then subsequently discovered Mendel's earlier work.

Figure 2.4

One of Mendel's experiments with peas differing in two traits.

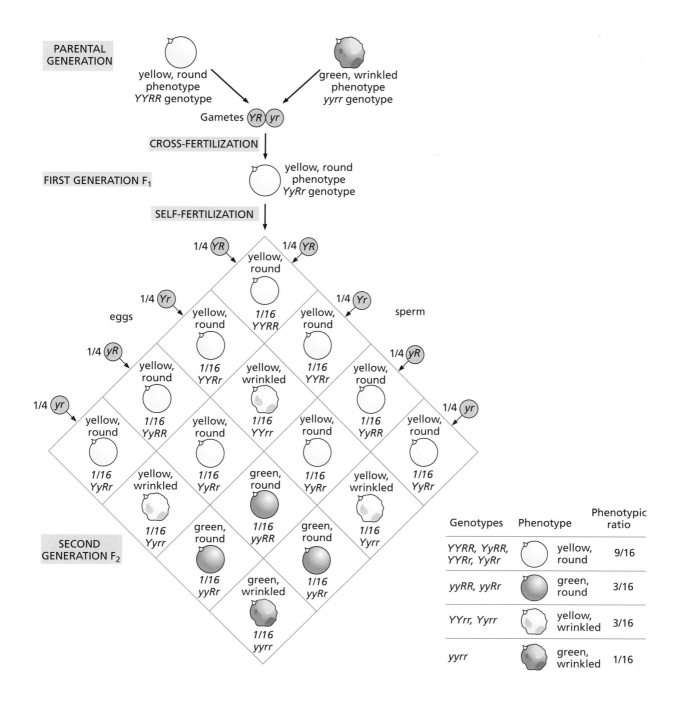

THOUGHT QUESTIONS

1 Mendel's experiments distinguished between two alternative hypotheses: either traits blend, so that the offspring have traits intermediate between those of their parents, or traits are inherited as discrete particles that do not blend. Which hypothesis is favored by Mendel's results? Did he disprove one hypothesis? Did he prove one hypothesis?

2 An experimenter cross-pollinates flowers of tall pea plants with pollen from flowers from short pea plants. She harvests the seeds and plants them, and all of the plants that grow in the F_1 generation are as tall as the tall parent plants. The F_1 plants produce flowers, each containing many sperm and many eggs. Each sperm and each egg carries only one allele of each gene. What are the possible genotypes for plant height among the sperm from the F_1 plants? What are the genotypes among the eggs? What fractions of the gametes possess each genotype? Would the distribution of types of gametes in the F_1 plants be the same or different if they were produced by cross-pollinating flowers of short pea plants with pollen from tall pea plants?

3 To study two traits at a time, Mendel first crossed one line of plants that bred true for two traits (round and yellow) with another line that bred true for different alleles of the same two genes. F_1 flowers were then self-pollinated and the F_2 peas produced as before. If you counted 16 of the F_2 peas, how many would you expect to be the doubly dominant phenotype, round and yellow? If you counted 1600 F_2 peas, how many round and yellow ones would you expect? Might your actual results deviate from these expectations? Comment on the difference between large and small sample sizes.

4 Why are certain traits studied in some species and not in others? Why were pea plants a good choice for Mendel's experiments?

The Chromosomal Basis of Inheritance Explains Mendel's Hypotheses

Notice that some of Mendel's assumptions raise questions that Mendel himself did not answer:

1. Where are the genes located?
2. Why do the genes exist in pairs?
3. Why do different traits assort independently?
4. What are the genes made of?

Figure 2.5

Structure of a nucleated cell. This particular cell is from an onion root tip.

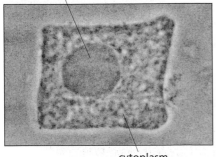

nucleus

cytoplasm

Answers to the first three of these questions were suggested by a young American scientist, Walter Sutton, who read about the rediscovery of Mendel's work in 1900. By this time, it was already well known that the cells that make up the bodies of plants and animals contain a central portion called the **nucleus** and a surrounding portion called the **cytoplasm** (Figure 2.5; for more on cell structure see Chapters 6, 10, and 12). Plants and animals are able to develop and grow because their cells divide. One cell divides to become two cells. Scientists who looked at dividing cells through their microscopes saw that the division of the cytoplasm is a very simple affair but that the dividing nucleus undergoes a complex rearrangement of the rod-shaped bodies called **chromosomes**. Chromosomes cannot usually be seen, except when cells divide. When chromosomes are visible during cell division, differences in their structure can be seen. Each chromosome has a narrow constriction (the centromere) that divides it into two portions, called the 'long arm' and the 'short arm'. By measuring the lengths of these arms, we can distinguish different chromosomes from one another.

The number of different chromosomes in a gamete is called the **haploid** number of chromosomes (*N*), with one of each type. All other body cells, called **somatic cells**, have twice as many chromosomes, called the **diploid** number of chromosomes (*2N*), with two of each size and shape (Figure 2.6). In diploid cells, each chromosome has a partner that matches it in overall length, in the lengths of the long and short arms, and in other features. Each pair of chromosomes that looks alike is called a **homologous pair**, and the chromosomes of diploid cells characteristically occur in such pairs. The two chromosomes in a homologous pair are physically separate (not attached), but they do pair up with one another during meiosis. Though they look alike under a microscope and carry the same genes, they often carry different alleles, and thus different hereditary information. Gametes, which are haploid cells, have one chromosome from each homologous pair.

Sutton noticed that eggs in most species are many times larger than sperm because they have a greater amount of cytoplasm (see Figure 9.7, p. 298). The nuclei of egg and sperm are approximately equal in size, and these nuclei fuse during fertilization. From these facts, Sutton reasoned as follows:

1. The genes are probably in the nucleus, not the cytoplasm, because the nucleus divides carefully and exactly, whereas the cytoplasm divides inexactly. Also, if genes were in the cytoplasm, the larger amount of cytoplasm in the egg would lead one to expect the egg's contribution always to be much greater than the sperm's, contrary to the observation that parental contributions to heredity are usually equal.

2. No other structures in cells, apart from chromosomes, are known to exist singly in gametes and twice as numerously in somatic cells. The known arrangement and movement of chromosomes exactly parallels what Mendel had postulated for genes. Genes must be located on chromosomes.

3. Mendel's genes assort independently because they are located on different chromosomes. However, there are only a limited number of chromosomes (4 pairs in fruit flies, 23 pairs in humans), but hundreds or thousands of genes. Sutton predicted that Mendel's law of independent assortment would apply only to genes located on different chromosomes. Genes located on the same chromosome would be inherited together as a unit, a phenomenon now known as linkage.

Sutton's idea that genes are located on chromosomes came to be called the chromosomal theory of inheritance. To understand the chromosomal theory of inheritance and Sutton's prediction that some pairs of traits would not follow Mendel's law of independent assortment, we must understand more about chromosome movements in dividing cells.

CONNECTIONS CHAPTERS 6, 9, 10, 12

Figure 2.6

Chromosomes in a gamete and a somatic cell of a species having two chromosomes in gametes (*N* = 2) and two pairs of chromosomes in somatic cells (*2N* = 4).

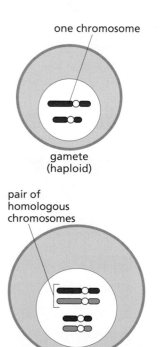

one chromosome

gamete (haploid)

pair of homologous chromosomes

somatic cell (diploid)

Mitosis

In mitosis, a somatic cell divides in two in a way that leaves each new cell with a diploid set of chromosomes, no more and no less. Each offspring cell that results from mitosis is thus genetically identical to the beginning cell.

The example used in Figure 2.6 shows a cell with two pairs of chromosomes, for a total diploid number of four. Prior to cell division, the cell enters a stage called **interphase**. The cell enlarges in size, increasing its cytoplasm. It also makes a copy of each of its chromosomes, so that the cell has twice its diploid number of chromosomes (eight chromosomes in our example). The relationship between duplicated chromosomes and homologous chromosomes is shown in Figure 2.7. As shown in Figure 2.8, during interphase none of the chromosomes are visible in the nucleus.

Mitosis then proceeds in four stages: prophase, metaphase, anaphase, and telophase, shown in Figure 2.8.

- With the start of **prophase**, the chromosomes shorten and thicken and thus become visible under a microscope. Each chromosome is now attached to its new duplicate, which was made during interphase. By the end of prophase, the membrane that surrounds the nucleus breaks down.

- During **metaphase**, each chromosome, attached to its duplicate, lines up along the center of the cell. The other chromosome of each type is also attached to its newly synthesized duplicate, and also goes to the midline of the cell. Homologous chromosomes are not in contact with one another, but are randomly spread along the midline, each attached to its duplicate.

- In **anaphase**, each chromosome separates from its attached duplicate, and the chromosome and its duplicate begin to move to opposite ends of the cell. The other chromosome of each type does the same, separating from its attached duplicate, and this homologous chromosome also moves to the opposite end of the cell from its duplicate.

- In **telophase** the chromosomes complete their move to the two ends of the elongated cell. Because each chromosome and its duplicate (made in interphase) have moved to opposite ends of the cell, each end of the cell now has the diploid number of chromosomes composed of two chromosomes of each homologous pair. At the end of telophase, a nuclear membrane reappears around each diploid set of chromosomes, forming two nuclei, one at each end of the cell, and completing mitosis.

After mitosis is finished, the cytoplasm divides between the two nuclei, finalizing the division of one diploid cell into two diploid cells.

Meiosis and sexual life cycles

Recall that pea plants reproduce sexually. In all species that reproduce sexually, two haploid gametes join to form a diploid cell called a **zygote** that can become a new individual organism. The gametes are produced in specialized cells in a process called meiosis.

Meiosis. Meiosis proceeds similarly to mitosis, but with the addition of a second round of cell division. This causes a different end result from

Figure 2.7

Comparison of a single chromosome, a duplicated chromosome, and a homologous pair of duplicated chromosomes.

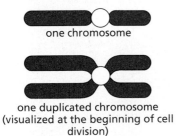

one chromosome

one duplicated chromosome
(visualized at the beginning of cell division)

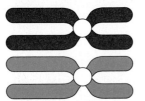

one pair of homologous chromosomes
with each of the chromosomes
duplicated

that in mitosis: in meiosis, four cells are produced, each with a haploid number of chromosomes.

As in mitosis, each duplicated chromosome thickens and becomes visible; then the nuclear membrane breaks down and the chromosomes line up along the midline of the cell. At this step, however, there is a major difference from the mitotic process. In mitosis, the homologous pairs of chromosomes come to the midline separately. In the first division of meiosis, as shown in Figure 2.9, the two homologous chromosomes, each with its attached duplicate, come together at the midline as quadruplicates. From each homologous pair, one chromosome (with its attached duplicate) then goes to each end of the cell. Nuclear membranes form and the cell separates into two offspring cells, completing the first

Figure 2.8

Mitosis. The cell divides in such a way that each of the two offspring cells contains the same number of paired chromosomes as the parent cell did. The drawings on the right show a cell like that in Figure 2.6, with a diploid number of $2N = 4$ chromosomes (two pairs). The photographs on the left, taken through a microscope, show an actual cell with more chromosomes.

INTERPHASE

The cell increases in size and each chromosome is duplicated, although they are not visible in the nucleus at this stage. The cell now has twice its diploid number.

nuclear membrane

PROPHASE

The duplicated chromosomes thicken and shorten, becoming visible under a microscope as attached pairs. The membrane surrounding the nucleus breaks down near the end of prophase.

chromosome attached to its duplicate

nuclear membrane fragments

METAPHASE

Each chromosome and its attached duplicate line up along the midline of the cell.

chromosomes separating from duplicates

ANAPHASE

Each chromosome separates from its duplicate; the chromosome and its duplicate move toward opposite ends of the cell.

diploid set of chromosomes

cytoplasm dividing

TELOPHASE

A complete diploid set of chromosomes arrives at each end of the cell, a nuclear membrane reassembles around each set, forming two nuclei, and the cytoplasm begins to separate in two.

nuclear envelope reassembling

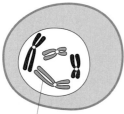

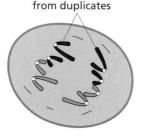

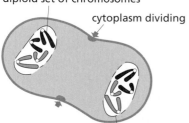

time = 0 min

time = 250 min

time = 279 min

time = 315 min

Figure 2.9

Meiosis. A cell with a diploid number of chromosomes ($2N = 4$ in this example) divides twice to produce four haploid gametes (here with two chromosomes each).

meiotic division. At this stage of meiosis, each new cell contains one chromosome of each homologous pair, and each chromosome remains attached to its new duplicate, synthesized at the beginning of meiosis. Each chromosome and its duplicate have come from only one of the members of a homologous pair. By contrast, at the end of mitosis, each cell has one copy of both chromosomes of each homologous pair. The first meiotic division is followed by a second division in each of the two new cells. The nuclear membranes break down once again and each chromosome now separates from its attached duplicate. Nuclear membranes form and the cells divide. The final result is four cells. Because there was no further replication of chromosomes between the first and second meiotic divisions, each of the four new cells, or gametes, ends up with the haploid number of chromosomes.

Meiosis, therefore, has three major differences from mitosis: (1) it has an additional cycle of cell division; (2) the end result is four cells each containing a haploid number of chromosomes; (3) during metaphase, the homologous chromosomes align together forming quadruplicates at the midline. This latter difference enables the additional feature of **crossing-over** to occur. One chromosome from each attached pair can exchange part of itself with the corresponding part on the homologous attached pair, as indicated by the exchange of red and blue segments in Figure 2.9. This phenomenon allowed the discovery of gene linkage, which is discussed below.

Sexual life cycles. After the haploid gametes are produced, they can be brought together by sexual reproduction, forming a diploid zygote, a process investigated by the Austrian zoologist Theodor Boveri and the American zoologist Ernest E. Just (see Figure 1.4). Figure 2.10 shows a sexual life cycle. The gametes fuse to form a zygote; then the zygote in most

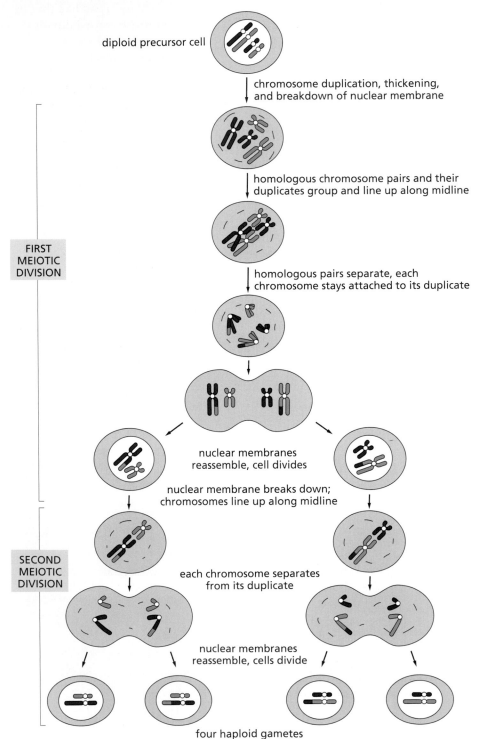

diploid precursor cell

chromosome duplication, thickening, and breakdown of nuclear membrane

homologous chromosome pairs and their duplicates group and line up along midline

homologous pairs separate, each chromosome stays attached to its duplicate

FIRST MEIOTIC DIVISION

nuclear membranes reassemble, cell divides

nuclear membrane breaks down; chromosomes line up along midline

SECOND MEIOTIC DIVISION

each chromosome separates from its duplicate

nuclear membranes reassemble, cells divide

four haploid gametes

species undergoes repeated mitosis to become a multicellular adult. Each somatic cell contains the full diploid set of chromosomes and thus all somatic cells are genetically identical. As the organism develops, different somatic cells specialize for different functions (see Chapter 12), although they all retain the full diploid set of chromosomes. Some of these cells specialize to undergo meiosis, producing new haploid gametes, and completing the sexual life cycle. Most, but not all, multicellular organisms have a sexual life cycle, alternating between haploid gametes and diploid somatic cells.

Gene linkage

Sutton predicted that independent assortment would not apply to all pairs of genes and that some genes would be found to be linked. Other investigators quickly confirmed his prediction. A British geneticist, William Bateson, described linked genes in a cross of varieties of garden peas. Other investigators soon discovered similar examples in fruit flies (*Drosophila*), corn (*Zea mays*), and other species, showing that Mendel's laws and Sutton's theory were not unique to peas or to plants. Figure 2.11 shows a cross revealing that two genes are linked in corn. Plants of genotype *CCSS* (colored, full seeds) crossed with *ccss* plants (colorless, shrunken seeds) produced all seeds of the colored, full phenotype among the F_1. In this experiment, the F_1 plants (*CcSs* heterozygotes) were not self-fertilized as in Mendel's experiments such as in Figures 2.3 and 2.4. Rather, the F_1 heterozygotes were fertilized by plants of the doubly recessive *ccss* genotype. This type of cross, in which an F_1 is crossed with one of the parental types, is called a backcross. In the backcross offspring, most of the seeds that were colored were also full, and most of the seeds that were colorless were also the shrunken phenotype. The genes for these two traits were said to be linked. That they were linked, rather than independently assorting, could be explained by assuming that the genes for the two traits were on the same chromosome.

Figure 2.10

Sexual life cycles. In sexual reproduction, haploid gametes join by fertilization to form a new diploid individual with one of each pair of homologous chromosomes coming from each parent. In multicelled organisms the diploid zygote divides by mitosis to form the adult organism. Each of the somatic (body) cells contains a set of chromosomes the same as that in the zygote. In a male organism, meiosis produces sperm, as shown. In a female organism, meiosis produces eggs.

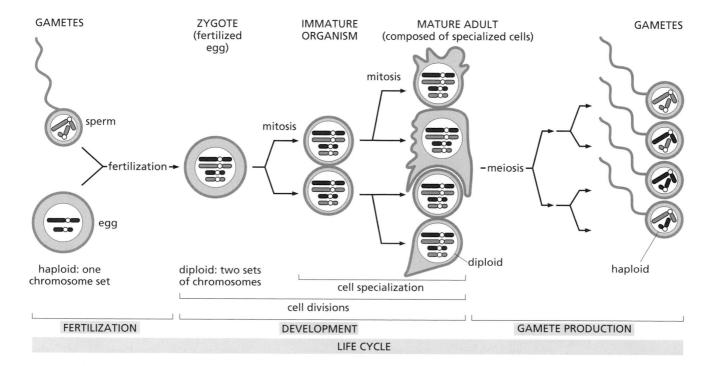

Most of the progeny in Figure 2.11 show the parental linked phenotypes. However, notice also in Figure 2.11 that small numbers of the backcross progeny were not of the colored and full or colorless and shrunken seed phenotypes but instead had colored and shrunken or colorless and full seeds. These atypical plants had new (nonparental) combinations of the phenotypes for the traits; the underlying recombinant genotypes were inferred to have arisen from the process of crossing-over in which chromosomes break and recombine by exchanging pieces. Some microscopists thought they had observed X-shaped arrangements of the chromosomes during meiosis (see Figure 2.11) that looked like crossing-over, but many scientists were unsure.

Confirmation of the chromosomal theory

Sutton's theory that genes are located on chromosomes had to wait three decades for confirmation by other researchers in genetics. In 1931, Harriet Creighton and Barbara McClintock confirmed the chromosomal theory of inheritance in corn; later that year Curt Stern observed the same thing in fruit flies. Creighton and McClintock used plants whose chromosomes had structural abnormalities on either end, enabling them to recognize the chromosomes under the microscope. What they were able to demonstrate was that *genetic recombination* (the rearranging of genes) *was always accompanied by crossing-over* (the rearranging of chromosomes). McClintock went on to discover genes that move from place to place, the so-called transposable elements or jumping genes, a discovery for which she later received the Nobel Prize.

The frequency of recombination between linked genes is very roughly a measure of the distance between them along the chromosome. Recombination between closely linked genes (those close together on a chromosome) is a rare event, while recombination between genes farther apart is more frequent. By making crosses between individuals having different alleles for pairs of linked genes, geneticists were able to determine the linear arrangement of many genes and the approximate distances between genes on the chromosomes of many species.

An interesting footnote to Mendel's work was provided in 1936 by the British geneticist R.A. Fisher, who noticed that garden peas have seven pairs of chromosomes. Mendel had picked seven traits that assorted independently because each is on a different chromosome! Because the probability of this

Figure 2.11

A cross between pure-line corn plants having different alleles for two linked genes.

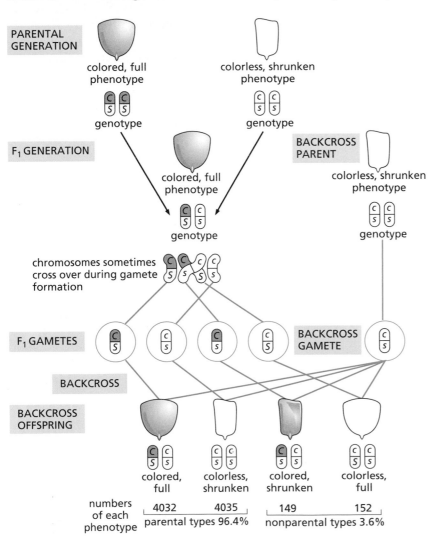

numbers of each phenotype	4032	4035	149	152
	parental types 96.4%		nonparental types 3.6%	

occurring by chance is extremely remote, Fisher concluded that Mendel may have studied many more traits and only reported the results for the seven independently assorting traits whose inheritance he could understand (see Figure 2.1).

THOUGHT QUESTIONS

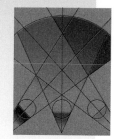

1 In a cross between pea plants of genotype *YYRR* (yellow, round seeds) and *yyrr* (green, wrinkled seeds), the F_1 plants are all *YyRr*. Make a series of large drawings showing the movements during mitosis of the chromosomes carrying these genes in cells of an F_1 heterozygous plant.

2 For the same F_1 heterozygous plants (*YyRr*) described in question 1, make a series of drawings showing the way in which the chromosomes separate in the first division of meiosis and in the second division of meiosis. Label each diagram with the symbols *Y*, *y*, *R*, and *r* to show how all four types of gametes originate.

3 Would you expect the cross *YYRR × yyrr* to give the same results as *YYrr × yyRR*? Why or why not?

Genes Carried on Sex Chromosomes Determine Sex and Sex-linked Traits

Not long after Sutton proposed that chromosomes carry genes, chromosomes and their abnormalities were studied in fruit flies and in people. Much has been learned about human genetics from the study of chromosomes. In particular, some unusual conditions are associated with alterations of the chromosomes.

During mitosis and meiosis, the double-helical DNA becomes further coiled around supporting proteins and then coiled around itself until the structure becomes thick enough to be visible under a light microscope. This structure is generally what we think of as a chromosome. During mitosis, when the chromosomes are visible, cells can be squashed onto a glass slide. Photographs of the chromosomes can be made through a microscope and the photographs can then be cut up. Using chromosome lengths and banding patterns, geneticists can line up the photos, putting the homologous chromosome pairs together. Such an arrangement is called a **karyotype**. The karyotypes of a female human and a male human are shown in Figure 2.12. The chromosomes of nearly every person can be arranged in a karyotype similar to one of those shown in Figure 2.12, with

Figure 2.12

Human female and male karyotypes.

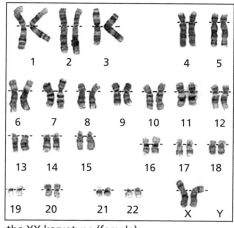

the XX karyotype (female)

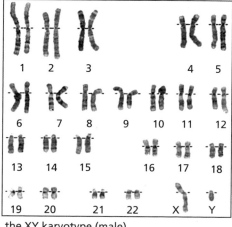

the XY karyotype (male)

46 chromosomes arranged in 23 pairs. Of the 23 pairs of chromosomes, 22 pairs are the same in both sexes (the **autosomal chromosomes**). The two chromosomes in each of these pairs are homologous, meaning that they carry the same set of genes (although possibly two different alleles of any gene). The 23rd pair of chromosomes (labeled X and Y in Figure 2.12) are called the **sex chromosomes** because they differ in males and females and have a role in determining a person's sex. The X and Y chromosomes are only partly homologous: they do pair during meiosis, but only some of their genes are present on both the X and Y chromosomes.

Sex determination

Human females typically have two similar sex chromosomes, symbolized as XX. Human males typically have one X chromosome and one different sex chromosome, the Y chromosome, and thus are symbolized as XY (see Figure 2.12).

Not all human females are XX and not all human males are XY. There are unusual situations in which a cross-over during meiosis between an X and a Y chromosome is followed by an exchange of chromosome pieces. Approximately 1 in 20,000 normal males is chromosomally XX, but one of his X chromosomes contains a small piece of the Y chromosome. About the same frequency of normal females are chromosomally XY but are *missing* the same small piece of Y chromosome. One such XY female had 99.8% of the Y chromosome, indicating that a male-determining factor, or testis-determining factor, was located in the 0.2% portion of the Y chromosome that she did not have. Examination of this 0.2% portion of the Y chromosome led to the identification of a gene now called *SRY*, hypothesized to induce development as a male.

The *SRY* gene helps determine sex by producing a protein, the SRY protein, that allows the production of another protein, the testis-determining factor, that converts progesterone into testosterone (Figure 2.13). Both progesterone and testosterone are hormones, small molecules by means of which cells communicate with one another. Testosterone acts on cells to induce the development of male organs in embryos. If progesterone is not converted to testosterone, it is instead converted to estrogen, which triggers the development of female organs. Embryos with an *SRY* gene thus become males, and embryos without an *SRY* gene become females. However, there are also case of *SRY*-negative individuals who are phenotypically male, demonstrating that other genes in addition to *SRY* are involved in sex determination.

Sex-linked traits

Very few genes are, like *SRY*, located on the Y chromosome. Many more genes are located on the X chromosome. Genes that are on the X chromosome and not on the Y chromosome are said to be **sex-linked**. Females can be either homozygous or heterozygous for sex-linked traits because they have two X chromosomes and therefore two alleles of every sex-linked gene. If the allele for a trait is recessive then a heterozygous woman has the dominant phenotype but is said to be a carrier of the trait. Because a male has only a single X chromosome, he has only a single allele for each sex-linked trait, and this allele determines his phenotype for that trait. The inheritance of a recessive sex-linked trait,

Figure 2.13

The relationship (simplified) between the *SRY* gene and sexual development in humans. TDF, testis-determining factor.

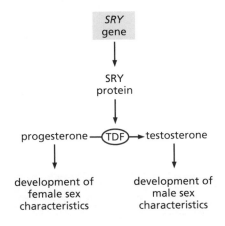

red–green colorblindness, is shown in Figure 2.14. If a woman is heterozygous for this trait, she shows the dominant phenotype and is not colorblind. However, a male who carries the recessive allele for colorblindness on his X chromosome will be colorblind. Note that a single sex-linked allele is phenotypically displayed in a male regardless of whether it is dominant or recessive.

Females generally possess two X chromosomes, but in a given cell only one of them has active genes that make a product or express a phenotype. Females thus express one phenotype (from their mother's X chromosome) in some cells and another phenotype (from their father's X chromosome) in other cells. This expression of two phenotypes at the cellular level is called **mosaicism**. All females who are heterozygous are mosaics for X-linked genes. For example, in a female heterozygous for an X-linked allele for colorblindness, patches of cells within the retina of the eye (see Chapter 13, p. 478) cannot respond to color. Other patches of cells, which express the normal allele on the other X chromosome, respond normally.

Chromosomal variation

Most humans have 46 chromosomes, consisting of a pair of sex chromosomes and 22 other pairs of chromosomes. Other chromosomal patterns have multiple consequences, called syndromes, usually named after the physicians who identified them. Several such variations of the sex chromosome number are known. For example, the XXY chromosomal type (Figure 2.15A) results in **Klinefelter syndrome**; persons with this condition have male phenotypes but are sterile. Some of the symptoms of Klinefelter syndrome can be successfully treated with hormones. In contrast, **Turner syndrome** results from the XO chromosomal type, in which only one X chromosome is present, the O representing its missing partner (Figure 2.15B). Persons with Turner syndrome develop as females; however, their ovaries do not produce female hormones. Puberty does not take place and gametes do not develop, resulting in infertility. The infertility cannot be overcome at present, but the other symptoms of Turner syndrome can now be treated hormonally with much success.

Turner and Klinefelter syndromes are believed to result from the same cause, an abnormal meiosis of the sex chromosomes. In the abnormal meiotic division, the two sex chromosomes fail to separate, resulting in some egg cells' having two of the mother's X chromosomes and some having none. Abnormal separation of chromosomes during gamete production has in fact been observed, partly confirming the hypothesized series of events shown in Figure 2.16.

Figure 2.14

Inheritance of red–green colorblindness, a sex-linked recessive trait.

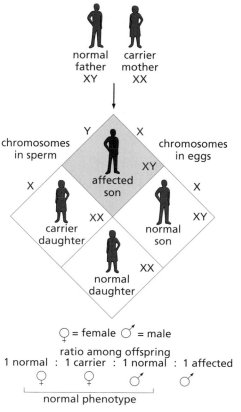

Figure 2.15

Two variations in human X and Y karyotypes.

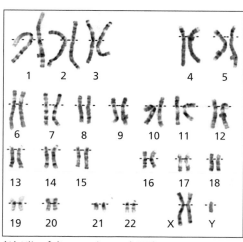

(A) Klinefelter syndrome (XXY)

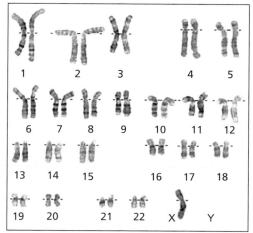

(B) Turner syndrome (XO)

Also supporting the hypothesis is the very rare XXX chromosomal abnormality; most XXX females are mentally retarded and sterile. The Y-only type of embryo, also predicted by the hypothesis, has never been observed, presumably because it dies at a very early stage of development. Abnormal chromosome separation can also take place during the formation of sperm cells, resulting in XY sperm and O sperm. When these fertilize a normal X egg, again either Klinefelter syndrome (XXY) or Turner syndrome (XO) can result.

There are also variations in number for chromosomes other than the sex chromosomes. The most common of these is associated with **Down syndrome**, marked by facial characteristics (including an epicanthic fold over the eyes), heart abnormalities, and a variable amount of mental retardation. Down syndrome usually results from an extra chromosome 21 (Figure 2.17). Other chromosome abnormalities are less common. For example, Patau syndrome (three copies of chromosome 13) results in severe mental retardation, a small head, extra fingers and toes, and usually death by one year of age. In addition to extra chromosomes, part or all of a chromosome can be missing. The *cri du chat* syndrome is caused by deletion of the short arm of chromosome 5 and results in a small head, a catlike cry, and mental retardation.

In addition to these changes in chromosome number, there are several kinds of large-scale changes involving chromosome fragments. Chromosome fragments may become duplicated (repeated); they may become attached at a new location, possibly on a different chromosome; or they may be lost entirely. A chromosome fragment may also be turned end-to-end and reinserted at its former location. Of these four types of chromosomal changes, end-to-end inversions are the most frequent, and have the most limited effects, while the other three types may result in nonviable phenotypes when the rearranged fragments of DNA are long.

Figure 2.16

Abnormal meiosis during egg production, showing how certain chromosomal variations may arise.

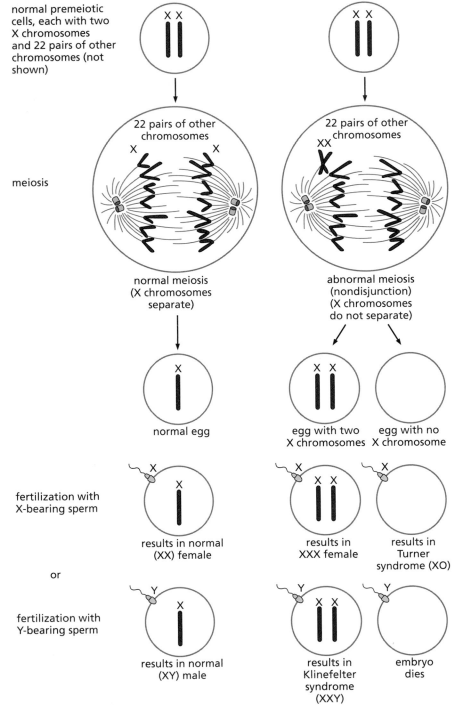

normal premeiotic cells, each with two X chromosomes and 22 pairs of other chromosomes (not shown)

meiosis

22 pairs of other chromosomes

22 pairs of other chromosomes

normal meiosis (X chromosomes separate)

abnormal meiosis (nondisjunction) (X chromosomes do not separate)

normal egg

egg with two X chromosomes

egg with no X chromosome

fertilization with X-bearing sperm

results in normal (XX) female

results in XXX female

results in Turner syndrome (XO)

or

fertilization with Y-bearing sperm

results in normal (XY) male

results in Klinefelter syndrome (XXY)

embryo dies

Social and ethical issues regarding sex determination

As we have seen, the determination of a person's sex is not always unambiguous. There is a strong societal expectation that each person should be categorized as either male or female, yet 17 out of 1000 people cannot be so clearly categorized. Some are people with XY chromosomes but female anatomy or XX chromosomes but male anatomy. Some people have partly male and partly female anatomy.

Realization that there are XX males and XY females forced the International Olympic Committee to reexamine the stipulation that only XX individuals could compete in female sporting events. If chromosome appearance is not sufficient to determine which individuals are male, should the presence of the *SRY* gene or the hormone testosterone be used as the test? Either turns out to be problematical. Females also have testosterone, although generally in smaller amounts than males. Also, there are rare XY individuals who have the functional *SRY* gene and produce male concentrations of testosterone but are nevertheless phenotypically female because they lack a functional allele of a different gene, the gene for a protein that allows cells to respond to testosterone. Cells cannot respond to a hormone during development or during adult life unless they possess receptor molecules to bind that hormone. The International Olympic Committee now uses the presence or absence of the functional *SRY* gene to decide the sex of Olympic athletes, but clearly sex is not determined by one single gene. Other genes, including genes such as the testosterone receptor gene on chromosomes other than the sex chromosomes also have a role in sexual development. Thus, even though we refer to X and Y as the sex chromosomes, many genes on many other chromosomes are also involved.

Clearly male and female are not either/or categories. There are persons who fall between, either because of chromosomal variation or variation in alleles at particular genes such as *SRY* or the testosterone-receptor gene. All of these different variations are called intersex. In many cases, babies born with ambiguous genitals have been treated, either with

Figure 2.17

Down syndrome.

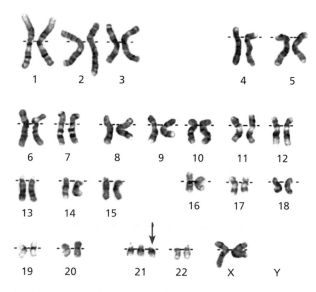

The karyotype with an extra chromosome 21 associated with most cases of Down syndrome

A child with Down syndrome

hormones or surgically, to make them conform to one sex or the other. Sometimes this has been done without the informed consent of the parents, raising further ethical issues.

People with variations in the relation between their karyotype and their phenotype are not defective individuals; they just do not fit a previously held view of what determines the sex of an individual. More recently, the research of Anne Fausto-Sterling and others has been calling into question the practice of assigning everyone to one of two categories of male or female. Why should categories that are essentially social constructions (derived by society) be placed over the much greater range of variation that exists in biological reality?

THOUGHT QUESTIONS

1 Sports officials have repeatedly asked female athletes to submit to testing to confirm their femaleness, but comparable proof is seldom demanded of males. Why do you think this disparity exists? Can you think of other ways besides sex in which athletes might be classified? Could we use age as a criterion? Could we use fat-to-muscle ratios?

2 Individuals with Klinefelter syndrome or Turner syndrome are fully functional aside from their infertility, yet many are 'treated' with hormones to give them a less indeterminate appearance of secondary sexual characteristics such as breasts. Why

do humans tend to think of such differences as 'abnormalities' needing 'treatment' rather than as normal variation within a trait?

3 Should the sexual phenotype include more than two categories, categories other than male and female? If it did, would a person who was intersex want to reveal that on any of the many forms where we are asked to declare which sex we are? Why or why not?

4 Are biological categories neutral and objective? That is, can they be free of the values placed on them by society?

The Molecular Basis of Inheritance Further Explains Mendel's Hypotheses

We now address the last of the four questions posed earlier. What are genes made of? In other words, what chemical substance transfers genotypes from parents to offspring?

The story begins with a curious experiment carried out in 1928 by Frederick Griffith, a U.S. Army medical officer attempting to develop a vaccine against pneumonia. Griffith worked with two strains of bacteria that differed in their outer coats. Strain S had an outer coat that gave a smooth appearance when masses of bacteria (called colonies) were grown on agar in dishes. The smooth-colony bacteria were virulent, which means that the bacteria cause a disease (pneumonia in this case). Strain R of the same bacterial species lacked the outer coat, which gave the colonies of this strain a rough appearance, and they were nonvirulent. (S stands for 'smooth,' R for 'rough.') When strain S was injected into mice, all the mice died of pneumonia. Strain R, when injected, did not kill mice, nor did bacteria of strain S that had been killed by heat. The surprising result, shown in Figure 2.18, was that a mixture of live bacteria of strain R and heat-killed bacteria of strain S did kill mice. Furthermore, living S bacteria were isolated from all mice that died this way. Griffith interpreted this experiment as showing that something in the dead S bacteria had somehow transformed the living R bacteria into virulent S bacteria, and this type of change came to be known as **transformation**. The bacteria had been altered genetically, not just

Figure 2.18

Griffith's experiment demonstrating hereditary transformation in bacteria.

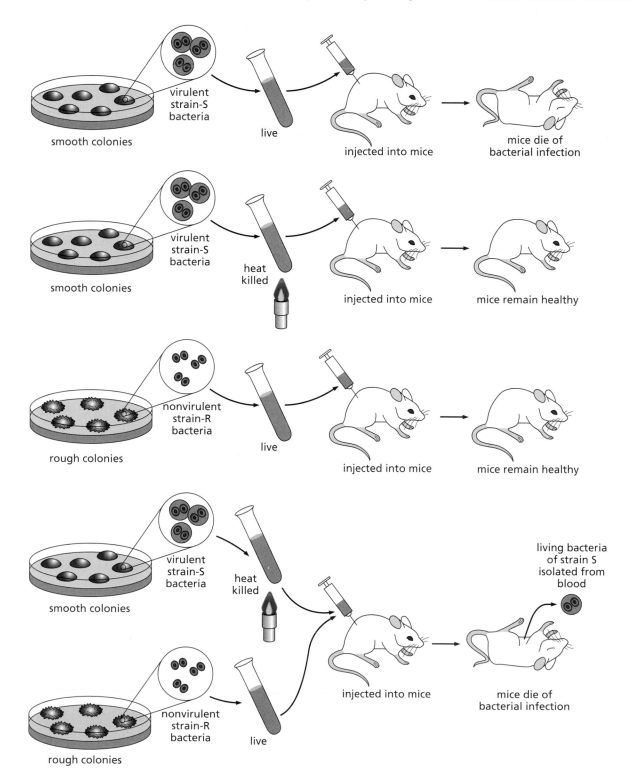

phenotypically. A change that was only phenotypic would not be passed on to future generations, but Griffith demonstrated that descendents of the transformed bacteria were also of strain S and continued to kill mice.

What Griffith had done in his experiment was transfer a genetic trait from one bacterial strain to another. What was the chemical substance that had been transferred? Griffith's use of bacteria as an experimental species and the unequivocal evidence that genetic material had been transferred in his experiment is significant because it was the background for work two decades later that demonstrated what the chemical substance is.

This section describes the research that revealed the chemical substance of the gene, the composition of this substance, and how it is replicated. We also learn how the chemical substance directs the synthesis of proteins and how changes in the substance can produce changes in proteins.

DNA and genetic transformation

From the beginning of the twentieth century into the 1940s, most researchers thought that proteins were the most likely candidates to be the chemical substance of genes. They thought this because proteins were known to be complex and varied, while most other molecules were thought not to be. In an attempt to discover whether protein was indeed the chemical substance, three bacteriologists, Oswald Avery, Colin MacLeod, and Maclyn McCarty, in 1944 conducted a chemical study of the bacteria that Griffith had injected into his experimental mice. First, they were able to show that a chemical extract of heat-killed S bacteria transformed strain R into strain S. They then separated the strain-S extract into different fractions, each containing different types of chemical molecules. They found that the fraction containing the **nucleic acids** transformed strain R into strain S, but that the protein fraction did not.

Finally, they distinguished between the two major types of nucleic acids (DNA and RNA) by using enzymes. **Enzymes** are biological molecules (nearly always proteins) capable of speeding up chemical reactions without themselves getting used up in those reactions. Enzymes control many biological processes, and the action of most enzymes is very restricted in that specific enzymes act on specific types of molecules. The enzyme deoxyribonuclease (DNase) specifically breaks down DNA. DNase treatment destroyed the ability of the strain-S extract to transform the strain-R bacteria, demonstrating that the chemical that carries the genetic material is, in fact, **deoxyribonucleic acid (DNA)**. On the other hand, the enzyme ribonuclease, which breaks down RNA, had no effect on the ability to transform; **ribonucleic acid (RNA)** is therefore not the genetic material.

The discovery of Avery and his co-workers did not get the attention it deserved. Doubters still remained. It took about a decade for many scientists to accept that DNA is the genetic material. Experimental work on the problem had meanwhile shifted from using bacteria to using viruses.

In 1952, two American virologists, Alfred Hershey and Martha Chase, published the results of a landmark experiment that confirmed the finding that DNA, not protein, is the genetic material. For their experiment, Hershey and Chase used a virus that infects bacteria and reproduces within them. They infected bacteria with the viruses and studied the viral offspring. It was known that this virus consisted only of protein and DNA. Was it the protein or the DNA that carries the genetic material?

In preparation for their experiment, Hershey and Chase grew some viruses in a medium containing radioactive phosphorus (^{32}P) and others in a medium containing radioactive sulfur (^{35}S), atoms that the virus needs to make new viruses. Because DNA contains phosphorus but protein does not, the new viruses grown with ^{32}P had radioactive phosphorus in their DNA but no radioactivity in their protein. In contrast, the proteins, but not the DNA, of viruses grown with ^{35}S were radioactive because proteins contain sulfur but DNA does not. Because radioactivity is easily detected, material prepared in this way is said to be radioactively labeled. Hershey and Chase applied the radioactive labels so they would be able to 'see' what happened to the viral DNA and protein when the viruses infected bacteria.

Hershey and Chase exposed *Escherichia coli* bacteria to the radioactively labeled viruses for long enough to permit the viruses to infect the bacteria. Part of the virus enters the cells that they infect and directs the replication of more viruses, producing thousands of new virus particles and eventually killing the bacteria and breaking them open to release the viruses (Figure 2.19A). Was the injected material that was carrying the

Figure 2.19

The Hershey–Chase experiment. This experiment confirmed DNA as the genetic material.

(A) Pattern of viral infection of *E. coli* bacteria.

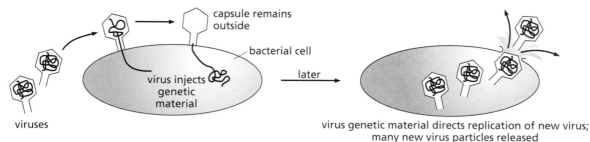

(B) Viruses grown with radioactive sulfur.

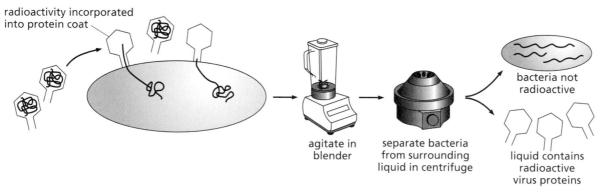

(C) Viruses grown with radioactive phosphorus.

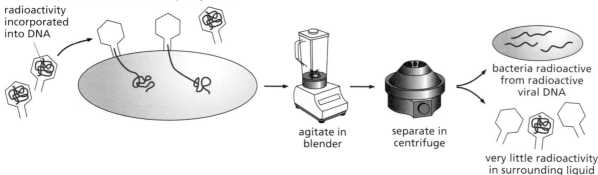

viral genotype DNA or protein? Hershey and Chase devised a way to interrupt the viral cycle after the infection period by using a kitchen blender to knock the attached virus capsules off the bacterial surfaces. These detached capsules could easily be separated from the bacteria by spinning the mixture in a centrifuge. Viral material that had been injected into the bacteria continued the process of viral reproduction, eventually rupturing and killing the bacteria. When ^{35}S-labeled viruses were used, the radioactive proteins remained outside the bacteria (Figure 2.19B); the viruses eventually released were not radioactive so they had not used radioactive protein to replicate themselves. However, when the ^{32}P-labeled viruses were used, the radioactive DNA entered the bacteria, making the bacteria radioactive. The viral offspring released when these bacteria broke open were also radioactive (Figure 2.19C), showing that they had used some of the radioactive DNA in their reproduction.

This experiment upheld the hypothesis that the genetic material of the virus was DNA. It falsified the hypothesis that the viral genetic material was made of protein. DNA had been identified as the genetic material, but its composition and structure were still a mystery.

The structure of DNA

Chemical breakdown of DNA into its parts showed that it was made of phosphate groups, deoxyribose (a sugar), and four nitrogen-containing bases called adenine (A), guanine (G), thymine (T), and cytosine (C) (Figure 2.20A). Biochemists soon realized that the deoxyribose could form a middle link between the phosphate groups and the nitrogenous bases, creating units called **nucleotides** (Figure 2.20B). Beyond this, the structure of DNA was unclear, and it was not at all obvious how the structure could give DNA the ability to carry genetic information.

Chargaff's rules. Some researchers suggested that the order of bases in DNA repeated regularly: AGTCAGTCAGTC.... If this were true, then the amounts of the four nitrogenous bases should be equal: there should be 25% of each of the bases in a DNA molecule. To test this hypothesis, biochemist Erwin Chargaff of Columbia University took DNA from various sources, broke it down using enzymes, and measured the relative amounts of the four nitrogenous bases. His findings were as follows.

1. The proportions of the four nitrogenous bases are constant for all cell types within a species. For example, all human cells contain about 31% adenine (A), 19% guanine (G), 31% thymine (T), and 19% cytosine (C), regardless of whether the DNA is from brain cells, liver cells, kidney cells, or skin cells.

2. Although the proportions are constant within a species, they differ from one species to another. All humans, for example, have the same proportions of the four bases. Proportions are different in rats and in bread molds, but all rats have the same proportions as one another and so do all bread molds.

3. The most unexpected finding, and the hardest to explain, was that the proportion of adenine was always the same as the proportion of thymine (within the limits of experimental error), and the levels of cytosine and guanine were also equal. These findings (symbolized as A = T and G = C) became known as Chargaff's rules.

Figure 2.20

The nucleotides of DNA.

(A) Two of DNA's nitrogenous bases, adenine (A) and guanine (G), are larger than the other two, thymine (T) and cytosine (C).

(B) With the addition of deoxyribose (a five-sided sugar molecule) and a phosphate group, each of these nitrogenous bases forms a nucleotide.

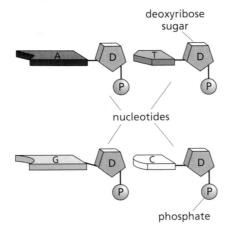

The three-dimensional structure of DNA. In 1953 James Watson and Francis Crick, two geneticists working in Cambridge, England, proposed a structure for DNA that explains Chargaff's rules and also explains how DNA carries genetic information. They did this with the help of some X-ray diffraction information obtained from the Cambridge laboratory of biochemist Maurice H.F. Wilkins, with whom they later shared a Nobel Prize. The data from Wilkins's laboratory were gathered and interpreted by Rosalind Franklin, a biochemist, whose contribution never received the recognition it deserved (see our Web site, under Resources: Franklin).

The X-ray diffraction information suggested certain dimensions and distances for the repetition of structures within the DNA molecule. From this information, Watson and Crick constructed the model of DNA structure summarized in the following list:

1. Each phosphate in a DNA molecule is attached to a deoxyribose sugar, which in turn is attached to a nitrogenous base. The three parts together constitute a nucleotide (see Figure 2.20B).

2. The phosphate group of one nucleotide is also connected to the deoxyribose sugar of the next nucleotide. The alternation of phosphates and sugar units thus forms a backbone that holds the nucleotides together in a strand, with the nitrogenous bases pointing inward (Figure 2.21A).

3. Each strand is a linear sequence of bases (it does not branch) that, because of the angles of the chemical bonds, is twisted in the shape of a corkscrew (a helix).

Figure 2.21

The three-dimensional structure of DNA.

(A) Thousands of nucleotides are strung together by a sugar-phosphate backbone.

(B) Two strands of DNA twist around one another to form a double helix. A straightened portion of this double helix resembles a ladder with the paired (complementary) bases forming the rungs.

(C) Two nucleotide sequences running in opposite directions pair with one another, with each adenine (A) pairing with a thymine (T), and each guanine (G) pairing with a cytosine (C).

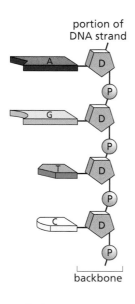

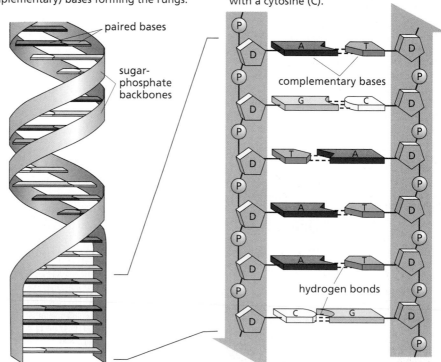

4. DNA has two strands wound around each other, forming a double helix, with the bases arranged in the interior, like steps in a spiral staircase (Figure 2.21B).

5. The strands run in opposite directions and are so arranged that an adenine on one strand is always paired with a thymine on the other strand, and vice versa. Also, cytosine on one strand is always paired with guanine on the other strand, and vice versa (Figure 2.21C). These pairings of matching (complementary) bases explain Chargaff's rules.

6. C and G fit together and T and A fit together because their shapes are complementary (see Figure 2.20). Because of these base pairings, each strand contains all the information necessary to determine the structure of the complementary DNA strand.

DNA replication

Watson and Crick's model for the structure of DNA led quickly to an understanding of the mechanism of **replication**, the process by which the cell makes copies of the DNA molecules. Before DNA replication, the two strands of the double helix unwind and separate from each other, as shown in Figure 2.22A. Notice the orientation of the 5-sided deoxyribose molecules in the orange-colored backbone, which shows that the two

Figure 2.22

DNA replication. The two strands run in opposite directions, as denoted by the arrowheads on the orange (old) backbones. The direction of synthesis is indicated by the arrowheads on the red (new) backbones. Note that either strand contains all the information needed to synthesize the other (complementary) strand.

(A) UNWINDING
The two strands of the DNA double helix separate.

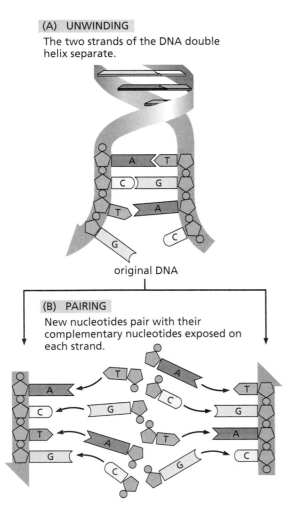

original DNA

(B) PAIRING
New nucleotides pair with their complementary nucleotides exposed on each strand.

(C) JOINING
Each new row of bases is linked into a continuous strand by joining adjacent sugars and phosphates. Each double helix contains one new strand and one original strand.

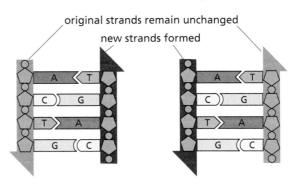

original strands remain unchanged
new strands formed

(D) THE REPLICATION FORK

early step   later step
replication fork moves

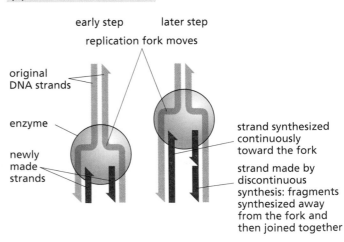

original DNA strands

enzyme

newly made strands

strand synthesized continuously toward the fork

strand made by discontinuous synthesis: fragments synthesized away from the fork and then joined together

strands run in opposite directions. After separating, the strands are bound by an enzyme (**DNA polymerase**), which actually begins the replication process. New strands of DNA are synthesized one nucleotide at a time, with one of the existing strands serving as a template (pattern to be copied). If the next unmatched base on the template is adenine (A), then a thymine (T) nucleotide (adenine's complementary base) is added to the growing new strand opposite the adenine. In like manner, G is added opposite C, C opposite G, and A opposite T (Figure 2.22B). The other existing strand is simultaneously acting as a second template and other complementary bases pair with it. The backbone of the new strand is formed by joining (bonding) the phosphate group of the incoming nucleotide to the deoxyribose sugar of the previous nucleotide (Figure 2.22C).

Recall that the two originally existing DNA strands run in opposite directions. Consequently, synthesis on the two template strands proceeds in opposite directions. One new strand of DNA is thus synthesized continuously, with the direction of synthesis running towards the replication fork, the place where the two original strands are coming apart. The other strand, however, is synthesized in the opposite direction, away from the replication fork. This strand must be synthesized in short fragments that are later joined together (Figure 2.22D).

THOUGHT QUESTIONS

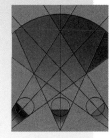

1 Before either mitosis or meiosis, the DNA in each chromosome is replicated, forming two identical chromosomes that are attached as a pair at the beginning of mitosis or meiosis. Does one of these contain the old DNA and the other the newly replicated DNA? Or does each contain some of the old DNA and some of the newly replicated DNA?

2 Is the DNA sequence in the two chromosomes of an attached pair identical? Is the DNA sequence in the two chromosomes of a homologous pair identical?

3 DNA is a double helix, two complementary strands that wind around each other. A particular gene is on one of the two strands. A different gene at a different location may be on the same strand or on the other strand. In transcription, the strand with the gene acts as a template for mRNA synthesis. What happens on the other strand during transcription? What happens on the other strand during DNA replication?

Concluding Remarks

In this chapter we have seen how scientists, beginning with Mendel, used observation and experimentation to understand the patterns of inheritance of simple, either/or traits. The same rules that work for pea plants work for other species, including humans. The units that assort and segregate in inheritance have come to be known as genes. Genes are located on chromosomes, a hypothesis that was first suggested because the numbers, locations, and movements of chromosomes could explain the

observed patterns of inheritance. Genes were later found to be composed of DNA, a molecule that consists of long chains of nucleotides. The double-stranded structure of DNA accounts for its ability to be replicated accurately. These landmark discoveries are summarized on our Web site, under Resources: Landmarks in genetics. Scientists now realize that no gene works independently of its cellular environment and that phenotypes for most traits are modifiable. Many traits in many species follow much more complex patterns of inheritance than the simple Mendelian either/or traits that we have seen in this chapter. In the next chapter we discuss some human traits that can be described as simple Mendelian traits, and many others whose inheritance is much more complex.

Chapter Summary

- The study of those aspects of biological traits that are inherited is called **genetics**.

- Hereditary information is carried in the form of **genes**, which are parts of **chromosomes**.

- An **allele** is a variant of a gene. **Dominant** alleles show up in the **phenotype** when either **homozygous** or **heterozygous**. **Recessive** alleles are expressed only in the phenotype when they are homozygous. A **genotype** is the sum of an organism's alleles.

- Plant and animal cells consist of two regions: a central portion called the **nucleus**, which contains the chromosomes, and a surrounding portion called the **cytoplasm**.

- Most cells in sexually reproducing species have the **diploid** number of chromosomes, and the chromosomes exist as **homologous pairs**. The gametes are an exception because they have the **haploid** number, including only one chromosome of each pair.

- In sexual reproduction, gametes fuse to form a **zygote** with the diploid number of chromosomes.

- Chromosomes are replicated and separated in **mitosis** during cell division, maintaining the diploid number and the full genotype.

- During **gamete** formation, **meiosis** halves the chromosome number to the haploid value and results in **segregation** of the alleles into different gametes.

- When two genes are located on different chromosomes, their alleles segregate independently during meiosis, undergoing **independent assortment**. When two genes are on the same chromosome, their alleles show **linkage**, staying together in meiosis unless crossing-over occurs.

- A **karyotype** is the full set of chromosomes from a cell, photographed during mitosis and arranged in homologous pairs by size, shape and banding patterns.

- One pair of human chromosomes differs between males and females; the chromosomes of this pair are called **sex chromosomes**.

- Sexual development in humans is coded for by many genes, including a gene on the Y chromosome.

- Genes that are on the X chromosome and not on the Y chromosome are called **sex-linked genes**.

- **Deoxyribonucleic acid (DNA)** and **ribonucleic acid (RNA)** are two types of **nucleic acid** molecules.

- Each chromosome contains two very long spiral molecules of **DNA,** wound around each other, forming a double helix. The two strands are composed of complementary, not identical, nucleic acids. Genes are segments of one strand of the double stranded DNA.

- Prior to mitosis, DNA undergoes **replication** to produce two identical double helices. Replication requires **enzymes,** biological molecules that speed up chemical reactions but are not themselves changed in the reaction. The replication enzymes move along each DNA strand separately. On each DNA strand, the enzymes facilitate the addition of one nucleic acid at a time to produce a new complementary strand.

CONNECTIONS TO OTHER CHAPTERS

Chapter 1	Mendel gave genetics its first paradigm; the structure of DNA is the basis for the current paradigm.
Chapter 3	Genes result in phenotypes by being the blueprint for synthesis of proteins. The same laws of inheritance that apply to pea plants apply to humans. Many diseases have a genetic component.
Chapter 4	Genetic engineering requires a thorough understanding of genes and their structure.
Chapter 5	Genetic changes supply the raw material for evolutionary change. Evolution takes place whenever the frequencies of different alleles change.
Chapter 6	The structure of genes and the processes of gene transmission across the generations are the same, or very nearly the same, in all known organisms.
Chapter 7	Human populations differ in the frequencies of many alleles.
Chapter 8	Mating patterns and sexual strategies can alter the frequencies of different alleles in populations.
Chapter 18	Conserving genetic diversity is an important aspect of protecting biodiversity.

PRACTICE QUESTIONS

1. A pea plant that is homozygous tall (*TT*) produces male gametes of what genotype? What is the genotype of the female gametes it produces?

2. If pollen from a homozygous tall pea plant fertilizes eggs from a homozygous short pea plant, what are the genotype(s) of the F_1 offspring? What are their phenotype(s)? What are the F_1 genotype(s) and phenotype(s) if the pollen is from a homozygous short plant and the eggs are from a homozygous tall plant?

3. If two F_1 plants from the previous question were crossed, what genotypes would be found among

the F_2 plants? In what proportion would they be found?

4. What are the offspring genotype(s) and phenotype(s) if pollen from the F_1 in question 2 fertilizes eggs from a homozygous short pea plant?

5. Is a homozygous tall pea plant taller if it is grown in nutrient-rich soil than if it is grown in nutrient-poor soil? What about a heterozygous tall pea plant?

6. What is the height of the F_1 plants when plants that are homozygous for both tall height and yellow peas are crossed with plants that are

homozygous for both short height and green peas ? What is their pea color?

7. If the F_1 plants from question 6 are self-fertilized, what are the genotypes, phenotypes, and phenotypic ratios of the F_2 plants?

8. If plants that are heterozygous for tall height and yellow peas are fertilized with pollen from plants homozygous for short height and green peas, what are the genotypes, phenotypes, and phenotypic ratios of the resulting offspring?

9. If V represents the allele for violet-red color (dominant) and v represents the allele for white (recessive), what genotypes of offspring would be produced in each of the following crosses, and in what proportions?

    a. $Vv \times vv$

    b. $Vv \times Vv$

    c. $vv \times vv$

    d. $VV \times VV$

    e. $Vv \times VV$

10. If N is the number of structurally different chromosomes in a mammalian species, how many chromosomes does a liver or skin cell from this species have before it enters mitosis? How many chromosomes does one of the offspring cells have when mitosis is finished?

11. If N is the number of structurally different chromosomes in a reptilian species, how many chromosomes does a cell of this species have before it enters meiosis? How many chromosomes does one of the offspring cells have when meiosis is finished?

12. How many of the two attached chromosomes go to each end of the cell during anaphase of mitosis? How many of the two attached chromosomes go to each end of the cell during the first cell division of meiosis?

13. Do homologous pairs of chromosomes line up together on the midline of the cell during

metaphase of mitosis? Do homologous pairs of chromosomes line up together on the midline of the cell during the first cell division of meiosis?

14. How many genes undergo DNA replication before mitosis? How many genes undergo DNA replication before meiosis?

15. If a diploid cell containing 10 chromosomes divides by mitosis, how many centromeres are present:

    a. in the interphase cell?

    b. at prophase?

    c. at anaphase?

    d. in each daughter cell?

16. If one strand of DNA contains the sequence TAGGCA,

    a. what is the opposite (complementary) DNA sequence?

    b. what mRNA sequence would be formed by this DNA sequence?

    c. how many codons are contained in this portion of mRNA?

    d. what anticodons of transfer RNA would bind to these codons?

17. In Griffith's experiment, which of the results were most unexpected? Why?

18. Specify what type of individual would be formed from a human zygote containing each of the following:

    a. 46 chromosomes, including two X

    b. 47 chromosomes, including two X and a Y

    c. 45 chromosomes, including one X and no Y

    d. 46 chromosomes, including one X and one Y

    e. 47 chromosomes, including two X and three copies of chromosome number 21

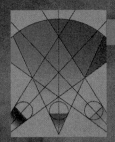

Issues

- How did molecular genetics grow out of Mendel's hypothesis? How do they help explain each other?

- How can we use the study of genetics to fight disease?

- Are there any inherent dangers of genetic research that we should be aware of?

- Is every genetic variant a defect?

- What are my genetic risks? What are my baby's risks? How can I find out?

Biological Concepts

- Molecular genetics (RNA; transcription; translation; gene expression; DNA markers and genes)

- Human health and disease (human genetic diseases; disease predisposition; genetic testing)

3 Human Genetics

A nervous couple sits in the waiting room, anxiously anticipating the results of a test. Their last child had lived her short life in almost constant pain, and had died, blind, at age three, a victim of Tay–Sachs disease. The couple wants another child, but their previous experience was a heart-wrenching nightmare that they don't want to go through again. They are awaiting the results of an amniocentesis, a technique that you will read about later in this chapter. A doctor enters the room with good news: the enzyme that her technicians were testing for is present in the amniotic fluid. The mother-to-be is carrying a child who will not get Tay–Sachs disease. The couple can look forward to raising a healthy child in a happy home.

Scenes like the one just described are happening more often with each passing year. An increasing number of couples are undergoing medical procedures that did not exist when they themselves were born, seeking assurances that would have been unthinkable a mere 25 years ago. Tay–Sachs disease is one of a growing number of conditions that can now be diagnosed before birth. Along with physical characteristics, these conditions (traits) are passed on from one generation to the next.

Some traits follow the simple Mendelian patterns of inheritance that we studied in the previous chapter. Many more, however, follow much more complex patterns because they are governed by multiple genes, each contributing a small amount to the trait. How do scientists find which genes are associated with which trait? Molecular biology has greatly changed how scientists go about the search. It has also changed what we mean by the concepts of 'trait' or 'phenotype' and has led to many new uses for genetic information.

What do Genes do?

In Chapter 2 we saw how the concept of the gene developed, culminating with the identification in the 1950s of DNA as the molecule of which genes are made. This discovery opened up a new line of research to explain exactly how a genotype results in a phenotype. As we will now see, genes act as the template for the synthesis of proteins, which then contribute in various ways to the production of a phenotype. The transition from DNA to protein involves two steps: transcription of DNA to another nucleic acid called RNA, followed by translation of RNA to protein. Together these two steps are called **gene expression**.

Gene expression: transcription and translation of genes

The DNA containing the hereditary information is packaged in chromosomes. Each chromosome contains two very long strands of DNA (in humans the length of DNA in *each cell* is about 1 m or 3 feet). A gene is a segment of the DNA strand, that is, a subset of bases within the linear sequence of the whole DNA. Each person or pea plant (or any diploid individual) has two chromosomes of each type, and thus has two genes for each hereditary trait, one on each chromosome of a homologous pair. Within a species, there may be any number of possible alleles for each gene, but each individual can only have two, one on each chromosome of a homologous pair.

Genes (made of DNA) are expressed as phenotypes by providing information for the synthesis of RNA, which in turn provides information for the synthesis of **protein**. Often several proteins (and therefore several genes) interact to produce a phenotype. We can therefore compare the nucleic acids DNA and RNA to blueprints that contain instructions for building proteins.

RNA (Figure 3.1) differs from DNA in that (1) the backbone contains the sugar ribose rather than deoxyribose, hence the name ribonucleic acid; (2) it is mostly single-stranded; and (3) the nitrogen-containing bases include uracil (U) instead of thymine (the other three nitrogen bases—adenine, guanine, and cytosine—are the same). In RNA, C pairs with G and A pairs with U. Genes undergo **transcription** into RNA, meaning that information is transferred from DNA to RNA, still within the language of nucleic acids with their linear sequence of bases. *The linear sequence of nucleotides in DNA determines the linear sequence of nucleotides in RNA.* Some gene products stop here, as special types of RNA that have functions described later. The product of transcription for most genes is **messenger RNA (mRNA)**, which then goes through a second information transfer (**translation**), in which the information is changed into another language, the language of amino acids. *The linear sequence of nucleotides in messenger RNA determines the linear sequence of amino acids in the protein.*

Information flows from DNA to RNA to protein. This summary statement has been called the central dogma of molecular biology (Figure 3.2A). As we see in Chapters 12 and 16, this central dogma has been considerably modified by the finding that information flow is not only in one direction. Proteins can affect the transcription of DNA and translation of RNA, and in some viruses RNA can be a template for DNA synthesis. Moreover, one protein is not always the product of a single gene. Our current concepts of information flow are more accurately represented as a network (Figure 3.2B).

Transcription—DNA to RNA. During transcription, a portion of DNA is used as a template to make a single-stranded mRNA. There are several differences between transcription and DNA replication (see Figure 2.22, p. 58). In DNA replication, the whole length of both DNA strands is copied to make two new strands of DNA. In transcription, a small, discrete part of a single DNA strand is the template for the synthesis of RNA. The portions of a DNA strand that contain the necessary information to make different proteins are the genes. A particular gene is transcribed from only one of the DNA strands but, at a different place on the same chromosome, a different gene may be transcribed from the other DNA strand. Transcription begins when an enzyme (**RNA polymerase**) combines with a short DNA

Figure 3.1

The structure of RNA. The molecule as a whole is usually single-stranded, but short portions of some RNA molecules can base-pair with other portions of the same molecule.

a single strand of RNA

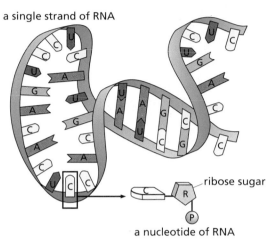

ribose sugar

a nucleotide of RNA

Figure 3.2

Changing concepts of the flow of genetic information.

(A) The central dogma of molecular biology, as it was understood in the 1960s: information flows from DNA to RNA and then to protein.

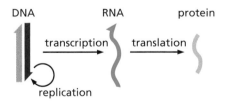

DNA   RNA   protein

transcription   translation

replication

(B) Many exceptions are now known. For example, some viruses, including the one that causes AIDS, have a 'reverse transcription' process in which RNA is used to make DNA. Also, many proteins influence the timing and amount of gene transcription and RNA translation.

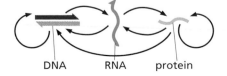

DNA   RNA   protein

CONNECTIONS CHAPTER 12

sequence marking the beginning of a gene (Figure 3.3). The enzyme causes the DNA of the gene to unwind, allowing RNA bases to pair with the nucleotides of the gene. Nothing happens on the other DNA strand. The RNA nucleotides become bonded together to form the backbone RNA strand. Finally the RNA strand comes off its DNA template, allowing the DNA to twist back into its double helical shape.

The product of transcription is usually mRNA, which carries the information for protein synthesis. Two other forms of RNA that function in protein synthesis, transfer RNA and ribosomal RNA, are also transcribed from DNA, but are not themselves translated into proteins. Several proteins are known that can either inhibit or enhance transcription, providing a means by which the amount of RNA being made can be controlled (see Figure 12.5, p. 420).

Translation—RNA to protein. Transcription to mRNA is followed by translation, during which the mRNA sequence of nitrogenous bases is

Figure 3.3

The transcription of DNA into RNA.

STEP 1

Enzyme binds to a DNA strand at the beginning of a gene.

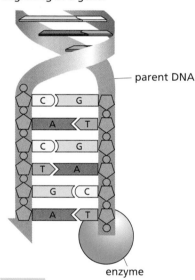

parent DNA

enzyme

STEP 2

Enzyme unwinds a portion of the double helix, separating the strands locally. RNA nucleotides pair one at a time with the complementary nucleotides on *one* DNA strand.

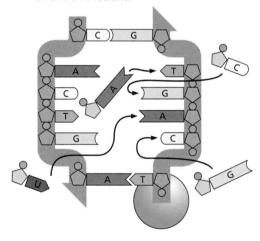

STEP 3

Other enzymes join the new RNA nucleotides to form a continuous RNA strand, complementary to one of the DNA strands.

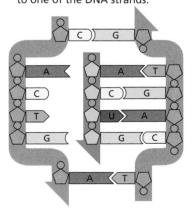

STEP 4

The RNA molecule separates, and the DNA strands pair once more.

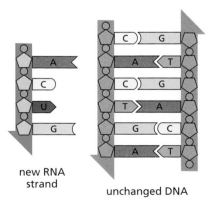

new RNA strand

unchanged DNA

translated into a sequence of amino acids that make up a protein chain. Translation uses groups of three successive nitrogenous bases on the mRNA as coding units, or **codons**. Each codon corresponds to one amino acid.

Translation uses all three forms of RNA: mRNA as the template, transfer RNA to match a codon with an amino acid, and ribosomal RNA to form the 'scaffold' on which the process takes place. Each mRNA codon pairs with a complementary three-base sequence called an **anticodon**, which is part of the transfer RNA (tRNA) molecule (Figure 3.4). Each tRNA molecule has a specific anticodon and carries a specific amino acid molecule that it can transfer to the growing protein chain during protein synthesis, but this addition only takes place if the anti-codon matches the next mRNA codon.

The mRNA and tRNAs are held in the proper relation by a particle called a **ribosome**, containing both protein and ribosomal RNA (rRNA) (Figure 3.4). As each mRNA codon is bound by its complementary tRNA anticodon, one amino acid is added at a time to the growing protein. The next three nucleotides on the mRNA form the next codon, ready for the corresponding tRNA anticodon to bind. The process repeats until the end of the mRNA is reached and the protein is complete (Figure 3.5).

Proteins and phenotypes. There are about 20 different amino acids that can be combined to make proteins. Proteins are linear, unbranched chains of these amino acids. After the chains have been synthesized by translation, they fold into complex shapes that determine their function. How they fold depends on the sequence of amino acids (for more on protein structure see Figure 10.5, p. 334).

The relationship between proteins and phenotypes is actually twofold. At the molecular level, it can be said that the protein itself is the phenotype for its gene. At the level of the organism (the level at which Mendel described visible phenotypes) proteins work together with other proteins and with other types of molecules to result in the phenotype. A phenotype can be altered by changes in the DNA.

Mutations

DNA occasionally undergoes permanent her-itable changes of its sequence of nucleotides. These changes are known as **mutations**. Some of these can result from mistakes during DNA replication, or they may be caused by chemical or physical agents (such as ultraviolet light and radiation—see Chapter 12). Most mistakes and damage are immediately fixed by various self-correction mechanisms and so do not persist.

There are two general types of mutation: (1) base substitution; (2) base insertion or deletion. The simplest kind of mutation is a single-base substitution (called a **point mutation**), such as the substitution of one nucleotide (A, G, C, T) for another. Because

CONNECTIONS
CHAPTERS 10, 12

Figure 3.4

The roles of transfer RNA (tRNA) and ribosomal RNA (rRNA) in translation. In the example shown here, the mRNA codon CUU matches the tRNA anticodon GAA, which corresponds to the amino acid carried by the tRNA, phenylalanine, abbreviated Phe. The ribosome containing rRNA holds the tRNA in place on the mRNA.

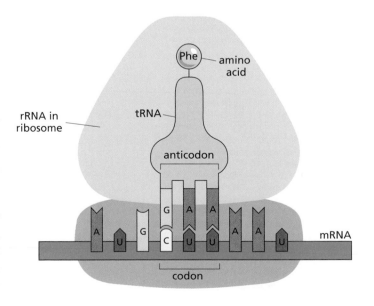

substitutions of this kind change the mRNA codon, they may result in the wrong amino acid being inserted into a protein chain (Figure 3.6A). The number of nucleotides has not changed; only one codon, and therefore at most one amino acid, is changed in the growing protein chain.

Larger changes occur when one or two extra nucleotides are inserted into a DNA sequence, or when one or two are deleted from the sequence. Mutations that change the number of nucleotides change all of the codons that follow the point of insertion or deletion. This is because the DNA or RNA code contains no 'commas' or other 'punctuation' to signify where a new codon starts. Each new codon is simply the next three bases after the previous codon, so the first codon determines the starting point for the second, and so on. If an extra base is inserted or a base is deleted, that codon and all codons that follow the mutation are changed or shifted over. (Hence this type of mutation is often called a **frameshift mutation**, and we say that the reading frame has shifted.) In the example shown in Figure 3.6B, a deletion of a G causes a change in mRNA codons from that point on, leading to the wrong amino acids' being added to a growing protein during translation. CGT in DNA is transcribed to the complementary GCU in mRNA, the codon for the amino acid alanine (Ala). By removing G, the DNA sequence becomes CTC, which is transcribed to the complementary mRNA codon GAG. GAG is the codon for the amino acid glutamine (Glu), not Ala. Note that the C that would have been part of the next codon is now part of the Glu codon. All succeeding codons are likewise shifted, leading to a protein with different codons and therefore a different amino acid chain from that coded by the unmutated DNA strand. Because so many amino acids are affected, most mutations of this kind result in nonfunctional proteins.

As mentioned earlier, protein function depends on the shape of the folded protein molecule. A substituted amino acid may alter the protein shape and therefore change or impair the protein's function. The more amino acids that are changed, the more likely it is that this change alters protein function. Phenotypic consequences of inserting a

Figure 3.5

Translation of a nucleic acid sequence into a protein. Each group of three bases in messenger RNA (mRNA) serves as a codon to determine what amino acid is to be inserted next into the protein sequence. The ribosome holds the mRNA while tRNAs bring successive amino acids to the growing protein chain.

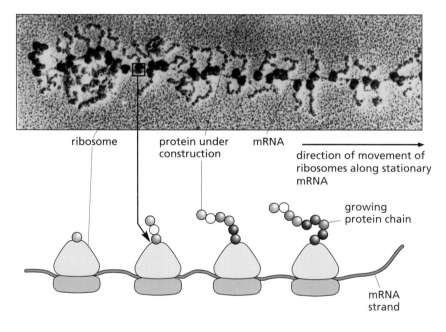

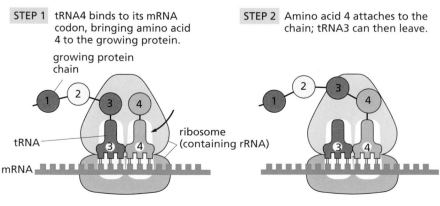

STEP 1 tRNA4 binds to its mRNA codon, bringing amino acid 4 to the growing protein.

STEP 2 Amino acid 4 attaches to the chain; tRNA3 can then leave.

changed amino acid run the gamut from those that are undetectable to those that are fatal, but changes that affect many amino acids at once are more often harmful or fatal.

Mutations that persist in somatic (body) cells can cause problems in the individual carrying the affected genetic material (see Chapter 12), but the change will not be passed on to the next generation. Only mutations in cells that give rise to gametes may be passed on to future generations and thus have a role in evolution. We owe the rich diversity of genes in the world's species to the creation of different alleles, which originate as mutations. Mutations are also useful to the scientist because they serve as essential tools in the laboratory.

Figure 3.6

Examples of two types of mutations. Glu, Leu, etc. are abbreviations for different amino acids.

(A) Single nucleotide substitution. When, for example, a C is substituted for a G in the DNA strand, the mRNA codon matches with the tRNA carrying the amino acid glycine rather than the tRNA carrying the amino acid alanine.

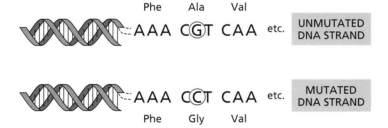

(B) Change in the number of nucleotides. When, for example, a G is deleted from the DNA strand, the codon that had the G and all subsequent codons are misread and different amino acids are placed in the chain.

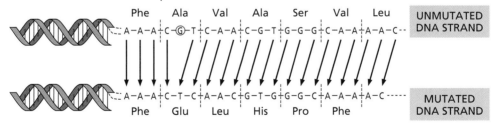

THOUGHT QUESTIONS

1 DNA has been called the master molecule because it controls (or determines) RNA sequences, which control protein sequences, which (as enzymes) control the cell's other activities. Is the term 'master molecule' an accurate description? Does the language of control (e.g., DNA controls the type of RNA produced) say more about the molecules or about the scientists? Does the use of a word such as 'master' suggest a hierarchical approach in which information flows in one direction only?

2 The term 'mutation' often has a negative connotation. Is mutation always 'bad'?

Some Diseases and Disease Predispositions Are Inherited

Figure 3.7

An example of simple Mendelian inheritance in humans. Albinism, a recessive trait, arises in most cases from matings between heterozygotes, although it could also arise from matings between a heterozygous person and someone who is homozygous recessive.

Early in the twentieth century, pioneering geneticists discovered that Mendel's rules, formulated on the basis of experiments with pea plants, could also explain the inheritance of many human traits. For example, albinism is a total lack of melanin pigment in the skin, eyes, hair, and the body's internal organs, and the inheritance of albinism follows Mendel's rules. The skin of albinos is white and their hair is white as well. One of the normal functions of melanin is to block ultraviolet light, and albinos sunburn easily and are very sensitive to bright lights. All geographical races of humans have albino individuals, as do many other species.

The inheritance of albinism is shown in Figure 3.7. A recessive allele is responsible for this condition and thus it can be transmitted without

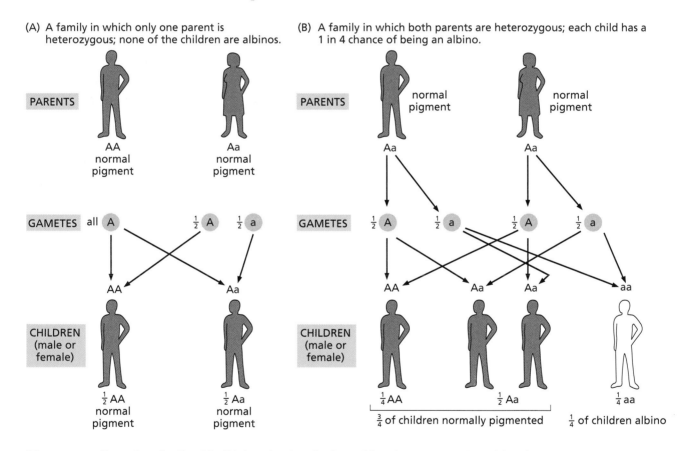

(A) A family in which only one parent is heterozygous; none of the children are albinos.

(B) A family in which both parents are heterozygous; each child has a 1 in 4 chance of being an albino.

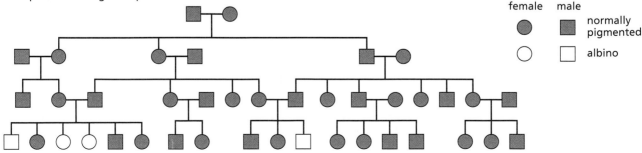

(C) Human pedigree for a family with albinism; the short horizontal lines between a male and female represent matings that produced the children in the row below.

detection through many successive generations of normally pigmented individuals (Figure 3.7A and C). However, matings between heterozygous individuals may produce albino children (Figure 3.7B). The likelihood that both parents are heterozygotes is increased if mates are chosen from among related persons such as cousins (see Figure 3.7C). This is because the same rare recessive alleles are more likely to be present in other family members than they are in the general public. Diagrams giving the pattern of mating and descent, as in Figure 3.7C, are called **pedigrees**.

Shortly after the rediscovery of Mendel's laws, an English physician, Archibald Garrod, made an important discovery: Mendel's laws applied not only to visible characteristics such as eye colors and albinism, but also to certain medical conditions. Garrod's identification of the genetic basis of a condition called alkaptonuria is described later in this section.

Identifying genetic causes for traits

Before searching for genes that cause a particular disease or trait, we first need to know whether there is any basis for thinking that the disease is inherited. As we will see, some diseases are the direct result of particular gene mutations. In other diseases, genetic effects are indirect and contribute to disease susceptibility, the likelihood that a person will get a disease. Susceptibility to many diseases seems to have a hereditary component. However, certain disease susceptibilities and many nondisease traits are the result of the interaction of multiple genes, not the result of mutations in single genes.

Several kinds of studies are used in answering the question of whether a disease or other trait is inherited.

Pedigrees. Geneticists have several ways of studying human hereditary traits. One of the most basic methods is to present the available data in pedigrees, as in Figure 3.7C. Pedigrees are most useful when they span many generations and hundreds of people, or when separate pedigrees are available for hundreds of different families. The study of pedigrees can help to identify whether a condition is inherited, and permit us to determine which traits are dominant, which are recessive, and which have a more complex genetic basis. If the genetic basis of a trait is complex, or not fully known, then pedigrees can also help in an empirical determination of risks, including medical risks. For example, a child has a greatly increased risk of having insulin-dependent diabetes mellitus if one or both of the child's parents has the disease. The term **risk** has a precise statistical meaning: it is the probability that a particular condition will occur or that a particular condition will be inherited.

Studies of twins and of adopted children. Studies of twins are sometimes useful in suggesting the extent to which the presence of a trait can be explained genetically, rather than environmentally. In such studies, numerous twin pairs are located in which at least one twin has the condition being investigated. For pairs of this kind, the frequency with which the other twin has the condition is studied. This frequency is called the **rate of concordance**. For the vast majority of traits, studies on twins find that a mixture of both heredity and environment is involved. Traits under strong genetic control usually have a higher frequency of both twins sharing a trait when they are monozygous twins (identical twins, derived from a single fertilized egg) than when they are dizygous twins (fraternal twins, derived from two separate eggs). In contrast, traits with mainly environmental causes have similar rates for both types of twins.

Adoption studies can also provide important clues about whether a particular trait is heritable: if adopted children show a higher rate of concordance with their birth parents than with their adoptive parents, then the hypothesis of a genetic cause is made stronger. Some researchers have, however, criticized this type of study because adoption agencies do not place children at random but purposely try to match children with adopting parents whose backgrounds are similar to the backgrounds of the children's birth parents. This practice introduces a bias that raises the concordance rates between adopted children and the parents who adopted them. Another complication is that many children are adopted by relatives. This makes it very difficult to sort out which similarities are environmental and which are genetic.

Linkage studies. Once evidence has been found that there is a genetic basis for a trait, linkage studies can help locate the relevant gene or genes. To do this, we first need a set of **DNA markers**, pieces of chromosomes that are visibly different under the microscope or short sequences of DNA that can be revealed by the molecular techniques discussed later in this chapter. If we can locate a DNA marker whose pattern of inheritance is the same as the pattern of inheritance of the trait, then we can conclude that a gene associated with the trait is located near the marker. One problem is that we may at first not know where to look, so we need many markers, scattered across all the chromosomes. Also, large pedigrees are needed to carry out this type of analysis. After a linkage between DNA markers and a trait has been found, other molecular techniques are used to close in on the actual gene. Finally, after the gene has been located, its full sequence can be determined.

In the past, genes that did not assort independently were found to be on the same chromosome and we say they are linked. The frequency of crossing-over was used as a measure of the distance between genes (see Chapter 2, pp. 45–46). If linkage to a visible chromosomal abnormality could be established, then the group of genes could be assigned to a particular chromosome. These classical genetic techniques were developed in other species, using hundreds or thousands of offspring in each generation to assess cross-over frequencies. These techniques could not be used on humans because humans have such small family sizes and such a long generation time, and because humans cannot be bred for experimental purposes. Before 1980, very few human genes had been mapped to their chromosomal locations.

DNA markers. Several marker systems have now been discovered for studying human DNA and human genetics. DNA contains, in addition to genes, non-coding regions that vary from one individual to another in their length or their sequence. Many of the markers are in these non-coding regions of DNA. The first of these DNA marker systems was discovered in 1980, and was called restriction-fragment length polymorphisms (abbreviated RFLPs and pronounced 'riflips'). Restriction fragments are short pieces of DNA produced by cutting the DNA with specific enzymes called restriction enzymes, which we will see more about in Chapter 4. These fragments are useful markers because they are different lengths in different people, and therefore one person's DNA can be distinguished from another's. There are only a limited number of RFLPs in the human genome, however. More recently, other DNA markers have been discovered that occur with greater frequency throughout the genome than do RFLPs. Each of these is a short, unique DNA sequence with a known location in the genome. Each DNA marker sequence exists in different alleles; that is, the sequence varies slightly from one person to another, again allowing one person's DNA to be distinguished from another's. For some DNA markers the same sequence is repeated a variable number of times (microsatellite markers). For other DNA markers some nucleotides differ from one allele to another. These include expressed sequence tags, small portions of genes that vary from one person to another, and single-nucleotide polymorphisms, sequences located either in genes or in noncoding DNA regions that differ by a single nucleotide.

Each different marker can be detected by a specific **DNA probe**, a piece of DNA with a sequence complementary to the marker sequence. When a radioactive DNA probe is added to some DNA, the DNA picks up the radioactivity if it contains the marker sequences that can pair with the probe. DNA probes cause only those fragments to show up that have sequences complementary to the probe sequence. For markers such as microsatellites or RFLPs that vary by length, the alleles can be distinguished by length after separation by electrophoresis (Figure 3.8). For markers that differ by sequence, a different DNA probe is required for

Figure 3.8

Microsatellite DNA marker alleles can be distinguished by the distance they travel during electrophoresis.

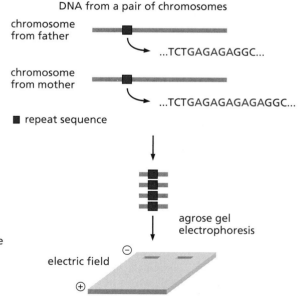

(A) DNA CONTAINING A MICROSATELLITE

The microsatellite differs in length depending on the number of repeats that exist within it. In this example, the allele from the father is shorter because it has fewer repeats than the allele from the mother, which is longer because it has more repeats.

(B) AMPLIFICATION OF MICROSATELLITE BY PCR

(Using techniques shown in Figure 3.12.)

(C) SEPARATION BY ELECTROPHORESIS

The mixture of microsatellite PCR products is placed on a gel and exposed to an electric field. Because DNA has a negative charge, the pieces move toward the electrode of positive charge. In the time that the current is on, smaller pieces travel farther through the gel than the larger ones do.

(D) DETECTION WITH A PROBE

None of the pieces can be seen; however, they can be detected with a variable-repeat probe tagged radioactively or chemically (bands shown in color). The probe is a small piece of DNA with a sequence complementary to the sequence of that variable repeat, so the probe will bind to those pieces of DNA containing that variable repeat. The probe thus does two things: it identifies pieces with that specific repeat and it allows scientists to determine whether the sequence is repeated a few times (to give a short DNA piece that travels farther) or many times (to give a long piece that travels a shorter distance within the gel).

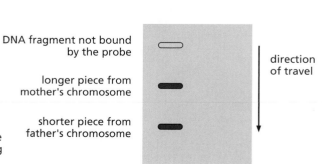

each allele. These DNA probes are clustered onto a solid surface called a DNA chip, or **microarray**. A fluorescently labeled sample of DNA is added. If the DNA sample contains the complement to the DNA probe, it binds and a microscope attached to a computer detects its fluorescence. One million DNA probes can be placed on a one square centimeter microarray, so that two alleles of each of a half million markers can be tested at the same time.

Such marker sequences have been identified throughout the DNA of humans and their chromosomal locations and genetic map positions have been established. Geneticists use all of these different types of markers in developing linkage maps. Because the chromosomal locations of the markers are known, geneticists can determine the positions of presumed genes located near the markers by finding a pattern of linkage between the trait and the marker. DNA samples are collected from members of a family with a pedigree in which the trait appears. Molecular geneticists then search the DNA samples for a marker, among the thousands of markers known, that is inherited within the family in the same pattern as the trait of interest. If all the individuals within a large pedigree who have a particular trait have a specific DNA marker, and all those without the trait do not have the marker, the gene for the trait is presumed to be located near that marker (Figure 3.9A). For the other markers that are not in the vicinity of a gene linked to the trait, the

Figure 3.9

Using DNA markers to establish a linkage between a DNA region and an inherited phenotype.

(A) Marker 1, detected by probe 1, is linked to the trait

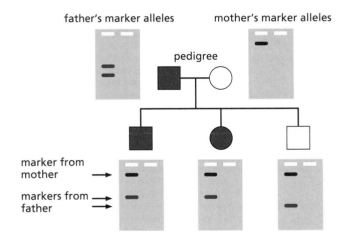

○ unaffected female
□ unaffected male
■ ● affected individual

The father has the trait and is heterozygous at marker 1. The mother does not have the trait and is homozygous at marker 1; the mother's marker allele has a greater number of repeats than either of the marker alleles from the father's DNA.

The children who have the trait have one of their father's length of microsatellite marker. The child without the trait does not. The marker lengths detected with probe 1 thus follow the pattern of inheritance of the trait, indicating that an associated gene may be nearby.

(B) Probe 2 detects a different microsatellite marker, and shows that it is not linked to the trait

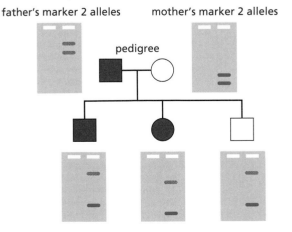

A child with the trait and a child without the trait have the same band pattern for the marker detected by probe 2. The marker detected by probe 2 is therefore not near any gene associated with the trait.

mother's and father's markers can be found in the children, but the pattern of bands does not follow the pattern of inheritance of the trait (Figure 3.9B). In this way, an increasing number of presumed gene locations are being discovered at an accelerating rate.

Identifying a specific gene as the cause of a trait. It is important to understand that DNA markers are not the genes themselves. In other words the linkage indicates the approximate *location* of a gene of interest. It does not tell us what the gene is, or what its function is, or what alleles are associated with disease or nondisease.

Often when such a location has been identified, news reports are published claiming that 'the gene for X' has been discovered, but no gene has even been investigated, only a DNA region that maps with the trait. When a linkage area has been identified, sequence databases can be consulted to see what genes are known in this area of the chromosome. If genes are known, their protein products can sometimes be deduced from the nucleotide sequence and those proteins can be further investigated for their connection with the trait or disease. Often, however, the genes within a linkage area have not yet been identified, and neither their protein product nor their function is known.

The first gene to be located by using DNA molecular markers was the gene for Duchenne's muscular dystrophy, which is located on the X chromosome. Other diseases for which linkage areas were located and identified during the 1980s and 1990s include Huntington's disease (chromosome 4), cystic fibrosis (chromosome 7), Alzheimer's disease (chromosome 21), one form of colon cancer (chromosome 2), and two forms of manic depression (chromosome 11 and the X chromosome). Of these linkage areas, genes have so far been found for muscular dystrophy, Huntington's disease, and cystic fibrosis. Among these diseases, only cystic fibrosis seems to be inherited on the basis of single-gene Mendelian genetics. Even in cystic fibrosis, how the gene product produces the disease is not fully understood.

Some hereditary diseases associated with known genes

Some human diseases that follow simple Mendelian genetics were identified early in the twentieth century. Genes for some other diseases have been found recently by using DNA markers. In this section we consider some hereditary diseases whose underlying genes are known. In some cases the mechanism by which the gene mutation produces the disease is known, but for others, although the full DNA sequence of the gene and its mutations may be known, the mechanism by which these result in disease is not.

Alkaptonuria. Alkaptonuria is a rare condition in which a patient's face and ears may be discolored and in which their urine turns black upon exposure to air. Archibald Garrod, mentioned earlier in this chapter, tested the urine of these patients and discovered that the color is caused by an acid. We now know that this substance, homogentisic acid, is formed in the course of breaking down the amino acid tyrosine. In most individuals, the homogentisic acid can be broken down harmlessly with the help of an enzyme. However, in patients with alkaptonuria, the necessary enzyme is missing or defective. Garrod realized that an error in an important biochemical (metabolic) process was responsible, and he

called this type of condition an "inborn error of metabolism". He studied the families of individuals with alkaptonuria and two other such conditions, including albinism, and found a common pattern: each of these inborn errors of metabolism was inherited as a simple Mendelian trait, and in each case the lack of a functional enzyme was recessive. Many other inborn errors of metabolism have since been discovered and their biochemical defects identified. Each of these inborn errors is caused by a recessive allele, the product of a DNA mutation that, when transcribed and translated to its protein product, changes a functional enzyme into a nonfunctional one. Alkaptonuria, albinism, and the condition called phenylketonuria, which is described next, all arise from errors in a series of closely related metabolic pathways (Figure 3.10).

Phenylketonuria. Phenylketonuria (PKU) is a genetically controlled defect in amino acid metabolism. The amino acid phenylalanine, which is present in most proteins, is normally converted by an enzyme into another amino acid, tyrosine; the tyrosine is then broken down by the pathway shown in Figure 3.10. A defect in the enzyme that usually con-

Figure 3.10

Biochemical pathways for three inborn errors of metabolism: phenylketonuria, alkaptonuria, and albinism.

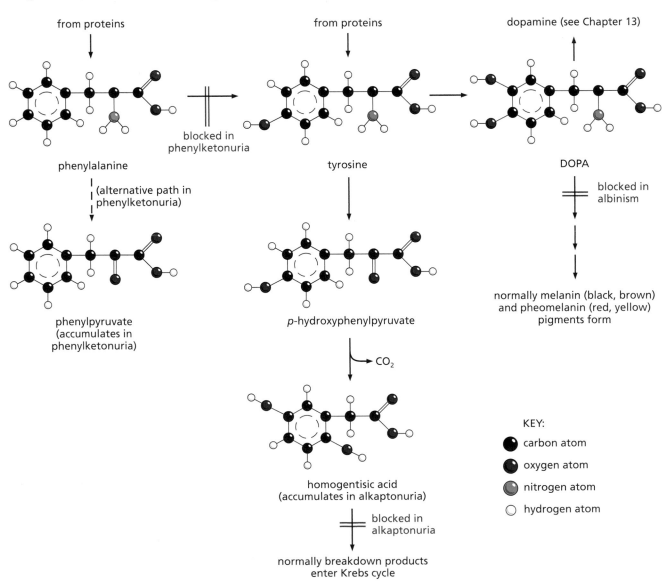

verts phenylalanine to tyrosine causes the phenylalanine to be processed by an alternative pathway. A product of this alternative pathway accumulates in the blood and in all cells, acting as a poison that causes most of the debilitating symptoms of the disease: insufficient development of the insulating layer (myelin) around nerve cells, uncoordinated and hyperactive muscle movements, mental retardation, defective tooth enamel, retarded bone growth, and a life expectancy of 30 years or less. Thus a change in one gene (and one enzyme) can have many phenotypic consequences throughout the body.

Fortunately for people carrying this genetic defect, a simple test for its presence exists. If phenylketonuria is detected at birth or earlier, it is possible to avoid the symptoms of the disease by greatly limiting those foods that contain phenylalanine (nearly all proteins, including breast milk), and diet foods and soda containing the artificial sweetener aspartame, which is metabolized to phenylalanine. Small amounts of phenylalanine are essential in protein synthesis (see Chapter 10), but the diet must be carefully monitored to guard against the larger amounts of phenylalanine whose breakdown products would be in toxic amounts that cause the disease symptoms.

CONNECTIONS
CHAPTER 10

Alkaptonuria and phenylketonuria are inherited via single recessive alleles that follow Mendel's laws. Yet these traits are medical conditions, rather than visible traits like those studied by Mendel. New findings thus broadened the concept of a 'trait.' The concept has been broadened further by the discovery of a genetic basis for diseases that are not just present or absent but that show a range of severity in different people.

Duchenne's muscular dystrophy. Duchenne's muscular dystrophy is a sex-linked genetic disorder that causes muscles to become weak and nonfunctional. In most cases, the inability of the muscles of the diaphragm to keep the patient breathing leads to death during the teenage years or in the early twenties. After the gene responsible for this disease was located by linkage studies using DNA markers, its protein product, dystrophin, was found. So far, this discovery has not led to new therapies for the disease, but the research on dystrophin is greatly adding to our understanding of normal muscle contraction, and it is hoped that this knowledge will lead to effective treatments.

Cystic fibrosis. Cystic fibrosis is the most common genetic disease among the white population in the United States and much of western Europe. Cystic fibrosis is characterized by thickened fluids, especially the fluids that line the lungs. Normally, as these fluids are cleared from the lungs, they remove respiratory bacteria, preventing infections. The thicker-than-normal fluids of cystic fibrosis are not cleared from the lungs as they should be, so people with cystic fibrosis have breathing difficulties and frequent lung infections.

Cystic fibrosis was mapped by linkage studies to a region on chromosome 7 and later a gene was identified. The product of that gene is a membrane transporter, a protein that carries ions, chloride in this case, into or out of a cell. A very large number of different mutations have now been found in the gene for this transporter protein. Some mutations lead to few or no symptoms; some are associated with more severe disease. However, no particular mutation is completely predictive of disease

onset or severity. Although identification of the gene and its product has increased our understanding of the disease mechanism, it is still not known how the mutations lead to the many symptoms of the disease.

Huntington's disease. Huntington's disease (or Huntington's chorea) is a neurological disorder that typically begins in middle age with uncontrollable spasms or twitches of the hands or feet (see Chapter 13, p. 474). As the disease progresses, the spasms become more pronounced, and the patient gradually loses conscious control of all motor functions and of mental processes. The disease progresses slowly, but is invariably fatal. American song writer and balladeer Woody Guthrie died of Huntington's disease. Although it is always lethal, Huntington's disease does not appear until after its victims have lived through their prime reproductive years, during which they may have passed the mutation to their children. Studies of family trees show that Huntington's disease is inherited as a dominant trait. The gene responsible for the disease was located on chromosome 4 in 1983. This gene was fully isolated and identified in 1993, but its normal function remains unknown.

Subsequent studies have shown that Huntington's disease is associated with a different type of DNA mutation, called a dynamic mutation. Most mutations are rare, stable, and inherited unchanged from one parent or the other. In dynamic mutations, a three-nucleotide segment is repeated many times, increasing in number during mitosis in different tissues. The number of copies can also change during meiosis, so a child may inherit more copies of the repeat than is present on either parent's chromosome.

The allele that causes Huntington's disease differs from the other alleles of the gene in having many extra repetitions of the three-nucleotide sequence AGC. Persons with fewer than 30 repeats are unlikely to get the disease. Persons with more than 38 repeats are almost certain to get the disease, with the age of onset being younger when there are more than 50 repeats. A test for the allele responsible for Huntington's disease has since been devised, but there is a problem in interpreting it because there is an overlap in the number of repeats associated with a particular outcome. The 'normal' range in number of repeats is 9–37, but some people with as few as 30 repeats have become ill. Because the number of repeats can increase during meiosis, people with 30–37 repeats may pass the disease to their offspring.

Huntington's is one of several diseases, called trinucleotide repeat diseases, associated with dynamic mutations. Another is fragile X syndrome, the most common type of hereditary mental retardation, in which the repeat is CCG in a gene on the X chromosome. The gene has been identified but its normal function remains unknown.

Genes increasing susceptibility to disease. The transmission of human traits controlled by single genes follows the rules that Mendel developed for peas. The situation is more complex when we study phenotypic traits such as height or skin color that are controlled by many genes at once and that are also influenced by environmental variables such as nutrition.

Genes have now been found that do not lead directly to a trait (such as a disease) but increase the probability that some trait or disease will develop. Having a particular allele of these genes does not mean that a person is certain to get a particular disease, only that the likelihood of their doing so is higher than that in people with other alleles. These genes, like all genes, code for proteins. The proteins associated with susceptibility are often regulatory proteins that change the body's response to some environmental factor. The genes have sometimes been called susceptibility genes, although 'genes associated with increased susceptibility' is a more accurate description. Those associated with predisposition to cancer, discussed in Chapter 12, are examples. As we mentioned earlier for other multigene traits, we must be cautious not to think of these 'susceptibility genes' as simple, either/or, Mendelian alleles.

In many cases all we know at present is a statistical association between some traits and some DNA markers; the genes themselves have not yet been identified. Traits such as bone density or obesity in mice, and behavioral traits such as ethanol consumption by rats, have been found to each be associated with several DNA markers, but no genes have yet been identified. Despite this, news stories have overplayed such statistical associations, referring, for example, to the 'gene for obesity.' Such terms are premature and surely oversimplified. Each of these traits is influenced by multiple genes, so we cannot think of them as we think of Mendelian either/or traits. These traits vary along a continuum and are not just present or absent like those studied by Mendel. Thus the concept of a single gene determining a single trait does not apply. A single gene codes for a single protein, but the amount of that protein produced depends on what other genes and proteins are present. It is the interaction of them all that results in the trait.

THOUGHT QUESTIONS

1 What is the difference between a genetic marker and a gene?

2 Is research on genetic diseases important for society even if it does not lead to new methods of treatment?

3 With our thinking influenced by genetic diseases like PKU or Huntington's disease, we have been accustomed to thinking of 'mutations' as 'defects.' Not every mutation is a defect, however. With genetics research now turning to the identification of disease susceptibility, it is likely that we will all turn out to have a hereditary predisposition to something. Do new research directions necessitate reconceptualizing the term mutation as 'variation,' rather than 'defect'?

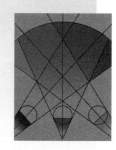

Genetic Information Can Be Used or Misused in Various Ways

After a genetic basis has been identified for a particular trait, what happens next depends in a large part on the values that individuals and society place on that trait. We have considered in this chapter many traits that at least some people consider undesirable, although not all of them impair health or longevity. In many cases, but not others, we can identify a specific gene that fails to make a functional protein of some sort. These are often called genetic defects, a category that includes all inborn errors of metabolism such as those shown in Figure 3.10. The term genetic defect thus means that a specific allele and its product are defective; it does not mean that the person bearing the gene is defective. For this reason, many people prefer terms such as genetic disease or genetic condition, rather than genetic defect.

Humans can deal with hereditary conditions and hereditary risks in many ways. We can conveniently describe four broad categories of response: gathering and sharing information through genetic testing and counseling, changing individual genotypes, changing the gene pool at the population level, and changing the balance between genetic and environmental factors. Many of the methods in these four categories, which we consider below in turn, raise important ethical questions. The ethical questions can be summarized as follows.

- Who decides who should be tested?
- Who has access to the results of the test?
- What are the responsibilities of a person who carries a gene for a hereditary disease?
- Do we have a responsibility to maintain genetic diversity?
- Who determines what traits (if any) are called 'defects'?

Think about these ethical issues as you read the rest of the chapter.

Genetic testing and counseling

Advances in medical genetics have led to better ways of detecting genetic diseases and to ways of detecting them earlier. Identification of chromosomal variations, mutated alleles of genes, or products of mutated alleles may allow the detection of a disease at the earliest possible stage.

Prenatal detection of genetic conditions. Some conditions can be detected before birth, *in utero* (literally, 'in the womb'). From conception until about the eighth week of pregnancy, a pregnant woman is carrying an embryo; from the eighth week until birth, the term fetus is used rather than embryo. In the technique of amniocentesis (Figure 3.11A), a small amount of fluid (amniotic fluid) is withdrawn from the sac in which the fetus is developing in the mother's uterus (see Chapter 9, pp. 302–303). The fluid itself is analyzed for the presence or absence of certain enzymes

Figure 3.11

Techniques for prenatal detection of genetic conditions.

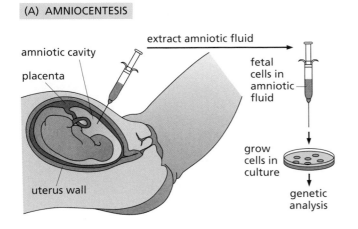

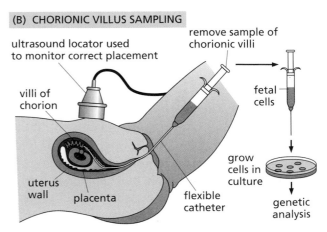

that might indicate a genetic defect in the fetus. Also, amniotic fluid usually contains cells that have been shed from the surface of the fetus. Growing the cells in the laboratory can reveal additional information.

Instead of amniocentesis, chorionic villus sampling is sometimes used for prenatal detection. This technique is a type of biopsy (removal of living tissue for examination) of the placenta (Figure 3.11B). The placenta is the structure by which the fetus attaches to the wall of the uterus (see Chapter 9, pp. 300–301 and Chapter 14, pp. 530–531). Because the part of the placenta biopsied is tissue derived from the fetus, not from the mother, the cells sampled by this technique are fetal cells. Chorionic villus sampling can be performed earlier in pregnancy than amniocentesis can. Certain low but nonzero risks are associated with amniocentesis or chorionic villus sampling, including a risk of mechanical injury to the growing fetus and a risk that the pregnancy will be prematurely terminated. Because of these risks and other reasons, these tests are not performed routinely on every expectant mother.

After fetal cells have been obtained by either amniocentesis or chorionic villus sampling, chromosomes from these cells are analyzed for evidence of Down, Turner, or Klinefelter syndromes. Also, if there is a reason to study a particular gene, its sequence can be determined by comparing it with a known sequence for that gene. This cannot be done on the minute amounts of DNA that exist in the cells unless these amounts are first increased, or amplified. Amplification of DNA is accomplished by the **polymerase chain reaction** (**PCR**) (Figure 3.12). The polymerase chain reaction is often used to detect genetic conditions by using DNA from eight-cell embryos before implantation. These embryos are derived from *in vitro* fertilization (literally 'in glass'), meaning that the fertilization of the egg by the sperm took place in laboratory glassware rather than inside the body (*in vivo*). One of the eight cells can be removed for genetic testing and the other seven can be implanted into a woman's uterus to grow to term.

Several dozen genetic diseases are now detectable through prenatal tests, including Tay–Sachs disease, cystic fibrosis, and phenylketonuria.

CONNECTIONS
CHAPTERS 9, 14

Figure 3.12

The polymerase chain reaction (PCR).

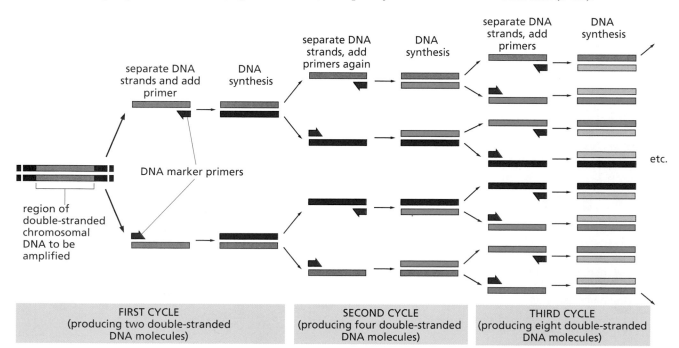

region of double-stranded chromosomal DNA to be amplified

DNA marker primers

separate DNA strands and add primer — DNA synthesis

separate DNA strands, add primers again — DNA synthesis

separate DNA strands, add primers — DNA synthesis

etc.

FIRST CYCLE
(producing two double-stranded DNA molecules)

SECOND CYCLE
(producing four double-stranded DNA molecules)

THIRD CYCLE
(producing eight double-stranded DNA molecules)

Testing newborns or adults. Other tests are done on newborns or on adults. Tests that are simple and inexpensive can be used for mass screening. For example, many hospitals routinely screen all infants at birth for phenylketonuria (by a blood test to detect the amino acid phenylalanine, not by a DNA test). Such screening is considered ethical because it is done on all infants and it is a clear benefit to the infant for the information to be known. Tests that detect heterozygosity for defective alleles (e.g., testing for carriers of the allele causing sickle-cell anemia; see Chapter 7) can be performed on adults before they become parents. Those undergoing this type of screening must first give their **informed consent**. They must sign a form stating that they understand the nature of the test, the possible outcomes (including the conditions that the test can detect and the likelihood that the genotype will result in a disease phenotype), the possible risks of the procedure, and the possible benefits.

People in these situations often consult a genetic counselor to help them understand the test and the risks before giving their consent. For example, they can be advised that if they are identified as being heterozygous and they have children with a person also heterozygous for the same allele, each of their offspring has a 25% chance of being homozygous for the recessive condition. (To understand why, refer back to Figure 3.7B.) The extent to which homozygosity predicts disease severity differs with the disease.

Who should be tested? Genetic testing is expensive and it would not be reasonable to test everyone for all genetic disorders. So an effort is made to identify persons at higher risk of having certain defective alleles. Figure 3.13 shows two ways in which risk is estimated. First, family history identifies persons at higher risk and may prompt testing because having a family history of a genetic condition increases the probability that a family member carries the defective allele (see Figure 3.13A). However, because recessive traits show up phenotypically only when the genotype is homozygous or sex-linked, a recessive trait may not show up in a pedigree and a person may not know the family history. A second way of estimating risk is by the population frequency of a trait (that is, how many people have the trait in a given population). The likelihood of carrying a

Figure 3.13

Two ways of assessing risk for a recessive trait with single-gene, simple Mendelian inheritance.

(A) FAMILY PEDIGREE

Each child of two parents heterozygous for a recessive trait has a 25% probability of being homozygous recessive

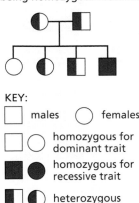

(B) POPULATION FREQUENCY

If 1% of a population expresses a trait known to be recessive (meaning that those who express the trait are assumed to be homozygous), 18% can be assumed to be heterozygous. Out of 100 individuals shown here, 1 is homozygous for the recessive trait and 18 are heterozygous.

recessive allele for a particular condition is higher for a person from a population in which the allele is more frequent (Figure 3.13B). Diagnostic testing for a particular genetic trait is sometimes recommended specifically to persons from those populations or ethnic groups known to have a greater prevalence of the trait. When the frequency of an allele is higher in a particular group, the probability of having an offspring homozygous for the trait is higher if both parents are from the same group than if one marries outside the group.

Examples of genetic testing for recessive alleles because of within-group risk are the following.

- Many African Americans now seek testing to see whether they carry a sickle-cell allele because the frequency of sickle-cell anemia is higher among African Americans (see Chapter 7, pp. 228–231).

- People of Mediterranean or southeast Asian descent may seek testing to see whether they carry an allele for thalassemia (see Chapter 7, p. 232) because the frequency of this disorder is higher in these groups.

- Ashkenazi Jews (those of eastern European descent) commonly seek testing for the recessive allele that causes Tay–Sachs disease, a fatal disorder of brain chemistry, because the frequency of the disease is higher in their group.

- People of western European (especially Irish) descent may seek testing for mutations in the membrane transporter gene responsible for cystic fibrosis because the frequency of the disease is higher in this group.

Each of these four diseases is a single-gene trait with recessive inheritance. There may be more than one defective allele for a disease and a range of disease severity depending on the exact mutation in the allele; cystic fibrosis is an example. Therefore, determining by genetic testing that a person is homozygous recessive most often does not tell you the severity of future disease, or even whether disease will actually develop.

People outside any higher-risk group can also inherit each of these diseases but their probability of doing so is lower simply because the frequency of the recessive allele is lower in their groups. These traits are rare even in the higher-risk groups where they are 'more frequent,' far more rare than the 1% shown in Figure 3.13B. Therefore most people, even those in a higher-risk group, are not heterozygous carriers of these rare recessive traits. (The frequency of heterozygous carriers can be calculated from the frequency of the recessive trait by a method we discuss in Chapter 7, pp. 222–223.)

Genetic testing of this sort should be done only on a voluntary, informed-consent basis and, in general, only when it is of potential benefit to those being tested or to their children. Community leaders of various ethnic groups, including many clergy, have helped to organize genetic testing programs and have encouraged people to participate.

Using information from genetic tests. When a genotype for a disease is detected, the decision about what to do is left up to the person, or to his or her parents if the person is a child. A patient's decision should be based on a clear knowledge of the possible choices, their consequences, and the extent to which an outcome can or cannot be predicted by the test. Genetic counselors help people to understand these choices, but the code of ethical conduct of genetic counselors prohibits them from making a decision on behalf of their clients: clients could rightfully resent any counselor who has pressured them into a decision.

Some decisions that must be made after genetic testing are difficult for the people making them. Couples who know they are at risk of bearing children with a genetic disease may decide to adopt children instead. In other cases, knowledge of a genetic condition permits medical intervention at the earliest possible stages, when chances of successful treatment may be better. For conditions that cannot be treated, some couples may choose to abort the fetus bearing the genetic defect. However, people committed to a pro-life position believe that the potential benefits to those being tested or to any future children can never justify what they view as the murder of a fetus.

Genetic testing has already led to some highly inventive mixtures of tradition and modern technology. The Hasidic Jews of Brooklyn, New York, who are mostly descended from the Ashkenazi Jews of Eastern Europe, have a relatively high population frequency of the gene for Tay–Sachs disease. Marriages are traditionally arranged within the Hasidic community, and marriages outside the community are rare; this pattern generally increases the rate at which recessive alleles come together and produce recessive phenotypes. Because they are ethically opposed to all abortions, the Hasidim do not permit genetic testing *in utero*. The availability of a test that detects Tay–Sachs heterozygotes has, however, allowed the Hasidic community to set up a computerized registry under their strict control. Testing of all persons within the community is encouraged, and the results are entered into the registry under a code number that guarantees confidentiality. The registry permits the traditional matchmakers to check potential couples before proposing a match; if both partners are carriers for Tay–Sachs disease, the matchmaker is warned of this fact and the match is never made. Before this registry was set up in 1984, the Kingsbrook Jewish Medical Center in Brooklyn, which serves the Hasidic community, had an average of 13 Tay–Sachs children under treatment at any one time; after just 5 years, the number of Tay–Sachs children under treatment in the hospital dropped to two or three.

The ethics of genetic testing. Genetic testing is sometimes a mixed blessing. If a genetic defect can either be cured or phenotypically suppressed, or if heterozygote detection permits at-risk couples to decide against having children, then a genetic test can be justified on the grounds that it relieves future suffering. However, most genetic defects cannot be cured. What is the point of testing a person for a condition such as Huntington's disease that can be neither controlled nor cured? One reason is that it permits people who carry a genetic condition to decide whether or not to have children. Will a person who tests positive for such a genetic disease be denied insurance or employment on the basis of the test results? Will a woman choose to abort a fetus if a genetic disease is detected *in utero*? Box 3.1 examines some of the ethical questions that arise in connection with various forms of prenatal and at-birth testing.

Another ethical issue concerns the use of prenatal screening not for the purpose of detecting a disease-associated allele but to find out

BOX 3.1 Ethical Issues in Medical Decision-making Regarding Genetic Testing

Should society influence the private decisions of individuals? To what extent do (or should) financial considerations limit the choices available? Suppose a child is born with a birth defect or other congenital condition. Is it ever ethical to withhold treatment? (Similar ethical issues are raised by conditions resulting from injuries, infectious diseases, poor maternal nutrition, or other causes.) What if the same disease is diagnosed in a fetus *in utero*—is it ethical to abort the fetus? The decision to abort a fetus or to withhold treatment from a child with a genetic disease raises important ethical questions. Here are some questions to consider.

1. Tay–Sachs disease is a genetically controlled disease whose victims are in constant pain and never survive beyond about 4 years of age. Does it make sense to spend thousands of dollars on the medical care of a child who has no chance of living beyond age 4, or even of enjoying those few years free from pain? Would it make a difference if a few people with the disease were capable of surviving? What if we were dealing with a disease that people could survive, but only with some disability?

2. When genetic testing has been done, people often meet with a genetic counselor to have the results of the test explained to them. The code of ethics of genetic counselors includes the ethic that they give information about 'health risks' in a non-directive and 'value-neutral' way. Is this possible? In what ways does a person's concepts of 'health' influence how they understand information about health risks?

3. The involvement of third-party insurance policies raises more issues. Should insurance policies pay for genetic testing? Should insurance policies pay medical expenses for genetic diseases that could have been avoided after screening? Some insurance policies will pay for medical treatment, but not for the testing that might have avoided the need for the treatment. Do you think insurance policies should cover genetic screening?

4. Should genetic screening be covered for certain ethnic groups but not others, just because the risks differ? For example, thalassemia is more prevalent among Italians, Greeks, and certain southeast Asians; should insurance cover testing for this condition in a person of Italian descent, but not in a person of English or Danish descent? Or in the instances common in the United States, in which descent is either mixed or unknown?

5. If a genetic disease is detected during pregnancy, should insurance policies cover termination of the pregnancy if desired by the parents? If parents elect not to terminate a pregnancy, and a child is born with a genetic disease, should insurance policies cover any specialized medical care that might be necessary? Should insurance companies be allowed to deny coverage or increase premiums if a genetic disease is discovered?

6. Screening for some inherited diseases such as PKU are done on all newborns. Informed consent is not gathered as the screening is universal and diet can prevent the harmful effects of the mutation. As tests are developed for other genetic diseases, should screening for these become routine? Do parents have the right to refuse to be told the results of such tests? What if there is no cure for the condition detected?

7. What role might science have in answering questions of medical ethics that are related to genetics? For example, can science help in assessing the benefits and risks of genetic screening? How reliable are the genetic tests? (For example, the test to detect a cystic fibrosis allele is currently less than 90% accurate.) To what extent does the detection of an allele predict a harmful phenotype? What constitutes sufficient evidence that a condition is genetically determined? What are the limits of the ability of science to contribute answers to these questions of medical ethics?

whether the fetus is a boy or a girl, something easily determined from examining the chromosomes. Will couples use this technology to select the sex of their offspring? This already happens in India, where abortion is legal and determination of the sex of the fetus by ultrasound is widely available to those who can afford (or can borrow) a fee of a few hundred dollars. A 1988 study of 8000 abortions in India's clinics showed that 7997 were female and only 3 were male. In the United States, clinics offering prenatal genetic testing have found that over one-fourth of the couples who come to them are motivated by the possibility of choosing their baby's sex.

As genetic testing becomes more common, it is inevitable that test results will occasionally be misused. In one case, school officials were told that a child needed to be kept on a special diet because he had phenylketonuria (PKU). Although he was functioning normally, the child was placed in a class for the learning disabled. The school officials apparently knew that PKU could cause mental retardation, but were unaware that this outcome could be averted by the special diet. As this case shows, the misuse of genetic information may result from ignorance.

Discrimination by employers or insurers represents another possible misuse of genetic information. Employers may refuse to hire or promote, or insurance companies may refuse to insure, persons who have or who are suspected of having a genetic condition. In some cases, benefits have been denied to heterozygous carriers of recessive conditions or to persons at risk for other reasons whose phenotype was unaffected. The practice is still uncommon, but it is growing and is likely to continue to grow as more and more genetic information becomes available through medical testing. One of the best safeguards against this kind of discrimination is a strict adherence to rules governing the confidentiality of medical records and other personal information held by health care providers and medical testing laboratories. Recent rulings state that the Americans with Disabilities Act protects people from discrimination on the basis of their genetic profile.

The language used by geneticists and genetic counselors may be misleading when only those persons who are homozygous for the nondisease allele are reported as 'normal.' Heterozygotes are called 'carriers' although phenotypically they show no disease. In Figure 3.13A, for example, only the child on the far left of the diagram would be reported as normal, even though 75% of the children and both of the parents are phenotypically normal with no sign of disease.

Altering individual genotypes

Some rare genetic traits impair health. In the future, it may become possible to correct certain genetic defects by direct alteration of the individual genotype. A more realistic possibility is a form of gene splicing in which the functional allele is added to the DNA of persons with defective alleles, a practice commonly referred to as gene therapy. A new gene is inserted into a cell that also continues to carry the mutated allele. We will cover the details of gene therapy in the next chapter.

Altering the gene pool of populations

Some people have proposed that, instead of treating people one at a time, we should alter the genetic makeup of populations (the entire gene pool) by changing the frequencies of certain genotypes. One difference between this approach and the approaches already described has to do with who is perceived to reap the benefits. Genetic testing, counseling, and the altering of individual genotypes are justified in terms of the pain and suffering that may be spared to individuals. In contrast, all attempts to alter the gene pool carry with them notions of harm or benefit to society rather than to the individual.

Positive eugenics. The altering of the gene pool through selection is called **eugenics**, from the Greek words meaning 'good birth.' This idea is not at all new: Plato's *Republic* (book 5) suggests that the best and healthiest individuals of both sexes be selected to be the parents of the next generation, much as we breed our horses and cattle. Plato's type of eugenics is called **positive eugenics**, meaning an attempt to alter the gene pool by selectively *increasing* the genetic contributions of certain chosen individuals or genotypes. Positive eugenics was also proposed in the twentieth century by the Nobel Prize-winning geneticist H.J. Muller, who advocated setting up sperm banks to which selected male donors would contribute. Muller thought that women would eagerly seek artificial insemination with these sperm in the hopes of producing genetically superior children. Several entrepreneurs have established sperm banks (and a smaller number of egg banks) offering to infertile couples (and others) the gametes of people thought to carry desirable traits. The system is not regulated, however, by any public agency, and many sperm banks and egg banks seem to be motivated more by profit than by any desire to change the gene pool. In 1999, a photographer in California began advertising the eggs of several fashion models, offering them at auction to the highest bidder over the Internet.

Ethical and other questions raised by positive eugenics usually center on the lack of an agreed standard for human excellence. The traits most often discussed by those who favor eugenics are intelligence and athletic ability. However, these traits are genetically complex and are highly influenced by education, training, and other environmental variables. Studies attempting to demonstrate a genetic influence on these and other traits were in many cases poorly done, leading many scientists to doubt the existence of any reliable evidence concerning the genetic control of human intelligence and other complex traits.

The complexity of the human genotype raises other issues. What if Einstein had been heterozygous for some genetic disease? If a society wanted to use his germ cells to breed people of superior intelligence, they would also unwittingly be selecting whatever other traits he happened to possess, possibly including a genetic defect in the process. Suppose a person inherited the manic-depressive disorder of Robert Schumann or Vincent Van Gogh, instead of their creative talents? What liability or what responsibility would a sperm bank face if a descendant were born with a genetic defect? What constitutes 'superiority' in an individual, and who should have the power to make such choices?

Negative eugenics. Most discussions of eugenics have centered on **negative eugenics**, the prevention of reproduction among people thought to be genetically defective or inferior. Founded by Francis Galton (1822–1911), a cousin of Charles Darwin, the modern eugenics movement has generally tended to emphasize negative measures. Galton and his supporters were very much interested in measuring intelligence, and they developed some of the early versions of what we now call IQ tests. Through the use of these and other tests, supporters of eugenics have long sought scientific respectability for their attempts to label certain people as genetically defective or inferior.

The Nazis instituted a program of negative eugenics in Germany, beginning with the forced sterilization of mental 'defectives,' deaf people, homosexuals, and others. The eugenics program soon grew into a program for the mass killing of all those millions who did not belong to Hitler's 'master race.' By 1945, the Nazis had killed millions in the name of racial purity and Aryan superiority. The Nazis also practised positive eugenics by encouraging German women with certain traits to have more children.

In the United States, the eugenics movement started as a series of attempts to identify, segregate, and sterilize mental 'defectives.' The movement soon found allies among racists and especially among those who sought to curb the new waves of immigration during the period from about 1890 to 1920. During the 1890s, one Kansas doctor sterilized 44 boys and 14 girls at the Kansas State Home for the Feeble-Minded, while Connecticut passed a law prohibiting marriage or sexual relations between any two people 'either of whom is epileptic, or imbecile, or feeble-minded.' A 1907 Indiana law required the sterilization of 'confessed criminals, idiots, imbeciles, and rapists in state institutions when recommended by a board of experts.' Fifteen other states passed similar laws, as often for punitive as for eugenic reasons. From 1909 to 1929, 6255 people were sterilized under such laws in California alone.

The writings of the American eugenicists became increasingly racist and anti-immigrationist in tone during this period. One eugenicist, for example, wrote in 1910 that "the same arguments which induce us to segregate criminals and feebleminded and thus prevent breeding apply to excluding from our borders individuals whose multiplying here is likely to lower the average [intelligence] of our people." In the 1960s, H.J. Muller wrote several articles warning against the practice of protecting and extending the lives of the 'genetically unfit,' those whom natural selection would tend to eliminate from the population. According to Muller, our medical intervention would only perpetuate genetic defects in our gene pool. Muller spoke pessimistically of a population divided into two groups, one so enfeebled from genetic defects that their very lives had to be sustained by extraordinary means, and the other group consisting of phenotypically normal people who had to devote their entire lives to the care and sustenance of the first group. Muller's views have not been substantiated by any evidence.

Biological objections to eugenics. Biological arguments against negative eugenics are based on the realization that eugenic measures could be expected to produce only small changes at great cost. Most known genetic defects are both rare and recessive, and selection against rare, recessive traits can only proceed very slowly no matter what the circumstances. As the trait gets increasingly rare, selection against it becomes increasingly ineffective. For example, the gene for albinism has a frequency of about 1 in 2000 in many human populations. If a eugenic dictator ordered all albinos to be killed or sterilized, theoretical calculations show that it would require about 2000 generations (about 50,000 years) of constant vigilance just to reduce the frequency of this trait to half of its present value. The reason why the process works so slowly is that most individuals carrying the gene for a rare, recessive trait are heterozygous and their phenotype does not reveal the presence of the gene. In contrast, modern techniques that allow the detection of the gene in heterozygous form would greatly increase the effectiveness (and hence the dangers) of negative eugenic measures.

For characteristics such as height or IQ, which are controlled by many genes and are influenced strongly by environmental factors, estimates are that eugenic selection would be so slow as to be barely perceptible. One geneticist calculated that it would take about 400 years of constant, unrelenting and totally efficient selection to raise IQs by about 4 points; the same improvement could be achieved through education in as little as 4 years, and with far less cost. This topic is addressed in more detail in Chapter 7.

Finally, eugenic measures can at best address only a small percentage of undesirable conditions, because most physical disabilities and medical conditions result from accidents, from infectious illnesses, or from exposure to toxic substances in the environment, not from inherited genetic makeup, and eugenic measures are powerless to alter these nongenetic causes. Genetic conditions may also result from new mutations, rather than from the inheritance of defective genes, and eugenic measures have no capacity to eliminate newly mutated genes or to depress the frequency of any gene below the mutation rate.

There is no biological basis for the claims of any eugenics movement that their methods could in any way improve humankind other than at great cost. The risks of negative eugenics are especially great, and include the possibility of genocide—the attempted extermination of a race or ethnic group. In addition there are no biological benefits. We are coming to know that population health and stability depends on genetic diversity; thus, narrowing a gene pool eugenically makes the population more vulnerable to infectious disease. Infectious disease is a far more widespread and common cause of sickness and death than is genetic disease.

Changing the balance between genetic and environmental factors

Although many traits are inherited, most are also influenced by the environment. Phenotypes result not just from genes but from the interactions between genes and their environment. In addition, far more disability is caused entirely environmentally through accidents and illnesses than is caused genetically. Even for most conditions that are caused genetically, elimination of an allele from the population is hardly the only option. Most genetic conditions can be modified or accommodated in several different ways.

Euphenics. **Euphenics** (literally, 'good appearance') includes all those techniques that either modify genetic expression or alter the phenotype to produce a modified phenotype. Plastic surgery, such as to repair body parts, is a form of medical intervention that alters the phenotype. Other examples include the installation of pacemakers in defective hearts, the giving of insulin to diabetics, and the dietary control of phenylketonuria. Although the genes remain unchanged, euphenics modifies or compensates for their phenotypic expression is in such a way that they no longer cause harm.

Many leading geneticists have argued, as an alternative to eugenics, that there is nothing wrong with altering the phenotype or the environment so that formerly disabling genotypes are no longer so harmful or debilitating. A leading advocate of this viewpoint was Theodosius Dobzhansky (1893–1975), who favored measures to permit people with hereditary 'defects' to overcome their handicaps and become phenotypic copies (phenocopies) of normal, healthy human beings. Once phenotypes could be controlled culturally, said Dobzhansky, the presence of formerly defective genotypes would cease to be the subject of any great concern. Euphenic intervention is already common practice for a number of genetic conditions. As our ability to modify phenotypes increases (e.g., with advances in corrective surgery), this type of practice is likely to become more common.

Euthenics. Another type of intervention is called **euthenics**. In this form of intervention, both genotype and phenotype remain unchanged, but the environment is modified or manipulated so that the phenotype is no longer as disabling as before. (In euphenics, by contrast, the phenotype is modified.) Examples of euthenic measures include canes, crutches, wheelchairs, and wheelchair ramps for those who cannot walk unaided, guide dogs and Braille for the sight-impaired, eyeglasses for the nearsighted, and so on. Conditions that are improved or assisted by euthenics may be either genetic or not.

Most people with disabilities support research that would prevent the recurrence of their condition in other people, especially if pain or paralysis are involved. However, medical research is expensive, its results are uncertain, and its benefits may take many years to become widely available. Euthenic measures are often less expensive and more quickly made available once they have been developed. Many of the people who use euthenic devices feel that they would be better served by simple improvements in the devices (e.g., better wheelchairs) than they would be if medical research were our only emphasis.

Eupsychics. Many people with uncommon traits or conditions (whether genetic or not) feel that they are best served by being accepted as they are and do not necessarily want to be 'cured.' (For an example, see our Web site, under Resources: Deafness.) Social and behavioral measures, or **eupsychics**, may lessen the impact of or compensate for disabling conditions. Included are the special education of handicapped individuals, mainstreaming (education of the handicapped in a regular public school setting), and the education and social conditioning of nonhandicapped members of society so that they will better understand and accommodate the needs of all citizens.

Genetic research seeks to understand the molecular mechanisms underlying normal physiology and health, but genetic testing and counseling emphasize diseases. The assumptions underlying genetic testing have at times included viewing variations as defects and desiring to apply to humans some arbitrary standard of perfection. Society may be better

served by an emphasis on the abilities, rather than the disabilities, of each individual. All individuals should be encouraged to develop their talents and abilities to the fullest. Whenever a person is discouraged from trying to develop a certain skill, ability, or talent, both the individual and the society are the losers in the long run.

THOUGHT QUESTIONS

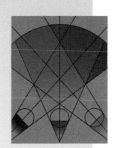

1 Are all 'birth defects' genetic defects? Do the same ethical questions regarding diagnosis and counseling apply to both genetic and nongenetic traits?

2 In some hospitals, screening for phenylketonuria is often performed on all infants. Does this violate the principle of informed consent? Is this practice ethical or not? How is it commonly justified? Do you feel that the justification is adequate?

3 Some groups opposed to abortions have also begun to object to certain kinds of genetic testing. What good, they ask, can come from knowing that a fetus suffers from a particular genetic or chromosomal defect if the parents are opposed to abortion of the fetus on religious or similar grounds? For such situations, discuss the costs, benefits, and ethical status of genetic testing. Does it matter what kind of testing is performed? Does it matter what genetic or chromosomal defect is being tested for?

4 Do unborn children have a 'right' to inherit an unmanipulated set of genes? Do they have a right to inherit 'corrected' genes if such a possibility exists? What kind of informed consent can we expect on behalf of unborn generations? Can a person make decisions that affect the genotype of all of his or her progeny? Do we need safeguards to protect future generations against the selfish interests of the present generation?

5 In what sense is 'positive' eugenics positive? In what sense is it negative? Should human populations be bred as we breed domesticated animals?

6 If public funds will be spent for the care and treatment of a person with a genetic condition, does that alter the ethical balance between the rights of the individual and the rights of society? If a euphenic measure is available, should public funds be used to 'correct' the condition? Should the person have the right to refuse such treatment? Should public funds be withheld from a person who refuses such euphenic measures?

7 How much access to genetic information about a subscriber or employee *should* an insurance company or an employer have? How much access *do* they have now?

Concluding Remarks

Through the processes of transcription and translation, genes get expressed as proteins, and proteins lead ultimately to phenotypes. As we have seen in many contexts in this chapter, however, most of human genetics is more complex than the either/or traits that Mendel studied in pea plants. It is not that Mendel's laws do not apply to humans; they do. There are some human conditions that do follow simple Mendelian inheritance with a mutation in a single gene leading to the trait. Rather, since Mendel's time, we have learned that the inheritance of most traits in all organisms, not just in humans, is more complex than Mendel and scientists early in the twentieth century envisioned. Many traits, including sex determination and disease susceptibility, are influenced by multiple genes. The expression of the genotype as a phenotype is also influenced by the environment. Most biologists today would agree that phenotypes are far more malleable than was assumed in the past, contradicting earlier ideas of biological determinism that assumed that genotypes solely determined phenotypes. In addition, phenotypes can be modified by technologies and other aspects of cultures.

We are entering an era in which people will be able to find out a lot about their own genotype and the genotypes of their children. Identifying a genetic predisposition to a chronic disease may allow a person to make healthier choices about his or her diet and lifestyle. As we saw with the couple in the introduction to this chapter, people will have the power to find out whether a fetus carries a genotype for a fatal illness. This new knowledge will give people choices that they have not had in the past, but none of those choices are likely to be easy.

Chapter Summary

- **DNA**, a **nucleic acid**, is the template for the synthesis of another nucleic acid, **RNA**, in a process called **transcription**.

- **mRNA** is the template for protein synthesis in a process called **translation**. Three mRNA bases form a **codon**, directing the addition of one amino acid to a **protein**.

- Changes in the DNA sequence (**mutations**) are reflected as changes in the mRNA sequence that usually result in changes in the amino acid sequence of proteins. Such changes may affect the folded shape and therefore the function of the proteins. Gene mutations create new alleles.

- **Phenotypes** result from the activity of proteins, often of several proteins acting together.

- Human genes are now being identified at an increasingly rapid pace through **pedigrees** and linkage with DNA markers.

- Many genetic diseases result from alleles that code for nonfunctional proteins. Some of these are inherited as simple Mendelian alleles; others show more complex patterns of inheritance.

- **Risk** is the probability of a condition's occurring; some genotypes are associated with increased risk.

- Genetic diseases can be diagnosed prenatally by amniocentesis, by chorionic villus sampling, and by other techniques, including those that use the **polymerase chain reaction** (**PCR**) to amplify DNA sequences into many copies. Testing for genetic diseases can also be done on children or adults. In any case, testing requires prior **informed consent**.

- Altering the gene pool at the population level is called **eugenics**.

CONNECTIONS TO OTHER CHAPTERS

Chapter 1	Molecular genetics is forcing a change in the scientific paradigm that includes our concepts of phenotypes and traits.
Chapter 1	Manipulating human heredity has raised several ethical concerns.
Chapter 2	The same basic laws of genetics apply to humans as to other animals and to plants.
Chapter 4	Human genes are often very similar to those from other species, and studying these relationships can tell us about species' evolutionary history.
Chapter 5	Evolution takes place whenever allelic frequencies change.
Chapter 7	Human populations differ in the frequencies of many alleles.
Chapter 8	Mating patterns and sexual strategies can alter allelic frequencies in populations.
Chapter 9	Population control seeks to manage the size of populations; eugenics, in contrast, seeks to alter the gene pool.
Chapter 11	Genetic engineering can be used to improve the traits of commercially important plant species.
Chapter 12	Predispositions for some cancers are hereditary.
Chapter 13	Huntington's disease is one of several brain disorders for which a genetic basis has been identified.
Chapter 18	Conserving genetic diversity is an important aspect of protecting biodiversity.

PRACTICE QUESTIONS

1. When a man and a woman who are both albino have children, what percentage of their children will be albino?

2. Are DNA markers genes?

3. A particular disease is suspected of being a genetic disease. A person with the disease tests positive for 15 DNA markers, each detected with a specific DNA probe, and negative for 20 other markers. What would be the next step in trying to determine which of the markers might be located near a gene responsible for the disease?

4. What is the risk of a child's having a recessive genetic trait when both parents are from a population in which the frequency of the recessive allele is 1 in 1000 (0.1%)? What is the risk when one parent is from a population with an allelic frequency of 0.1% and the other parent is from one in which the frequency is 1 in 100,000 (0.001%)?

5. How many copies of a DNA fragment are synthesized in 20 rounds of amplification by PCR (assuming that each step works correctly)? How many will be synthesized in 30 rounds?

6. How many different genes are mutated in cystic fibrosis? Do all people with cystic fibrosis have the same mutation?

7. A point mutation that substitutes a single base for another changes that codon but not the succeeding ones. If two bases that are next to each other are replaced by two different bases, how many codons will be altered? On what might your answer depend?

Issues

- Does genetic engineering fundamentally change the biology of an organism?
- Does gene therapy work?
- When should gene therapy be used? When should it not be used?
- Do DNA tests positively identify individuals?
- Why does the U.S. government fund the Human Genome Project?
- What benefits have been derived from the Human Genome Project?
- How could the results of the Human Genome Project be misused? How can we guard against such misuse?

Biological Concepts

- Biotechnology (The Human Genome Project; genetic engineering)
- Molecular biology (genomics; bioinformatics)
- Structure–function relationships (proteomics)

Chapter Outline

Genetic Engineering Changes the Way That Genes Are Transferred

Methods of genetic engineering

Genetically engineered insulin

Gene therapy

Molecular Techniques Have Led to New Uses for Genetic Information

The first DNA marker: restriction-fragment length polymorphisms

Using DNA markers to identify individuals

Using DNA testing in historical controversies

The Human Genome Project Has Changed Biology

Sequencing the human genome

The human genome draft sequence

Mapping the human genome

Some ethical and legal issues

Genomics Is a New Field of Biology Developed as a Result of the Human Genome Project

Bioinformatics

Comparative genomics

Functional genomics

Proteomics

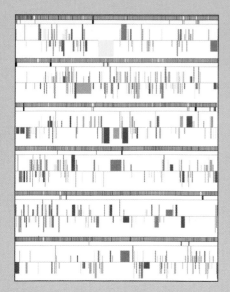

4 Genetic Engineering and Genomics

As a result of information published in 2001, humans now know more about themselves, at least at the molecular level, than they ever have before. This watershed date marked the publication of the draft of the nucleotide sequence of all of the DNA in human chromosomes. Along the way, a complete map of the location of these nucleotide sequences on the chromosomes was also produced. All of this information is stored in an enormous database that is publicly available for use by any scientist in the world. A tremendous amount of basic molecular biology has been discovered in the course of the Human Genome Project that produced this database. As a tool for biological research, this database potentially offers new ways of studying everything else in biology. In addition, the project has spawned many practical advances in biotechnology and genetic engineering.

Genetic Engineering Changes the Way That Genes Are Transferred

Genetic engineering is the direct alteration of individual genotypes. It is also called recombinant DNA technology or gene splicing, terms which are used interchangeably. Human genes can be inserted into human cells for therapeutic purposes (gene therapy, p. 100). In addition, because all species carry their genetic information in DNA and use the same genetic code, genes can be moved from one species to another. The uses of genetic engineering in plants are discussed in Chapter 11. Here we see some of the applications of genetic engineering for human medicine.

Methods of genetic engineering

Whether the 'engineered' gene is one from the same species or a different species, the techniques are much the same. All these technologies depend on being able to cut and reassemble the genetic material in predictable ways. This is possible owing to the discovery of special enzymes called restriction enzymes.

Restriction enzymes. **Restriction enzymes** are enzymes used to cut DNA at specific sites. There are several hundred restriction enzymes currently known and each cuts DNA at a different nucleotide sequence; these target sites are generally about four to eight nucleotides long (Figure 4.1). Each of these restriction enzymes is a normal product of a particular bacterial species, and most are named after the bacteria from which they are derived. Thus, in Figure 4.1, *Hae*III is an enzyme from the bacteria *Haemophilus aegypticus* and *Eco*RI is from *Escherichia coli*. They are called restriction enzymes because their normal function within the bacteria is to restrict the uptake of DNA from another bacterial species. Each species' restriction enzyme cuts the DNA from other species, but not its own, because its own DNA does not contain the nucleotide sequence that is the target site for its own enzyme.

Several other enzymes are known that can break apart a DNA molecule, but an enzyme that acts indiscriminately is of little use in genetic engineering. Restriction enzymes act specifically. Each restriction enzyme generally cuts a sample of DNA in several places, wherever the DNA contains a particular sequence of bases that the enzyme recognizes, forming a series of pieces (called restriction fragments). A given restriction enzyme mixed with the same sequence of DNA always produces the same number of fragments. The length of the pieces may vary if there are variable repeat sequences, for example, but the number of pieces and the places cut are always the same. Before the discovery of restriction enzymes, breaking chromosomal DNA into pieces was done mechanically, producing different numbers of pieces every time the procedure was done, making the results of DNA techniques impossible to reproduce from one experiment to the next. Because restriction enzymes always cut at the same sites, they can be used in genetic engineering.

Restriction enzymes in genetic engineering. The first step in inserting a gene for genetic engineering is to isolate the gene in question. This is carried out by using a restriction enzyme to snip out the desired segment of DNA. Each restriction enzyme cuts the DNA at specific places, defined by their DNA sequences. The most useful restriction enzymes are those that cut the two DNA strands at locations that are not directly across from each other, producing short sequences of single-stranded DNA known as sticky ends (see Figure 4.1). For example, the commonly used restriction enzyme *Eco*RI always targets the sequence GAATTC, cutting it between G and AATTC, breaking the two-stranded sequence into fragments that have sticky ends. The ends are called 'sticky' because they can stick together spontaneously with another molecule containing complementary sticky ends. In fragments cut with *Eco*RI, the single-stranded AATT sequences can pair with one another, stick together, and then be joined permanently. (An enzyme such as *Hae*III that cuts at sites directly across from each other forms 'blunt', rather than sticky, ends, as shown in Figure 4.1.)

If a particular restriction enzyme produces sticky ends, all fragments cut with that enzyme will have sticky ends that match one another. Thus, a fragment can be joined to any other fragment cut with the same enzyme. This makes it possible to use restriction enzymes to cut a DNA sequence and insert a functional gene with matching sticky ends.

Figure 4.1

Restriction enzymes. The nucleotide sequences recognized and cut by the restriction enzymes *Hae*III and *Eco*RI are shown.

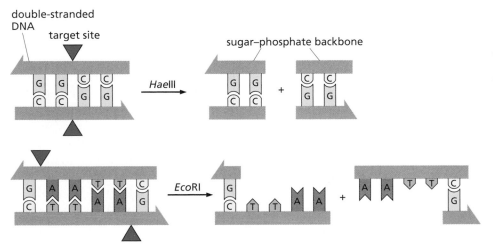

The *Hae*III target site is four bases long and the enzyme cuts the DNA strands at sites directly across from each other, leaving double-stranded ('blunt') ends.

The *Eco*RI target is six bases long and it cuts the DNA between G and AATTC. The sites on the two strands are not directly across from each other, leaving short single-stranded ('sticky') ends.

Restriction enzymes that produce blunt ends are useful in other ways, but are not useful for genetic engineering because the fragments cannot be put back together.

Cutting an entire chromosome with a restriction enzyme produces many fragments, only one of which contains the gene to be isolated. A **DNA probe** specific for the gene will isolate the fragment containing the gene of interest. As we have seen before, such a probe is a complementary DNA strand that carries a radioactive or chemical tag. The probe allows geneticists to isolate the labeled sequences, and then separate the desired genes from the DNA probes that pair with them.

A functional gene isolated in this way can then be inserted into another piece of DNA. The target DNA is cut with the same restriction enzyme, so sticky ends complementary to the fragment are available, and the gene can be incorporated permanently. So far, most genetic engineering of human genes has involved the introduction of these human genes into bacteria. The reasons for this are largely practical: many human gene products are useful in medicine but are more readily produced in large amounts inside genetically engineered bacteria than inside people. For example, the hormone somatostatin, also called growth hormone, is highly valued for the treatment of certain types of dwarfism. The hormone is, however, difficult to obtain from human sources (the traditional way is to extract it from the pituitary glands of dozens of cadavers) and is therefore very expensive. Insulin, the hormone needed by diabetics, is another example of a human gene product. Both of these hormones could be obtained from sheep or pigs or other animals, but the animal hormones are not as active in humans as the human hormones, and some patients are allergic to hormones obtained from other species. Genetic engineering provides a cost-effective way of manufacturing large amounts of these human hormones in bacteria.

Genetically engineered insulin

Human insulin was the first commercially produced genetically engineered product. The initial step is to grow human cells in tissue culture. Tissue culture is a procedure in which cells that have been removed from an organism are grown in a dish of nutrient-rich medium kept at body temperature in an incubator. After a sufficient number of cells have grown, DNA extracted from the cell nuclei is then exposed to a restriction enzyme that cuts the DNA into desired fragments. One fragment contains the human gene for insulin, which can be isolated using a DNA probe.

The same restriction enzyme is used on nonchromosomal DNA molecules, called **plasmids**. Bacteria have a single chromosome in the form of a closed loop. Many also have a number of plasmids, short circular DNA pieces that are separate from the bacterial chromosome (Figure 4.2). Plasmids are used in genetic engineering because, being short, they have fewer sites at which a given restriction enzyme can cut. Cutting a DNA sequence in the plasmid with the same restriction enzyme that was used on the human DNA creates sticky ends that match the DNA fragment taken from the human cell. This allows incorporation of the human gene for insulin into the bacterial plasmid. The bacteria are then treated so that they take up the engineered plasmid. In most cases, the plasmid also contains another DNA sequence that can be used to

Figure 4.2

A bacterial cell showing its single chromosome and one plasmid.

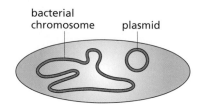

bacterial chromosome   plasmid

select the bacteria that have incorporated an engineered plasmid. For example, the plasmid might contain the gene for an enzyme that gives the bacteria resistance to a common antibiotic; the antibiotic can then be used to select the bacteria that have incorporated this gene while killing the majority that are still susceptible. The procedures sound easy and straightforward, but each step of the process is technically difficult and only a small proportion of the attempts succeed.

The genetically altered bacterium can now be cloned, that is, allowed to multiply asexually, which produces vast numbers of genetically identical copies of itself and its engineered plasmid. The resultant bacteria then transcribe and translate the human gene to produce human insulin (Figure 4.3). The human insulin extracted from these bacteria, called recombinant human insulin, can be given to diabetic patients.

Figure 4.3

Production of genetically engineered insulin.

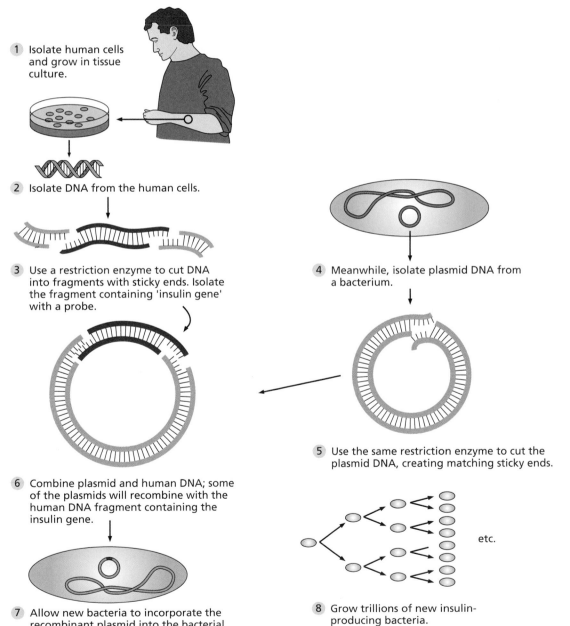

1 Isolate human cells and grow in tissue culture.

2 Isolate DNA from the human cells.

3 Use a restriction enzyme to cut DNA into fragments with sticky ends. Isolate the fragment containing 'insulin gene' with a probe.

4 Meanwhile, isolate plasmid DNA from a bacterium.

5 Use the same restriction enzyme to cut the plasmid DNA, creating matching sticky ends.

6 Combine plasmid and human DNA; some of the plasmids will recombine with the human DNA fragment containing the insulin gene.

etc.

7 Allow new bacteria to incorporate the recombinant plasmid into the bacterial cell, then screen bacteria to find the ones that have incorporated the human gene for insulin.

8 Grow trillions of new insulin-producing bacteria.

Gene therapy

Instead of growing human insulin in bacteria (see Figure 4.3), genetic engineering could theoretically be used to introduce the insulin gene into human cells that do not possess a functional copy. (That would still not cure diabetes unless these cells were also capable of appropriately increasing or decreasing their output of insulin according to conditions.) This type of genetic engineering is called **gene therapy**, the introduction of genetically engineered cells into an individual for therapeutic purposes.

Treatment for hereditary immune deficiency. Human gene therapy has been used successfully to treat severe combined immune deficiency syndrome (SCIDS), a severe and usually fatal disease in which a child is born without a functional immune system. Unable to fight infections, these children will die from the slightest minor childhood disease unless they are raised in total isolation: the 'boy [or girl] in a bubble' treatment. The enzyme that controls one form of SCIDS has been identified; it is called adenosine deaminase (ADA) and its gene is located on chromosome 20. A rare homozygous recessive condition results in a deficiency of this enzyme, which in turn causes the disease.

Gene therapy for this condition consists of the following procedural steps, shown in Figure 4.4.

1. Normal human cells are isolated. The cells most often used are T lymphocytes, a type of blood cell that is easy to obtain from blood and easy to grow in tissue culture.

2. The isolated cells are grown in tissue culture.

3. The DNA from these cells is isolated.

4. A restriction enzyme is used to cut the DNA into fragments with sticky ends; one will contain the functional gene for ADA. A probe with a complementary DNA sequence is then used to isolate and identify the fragments bearing the gene.

5. The same restriction enzyme is used to create matching sticky ends in viral DNA isolated from a virus known as LASN. This virus was chosen because it can be used as a **vector**: it can transfer the gene into the desired human cells—the host. (Other vector viruses have also been used; each virus type varies in the size of DNA fragment that can be inserted and the type of cell that it can enter.)

6. The viral DNA is then mixed with the human DNA fragments and allowed to combine with them.

7. The virus is allowed to reassemble itself; it is then ready for further use.

8. Blood is drawn from the patient to be treated and T lymphocytes are isolated from this blood. These lymphocytes, like all of the other cells from this person, are ADA-deficient because they do not possess a functional ADA allele.

9. The virus is now used as a vector to transfer the functional gene. The virus must get the gene not only into the lymphocyte but also into its nucleus. The gene must incorporate into the cell's DNA in a location where it will be transcribed and where it does not break up some other necessary gene sequence.

10. The lymphocytes are tested to see which ones are able to produce a functional ADA enzyme, showing that they have successfully incorporated the functional ADA allele.

11. The genetically engineered lymphocytes are injected into the patient, where they are expected to outgrow the genetically defective lymphocytes because the ADA-deficient cells do not divide as fast as cells with the ADA enzyme.

Figure 4.4

An example of gene therapy showing the transfer of the human gene responsible for adenosine deaminase (ADA).

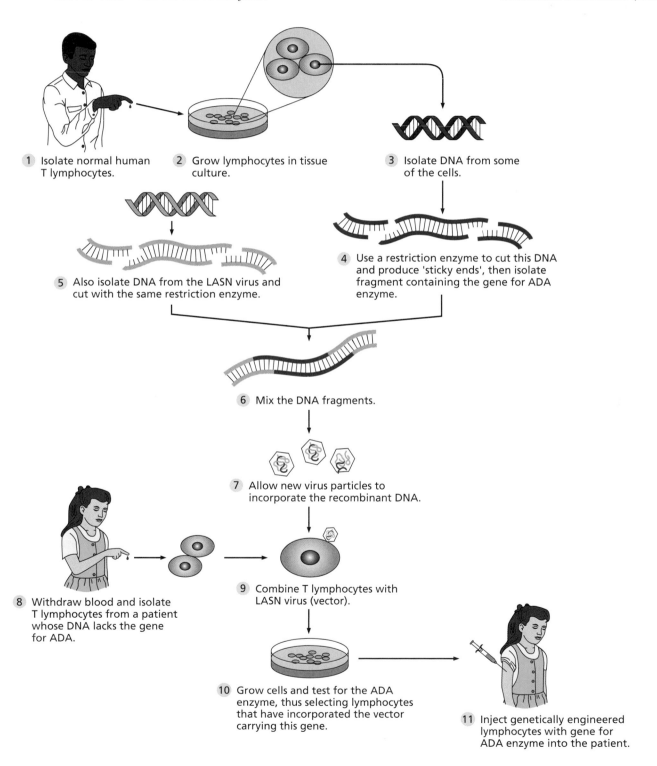

1 Isolate normal human T lymphocytes.

2 Grow lymphocytes in tissue culture.

3 Isolate DNA from some of the cells.

5 Also isolate DNA from the LASN virus and cut with the same restriction enzyme.

4 Use a restriction enzyme to cut this DNA and produce 'sticky ends', then isolate fragment containing the gene for ADA enzyme.

6 Mix the DNA fragments.

7 Allow new virus particles to incorporate the recombinant DNA.

8 Withdraw blood and isolate T lymphocytes from a patient whose DNA lacks the gene for ADA.

9 Combine T lymphocytes with LASN virus (vector).

10 Grow cells and test for the ADA enzyme, thus selecting lymphocytes that have incorporated the vector carrying this gene.

11 Inject genetically engineered lymphocytes with gene for ADA enzyme into the patient.

Technical difficulties in gene therapy are numerous. Transferring large pieces of DNA into cells is difficult (most genes are large). Inserting a gene in a location in the DNA where its protein product will be transcribed and translated in a normal way is far more difficult.

The gene therapy described above provides a functional gene that is transcribed and translated by the body cells, producing the missing enzyme in lymphocytes. Because lymphocytes are not the only cells that need the ADA enzyme, the patient must also receive injections of the ADA enzyme coupled to a molecule that permits it to enter cells. (This last step might not be necessary for the treatment of other enzyme defects.) The enzyme controls the symptoms of the disease, but it is not a cure because the underlying disease is still present. Gene therapy for ADA was first successfully used on a 4-year-old girl in 1990. A second patient, a 9-year-old girl, began receiving treatments in 1991. Both patients are being closely monitored, and their immune systems are now working properly. However, because the genetically engineered cells are mature lymphocytes, which have only a limited lifetime, repeated injections of genetically engineered cells are needed.

To get around this problem, in the hope of bringing about a more lasting cure, some Italian researchers have tried using both genetically engineered lymphocytes (as described above) and genetically engineered bone marrow stem cells. Stem cells divide to form all the developed types of blood cells (see Chapter 12) and they maintain this ability throughout life. Therefore, after repaired lymphocytes die off, stem cells with repaired DNA could divide to provide new, ADA-functional lymphocytes, possibly for the lifetime of the individual. This type of therapy was begun on a 5-year-old boy in 1992, and since then several other children have received this treatment.

Questions of safety and ethics. There are legitimate safety concerns with human gene therapy. For example, any virus used as a vector must be capable of entering human cells. Might such a virus cause a disease of its own? To preclude this possibility, the viruses used in human gene therapy have been from viral strains with genetic defects that render them incapable of reproducing and spreading to other cells. Might random insertion into the host DNA destroy some other gene? Methods are being developed for directing the insertion location, but it is still largely a random event. In 1999, gene therapy clinical trials were halted in the United States when an 18-year-old boy died after receiving a viral vector for gene therapy for a metabolic disease. The reasons for his death were not apparent, so clinical studies were halted until issues of safety could be addressed. The boy's father has testified at a U.S. Senate hearing that the boy and his family were not fully informed of the dangers of the experiment. Others have raised ethical objections to the use of the term 'gene therapy' in clinical trials when most of the experiments that have been done so far have not been designed to cure any condition, only to alleviate symptoms (or to test the safety of the procedure itself).

Gene therapy also raises other ethical concerns. New recombinant DNA procedures are very expensive to develop. This raises ethical issues of fairness: will the benefits of genetic engineering be available only to those who can afford them? Should government programs provide them through Medicare and Medicaid? Should insurance cover their use? How can society's health care resources best be distributed? If medical resources are limited, should an expensive procedure used on one person take up needed resources that could cover inexpensive treatments of other diseases for many people? These particular questions are not unique to genetic engineering; they apply to any expensive form of medical treatment.

Genetic engineering may someday become commonplace in human cells. In theory, gene therapy could be practised either on somatic cells or on gametes. If it were performed on somatic cells, the effects of the gene therapy would last as much as a lifetime, but no longer. For example, insertion of the functional allele for insulin into the pancreatic cells of patients with diabetes might cure them of the disease, but they would still pass on the defective alleles to their children. A general consensus has been reached that using gene therapy on somatic cells has an ethical value if it is used for the purpose of treating a serious disease.

If successful gene therapy is performed on germ cells, then the genetic defect will be cured in the future generations derived from those germ cells. In addition to all the ethical questions raised earlier, gene therapy on germ cells raises many additional ethical questions. Most medical ethicists today advise caution and waiting in the case of germ-cell gene therapy on humans until we have more experience with gene therapy on somatic cells or in other species.

THOUGHT QUESTIONS

1 The use of growth hormone for the treatment of shortness (not dwarfism) in otherwise healthy children is controversial, but its testing for this purpose was approved in 1993 by the Food and Drug Administration. When does a phenotypic condition unwanted by its bearer become a disease to be treated? Who decides? Should the use of human growth factor produced by engineered bacteria to increase someone's height be allowed? Is this simply another form of cosmetic surgery, similar to breast implants or face-lifts?

2 If a person dissatisfied with his or her phenotype suffers from lack of self-esteem on that account, does the lack of self-esteem justify a procedure to correct the phenotype? (This same argument is raised to justify traditional forms of cosmetic surgery.) Do parents have the right to anticipate for a child what the future effects on self-esteem will be with and without corrective procedures? For a phenotype such as height that develops over a period of years, at what age is it appropriate (if ever) to evaluate the phenotype and decide upon corrective measures?

3 A procedure such as gene therapy is expensive. Who should pay for it? Is gene therapy a limited resource? Does giving gene therapy to one patient thereby deprive another of medical care?

Molecular Techniques Have Led to New Uses for Genetic Information

Molecular biology is an interdisciplinary field that focuses on DNA. Although there are many other kinds of molecules, molecular biologists are concerned mostly with DNA. Molecular biology techniques can tell us a lot about human genetics, and several marker systems have now been discovered for studying human DNA. The first of these marker systems, restriction-fragment length polymorphisms, is described here. More recently other markers, with names such as expressed sequence tags, microsatellites, and single-nucleotide polymorphisms, have been discovered.

Each person has a unique DNA sequence. If it were practical to sequence a person's whole genome, his or her DNA could definitively identify a person. The human genome is far too long for it to be useful for such identification, but the DNA marker techniques that have been so useful in mapping gene regions have also proved useful in distinguishing, with a high probability, any person from another except for identical twins. Two frequent uses of this technique are in the identification of suspects in police investigations and in disputes over paternity.

The first DNA marker: restriction–fragment length polymorphisms

In 1980 a new mapping technique was devised that could readily be used in human studies, as well as in studies on other species. DNA contains, in addition to genes, noncoding regions that vary in length from one individual to another. Short sequences of nucleotides, 3–30 bases long, are repeated over and over anywhere from 20 to 100 times. These are called short tandem repeats. Several thousand different such repeats are now known in humans, each with a unique sequence not found elsewhere in the genome. When DNA containing variable numbers of repeats is cut with a restriction enzyme, fragments of DNA of various lengths are produced (Figure 4.5A). Variations (also called polymorphisms) in the lengths of the fragments produced with restriction enzymes are known as **restriction-fragment length polymorphisms**, or **RFLPs** (pronounced "riflips"). The fragments of different lengths are separated by a technique called electrophoresis (Figure 4.5B). As we saw in Chapter 3 (Figure 3.8, p. 73), because DNA carries an electric charge it moves in an electric field. When a DNA sample that has been cut into fragments is loaded onto a gel and electric current is applied, the fragments move. The gel material retards the movement of the fragments somewhat, and the larger the fragment, the more its movement is retarded by the gel. In the time that the electric current is on, smaller fragments will therefore move farther than large fragments. Because the nucleotide sequence of each short tandem repeat is unique, each can be detected by a specific probe, a piece of DNA with a sequence complementary to the repeat sequence (Figure 4.5C). Probes are specific and cause only those fragments to show up that have sequences complementary to the probe sequence.

Using DNA markers to identify individuals

Using the same DNA marker techniques that we saw above, geneticists can compare DNA samples from different persons. The samples are cut with restriction enzymes. Pieces are separated according to size by electrophoresis and then transferred to a paper material. Radioactively labeled probes complementary to known DNA sequences are then used to detect the fragments containing particular variable repeats. These fragments appear as bands, with their location indicating the fragment length. Several probes can be used at once so that many bands show up, not just one or two as in the example shown in Figure 4.5, in which just one probe was used.

Bands at the same position indicate fragments of the same length in samples being compared. If the band patterns are not the same, then it can be stated with certainty that two samples did not come from the same person. In the example from a criminal investigation shown in Figure 4.6, person 1 can be eliminated as a suspect because the band pattern from the evidence is not the same as that from sample 1. The reverse is not true, however; band patterns that are the same are not an absolute guarantee that the samples came from the same individual. What are being visualized are chunks of DNA of variable lengths, not the DNA

Figure 4.5

Restriction-fragment length polymorphisms (RFLPs).

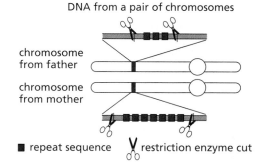

DNA from a pair of chromosomes

chromosome from father

chromosome from mother

■ repeat sequence   restriction enzyme cut

(A) CUTTING DNA WITH RESTRICTION ENZYMES

The pieces differ in length depending on the number of repeats that exist within a piece. In this example, the piece from the father is shorter because it has fewer repeats than the piece from the mother, which is longer because it has more repeats.

(B) SEPARATION BY ELECTROPHORESIS

The mixture of pieces is placed on a gel and exposed to an electric field. Because DNA has a negative charge, the pieces move toward the positive electrode. In the time that the current is on, smaller pieces travel farther through the gel than the larger ones do. None of these pieces is visible yet.

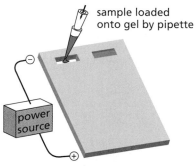

sample loaded onto gel by pipette

power source

(C) DETECTION WITH A PROBE

None of the pieces can be seen; however, they can be detected with a variable-repeat probe tagged radioactively or chemically (bands shown in color). The probe is a small piece of DNA with a sequence complementary to the sequence of that variable repeat, so the probe will bind to those pieces of DNA containing that variable repeat. The probe thus does two things: it identifies pieces with that specific repeat and it indicates whether the sequence is repeated a few times (to give a short DNA piece) or many times (to give a long piece). Other probes will find other sequences that are repeated in other chromosomal locations.

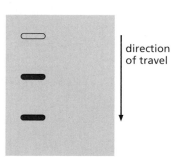

DNA fragment not bound by the probe

longer piece from mother's chromosome

shorter piece from father's chromosome

direction of travel

Figure 4.6

Forensic DNA technology. In this example, the evidence sample shows the same pattern of bands as DNA from suspect 2. There is therefore a high probability that the DNA in the evidence is from that suspect. The person from whom sample 1 was taken can be eliminated as a suspect.

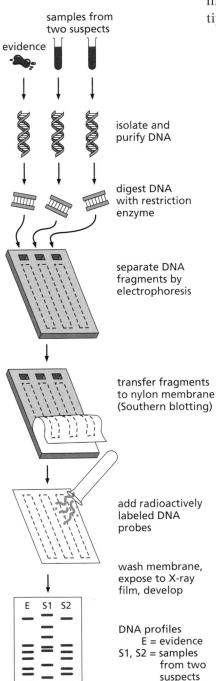

samples from two suspects

evidence

isolate and purify DNA

digest DNA with restriction enzyme

separate DNA fragments by electrophoresis

transfer fragments to nylon membrane (Southern blotting)

add radioactively labeled DNA probes

wash membrane, expose to X-ray film, develop

DNA profiles
E = evidence
S1, S2 = samples from two suspects

sequences of the chunks. A score is calculated that indicates how likely it is that a randomly chosen person, other than the one tested, could have the same band pattern.

The likelihood that another, randomly selected person could have the same banding pattern is made very small in two ways. First, the DNA probes selected are those that pick up specific DNA markers that are rare in a given population. Also, several DNA probes are used, one after another, to produce a composite banding pattern. The probability that the bands produced with just one DNA probe are the same for two people is equal to the frequency of that DNA marker in the population. If more than one DNA probe is used, the probability of both band patterns' matching is equal to the population frequency of the first DNA marker multiplied by the population frequency of the second, and so on for multiple DNA probes and markers.

There are many ways in which the banding pattern can yield flawed or ambiguous results if samples are not properly processed. In samples from crime scenes, there is often DNA from mixed sources, including DNA from several people and from bacteria or fungi. Protein material in the sample may slow the movement of a restriction fragment in the electrophoresis, making the DNA fragment appear as though it were larger than it is. Other chemicals in the samples, such as the dyes in cloth, can interfere with the restriction enzymes cutting the DNA. However, when the tests are done properly and with the proper controls, they can be very reliable. In addition to linking suspects to material taken from crime scenes, the methods can be used to settle questions of disputed parentage. The methods can also be used to identify the dead when an intact corpse is not available, as in the aftermath of the terrorist attacks in the United States on 11 September, 2001.

Using DNA testing in historical controversies

An unusual use of this technique helped shed new light on a historical controversy involving Thomas Jefferson, the third president of the United States. DNA markers were used to investigate whether Thomas Jefferson could have been the father of children borne by one of his slaves, Sally Hemings. Two oral traditions exist: descendants of Hemings's sons, Eston Hemings Jefferson and Thomas Woodson, believe that Jefferson was their ancestor, while descendants of Jefferson's sister believe that one of her children, Jefferson's nephew, fathered Sally Hemings's later children. Researchers compared Y chromosomal DNA from descendants of two of Sally Hemings's sons with DNA from descendants of one of Thomas Jefferson's uncles. No Y chromosomal DNA was available from Thomas Jefferson's direct descendants because he had no sons who survived to have children.

The DNA data show that a set of 19 markers (collectively called the haplotype) is shared by all five of the descendants of Jefferson's uncle who were tested and by the descendants of Eston Hemings Jefferson. The haplotype is not shared by descendants of Hemings's other son, Thomas Woodson, or by the descendants of Jefferson's nephew, nor was it found in almost 1900 unrelated men. Thus, Jefferson may definitively be ruled out as the father of Thomas Woodson.

In the case of the positive match, however, the evidence supports, but does not prove, the idea that Thomas Jefferson could have been Eston Hemings Jefferson's father. As we explained earlier, positive matches indicate probabilities, not definite identity. The researchers state that because "the frequency of the Jefferson haplotype is less than 0.1%," their results are "at least 100 times more likely if the president was the father of Eston Hemings Jefferson than if someone unrelated was the father." They also state that they "cannot completely rule out other explanations of our findings," but that "in the absence of historical evidence to support such possibilities, we consider them to be unlikely." Interestingly, although the authors are very precise in the text of their article, the title, "Jefferson fathered slave's last child," overstates their results (E.A. Foster et al. *Nature* 396: 27, 1998).

THOUGHT QUESTIONS

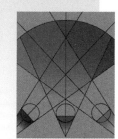

1 Thomas Jefferson had daughters who survived to have children. Why was the DNA of their descendants not used in the study to determine the paternity of Eston Hemings Jefferson and Thomas Woodson?

2 The authors of the Jefferson study state that they "cannot completely rule out other explanations of our findings." What other explanations are biologically possible?

3 Think about the study done on DNA from descendants of Jefferson's family and Sally Hemings's sons. Why is the title of the study, "Jefferson fathered slave's last child," an overstatement of the results?

4 In the study on Jefferson's descendants, why did the researchers test DNA at 19 DNA marker sites, rather than just at one or two sites?

The Human Genome Project Has Changed Biology

The complete genetic material of an entire organism is known as its **genome**. In 1986, scientists proposed a project to make a genetic map, or catalogue, of a prototypical human, including the chromosomal location of all human genes and the complete DNA sequence of the genome. Many scientists and physicians think that many medical and other benefits could flow from knowing the location and sequence of all the genes. Such knowledge would facilitate locating genes that are associated with diseases or disease susceptibility. It will also make possible the development of drugs that are much more specifically tailored to block particular molecules. This effort became known as the Human Genome Project.

The Human Genome Project was funded by the U.S. Congress to begin work in the fall of 1989, and James Watson, co-discoverer of the double-helical structure of DNA, was appointed as the first director. Watson stated his belief that the Human Genome Project would tell us what it means to be human.

It should be noted, however, that although we talk of *the* human genome sequence, the DNA sequence of each person is unique. There is no one DNA sequence that is representative of every human, just as no one person could be said to represent all humans in any other method of describing people. It is estimated that one person differs from another in about 0.1% of the 3 billion base pairs in the human genome. People share the same genes but the nucleotide sequences of those genes vary in different alleles.

Sequencing the human genome

One of the stated goals of the Human Genome Project was to determine the human DNA sequence. When we read in the newspaper or hear on television about a genome being sequenced, what does this mean? The 'sequence' of DNA is the order in which the four nucleotide bases (see Chapter 2, p. 56) appear from one end of the DNA molecule to the other. Because DNA is an unbranched molecule, the sequence of bases can be 'read' from one end to the other.

Determining the order of nucleotides by using fluorescent dyes. Because the amount of DNA in even one chromosome is enormous, it is not practical to work with the whole length of a chromosome in determining sequences. The maximum size of pieces that can be sequenced is currently about 500–700 bases long. The chromosomes are therefore separated and each is cut into overlapping pieces with restriction enzymes. Each piece is inserted into a plasmid which enters a bacterium. The bacteria then divide repeatedly and make large quantities of one piece at a time, as we saw on p. 98 for bacterial production of human insulin.

The nucleotide sequence of each of the pieces can then be determined using an established method (called the di-deoxy method) based on DNA synthesis. The DNA is used as a template for synthesis of new DNA strands in a test tube, as outlined in Figure 4.7. The overall result is the production of a series of smaller pieces, each piece one nucleotide longer than the next. Each of the small pieces is then separated by electrophoresis. The pieces are made visible with a fluorescent dye, a different color used for each of the four nucleotides. Unlike the specific probes used with DNA markers, fluorescent dyes make all of the pieces visible that end in that nucleotide. The sequence of bases in the DNA fragment can thus be read from the gel: the base found at the end of the shortest piece is first (traveled farthest in the gel), followed by the base found at the end of the next longer piece (traveled the second farthest in the gel), and so forth.

Mistakes can occur in either copying or sequencing, and repeating the process does not always give the same answer, so the technique must be repeated several times by different laboratories until a consensus sequence is established. After the sequence of each piece has been determined, the pieces must be arranged in their original order to get the overall sequence. Remember: this sequence analysis has been carried out on only one fragment of a chromosome at a time. The next challenge is to piece together the sequenced fragments, which is part of the mapping procedure discussed below.

The non-coding DNA. Most of the human chromosomal DNA does not code for genes, however, and the Human Genome Project included the sequencing of these non-coding regions. The non-gene DNA consists of 'spacer' sequences that are never transcribed, and other kinds of sequences that are transcribed but never translated. The function of most of these non-gene sequences is currently unknown, and the wisdom of spending an estimated $15 billion on their sequencing is a question on which opinion, even among scientists, differs widely. These non-coding regions, however, have turned out to be the locations of many of the DNA markers discussed earlier, which have allowed us to find where specific

Figure 4.7

Discovering the nucleotide sequence of a piece of DNA.

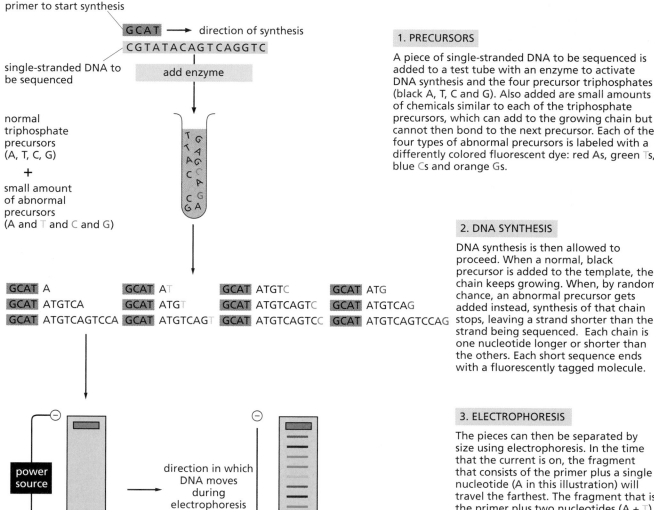

1. PRECURSORS

A piece of single-stranded DNA to be sequenced is added to a test tube with an enzyme to activate DNA synthesis and the four precursor triphosphates (black A, T, C and G). Also added are small amounts of chemicals similar to each of the triphosphate precursors, which can add to the growing chain but cannot then bond to the next precursor. Each of the four types of abnormal precursors is labeled with a differently colored fluorescent dye: red As, green Ts, blue Cs and orange Gs.

2. DNA SYNTHESIS

DNA synthesis is then allowed to proceed. When a normal, black precursor is added to the template, the chain keeps growing. When, by random chance, an abnormal precursor gets added instead, synthesis of that chain stops, leaving a strand shorter than the strand being sequenced. Each chain is one nucleotide longer or shorter than the others. Each short sequence ends with a fluorescently tagged molecule.

3. ELECTROPHORESIS

The pieces can then be separated by size using electrophoresis. In the time that the current is on, the fragment that consists of the primer plus a single nucleotide (A in this illustration) will travel the farthest. The fragment that is the primer plus two nucleotides (A + T) will travel not quite as far, and so forth.

4. READING THE SEQUENCE

A fluorescence detector reads each band of the gel, detecting the color of the dye labeling that band.

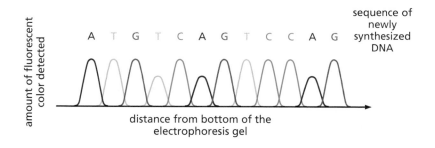

genes are located. Other scientists suggest that these non-coding regions will also turn out to be important for other reasons. For example, the non-coding regions are the binding sites for proteins, such as the SRY protein (see Chapter 2, p. 48), that regulate DNA folding, and thus regulate when a gene is transcribed.

The human genome draft sequence

In February 2001 two groups simultaneously announced completion of a draft of the sequence of the human genome. One group, the International Human Genome Sequencing Consortium, involving laboratories from the United States, Britain, Japan, France, Germany, and China, published their results in *Nature* (409: 860). The other group, a biotechnology company called Celera Genomics, published their results on the same day in *Science* (291: 1304). The draft covers about 94% of the estimated 3 billion bases in the complete genome. Of those 3 billion bases, 1 billion have been sequenced to completion, including all of those on the smallest paired chromosomes, chromosomes 21 and 22. The other 2 billion bases contain gaps and areas where different efforts at sequencing have resulted in different answers.

Completion of the draft sequence supported some previously established hypotheses, but also produced some surprises. Some key results are:

1. *About 95% of the human genome represents non-coding DNA, a large proportion of which is composed of repetitive sequences.* Less than 5% of the human genome is composed of genes, sequences that code for RNAs or proteins. It has been known for a while that the complexity of an organism does not correlate with the size of its genome. Much of the excess size is due to these non-coding, repeat sequences. Detailed knowledge of these sequences is opening up a new resource for studying evolution. These sequences can be likened to living fossils carried within each of us. They are already used in population genetic studies examining the migrations of human populations.

2. *The actual number of genes is smaller than previously estimated.* In humans it is difficult to predict which sequences represent genes, for reasons we discuss later. Thus, although the draft sequence of the human genome has been published, the number of genes remains unknown. The estimate of the number of genes is currently between 30,500 and 35,500. (Previous estimates had been between 50,000 and 100,000 genes.) The numbers of genes in the fruitfly (*Drosophila melanogaster*) and the roundworm (*Caenorhabditis elegans*) have been ascertained; comparisons reveal that humans are likely to have only twice as many genes as each of them.

3. *The protein products of many human genes remain unknown.* It has been found that many of the known genes can be translated in different ways to produce alternative protein variants from the same gene (see Figure 4.10, p. 117). Thus, although we have only twice as many genes as fruitflies, we may have five times as many different proteins.

4. *A very high percentage of our genes are not unique to humans but are closely similar to comparable genes from other species.* In fact, only 1% of human genes have no sequence similarity to any other organism. Our genes are similar to 46% of the genes in yeast, among the simplest organisms whose cells have a nucleus. Changes within genes over time provide clues to rates and paths of evolution.

5. *More than 200 human genes and their protein products have been found to have significant similarity to those in bacteria.* These genes are not found in intermediate organisms such as fruitflies, and one school of thought suggests that these genes jumped from bacteria to humans or vice versa.

6. *Mutation rates differ in different parts of the genome.* They are also higher in males than in females, although the reason for such a difference is not known.

7. *Within each gene, there is an average of 15 sites at which different individuals carry a different nucleotide, or at which the same individual may have a different nucleotide on each chromosome in a pair.* These variations, called single-nucleotide polymorphisms, are greatly expanding how many alleles we think are possible for different genes. In addition, these small changes may affect the physiology of the organism possessing them. Some of these polymorphisms are associated with disease; most are not, but are instead associated with small changes in protein function or regulation. Knowledge of such small-scale variations continues to challenge our concepts of terms such as 'heterozygous', 'dominant' and 'recessive', and 'allele'. It also makes it clear that there is no such thing as *the* human genome sequence. The genome sequence within each individual is unique.

In April of 2003, only two years after publication of the draft sequence, the sequence of the human genome was completed. Its publication in the journal *Nature* was timed to coincide with the fiftieth anniversary of Watson and Crick's article describing the double helical structure of DNA.

Mapping the human genome

Another goal of the Human Genome Project was to map the human genome. Mapping a species' genome means identifying the chromosomal location of each gene and the order of the genes relative to one another. Just determining the sequence of a piece of DNA does not tell you its location in the genome. The molecular techniques developed as part of the Human Genome Project have accelerated the mapping and identification of genes more generally.

One way to map a large piece of DNA is to cut the same long piece with two different restriction enzymes, derive the sequence of each of the pieces, then use computers to discover how the two sets of pieces overlap. Figure 4.8 shows how sequence data from overlapping fragments of DNA are used to derive the original order of the fragments. Figure 4.8A shows two sets of fragments of DNA produced by cutting a DNA sample with different restriction enzymes. The first restriction enzyme cut the

DNA into six pieces only; the second resulted in eight pieces. The bases in the sequences of each of the eight pieces can be lined up to match the bases in the six pieces. Can you see how you would use this idea to determine the order that the six pieces had originally been in? Now turn the page and look at Figure 4.8B.

In our example the largest piece contains 40 bases. Actual DNA pieces for sequencing are around 500 bases in length. Because the pieces are so much longer and there are so many of them, computers are needed to line up the overlaps. The accuracy of the method increases with the length of the overlapping region. The longer the sequence of the overlap between two pieces, the higher the probability that the sequence will appear only once in the genome, allowing the unambiguous assignment of the position of the two pieces relative to each other.

Celera used this approach first in 1995 with the complete sequencing of the genome of the bacteria *Haemophilus influenzae*. The same approach was used successfully on the genomes of the 599 viruses, 31 eubacteria, and 7 archaebacteria that were sequenced between 1995 and 2002. They believe that the same approach will work for mapping the human genome.

But there are obstacles to applying this approach to mapping the human genome. One obstacle is size; the human genome is about 25 times larger than any previously sequenced genome, although it is far from being the largest genome known. (One species of single-celled amoeba has a genome 200 times larger than humans!) Another obstacle to accurate reassembly is the fact that much of the non-coding DNA in the human genome is composed of repeated sequences of nucleotides. This enormously complicates the job of putting pieces into unambiguous order. Species whose genomes had previously been sequenced do not contain these repeats, so it was much easier to determine which piece went where in these genomes.

The International Human Genome Sequencing Consortium therefore used DNA markers in addition to sequence overlap to map the locations of the pieces. In the technique used by the Consortium, the total DNA in the genome was split into 29,298 overlapping large fragments with a variety of restriction enzymes. Each large piece was further split into pieces of a size that could be sequenced. Sequencing of the small pieces has been proceeding at the same time as the mapping of the large fragments, and one advantage of this approach is that different laboratories can be simulta-

Figure 4.8(A)

Combining the sequences of small pieces into the sequence of the original whole chromosome. Here are the fragments of a sequence cut with two different enzymes. Can you piece them together to reconstruct the complete sequence? Don't turn the page until you've tried it!

In this example, a DNA sequence of 150 bases is cut with two different restriction enzymes, producing the following fragments, each of which has been sequenced.

Fragments from the first restriction enzyme:

GGTCGGCTATGTAACGAGTTGCC
TCTTGTTCCTAGCTTGTCAACCGGGGATGAATGTTTACTG
CACGCGGACCGTCGGTTCAT
GTCGCAGAGCCTATTGCGAGAAGT
GCCCACCTT
TTATTGAGTTGATGCTCGACGTAGCCAGACTTAA

Fragments from the second restriction enzyme:

ACCGGGGATGAATGTTTACTGGTCGCAGAG
CCTATTGCGAGAAGTGGTCGGCTA
CTTGTCA
TGATGCTCGACGT
CGTCGGTTCAT
AGCCAGACTTAACACGCGGAC
TGTAACGAGTTGCCGCCCACCTTTTATTGAGT
TCTTGTTCCTAG

Try to piece these fragmentary sequences together and determine the entire sequence of 150 bases, before you turn the page.

neously working on different pieces of the puzzle. Indeed, the location of each of the large fragments within the genome has now been mapped and the map is publicly available. Mapping of all of the small pieces is still proceeding.

Because Celera started with all small pieces, the Consortium maintains that Celera will not be able to reassemble the sequences of their small pieces without referring to the publicly available data posted by the Consortium. Celera maintains that because the Consortium map and sequence data are publicly available, Celera should use it to help assemble their small pieces more quickly. Why continue to insist on the slow way, when those data can now be used in a more rapid way?

The Consortium requires rapid, public disclosure of all data. Their decision to publish a draft sequence as fast as possible was driven, in their words, by "concerns about commercial plans to generate proprietary databases of human sequences that might be subject to undesirable restrictions on use" (*Nature* 409: 863). These worries have to do with the stated intentions of Celera Genomics to require others to pay for access to their databases.

Some ethical and legal issues

Many of the issues already covered in Chapter 3 regarding genetic testing will become more commonplace as molecular genetics continues to change medicine. How does an individual's right to privacy balance against family members' desire to know the results of genetic tests or an insurance carrier's or employer's desire to cover or to hire only employees who will remain healthy? How does an individual's desire to control their own reproduction balance against possible eugenic aims of society or against further stigmatization of disabled people? How can genetic counseling be value-free while providing education about genetics and not just about the testing procedure itself?

When the Human Genome Project was funded, scientists saw the need for examination of the ethical, legal, and social issues (anticipated and unanticipated) that would be raised by the research. One percent of the funding was set aside for this effort. The issues just mentioned are among those being studied, but there are many others. Social workers, anthropologists, ecologists, ethicists, and others are working together to examine the issues raised by the study of genetic variation in human populations and by the integration of genetic information into health care as well as into non-clinical settings. Others are studying the ways in which socioeconomic factors, race, and ethnicity influence people's understanding, interpretation, and use of genetic information. Simultaneously, new genetic information continues to change our concepts of race and ethnicity (see Chapter 7). Others are examining how genetic knowledge and concepts interact with different philosophical and theological traditions. Many of the working groups have composed reports with their answers to many of these questions and their guidelines for the use of genetic information. These reports are available at the Web site www.genome.gov.

In addition, the data derived from the Human Genome Project raise questions of ownership and patent rights. Who owns the human genome

CONNECTIONS
CHAPTERS 3, 7

or the sequence of any particular gene? If a researcher localizes a gene to a particular chromosome, can that researcher patent the information? Can a gene sequence be copyrighted in the manner of a book? Can the genes themselves be patented? Certain biotechnology companies stand to profit greatly from the marketing of gene sequences, tests for gene sequences, or cures for various genetic diseases, but the sharing of information on gene sequences seems at first glance to threaten their competitive position. Several corporations intend to determine as many gene sequences as possible and then copyright them and sell the information at a profit. Other scientists think that the human genome should be public information, and that scientists should share this information cooperatively, particularly if public money in the form of research grants has been used in production of the knowledge. A middle ground is developing, wherein most sequences are posted in data banks with public access, but fees are sometimes charged for that access.

Figure 4.8(B)

Here is the complete sequence of 150 bases. Geneticists usually work with hundreds of fragments at once, each of them longer than this entire sequence, so the task of piecing them together is much more difficult.

When the two sets of fragments are lined up in this way, the order of the bases in the first row is the same as the order of the bases in the second row.

and the complete sequence is therefore as follows:

```
TCTTGTTCCTAGCTTGTCAACCGGGGATGAATGTTTACTGGTCGCAGAGC-
TCTTGTTCCTAGCTTGTCAACCGGGGATGAATGTTTACTGGTCGCAGAGC-
TCTTGTTCCTAGCTTGTCAACCGGGGATGAATGTTTACTGGTCGCAGAGC-

CTATTGCGAGAAGTGGTCGGCTATGTAACGAGTTGCCGCCCACCTTTTAT-
CTATTGCGAGAAGTGGTCGGCTATGTAACGAGTTGCCGCCCACCTTTTAT-
CTATTGCGAGAAGTGGTCGGCTATGTAACGAGTTGCCGCCCACCTTTTAT-

TGAGTTGATGCTCGACGTAGCCAGACTTAACACGCGGACCGTCGGTTCAT    deduced sequence
TGAGTTGATGCTCGACGTAGCCAGACTTAACACGCGGACCGTCGGTTCAT    fragments from first enzyme
TGAGTTGATGCTCGACGTAGCCAGACTTAACACGCGGACCGTCGGTTCAT    fragments from second enzyme
```

THOUGHT QUESTIONS

1 To what extent do you agree with Watson's statement that sequencing the human genome will tell us what it means to be human? Suppose you knew the exact gene sequence of part or all of your genome; what would you really know about yourself?

2 If only stretches of DNA 500–700 bases long can be sequenced at a time, how many of these small sections of DNA must be sequenced to determine the sequence of the entire human genome? (Think also about the overlaps required to piece the sequences together; assume an average of 10% overlap.)

3 Will the DNA sequence of the human genome tell us what traits are controlled by each part of the sequence? Will it tell us which sequences represent genes and which sequences represent spacers?

4 If you have a certain rare genetic condition, and scientists use cell samples from your body to determine the gene's DNA sequence, what rights (if any) does this give you to the information? Do the scientists have the right to publish your gene sequence, or any part of it? Is it an invasion of your privacy? Can the scientists sell the information? If they do, are you entitled to a share of the profits?

Genomics Is a New Field of Biology Developed as a Result of the Human Genome Project

The Human Genome Project also funded the sequencing of the genomes of many other species. This may seem odd at first because the name of the project specifies the *human* genome, but there were several reasons for including these other species. The study of the genomes of species has become an entire new area of biology called **genomics**. This field has arisen to help unfold the mysteries of human genes now that the sequences and mapping are nearing completion. One focus of genomics is the identification of individual human genes. The combination of molecular biology and computer science that has been necessary to navigate through the tremendous amounts of data produced by the various genome projects is called **bioinformatics**.

Bioinformatics

Just as the NASA space program led to many unexpected 'spin-off' technologies in the 1960s and 1970s, the Human Genome Project is doing so as well, with new computer technologies and genetic engineering having wide applications outside genetics. DNA sequencing and mapping would not have been practical before the advent of large computers. Although the techniques for determining sequences of short pieces of DNA are rather simple (see Figure 4.7), finding the overlaps that indicate how the small sequenced pieces were originally arranged (see Figure 4.8) requires massive computer power. Then, when the longer sequences have been determined, storing the data has necessitated the development of larger and larger computer databases and new methods for searching them. Genomics requires the development of new types of computer software. The need for people who are trained in both molecular biology and computer science who can work with these data has made bioinformatics a fast-growing new field of employment.

One research project within bioinformatics has been the development of computer programs to locate genes within a genome. In the past, as we have seen, scientists worked from a trait, back to finding a gene. Now that the genome sequence is complete for many species and nearly complete for humans, the method of gene discovery has changed. Now people are examining the sequence data itself and trying to determine which parts may be genes, without any prior knowledge of a trait or a function for those genes. Many such genes have already been found in bacteria and yeast, and they are referred to as "orphan genes" because, at the time of their discovery, no function was known. (The later identification of their function is part of the research program of functional genomics, described later.)

Within bioinformatics, people are programming computers to scan the sequence data to locate genes, meaning areas that code for RNAs and proteins. To do so, programmers must discern 'rules' of the genetic code: what characteristics of a sequence distinguish a coding region from a non-coding region? The computer search for genes within sequences is called gene scanning.

Gene scanning in different organisms. Interestingly, most genes start with the codon ATG and end with one of three 'stop codons': TAA, TAG, or TGA. If the nucleotides A, T, G and C were distributed randomly, each of the stop codon triplets would be expected to occur on average every 4^3 or 64 bases. But nucleotides are not distributed randomly within genes; they are retained in a non-random pattern as a result of evolution because they code for a product conferring advantage to the organism. In bacteria, genes are typically 300–500 codons long, are contiguous, and do not overlap. In addition, bacteria have very little non-coding DNA. These factors make gene scanning in bacteria relatively easy. A computer can scan the sequences that follow any ATG and find those areas where the next stop codon occurs a few hundred bases further along.

Gene scanning is much more difficult in other organisms, namely the nucleated organisms (eucaryotic organisms; see Chapter 5). In contrast to bacterial species, they have long, non-coding stretches of nucleotides (called **introns**) dispersed among much shorter regions that correspond to codons. While the coding regions (called **exons**) are roughly the same lengths in different species, the size of the non-coding introns is much greater in humans than in other species (Figure 4.9).

Within the human genome, less than 5% is located within genes; furthermore, within these human genes, only about 5% of the nucleotides comprise coding sequences. This makes it difficult to use raw sequence data to predict which nucleotide regions represent genes. Thus, gene-scanning programs are continuing to be refined to include up-to-date knowledge about the characteristics of the 'departures from randomness' within genes in species other than bacteria. One such departure from randomness is called 'codon bias:' not all codons are used equally by a given species. For example, the amino acid alanine can be coded by four different codons in humans; furthermore, three of those are used much more frequently than the fourth. In a non-coding region of the genome, all of the codons have an equal probability of being represented, but in a coding region the one codon is present less frequently.

The presence of non-coding regions within genes is clearly a complication for gene scanning. Of what benefit could it be to an organism for its genes to be interrupted by such non-coding regions? It is these non-coding stretches that have allowed the shorter coding segments to recombine to form new genes. This provides a mechanism for rapid genetic change (more rapid than by mutation). New genes are produced by the novel assembly of parts. There is another way in which the division of genes into many coding regions is adaptive, and that is in providing a mechanism by which slightly different versions of a protein can be made in different tissues, adapted to the cellular environment and function of that tissue. An example is shown in Figure 4.10. The

Figure 4.9

Single strands of DNA showing the differences in gene structures in bacteria compared with eucaryotic cells. (A) Bacterial genes contain only coding regions; that is, the DNA is all transcribed to mRNA. (B) In eucaryotic cells non-coding regions that are not transcribed are located within the coding regions of genes. (C) In humans (a eucaryotic species) the amount of non-coding DNA is much greater than the amount of DNA that codes for a protein product.

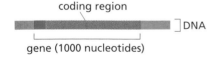

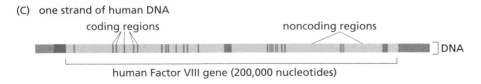

human gene for a protein called α-tropomyosin contains many coding regions scattered among non-coding regions. This gene can be transcribed to mRNA in different ways, so that in the cells of one tissue one set of coding regions is used, and in the cells of another tissue, a different set of coding regions is transcribed. This results in different mRNAs in the different cells, and therefore in slightly different proteins after translation. Each protein is still α-tropomyosin, but with a slightly different amino acid sequence and therefore a slightly different functional capability.

Although scientists think these large non-coding regions within genes are adaptive for the organism, they do present a significant obstacle to identifying genes by gene scanning. In fact, it does not appear that gene-scanning programs alone will be able to identify all of the genes in a eucaryotic genome. Hence, the Human Genome Project also funded work on the genomes of other species, so that human genes could be located by comparison with the genes of other species, a field now known as comparative genomics.

Figure 4.10

Within human genes all of the nucleotides are transcribed into RNA but only some of the RNA nucleotides are translated into protein.

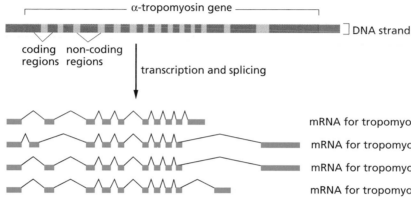

Comparative genomics

When scientists compare sequences of genes from one species with those from another, they are working in the field of comparative genomics. The size of the genome of many species has been determined. As we saw earlier, the overall size does not always correlate with the complexity of the organism. This is due to the very great differences in the amount of non-coding DNA in various genomes, so that overall size does not correlate with the numbers of genes present.

As we have just discussed, genes are much easier to identify in some species than in others. Once a gene has been identified and its sequence determined in one species, there is often enough sequence similarity for its counterpart gene (or genes) to be located in other species. This is the major reason why other species' genomes were also examined as part of the Human Genome Project. Another reason was that sequencing the genomes of other species allowed scientists to develop the technology that was later used to analyze human genome sequences.

Many human genes have been located by their similarity to yeast genes. A yeast cell, like a human cell, has a nucleus and many of its genes have remained very similar to the counterpart genes in humans. Animals are even more similar and one animal that is proving to be quite useful in comparative genomics is the pufferfish, *Fugu rubripes*. Its genome is only one-seventh the size of the human genome, yet it is estimated to have the same number of genes. Because of its small genome size, gene location is much easier in pufferfish, and may subsequently allow mapping of the human counterpart genes. The mouse genome is almost complete, and

many more human genes will be found by comparison with those in the mouse. Many known genes in mice are located in the same order on their chromosomes as they are on human chromosomes, and this correspondence is extremely helpful in mapping genes. The mouse genome, however, is even larger than the human genome, so the problems of working with a large genome still pertain. See our Web site for information on the sizes of the genomes of various species (under Resources: Genome sizes).

Aside from its usefulness in locating human genes, comparative genomics has produced new data for evolutionary biologists. Species that have a common ancestor have more genes and more nucleotide sequences in common than species that do not. Unfortunately, the scientists working on a particular species have often independently devised the database for each species' genome. Consequently, another goal of bioinformatics is to devise ways of making the different databases compatible and interactive, thereby facilitating comparative genomics.

In addition to finding similarities between species, comparative genomics has led to the realization that within a species there are groups of genes that share large portions of their sequences. These 'gene families' are presumed to have evolved from a common ancestral gene. Finding one gene in the family enables the others to be located, and most often the different protein products of the family members to be identified. For example, the human hormones oxytocin and vasopressin (both proteins) belong to the same gene family, and they have very similar amino acid sequences and genes that code for them. The same is true of the oxygen-carrying proteins hemoglobin and myoglobin.

Functional genomics

In Chapter 3 we described how Archibald Garrod and other scientists studied "inborn errors of metabolism," disease conditions caused by changes in biochemical pathways. The study of similar changes in bacteria or yeast have often led to the discovery of entire chains of biochemical reactions. In the past, scientists looking for the molecules involved in such a biochemical pathway, would start with a trait and work backwards to a protein. Pedigrees such as we saw in Chapter 3 would be linked with different forms of a protein. After purifying the protein and discovering its amino acid sequence, its gene sequence could be inferred. Gene sequencing turns this whole process around. Genes are found by linkage to DNA markers, and only later is the protein product found. However, finding a gene, mapping its location, sequencing it, and even deriving the amino acid sequence of its protein product, will not tell us its function.

New sequences can be compared with those whose function is already known. This is the field of functional genomics. Species that can be easily manipulated experimentally have been most useful in discovering gene functions. The zebrafish is a vertebrate that reproduces rapidly, and many of its internal structures are visible in the living fish because overlying structures are transparent. For these reasons, zebrafish have become an experimental species of great interest to scientists working on the genetics of development. An even simpler species, the yeast *Saccharomyces cerevisiae*, has been found to share many genes with humans. Gene functions that were discovered in yeast have proved to

have parallels with disease-associated gene mutations in humans. For some examples see our Web site, under Resources: Yeast genes. The functions of the yeast genes are discovered by several methods. One is to examine which genes are transcribed to mRNA when the yeast undergoes a particular response or function; another is to inactivate (mutate) a gene and see what effect this has. Once the sequence of a gene is known, it is relatively easy to mutate it by manipulating the DNA causing a change in the protein product, which is now not functional. The opposite approach can also be used: extra copies of the gene can be inserted and observations made of the change in function under different environmental conditions. These approaches are not confined to yeast, but are also done to discover gene functions in mice and other species.

Earlier in this chapter we saw how a gene could be added to a genome, using a vector. In Figure 4.4 we saw how the functional gene for ADA was added to the genome in human cells. This method adds a gene, but in an unpredictable location within the genome. The non-functional gene is still present, and indeed one of the possible dangers of the technique is that the new gene may get added in a place that disrupts some other gene. More recently, methods have been developed for changing the sequence of a specific gene. In theory, this technique could be used to repair a non-functional gene, to mutate a gene in a specific way, or to disrupt a functional gene. A vector is used to carry into the cell a piece of DNA partly complementary to the gene to be altered. The introduced double-stranded DNA becomes substituted for the gene region as a result of crossing-over at two sites where the insert and the gene have the same sequences (Figure 4.11). If the inserted DNA is non-functional, as shown in this example, the normal gene is disrupted. The effect of deletion of that gene's protein can then be studied in the offspring cells (yeast or tissue culture of human cells). If the gene disruption is carried out on a cell from a very early stage of development, an entire organism can develop that is lacking the gene and its protein product. (This topic, part of stem cell research, is covered in more detail in Chapter 12.) Mutated mice with particular genes nonfunctional or 'knocked out', or mice with overexpressed genes produced by the insertion of additional copies of a functional gene, have led to important clues to the functions of human genes.

Families of genes have been found within species that have structurally related protein products but very often quite different functions. This has led scientists to realize that gene duplication and mutation can occur first, and that new functions can follow. Of course, a gene that is present in just a single copy cannot change to a new form (possibly with a new function) without giving up its original form and function. A duplicated gene, however, can undergo changes in one copy (possibly evolving new functions) while the other copy remains unchanged.

Proteomics

Just as the complete DNA sequence of an organism is its genome, the complete protein content of that organism is its proteome. **Proteomics** is the study of how the protein content changes over time in a cell and in an organism, how it differs in different tissues,

Figure 4.11

In human genes, different combinations of coding regions can be transcribed to produce different mRNAs in different tissues.

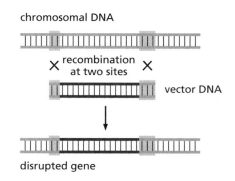

chromosomal DNA

× recombination at two sites ×

vector DNA

disrupted gene

and how it relates to the health and function of the organism. Proteins are synthesized as a result of transcription of genes and translation of mRNA, as we saw in Chapter 3. However, there are further modifications to a protein after it has been translated that affect both its activity and its concentration. No protein stays in a cell forever; all are degraded and removed. We will see more about these aspects of protein function in cells in Chapter 12.

CONNECTIONS ▷ CHAPTERS 3, 12, 14

Knowing the DNA sequence of genes has hastened the discovery of the amino acid sequence of proteins. Computer programs use the known energies and bond angles of chemical bonds to turn amino acid sequence data into molecular models of the three-dimensional shape of a protein or portion of a protein. Having the ability to visualize these shapes by computer graphics has led to new strategies for the design of medicines. In the past, natural products and synthetic compounds were randomly tested in functional assays to see which would work for a particular need. Now small molecules can be designed to exactly fit a critical enzymatic site of a protein. Once the molecule has been designed by computer simulation, medicinal chemists then synthesize it and biologists test to see whether it has the desired outcome of blocking the protein's function. The action of such drugs is far more specific, and the drug will therefore have fewer side-effects than traditionally developed drugs, for reasons we will study in Chapter 14.

To synthesize a protein with even a slightly different structure can be very difficult and costly. However, once the sequence for the gene coding for that protein is known, it becomes relatively easy to modify the protein by changing the sequence of its gene. Roughly the same technique as that in Figure 4.11 is used, but the inserted piece of DNA differs from the normal piece by only a few nucleotides. Such modifications can, for example, lead to the development of proteins that are stable under a wider variety of conditions. These proteins find a variety of industrial applications. Stain removers in laundry detergents, altered enzymes for food processing, and cleanup of pollution, are just a few examples.

Rather than studying one protein change at a time, proteomics also has another goal: to study all of a cell's proteins in the aggregate. Such a goal has been unattainable in the past, and is a big factor explaining why reductionism (reducing a problem to its simplest form) has been a widespread experimental approach in biology. Proteomics is in its infancy, but promises to be a much more integrative approach.

THOUGHT QUESTIONS

1 In what ways are humans poor subjects for genetic research? In what ways are humans good subjects? Which of your reasons are purely biological, and which have ethical components?

2 Why are certain traits studied in some species and not others?

3 Will genomics allow the findings in one species to be applied in other species? Why or why not?

Concluding Remarks

As genomics has discovered genes with useful properties within one species, genetic engineering has given us the tools to transfer those genes into another species. The Human Genome Project has discovered that transfer of genes from one species to another does also occur in nature. Viral and bacterial genes are found in the human genome, for example. Because we have almost always studied the effect of one gene or one protein at a time, transferring a gene into a new genomic environment may lead to different results from those that we expect. As we develop the tools to alter genomes, proteomics may give us the ways to study the effects of such changes throughout the cell. We also need to be mindful of effects at the level of the whole organism and effects of genetic engineering on ecosystems as well, which we will explore further in Chapter 11.

Chapter Summary

- **Restriction enzymes** cut DNA into fragments with known sequences at their ends. Restriction enzymes that produce fragments with single-stranded sticky ends are used in genetic engineering to splice new genes into genomes.

- Variations in the lengths of these fragments are called **restriction-fragment length polymorphisms** or **RFLPs**. RFLPs have helped in finding the location of many genes, as well as in the identification of individuals and in genetic engineering.

- **Genetic engineering** consists of inserting functional genes into cells, thereby altering the cell's genotype. The recipient cells may be bacterial cells that may then acquire the ability to make certain human proteins, or they may be human cells that acquire a functional allele and are injected into a patient as **gene therapy**.

- Bacterial **plasmids** are used to carry genes into a new species.

- A **genome** is the total genetic information carried by a particular organism. The Human Genome Project has now produced a draft sequence and map of the human genome.

- DNA markers of various kinds have allowed the mapping of genes to locations within the genome. Markers also allow the identification of individuals.

- **Genomics** is the study of the genome, either the comparison of genomes of different species or as a method of discovering gene functions.

- **Bioinformatics** combines computer science and molecular biology in the analysis of genomes and the identification of genes within a genome.

- **Proteomics** is the study of all of the proteins present within a cell.

CONNECTIONS TO OTHER CHAPTERS

PRACTICE QUESTIONS

1. If one individual human differs from another in 0.1% of the genome, how many bases are different?

2. In the following stretch of DNA, how many fragments will result from digestion with the *Hae*III restriction enzyme shown in Figure 4.1? How many will result from digestion with *Eco*RI?

strand 1

A T C C G T A G G C C T A A C C A T C C T A G T G C
T A G G C A T C C G G A T T G G T A G G A T C A C G

strand 2

3. Why are restriction enzymes that produce fragments with 'sticky ends' more useful in genetic engineering than restriction enzymes that produce fragments with 'blunt ends?'

4. Could the following sequence be used as an insert into genomic DNA? Why or why not?

strand 1

A A G C T T A A C G G A T T A G C A A G C
    C G A A T T G C C T A A T C G T T C G A A

strand 2

5. Could the following sequence be used as an insert into genomic DNA? Why or why not?

strand 1

A A G C U U A A C G G A U U A G C A A G C
    C G A A U U G C C U A A U C G U U C G A A

strand 2

6. When a plasmid is being cut with a restriction enzyme in preparation for inserting a DNA fragment, the plasmid needs to be cut with the same restriction enzyme as was used to make the DNA fragment. Why?

7. Can DNA marker band patterns be used to identify maternity, as well as paternity?

8. Can DNA marker testing be used to identify individual organisms in other species besides humans?

Issues

- Did life evolve?
- Is science compatible with religion?
- How does evolution relate to genetics?
- How do new species originate?
- Is life still evolving?
- Will life continue to evolve?
- Is the science of evolution static or changing?

Biological Concepts

- Evolution (descent with modification, natural selection)
- Fossils and geologic time
- Environmental influences on species
- Adaptation
- Form and function
- Species and speciation
- Origin of life

Chapter Outline

The Darwinian Paradigm Reorganized Biological Thought

Pre-Darwinian thought
The development of Darwin's ideas
Descent with modification
Natural selection

A Great Deal of Evidence Supports Darwin's Ideas

Mimicry
Industrial melanism
Evidence for branching descent
Further evidence from the fossil record
Post-Darwinian thought

Creationists Challenge Evolutionary Thought

Bible-based creationism
Intelligent design
Reconciling science and religion

Species Are Central to the Modern Evolutionary Paradigm

Populations and species
How new species originate

Life on Earth Originated by Natural Processes and Continues to Evolve

Evolution as an ongoing process

5

Evolution

5 Evolution

*A*sk any biologist to name the most important unifying concepts in biology, and the theory of evolution is likely to be high on the list. As geneticist Theodosius Dobzhansky explained, "nothing in biology makes sense, except in the light of evolution." However, many people in the United States are unaware of the importance of evolution as a unifying concept: public opinion surveys reveal that 25–40% of Americans either do not believe in evolution or think that evidence for it is lacking. (The percentage varies depending on how the question is worded.) In this chapter we examine both the theory of evolution and the opposition to it.

As explained in Chapter 1, scientists use the word 'theory' for a coherent cluster of hypotheses that has withstood many years of testing. In this sense, evolution is a thoroughly tested theory that has withstood nearly a century and a half of rigorous testing. Scientific evidence for evolution is as abundant as, and considerably more varied than, the evidence for nearly any other scientific idea. To refer to evolution as "just a theory" is thus a grave misunderstanding of both scientific theories in general and evolutionary theory in particular. When physicists speak of the atomic theory or the theory of relativity, or when medical professionals speak of the germ theory of disease, they are speaking of great unifying principles. These principles are now well established, but they have withstood repeated testing for somewhat fewer years than the theory of evolution has. Educated people no longer doubt the existence of atoms or of germs, and nobody refers to any of these concepts as 'just a theory.' In the way that the atomic theory is a unifying principle for much of physics and chemistry, the theory of evolution is a unifying principle for all of the biological sciences.

The Darwinian Paradigm Reorganized Biological Thought

Arguably the most influential biology book of all time was published in 1859. *On the Origin of Species by Means of Natural Selection*, written by the English naturalist Charles Darwin (1809–1882), contains at least two major hypotheses and numerous smaller ones, along with an array of evidence that Darwin had already used to test these hypotheses. Both hypotheses deal with **evolution**, the process of lasting change among biological populations. Together, these hypotheses offer explanations for the origins and relationships of organisms, the great diversity of life on Earth, the similarities and differences among species, and the adaptations of organisms to their surroundings.

The first major hypothesis, **branching descent**, is that species alive today came from species that lived in earlier times and that the lines of descent form a branched pattern resembling a tree (Figure 5.1). Darwin used this hypothesis, which he called "descent with modification," to explain similarities among groups of related species as resulting from common inheritance. The second major hypothesis is that parents having genotypes that favor survival and reproduction leave more offspring, on average, than parents having less favorable genotypes for the same traits in a given environment. Darwin called this process **natural selection**, and he hypothesized that major changes within lines of descent had been brought about by this process. Both of these two hypotheses are falsifiable, and they have been tested hundreds if not thousands of times, without being falsified, since Darwin first proposed them in 1859. Darwin's two hypotheses made sense of several previously noticed but unexplained regularities in anatomy, classification, and geographic distribution. As both a unifying theory and a stimulus to further research, Darwin's *Origin of Species* fits the concept of a scientific paradigm expounded by Thomas Kuhn and explained in Chapter 1 (p. 13). Modern evolutionary thought is still largely based on Darwin's paradigm of branching descent and natural selection, expanded to include the findings of genetics.

Pre-Darwinian thought

Darwin's evolutionary theory was not the first. An earlier theory had been proposed by the French zoologist Jean-Baptiste Lamarck in 1809. Lamarck believed in what he called *la marche de la nature* (the parade of nature), a single straight line of evolutionary progress. This idea was based in part on the earlier idea of a scale of nature, called 'scala naturae' in Latin or 'chain of being' in English, an idea described on our Web site (under Resources: Scala naturae).

Lamarck also noticed that species were adapted to local environments. An **adaptation** is any feature that enables a species to survive in circumstances in which it could not survive as well without the adaptation. Adaptations had been observed since ancient times, but scientists of Lamarck's generation were among the first to propose hypotheses to explain adaptation. Along with several contemporaries, Lamarck was an environmental determinist, meaning that he believed in the almost limitless ability of adaptation to mold species to their environments and achieve a perfect match. Each environmental determinist favored a different explanation for adaptation. Lamarck's own explanation was based on the strengthening or increase in size of body parts through repeated use, or their weakening or decrease in size through disuse. Lamarck thought that such changes, acquired during the life of an individual, would be passed on to the next generation, but we now know that these **acquired characteristics** are not inherited and do not contribute to evolution. Other scientists, including Darwin, recognized adaptation to the local environment as an important phenomenon. However, Darwin differed from the determinists in seeing important limitations to the ability of adaptation to modify species.

Figure 5.1

The pattern of branching descent. Living species in the top row are descended from the ancestors below them. The red circle represents the common ancestor to all other circles, and the red square is likewise ancestral to all the squares. The red hexagonal shape at the bottom is ancestral to all species shown in this family tree. In a classification, all the squares would be placed in one group and all the circles in another.

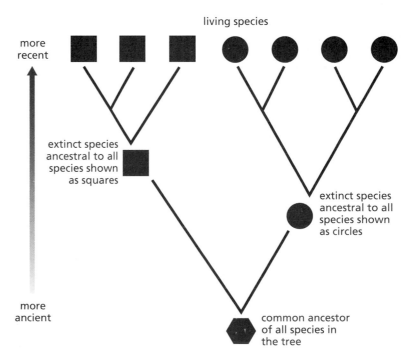

British naturalists had quite different explanations for adaptation. The Natural Theology movement, led by the Reverend William Paley, sought to prove the existence of God by examining the natural world for evidence of perfection. By careful examination and description, British scientists found case after case of organisms with anatomical structures so well constructed, so harmoniously combined with one another, and so well suited in every detail to the functions that they served that one could only marvel at the degree of perfection achieved. Such harmony, design, and detail, they argued, could only have come from God. Paley offered well-planned adaptation as proof of God's existence: "The marks of *design* are too strong to be gotten over. Design must have a designer. That designer must have been a person. That person is God." (Paley, *Natural Theology*, end of Chapter 23; page numbers vary among many editions.) In a nation in which many clergymen were also amateur scientists, it became fashionable to dissect organisms down to the smallest detail, all the better to marvel at the wondrously detailed perfection of God's design. A large series of intricate and sometimes amazing adaptations were thus described, which Darwin would later use as examples to argue for an evolutionary explanation based on natural selection.

The publication of *On the Origin of Species* challenged many scientific ideas, including those of Lamarck and Paley, and it thus caused controversy among scientists. Some people felt that it also challenged social and religious views that had been taught for centuries. Today, there are still people who are antievolutionists and we discuss their ideas later in this chapter.

The development of Darwin's ideas

From 1831 to 1836, Charles Darwin traveled around the world aboard the ship H.M.S. *Beagle*. His observations in South America convinced him that the animals and plants of that continent are vastly different from those inhabiting comparable environments in Africa or Australia. For example, all South American rodents are relatives of the guinea-pig and chinchilla, a group found on no other continent. South America also had llamas, anteaters, monkeys, parrots, and numerous other groups of animals, each with many species inhabiting different environments throughout the continent, but different from comparable species elsewhere (Figure 5.2). This was definitely not what Darwin had expected! Environmental determinist theories such as Lamarck's had led Darwin to expect that regions in South America and Africa that were similar in climate would have many of the same species. Instead, he found that most of the species inhabiting South America had close relatives living elsewhere on the continent under strikingly different climatic conditions. They had no relationship, however, to species living in parts of Africa or Australia with similar climates. The animals inhabiting islands near South America were related to species living on the South American continent. Fossilized remains showed that extinct South American animals were related to living South American species. "We see in these facts some deep organic bond, prevailing throughout space and time, over the same areas of land and water, and independent of their physical conditions. The naturalist must feel little curiosity, who is not led to inquire what this bond is. This bond, on my theory, is simply inheritance, that cause which alone, as far as we positively know, produces organisms quite like, or...nearly like each other." (Darwin. *Origin of Species*, 1859, p. 350.)

The Galapagos Islands. The Galapagos Islands are a series of small volcanic islands in the Pacific Ocean west of Ecuador. Darwin's visit to these islands proved especially enlightening to him. In this archipelago, a very limited assortment of animals greeted him: no native mammals or amphibians were present; instead there were several species of large tortoises and a species of crab-eating lizard. Most striking were the land birds, now often called 'Darwin's finches' (phylum Chordata, class Aves, order Passeres): a cluster of more than a dozen closely related species, each living on only one or a few islands (Figure 5.3). The tortoises also differed from island to island, despite the clear similarities of climate throughout the archipelago. Darwin hypothesized that each species cluster had arisen through a series of modifications from a single species that had originally colonized the islands. The islands, Darwin noted, were similar to the equally volcanic and equally tropical Cape Verde Islands in the Atlantic Ocean west of Senegal, which Darwin had also visited, but the inhabitants were altogether different. Darwin concluded that the Galapagos had received its animal colonists (including the finches) from South America, while the Cape Verde Islands had received theirs from Africa, so in each example the closest relatives were found on

Figure 5.2

An assortment of South American mammals. These species are very different from the mammals found on other continents, even where climates are similar.

chinchilla

tapir

tree sloth

spider monkey

agouti

coatimundi

paca

guanaco

armadillo

giant anteater

kinkajou

Figure 5.3

Some of Darwin's finches from the Galapagos Islands.

Medium ground finch (*Geospiza fortis*)

Large cactus finch (*Geospiza conirostris*)

Warbler finch (*Certhidea olivacea*)

the nearest continent, not on geologically similar or climatically similar but distant islands. Geographic proximity, in other words, was often more important than climate or other environmental variables in influencing which species occurred in a particular place.

Descent with modification

As the result of his studies of species distribution on continents and on islands, Darwin concluded that each group of colonists had given rise to a cluster of related species through a process of branching descent. He called this process "descent with modification," and he emphasized that each species in the cluster had been differently modified from a common starting point. Darwin was the first evolutionary theorist to emphasize that clusters of related species indicated a branching pattern of descent, a series of treelike branchings in which species correspond to the finest twigs, groups of species to the branches from which these twigs arise, larger groups to larger branches, and so forth (see Figure 5.1). In this diagram, each branch point represents a time of species formation and genetic divergence, and the base of the tree represents the common ancestor of all the species that arose from it. Darwin used a very similar treelike diagram, the only illustration in his book.

Darwin found many large groups of related animal species inhabiting each continent. These groups were unrelated to the very different groups inhabiting similar climates on other land masses. Several large land areas had flightless birds, but they differed strikingly from one continent or island to the next: rheas in South America, kiwis and extinct moas in New Zealand, emus and cassowaries in Australia, extinct elephant birds on Madagascar, and ostriches in Africa. Each land mass had its own distinct type of flightless bird, although they all lived in regions of similar climate. Theories of environmental determinism (such as Lamarck's) could not explain these differences, nor could theories of divine creation explain why God had seen fit to create half a dozen distinct types of flightless birds where one might have sufficed.

Before Darwin's time, biological classifications had already taken their modern hierarchical form, as described in Chapter 6, and as illustrated in Figure 6.1. Darwin explained this hierarchy as the natural result of branching descent with modification, a process that produces the similarities and differences that biologists have used in classifying organisms.

Natural selection

When Darwin returned to England, he began reading about the ways in which species could be modified. How, he wanted to know, could a single colonizing species produce a whole cluster of related species on a group of islands? To help find clues to answer this question, Darwin contemplated the results of plant and animal breeding in Britain. During the preceding hundred years, British breeders had produced many new varieties of plants, such as roses and apples, and animals, including dogs, sheep, and pigeons, by careful breeding practices. Through these same practices, the breeders had greatly improved wool yields in sheep, and milk yields in cattle. By methodically selecting the individuals in each generation with the most desired traits and breeding these individuals with each other, the breeders had modified a number of domestic species through a process that Darwin called **artificial selection**. This process simply took advantage of the natural variation that was present in each species, yet it produced breeds that were strikingly different from their ancestors. Darwin remarked that some of the domestic varieties of pigeons or dogs differed from one another as much as did natural species, despite the fact that the domestic varieties had been produced within a short time from a known group of common ancestors. Could a similar process be at work in nature?

At about the same time, Darwin read Thomas Robert Malthus' *Essay on Population* (see Chapter 9, p. 287). Malthus emphasized that, in the natural world, each species produces more young than are necessary to maintain its numbers. This overproduction is followed in each generation by the premature death of many individuals and the survival of only a few. When Darwin compared this process with the actions of the animal breeders, he concluded that nature was slowly bringing about change in each species. Individuals varied in every species, and those that died young in each generation differed from those that survived to maturity and mated to produce the next generation. In this 'struggle for existence,' he hypothesized that

> ...individuals having any advantage, however slight, over others, would have the best chance of surviving and of procreating their kind.... On the other hand, we may feel sure that any variation in the least degree injurious would be rigidly destroyed. This preservation of favourable variations and the rejection of injurious variations, I call Natural Selection.... Natural selection...is a process incessantly ready for action, and is as immeasurably superior to man's feeble efforts, as the works of Nature are to those of Art. (Darwin. *Origin of Species*, 1859, pp. 61, 81.)

CONNECTIONS
CHAPTERS 2, 9

All modern descriptions of natural selection are stated in terms of the concepts of genetics outlined in Chapter 2. New genotypes originate by mutation and by recombination, both of which act prior to any selection. Darwin, of course, knew nothing of mutations or of modern genetics, but he did realize that heritable variation had to come first and that "any variation which is not heritable is unimportant to us."

Natural selection may be defined as consistent differences in what Darwin called "success in leaving progeny," meaning the proportion of offspring that different genotypes leave to future generations. The relative number of viable individuals that each genotype contributes to the next generation is called its **fitness**, and *natural selection favors any trait that increases fitness*. Darwin's theory of natural selection is the basis for all modern explanations of adaptation.

Many agents of natural selection operate in nature. Often, the selecting agents are predators. Selection by predators may be convenient to study, but many other agents of selection are known. Any cause of death contributes to natural selection if it reduces the opportunity for reproduction and if some genotypes are more likely to die. Some genotypes may be more susceptible to particular diseases or parasites, and die in greater numbers from these causes, while other genotypes might be more resistant and thus survive more readily. Starvation and weather-related extremes of cold, dryness, or precipitation may also be agents of selection if some genotypes can survive these conditions better than others. These and other causes of mortality are all agents of selection if there are differences in the death rates for different genotypes.

Not every agent of selection causes death, however. Natural selection also favors those genotypes that reproduce more and leave more offspring. A special type of selection, called **sexual selection**, operates on the basis of success (or lack of success) in attracting a mate and reproducing. For example, animals of many species attract their mates with mating calls (such as bird songs), visual displays (as in peacocks; see Figure 8.2, p. 253), or special odors (as in silkworm moths or many other invertebrate animals). Individuals that do not perform well enough to attract a mate may live long lives but leave no offspring.

THOUGHT QUESTIONS

1 Darwin did not initially use the expression "survival of the fittest." The phrase was first used by the British social philosopher Herbert Spencer, and was popularized especially by those who saw Darwinism as a license for unbridled, cutthroat competition in the era of the 'robber barons.' (Darwin did adopt the expression in his later editions.) Do you think that this expression accurately reflects Darwinian thought regarding the animal and plant kingdoms?

In what ways does it not?

A Great Deal of Evidence Supports Darwin's Ideas

Darwin had proposed two major hypotheses: branching descent ("descent with modification") acting over long periods of time, and natural selection as a mechanism that explained how evolutionary change takes place. In the years since Darwin proposed these two hypotheses, many scientists have used scientific methods to conduct thousands of tests of both hypotheses. The results of these tests have yielded much evidence that supports the hypotheses, and none that falsifies them.

Mimicry

One of the earliest tests of Darwin's hypothesis of natural selection involved the phenomenon of **mimicry**, in which one species of organisms deceptively resembles another. In one type of mimicry, a distasteful or dangerous prey species, called the model, gives a very unpleasant and memorable experience to any predator that attempts to eat it. Predators always avoid the model after such an unpleasant experience. A palatable prey species, the mimic, secures an advantage if it resembles the model enough to fool predators into avoiding it as well.

Selection by predators explains mimicry rather easily. Any slight resemblance that might cause a predator to avoid the mimic as well as its model is favored by selection and passed on to future generations of the mimic species, while individuals not protected in this way would be eaten in greater numbers. Predator species differ greatly in their abilities to distinguish among prey species, so a resemblance that fools one predator might not fool another. Any advantage that increases the number of predators fooled is favored by selection, causing closer and closer resemblance to evolve with the passage of time.

Sometimes several species that resemble each other are all distasteful to predators. Predators learn to avoid distasteful species, but a certain number of prey individuals are killed for each predator individual that learns its lesson. Without mimicry, each prey species must sustain this loss separately. Mimicry allows predators to learn the lesson with fewer individuals of each prey species dying in the process. The mimicry therefore benefits each prey species and is thus favored by natural selection.

Mimicry often varies geographically. Some wide-ranging tropical species mimic different model species in different geographic areas. The deceptive resemblance is always to a species living in the same area, never to a far-away species. Environmentalist theories such as Lamarck's had no way to account for the evolution of mimicry, and the patterns of geographic variation could not be explained by either environmentalist theories or by Paley's natural theology. Natural selection, however, explains the variation as resulting from selection by predators.

In a well-known case of mimicry, the model is the monarch butterfly, a distasteful species that feeds on milkweed plants. An unrelated species, the viceroy, is similar in superficial appearance, and is thus avoided by many predators (Figure 5.4); some of these predators may find the viceroy distasteful as well.

Industrial melanism

The power of natural selection is also demonstrated by a phenomenon called **industrial melanism**, when darker colors evolve in areas polluted by industrial soot in species that are usually light in color elsewhere. In the British Isles, a species known as the peppered moth (*Biston betularia*) (phylum Arthropoda, class Insecta, order Lepidoptera) had long been recognized by an overall light gray coloration with a salt-and-pepper pattern of irregular spots. A black variety of this species was discovered in the 1890s. The black moths increased until they came to outnumber the original forms in some localities (Figure 5.5). The British naturalists E.B. Ford and H.B.D. Kettlewell studied these moths for several decades from about 1944 onward. Downwind from the major industrial areas, the woods had become polluted with black soot that killed the lichens grow-

Figure 5.4

An example of mimicry. (A) *Limenitis arthemis*, a nonmimic relative of (B) *Limenitis archippus*, the viceroy. The viceroy resembles the unrelated monarch butterfly (C, *Danaus plexippus*), the model. The monarch is avoided by predators after just a single unpleasant experience (D, E). The warning color pattern of the monarch helps predators learn to avoid it; the viceroy is protected because its color pattern mimics that of the monarch.

(A) Butterfly closely related to the species from which the viceroy evolved

(B) Viceroy

(C) Monarch

(D) Blue jay eating monarch

(E) Blue jay vomiting after eating monarch

ing on the tree trunks. Most of the moths living on the darkened tree trunks in these regions were black. However, where the woods were untouched by the industrial soot, the tree trunks were still covered with lichens and the moths had kept the light-colored pattern. Ford and Kettlewell hypothesized that the moths resembling their backgrounds would be camouflaged and thus harder for predators to see. To test this hypothesis, they pinned both light and dark moths on dark tree trunks in polluted woods, and they also pinned both types on lichen-covered tree trunks in unpolluted woods. They observed that birds ate more of the dark moths in the unpolluted woods (favoring the survival of the light-colored pattern), but birds in the polluted woods ate more of the light-colored moths, not the dark ones. These observations and the geographical patterns of variation (see Figure 5.5) were easily explained in terms of natural selection by predators. In addition, since the experiments were first conducted, laws to control smokestack emissions and other forms of pollution were passed and enforced, and many of the woods affected by pollution have returned to their former state. In these woods, the lichens have returned to the tree trunks, and most of the moths in these places again have the original light-colored pattern.

Industrial melanism in insects demonstrates that species can change in response to changes in the environment and were not created with permanent, unalterable traits. Lamarckian mechanisms fail to explain industrial melanism because, in these species, the adult colors do not change once they are formed, and there is nothing that individuals can do that would alter their color. The experiments with birds as predators clearly show natural selection at work.

Evidence for branching descent

Darwin's contemporaries immediately recognized that his concept of "descent with modification" could be used to make sense out of a variety of observations not easily explained by other means. The branching pattern of descent explained the formation of groups or clusters of related species in particular geographic areas. Moreover, it explained the arrangement of these groups into a hierarchy of smaller groups within larger groups. Biologists before Darwin had been making classifications this way for about a century, but it was his theory of branching descent that explained why this type of classification made sense. Darwin predicted that biological classifications would increasingly become genealogies (that is, maps of descent similar to Figure 5.1) as more and more details about the evolution of each group of organisms became known. Darwin's prediction came true as scientists found more and more evidence showing that relationships among species arise in branching patterns of descent. Evidence for such relationships comes from the comparative study of the anatomy, biochemistry, physiology, and embryology of different species.

Figure 5.5

Industrial melanism in peppered moths in the British Isles.

Geographic variation in the frequency of melanic moths in the 1950s, which reached as high as 100% in polluted localities downwind from major industrial centers.

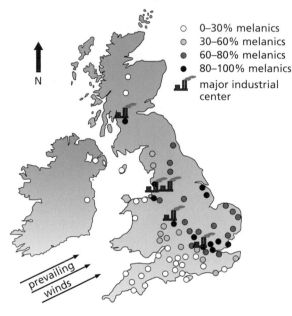

N

○ 0–30% melanics
◔ 30–60% melanics
◉ 60–80% melanics
● 80–100% melanics
⌐ major industrial center

prevailing winds

The melanic (black) variety and the original 'peppered' variety (below the right wing-tip of the melanic moth) on a light, lichen-covered tree trunk.

The same two varieties on a dark, soot-covered tree trunk.

Homologies. The construction of family trees is based in large measure on the study of shared structures or gene sequences. Under Darwin's paradigm, shared similarities are evidence that the organisms in question share a common ancestry. In a sense, a shared similarity is a falsifiable hypothesis that the several species sharing it are related to one another by descent. By itself, one such similarity reveals very little, but a large number of similarities that fit together into a consistent pattern strongly suggest shared ancestry. When the evidence for shared ancestry is compelling, the similarity is called **homology**.

Darwin noted the similarities among the forelimbs of mammals: "what can be more curious than that the hand of a man, formed for grasping, that of a mole for digging, the leg of the horse, the paddle of the porpoise, and the wing of the bat, should all be constructed on the same pattern, and should include the same bones, in the same relative positions?" (Figure 5.6). Darwin wondered why similar leg bone structures appeared in the wings and legs of a bat, used as they are for such totally different purposes. Among the Crustacea (barnacles, crabs, shrimp, etc.), most species have a thorax region with eight pairs of leglike appendages

Figure 5.6

Homologies among mammalian forelimbs adapted to different functions.

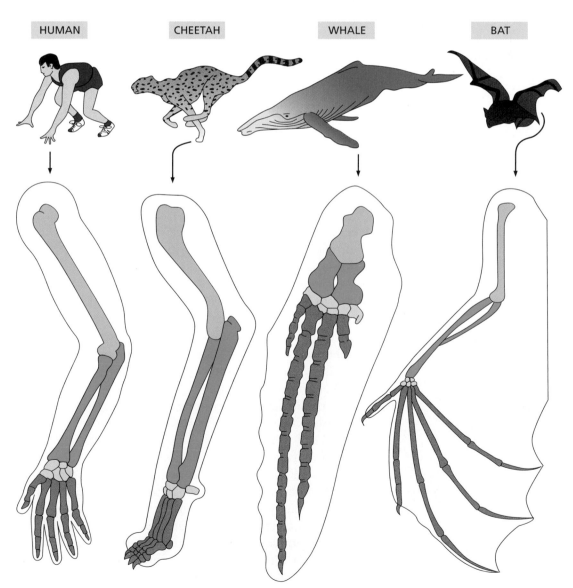

used for locomotion, but the group that includes lobsters and crabs has the first three of these pairs modified into accessory mouthparts, leaving only five pairs of walking legs. Why, asks Darwin, should a crustacean that has more mouthparts have correspondingly fewer legs, or why should those with more legs have fewer mouthparts? Darwin's answer is that all these structures arose by modification of the same type of repeated part. Crustacean mouthparts and legs are derived from a common set of leglike appendages. When the structure of these appendages varied (through mutation and other causes), natural selection favored different structures for different uses. Individuals with better-functioning mouthparts near the mouth, or with better-functioning legs near the center of gravity, left more offspring, and the proportions of the responsible genotypes increased in each population. As a consequence, when more appendages were used around the mouth as food-handling structures, there were fewer appendages remaining to be used as legs. An omnipotent God, however, could have added mouthparts without taking away legs (without being limited by the total number of appendages), leaving Darwin to declare, "How inexplicable are these facts on the ordinary view of creation!" (Darwin. *Origin of Species*, 1859, p. 437.)

Vestigial structures. Structures whose function has been lost in the course of evolution tend to diminish in size. Often, they persist as small, functionless remnants, called **vestigial structures.** A good human example is the coccyx, a set of two or three vestigial tail bones at the base of the spinal column, homologous to the tails of other mammals. The Darwinian paradigm of natural selection and branching descent explains these vestigial structures as the remnants of structures that had once been functional. Neither Lamarck nor the creationists had any explanation for the presence of vestigial structures, and certainly not for the homologies of vestigial structures in many species with their functional counterparts in related species.

Convergence. Similarities that result from common ancestry (that is, true homologies) should also be similar at a smaller level of detail, and even similar in embryological derivation, meaning that they should grow from the same source tissues. A hypothesis of homology can thus be falsified if two similar structures turn out to be dissimilar in detailed construction or in embryological derivation. There are also cases in which several hypotheses of homology are in conflict because they require different patterns of relationship for different characters. In such cases, evolutionists examine all the similarities more closely and repeatedly to see whether a reinterpretation is possible for one set of similarities.

Convergence is an evolutionary phenomenon in which similar adaptations evolve independently in lineages not closely related. Similarities for which the hypothesis of homology is falsified by more careful scrutiny are often reinterpreted as convergent adaptations, meaning structures that evolved independently in unrelated lineages. Resemblance resulting from convergent adaptation is called **analogy.** Distinguishing homology from analogy is an ongoing aim of evolutionary classification. For example, the wings of bats and insects are analogous, rather than homologous, structures. They are constructed in different ways and from different materials, and their common shapes (which they also share with airplane wings) reflect adaptation to the aerodynamic requirements of flying. Although bat wings are not homologous to insect wings, they are homologous to human arms, whale flippers, and the front legs of horses and elephants. These all have similar bones, muscles, and other parts in similar positions despite their very different shapes and uses, while insect wings have no bones and their muscles are very differently located.

Cephalopods as an example. One frequent test of the hypothesis of branching descent is to identify a group of organisms that share some particular character, such as an anatomical or biochemical peculiarity. The general hypothesis of branching descent then gives rise to a more specific hypothesis, that these organisms all share a common descent from a common ancestor. An example of this type of reasoning can be illustrated by the Cephalopoda, a group of mollusks that includes the squids, octopuses, and their relatives. All cephalopods can be recognized by the presence of a well-developed head and a mantle cavity beneath (Figure 5.7). The mantle cavity contains the gills, the anus, and certain other anatomical structures. Other mollusks have mantle cavities, but only in the Cephalopoda is the mantle cavity located beneath the head and prolonged into a nozzlelike opening known as the hyponome. Knowing this, we can formulate the specific hypothesis that all cephalopods share a common descent.

If our hypothesis is true, then we should be able to find, as evidence, some additional similarities among cephalopods not shared with other mollusks. Such similarities do exist: all cephalopods have beaklike jaws at the front of the mouth, and a muscular part, called the foot in other mollusks, subdivided into a series of tentacles (see Figure 5.7). Moreover,

Figure 5.7

Three living types of cephalopod mollusks (kingdom Animalia, phylum Mollusca, class Cephalopoda): the cuttlefish, the squid, and the chambered nautilus.

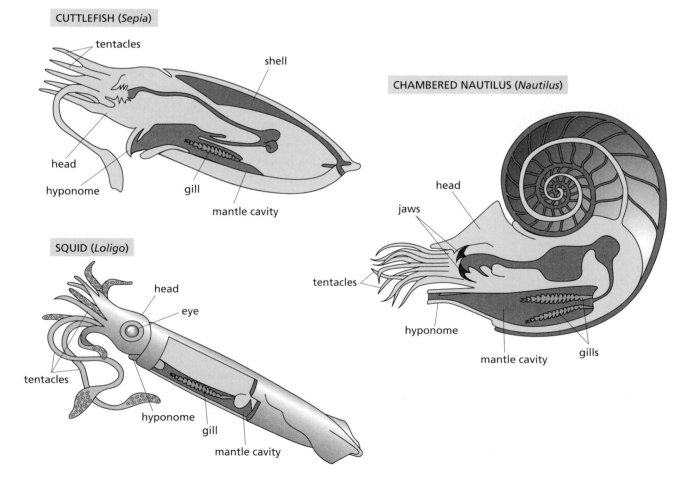

all cephalopods have an ink gland that secretes a very dark, inky fluid. When a squid or octopus feels threatened by a predator, it releases this fluid into its mantle cavity and quickly squirts the contents of the mantle cavity forward through its nozzlelike hyponome. This action hides the animal and propels it backwards, in a direction not expected by its predator. The predator's attention is meanwhile held by the puff of inky fluid. By the time the color dissipates, the squid or octopus has vanished. All members of the Cephalopoda have this elaborate and unusual escape mechanism, including squids, cuttlefishes, octopuses, and the chambered nautilus. The hypothesis of a common descent for all the Cephalopoda is thus consistent with the known data, meaning that the hypothesis has been tested and not falsified.

Further comparisons among species. Since Darwin's time, many additional types of similarities have been discovered among organisms. The comparative study of embryonic development has resulted in the discovery of many new similarities among distantly related species, some of which are described in Chapter 6. Comparative genomics (Chapter 4) and comparative studies in biochemistry have revealed the detailed structure of protein chains, DNA and RNA sequences, and other large molecules of biological interest. As with anatomical similarities, similarities in biochemistry or in embryology group related species together, and small groups are contained within larger groups at many hierarchical levels. Each new type of similarity has brought new evidence of branching descent with modification: in most cases, the groups established by older methods are reaffirmed when newer methods result in the same groupings. In a few cases, new groupings are discovered, and sometimes these are later corroborated by further evidence such as new fossil discoveries.

On the basis of anatomical, embryological, and biochemical comparisons, hypotheses of common descent have been tested and confirmed for cephalopod mollusks and for many other groups of animals and plants. This increases our confidence in the larger hypothesis that all species of organisms have evolved from earlier species in patterns of branching descent. The many facts of comparative anatomy, comparative physiology, embryology, biogeography, and animal classification are all consistent with the hypothesis that modern species have evolved from ancestors that lived in the remote past, and they make little sense otherwise.

Since Charles Darwin published his evolutionary ideas in 1859, thousands of tests have been made of his twin hypotheses of branching descent and natural selection. Because these thousands of tests have failed to falsify either hypothesis, both now qualify as scientific theories that enjoy widespread support. The Darwinian paradigm continues to this day as a major guide to scientific research.

Further evidence from the fossil record

The history of life on Earth is measured on a time scale encompassing billions of years. This geological time scale (Figure 5.8) was first established by studying **fossils**, the remains and other evidence of past life forms. Most fossils are formed from the burial of plants or animals in sediment. The soft parts of these organisms are often consumed or decomposed, but they may leave imprints in soft sediments if buried rapidly. Scientists had recognized since 1555 that most fossils were the remnants of species no longer living, thus providing clear-cut evidence for extinction. Comparisons of fossils with living species provide evidence of change over time and thus have a role in supporting the theory of descent with modification and the concept of evolution more generally.

Stratigraphy. The geological time scale was first established through the study of layered rocks (**stratigraphy**). One of the first principles established in the study of these rocks was that when rock layers have not been drastically disturbed, the oldest layers are on the bottom and successively newer layers are on top of them. Using this principle, geologists can identify the rock formations in a particular place as part of a local sequence, arranged chronologically from bottom to top.

Figure 5.8

The geological time scale.

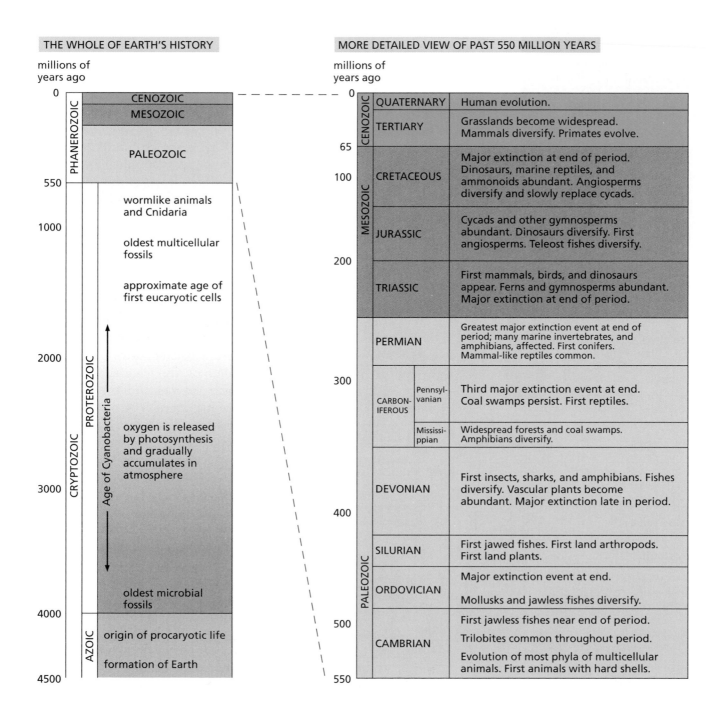

Local sequences from different places can be matched with one another in several ways, but the most reliable of these proved to be the study of their fossil contents. Two rock formations are judged to be from the same time period if they contain many of the same fossil species (the principle of **correlation by fossils**). The rocks do not need to be similar in composition or rock type—one can be a limestone and the other a shale—but if their fossil assemblages are similar, they are judged to be equally old. A single species of fossil is never sufficient; several fossil species are needed, and they must occur together with some consistency. Using this technique, paleontologists (scientists who study fossils) have been able to match up formations of the same age from many different localities, enabling them to assemble the world's various local sequences into a 'standard' worldwide sequence, which is the basis for the complete sequence of time periods shown in Figure 5.8. The dates assigned to these periods are determined by measuring the rate of radioactive decay in certain rocks.

Family trees. The age of a fossil, by itself, tells us very little about its place in any family tree. The relative ages of fossils only begin to have meaning when we study a group of organisms represented by many fossils. The family tree or genealogy of any group, called its **phylogeny**, fits into a pattern like that shown in Figure 5.1. Any such family tree is a hypothesis that biologists use to explain how the anatomical and other characteristics of each species are related, which leads to the classification of the group as a whole. In any family tree, the known fossils must fit into a consistent framework.

Figure 5.9

Family tree of the class Cephalopoda (phylum Mollusca), showing branching descent over time. Horizontal width represents number of species in each group; vertical distance represents time.

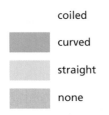

coiled

curved

straight

For example, there are many fossils of cephalopod mollusks, permitting further tests of the hypothesis of a common descent for all the Cephalopoda. Living and extinct cephalopods can be arranged into a family tree consistent with our knowledge of the characteristics of each species and the relations among them (Figure 5.9). Differences among the living cephalopods can be explained with reference to this family tree. The chambered nautilus is very different from other living cephalopods because it is fully housed within a coiled shell and has four gills, while the squids and octopuses have only two gills and a very small, reduced shell or else none at all. One would therefore imagine a family tree in which octopuses and squids have a common ancestor that the chambered nautilus does not share. The fossil

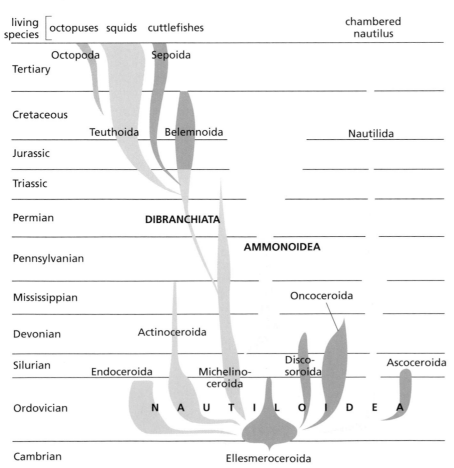

Cephalopoda conform to these expectations. The group of cephalopods with the oldest fossil record is the nautiloids, of which the chambered nautilus is the only living remnant. A second group of cephalopods, called the ammonoids (see Chapter 18, p. 351), flourished in Mesozoic times, during the age of dinosaurs. A small, third group had an internal shell that became reduced in size. When the ammonoids became extinct, this third group, the Dibranchiata, persisted and is represented today by the squids and octopuses. Thus, the fossil record of the cephalopod mollusks, including both the anatomy and age relationships of fossil forms, confirms the relationships hypothesized on the basis of the anatomy of the living forms.

The fossil record has repeatedly confirmed hypotheses of descent for particular living species. For example, Thomas Henry Huxley, one of Darwin's early supporters, studied the anatomy of birds and declared them to be "glorified reptiles." The interpretation of birds as descendants of the reptiles was strengthened by the discovery of *Archaeopteryx*, a fossil with many birdlike and also many reptilian features. Among these were a long tail, simple ribs, a simple breastbone, and a skull with a small brain and tooth-bearing jaws (Figure 5.10). Despite these reptilian features, *Archaeopteryx* had well-developed feathers and was probably capable of sustained flight. The discovery of transitional forms like *Archaeopteryx*, coupled with the fossils of other early birds and of feathered dinosaurs close to bird ancestry, strengthens our confidence in the hypothesis that birds evolved from reptiles. Other transitional forms are known, such as those between older and more modern bony fishes, between fishes and amphibians, between reptiles and mammals, and between terrestrial mammals and whales. Instead of being exactly intermediate in each trait, transitional forms usually exhibit a mix of some innovative characteristics and some ancestral characteristics.

Post-Darwinian thought

One of the hallmarks of science is that hypotheses are subjected to rigorous and repeated testing. Darwin's hypotheses have been thoroughly and repeatedly tested for nearly a century and a half, and the general outlines of branching descent and natural selection have been repeatedly and consistently confirmed. A second hallmark of science is that theories are extended and modified as new data are discovered. Here again, Darwin's ideas have been extended and supplemented by newer findings. Many additional details are now known, none of which contradict the basic concepts of natural selection and branching descent. A third hallmark of a scientific theory is that it acts as a spur to further research, and Darwin's two theories have stimulated more scientific research than just about any other theory in the history of biology, with the possible exception of Mendel's theories in genetics. Evolution guides our thinking in nearly every field of biology, which is why "nothing in biology makes sense, except in the light of evolution."

During the period 1860–1940, scientists who doubted the effectiveness of natural selection proposed many other hypotheses to explain evolutionary change. Our modern theory of mutations originated from one such hypothesis. In Czarist Russia, scientists of nearly every political stripe (from conservative monarchists to socialists and anarchists) found the British idea of competition very distasteful. They were therefore reluctant to embrace the concept of natural selection, which they felt was based on a model of competition. (Darwin had emphasized that he meant competition in a "large and metaphorical sense," but his Russian readers still found the similarity with capitalist economics distasteful.) As an alternative, the anarchist Petr Kropotkin, and the novelist and pacifist Leo Tolstoy, developed theories of "mutual aid" or **mutualism** as an important evolutionary mechanism. According to this view, organisms succeed (and leave more offspring) if they cooperate with one another instead of competing, as among social insects (see Chapter 8). Modern biologists now view mutualism as an adaptive interaction between species that may evolve as the result of natural selection. Natural selection favors any change that increases reproductive success, and this frequently includes changes that benefit other species directly. The theory of mutualism has thus been accepted into the mainstream of Darwinian evolutionary thought, and is no longer viewed as something incompatible with natural selection. Darwin himself gave several examples of cooperative interactions between species.

Although Darwin's theories of natural selection and branching descent continue to guide biological research to this day, the early 1940s saw the expansion of the evolutionary paradigm called the modern synthesis. Dar-

Figure 5.10

The early bird *Archaeopteryx*, compared with a modern-day pigeon. Modern birds lack teeth, and evolution has enlarged the braincase and strengthened other parts (wing, ribs, breastbone, pelvis, tail) highlighted here.

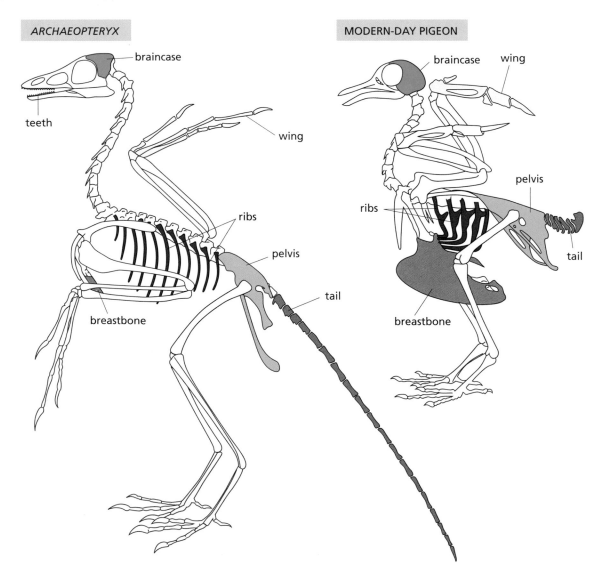

win's ideas are retained in this expanded paradigm, but the findings of genetics are also incorporated and are used to explain the source of heritable variation. Natural selection is well documented as an important cause of evolutionary change, but it is by no means the only cause. Chance alone (accidents of sampling which individuals die, which live, and which reproduce) can cause erratic changes in the allele frequencies of natural populations, especially small ones. This phenomenon, called **genetic drift**, is discussed further in Chapter 7. The changes produced by genetic drift are usually nonadaptive, and they increase the chances that a small population will die out. Later in this chapter we will also discuss the importance of geographic isolation, a nonselective force that sets up conditions that bring about the evolution of new species.

CONNECTIONS CHAPTERS 7, 8

Beginning in the 1970s, Niles Eldredge and Stephen Jay Gould advocated a theory that they viewed as alternative to Darwinian thought. Darwin had frequently emphasized that evolutionary change was gradual, but Eldredge and Gould claimed instead that species remain static for long periods and then change abruptly. The new species begins, they said, as a small, isolated population on the geographic periphery of the original species. The small size of the isolated population allows it to undergo rapid change, producing a new species. Once the new species becomes successful, its numbers and geographic range may increase to the point where it invades the geographic range of the original species from which it evolved. If the new species successfully outcompetes the original one, the original one may become extinct. What we often see in the fossil record is the abrupt replacement of one species by another rather than a gradual change. Gould always claimed that this **punctuated equilibrium** theory is an alternative to Darwin's gradualism, but certain other evolutionists (such as Ernst Mayr) view the two theories as fully compatible.

THOUGHT QUESTIONS

1 For a family tree such as that shown in Figure 5.9, what kinds of fossil evidence (be specific) would falsify the descent pattern shown? What kinds of evidence would cause paleontologists to modify the family tree but continue to believe in a process of descent with modification? What kinds of evidence would falsify the hypothesis of descent with modification?

2 One of the Galapagos finches studied by Darwin has woodpeckerlike habits and certain woodpeckerlike features: it braces itself on vertical tree trunks with stiff tail feathers in the manner of true woodpeckers and drills holes for insects with a chisel-like bill. However, it lacks the long, barbed tongue with which true

woodpeckers spear insects; it uses cactus thorns for this purpose instead. How would Lamarck have accounted for this set of adaptations? How would Paley? How would Darwin? Which of these explanations accounts for the absence of the barbed tongue in the woodpecker finch? How would each hypothesis account for the absence of true woodpeckers on the Galapagos Islands?

3 Is the study of evolution static or changing? Find some recent news articles dealing with new fossil discoveries or other new findings that deal with evolution.

4 Explain antibiotic resistance in bacteria by using Darwin's concept of natural selection.

Creationists Challenge Evolutionary Thought

Opposition to the idea that life evolves has come from various quarters. Many opponents of evolution have been creationists, people who believe in the fully formed creation of species by God. In this section we discuss a variety of creationist ideas, along with the creationist opposition to evolution.

Creationists, by definition, believe that God created biological species. The majority of creationists believe that God created species much as we see them today, and that they did not evolve. Creationists are usually devout believers and most of them are Christian, but beyond these similarities creationists do not always share all of the same beliefs. Some creationists have been practising scientists who conduct research and follow scientific methodology, while others are strongly antiscience and may even seek the destruction of science and of scientific institutions.

Three major groups of creationists stand out:

1. Bible-based creationists, who insist on the biblical account of creation. These creationists work outside science and reject any scientific theory that conflicts with scripture; some of them are openly hostile to science.

2. Intelligent-design creationists, who try to work within the framework of science to find evidence of design in nature. They claim that biological systems are so complex and so well adapted to their functions that only an intelligent (and benevolent) designer could have made them.

3. Theistic evolutionists, who believe that God created the universe and all life, but that species evolved after that time and that evolution is one of God's creative processes. Several practising scientists and various clergy adhere to this view.

Bible-based creationism

In the United States, most creationists have based their beliefs on a literal interpretation of the Bible. Believing in the inerrant truth of their ideas, these creationists reject all science and all scientific evidence that contradicts their beliefs. Some of them are openly hostile to science. Almost all of these creationists are Protestant Christians, but they represent a small minority within the Protestant tradition. The large, established denominations accept the evidence for evolution as fully compatible with their religious beliefs.

Various shades of opinion divide the Bible-based creationists into separate groups. One group, the 'Young Earth' creationists, insists that the six days of creation mentioned in Genesis were each 24 hours in length. Another group, the 'Day–Age' creationists, seeks to reconcile science with biblical accounts by proposing that the six days of creation should be interpreted as six ages of indefinite duration. (The Hebrew word 'yom' is often used this way elsewhere in the Bible.) Many mainstream clergy of various faiths subscribe to this view. Somewhat similar in their views are the 'Gap' creationists, who reconcile biblical with geological time scales by claiming that a long, indefinite gap of time intervened between the events described in Genesis chapter 1 and Genesis chapter 2.

Early fundamentalism. In the early twentieth century, most opposition to evolution came from certain Protestants, mostly in the United States, who declared that evolution conflicted with the account of creation given in the Bible. These people founded a number of societies, including the Society for Christian Fundamentals (the origin of the term *fundamentalist*). The fundamentalists persuaded several state legislatures to pass laws restricting or forbidding the teaching of evolution in schools. Some of these state laws remained until the 1960s.

In 1925, a famous court case was brought in Tennessee by the fledgling American Civil Liberties Union. A teacher, John H. Scopes, was arrested for reading a passage about evolution to his high school class. The trial attracted worldwide attention. Scopes lost and was assessed a $100 fine. Upon appeal, the case was thrown out because of the way in which the fine had been assessed; the merits of the case were never really debated. The Scopes trial did, however, have a chilling effect on the textbook publishing industry: books that mentioned evolution were revised to take the subject out, and most high school biology textbooks published in the United States between 1925 and 1960 made only the barest reference, if any, to Charles Darwin or any of his theories.

Creationism in recent decades. The Soviet launch of the Earth-orbiting satellite Sputnik in 1958 set off a wave of self-examination in American education. Groups of college and university scientists began examining high school curricula with renewed vigor, and several new high school science textbooks were written. Most of the new biology texts emphasized evolution, or at least gave it prominent mention.

Alarmed in part by the new textbooks, a new generation of creationists began a series of attacks on the teaching of evolution. These new creationists tried to portray themselves as scientists, calling their new approach 'creation science' even though they never conducted experiments or tested hypotheses. Instead of making their studies falsifiable, the new creationists claimed that they held the absolute truth:

> Biblical revelation is absolutely authoritative.... There is not the slightest possibility that the *facts* of science can contradict the Bible and, therefore, there is no need to fear that a truly scientific comparison...can ever yield a verdict in favor of evolution.... The processes of creation...are no longer in operation today, and are therefore not accessible for scientific measurement and study. (H.M. Morris (Ed.). *Scientific Creationism*. San Diego: Creation-Life Publishers, 1974, pp. 15–16 and 104.)

> We do not know how the Creator created, what processes He used, for He used processes which are not now operating anywhere in the natural universe.... We cannot discover by scientific investigations anything about the creation process used by the Creator. (D.T. Gish. *Evolution: The Fossils Say No!*. San Diego: Creation-Life Publishers, 1978, p. 40.)

In contrast, Darwin knew that his theories were—rightly—subject to empirical testing and possible falsification (see the quotation on page 147).

Some creationist writings contain faulty explanations of scientific concepts. One such misinterpretation uses the second law of thermodynamics. According to this law, a closed system (one in which energy neither leaves nor enters) can only change in one direction, to that of less order and greater randomness. Thus, a building may crumble into a pile of stones, but a pile of stones cannot be made into a building without the expenditure of energy. Creationists have claimed that this law precludes the possibility of anything complex ever evolving from something simpler. The second law of thermodynamics does not, however, rule out the building up of complexity; rather, it states that making something complex out of something simple requires an input of energy. The second law of thermodynamics *does* apply to all biological processes. If the Earth were a thermodynamically closed system, life itself would soon cease. However, the Earth is not a thermodynamically closed system because energy is constantly being received from the Sun, and this energy allows life to persist and evolve. Creationist claims on this point may have originated as an innocent error, but the point has been so well refuted that its continued use can only be a deliberate misrepresentation that lies outside the bounds of science or of honest debate.

In the 1960s, because many of the laws in the United States forbidding the teaching of evolution had been declared unconstitutional, one group of creationists, led by Henry Morris, Duane Gish, and John Slusher, decided on a new approach. Evolution could be taught in the schools, they argued, but only if 'creation science' was taught along with it and given equal time. (The concept of 'equal time,' was originally a measure to ensure fairness in political campaigns.) A few state legislatures passed laws inspired by this new group of creationists. An Arkansas law known as the Balanced Treatment Act (Public Law 590) was finally declared unconstitutional in 1981, and a similar Louisiana law was declared unconstitutional a few years later. Interestingly, in the challenges to these laws, the scientific issues *were* raised in court, and prominent scientists were called upon to testify. Specifically, in the Arkansas and Louisiana cases, the U.S. Court of Appeals was asked to rule on what is scientific and what is not. The court finally ruled that evolution is a scientific theory and may be taught, whereas 'creation science' is not science at all because it involves no testing of hypotheses and because its truths are considered to be absolute rather than provisional. Instead, 'creation science' was found to be a religion, or to include so many religious concepts (creation by God, Noah's flood, original sin, redemption, and so forth) that it could not be taught in a public school without violating the U.S. Constitution's historic separation of church and state.

In the 1990s, a new group of creationists emerged, advocating the view that modern science, particularly evolution, is the basis for the materialistic philosophy that they claim is responsible for all that is wrong in today's society. The avowed aim of this group, called the 'Wedge group', is to destroy all of science, and evolution in particular, by driving in a thin 'wedge' and then continuing to drive it in deeper and deeper until the body of science is split asunder. Small but well-financed, this group is the guiding force behind the Discovery Institute and the Center for the Renewal of Science and Culture (CRSC), both of which support intelligent-design creationism.

Creationists continue to exert influence today. Despite state laws that have been declared unconstitutional, creationists continue to pressure local school boards and state education departments to support their approach. These efforts have sometimes been successful. In 1999, the Board of Education in Kansas approved, at the urging of creationists, a statewide science curriculum that included no mention of evolution. They also approved a statewide program for testing scientific knowledge and understanding, but decided that an understanding of evolution should not be part of this testing program. Two years later, after two of the officials who had voted for this curriculum had been voted out of office and a third had left voluntarily, the Board of Education reversed its earlier decision and restored evolution to the science curriculum in Kansas. This kind of opposition to the teaching of evolution is largely an American phenomenon. Biologists in most countries other than the United States have not faced similar opposition.

Intelligent design

In the eighteenth and nineteenth centuries, many creationists were also scientists who proposed and tested hypotheses. For example, Reverend William Paley and his supporters proposed that biological adaptations were the work of a benevolent God. In 1996, American biochemist Michael Behe resurrected Paley's preevolutionary arguments and revised them in the new language of cell biology and biochemistry.

Paley's Natural Theology. Paley sought to prove the existence of God by examining the natural world for evidence of perfection in design. The anatomical structures examined by Paley and his supporters were so well suited to the functions that they served and were, in his view, so perfectly designed that they could only have come from God. Paley's school of Natural Theology was very influential in Britain in the early nineteenth century, and the young Charles Darwin was educated in its lessons.

Paley and his supporters had paid much attention to complex organs such as the human eye. The eye, they pointed out, was composed of many parts, each exquisitely fashioned to match the characteristics of the other parts. What use would the lens be without the retina, or the retina without a transparent cornea? An eye, they argued, would be of no use until all its parts were present; thus it could never have evolved in a series of small steps, but must have been created, all at once, by God.

Paley pointed to the structure of the heart in human fetuses as containing features that adaptation to the local environment could not account for. In adult mammals, including humans, the blood on the left side of the heart is kept separate from the blood on the right side of the heart (see Chapter 10, pp. 353–354). In fetal mammals, the blood runs across the heart from the right side to the left, bypassing the lungs, which are collapsed and nonfunctional before birth. As the blood enters the left side of the heart, it passes beneath a flap that is sticky on one side. When the baby is born, its lungs fill, and blood flows through them. The blood returning to the heart from the lungs now builds up sufficient pressure on the left side that the flap closes. Because it is sticky on one side, it seals shut. No amount of adaptation to the environment, said Paley, could endow a fetus with a valve that was sticky on one side so that it would seal shut at birth. Only a power with foresight could have realized that the fetus would need a heart whose pattern of blood flow would change at birth, and thus designed the sticky valve. Paley attributed the

CONNECTIONS
CHAPTER 13

foresight to God, and he insisted that no other hypothesis could explain such an adaptation to future conditions.

What is most interesting is that Paley and his many supporters understood the nature of science and used the methods of science to argue their case. Paley in particular sought scientific proof of God's existence and benevolence by arguing that no other hypothesis could explain the evidence as well. This example shows that good science is certainly compatible with a belief in God or a rejection of evolution. In fact, the best scientists of the period from 1700 to 1859 were, with few exceptions, devout men who rejected the pre-Darwinian ideas of evolution on scientific grounds.

Darwin's response. Darwin was quite familiar with Paley's arguments, and he offered evolutionary explanations for many of the intricate and marvelous adaptations that Paley's supporters had described. In each case, Darwin argued that the hypothesis of natural selection could account for the adaptation as well as the hypothesis of God's design.

To counter Paley's argument about complex organs such as the eye, Darwin pointed out that the eyes of various invertebrates can be arranged into a series of gradations, ranging in complexity from "an optic nerve merely coated with pigment" to the elaborate visual structures of squids, approaching those of vertebrates in form and complexity. A large range of variation in the complexity of visual structures is found within a single group of organisms, the Arthropoda, which includes barnacles, shrimps, crabs, spiders, millipedes, and insects. All the visual structures, regardless of their degree of complexity, are fully functional adaptations, advantageous to their possessors. It would therefore be quite reasonable, argued Darwin, to imagine each more complex structure to have evolved from one of the simpler structures found in related animals. Eyes, in other words, could have evolved through a series of small gradations.

As an additional argument against Paley, Darwin also pointed out several adaptations that were *less* than perfect, or that seemed to be 'making do' with the materials at hand. The gills in barnacles are modified from a brooding pouch that once held the eggs. The milk glands of mammals are modified sweat glands. The giant panda, evolved from an ancestor that had lost the mobility of its thumb, developed a new thumb-like structure made from a little-used wrist bone. (This last example was not known in Darwin's time, but fits well into his argument.) These many adaptations seem more easily explained by natural selection than by God's design because the design is imperfect and God could presumably have 'done better.' Natural selection is limited to the use of the materials at hand, and then only if there is variation; an omnipotent God could have made barnacle gills from entirely new material without taking away the brood pouches, and could have given pandas a true thumb instead of modifying a wrist bone. Darwin and his supporters used examples like these to show that the evolutionary explanation fitted the available evidence better than Paley's explanation of divine planning. For example, natural selection perpetuates only those hearts whose flaps seal properly at birth.

'Irreducible complexity.' With today's knowledge of cell biology and biochemistry has come a return to Paley's argument from design at the molecular level. The major proponent of this argument is biochemist Michael Behe. He begins with the claim that every living cell contains many sophisticated molecular systems that he calls "irreducibly complex." An irreducibly complex system, according to Behe, is any system that is nonfunctional unless all of its parts are present and functional. Behe's argument, which echoes Paley's, is that no irreducibly complex system could evolve by small, piecemeal steps. According to this creationist argument, natural selection can only improve upon a functioning system, so could never create a system that requires many parts in order to function at all. Thus, if a complex system cannot function without a minumum of five components, then natural selection could never bring about the evolution of a second component when only one existed, or a third component when only two existed, because none of these changes would improve anything if the system remained nonfunctional with fewer than all five of the required components. Paley had earlier made the same argument with regard to the several parts of the eye, as did the British zoologist St. George Mivart in Darwin's time. Darwin himself realized the power of this argument, for he wrote:

> If it could be demonstrated that any complex organ existed, which could not possibly have been formed by numerous successive, slight modifications, my theory would absolutely break down. (Darwin. *Origin of Species*, 1859, p. 189.)

There are at least two responses to counter the argument of irreducible complexity. One is to show that the system is not, in fact, irreducibly complex, and that a partial system with only one or a few of its components does function in some capacity and represents an improvement over the same system with fewer components or none at all. If one component is an improvement over none, and two are an improvement over one, then the entire system can evolve piecemeal, step by step, because each step is an improvement over the previous ones, and natural selection favors each small, successive change.

The other response to irreducible complexity is to recognize the role played by changes in function. A system may be incapable of its present function unless fully formed, and thus be described as irreducibly complex. However, the system, or some of its parts, may originally have served a different function, and thus could have evolved by a series of small steps as long as each step improved some ability to serve some function. This can be illustrated by the evolution of insect wings, which developed from external folds of the thoracic wall. By building models of insects with no folds, tiny folds, medium-size folds, and folds large enough to function as wings, scientists were able to show that an increase in the size of the fold from none to small or from small to medium would hardly have improved flying ability. Natural selection would probably not have been able to bring about the early increases in the size of the folds, based upon their usefulness in flying. On the other hand, the function of the folds in cooling the body was also considered. Muscular activity generates heat, and an animal would be in danger of cooking its own tissues if it exercised vigorously without somehow dissipating heat. The efficiency of the flaps in cooling the body also varies with size, as shown in Figure 5.11. Most of the improvement comes in the smaller sizes, with medium flaps dissipating more heat than small ones, which in turn dissipate more heat than none

Figure 5.11

The evolution of insect wings. The efficiency of thoracic folds in primitive insectlike arthropods was measured according to two criteria: efficiency in cooling the body down by dissipating heat, and efficiency in airborne locomotion by adding to downward air resistance and to lift. Up to a certain size, increments in the size of the folds improved cooling ability but had little effect on locomotion. Thus, early increases in fitness among small to moderate wing sizes depended on improved cooling; however, later increases in fitness depended more on improvements in flying ability.

at all. Large flaps, on the other hand, are scarcely any more efficient than medium-size flaps in their cooling ability. Thus, the early stages in the evolution of the wing flaps are thought to have been selectively favored because they improved the body's ability to exercise more without overheating. Only after the flaps had reached a certain medium size, their function as wings became more important than their function in dissipating heat. Thus, the early stages were selectively improved because they helped dissipate heat, while the later stages were selectively improved because they functioned as wings.

As this example shows, the early stages in the evolution of a structure may have been useful for a totally different function than the one they now serve. Half-built structures, or systems with only a few components, may have improved the ability of their possessors to pass on their genes even without fulfilling their present function. Many structures are now known to have changed their function in the course of evolution.

With these evolutionary counter-arguments in mind, we can now examine Michael Behe's claims of irreducible complexity for several systems that function within cells. All of them could have evolved gradually, step by step, despite Behe's insistence to the contrary. For example, the clotting of blood is a multi-step process that Behe argues is "irreducibly complex" because none of it would work unless all of it were present. In fact, blood can clot upon exposure to air, and the many chemicals that improve clotting ability could certainly have evolved one at a time, each representing a piecemeal improvement. Natural selection would favor the evolution of any protein or other compound that aided in the clotting process and reduced the chances of bleeding to death. Because blood chemistry does not fossilize, we have no proof of how blood clotting evolved, but a gradual evolution is certainly plausible.

Another of Behe's examples discusses complement proteins and antibody production. A somewhat detailed explanation of how this system works, and how various parts of the system can and do function apart from the rest, can be found on our Web site (under Resources: Complement). If parts of the system are useful without the remainder (and, in fact, functional in species known to lack the complete system), then they certainly could have evolved in small steps by natural selection.

Conclusions. It is important to note that Behe has conducted no research and provided no scientific evidence to support his claims of irreducible complexity or to test any other hypothesis. Perhaps we should describe his claims as philosophical rather than scientific. One important measure of a scientific theory is the amount of research that it

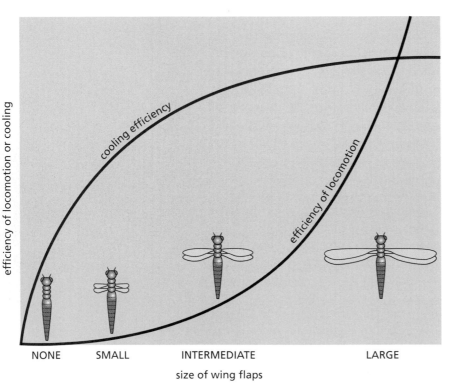

stimulates. Behe's concept of irreducible complexity has stimulated no research that supports any of his claims, but many arguments have been offered to show that the systems that Behe discusses are not, in fact, irreducibly complex.

All of the systems that Behe claims to be irreducibly complex can be explained as the products of gradual, step-by-step evolution, especially if changes in function are considered. Various biologists have examined Behe's claims and none, to our knowledge, support them. Of the systems that Behe describes, none withstands scrutiny as an argument against evolution.

Despite the many criticisms that have been raised against Behe's arguments, the state of Ohio in 2002 seriously considered a proposal to add intelligent design to the science curriculum as an alternative to evolution by natural selection, even though there is no evidence (produced by hypothesis testing) to support the idea. Irreducible complexity is not a scientific theory, and does not qualify as science in the minds of most scientists.

Reconciling science and religion

A majority of scientists are religious, and a majority of devout people of all religions also accept scientific findings. There are many ways of reconciling religious and scientific viewpoints, and a majority of theological seminaries of all faiths teach that science and religion are fully compatible. The following are examples of the ways in which some people have reconciled religious beliefs and science.

René Descartes was the originator of a dualistic philosophy that separates science and religion as operating in different spheres. In this view, science informs us about the physical world, including the human body, while religion informs us about the spiritual world, including both God and the human soul. Questions about the body can be answered by science, while questions about the soul or about God can be answered only theologically. A separation between science and religion, based on this dualism, has become the official view of the Roman Catholic Church.

The Protestant theologian Reinhold Niebuhr defines religion as the study of the "ultimately unknowable." In this view, advances in science have expanded the frontiers of knowledge—the study of what is knowable, but religion is the study of what remains—the ultimately unknowable. Thus, religion and science operate in separate spheres, and there is no possibility of an incompatibility between them.

Various scientists have expressed the view that God should be excluded from scientific theories whenever it is possible to do so. One such scientist was the early nineteenth century French astronomer and mathematician Pierre Simon LaPlace, one of the authors of the idea that galaxies and solar systems form from swirling masses as the result of natural gravitational forces. When he published his book on this 'nebular hypothesis,' he presented a copy to the emperor Napoleon, who asked him why he had not mentioned God in his book. LaPlace replied, "I have no need of that hypothesis." A similar attitude caused the British geologist Charles Lyell to exclude all miracles from his geological theories.

According to some twentieth-century versions of a theory called 'operationalism,' God's presence in certain scientific explanations may not be needed. Thus, the statement, "The Grand Canyon was formed by

the action of running water over long periods of time" is indistinguishable from the statement, "God formed the Grand Canyon by the action of running water over long periods of time." Any evidence that could support either of these statements would also support the other, and any evidence against either statement would also be evidence against the other. The two statements are operationally equivalent (or indistinguishable) because no evidence could possibly distinguish between them. According to this view, God is not a necessary part of the explanation, in line with the view that LaPlace had expressed earlier.

A number of scientists and religious thinkers have reconciled science with their religious beliefs by accepting the findings of science as an explanation of the ways that God operates. God created the world along with the natural laws that govern it, and science attempts to discover these natural laws. Isaac Newton, William Paley, and Albert Einstein all expressed views along these lines. One version of this approach is that God set natural laws in place but then withdrew to allow the universe to unfold according to the workings of these natural laws. Another version is that God occasionally intervened to set things right by making exceptions to natural laws. Einstein, who favored the non-interventionist interpretation, ridiculed this second approach in his statement, "I can't believe that God plays dice with the world."

Theistic evolution represents an attempt along these lines to reconcile evolution with a creationist viewpoint, either with or without divine intervention. The Jesuit philosopher and paleontologist Pierre Teilhard de Chardin believed that evolution, including human evolution, was part of God's method of creation in accordance with natural law. Charles Lyell, a geologist who had inspired Darwin's early thinking, and Alfred Russell Wallace, a naturalist who discovered natural selection independently of Darwin, both came late in life to the belief that evolution was the consequence of natural laws, but that divine intervention had been necessary to bring about the evolution of human beings. Most scientists, however, see no need for any such exceptions to explain human evolution.

THOUGHT QUESTIONS

1 In what ways did William Paley use scientific evidence? Did he use testable hypotheses? Which of today's creationists use falsifiable hypotheses to support their claims?

2 How much time should be devoted in science classes to alternative explanations or theories that have been tested and rejected? Should time be given to explanations that are not testable? Should all explanations be given equal time? How much (if any) of a science curriculum would you devote to divine creation as an alternative to evolution? To astrology as an alternative to astronomy? To the theory that disease is caused by demons or evil spirits?

3 Does the teaching of unpopular or rejected theories encourage students to think critically? Does it encourage attitudes of fairness? Does it increase students' understanding of what science is and how science works?

4 Do you think that the concept of intelligent design should be taught in high schools as an alternative to evolution by natural selection? Why or why not?

Species Are Central to the Modern Evolutionary Paradigm

The evolutionary paradigm known as the modern synthesis was based largely on the fusion of genetics with Darwinian thought. The corner-stone of the modern synthesis paradigm is a theory of speciation, the process by which one species branches into two species.

Populations and species

A biological **population** consists of those individuals within a species that can mate with one another in nature. If we look backward in time, we realize that any two individuals in a population share at least some of their alleles because of common descent. If we look into the future, we see that any two opposite-sex individuals in a population are potential mates. Membership in a population is determined by descent and by the capacity to interbreed.

Biological populations within a species may exchange hereditary information (alleles) with one another. The combining of genetic information from different individuals or the exchange of genetic information between populations is called **interbreeding**. The existence of biological barriers to such exchange is called **reproductive isolation**. Interbreeding between populations of the same species takes place when members of different populations mate and produce offspring; reproductive isolation inhibits such matings to varying degrees.

Species are defined as *reproductively isolated groups of interbreeding natural populations*. There are several points to note in this definition. Physical characteristics (morphology) are not part of the definition of species; species are defined by breeding patterns instead. Populations belonging to the same species will interbreed whenever conditions allow them to. Populations belonging to different species are reproductively isolated from one another and will thus not interbreed. Any biological mechanism that hinders the interbreeding of these populations is called a **reproductive isolating mechanism**, as explained below. Species are composed of natural populations, not of isolated individuals. Thus, the mating behavior of individuals in captivity can only serve as *indirect evidence* of whether *natural populations* would interbreed under natural conditions.

The many reproductive isolating mechanisms fall into two categories: those that prevent mating and those that interfere with development after mating has occurred. Mating is prevented when potential mates never encounter each other, possibly because they live in different habitats, or because they are active at different times of day or in different seasons, or because they are not physiologically capable of reproduction at the same time. Figure 5.12 shows that wood frogs

Figure 5.12

Reproductive isolation of several frog species by season of mating, an ecological means of preventing mating between species.

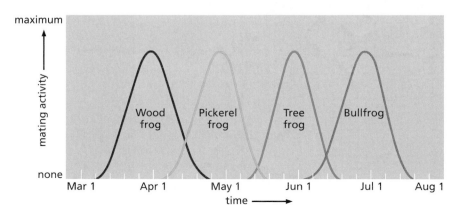

are fully isolated ecologically from tree frogs and bullfrogs by breeding at different seasons; they are partly isolated from pickerel frogs because the breeding seasons overlap only slightly. Mating can also be prevented by differences in behavior, allowing potential mates with different courtship rituals to live together in the same place without mating. For example, different species of fireflies (phylum Arthropoda, class Insecta, order Coleoptera, family Lampyridae) use different flashing patterns and flight patterns (Figure 5.13) as mating signals. In addition, insects and some other animals have hardened and inflexible sexual parts (genitalia); mating of these animals requires a 'lock and key' fit, and mating is prevented if the parts do not fit together properly.

There are other isolating mechanisms in which mating occurs but the offspring do not develop. In animals, sperm from a male of another species may die before fertilization takes place. In plants, the pollen may fail to germinate on the flowers of another species. If a mating takes place between species, the fertilized egg may die after fertilization. Incompatible chromosomes may disrupt cell divisions and developmental rearrangements, leaving the embryo or larva to die. Alternatively, hybrid individuals may live for a while but not reach reproductive age, or they may be sterile. For example, a mule is a sterile hybrid between a horse and a donkey. The sterility of mules keeps the gene pools of horses and donkeys separate, so they remain separate species.

How new species originate

To explain how a new biological species has come into existence, we need to explain how it has become reproductively isolated from closely related species. The origin of a species is thus the origin of one or more reproductive isolating mechanisms.

In the vast majority of cases, new species have come into existence through a process of **speciation** that includes a period of **geographic isolation** in which populations are separated by some sort of barrier such as a mountain range or simply an uninhabited area that the organisms do not cross. The essence of the theory is that reproductive isolating mechanisms originate during times when such barriers separate populations geographically. Geographic isolation is not by itself considered to be a reproductive isolating mechanism; rather, it sets up the conditions under which the separated populations may evolve along different lines, resulting in reproductive isolation.

What happens depends in part on the length of time for which the populations are geographically isolated—more time allows more chances for reproductive isolating mechanisms to evolve. Another factor is that natural selection must favor different traits on the two sides of the geographic barrier. That is, conditions must be different enough for one set of traits to increase fitness in one locale and for a different set of traits to increase fitness in another

Figure 5.13

Flashing patterns used as mating signals by different species of fireflies. The species 1 to 9 are reproductively isolated from one another by the behavioral differences shown in these patterns. Details in this form of behavioral isolation include the duration of each flash, the number of repetitions, and the location of the insect when it flashes. A firefly will respond only to the flashing pattern of its own species.

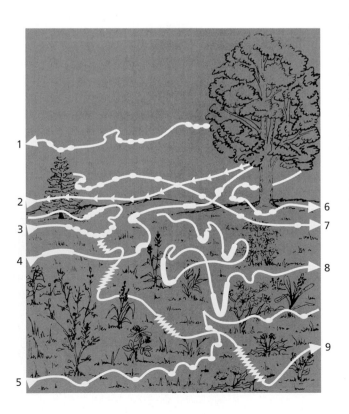

locale. If the populations on opposite sides of the barrier are selected differently for a long enough period, then one or more reproductive isolating mechanisms may evolve between the two groups of populations and separate them into different species (Figure 5.14). If the populations later come into geographical contact again, the reproductive isolating mechanisms that have evolved during their separation will keep them genetically separate as two species. For example, frog or cricket populations isolated on opposite sides of a mountain chain or a large body of water may develop different mating calls. Because the animals respond only to the mating calls of their own population, the two populations will be reproductively isolated and thus become separate species.

The geographic theory of speciation predicts that examples of incomplete speciation may be discovered. If two populations are separated for a very long time (or if selective forces on opposite sides of a barrier differ greatly), then the populations are likely to split into two species. If the separation is brief, then speciation is unlikely. These two situations lie at opposite ends of a continuum. Somewhere along this continuum lies the situation in which populations have been separated by a geographic barrier long enough for reproductive isolation to begin evolving, but not yet long enough for the reproductive isolation to be perfected. Partial or imperfect reproductive isolation between two populations would lessen the chances of interbreeding between them, but not prohibit it entirely. Such situations have indeed been found, for example, among the South American fruitflies known as *Drosophila paulistorum*. Crosses between divergent populations of *Drosophila paulistorum* produce fertile hybrid females but sterile hybrid males. The geneticists studying these flies referred to them as "a cluster of species *in statu nascendi*" (in the process of being born).

People intuitively group similar species together and give names to many collective groups: birds, snakes, insects, pines, orchids, and so forth. Biologists organize these collective groups into a classification that reflects the degree of evolutionary relatedness among species, using methods described in Chapter 6.

Figure 5.14

Geographic speciation: the evolution of reproductive isolation during geographic isolation. Genetically variable populations that spread geographically can develop locally different populations that are capable of interbreeding with one another initially. If the populations are separated for a long enough time by a barrier such as a mountain range or a deep canyon, they may develop differences that prevent interbreeding even after contact is resumed.

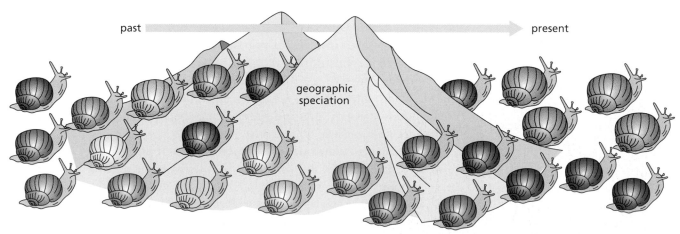

past → present

geographic speciation

| Initial population has lots of genetic variation | Mountain range arises, separating population into two groups | Environment becomes different on the two sides | Two populations diverge as mutation and selection fit organisms to environment | When populations come into contact again, reproductive isolating mechanisms keep species genetically separate |

THOUGHT QUESTIONS

1 What kinds of reproductive isolating mechanisms might prevent related species of antelopes from interbreeding? Answer the same question for related species of birds, related species of trees, and related species of butterflies.

2 Several kinds of organisms reproduce asexually, producing offspring without combining gametes from two parents. Can asexually reproducing organisms belong to species? Can the definition of species be modified to apply to asexually as well as sexually reproducing organisms?

Life on Earth Originated by Natural Processes and Continues to Evolve

In addition to explaining how species change and how new species arise, modern evolutionary theory also accounts for the origins of life on Earth. The origin of life, the early history of life on Earth, and the effects of life on Earth's atmosphere will all be discussed in Chapter 19.

Evolution as an ongoing process

Evolution is a process that takes place within species as well as between species, and the process continues in the present as it has in the past. The evolutionary changes in populations and the changes that create new species can be studied as they occur. Within the twentieth century, the peppered moths of some locations in England changed from predominantly light-colored to almost all dark and back again. In one species of Galapagos ground finches, *Geospiza fortis*, the average bill size changes back and forth. Small-beaked birds that eat soft seeds survive and produce the most offspring in years when rainfall is adequate, but birds with larger beaks are at an advantage in drought years because they can open large, tough old seeds. The average bill size of birds within the population thus increases in drought years and decreases in wet years. In fruit-flies, different chromosomal variations (inversions) are favored in different seasons. We see that evolution responds adaptively to fluctuating environmental conditions. Different alleles are selected by different environmental conditions at different times because their phenotypes are more adaptive in those conditions.

Selection also continues to operate in human populations and in bacteria. For example, infant mortality is much higher among babies born under about 3 kg (7 lb) in weight, even with all that modern medicine can offer. Natural selection thus favors birth weights close to this optimum value. Selection also favors certain human genotypes in certain environments (Chapter 7) and during epidemics (Chapter 17). The use of antibiotics has favored the evolution of resistance to these drugs among bacteria (Chapter 17).

Concluding Remarks

Considerable evidence now shows that evolution has taken place in the past and that organisms continue to evolve today, though often slowly. The ways in which species resemble one another and are related to one another reflect branching patterns of descent. Evolutionary change is brought about by natural selection, a process that operates whenever some genotypes leave more offspring than others. Species are reproductively isolated from one another, and the splitting of a species therefore requires the evolution of a new reproductive isolating mechanism. All species, including humans, arose by speciation and are products of evolution. Our attempts to classify the resulting diversity of species are explained in Chapter 6.

Chapter Summary

- **Evolution** is the central, unifying concept of biology.
- Darwin's major contributions included his theories of **branching descent** ("descent with modification") and **natural selection**.
- Only inherited traits contribute to evolution and bring about **adaptation**; acquired characteristics do not.
- **Evolution** operates through **natural selection**: there is heritable variation in all **species**, and different genotypes differ in **fitness** by leaving different numbers of surviving offspring.
- Forces of **natural selection** include predators, disease, and **sexual selection**.
- **Mimicry** is easily explained by natural selection but not by any alternative hypothesis.
- Branching descent with modification accounts for **homology** between species. **Fossils** provide important evidence for evolution, as does the comparative study of anatomy, biochemistry, and embryological development.
- The modern synthesis combines Mendelian genetics with Darwinian evolution. It describes the evolution of genes and phenotypes in populations, and it includes a theory for the formation of species through geographic isolation.
- **Speciation** occurs through the build-up of genetic differences between **populations** arising primarily during times of geographic isolation. Over time, this results in **reproductive isolation**, which prevents **interbreeding** between species.
- **Evolution** continues today in all species. In many cases, we can detect ongoing change from year to year.

CONNECTIONS TO OTHER CHAPTERS

Chapter 1	Darwinian evolution and modern evolutionary theory are both good examples of successful paradigms.
Chapter 1	Presenting creationist ideas in school classrooms raises several social policy issues.
Chapter 2	Gene mutations provide the raw material for evolution.
Chapter 4	Comparative genomics reveals evolutionary patterns of descent.
Chapter 6	Branching descent and other evolutionary processes have produced a great diversity of species that have been described and classified, and many others that await discovery and description.
Chapter 7	Differences have evolved and continue to evolve both within and among human populations.
Chapter 8	Social behavior and reproductive strategies are, in part, products of evolution.
Chapter 9	Successful species may increase so rapidly in numbers that they outstrip the available resources.
Chapter 11	Plant characteristics resulting from evolution include the presence of chloroplasts and vascular tissues.
Chapter 13	Differences in brain anatomy in different species provide good evidence of evolution.
Chapter 16	Viruses and other microorganisms may evolve disease-causing strains, as well as strains resistant to certain medicines.
Chapter 17	Bacteria often evolve antibiotic resistance through natural selection.
Chapter 18	Speciation increases biodiversity, whereas extinction diminishes biodiversity.
Chapter 19	The evolution of life has changed the entire Earth, including the atmosphere and all habitats.

PRACTICE QUESTIONS

1. Match the ideas in the first list with the people in the second. One name needs to be used twice.

   a. Evolution is a branching process.

   b. Adaptations should be studied carefully as a way of understanding God's creation.

   c. Evolution should never be taught.

   d. New species originate by a process that includes geographic isolation.

   e. Adaptation occurs by use and disuse.

   f. Organisms with successful adaptations will be perpetuated, whereas those with unfavorable characters will die out.

   g. Evolution and creation science should be given equal time in science classes.

   i. Creationist supporters of the 'balanced treatment act'

   ii. Creationists of the period 1890–1940

   iii. Charles Darwin

   iv. Jean Baptiste Lamarck

   v. William Paley

   vi. Modern evolutionary biologists

2. The Bahamas are a group of islands in the Atlantic, made mostly of coral fragments. The closest mainland is North America, but political ties are to Great Britain. According to Darwin's reasoning, the birds and other species living on these islands should have their closest relatives in:

   a. other islands of similar composition in the Pacific

   b. islands such as the Canary Islands, in the Atlantic at a similar latitude

   c. North America

   d. England

3. Which theory had no way of explaining the sticky flap in the fetal heart?

   a. Darwin's

   b. Paley's

   c. Lamarck's

4. In mimicry, the mimics and their models always:

   a. live in similar climates, although they may be far away

   b. live close together

   c. taste the same to predators

   d. are camouflaged to resemble their backgrounds

5. Which of these is NOT considered a reproductive isolating mechanism?

   a. two geographically separated species

   b. two species breeding in different seasons

   c. two species that produce infertile hybrids when they mate

   d. two species with different mating calls

   e. two species whose external genitalia cannot fit together

6. Give a clear definition of the term *species*.

7. What is the basic argument used by supporters of intelligent design? What kinds of evidence can be used against this argument?

Issues

- Why is life so diverse?
- Why have some groups proliferated into so many species?
- Why is classification important?
- How do our classifications reflect evolution?
- How are classifications established?
- Did humans evolve? How and why? Where and when?

Biological Concepts

- Evolution (adaptation, descent with modification)
- Form and function
- Species and speciation
- Products of evolutionary change (phylogenetic classification, biodiversity)
- Types of cells (procaryotic, eucaryotic)
- Hierarchy of organization
- Classification of organisms (three domains and six kingdoms)
- Sexual recombination
- Major groups of organisms
- Plant structure and adaptations (vascular tissues, seeds, flowers, etc.)
- Animal structure and adaptations (bilateral symmetry, sense organs, head, assembly-line digestion, body cavities, segmentation, etc.)

Chapter Outline

Why Classification Is Important

"All those names"

Taxonomic theory

Modern Classifications Recognize a Great Difference Between Procaryotic and Eucaryotic Cells

Procaryotic cells

Eucaryotic cells

Endosymbiosis and the evolution of eucaryotes

Six Kingdoms of Organisms Are Included in Three Domains

Domain and kingdom Archaea

Domain and kingdom Eubacteria

Domain Eucarya

Kingdom Protista

Kingdom Plantae

Kingdom Mycota

Kingdom Animalia

Humans Are Products of Evolution

Our primate heritage

Early hominids

The genus *Homo*

6 Classifying Nature

*L*ife on Earth is extremely diverse, with an estimated 10 million species or more. Why is life so diverse and so prolific? How do we describe this diversity? How can we begin to understand it? As Charles Darwin observed, every living species has so great a tendency to reproduce that, if left unchecked, there would soon be no standing room left on Earth for all its progeny. As we saw in Chapter 5, species can also undergo speciation: splitting and thus creating additional species. This process has been going on so long and so frequently as to have produced the many millions of species alive today, plus an even greater number of species that have become extinct in the past. In this chapter we will point out certain innovations that greatly increased the ability of biological species to succeed in life and to speciate further.

Why Classification Is Important

In order to describe and understand life's diversity, we need a system of classification. People intuitively group all insects together and all birds together. All people have common names for collective groups of similar species: birds, snakes, insects, pines, orchids, and so forth. Some collective groups, such as beetles, are contained within larger groups, such as insects. Biologists organize these collective groups into a **classification**, an arrangement of larger groups that are subdivided into smaller groups, each reflecting their degree of evolutionary relatedness. Classifications help us to catalog and describe life's diversity, which is the first step on the road to understanding this great diversity.

Any collective group of similar organisms, such as insects or orchids, is called a **taxon**, and **taxonomy** is the study of how these taxa (plural of *taxon*) are recognized and how classifications are made. In a biological classification, species are grouped into successively more inclusive groups, or 'higher taxa': related species are grouped into genera (singular, **genus**), related genera into **families**, related families into **orders**, related orders into **classes**, related classes into **phyla**, and related phyla into **kingdoms**, such as the animal or plant kingdoms. All these are arranged as groups within groups, with the less inclusive (smaller) groups sharing more characters and the more inclusive (larger) groups sharing fewer characters. Thus, species within a genus have more characters in common than do families within an order.

For example, human beings constitute the species *Homo sapiens*. Figure 6.1 shows, beginning on the right, that *Homo sapiens* is grouped together with *Homo erectus* and certain other fossil species into the genus *Homo*. (A genus always has a one-word name that is capitalized; a species has a two-word name in which the first word is the name of the genus; after the two-word name has been introduced, the genus may subsequently be abbreviated, for example, *H. sapiens*.) The genus *Homo* is grouped together with the extinct genera *Australopithecus* and a few others into the family Hominidae. This family is included in the order

Primates, which also includes apes (Pondigae), monkeys (Cebidae and Cercopithecidae), and lemurs (Lemuridae). The primates are grouped together with rodents (Rodentia), carnivores (Carnivora), bats (Chiroptera), whales (Cetacea), and over two dozen other orders into the class Mammalia, including all warm-blooded animals with hair or fur that feed milk to their young. Mammals are one of several classes in the phylum Chordata, a group that includes all vertebrates (animals with backbones) and a few aquatic relatives such as the sea squirts and amphioxus. The Chordata and several dozen other phyla are together placed in the animal kingdom (Animalia). Animals are one of the several kingdoms to be discussed later in this chapter.

"All those names"

Communication among scientists requires a common vocabulary. Even hunters, gatherers, farmers, and other nonscientists need names for all the types of organisms familiar to them. However, in a linguistically diverse world, different people speak many different languages. Many years ago, the scientists of Europe discovered that they could best communicate with scientists in other countries by using Latin names. Thus, while the common names of animal and plant species differ from one language to another (e.g., dog, hund, cao, perro, chien), the scientific name, *Canis*, is the same for scientists the world over. Moreover, while modern languages change (modern English differs from Shakespeare's and even more from Chaucer's), scientific names are based mostly on Latin and Greek roots that do not change over time.

The modern hierarchical form of classification, which we have just described, originated with the eighteenth-century Swedish naturalist Carl von Linné, who wrote under the name Linnaeus. Linnaeus also began a system in which each species has a two-word name, such as *Homo sapiens*. The use of two words for species names allows us to recognize millions of species with far fewer names. Thus, *Canis familiaris* is

Figure 6.1

The place of the species *Homo sapiens* in the classification of organisms. Reading from right to left shows the increasingly more inclusive taxa to which our species belongs. Reading from left to right focuses on taxa of increasingly narrow scope.

KINGDOMS	PHYLA	CLASSES	ORDERS	FAMILIES	GENERA	SPECIES
	Phylum Porifera					
Kingdom Archaea	Phylum Cnidaria	Class Agnatha	Order Monotremata			
Kingdom Eubacteria	Phylum Platyhelminthes	Class Chondrichthyes	Order Marsupialia	Family Cebidae		
Kingdom Protista	Phylum Echinodermata	Class Osteichthyes	Order Chiroptera	Family Lemuridae		
Kingdom Animalia	Phylum Chordata	Class Mammalia	Order Primates	Family Hominidae	Genus *Homo*	*Homo sapiens*
Kingdom Mycota	Phylum Mollusca	Class Aves	Order Rodentia	Family Pongidae	Genus *Australopithecus*	*Homo erectus*
Kingdom Plantae	Phylum Annelida	Class Reptilia	Order Carnivora	several other families	a few more genera	*Homo habilis*
	Phylum Arthropoda	Class Amphibia	Order Artiodactyla			
	many other phyla	several other classes	many other orders			

the domestic dog, *Canis lupus* is the wolf, *Canis latrans* is the coyote, and *Canis aureus* is the jackal. Any person encountering any of these names for the first time immediately recognizes that they are all related species placed in the same genus. The name of each family is always based on the name of a well-known genus (e.g., Canidae for the dog family, comprising *Canis* and its relatives).

Larger collective groupings, such as orders and classes, also have names. Most of these names are descriptive of some important characteristic of the group. Thus, the class Mammalia (from the Latin word *mamma*, meaning a breast) includes all species whose young are nourished with their mother's milk, and the class Amphibia (from Greek roots *amphi* and *bio*, literally meaning 'both lives') contains frogs, salamanders, and other animals that live in water as gill-breathing larvae before they transform into lung-breathing adults. The names, in other words, have a meaning that relates to the organisms and that makes the name easier to learn and remember, at least when you only learn a few at a time. Unfortunately for many beginners, there are many names to know, and students who see them all at once may be overwhelmed by the sheer number of unfamiliar terms. We suggest that you begin by simply learning to recognize the same name on repeated encounters—it is, after all, just a name. If you learn the meaning of the name (which often means learning its Latin or Greek roots), it may help you to associate it with an important characteristic of the group. Remember that a classification is a tool designed to make communication easier.

Taxonomic theory

It is easy for students to look upon classifications as fixed and unchanging, but this is a false impression. Each classification is really just a hypothesis about how best to describe the variation among the organisms being discussed. If you read several accounts of the classification of the same group of organisms, you will probably find that not all authorities follow exactly the same classification. How, then, are classifications made? Sometimes, one person may become an expert on a particular group of organisms and then everyone accepts their classification as authoritative, but this situation is uncommon. More often there are multiple researchers, each following a slightly different classification, and each attempting to attract followers to their way of thinking. So how do scientists determine that one classification is preferable to another for the same group of organisms?

One important goal of classification is to summarize and communicate what we know about groups of organisms (taxa). Thus, one criterion of a good classification is that it should aid in describing variation among taxa, summarizing both their differences and their similarities. Early classification systems were based on common physical structures, and members of each group were therefore expected to share visible characteristics. Many familiar groups recognized by ordinary people were given formal names and recognized in the classification, for example, birds—class Aves. Fishes, insects, clams, ferns, and orchids were also recognized in many early classifications; some of these groups are still recognized today. One reason that these taxa are useful is that there are many shared similarities that unite all their members and that distinguish these taxa from related taxa. However, other groupings, such as 'worms,' were shown to be heterogeneous when the included organisms were studied more closely; such unnatural groups are usually abandoned.

Another goal of classification is to describe evolution. In order to describe all the different descendants of some ancient species, we clearly need some collective name for the group. Such a group is called a **clade** (from a Greek word, meaning branch), and the study of branching patterns in evolution, from common ancestors to their descendants, is therefore called **cladistics**. On a family tree such as the one shown in Box 6.1 or in Figure 5.1, a clade consists of any branching point, representing an ancestor, and all the lines branching from it above, representing its descendants. In traditional practice, the taxa and their groupings were decided upon first, and family trees were then offered as explanatory hypotheses that justified the classification. Cladistics reverses this order, deriving the family tree first and then basing the classification on the tree.

Biologists who follow cladistics develop family trees and classifications in which each larger taxon corresponds as closely as possible to a clade. After years of intensive study of a large and varied group of organisms, they make lists of important anatomical features and other characters that they think will help in classification. Such characters may include the occurrence of particular plant pigments; reproductive adaptations such as seeds or hard-shelled eggs; the presence of openings in the skull; the number and shape of teeth; the presence of certain glands and their secretions; or the anatomical structure of the limbs and other body parts. Often, characters present only in larvae or in embryonic life stages are used. After many such characters are listed for the species being studied, the biologist will deduce (sometimes with the aid of a computer program) the family tree that can explain the evolution of these characters with the smallest number of evolutionary changes. That is, if five species all possess a certain shared character, five changes would be required if each acquired the character independently, but only one change would be required if the character were acquired by an ancestor common to all five. From information of this kind, the five species are assigned to a clade, meaning a common branch of the family tree. This approach usually produces a consistent pattern, with smaller clades nested within larger clades. If, however, one character does not fit consistently with the others (that is, if it suggests clades that cannot be nested with the clades suggested by other characters), then that character tends to be discarded as misleading and the family tree is then based on the remaining characters. Once a family tree has been established, a classification is then drawn up, based on this family tree, by giving a name (e.g., Insecta or Diptera) and a rank (e.g. class or order) to each clade. As an example, Box 6.1 shows how a family tree and classification of land vertebrates may be derived.

Molecular biology (including genomics) provides further evidence that can be used to construct classifications. First, the DNA sequences or protein sequences of various species are compared. Older molecular approaches simply compared all pairs of species for the overall similarity of their molecular sequences and grouped species into a classification based on the percentage of overall similarity (or percentage homology).

BOX 6.1   Evolution and Classification of the Land Vertebrates

The following example presents a simplified version of cladistics, the procedure commonly used to formulate a family tree and base a classification upon it. The organisms in this example are the land vertebrates, including amphibians (frogs, salamanders, etc.), turtles, lepidosaurs (lizards and snakes), crocodilians, birds, monotremes (egg-laying mammals such as the platypus), marsupials (pouched mammals), and placental mammals (those retaining the young inside the uterus throughout gestation). The groups (taxa) being classified are compared with an 'outgroup' of organisms outside the assemblage being classified, but related to it. Characters present in this 'outgroup' are presumed to be primitive (or ancestral) for the assemblage in question. In this example, lungfish are the outgroup.

First, a character matrix is prepared as follows:

TAXON / CHARACTERS	Lungfish	Amphibians	Turtles	Lizards and snakes	Crocodilians	Birds	Monotremes	Marsupials	Placental mammals
1 egg with amnion	no	no	yes	yes	yes	yes	yes	yes	yes
2 hair	no	no	no	no	no	no	yes	yes	yes
3 warm blooded	no	no	no	no	no	yes	yes	yes	yes
4 single centrale bone	no	no	yes	yes	yes	yes	no	no	no
5 two-windowed skull	no	no	no	yes	yes	yes	no	no	no
6 shell	no	no	yes	no	no	no	no	no	no
7 opening in lower jaw	no	no	no	no	yes	yes	no	no	no
8 slit-shaped anus	no	no	no	yes	no	no	no	no	no
9 feathers	no	no	no	no	no	yes	no	no	no
10 major arteries connected	no	no	no	no	yes	no	no	no	no
11 live birth	no	no	no	seldom	no	no	no	yes	yes
12 superficial jaw-opening muscle	no	no	no	no	no	no	yes	no	no
13 placenta	no	no	no	seldom	no	no	no	no	yes
14 pouch	no	no	no	no	no	no	no	usually	no

In this matrix, each taxon being classified is listed as either having or lacking each of the character traits used in the study. In our example, we are only using 14 characters; professional studies usually use many more than this.

Our characters are as follows:

1. An egg containing certain internal membranes (amnion, etc.), capable of being laid on land

2. Hair or fur

3. Maintenance of a constant, warm body temperature by using heat generated through physiological activity (endothermy, commonly called "warm blooded")

4. A single centrale bone in the ankle

5. A skull with two window-like openings behind the eye

6. A shell enclosing most of the body

7. A window-like opening in the lower jaw

8. A transverse, slit-shaped anus

9. Feathers

10. An opening between the major arteries as they exit from the heart, permitting the blood to be diverted right or left as needed to balance pressure when the animal dives

11. Giving birth to live young

12. A jaw-opening muscle belonging to a superficial layer encircling the neck, and used instead of a deeper muscle beneath the jaws

13. A placenta formed by two embryonic membranes (chorion and allantois) and attaching to the inner wall of the uterus during gestation

14. A pouch in which the young is nursed following its birth

For this matrix of characters, the family tree shown here is the simplest one possible, meaning that it requires the fewest evolutionary changes. Each labeled transition on the tree marks the evolution of a trait shared by all taxa on all of the branches to the right of that point. The family tree can be constructed from the character matrix by hand or by a computer program, using the following basic protocol:

a. Allow all the taxa sharing a derived trait (a 'yes' in the above matrix) to form a single branch of the tree. Allow the various branches all to be nested in one another. For example, the land-based egg (character 1) is presumed to have evolved only once, in a common ancestor of all the animals possessing this trait, that is, all of the animals to the right of the branch labeled character 1. If all of the characters evolved only once, then the tree is finished.

b. Where the first strategy does not result in a tree with each character showing up on only one branch of the proposed tree, we choose the simplest family tree as being the most likely. When a few characters do not fit in consistently it may indicate that the character trait evolved but then was lost by some, but not all, species in a group, or that the character evolved more than once (by convergence). When this happens, select the tree that requires the fewest evolutionary changes, minimizing the number of character traits for which we must assume multiple origins or acquisition and subsequent loss. In the family tree shown here, warm body temperature (3) is shown as having evolved independently in birds and in mammals. A different tree would result in many more traits for which multiple origins must be assumed.

The tree formed by this protocol allows us to recognize the clades (branches of the family tree) that form the basis for the classification. Some of the taxonomic groupings supported by this particular family tree are shown as brackets on the right of the diagram. 'Reptiles' are in quotation marks because they are a group that does not correspond to a clade in this arrangement.

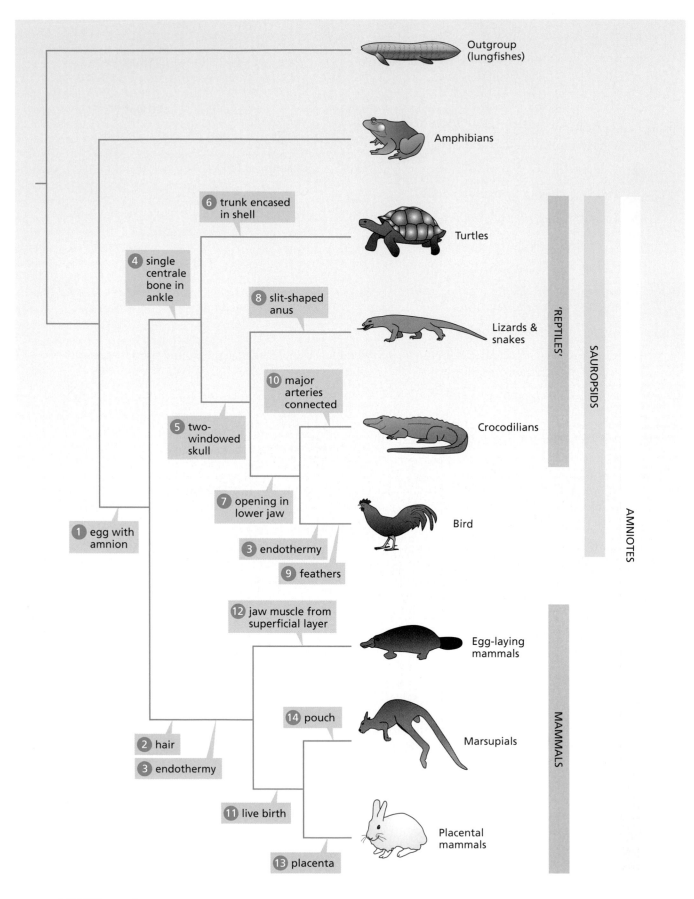

The newer approaches use the methods of cladistics to construct family trees with the smallest number of evolutionary changes needed to explain the patterns of variation, as explained earlier.

We have said that one goal of classification is describing groups with many physical characters in common, and that another goal is describing evolutionary history. Thankfully, these two goals usually result in very similar classifications. When they do not, more investigation is needed to determine the cause. If coherent groups, sharing many characters in common, do not correspond to clades, then an explanation is needed. Perhaps two unrelated groups evolved similar adaptations by convergence (see Chapter 5), developing many resemblances by analogy. Another possibility is parallel evolution, in which similar trends occurred independently several times among closely related groups. Perhaps some investigators were mistaken in describing or evaluating characters, for example, by misinterpreting one bone as a different bone, or by failing to recognize that a structure that appeared to be the same in two species had different embryological origins and was therefore a different structure in each case. One thing that is certain is that taxonomic disagreements often serve as a spur to further research when biologists attempt to discover the reasons behind the disagreement. After further research, classifications often change. For instance, the barnacles that grow on rocks or on ships were once thought to be related to clams or other filter-feeding animals with hard shells, but a study of their embryology revealed that they were built from the same parts as lobsters, shrimp, and other members of the class Crustacea, with which they are now classified.

CONNECTIONS
CHAPTERS 1, 5

THOUGHT QUESTIONS

1 The class Mammalia includes whales and bats along with the more numerous terrestrial species (rodents, monkeys, cats, bears, deer, elephants, humans, and many others). Why is this grouping a useful one? Think of as many reasons as you can.

2 Think of a large group of organisms that you know something about. How would a biologist decide which of several possible classifications is best for this group? What would she or he look for? What kind of evidence is relevant? (These are among the most basic questions of taxonomy.)

Modern Classifications Recognize a Great Difference Between Procaryotic and Eucaryotic Cells

In Chapter 1 we introduced the cell as the unit of organization for living things. A great gulf separates the major types of organisms on the basis of the structure of their cells. Bacteria and certain other organisms have a simple cellular structure with no internal compartments. In contrast, animals, plants, fungi, and protists have more complex cells with internal compartments and a true nucleus bounded by a membrane. Every modern classification of organisms gives prominence to this fundamental distinction of cell types.

Procaryotic cells

The first organisms to evolve were simple cells with no internal compartments and thus no nucleus. A membrane called the plasma membrane formed an outer boundary of the cell and kept its contents inside. Simple cells of this type are called **procaryotic** cells (Box 6.2). Procaryotic cells lack most of the complex internal structures possessed by more advanced (eucaryotic) cells. Procaryotic cells have a single chromosome, consisting of only nucleic acid without any protein. The DNA double helix of this chromosome is joined end to end in a circular form resembling a closed necklace. The region of the cell containing this procaryotic chromosome is not surrounded by a membrane or set apart from the rest of the cell in any other way. Many procaryotes also have fragments of DNA, called **plasmids**, which can detach from the chromosome, lead an independent existence for a long while, and then reincorporate into the chromosome (see Chapter 4, pp. 98–99 and Chapter 11, p. 400).

Bacteria are procaryotic cells, and a majority of procaryotes are bacteria. Two other groups of organisms that are procaryotic cells are the blue-green photosynthetic organisms (Cyanobacteria) and the very primitive Archaea.

Eucaryotic cells

Plants, animals, fungi, and protists are examples of organisms composed of **eucaryotic** cells (see Box 6.2*)*. Eucaryotic cells have various internal parts, called **organelles**, which are bounded by intracellular membranes separating the various functions of the cell into different compartments. A defining characteristic of eucaryotic cells is the presence of a nucleus surrounded by a double membrane called the nuclear envelope (the name *eucaryotic* means 'true nucleus'). The nucleus contains rod-shaped chromosomes composed of DNA and proteins. Many types of cell organelles are shown and explained in Box 6.2. Eucaryotic cells also have an internal network of protein filaments called a **cytoskeleton**. These filaments determine the shape of the cell and keep many organelles in their positions. Contraction of these filaments may help to move the whole cell. Keep in mind, however, that there are a vast number of variations on these cell types, both among species and among the various specialized cells within multicelled organisms; there is probably no actual cell that exactly matches the accompanying diagrams.

CONNECTIONS
CHAPTERS 4, 5, 11

Some eucaryotic organisms are single-celled (unicellular) and others are multicellular. All procaryotic organisms are unicellular.

Endosymbiosis and the evolution of eucaryotes

In Chapter 5 we outlined some of the evidence for the origin of life under hydrogen-rich (reducing) conditions. Chemical fossils left by the earliest organisms provide evidence for the presence of chlorophyll (and thus photosynthesis) during the time when the atmosphere changed to its present oxygen-rich composition. Simple structural fossils of the earliest organisms, between 4.0 and 4.2 billion (4.0–4.2 × 10^9) years old, show that they were single cells about the size of modern procaryotic cells, or roughly one-tenth the size of eucaryotic cells. Larger cells, comparable in size to modern eucaryotic cells, do not appear in the fossil record until about 1.5 billion years ago. But how did eucaryotic cells originate?

From procaryotes to eucaryotes. According to a theory first championed by the American cell biologist Lynn Margulis in the 1970s, eucaryotic cells arose from procaryotic cells by a process called **endosymbiosis** (literally meaning 'living together inside'). According to this theory, large procaryotic cells incapable of performing certain energy-producing chemical reactions (those of the Krebs cycle, described in Chapter 10, pp. 349–351) engulfed smaller procaryotic cells able to carry out these reactions. The larger (host) cells could obtain energy by digesting the smaller cells, but they could obtain even more energy if they allowed the smaller cells to go on living inside them and then used the products of the energy-producing reactions. In this situation, host cells that allowed the smaller cells to persist were favored by natural selection over host cells that digested the smaller cells. Over time, the smaller cells became energy-producing cellular organelles called **mitochondria** (Figure 6.2).

Eucaryotic cells that are capable of photosynthesis have additional organelles called **chloroplasts**. The pigments that carry out photosynthesis are contained within these organelles. Chloroplasts are believed to have evolved by a process similar to that described above for mitochondria. In this case, however, the smaller cells were cyanobacteria capable of photosynthesis. The larger cells achieved greater growth potential by harboring these smaller cells rather than digesting them. The large cells containing photosynthetic organelles did better and reproduced in greater numbers than similar cells without chloroplasts, and cells with these organelles ultimately persisted while many others without chloroplasts died out.

In support of the theory of endosymbiosis is the fact that both chloroplasts and mitochondria have their own types of membranes and their own DNA, separate and different from those of the eucaryotic host cells that contain them, but similar to the membranes and DNA of procaryotic organisms. The presence of chloroplasts or other plastids is used in this book (and many others) as the defining attribute that determines the boundaries of the plant kingdom, which we discuss later.

Figure 6.2

The origin of eucaryotic cells according to the widely accepted theory of endosymbiosis.

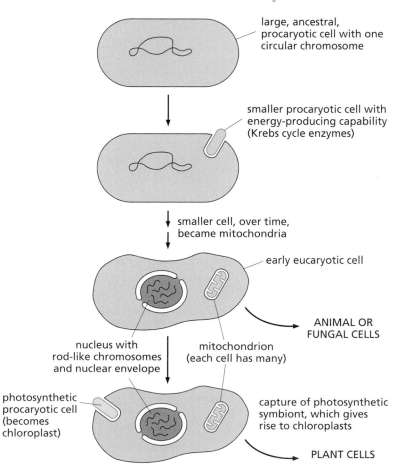

large, ancestral, procaryotic cell with one circular chromosome

smaller procaryotic cell with energy-producing capability (Krebs cycle enzymes)

smaller cell, over time, became mitochondria

early eucaryotic cell

ANIMAL OR FUNGAL CELLS

nucleus with rod-like chromosomes and nuclear envelope

mitochondrion (each cell has many)

photosynthetic procaryotic cell (becomes chloroplast)

capture of photosynthetic symbiont, which gives rise to chloroplasts

PLANT CELLS

CONNECTIONS CHAPTER 10

BOX 6.2 Procaryotic and Eucaryotic Cells Compared

The great differences between procaryotic (bacterial) cells and eucaryotic cells (including both animal and plant cells) are shown in the accompanying drawings and also in chart form.

PROCARYOTIC BACTERIAL CELL

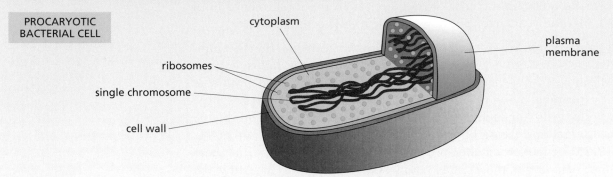

cytoplasm

ribosomes

single chromosome

cell wall

plasma membrane

EUCARYOTIC ANIMAL CELL

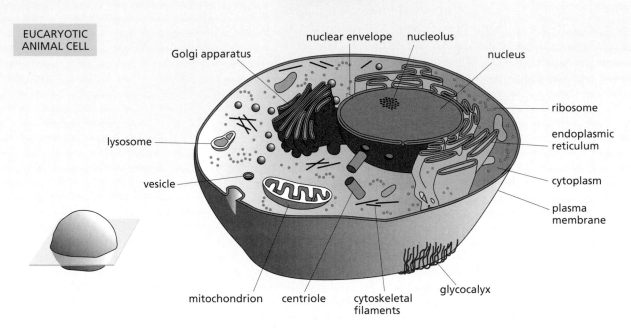

Golgi apparatus

nuclear envelope

nucleolus

nucleus

lysosome

vesicle

ribosome

endoplasmic reticulum

cytoplasm

plasma membrane

mitochondrion

centriole

cytoskeletal filaments

glycocalyx

EUCARYOTIC PLANT CELL

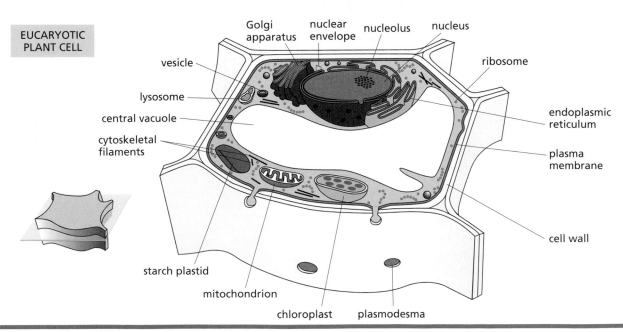

Golgi apparatus

nuclear envelope

nucleolus

nucleus

vesicle

lysosome

central vacuole

cytoskeletal filaments

ribosome

endoplasmic reticulum

plasma membrane

cell wall

starch plastid

mitochondrion

chloroplast

plasmodesma

structure	function	present in procaryotic cells	present in eucaryotic cells	
			plant cells	animal cells
plasma membrane	protection; communication; regulates passage of materials	✓	✓	✓
DNA	contains genetic information	✓	✓	✓
nuclear envelope	surrounds genetic material		✓	✓
several linear chromosomes	contain genes that govern cell structure and activity		✓	✓
cytoplasm	gel-like interior of cell	✓	✓	✓
cytoskeleton	aids in cell and organelle movement and in maintaining cell shape		✓	✓
endoplasmic reticulum	transport and processing of many proteins		✓	✓
Golgi apparatus	adds sugar group to proteins and packages them into vesicles		✓	✓
ribosomes	protein synthesis (translation) along mRNA	✓	✓	✓
lysosomes	contain enzymes; aid in cell digestion; have a role in programmed cell death		✓	✓
mitochondria	provide cellular energy		✓	✓
chloroplasts and other plastids	capture sunlight; produce energy for cell		✓	
central vacuole	maintains cell shape; stores materials and water		✓	
flagella (whiplike appendages)	cell movement	simple propeller type	complex undulating type	complex undulating type
cilia (hairlike appendages)	cell movement; present only in certain types of cells			✓
cell wall	protects cell; maintains cell shape	✓	✓	
glycocalyx	surrounds and protects cell			✓
pili (hairlike appendages)	mating; adherence	✓		
plasmodesmata	cell-to-cell communication		✓	

Eucaryotic diversity. As we have seen, evolutionary change occurs when natural selection acts on genetic diversity. Species evolve when new reproductive isolating mechanisms arise. Eucaryotic organisms such as animals and plants have speciated very often, so that most of the known species alive today are eucaryotic. In this section we examine the evolutionary advances that have taken place among eucaryotic organisms.

Crucial to the success of eucaryotic organisms are a number of important evolutionary advances. Of these, **sexual reproduction** and multicellularity evolved early, and perhaps repeatedly, among organisms whose bodies were still small and simple. The vast majority of eucaryotic organisms reproduce sexually, meaning that new individuals are formed only after a process of genetic recombination (see Chapter 2). In most cases, this means that new individuals derive their genes from two parents. Sexual recombination may initially have evolved because it was a simple yet efficient way of generating new gene combinations. If an organism produces a thousand offspring by **asexual reproduction**, all of them will, in the absence of a mutation, be genetically identical to the parent. Even if a mutation occurs, the offspring will still be very similar to the parents. Sexual recombination changes this. If an organism produces a thousand offspring through sexual recombination, they will differ genetically from the parent and also amongst themselves. Given the uncertainties of future conditions, the chances of a few offspring having better (i.e. more highly adapted) gene combinations than the parent are greatly increased if reproduction is sexual. Also, among the many offspring, more than one favorable combination of genes may arise, and these may eventually result in different kinds of organisms (i.e. different species). Thus, in eucaryotic organisms, sexual recombination supplies a mechanism to generate increased diversity, and it is therefore not surprising that there are so many different species of eucaryotic, sexually reproducing organisms in the world today.

Multicellularity probably evolved more than once independently. One route by which multicellularity evolved was by way of colonial organization. Single-celled organisms could in some cases maintain hospitable conditions better if they clumped together into small colonies. The aggregation into colonies would certainly reduce the surface area of each cell exposed to the outside environment because much of its surface would be in contact with its neighbors. The reduced surface area would aid in maintaining homeostasis. At first, all the cells in such a colony would be the same, and each would carry out all life processes, as before. As colonies grew larger, however, some interior cells began to lose contact with the outside environment altogether, and this began a simple division of labor between surface cells and interior cells. The interior cells were now in a more protected position, but they could no longer meet their own needs without having nutrients supplied to them or wastes taken away by other cells. The surface cells had to help supply nutrients to the interior cells and remove their wastes. In addition, certain functions (such as feeding or defense) could in most cases be carried out more efficiently by surface cells, while other functions (such as reproduction) could often be carried out better by the interior cells. Over time, the functions of cells in different parts of the organism became increasingly different from one another, and complexity thus increased in most cases.

Six Kingdoms of Organisms Are Included in Three Domains

The largest taxa of all have traditionally been called kingdoms. The animal and plant kingdoms have been recognized for centuries, and early classifications of organisms recognized only these two kingdoms. Plants were distinguished as being nonmotile organisms with rigid cell walls, capable of using sunlight as a source of energy. Animals, in contrast, were recognized for their ability to move, their lack of cell walls, and their inability to derive energy directly from sunlight. This two-kingdom classification continued in use despite the discovery of animals that do not move and other organisms, such as bacteria or fungi, that do not fit well into either the plant or the animal kingdom.

Advances in our knowledge made possible by electron microscopy, particularly by the discovery of the profound structural differences between procaryotic and eucaryotic cells, led to major classification changes. A five-kingdom classification system that became widely followed was first proposed in 1963.

Organisms did not change, but our arrangement of them changed because classifications are socially constructed, that is, devised by humans and agreed upon as a matter of social convention. To say that a classification scheme is socially constructed does not mean that the process of establishing a classification scheme is arbitrary or that all schemes are equally valid. Classifications are now usually understood as hypotheses about how organisms are related by patterns of descent. As more and more knowledge is accumulated about organisms, that knowledge is used to test the hypotheses and to replace rejected hypotheses with new ones. The division of living things into five kingdoms was simply a widely accepted theory, not an unchanging set of facts. Indeed, most biologists now add a sixth kingdom, the Archaebacteria or Archaea, more recently discovered, but evolutionarily ancient. The Archaea, the bacteria, and the cyanobacteria are the only procaryotic organisms. We use the six-kingdom classification in this book.

Even more recently, direct comparison of DNA sequences has shown that the Archaea and Eubacteria differ greatly from one another, even more than the Eubacteria differ from the various eucaryotes. Accordingly,

Figure 6.3

A family tree based on comparison of DNA (or RNA) sequences, showing the arrangement of organisms into three domains.

many biologists now sort the six kingdoms into three 'domains': one for the Archaea, one for the Eubacteria (including both the Bacteria and the Cyanobacteria), and a third for all organisms with eucaryotic cells—the Eucarya (Figure 6.3).

The kingdoms of organisms now recognized by biologists are listed in Box 6.3, along with a family tree showing how these kingdoms are thought to be related (see also Figure 6.3). We now proceed to examine

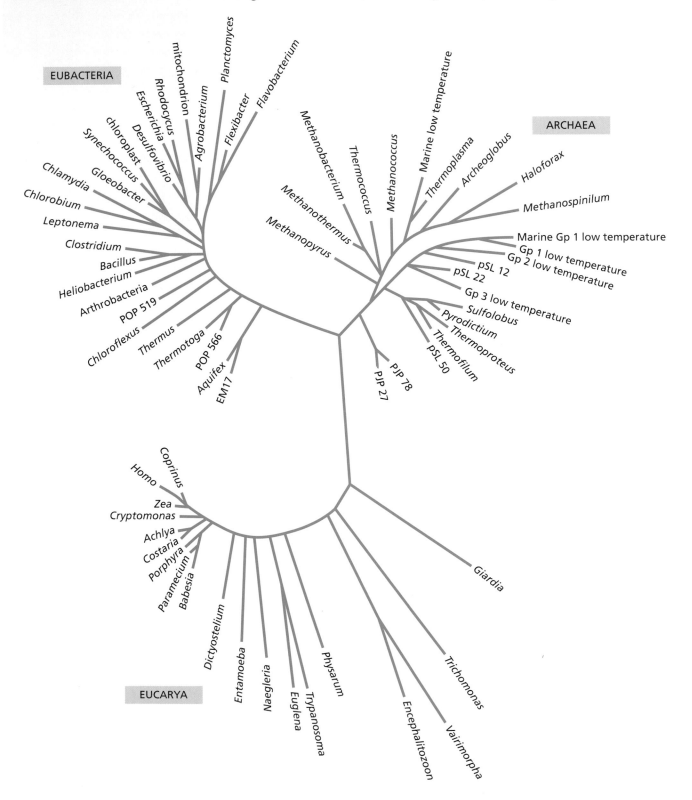

the six kingdoms, which are distinguished from one another by details of cell structure, development, nutrition, and overall morphology. A classification of organisms appears on our Web site (under Resources: Classification), and students are urged to consult this classification repeatedly as they read the remainder of the chapter.

Domain and kingdom Archaea

The domain Archaea contains a single kingdom, called either Archaea or Archaebacteria. The Archaea are one of the two procaryotic kingdoms and are characterized by the ability to live only in very special environments. Some members of the Archaea, the methane producers, live in oxygen-free environments, such as inside cows' guts, where, in the course of consuming energy, they perform chemical reactions that produce methane gas (CH_4) as a by-product. Others live in extreme environments once thought incompatible with life, such as areas of high salt concentrations, the edges of thermal vents in the ocean floor, or in hot springs. Scientists reasoned that adaptation for such extreme environments must require very different enzymes from those previously known, which proved to be so. One enzyme that can operate on DNA at very high temperatures is now used in the polymerase chain reaction, central to much work in biotechnology (see Chapter 3, p. 81).

Domain and kingdom Eubacteria

The domain Eubacteria, like the Archaea, contains but a single kingdom. This second procaryotic kingdom, now called Eubacteria, includes the more commonly known bacteria and the blue-green cyanobacteria. There is a great deal of diversity among bacteria. There are a handful of cell shapes: spherical, rodlike, gently curved (bananalike), and even spiral (like a corkscrew) (Figure 6.4). There are also differences in the structure of the cell wall, as revealed by commonly used staining techniques such as the Gram stain. An even greater diversity exists at the biochemical level. Some bacteria can tolerate oxygen in their environment and others cannot. Some bacteria can metabolize certain sugars and others cannot; the cyanobacteria can use cellular pigments to synthesize sugars

Figure 6.4

Bacterial diversity:
(A) in shape; (B) in structure of cell walls.

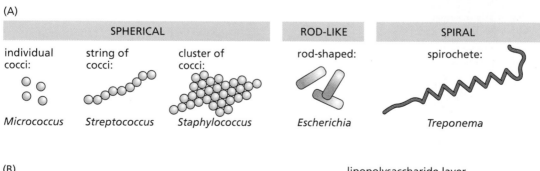

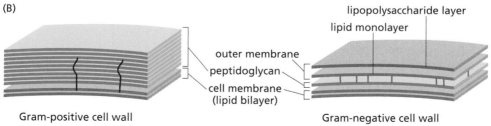

BOX 6.3 The Six Kingdoms of Organisms

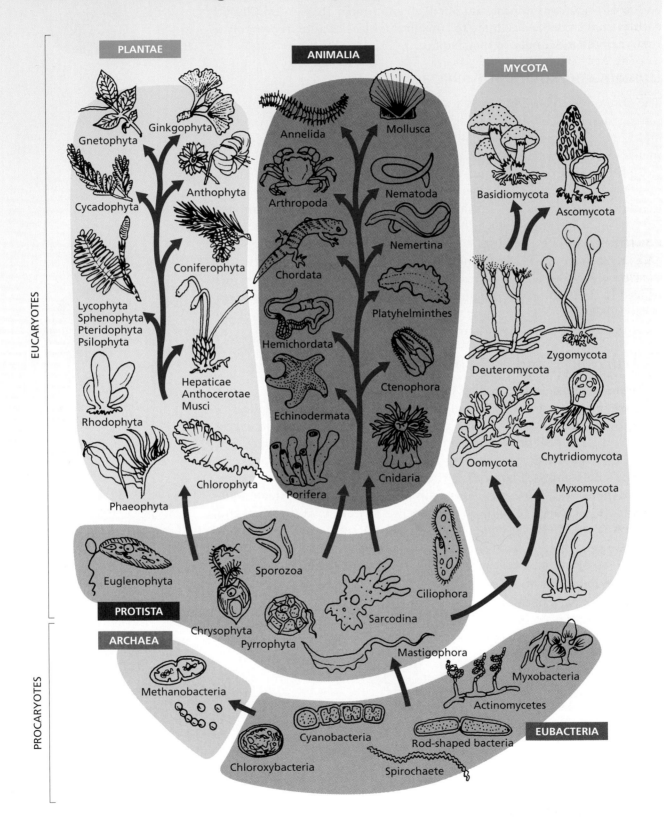

The placement of certain organisms differs among experts. In particular, the algae are sometimes included with plants, sometimes with protists, and sometimes divided between these two kingdoms. For further details, see the classification on this book's Web site, under Resources: Classification.

ARCHAEA

A small group of organisms, some of them adapted to extremely hot environments and many producing methane as a product of metabolism. Nucleic acid sequences of these organisms show them to be only distantly related to Eubacteria, with which they share the procaryotic type of cell structure shown in Box 6.2.

EUBACTERIA

The vast majority of procaryotic organisms, including the typical bacteria and Cyanobacteria (blue-green bacteria). No well-defined nucleus or nuclear envelope, nor any type of organelle (such as mitochondria or endoplasmic reticulum) that requires internal membranes.

PROTISTA

Eucaryotic unicells without plastids or cell walls. Various adaptations for locomotion may be present (cilia in one group, whiplike flagella in another group, protoplasmic extensions called 'pseudopods' in the largest group), but one group lacks motility and resembles the fungi in reproducing by spores. Some authorities list the algae here rather than among the plants. This and the remaining three kingdoms all have eucaryotic cells, as explained in Box 6.2.

PLANTAE

Eucaryotic organisms with plastids, including various algae, mosses, liverworts, ferns and fern allies, conifers and a vast array of flowering plants from buttercups to orchids and from grasses to trees. Most plants have nonmotile life stages and cells surrounded by cell walls, whose presence strengthens plant tissues. Nearly all possess chlorophyll a and are capable of carrying out photosynthesis using sunlight.

FUNGI OR MYCOTA

Nonphotosynthetic eucaryotic organisms with cell walls and absorptive nutrition, reproducing by means of spores. Includes slime molds, yeasts, mushrooms, and various other forms.

ANIMALIA

Eucaryotic organisms without plastids, usually possessing a life stage with at least some locomotor capabilities, and developing by means of an embryonic stage consisting of a hollow ball of cells (blastula). No chlorophyll or photosynthesis; no cell walls.

with the aid of sunlight (a process called photosynthesis), while most bacteria cannot. Some bacteria can move with the use of a simple flagellum, while others cannot. Many bacteria are free-living (in soil, for example), but many other species live only in a narrow range of host species. For example, most bacteria that use dogs as hosts cannot live in humans.

Molecular genetics is now providing new information, revealing that species diversity is immense within both the Eubacteria and the Archaea. Although only 4500 procaryotic species have thus far been characterized and described, it is now estimated that millions exist.

We often think of bacteria as the 'germs' that cause disease, as many of them do. However, a far greater number of bacterial species are beneficial to humans, to other organisms, or to entire ecosystems. For example, bacteria decompose dead material into chemical forms that other organisms can then use to sustain life. Without such decomposition by bacteria, all other forms of life would soon cease. Certain bacteria (and a few cyanobacteria) are also important in the reactions of the nitrogen cycle (see Chapter 11, pp. 374–375), which also sustains nearly all other species on Earth. Biotechnology uses bacterial plasmids and bacterial enzymes, and many industrial processes, including the making of cheese, yogurt, and sauerkraut, depend on chemical reactions performed by bacteria. We will discuss bacterial diversity further in Chapter 17.

CONNECTIONS CHAPTERS 11, 17

All members of the Eubacteria, whether bacteria or cyanobacteria, are single-celled organisms, but they often grow in colonies or filaments of many individual cells attached to a substrate or to one another. Some of these colonies have characteristics that differ from characteristics of single cells of the same species; they thus exhibit what may have been the first step in the evolution of multicellularity.

Domain Eucarya

All the remaining kingdoms have eucaryotic cells and are therefore placed in the domain Eucarya. The earliest eucaryotes, known as protists, remained small and in most cases unicellular. Multicellularity evolved independently several times and forms the basis for the division of the Eucarya into Protista and three additional kingdoms. An absorptive type of nutrition evolved in one group of eucaryotes, the fungi (kingdom Mycota). Although some fungi remained single-celled, most are now multicellular and carry out their absorptive nutrition with the aid of thin, absorptive filaments. Chloroplasts evolved in another group of eucaryotic organisms, and these organisms became plants (kingdom Plantae). Protective structures for the egg cells evolved among later plants, as did vascular tissue, seeds, and flowers. In yet another large group of eucaryotes, motility became increasingly developed, and multicellular animals (kingdom Animalia) evolved from this group. Many important innovations evolved later on among animals, including bilateral symmetry, body cavities, segmented body plans, and, in our own phylum, backbones.

Kingdom Protista

The earliest eucaryotes were simple, single-celled organisms. These species and their immediate descendants, those lacking the characteristics of the plant, animal, or fungal kingdoms, are placed in the kingdom

Protista (Figure 6.5). Among protists, mechanisms evolved to ensure that, when cells divided, all their chromosomes would be present in the offspring cells. Mitosis and meiosis (see Chapter 2) first evolved among the Protista, as did sexual recombination. By having haploid gametes that joined during fertilization to produce diploid fertilized eggs (**zygotes**), the eucaryotes became able to generate new genetic combinations, thus producing great genetic variation in every generation. Because variation is the raw material on which natural selection works, sexual recombination increased the rate of evolution among eucaryotes.

Eucaryotic organisms initially evolved in aquatic habitats; only much later did several eucaryotic groups independently colonize the land. Among the early eucaryotes, there must have been a great selective advantage in being able to move from place to place to find food or to escape from unfavorable conditions. Several different mechanisms for motility (movement) evolved, and the major kinds of protists are distinguished by these adaptations. The earliest protists had contractile protein filaments that allowed the cells to change shape and creep through their surroundings. One large group of protists, the phylum Sarcodina (containing *Amoeba* and its relatives), change their body shape to create movement. These protists move by sending out extensions called **pseudopods**; the flow of cytoplasm into the pseudopod determines the direction of movement. Another large group of protists, the phylum Mastigophora (or Flagellata), achieve motion through a whiplike structure (called a **flagellum**). A third group of protists move by means of numerous hairlike structures (called **cilia**). Representatives of these three groups of protists are shown in Figure 6.5. A fourth group of protists are nonmotile; this group, the Sporozoa, includes the parasites that cause malaria (see Chapter 7, pp. 226–227).

CONNECTIONS
CHAPTERS 2, 7

Kingdom Plantae

One of the great achievements of the early eucaryotic organisms was the acquisition of chloroplasts, which allowed these organisms to increase their energy production by photosynthesis. The simplest organisms possessing chloroplasts are called **algae**; all of them lack specialized tissues. Some experts place the single-celled algae among the kingdom Protista, and others place all the algae there, but many more experts include all algae in the plant kingdom, as we do in this book. The Plantae can then

Figure 6.5

Representatives from several phyla of the kingdom Protista.

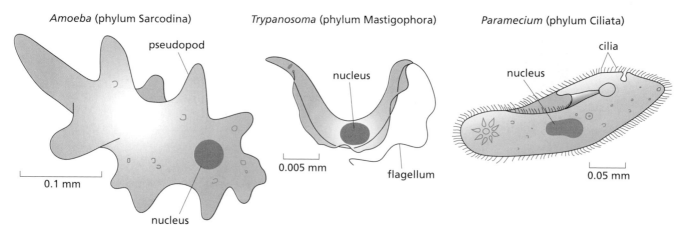

Amoeba (phylum Sarcodina)
pseudopod
0.1 mm
nucleus

Trypanosoma (phylum Mastigophora)
nucleus
0.005 mm
flagellum

Paramecium (phylum Ciliata)
cilia
nucleus
0.05 mm

be defined as eucaryotic organisms possessing chloroplasts, chlorophyll pigments, and a cell wall that commonly contains cellulose. (A few plant species have lost the pigments, but they still have all the other hallmarks of plants.)

Different groups of algae are distinguished by the types of photosynthetic pigments that they contain, by the chemicals that they use to store energy, and by various other adaptations. Representative algae of different groups are shown in Figure 6.6. One group of microscopic algae, called dinoflagellates, can occasionally exhibit sudden growth into huge oceanic blooms called 'red tide.' These algae produce a toxin that accumulates in shellfish that eat the algae, and this toxin can cause illness—even death—in people who eat contaminated shellfish.

The algae are ancient, aquatic plants, and they provide clues to the beginnings of multicellularity within the plant kingdom. Within several groups of algae, independently of one another, multicellular aggregations evolved. At first, these aggregations were just colonies of similar cells, similar to those formed by the green alga *Volvox* (see Figure 6.6). Then colonies began to function as multicellular single organisms. Cells located in different places within evolving organisms began to develop differently. Evolved differences between surface cells and those in the interior, or between cells near the top and the bottom of a former colony, allowed the organisms to take advantage of the differences in environment between the various locations.

Another group of ancient plants, the simple yet multicellular Bryophyta (mosses, liverworts, and hornworts), are important because they provide clues for the move from aquatic to terrestrial environments. The bryophytes have a layer of sterile, nonreproductive cells that surround and protect their egg cells; this adaptation permitted these plants to emerge from aquatic environments and colonize the land, although they still live in relatively moist environments. Most botanists believe that bryophytes evolved from green algae because important photosynthetic pigments and other characteristics are shared by both groups. Three species of bryophytes are shown in Figure 6.7.

Figure 6.6

Representative algae belonging to several phyla of the kingdom Plantae. Their different photosynthetic pigments are responsible for many different colors.

Dulse (*Palmaria*), a red alga (phylum Rhodophyta)

Ascophyllum, a brown alga (phylum Phaeophyta)

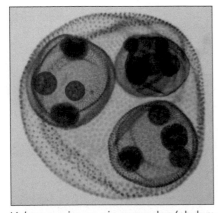

Volvox, a microscopic green alga (phylum Chlorophyta)

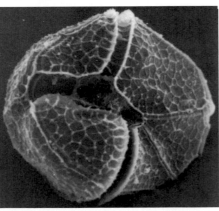

Peridinium, a one-celled dinoflagellate

Bryophytes do not have deep underground parts because all parts of the plant carry out photosynthesis and therefore need to be in the light. They cannot grow very tall because they lack vascular tissue that would conduct fluids and because they lack the roots and stems that would provide anchorage and support. They are therefore nonvascular plants.

Vascular plants with specialized tissues. Land plants were small at first and were restricted to moist habitats (such as the bryophytes discussed above), but some plants evolved vascular tissues, which allowed them to grow much taller, and these became the **vascular plants** (Tracheophyta). The most familiar and ecologically dominant plants are all vascular plants, and they include the largest and most conspicuous organisms in most terrestrial habitats. Vascular plants come in all shapes and sizes—a sample of this diversity is shown in Figure 6.8.

In algae and other simple plants, each cell carries out its own photosynthesis, absorbs its own nutrients, and gets rid of its own waste products. The increasing specialization of cells in vascular plants allows different parts of each plant to perform different functions efficiently. Groups of similar cells are organized into **tissues**, and groups of tissues are organized into **organs** such as leaves and roots. For example, each leaf is an organ, while each cell layer within a leaf is a tissue. The simplest plants containing separate types of tissues are the Bryophyta, but the diversity and complexity of tissue types increases dramatically among vascular plants. Photosynthesis is carried out principally in the leaves, and other plant parts are also specialized for particular functions. Roots are specialized for water absorption and fruits for reproduction and dispersal. The division of labor among different parts of the plant would not be possible without the specialization of plant tissues, particularly the vascular (conducting) tissues that efficiently transport materials from one part of the plant to another.

We will examine vascular plants further in Chapter 11.

Specializations of flowering plants. The most highly evolved vascular plants reproduce with the aid of seeds, which contain small diploid embryos capable of being dispersed to new locations away from the parent plant. These plants have branching roots and leaves with multiple veins. The largest and most diverse, as well as ecologically dominant, group of seed-producing plants is that of the flowering plants (Angiospermae or

CONNECTIONS   CHAPTER 11

Figure 6.7

Representative bryophytes (phylum Bryophyta). These are land plants with protective cells surrounding the egg, but their lack of vascular tissues keeps them small.

Conocephalum, a thalloid liverwort

Marchantia, another thalloid liverwort

Polytrichum, a moss

Anthophyta). The seeds of flowering plants develop within elaborate reproductive structures called flowers (see Figure 6.7). Eggs, each with a haploid set of chromosomes, are produced by the female part of the flower within the ovary. The male part of the flower produces haploid gametes (sperm) within pollen grains in the anthers. Pollination is the introduction of the pollen onto the stigma, the female receptive surface of the flower (Figure 6.9). A pollen tube grows from the pollen grain to the ovary, where the sperm fertilizes the egg, resulting in a diploid zygote. Many flowers are pollinated by wind, but a much larger number are pollinated by insects. The relations between flowering plants and the insects that pollinate them are often quite elaborate and are crucial to much of the diversity and also the evolutionary success of both flowering plants and insects (see Chapter 18, p. 645 and pp. 664–665).

After fertilization, the zygote undergoes cell division and becomes an embryo, and the structures surrounding the zygote mature into a seed. The seed-bearing structures in a flower ripen into fruits, defined as ripened ovaries that contain seeds. In addition to the seeds themselves, fruits often contain tissues that attract various animals by means of conspicuous colors, special odors, carbohydrate nutrients, or a combination of these. Animals that eat the fruits may disperse the seeds in their feces, often far from the parent plant. Seeds are also dispersed in other ways (Figure 6.10). Among the fruits that humans eat are many that we commonly recognize as fruits (e.g., apples, peaches, melons) and others that we do not always regard as fruits (e.g., nuts, grains, cucumbers, tomatoes, peppers, eggplant).

CONNECTIONS CHAPTER 18

Figure 6.8

An assortment of vascular plants (Tracheophyta), belonging to several groups.

Nephrolepis, a fern (phylum Pterophyta), showing reproductive structures on the underside of the leaf

Equisetum, a horsetail (phylum Arthrophyta)

Pinus, the white pine (phylum Coniferophyta)

Flowering plants or angiosperms (phylum Anthophyta)

Daisy (*Chrysanthemum*)

Trillium (*Trillium*)

Rose (*Rosa*)

Kingdom Mycota

Organisms of the kingdom Mycota are commonly known as fungi. These organisms typically live on dead or decaying organic matter that they absorb through threadlike extensions called **hyphae**. Fungi have mitochondria, but not chloroplasts, and so they do not carry out photosynthesis.

Typical fungi (subkingdom Eumycota) have cell walls and are non-motile, but a few primitive fungi (subkingdom Myxomycota) have motile stages in their life cycles. For example, the slime molds such as *Dictyostelium* (Figure 6.11) have multicellular reproductive stages that look like fungi and carry out absorptive nutrition, but at other times they live as motile, amoebalike individual cells. Thus, they are thought to resemble an early stage in the evolution of multicellularity. Yeasts, morels, and certain molds belong to the phylum Ascomycota, which includes both single-celled and hyphal growth forms. Most ascomycetes can reproduce either sexually or asexually. Mushrooms belong to the phylum Basidiomycota. Their feeding structure consists of a branched network of fine hyphae; the familiar mushrooms (composed of tightly packed hyphae) are their reproductive structures. All fungi reproduce with the aid of spores, which are minute haploid gametes that are not distinguishable as eggs or sperm. Fungi useful to humans include edible mushrooms, yeasts (used in brewing, baking, and biological research), and the mold *Penicillium*, the source of the antibiotic penicillin. Several types of fungi are shown in Figure 6.12.

Kingdom Animalia

Animals are multicellular organisms with eucaryotic cells and an embryonic life stage consisting of a hollow ball of cells called a **blastula**. Most animals have at least some motility during some stage of their life cycle.

As we have seen, there are many highly successful forms of life that are not animals. Even within the animal kingdom, most animals are very different from the group to which we belong. In fact, the majority of animals lack a stiffening backbone and are called **invertebrates**. The animal kingdom also includes a great diversity of life strategies, and each is biologically successful within the habitat that it occupies.

The animal kingdom is divided into about 30 phyla. Experts differ on the exact number of phyla because they are not in agreement about how to classify some animals. Some small phyla are not included in the following survey, but they are listed in the detailed classification found on this book's Web site, under Resources: Classification.

Minimal organization and the sponges. Multicellular organization in animals takes several forms. The simplest animals are sponges (phylum

Figure 6.9

Diagram of a complete flower, containing both male and female parts together. After fertilization and ripening, the ovary becomes a fruit. Variations in the number and structure of the parts shown here are among the most useful characters in plant classification. Some plants have incomplete flowers in which there are separate male flowers with undeveloped female parts and female flowers with undeveloped male parts.

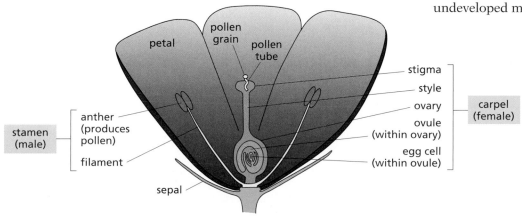

Figure 6.10

The dispersal of seeds in different kinds of fruits.

Porifera) (Figure 6.13). Various cell types are present in these aquatic animals, including wandering amoebalike cells, barrel-shaped cells with hollow interiors, and cells with a whiplike flagellum surrounded by a 'collar.' These cells, however, are not organized into different tissue layers,

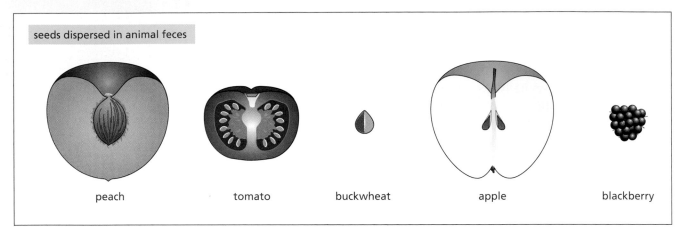

seeds dispersed in animal feces

peach   tomato   buckwheat   apple   blackberry

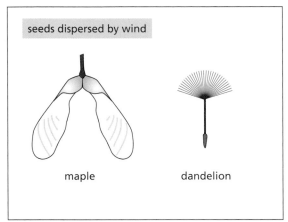

seeds dispersed by wind

maple   dandelion

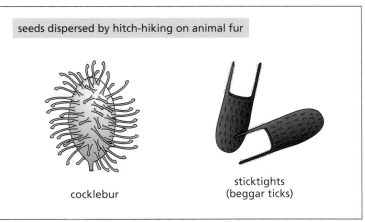

seeds dispersed by hitch-hiking on animal fur

cocklebur   sticktights (beggar ticks)

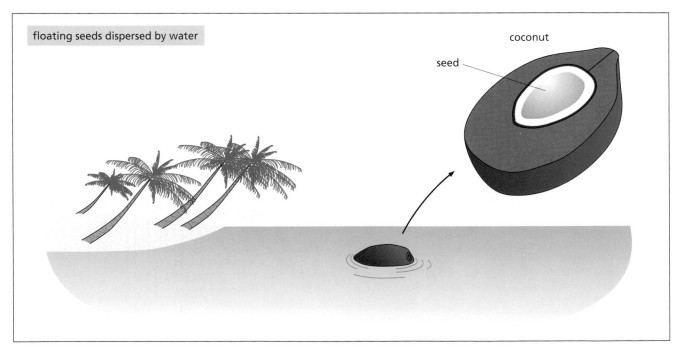

floating seeds dispersed by water

coconut

seed

Figure 6.11

Life cycle of the slime mold *Dictyostelium* (kingdom Mycota, subkingdom Myxomycota).

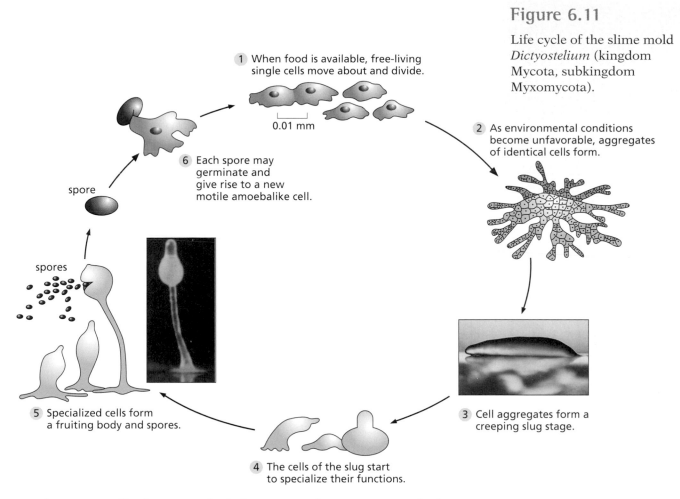

1 When food is available, free-living single cells move about and divide.

0.01 mm

2 As environmental conditions become unfavorable, aggregates of identical cells form.

6 Each spore may germinate and give rise to a new motile amoebalike cell.

spore

spores

5 Specialized cells form a fruiting body and spores.

3 Cell aggregates form a creeping slug stage.

4 The cells of the slug start to specialize their functions.

as they are in all other animal phyla. Sponges have a variety of adaptations to deter predators that would otherwise feed upon them: all sponges have sharp, needlelike structures (spicules), and many sponges also secrete poisonous chemicals. The sponges that lacked these defenses disappeared long ago.

Tissue layers and the phylum Cnidaria. The simplest animals having cells organized into tissues are found in the phylum Cnidaria (Figure 6.13). Like the sponges, these are aquatic animals. In the Cnidaria and in the embryos of all other animals except sponges, one portion of the hollow blastula puckers inward and turns inside-out. The resultant cup-shaped structure, called a **gastrula**, contains two distinct layers of cells: the layer that has moved to the inside is called **endoderm**, and the layer that remains on the outside is called **ectoderm**. These structures are shown in Figure 6.14.

The ectoderm and endoderm form two distinct tissue types, the beginnings of the differentiation of multicellular animals into a variety of such tissues. As in plants, tissues are groups of similar cells that form sheets or other integrated structures, each specialized to perform a different function. We discuss tissues further in Chapter 12. In addition to the ectoderm and endoderm, the gastrula contains an endoderm-lined central cavity, open to the outside. The fact that all animals (except

CONNECTIONS
CHAPTER 12

sponges) go through such a gastrula stage in their development is strong evidence that they all share a common ancestry.

The two tissue layers are arranged in two basic body plans among the Cnidaria. One plan (called a **polyp**) has the central cavity opening upward. Cnidaria with this body plan usually grow attached to the ocean bottom or to other animals; many live in large colonies that we recognize as corals. The other body plan (called a **medusa**) has the central cavity opening downward. Cnidaria with this body plan float freely in the water, and most can contract portions of their body to control their movement. Because of the large amount of jellylike material that lies between the outer and inner layer of cells, most of these Cnidaria are popularly known as 'jellyfish.' The major subgroups of Cnidaria are distinguished on the basis of whether their life cycle includes only one of these body plans or both of them. Both cnidarian body plans have a series of tentacles surrounding the opening of their central cavity. These tentacles contain specialized stinging cells, which are used to defend the animal against predators.

Bilateral symmetry and the flatworms. Most sponges, and a few other types of animals that live attached to the ocean bottom, have irregular body shapes that show no symmetry. Other sponges, and all members of the Cnidaria, have a radially symmetrical body plan, that is, a body plan arranged in a circle. (If you look down on them from above, you will see the same anatomical details repeated over and over around the edge of this circle.) Regardless of body plan, a cnidarian has the same chance of finding something nutritious, or something dangerous, in any direction, and natural selection has therefore favored radial body plans among these animals.

The vast majority of animals, belonging to over two dozen phyla in the animal kingdom, are characterized by **bilateral symmetry**, meaning that their bodies can be divided by a central plane such that structures on the left side of this plane are mirror images of corresponding structures on the right side. Bilateral symmetry is believed to have evolved in animals as an adaptation that came along with forward movement. Imagine an animal that creeps along the ocean bottom in such a way that one end of its body is in a forward position. Movement would be made easier by a streamlined or elongated body. New discoveries, whether of

Figure 6.12

Some types of fungi
(kingdom Mycota,
subkingdom Eumycota).

A morel, *Morchella* (phylum
Ascomycota)

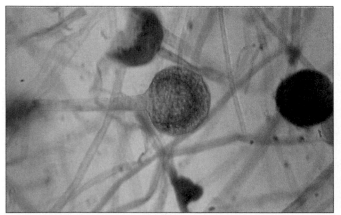

The black bread mold, *Rhizopus* (phylum Zygomycota)

A mushroom, the poison *Amanita* (phylum
Basidiomycota)

food or of danger, would be more likely to be made with the front end. Under these conditions, natural selection would favor the development of a front end with sense organs (eyes, feelers, taste organs, sound and motion detectors), feeding organs, and possibly also aggressive weapons.

An animal that creeps along the ocean bottom may also be expected to react differently to the water above than to the sediment below, and so natural selection would tend to favor organisms having structures on the top (dorsal) surface that differ from those on the underside (ventral). However, any structure or ability that is adaptive on the right side is equally adaptive on the left, and organisms would have no selective advantage if their right side differed from their left. As a result of selection under these conditions, body plans that are bilaterally symmetrical are common in the animal kingdom and present in many different phyla.

The simplest animals with bilateral symmetry are the flatworms of the phylum Platyhelminthes (Figure 6.15). They have somewhat elongated bodies, with sense organs concentrated at one end, which is recognizable as a head. The body is flattened, with broad upper and lower surfaces that in many species differ from one another in coloration and in other ways. The body plan is bilaterally symmetrical, with right and left halves of the body being mirror images of one another. Flatworms have a middle layer of tissue (called **mesoderm**) in addition to the ectoderm and endoderm (see Figure 6.14). All the animals yet to be described in this chapter have tissues derived from these three basic layers.

Assembly-line digestion and the roundworms. One way in which flatworms are similar to cnidarians is in their digestive system, which is just a sac with a single opening that serves as both entrance and exit. With this arrangement, much of what is discarded as waste is immediately taken in again as food, making the system very inefficient. A further inefficiency is that every region of the digestive tract, and every group of cells in the digestive lining, must be capable of performing the entire digestive process from beginning to end. With this arrangement, cells cannot specialize to carry out early or late steps of digestion.

Figure 6.13

Sponges (animals not organized into tissue layers) and cnidarians (animals with radial symmetry and with two tissue layers, an outer ectoderm and an inner endoderm).

Phylum Porifera

Sponge

Phylum Cnidaria

Hydra, a polyp

Jellyfish or medusa

Polyp of a sea anemone

Figure 6.14

Early stages in the embryology of animals. Adult sponges develop directly from a modified blastula stage. All other animals go through both blastula and gastrula stages. Members of the phylum Cnidaria form adult stages that still resemble gastrulas, whereas most other animals develop a middle layer (mesoderm) that gives rise to additional internal organs.

A more efficient arrangement, which first evolved in roundworms (phylum Nematoda) and several related phyla (Figure 6.15), is an assembly-line digestive tract with an entrance, the mouth, at the front end and an exit, the anus, at the hind end. With this arrangement, selection can

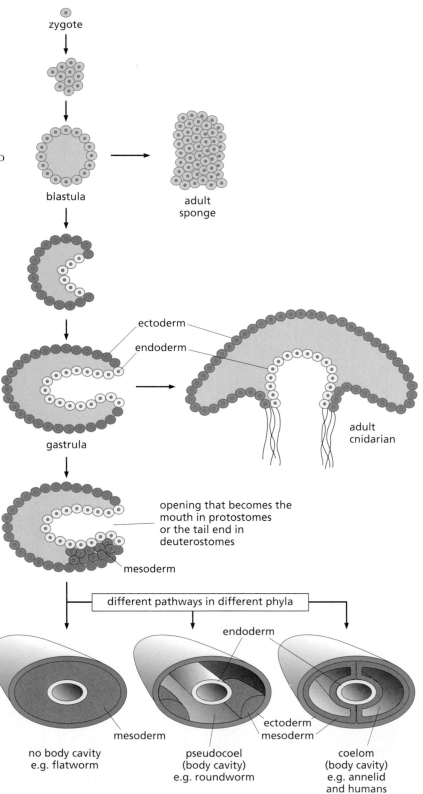

favor organisms in which the cells near the front end can perform the early stages of digestion more efficiently and those near the hind end are more efficient in completing the later stages. Certain parts of the digestive tract can now specialize in the processing of different types of nutrients, or of hard substances requiring mechanical break-up (see Chapter 10). Waste products are now discharged more efficiently because they are released from the hind end and left behind as the animal moves forward.

The evolution of body cavities. In the course of creeping forward along the ocean floor, some animals occasionally found reason to burrow into the bottom sediment. Burrowing is usually accomplished by a mechanical process of wedging the front of the body farther forward, then forcefully inflating part of the body to make it wider, then repeating the process. Forceful widening of the body thus alternates with forceful elongation, both in time and space. At any moment when the narrow portions of the body are pushing forward, the other parts of the body are widening to give the parts in front of them something against which to push.

This alternation of widening and elongating can be done much more efficiently if the body contains one or more fluid-filled cavities. Because fluids like water are not compressible, squeezing a fluid-filled bag (like a water balloon) in one place or in one direction causes it to bulge elsewhere. Thus, any fluid-filled cavity can be forcefully widened by contracting muscles running front to back, while the same cavity can be forcefully elongated by contracting muscles that encircle its girth.

In the course of evolution, various types of fluid-filled cavities of different constructions and different embryological derivations evolved in different phyla. These phyla are classified in part according to the nature of the body cavity and the cells lining its interior. Roundworms, horsehair worms, rotifers (see Figure 6.15), and a handful of other animal phyla are characterized by body cavities lined with cells derived from several embryonic layers, including endoderm. The remaining phyla described below all have body cavities entirely lined with mesoderm; such a body cavity is technically called a coelom (see Figure 6.14).

Figure 6.15

An assortment of invertebrate animals belonging to various phyla. Of the animals shown here, the flatworm (phylum Platyhelminthes) has no body cavity, while the roundworms (phylum Nematoda) and rotifers (phylum Rotifera) have a body cavity whose lining is made from several different embryonic tissue layers.

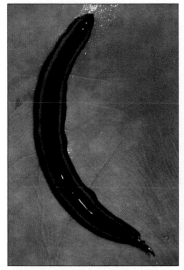

Giant flatworm (phylum Platyhelminthes)

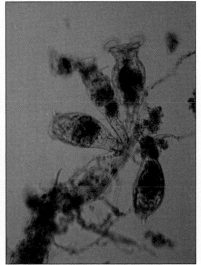

Rotifers (phylum Rotifera)

Roundworm (*Trichinella*) cyst in muscle tissue (phylum Nematoda)

Protostome phyla and the evolution of segmentation. The remaining animal phyla all have mesoderm-lined body cavities. They are separated into two large groups, the protostomes and deuterostomes, which evolved in different directions. The **protostomes** include the mollusks, annelids, arthropods, and several smaller groups (Figure 6.16). All of them share certain embryological similarities, such as the way in which their body cavity develops and the derivation of the mouth.

At some point in protostome evolution, the mesoderm and its body cavity became subdivided into a series of individual blocks or pouches (somites). These blocks of tissue were arranged from front to rear, setting the stage for the evolution of segmentation of the body. Some of these animals, such as the annelid worms, are thoroughly segmented in both larval and adult stages, but others, including many mollusks, have lost most of their segmentation as adults.

Animals of the phylum Annelida, of which the earthworm is a familiar example, are anatomically arranged as a series of repeated segments; all the body segments are similar to one another in size and in anatomical structure. Segmentation permits parts of the body to work as self-contained units, allowing rhythmic swimming or crawling motions. The worms crawl through soil or sediment by using rhythmic waves of muscle contraction squeezing against the fluid-filled body cavity of each segment. A few body segments elongate while the next few widen. Waves of elongation alternate with waves of widening of the body segments, and these alternating waves pass down the length of the body from the front end to the rear. The annelid worms have no legs, but they do have tiny bristles (called setae) that stick out of their sides and anchor the widened, nonmoving segments of the body to the surroundings, giving the elongated segments something to push against as they move forward.

The phylum Arthropoda is the largest and most diverse phylum of the entire animal kingdom. The Arthropoda include lobsters, crabs, shrimp (see Figure 6.16), barnacles, spiders, scorpions, centipedes, millipedes, and insects. Insects alone account for over two-thirds of the

Figure 6.16

Protostome animals with a true coelom.

Tree snail (phylum Mollusca)

Shrimp (phylum Arthropoda)

Copepod (phylum Arthropoda)

A water bear (phylum Tardigrada)

Tropical earthworm (phylum Annelida)

animal kingdom and over half of all living species on Earth. Arthropods evolved from annelid ancestors. The body segments of arthropods are fewer than in annelids, and these segments are more specialized, differing from one another in both size and anatomy. Most notably, the arthropods have a series of leglike structures that differ in most cases from segment to segment. Arthropods have a strong, protective outer coating (called an **exoskeleton**), making each segment rigid. The rigid segments are separated by flexible hinge regions. The legs are also arranged as a series of rigid segments separated by flexible, hinged joints, giving the phylum its name (from the Greek, *arthro* meaning 'hinged' or 'jointed,' plus *pod* meaning 'leg' or 'foot').

Animals of the phylum Mollusca are in most cases protected by a hard outer shell, secreted by a special layer called the mantle. Part of the mantle is retracted at the rear of the animal to form a mantle cavity that contains both an anus and gills that allow respiration in water. Anyone who has admired seashells has some idea of the tremendous variety of species of mollusks. In addition to the familiar snails, clams, and oysters, mollusks also include cephalopods such as the squid and octopus, in which the shell is hidden inside or has been lost entirely (see Figure 5.7, p. 135). The creeping movements of snails and the digging movements of clams are both very similar to the waves of contraction used by annelids.

Evolution of the deuterostome phyla. The remaining phyla of the animal kingdom are **deuterostomes**. All deuterostomes have body cavities completely lined with mesoderm, but they have evolved separately from the protostomes described above. One major difference is in the manner in which the body cavity usually develops; another difference lies in the embryologic formation of the mouth. In protostomes, the gastrula opens to the outside by an opening that becomes the future mouth. That same opening in deuterostomes ends up near the hind end of the animal, just above the anus, while the mouth develops as a secondary structure at the other end. (*Protostome* means 'first mouth,' while *deuterostome* means 'secondary mouth'.) Thus, in a very real sense, your head corresponds to an insect's hind end (they are homologous), and an insect's head corresponds to your rear.

The deuterostomes and protostomes evolved separately, but some convergent adaptations have appeared. For example, a form of segmentation of the muscles and certain other body systems evolved independently in both groups.

The deuterostomes include four phyla, all evolved from bilaterally symmetrical ancestors. All animals in these four phyla go through early (larval) stages of development that are bilaterally symmetrical, but a few become irregular (asymmetric) or radially symmetrical as adults. Two phyla, the echinoderms and chordates, are large groups.

The phylum Echinodermata includes sea stars (starfish), brittle stars (Figure 6.17), sea urchins, sand dollars, crinoids (sea lilies, Figure 6.17), and sea cucumbers. The living species of echinoderms show a fivefold (pentameral) symmetry as adults, but their larvae are bilaterally symmetrical. Their other characteristics include a bumpy or spiny skin protected by calcium carbonate deposits and a water-vascular system through which sea water circulates in a series of tubes.

Humans and other animals with backbones belong to the phylum Chordata (Figure 6.18). Also included in this phylum are several small and unfamiliar sea creatures such as the sea squirts (also called tunicates) and the small sea lancet or amphioxus. The common ancestors of echinoderms and chordates probably lived attached to the ocean bottom, either directly or by means of a stalk. Many extinct groups of echinoderms grew this way (crinoids still do; see Figure 6.17), but most living echinoderms and chordates have a free-living, unattached way of life. The transition from attached to free-living is best shown by the tunicates in the phylum Chordata (see Figure 6.18). These small animals generally spend their adult lives attached to a rocky bottom. Here they sit and pump water through a large basketlike structure (the pharynx) whose numerous slits strain the water through while suspended food particles collect on a sticky, ciliated surface coated with mucus. The attached, filter-feeding existence of adult tunicates was the ancestral way of life for deuterostomes. Larval tunicates, by contrast, are actively free-swimming animals that resemble tadpoles. They swim by means of a long tail that sweeps back and forth like the tail of a fish, and the muscle blocks and nerves of this tail are segmentally organized. Free-swimming members of the phylum Chordata, including all fishes, have more in common with the tunicate 'tadpole' than with the filter-feeding adult.

Although humans are obviously different from tunicates in many ways, as members of the Chordata they share major characteristics not found in any other group of organisms. Characteristics shared by all chordates include (1) a body axis containing a stiff, flexible rod (called a **notochord**), (2) a hollow nerve cord along the back, and (3) a series of openings called gill slits, located behind the mouth region. These three characteristics originate early in the embryo and are not retained in the adults of all species. For example, fish keep their gill slits throughout life and use them to breathe, but humans lose their gill slits long before birth, keeping only a few remnants here and there, such as the tube that connects the throat to the middle ear (see Figure 13.12, p. 480).

Animals with backbones (Vertebrata). Among the members of the phylum Chordata, the majority are backboned animals that make up the subphylum Vertebrata. The stiff, flexible rod found in other chordates is

CONNECTIONS
CHAPTER 13

Figure 6.17

Echinoderms
(deuterostomes of the
phylum Echinodermata).

Crinoid or sea lily

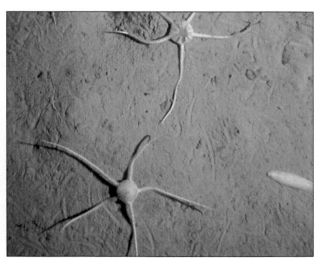

Brittle stars

functionally replaced in adult vertebrates by a backbone made of a series of individual bones or cartilages. Included in the vertebrates are four classes of fishes: (1) the jawless fishes; (2) the extinct, armored Placodermi; (3) the fishes with cartilage skeletons (including the sharks and rays); and (4) the fishes with skeletons of true bone (the group to which most fishes belong). There are more living species of bony fishes

Figure 6.18

Representatives of the phylum Chordata, another deuterostome phylum. All of these animals have embryonic notochords and gill slits.

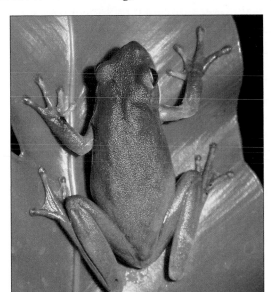

Tree frog (class Amphibia)

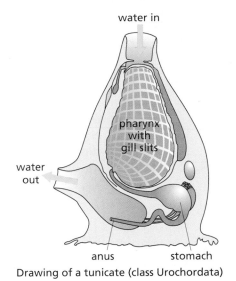

Drawing of a tunicate (class Urochordata)

Queen angle fish (class Osteichthyes)

Coral snake (class Reptilia)

Penguins (class Aves)

Kodiak brown bear (class Mammalia)

than of all the other vertebrate classes combined. All fishes are aquatic vertebrates that use gills to breathe and that swim by waving their hind end from side to side.

The four remaining vertebrate classes evolved, directly or indirectly, from the bony fishes, and thus all have bony skeletons. First of these are the amphibians (class Amphibia), which include the frogs, toads, and salamanders. These animals lay eggs in water, and the eggs develop into aquatic, gill-breathing larvae, commonly called tadpoles. After a while, the tadpoles undergo a rapid developmental change (metamorphosis) into adults that have legs and in most cases lungs. Fossil amphibians are known from the Devonian period to the present day.

Derived from the amphibians are the reptiles (class Reptilia), which include turtles, snakes, lizards, crocodiles, and many extinct species including dinosaurs. Unlike the amphibians, the reptiles have dry, scaly skin, and they lay their eggs on dry land (except for a few species that retain the egg inside the mother and give birth to live young). Fossil reptiles are known from the Pennsylvanian period to the present, but the Mesozoic era was populated by so many reptiles that it is often called the Age of Reptiles (see Figure 5.8, p. 137).

One group of reptiles, the Archosauria, included the dinosaurs and other dominant reptiles of the Mesozoic era. Derived from this group of reptiles are the birds (class Aves), distinguished by their possession of feathers. Most bird adaptations have to do with flying, including adaptations (such as hollow bones and the loss of one ovary) that lighten the body, and the high metabolism (and thus the high internal body temperature) that flying requires. Feathers do double duty as a flight surface and as insulation.

Another group of ancient reptiles had mammal-like features, and the class Mammalia, to which we belong, evolved from them. Mammals maintain a high and fairly constant body temperature, made possible by an insulating layer of hair or fur, supplemented in some cases by fat or blubber. A four-chambered mammalian heart prevents oxygen-rich blood from the lungs from mixing with oxygen-poor blood returning from other parts of the body. Also characteristic of mammals is the fact that they supply their young with milk, a secretion of the female's mammary glands. Mammals include kangaroos, shrews, monkeys, humans, bats, rats, squirrels, rabbits, whales, dogs, cats, bears, seals, elephants, horses, pigs, sheep, cattle, and many other species.

Humans are mammals because we share such mammalian characteristics as hair, a four-chambered heart, and the feeding of milk to our young. We are also chordates because we share in the embryonic gill slits and other characteristics that unite us with tunicates, amphibians, and other Chordata. We also share with all deuterostomes (chordates and echinoderms) the way our mouths develop embryologically. We share with many more phyla the anatomical structure and embryonic derivation of our body cavity, and with all animals the presence of motile cells and development from a blastula. We share the eucaryotic type of cell with four of the six kingdoms of organisms, putting us in the eucaryotic domain. The evolution of species over billions of years accounts for these patterns of shared characteristics.

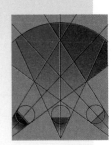
1 Why are algae sometimes considered protists? Why are they sometimes considered plants? How would you decide which is the better approach? If chloroplasts evolved only once, how would this affect your answer? What if chloroplasts evolved many times, independently?

2 Find four or more books on botany or general biology. List the phyla or divisions of the plant kingdom that each book recognizes. What similarities do you find? What differences do you find? How do you account for the differences?

3 Many bilaterally symmetrical animals have a long, thin 'wormlike' body shape. What advantages do you think such a body shape can confer? What problems do you think can arise from such a body shape?

Humans Are Products of Evolution

As we saw in the last section, humans are one species among many. At what point in evolutionary history did our ancestors evolve into something we could call 'human'? Answers to this question are reconstructed from fossils. The fossils help us to reconstruct our family tree, although there are frequent disagreements among scientists as to where a particular new fossil fits in.

Our primate heritage

Along with monkeys, apes, and lemurs, humans belong to the mammalian order Primates (Figure 6.19). We share many characteristics with other primates, but we did not evolve from any present-day species. Most adaptations shared by primates are related to the requirements of life in trees. Most primates live in trees today, and those that do not had ancestors that did. Nonprimate mammals whose ancestors never lived in trees do not share these adaptations. Primate characteristics directly related to the requirements of locomotion in trees are: (1) the independent and individual mobility of the fingers, (2) the ability of the thumb to oppose the action of the other fingers, and (3) the presence of friction ridges on the palm of the hand and the sole of the foot. Primates have also retained some primitive characteristics that many other mammals have lost in the course of their evolution, including the five-fingered hand, the collarbone (clavicle), and the ability to rotate the two bones of the forearm. Unlike those mammals that rely heavily on the sense of smell, primates rely heavily on vision. Primates have vision that merges images from both eyes to give three-dimensional information (binocular vision). This binocular vision is possible because the eyes came forward to the front of the skull during early primate evolution, so that the visual fields from the two eyes overlap. The portions of the brain related to vision are expanded in primates, especially when compared to those mammals whose eyes are to the sides of their heads and whose right and left visual fields are largely separate. Also, the outer surface of the brain (the cerebral cortex;

CONNECTIONS
CHAPTERS 8, 13

see Chapter 13) is more complex. The increased complexity of the primate brain is associated with an increased complexity of learned behavior (see Chapter 8). The reliance on learned behavior would be impossible without a lengthy period of very intensive parental care. Primates typically give birth to one offspring at a time. Primate nipples are restricted to a single pair in the chest region (other mammals have many pairs). Other mammals have two uteri, but in primates these are fused into a single uterus. Primates include lemurs, lorises, galagos, tarsiers, monkeys, apes, and humans. Among these, humans are most closely related to apes, but differ from all nonhuman primates in habitually walking upright.

Early hominids

In 1925, a fossilized child's skull was discovered in a cave near Taung, South Africa, and was named *Australopithecus africanus*. Although the skull had both apelike and human features, most experts treated it as just another ape. Additional fossils of *A. africanus* were discovered subsequently (Figure 6.20). These fossils included skulls with a very low opening for the emergence of the spinal cord from the base of the brain, showing that the skull balanced on top of an erect vertebral column. Parts of the foot, the pelvis, and the lower part of the vertebral column were also present, and these structures confirm that *Australopithecus* walked upright and was therefore more like humans than like apes. Direct evidence for upright walking comes from the discovery of a set of footprints at Laetoli, Tanzania, approximately 4.5 million years old. Primates that walk upright are placed in the family Hominidae and referred to as hominids.

Figure 6.19

Representative primates (phylum Chordata, class Mammalia, order Primates).

Ring-tailed lemur (*Lemur catta*)

Slow loris (*Nycticebus coucang*)

Squirrel monkey (*Saimiri sciureus*)

Mandrill (*Mandrillus sphinx*)

Chimpanzee (*Pan troglodytes*)

Scientists have since unearthed the remains of several other species of *Australopithecus* and related early hominids. The relationships among these hominids are shown in Figure 6.21. The oldest species is the recently discovered *Sahelanthropus tchadensis*, approximately 7 million years old. This earliest hominid had a very small brain and certain ape-like dental features. Another early hominid, *Australopithecus anamensis*, lived about 4 million years ago in Kenya. *A. anamensis* is thought to be the ancestor of all later species of *Australopithecus* and a close relative of the small hominid *Ardipithecus ramidus*. Another species, *Australopithecus afarensis*, about 3.5 million years old, is represented by the well-known skeleton known as Lucy, a female about 1.3 meters (slightly over 4 feet) in height. Enough of Lucy's skull is preserved to permit us to estimate the size of her brain in proportion to her body size, and these proportions are consistent with the hypotheses that *A. afarensis* was the common ancestor of the genus *Homo* and of several later species of *Australopithecus*. Two of these later species, *A. robustus* and *A. boisei*, were considerably larger than the better-known *Australopithecus africanus*, which lived from approximately 3.0 to 2.0 million years ago.

Two other early hominids are *Orrorin tugensis*, an early species estimated to be about 6 million years old, and *Kenyanthropus platyops*, a flat-faced species about 3 million years old. In both cases, the dates are uncertain and in dispute. Few fossils of either species are known, and these fossils are fragmentary. For these reasons, the exact relationship of either of these two species to the better-known hominids is unclear.

Of the species we have described, *Sahelanthropus tchadensis*, *Australopithecus anamensis*, and *Australopithecus afarensis* were probably along the line leading to *Homo*; the various other species were probably side branches of the family tree that died out without leaving any surviving descendants. The earliest *Australopithecus* came well before the earliest known *Homo* (about 4 million years ago, or half a million years after the Laetolil footprints), but later species of *Australopithecus* persisted side by side with *Homo*, at least in East Africa.

The genus *Homo*

Modern humans (*Homo sapiens*) and at least two extinct species are placed in the genus *Homo* (Figure 6.22). The oldest species of *Homo* was *Homo habilis*, which lived in East Africa from about 3.5 to 1.7 million years ago, coexisting with *Australopithecus boisei* and perhaps with other *Australopithecus* species. *H. habilis* had a brain that was small in absolute terms (about 400 cm^3, compared with 1200–1500 cm^3 for most modern humans), but the proportions of the brain to body size were more comparable to those of *Homo* than to those of *Australopithecus*.

Figure 6.20

Fossils of the genus *Australopithecus*.

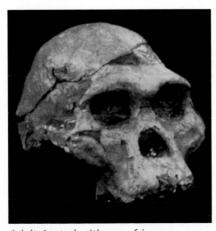

Adult *Australopithecus africanus* (Sterkfontein, South Africa)

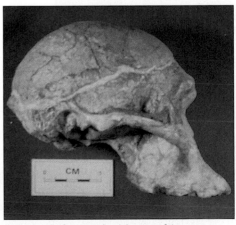

Side view of *Australopithecus africanus* (Sterkfontein, South Africa)

H. habilis has been found contemporaneously with certain types of tools, including simple stone tools. It is generally presumed that *H. habilis* was the maker of these tools.

A later species, *Homo erectus* (see Figure 6.22), is now known from fossils from about 1.5 million to 300,000 years old in China, Java, Europe, and several parts of Africa. A cave site at Choukoudian, China (near Beijing), has heat-fractured stones indicative of the use of fire. There is also evidence of round or oval tents supported by poles and held down along the margins by a circle of stones.

H. erectus was the ancestor of *Homo sapiens*, the species to which living humans belong. As *H. sapiens* evolved, tools became more sophisticated, and were in many cases mounted on wooden shafts. The *H. sapiens* that lived in Europe from about 150,000 to 50,000 years ago are called Neanderthals. Neanderthals hunted deer, horses, and even rhinoceroses and mammoths. Healed surgical wounds show that these skilled hunters took care of sick companions, set broken bones, and even performed simple brain surgery. They buried their dead and decorated the graves with flowers of preferred colors, mostly white or cream-colored. The decoration of graves is thought by several anthropologists to indicate a belief in an afterlife.

The more modern *H. sapiens* that replaced the Neanderthals were Upper Paleolithic people (including the Cro-Magnons) who lived from about 50,000 to 15,000 years ago. They had an even greater variety of tools, including fishhooks and harpoons. They hunted wooly mammoths and large herd animals. They also left records of their activities in the form of cave paintings, showing their interest in hunting and their understanding of both animal anatomy and physiology. By prominently drawing the heart and singling it out as a target, these hunters showed that they understood how vital this organ was. Their drawings of pregnant deer and of mating rituals show that they knew

Figure 6.21

A family tree of the family Hominidae. Dark orange areas show the known time range of species represented by fossils. Some dates are approximate.

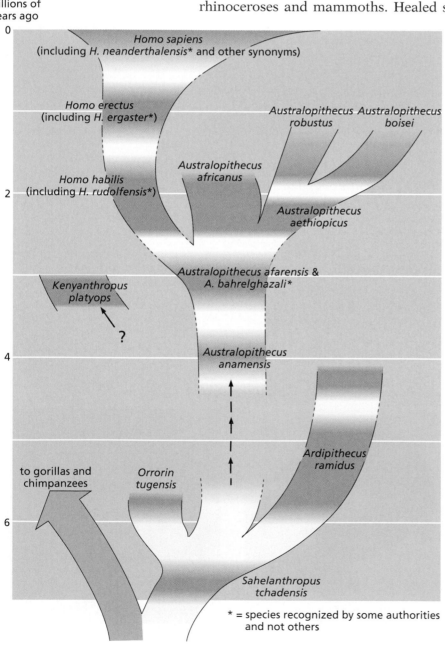

enough reproductive biology to understand the relationships between mating, birth, and subsequent herd sizes.

The discovery of agriculture ushered in a new phase of human history called the Neolithic. With the planting and harvesting of crops, humans began to settle down into villages, which later grew into towns and cities. Civilization has greatly changed the ways in which we live our lives. The rapid pace and power of cultural change leaves many people wondering whether biological evolution of *H. sapiens* has become a thing of the past. If we need to travel faster, the argument goes, our species tames horses or builds automobiles instead of evolving longer legs for faster running. Evolution by natural selection is much slower than cultural innovation. In this view, the future development of our species resides more in our technology than in our bodies.

No one questions that cultural changes in human beings have far outstripped biological ones as the most rapid and far-reaching changes taking place today. Cultural innovation spreads rapidly, in part because there are no species barriers to prevent transmission from one human group to another. (Language barriers and geographic barriers can always be crossed, especially in the age of television and jet travel.) The ease of travel and the global spread of people and their culture has brought us to an era in which there is no significant geographic isolation of human populations. Without geographic barriers, no reproductive isolating mechanisms will evolve, and all humans will remain one species. Cultural change is also more rapid than biological evolution because new inventions and other culturally acquired characteristics *are* inherited, although not genetically. Each generation inherits the stored knowledge of past generations (in libraries and museums, for example), along with tools (from tractors to telephones to satellites) and the technology needed to design and build new and better tools in the future.

Figure 6.22

Fossils of the genus *Homo*.

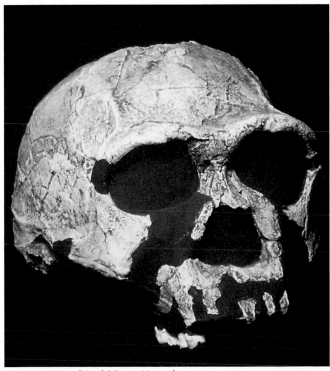

Homo erectus (Koobi Fora, Kenya)

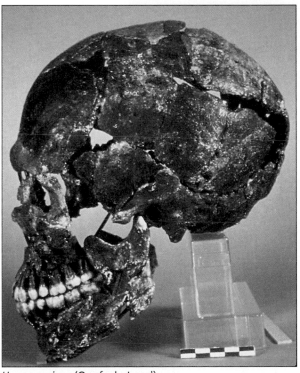

Homo sapiens (Quafzeh, Israel)

Although natural selection continues to take place, the environment, and therefore the traits favored by selection, have changed because of our own culture. Many of the selection forces that shaped human evolution in the past, including famines, epidemics, and predators, have been greatly diminished in modern times. Many traits that were once disadvantageous have become much less so. For example, poor eyesight is no longer an important barrier to survival and reproduction in societies that supply eyeglasses.

Human evolution has not stopped, however, because there continue to be situations in which the chance of survival or reproduction differs among people as a consequence of their genotypes. In our own species, genetic conditions such as cystic fibrosis, Tay–Sachs disease, muscular dystrophy, and others continue to cause numerous deaths before reproductive age, despite the best that medical technology has to offer. Other diseases that are generally survivable may reduce reproductive capacity, which decreases fitness. For example, chondrodystrophy is a rare disease, controlled by a dominant gene, in which the cartilage tissue turns bony at an early age, resulting in a form of dwarfism. Most chondrodystrophic dwarfs enjoy fairly normal health as adults, but have only about one-fifth as many children as their nondwarf siblings. Lowered reproductive rates are also found in diabetics. As these examples show, natural selection continues to affect the human species. Biological evolution thus continues to operate and to interact with cultural evolution in all human populations.

THOUGHT QUESTIONS

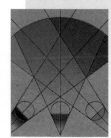

1 The Neanderthals were similar to us in many respects, though their skulls had a somewhat more 'rugged' appearance, with brow ridges and cheek bones protruding. How should we decide whether Neanderthals should be placed in their own species, separate from *Homo sapiens*?

2 In Europe, Upper Paleolithic culture replaced the culture of the earlier Neanderthal populations rather suddenly. Do you think that the replacement of one set of tools and traditions by another took place mostly by conquest, by intermarriage, or by some combination of the two? What evidence would you look for to test one hypothesis against the others?

3 Is the study of evolution static or changing? Find some recent news articles dealing with new fossil discoveries or other new findings that deal with evolution.

Concluding Remarks

The classification of organisms into species and higher taxa is an important way to summarize many of the results of evolution. Classifications cannot be static because our understanding of evolution keeps improving.

Biological evolution continues to act today in humans as it does in other species, although it now interacts with cultural evolution and technological revolution. Natural selection continues to operate by both differential mortality and differential reproduction, and continued selection will result in biological changes within all species, including humans. One frequent result of evolution within species is geographic variation. In the next chapter we examine some of the reasons for geographic variation within the human species.

Chapter Summary

- Scientists make **classifications** that group species together into **taxa** on the basis of their similarities and their evolutionary history.

- The theory behind classifications is called **taxonomy**. One important school of taxonomy, called cladistics, bases classifications on the sequence of branching points in family trees and on making taxa correspond to clades.

- Early cells were **procaryotic** and contained no internal membrane-bounded compartments or internal structural fibers. Procaryotic cells have no nucleus and only a single chromosome, containing DNA but no proteins.

- **Eucaryotic cells** contain a variety of internal organelles (including a membrane-bounded nucleus), internal structural fibers, and multiple chromosomes containing proteins as well as DNA.

- Highlights of early eucaryotic evolution include the origin of **organelles**, the acquisition of chloroplasts and mitochondria by **endosymbiosis**, the evolution of multicellularity, and the origin of sexual recombination of gametes.

- **Sexual reproduction**, which involves genetic recombination, has greatly increased the possibilities for generating new diversity among eucaryotic organisms.

- Most procaryotic and some eucaryotic organisms practice **asexual reproduction** in which no genetic recombination takes place.

- Evolutionary highlights within the plant kingdom include the origin of protective layers around egg cells and the origin of **tissues** including vascular tissues. **Vascular plants** have tissues that conduct water and allow plants to have different **organs** devoted to different functions in different parts of the plant. Vascular tissues also allow plants to grow tall. Seeds and fruits have permitted flowering plants to adapt to many habitats not previously available to simpler plants.

- Highlights of animal evolution include the origin of **bilateral symmetry**, the evolution of body cavities, the development of segmented body plans, and the evolution of the backbone.

- Early human fossils are placed in the genus *Australopithecus* and several related genera. The genus *Homo* includes various fossils and all living people.

CONNECTIONS TO OTHER CHAPTERS

Chapter 2	Differences in DNA structure help us to understand the relationships shown in our classifications.
Chapter 4	Different species carry different genomes, and these differences can be used in classification.
Chapter 5	Classifications reflect evolutionary history because differences among taxa are the results of branching evolution.
Chapter 7	Geographic variation within the human species was formerly understood as a classification based on 'races.' It is now understood to be based on the same evolutionary processes that operate within all species.
Chapter 8	Social behavior and reproductive strategies often vary among taxa and can be used in classification.
Chapter 10	Differences in metabolic pathways can be used in classification, especially among bacteria.
Chapter 11	Plant characteristics include the presence of chloroplasts responsible for photosynthesis. Plant classifications highlight the differences among plants, such as different photosynthetic pigments and the development of vascular tissues.
Chapter 12	The same developmental pathways that lead to tissue differentiation can also lead to cancer.
Chapter 13	Differences in brain anatomy among different species provide good evidence of evolution that can be used in classifications.
Chapter 14	The presence or absence of drug molecules in plants can be useful characteristics in their classification.
Chapter 16	Species close to us phylogenetically are susceptible to viruses similar to HIV.
Chapter 17	Newly emerging infections can reflect the evolutionary origin of new strains of pathogens.
Chapter 18	Biodiversity can best be understood by developing a comprehensive classification of all living species.

PRACTICE QUESTIONS

1. What are the major aims of classifications?

2. Name the six kingdoms of organisms currently recognized.

3. Name five organelles that eucaryotic cells possess but procaryotic cells do not.

4. Name three or four evolutionary advances among plants that you consider most important in their classification. Why was each important?

5. Why are arthropods, mollusks, and segmented worms grouped together? What do they have in common? What important differences characterize each of these phyla?

6. What selection pressures fostered the development of each of the following characteristics?

   a. radial symmetry

   b. bilateral symmetry

   c. complete 'assembly line' digestive tract

   d. body cavities

   e. segmentation

7. Why are vertebrates grouped together with echinoderms? What evidence exists for this arrangement?

8. What is the major anatomical difference between humans and apes? What is the major anatomical difference between the genus *Australopithecus* and the later members of the genus *Homo*?

9. Examine Box 6.1, and identify three taxa that correspond to clades.

Issues

- How can we describe and compare variation within and between populations?
- How is the study of population genetics related to human variation?
- Why do human populations differ biologically?
- Do human races exist?
- Is there a biological basis for the idea of race? Is biology the most accurate descriptor of race?
- Will changing biological concepts of race diminish racism? Why or why not?

Biological Concepts

- Populations and population ecology
- Population genetics (genetic variation, Hardy–Weinberg equilibrium, blood groups, genetic drift)
- Patterns of evolution (adaptation, physiology)
- Forces of evolutionary change (natural selection, environmental factors, communicable diseases, parasitism, human health)
- Gene action (molecular structure, genetic polymorphism)
- Scaling (body size and shape)

Chapter Outline

7 Human Variation

The human species is highly variable in every biological trait. Humans vary in their physiology, body proportions, skin color, and body chemicals. Many of these features influence susceptibility to disease and other forces of natural selection. Continued selection over time has produced adaptations of local populations to the environments in which they live. Much of human biological variation is geographic; that is, there are differences between population groups from different geographical areas. For example, northern European peoples differ in certain ways from those from eastern Africa, and those from Japan differ in some ways from those from the mountains of Peru. Between these populations, however, lie many other populations that fill in all degrees of variation between the populations we have named, and there is also a lot of variation within each of these groups.

Central to the study of human variation is the concept of a biological population, as defined in Chapter 5 (p. 151), and as explained again later. Both physical features and genotypes vary from one person to another within populations, but there is also a good deal of variation between human populations from different geographic areas as the result of evolutionary processes. How do populations come to differ from one another? How do alleles spread through populations? How do environmental factors such as infectious diseases influence the spread? Why are certain features more common in Arctic populations and other features more common in tropical populations? Why do we think of some of these variations as 'races'? These are some of the questions that are explored in this chapter.

There Is Biological Variation Both Within and Between Human Populations

All genetic traits in humans and other species vary considerably from one individual to another. Some of this variation consists of different alleles at each gene locus; other variation results from the interaction of genotypes with the environment. The simplest type of variation governs traits such as those discussed in Chapter 3 (pp. 75–77), in which an enzyme may either be functional or nonfunctional. The inheritance of these traits follows the patterns described in Chapter 2, which you may want to review at this time. In particular, be sure that you understand the meaning of dominant and recessive alleles and of homozygous and heterozygous genotypes. Many other traits, as we saw in Chapter 3, have a more complex genetic basis. In this section we examine how biological variation is described.

Continuous and discontinuous variation within populations

Many human traits vary over a range of values, with all intermediate values being possible; such variation is called **continuous variation**.

Continuously variable traits, such as height, can often be measured in an individual and expressed as a numerical value. Other traits that vary continuously, such as hair curliness or skin color, are seldom expressed numerically, although theoretically they could be.

Continuous variation can result from the cumulative effects of multiple genes, each of which by itself contributes a small effect. Dozens of known genes, perhaps even hundreds, influence height in one direction or another. If we make the simplifying assumption that these effects are independent of one another and that they add up, we can predict that a population of individuals will show a variation in height similar to the bell-shaped curve (**normal distribution**) of Figure 7.1. When we measure heights in any large population, we do in fact get a curve that closely matches this predicted curve. Many other continuous traits vary in much the same way as height. For most of these traits, a strong environmental component also exists. Height, for example, is strongly influenced by childhood nutrition as well as by genes. Environmental components of traits also contribute to the formation of a bell-shaped curve.

A numerical description of continuous variation in a population requires the use of statistical concepts such as average (**mean**) values. The average values are characteristic of the population as a whole, not of any individual member within the group. For a particular group of people, we can calculate an average height, weight, or head breadth, but these averages are just statistical abstractions—there are perfectly normal individuals that differ from the average, perhaps even greatly, as can be seen in Figure 7.1. Thus, the group average for a continuously variable trait tells us little about any individual. Also, whereas height can actually be measured (and average height computed), concepts such as 'tall' are relative: a height that is average in England may be considered tall in India or the Philippines.

Your individual traits result from both the genes that you inherited from your parents and the environmental factors to which you are exposed. What you inherit from your parents is a predisposition for a range of possible future variations in phenotype. For example, when a child is born, its exact height as an adult cannot be predicted, but if the mother and father are both significantly taller than average, it will, if it receives adequate nutrition, probably also be taller than average.

Discontinuous variation within a population is represented by traits that are either present or absent, with no intermediate values possible. Most of these traits have a simple genetic basis, so that someone's genotype may sometimes be deduced from their phenotypes and the phenotypes of their close relatives. Traits that vary discontinuously include blood groups and the presence or absence of conditions such as albinism or Tay–Sachs disease (see Chapter 3). A particular

Figure 7.1

Continuous variation in a single population: all intermediate values are possible.

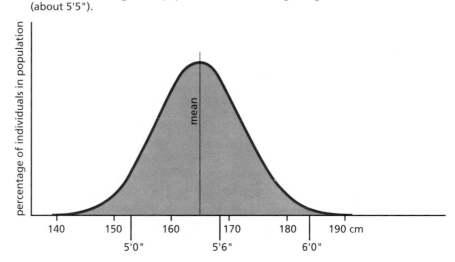

Distribution of height in a population whose average height is 165 cm (about 5'5").

percentage of individuals in population

mean

140   150   160   170   180   190 cm

5'0"   5'6"   6'0"

phenotype for such a trait is either present or not in a particular individual and is generally not altered by environmental influences.

To describe discontinuous variation in a population, we divide the number of people who have a particular phenotype by the total size of the population; the resulting fraction is the frequency of that phenotype. From these phenotypic frequencies, scientists can calculate the frequencies of the alleles responsible. These **allele frequencies** (originally called gene frequencies) are most easily studied for traits whose patterns of inheritance are known and simple. Like the average values of continuously variable traits, allele frequencies are characteristic of entire populations, not of individuals. All individuals have genotypes, but only populations can have allele frequencies.

Variation between populations

The study of genetic variation both within and between populations is called **population genetics**, and it includes the study of allele frequencies for discontinuous traits. The measuring of allele frequencies requires that the different genotypes, and the alleles responsible for them, can readily be distinguished from one another. It is for this reason that population geneticists often concentrate on those genes whose phenotypic effects are easy to tell apart. Most of those genes control discontinuously variable traits that are either present or not. Differences in the average values for traits that vary continuously are also of interest to population geneticists, but the study of these traits is more difficult because the phenotypes of continuously variable traits are often altered by environmental influences such as nutrition.

One of the central tenets of modern biology is that evolution can occur only if populations are genetically varied. However, biologists did not always think in terms of evolving and variable populations. For over 2000 years, biologists believed that species were constant, unvarying entities. Plato and Aristotle had declared that each species was designed according to an ideal form that they called an *eidos*, often translated as 'type' or 'archetype.' Biologists following this view developed the **morphological species concept**. Each species was described as having certain fixed and invariant physical characteristics (morphology). The whole 'type' of that species was believed to be a cluster of 'essential' characteristics inherited as a single unit.

Biologists now recognize that species are constantly evolving, largely as the result of natural selection working on the genetic variation that is present within populations (Chapter 5). The Human Genome Project (Chapter 4) has revealed that over 99.9% of the human genome is identical in all people. However, the remaining fraction of a percent varies geographically, meaning that populations from different locations differ from one another. We must have some clear way to describe this variation and to describe population groups.

To define a population or a larger group of populations, we could sort people by some physical trait, such as distinguishing between people who are tall, short, or average in height. For any trait that we could choose, much of the variation exists within each and every population. If we chose some other physical trait, such as eye color or hair curliness, we would find that each physical characteristic results in a different grouping of the same people. In addition, we find that groupings based exclusively and strictly on any single trait always group together people who are quite dissimilar in many other respects (especially on a worldwide basis). For these reasons primarily, biologists prefer not to base the definition of population groups on physical characteristics.

Instead of using physical characteristics to define populations, biologists use the term **population** to refer to all members of a species who live in a given area and therefore can interbreed with one another. Membership in a population is determined by geographical location and by mating behavior, not by physical characteristics. Populations that interbreed under natural conditions belong to the same species (Chapter 5). All humans are placed in a single species, *Homo sapiens*, because all of them have the capacity to mate with one another and produce fertile offspring. However, people in different geographical locations belong to different populations. Genetic variation within any population is usually less than in the species as a whole. In past centuries, geographic isolation kept many human populations more distinct than they are now with worldwide transportation and migration. Population boundaries are not the same as national boundaries. Several different populations may live in the same geographic area, especially if cultural factors have maintained their separateness and inhibited matings between them. Sometimes, these populations are distinguishable by their derivation from geographically separate earlier populations.

Human populations in different places differ from one another in many physical traits. The average Canadian is taller than the average Southeast Asian, and the average African has darker skin than the average European. For natural selection, however, the characteristics that matter the most are those with the greatest impact on health and disease (or life and death). For example, cystic fibrosis and skin cancer are more frequent among people of European descent, but people of African descent have a higher risk of sickle-cell anemia and are more susceptible to frostbite if exposed to very cold temperatures. Most discontinuously variable traits that are examined closely show differences in allele frequency from one human population to another. For continuously variable traits, the difference between the averages of two populations is much less than the variation within either population (Figure 7.2). For example, the average height in the United States is taller than in China, but many Americans are shorter than the Chinese average and many Chinese are taller than the American average.

Although it is easy to find human populations that differ from one another in both physical features (morphology) and genetic traits, it is usually very difficult to find sharp boundary

Figure 7.2

Continuous variation in two populations with different mean values.

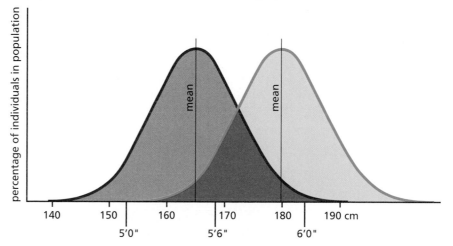

Distribution of height in two populations whose average values are 165 and 180 cm respectively. The variation within each population is greater than the difference between the average values of the two populations. Note that one of these populations is identical to the one shown in Figure 7.1.

lines dividing these populations from one another. If you were to walk from Asia to Europe and then to Africa, you would see populations differing only slightly, in most cases imperceptibly, from their neighbors, and you would meet representatives of the three largest population groups on Earth without finding any abrupt boundaries between them. Another way to say this is to say that variation between human populations is always continuous. This is so even when the trait in one individual is discontinuous. The population frequency of the allele responsible for the trait can vary continuously between zero (no one has the phenotype) and 100% (everyone has the phenotype). For discontinuous traits such as blood type, the allele frequencies of adjacent populations are generally close, just as is true for the average values of continuous traits, as seen in Figure 7.2.

Concepts of race

Humans have developed various ways of describing both themselves and the other human populations with which they have had contact. Biologists (who study all forms of life) and anthropologists (social scientists who study human populations and human cultures) have assisted in these descriptions by studying and measuring certain physical traits and allele frequencies. There are many ways in which human variation can be described, and there are many uses to which these descriptions have been put. One of the most problematic has been the attempt to separate people into different **races**. As we will soon see, there are various different meanings to this term, all of them different from the term 'population.' The term 'population' always describes smaller and more cohesive units than the term 'race.' No physical features are used in defining populations, but some race concepts have been based on physical features. In this section we describe four different concepts of race in the order in which they originated. The older concepts have not entirely died out; they have in many cases persisted side by side with the concepts that came later.

Races based on cultural characteristics. In the Bantu languages of Africa, the word for 'people' is *bantu*. Likewise, the Inuit word for 'people' is *inuit*. Every group of people has a name for itself and its members, and the name often means people or human. Names that people apply to other groups of people may simply be descriptive, but value judgments are often implied as well. In some instances, the value judgment implicit in the choice of name has been used to justify widespread abuses against the negatively labeled population. Such was the case when land and labor shortages resulted from large-scale cereal agriculture, a problem that arose independently in many places. A commonly developed solution to these shortages was to conquer neighboring people (the 'other') and confiscate their land. Slavery and several other systems of coercion were developed to secure the labor of conquered peoples. Slavery, oppression, and conquest all call upon the victorious people to practice certain atrocities on others that they would never tolerate within their own group. To justify these atrocities to themselves, and to protect their own members from practicing similar atrocities on one another, just about every conquering group has found it expedient to distinguish themselves from the 'other,' and furthermore to depict the conquered people as somehow inferior, subhuman, or deserving of their fate. Many of the groups that were culturally defined as races in the past are really language groups, cultural groups, or national groups that are hardly distinguishable on any biological basis from the group that traditionally oppressed them.

The imposition of social inequalities between 'Us' and 'Them' is now recognized as racism. **Racism** has many meanings, but all of them include the belief that some groups of people are better than others, and that it is somehow justified or proper for the more powerful group to subdue and oppress the less powerful. In most cases, the motivation to conquer and oppress others came first; the racist ideology came later.

The 'races' identified by the conquering group are socially constructed to serve the interests of the oppressors only. The distinctions and values of the oppressors are forcibly imposed on the oppressed, who are often taught to believe in their own inferiority. Most people now regard racism as unethical because it denies basic rights to many people and because it results in frequent crime, violence, and social conflict.

Separation based on race serves better the political and economic causes that have engendered it if the distinctions recognized are declared to be 'natural' and unchangeable, as opposed to characteristics that can easily be changed by education or religious conversion. Scientists belonging to racist societies have therefore sometimes attempted to 'prove' that the traits characteristic of another race have an inherited basis that cannot easily be changed, an assertion called biological or genetic **determinism** (or hereditarianism). Behind such assertions is the view that a group identity (an 'essence' or Platonic *eidos*) can be inherited, a view for which there is no basis in genetics. Anthropologist Eugenia Shanklin documents several instances in which scientists conducted 'scientific' studies to help 'prove' the values and prejudices of their own social group. In their genocidal campaigns of the 1940s, the Nazis exterminated many millions of Jews, gypsies, Slavs, and other groups, but not until they had declared each of them to be an inferior 'race.'

Racism and hereditarianism are not synonymous, but they often go together as attitudes shared by many of the same people. The supporters of eugenics (see Chapter 3) had many followers, including Nazis in Germany and anti-immigrationists in the United States. These followers sought ways to prove the inferiority, and especially the biologically unchangeable inferiority, of other people.

CONNECTIONS
CHAPTER 3

The other race concepts that we discuss later differ from this earliest concept, and resemble one another, in their avoidance of language, customs, and other cultural traits in the delineation of races. However, racism is not confined to those societies that embrace the cultural concept of race. Many biologists and anthropologists have pointed out that racism is also built into the the next concept, race delineated by body features.

The morphological or typological race concept. Biologists who study plant and animal species often describe the geographical variation within a species by subdividing the larger species into smaller and more compact subgroups, each of which is less variable than the species as a whole. These subgroups are generally called **subspecies**, but within our own species they are called races. To bring the study of human variation more in conformity with that of other species, scientists began to restrict their attention to characters that could be studied biologically and to exclude personality traits, languages, religions, and customs more influenced by culture than by biology.

Before the days of ocean-going vessels, most of the world's people had only a limited awareness of human variation on a worldwide scale. Each population, of course, knew about other populations nearby, but in most cases adjacent populations differed only slightly from one another. When trade extended over great distances, it usually did so in stages, so that none of the traders ever had to go more than a few hundred miles from home. The trade routes were also in most cases traditional, meaning that traders and migrants had generally come and gone over the same routes for centuries. This contributed to a gene flow or mixing of alleles that lessened the degree of difference between populations that would be noticed along the trade routes.

When explorers began to sail directly to other continents, they found people in other lands who differed more sharply from themselves in physical features. Many scientists subsequently became curious about the origin of these physical differences. Discussions of racial origins from about 1750 to 1940 tended to dwell on the origin of physical differences. A morphological definition of each race, based on physical features (morphology), was an outgrowth of the same thinking that had earlier resulted in a morphological species concept. At least initially, the major founders of this tradition were scientists who had no interest in oppressing the newly discovered peoples, so finding an excuse for racial oppression was less of a motive than was scientific curiosity. The emphasis was no longer on distinguishing only 'Us' from 'Them,' but on distinguishing among many different racial groups.

By the 1700s, biologists were actively describing and categorizing the variation in all living species. The eighteenth century naturalist Linnaeus (Carl von Linné) divided the biological world into kingdoms, classes, orders, genera, and species (see Chapter 6). He also divided humans into four subspecies: white Europeans, yellow Asians, black Africans, and red (native) Americans. The use of physical features such as skin color and hair texture to define subspecies was common among biologists using a morphological race concept. Other scientists in this same tradition recognized more races or fewer, but each race was always described on the basis of morphological characteristics such as skin color, hair color, curly or straight hair, and the occurrence of epicanthic folds of skin over the eyes.

Under the morphological concept of race, each race was defined by listing its common physical features as though they were invariant. For example, when describing a feature such as color, only one color was given, as if this color were invariant throughout the group and throughout time. This approach, which classified races on the basis of 'typical' or 'ideal' characteristics, ignoring variation, is called typology. Morphological definitions of race were always typological. Africans, for example, were declared to have black skins and curly hair, overlooking the fact that both skin color and hair form vary considerably from place to place within Africa and even within many African populations. All of the morphological characteristics were assumed to be inherited as a whole; a person was assumed to inherit a Platonic *eidos* (a 'type' or 'essence') for whiteness or redness, not just a white or red skin. Supporters of the typological concept of races were also supporters of a typological concept of species.

Years after morphological races had been defined, closer scrutiny revealed both variation within the morphological races and intergradation between them across their common boundaries. A few Europeans tried to save the morphological definitions by proposing that each race had originally been 'pure' and invariant, and that present-day variation within any population was the result of mixture with other races. One zoologist, Johann Blumenbach (1752–1840), divided up humans into American, Ethiopian, Caucasian, Mongolian, and Malayan races. He thought that each of these races was originally homogeneous (that is, 'pure'), and he named each after the place that he identified as its ancestral homeland. For example, white-skinned people are called 'Caucasian' because Blumenbach thought that this race originated in the Caucasus Mountains, east of the Black Sea.

There is no scientific support nowadays for the concept of originally pure races or for the concept of different ancestral centers of origin of different races; human populations have never been homogeneous and have always been quite variable. In some cases, however, Europeans and others who feared for the 'purity' of their own group sought to pass laws limiting contacts, especially sexual contacts, between the races that they recognized. Most of these laws were brutal but still ineffective in stopping what were viewed as interracial matings. There is no scientific basis for the belief that such matings are in any way harmful. On the contrary, variation within any species confers a long-term evolutionary advantage because it provides the raw material that natural selection can use to adjust to changing environmental conditions.

But hereditarian assumptions were even more strongly embedded in the morphological race concepts than they are in the culturally based race concepts. Lest one think that science has long since banished such attitudes in educated people, it is only necessary to point to the great storm of controversy that flourished over the subject of race and IQ in the 1970s. Arthur Jensen attempted to convince his readers that the mental abilities of African Americans were below those of other races and that these differences were fixed by heredity and unchangeable by educational means. A number of scientists, including Leon Kamin, Richard C. Lewontin, and Stephen Jay Gould, showed that his claims were unsupportable and based on fallacies and fabricated evidence. As recently as 1994, a book by Richard Herrnstein and Charles Murray once again brought up many of the same hereditarian arguments that had earlier been debunked (Box 7.1).

One of the strange ironies of a racist past is that many attempts at remediation, such as affirmative action, continue to require, at least for a time, the identification and naming of the same groups that were used previously for racially divisive purposes. Attempts to ensure fair and nondiscriminatory treatment for members of different socially recognized racial groups (in housing, employment, schooling, and so forth) require that we first identify and study the groups that we wish to compare. In this way, societies trying to overcome a history of racism find themselves using the very racial classifications of their racist past in order to redress the injustices of past generations.

Population genetics, clines, and race. Modern studies of human variation are based in large measure on genetics. Genetic variation between

BOX 7.1 Is Intelligence Heritable?

To address a question such as this, we must first define intelligence. Intelligence is not easily defined, but it includes the ability to reason and to learn new ideas and new forms of behavior, the measurement of which is far from simple. The biological bases for these abilities are likely to be multifaceted (Chapter 13), and genetic factors are likely to be the result of the interaction of many, many genes. Most discussions on the inheritance of human intelligence deal only with a single measure of this very complex trait, the IQ score, obtained from a test. IQ is not the same thing as intelligence and is at best an imperfect measure of mental abilities.

Also, to address this question, we must define the word 'heritable.' Heritability is defined in statistical terms as the proportion of the population's variation in some trait associated with genetic as opposed to environmental variation. Statistical association, or correlation, does not imply causation, and it certainly cannot be used to justify the claim that 'there is a gene for' the trait in question. One way to determine heritability of a trait in a domesticated species is to compare the variability of that trait in the population at large with the variability of the trait among highly inbred, genetically uniform individuals. Another way to determine heritability is to compare the variability that a trait exhibits at large with the variability of that trait among individuals raised in a standardized, experimentally controlled environment. Neither of these methods can be applied to humans, and the measures that are used to study humans are all indirect, complicated, and subject to criticism on technical grounds. For these reasons, there is no agreement on the heritability of any important human ability, including 'intelligence.' Moreover, because variation is a characteristic of populations and not of individuals, a term such as '60% heritable' would simply be a ratio of variation in one population to variation in another and would not tell us anything about the genotype or phenotype of any individual.

Numerous studies on IQ scores have shown the following:

- It is difficult to devise IQ tests that are free from cultural bias and from bias based on the language of the test, the gender and race of the test subjects, and the circumstances in which the test is administered.

- IQ scores seem to have both genetic and non-genetic components. Children's IQ scores correlate strongly with those of their parents. The IQ scores of adopted children usually agree more closely with their adoptive parents than with their birth parents, although studies on adopted children have been criticized for a variety of reasons (see Chapter 3, pp. 71–72).

- IQ scores can be greatly improved by environmental enrichment. They can also be adversely affected by poor nutrition, poor prenatal conditions, and a number of other environmental circumstances.

- Populations historically subject to discrimination, such as African Americans in the United States, Maoris in New Zealand, and Buraku-Min in Japan, have average IQ scores about 15 points below those of the surrounding majority populations. However, these lower average scores do not always persist in people who migrate elsewhere: descendants of Buraku-Min living in the United States have, on average, IQ scores on a par with those of other people of Japanese descent.

- In the United States, IQ scores of whites and also of blacks (African Americans) vary from state to state, in some cases more than the average 15-point difference between blacks and whites. Among African Americans born in the South but now living in the North, IQ scores vary in proportion to the number of years spent in northern school systems.

- Transracial adoption studies show that African American children adopted at birth and raised by white families had IQ scores close to (in fact, slightly higher than) the white average.

- Careful studies of matched samples in schools in Philadelphia failed to show significant average differences in IQ scores between black and white schoolchildren if differences in background were controlled. 'Matched samples' mean that children in the study were compared only with other children of comparable age, gender, family income level, parents' occupation, and similar variables.

Taken together, these data indicate that there is, at most, a small degree of heritability for IQ. They provide little support for the hereditarian claim that IQ is fixed and immutable, or that observed differences in scores cannot be diminished. They provide no support whatever for predicting any individual's IQ score on the basis of their inclusion in any group.

populations is continuous, and all boundaries between groups of populations are arbitrary. Even for traits that vary discontinuously and for which an allele frequency can be calculated for each population, geographic variation in allele frequencies is continuous between populations.

A continuous increase or decrease in the average value, allele frequency or phenotypic frequency of any one trait is called a **cline**, after a Greek word meaning 'slope' (as in words like 'incline' or 'recline'). Clines are an accurate (but lengthy) way of describing the geographic variation in each trait, one trait at a time, and each cline could be shown on a map. For a dozen characteristics, a dozen different maps would be needed, because the patterns of variation would in general not coincide.

The maps in Figure 7.3 show the clinal variation in the allele frequencies of three blood group alleles. Before such maps of allele frequencies can be drawn, local populations must first be identified and sampled. For example, blood groups must first be studied in many local populations; then the allele frequencies found in each geographic area can be drawn on maps such as those in Figure 7.3. From maps such as these, we learn that large continental areas usually show gradual clines. Thus, Figure 7.3A shows a gradual south-to-north increase in the frequency of allele A across North America, and Figure 7.3B shows a gradual west-to-east increase in the frequency of allele B across most of Eurasia. In geographic variation, clines of this sort are gradual, and boundaries of population groups are therefore arbitrary. Abrupt changes are uncommon, and when they do occur they generally coincide with geographic barriers that hinder both migration and gene flow. Examples of such barriers include the Sahara Desert, the Himalaya Mountains, and the Timor Sea north of Australia.

As can be seen in Figure 7.3, the frequencies of the blood group alleles A, B, and o vary greatly from one human population to another. The variations, however, do not necessarily coincide with other traits or with the groups recognized on the basis of morphology. Allele B, for example, reaches its highest frequency on mainland Asia, but is nearly absent from Native American populations or among Australian Aborigines. The frequency of allele A decreases from west to east across Asia and Europe. In Native American populations, allele A occurs mostly in Canada, and is mostly absent from indigenous Central or South American populations. The allele for blood group O has a frequency of 50% or more in most human populations, but its frequency approaches 100% in Native American populations south of the United States. African populations generally have all three of the alleles for ABO blood groups at levels close to worldwide averages.

Since the clinal variation concept was introduced in 1939, it has become customary to describe human variation by drawing maps of one cline after another. In addition to cline maps of phenotypes and allele frequencies, the techniques of molecular genetics (such as the DNA marker techniques described in Chapter 3, p. 72) are now being used to study clines at the molecular level. Clinal maps can also be drawn for continuous traits, in which case average values for the trait are calculated in each population. To describe the geographic variation in *Homo sapiens* or any other species, we could draw one map showing clinal variation in average body height, another showing variation in skin color or hair form, and so on. The population genetics approach encourages the scientific study and description of populations, including studies on the origins and former migrations of populations.

CONNECTIONS
CHAPTER 3

Figure 7.3

The clinal distribution of alleles for the ABO blood groups in indigenous populations of the world. Indigenous populations are those that have lived for hundreds or thousands of years in approximately the same region, to which they have had time to evolve adaptations. This includes Native Americans in the Western Hemisphere, Bantu and Xhoisan peoples in southern Africa, and Aborigines in Australia, but not the European colonists who came to these places after A.D. 1500.

After the Holocaust (1933–1945), the fledgling United Nations felt the need to refute many Nazi claims about race. The result was the 1948 *Statement on Race and Racism*, written by a committee that included several prominent anthropologists and geneticists. The statement, which has been revised several times since 1948, correctly pointed out that nations,

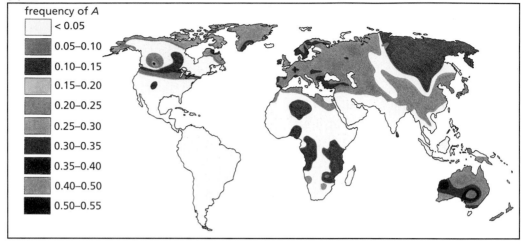

(A) The distribution of allele *A*

frequency of *A*
- < 0.05
- 0.05–0.10
- 0.10–0.15
- 0.15–0.20
- 0.20–0.25
- 0.25–0.30
- 0.30–0.35
- 0.35–0.40
- 0.40–0.50
- 0.50–0.55

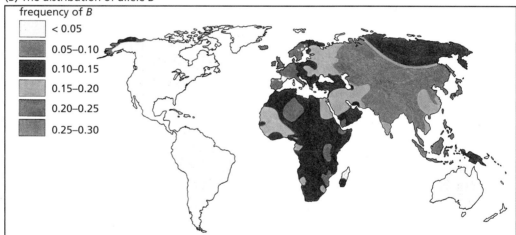

(B) The distribution of allele *B*

frequency of *B*
- < 0.05
- 0.05–0.10
- 0.10–0.15
- 0.15–0.20
- 0.20–0.25
- 0.25–0.30

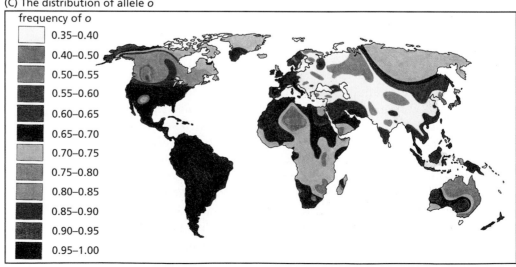

(C) The distribution of allele *o*

frequency of *o*
- 0.35–0.40
- 0.40–0.50
- 0.50–0.55
- 0.55–0.60
- 0.60–0.65
- 0.65–0.70
- 0.70–0.75
- 0.75–0.80
- 0.80–0.85
- 0.85–0.90
- 0.90–0.95
- 0.95–1.00

language groups, and religions have nothing to do with race, and that no group of people can claim any sort of superiority over another. The statement went further, however, to proclaim a new definition of race that replaced older, morphological definitions based on the inheritance of Platonic 'ideal types' with a new definition based on population genetics.

Under the population genetics definition, a **race** is a geographic subdivision of a species distinguished from others by the allele frequencies of a number of genes. A race could also be defined as a coherent group of populations possessing less genetic variation than the species as a whole. Either definition means that blood group frequencies are now considered more important than skin color in describing race, and that races are groups of similar populations whose boundaries are poorly defined. It also means that one cannot assign an individual to a race without first knowing what interbreeding population that individual belongs to. 'Race' is no longer a characteristic feature of any individual, because allele frequencies, like average phenotype values, characterize populations only, not individuals. Allele frequencies are consequences of population membership; they cannot be used to assign someone to a particular population or group. For this reason, racially discriminatory laws cannot and do not use population genetics; such laws rely invariably on the older morphological definitions or the still older social definitions of race.

Some writers maintain that racism is still contained in the population genetics race concept. They contend that studies that describe allele frequencies in geographic populations are merely reinscribing the racism of earlier concepts. Although far fewer people see racism in population genetics than in the earlier race concepts, some wish to go even further than the U.N. statement goes.

The 'no races' concept. Some scientists went still further in rejecting the heritage of the racist past: in the 1960s, led by the British anthropologist M.F. Ashley Montagu (who had earlier contributed to the U.N. definition), they declared that they would not recognize races at all. Among their arguments, one of the most compelling is that race concepts have always been misused by racists of the past and that the only way to rid the world of racism was to reject the entire concept of race. History is replete with examples of slavery, apartheid, discrimination, genocide, and warfare between racial groups. It is therefore easy to argue that the naming of races has in past generations done far more harm than good.

One stimulus to the 'no races' approach arises from the realization that there are no unique alleles or other genetic markers that could identify a person's race. Races, like populations, differ in the frequency of various alleles but do not have alleles that belong exclusively to that group. Even differences in allele frequencies have become less pronounced as a result of the great increase in international travel and migration that has occurred especially since World War II. To a certain extent, human populations have always mated with one another whenever there has been geographic contact between them; this is one reason why human population groups do not differ more than they do and why neighboring populations are so often similar. Since the advent of the jet age, frequent migrations have allowed more extensive contact and more opportunities for mating between people of different genetic backgrounds than ever existed before. Such matings have always occurred and always will; they even occur in societies that have tried to outlaw them. This type of mating will slowly but inevitably diminish the differences in the mix of alleles (the

gene pools) of populations, making it progressively more difficult to identify any significant differences between populations.

The study of human variation

All studies of human variation run the risk of being misused or misinterpreted by racists. Nevertheless, there are many good reasons for studying human variation, and this study serves as the basis for the entire field of 'human factors engineering.' To take a simple example, the design of a passenger compartment (for automobiles, aircraft, etc.) must accommodate a certain range in the size, sitting height, arm length, and other dimensions of its possible occupants. These and other accommodations must take into account the total range of human variation, including all races and both sexes. In airline cockpits and similar enclosures, controls should be both visible and reachable by persons of different sizes. Moreover, these features are often matters of safety as well as comfort. Vehicle seat belts and airbags, sports equipment, surgical equipment, wheelchairs and similar aids, boots, helmets, kitchen counters, telephone receivers, gas masks, toilets, and doorways all need to accommodate the range of dimensions of the human body. Variation in other human characteristics (breathing rates, sweating) must also be considered in the design of space suits, diving equipment, respiratory equipment for fire fighting, or protective clothing for other situations. Most of the variation relevant for human factors engineering is found within each population group, including variation by age and sex; variation between human populations is generally minor by comparison.

A further reason for studying genetic variation among human populations is that it can help us to understand evolution. Population genetics has helped us to recognize geographic patterns of disease resulting from natural selection acting on human populations. Studies of this kind can also help us to reconstruct the past history of particular human populations, or of the human species as a whole. In succeeding sections of this chapter we examine some of these studies.

THOUGHT QUESTIONS

1 Twentieth-century approaches to the description of human variation have in large measure been revolts against the earlier approaches. Against which of these earlier approaches was the 'no races' approach primarily directed? Against which earlier approach was the population genetics approach directed?

2 African Americans more often have high blood pressure and more often die from their first heart attacks than do white Americans. How would you decide whether this is the result of a difference in genes, in diets, in the availability of medical care, or in the lasting effects of discrimination in

U.S. society? If people in rural Africa seldom have heart attacks or high blood pressure, what possible hypotheses are falsified?

3 To produce research results of the kind referred to in Thought Question 2, one must have a way of assigning an individual to a population group. How does one determine a person's membership in a biological population? Is it sufficient to know that they live in a particular place? Will asking people to name the racial or ethnic group in which they claim membership (self-identification) produce biologically meaningful results?

Population Genetics Can Help Us to Understand Human Variation

The geographic variation shown in Figure 7.3 deals with human blood groups. We know a lot about the genetic basis of blood groups, and a person's blood group is easily determined, making blood groups good candidates for study by population geneticists. We now look in more detail at human blood groups and what their study has taught us about our own species.

Human blood groups and geography

In the days before reliable blood banks, blood transfusions were much riskier than they are today. Soldiers wounded in battle were generally treated in the field. If a transfusion was needed, it was done directly from the blood donor to a patient lying on an adjacent stretcher. Some transfusions were successful, but others resulted in death of the patient. Studies on the reasons for these different outcomes led to our knowledge of the existence of blood groups.

ABO blood groups. During the Crimean war (1854–1856), a British army surgeon kept careful records of which transfusions succeeded and which did not. From his notes he was able to identify several types of soldiers, including two types that he called A and B. Transfusions from type A to type A were nearly always successful, as were transfusions from type B to type B, but transfusions from A to B or B to A were always fatal. Also discovered at this time was a third blood type, O, which was initially called 'universal donor' because people with this blood type could give transfusions to anyone. These results were put to immediate practical use in treating battlefield injuries.

Karl Landsteiner, an Austrian pathologist who migrated to the United States, discovered the reason for these distinctions. Persons with blood type A make a carbohydrate of type A, which appears on the surfaces of their blood cells. Persons with blood type B make a carbohydrate of type B; persons with type AB make both type A and B carbohydrates; and persons with blood type O make neither of these carbohydrates. The A and B carbohydrates are also called antigens because they are capable of being recognized by the immune system (see Chapter 15). The immune system of each individual also makes antibodies against the blood group antigens that their own body does not make. In a person receiving a transfusion with incorrectly matched blood, these antibodies bind to the type A or B antigens, causing the blood cells to clump together within the blood vessels (Figure 7.4), often with fatal results. For explaining these immune reactions, Landsteiner received the Nobel Prize in 1930.

The A and B antigens allow all people to be classified into the four blood groups A, B, AB, and O. These blood groups are controlled by a gene that has three alleles: allele *A* is dominant and it contains information for producing antigen A (its phenotype); allele *B* is dominant and it contains information for producing antigen B (its phenotype); allele *o* is recessive and it functions as a 'place-holder' on the DNA but produces neither functional antigen. The *AA* and *Ao* genotypes both produce antigen A and are therefore assigned to blood group A. Likewise, both *BB* and *Bo* genotypes produce antigen B and result in the B blood type. Genotype *oo* produces neither A nor B antigens, which results in the O blood type (universal donor). Finally, genotype *AB* allows both alleles *A* and *B* to produce their respective antigens, resulting in the AB blood type. When they occur together, the *A* and *B* alleles are said to be **codominant** because the heterozygote shows both phenotypes.

For the purpose of matching blood donors and recipients, any person who shares your blood type is a good donor. It is therefore possible to collect blood in advance from many donors, sort the blood by blood type, and store it under refrigeration for use in an emergency. It is ironic that the doctor who developed this concept, an African American named Charles Drew (1904–1950), was denied its full benefits because many hospitals at the time kept separate blood banks for whites and nonwhite patients, a practice that has no biological foundation. Because the chemical composition of the allele products does not vary, type A antigen from an African American is identical to type A antigen from a Native American or from anyone else. A person with blood type A is therefore a good donor for almost any other person with blood type A.

Other human blood groups. Karl Landsteiner also discovered several other blood group systems that are totally independent of ABO. One such system, called the Rh system, actually has three genes located very close together on the same chromosome: the first gene has alleles *C* and *c*, the second has alleles *D* and *d*, and the third has alleles *E* and *e*. Unlike the ABO system, in which alleles are codominant, *c*, *d*, and *e* are recessive to *C*, *D*, and *E*. In all, there are eight phenotypic possibilities, of which phenotype cde (genotype *ccddee*, homozygous recessive for all three genes) is sometimes called Rh-negative and the others Rh-positive. The CDe phenotype is the most frequent phenotype in most populations, except in Africa south of the Sahara, where cDe predominates. The Rh-negative

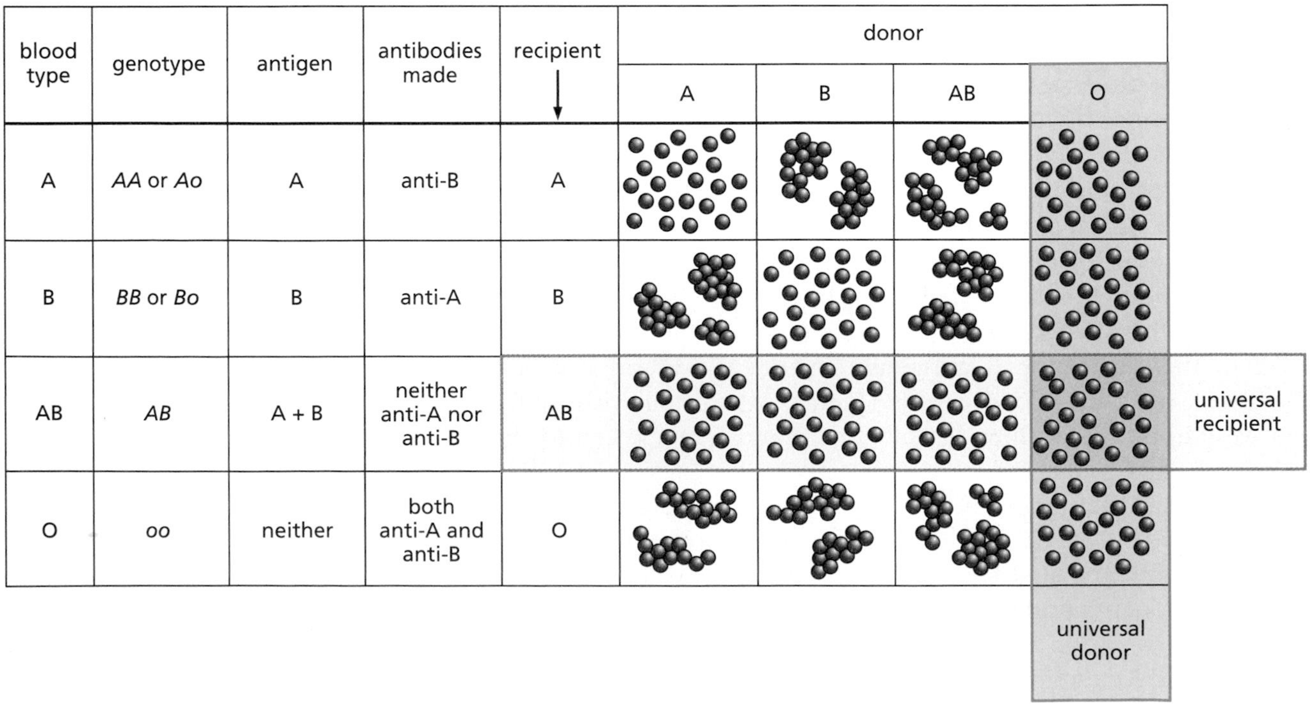

Figure 7.4

The human ABO blood groups. If a person of blood type A, who makes antibodies against blood type B, receives a transfusion of type B or AB blood, those antibodies cause the donated blood cells to clump together. Transfusion with matched blood or with blood type O (no A or B antigens) does not cause clumping.

blood type	genotype	antigen	antibodies made	recipient ↓	donor			
					A	B	AB	O
A	AA or Ao	A	anti-B	A				
B	BB or Bo	B	anti-A	B				
AB	AB	A + B	neither anti-A nor anti-B	AB				universal recipient
O	oo	neither	both anti-A and anti-B	O				

universal donor

phenotype cde is the second most common Rh phenotype in Europe and Africa, but is rare elsewhere.

Problems arise when a mother with the cde Rh-negative phenotype is pregnant with a baby who has a dominant *C* or *D* or *E* allele and is therefore Rh-positive. In this case, the mother makes antibodies against the C, D, or E antigens on the baby's blood cells, especially in response to the tearing of blood vessels during the process of birth. Because these antibodies are made at the end of pregnancy, they usually don't affect the first Rh-positive fetus that the mother carries. However, once these antibodies have been made, the mother's immune system attacks any subsequent pregnancy with an Rh-positive fetus, destroying many of the fetus' immature red blood cells, which can cause the death of the fetus (Figure 7.5). This problem can now be prevented by giving the Rh-negative mother gamma globulin (e.g., RhoGAM) at the time of the birth of any Rh-positive child; the globulin inhibits the formation of antibodies against Rh antigens, thereby protecting future pregnancies.

Separate from the ABO and Rh blood group systems are an MN system (with M most frequent among Native Americans and N among Australian Aborigines), a Duffy blood group system (with alleles *Fy*, *Fy^a*, and *Fy^b*), and many others.

Geographic variation in blood group frequencies. We saw earlier that the alleles for the ABO blood groups vary in frequency in different geographic locations (see Figure 7.3). Table 7.1 shows how the major geographic subgroups of *Homo sapiens* differ in the frequencies of various blood groups and other genetic traits. It is important to remember that

Figure 7.5

Rh incompatibility arising in an Rh-negative mother pregnant with an Rh-positive child.

FIRST Rh⁺ PREGNANCY	SOON AFTER BIRTH	SECOND Rh⁺ PREGNANCY
When an Rh-negative mother has her first Rh-positive pregnancy, C, D, or E antigens from the baby enter the mother's circulation during the detachment of the placenta following birth	The mother's immune system soon makes antibodies against Rh antigens C, D, or E	Antibodies made after the first pregnancy can endanger any subsequent Rh-positive fetus unless protective measures are taken

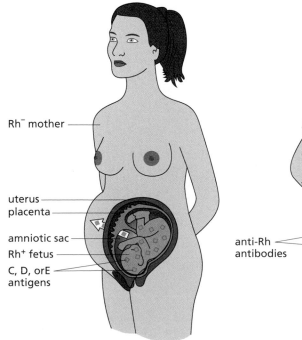
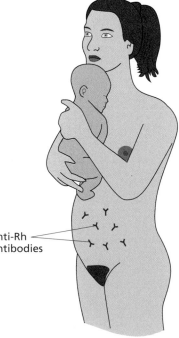
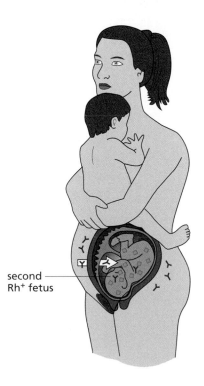

allele frequencies characterize populations only, not individuals. No blood group is unique to any population, so a person's blood type cannot identify them as a member of any population.

Frequencies of blood group alleles also vary on a smaller geographic scale. This is especially true among rural people who remain in their native villages or districts all their lives. The geneticist Luigi Cavalli-Sforza has documented variation in the ABO, MN, and Rh blood group frequencies from one locality to another across rural Italy. Similar results have been observed in rural populations in the valleys of Wales, in African Americans from city to city across the United States, and among the castes and tribes of a single province in India. These studies emphasize the hazards of assigning all people in a single country to a single population, especially when cultural barriers discourage random mating. However, populations that have become more mobile experience less of this microgeographic variation. As stated earlier, these are variations in allele frequencies and therefore can not be used to establish clear-cut boundaries between populations.

Isolated populations and genetic drift

In large, randomly mating populations in which selection and migration are not operating, the frequencies of the genotypes in the population tend to remain the same. This principle, which operates in all sexually reproducing species, is called the **Hardy–Weinberg principle**, and the predicted equilibrium is called the **Hardy–Weinberg equilibrium** (Box 7.2).

One of the criteria for a Hardy–Weinberg equilibrium is that the population be large. In small populations, allele frequencies tend to vary erratically, in unpredictable directions, from the expectations of the Hardy–Weinberg equilibrium. This phenomenon, called **genetic drift**, is

Table 7.1

Allele frequencies in the major geographic population groups.

AFRICAN POPULATIONS

Frequencies of blood group alleles *A*, *B*, and *o* in the ABO system and *M* and *N* in the MN system close to world averages; Rh blood group system with allelic combination *cDe* most frequent and *cde* second; allele *Fy* most common in the Duffy blood group system; allele P_1 more common than P_2 in the P blood group system; hemoglobin alleles Hb^S and Hb^C more frequent than in most other populations.

CAUCASIAN (EUROPEAN AND WEST ASIAN) POPULATIONS

Allele *A* in the ABO blood group system somewhat more frequent than in African or Asian populations; frequencies of *M* and *N* in the MN system close to world averages; Rh blood group system with allelic combination *CDe* most frequent and *cde* second; Duffy blood group system with allele Fy^a most frequent and Fy^b second; alleles P_1 and P_2 both common in P blood group system; alleles for G6PD deficiency and thalassemia more frequent than in most other population groups.

ASIAN POPULATIONS

High frequencies of allele *B* (and correspondingly less of allele *A*) in the ABO blood group system; *M* and *N* alleles at frequencies close to world averages; Rh blood group system with allelic combination *CDe* most frequent and *cde* rare or absent; Fy^a especially common in the Duffy blood group system; allele P_2 more common than P_1 in the P blood group system; some populations with high frequencies of alleles for thalassemia.

NATIVE AMERICAN (AMERINDIAN) POPULATIONS

Very high frequencies of allele *o* and virtually no *B* in the ABO blood group system; very high frequencies of allele *M* in the MN system; Rh blood groups with allelic combination *CDe* most frequent and *cde* rare or absent; allele P_2 more common than P_1 in the P blood group system; high frequencies of Di^a in the Diego blood group system.

AUSTRALIAN AND PACIFIC ISLAND POPULATIONS

Frequencies of blood group alleles *A*, *B*, and *o* in the ABO system close to world averages; high frequencies of allele *N* in the MN blood group system; Rh blood groups with allelic combination *CDe* most frequent and *cde* rare or absent.

defined as changes in allele frequencies in small to medium-sized populations due to chance alone.

The original model of genetic drift dealt with populations that remained small all the time, but other types of genetic drift were found to apply in particular situations. For example, if a large population became temporarily small and then large again, the random changes in allele frequencies that occurred when the population was small—the bottleneck—would be reflected in the allele frequencies of subsequent generations. This **bottleneck effect** is shown in Figure 7.6. Another type of genetic drift occurs if a small number of individuals become the founders of a new population. The allele frequencies in such a new population—whatever its subsequent size—will reflect the allele composition of this small group of founders, an influence known as the **founder effect**.

Several cases of genetic drift have been studied in isolated human populations. One well-studied example concerns the German Baptist Brethren, or Dunkers, a religious sect that originated in Germany during the Protestant Reformation. Forced to flee their native Germany, a few dozen Dunkers came to Pennsylvania in 1719 and started a colony that grew to several thousands and spread to Ohio, Indiana, and elsewhere. Because their strict religious code forbids marriage outside the group, they have remained a genetically distinct population.

Allele frequencies among the Dunkers have been influenced by genetic drift, particularly by the founder effect. If the Dunkers were a representative sample of seventeenth-century German populations, we would expect similar allele frequencies to those of present-day German populations derived from the same source. If, however, natural selection had changed the Dunker populations as the result of adaptations to their new location, then we would expect their allele frequencies to come closer to those of neighboring populations of rural Pennsylvania. Neither of these predictions is correct. Allele frequencies among the Dunkers differ from populations of both western Germany and rural Pennsylvania in a number of traits that have been studied. Blood group B, for example, hardly occurs at all among the Dunkers, although the frequency of the *B* allele is around 6–8% in most European-derived populations, including those of both Germany and Pennsylvania. Other genetically determined traits show similar patterns, including the nearly total absence of the *Fy^a* allele (from the Duffy blood group system) among Dunkers. The explanation that best agrees with the data is that the original founder population, known to have been made up of only a few dozen individuals, happened not to include anyone carrying *Fy^a* or

Figure 7.6

The bottleneck effect, a form of genetic drift that originates when populations are temporarily small.

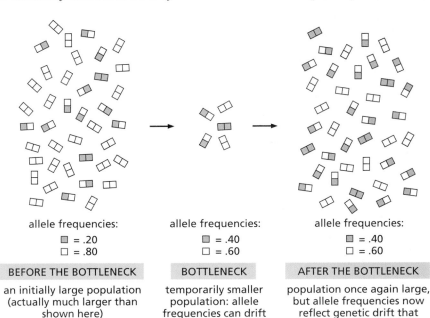

allele frequencies:
■ = .20
□ = .80

BEFORE THE BOTTLENECK
an initially large population (actually much larger than shown here)

allele frequencies:
■ = .40
□ = .60

BOTTLENECK
temporarily smaller population: allele frequencies can drift in random directions

allele frequencies:
■ = .40
□ = .60

AFTER THE BOTTLENECK
population once again large, but allele frequencies now reflect genetic drift that happened when the population was small

the allele for blood group B. Additional alleles may have been lost by genetic drift while the population remained small. The result was a population that derived its allele frequencies from the assortment of alleles that happened to be present in the founders. We can test this assumption by looking for the rare Dunkers who do possess an allele such as Fy^a. In every case that has been investigated, the occurrence of such an allele among the Dunkers can be traced to a person who joined the group as a religious convert within the last few generations.

Because they are genetically isolated, except for occasional religious conversions, the Dunkers have kept a unique combination of unusual allele frequencies. In the absence of blood group B, they resemble Native American populations; in the absence of Fy^a, they resemble

BOX 7.2 The Hardy–Weinberg Equilibrium

The Hardy–Weinberg principle can be stated as follows:

In a large, randomly mating population characterized by no immigration, no emigration, no unbalanced mutation, and no differential survival or reproduction (that is, no selection), *the frequencies of the alleles (genotypes) tend to remain the same.*

Allele frequencies are fractions of the total number of alleles present. If a population of 500 individuals (or 1000 alleles at a single genetic locus) contains 400 alleles of type A and 600 alleles of type a, then we say that the frequencies of the two alleles are 0.40 and 0.60 respectively, or 40% and 60% of the total number of alleles in the gene pool. At a given locus, the allele frequencies always add up to 1, or 100% of the population's gene pool.

Under the conditions specified in the Hardy–Weinberg principle, as stated above, there is a simple equilibrium of unchanging allele frequencies. Let us consider the case of a gene locus that contains two alleles, A and a. If the frequency of allele A is called p and the frequency of allele a is called q (where $p + q = 1$), then the equilibrium frequencies of all three diploid genotypes is given by the Hardy–Weinberg formula:

Genotypes	AA		Aa		aa	
Frequencies	p^2	$+$	$2pq$	$+$	q^2	$= 1$

This formula predicts that the frequency of the homozygous dominant genotype AA will be p^2, the frequency of the heterozygous genotype Aa will be $2pq$, and the frequency of the homozygous recessive genotype aa will be q^2.

To show that these equilibrium frequencies remain stable over successive generations and do not tend to change in either direction, consider the production of gametes in a population already at equilibrium. All of the gametes produced by the dominant homozygotes AA carry allele A, so the frequency of A gametes from AA homozygotes is p^2. Half of the gametes produced by the heterozygotes Aa also carry allele A, so the frequency of A gametes from heterozygotes is half of $2pq$, which equals pq. The total proportion of A gametes is thus $p^2 + pq$. We can now use simple algebra, separating out the common factor and then applying the equation $p + q = 1$ to calculate the frequency of A gametes:

Frequency of A gametes:

$$
\begin{aligned}
p^2 + pq &= p(p+q) \\
&= p(1) \\
&= p
\end{aligned}
$$

In similar fashion, the proportion of gametes carrying allele a is equal to pq (the other half of $2pq$) from the heterozygotes plus q^2 from the recessive homozygotes aa.

Frequency of a gametes:

$$
\begin{aligned}
pq + q^2 &= (p + q)q \\
&= (1)q \\
&= q
\end{aligned}
$$

So the frequency of A and a gametes corresponds to the frequency of A and a alleles.

African populations. In most traits, however, their derivation from a European source population is evident. These findings show that population resemblances based on a single blood group or gene system may often be misleading, and that distinctions among human populations, if used at all, should be based on a multiplicity of genetic traits.

The bottleneck effect has been used as a hypothesis to explain the near-total absence of blood group B among Native Americans and of cde (in the Rh blood groups) among Pacific Islanders. When the ancestors of these people first migrated from Asia, the random changes in allele frequency that occurred when the groups were small gave rise to distinct, isolated populations whose allele frequencies differed from those of the ancestral populations. Genetic drift of this kind would apply primarily to

Combining the gametes in all possible combinations (to simulate random mating) produces the following results:

Female gametes

	A	a	← Gametes
	p	q	← Frequencies
A / p	AA / p^2	Aa / pq	← Genotypes / ← Frequencies
a / q	Aa / pq	aa / q^2	

(Male gametes, shown on the left axis)

Taking the resulting genotypes from the chart above (and adding the two heterozygous combinations together), we obtain:

$$AA \quad\quad Aa \quad\quad aa$$
$$p^2 + 2pq + q^2 = 1$$

This is the same equation that we started with, which shows that the frequencies have not changed. It can also be shown that a population that does not start out at equilibrium will establish an equilibrium in a single generation of random mating.

Notice all the assumptions of the model: the population must be closed to both emigration and immigration, and there must be no unbalanced mutation and no selection. The population must be large enough to permit accurate statistical predictions, and the population members must mate at random. In reality, most natural populations are subject to mutation, selection, and nonrandom mating

(including inbreeding), and most usually experience emigration and immigration as well. The Hardy–Weinberg model, in other words, describes an idealized situation that is seldom realized in practice. The Hardy–Weinberg equilibrium is important to population genetics as an ideal situation with which real situations can be compared; if a population is *not* in Hardy–Weinberg equilibrium, one can ask why and then seek to measure the extent of the deviation from equilibrium. The same procedure is followed in other sciences as well. For example, 'freely falling bodies without air resistance' are an ideal situation in physics, and air resistance can be measured as a deviation from this ideal.

The Hardy–Weinberg equation is useful in estimating allele frequencies for traits controlled by a single gene. For example, if a population of 1000 has 960 individuals showing the dominant phenotype (such as normal pigmentation) and 40 displaying the recessive phenotype (such as albinism), then q^2, the proportion of homozygous recessive individuals, is equal to 40/1000, or 0.04. From this, we can calculate $q = \sqrt{0.04} = 0.2$.

From the fact that $p + q = 1$, we can calculate $p = 1 - q$. Substituting the value of 0.2 that we found for q gives us $p = 1 - 0.2$ or $p = 0.8$. Then the proportion of homozygous dominants in the population is $p^2 = (0.8)^2 = 0.64$ and the proportion of heterozygous individuals is $2pq = 2(0.8)(0.2) = 0.32$.

Figure 7.7

A family tree of human populations constructed on the basis of mitochondrial DNA sequences. 'Genetic distance' refers to the fraction of mitochondrial DNA sequence not shared by two populations, so that a fork at a genetic distance of 0.006 means that the populations share 99.4% of their mitochondrial DNA sequences.

groups of people, like the Polynesians or Native Americans, whose founder populations were initially small. The effects of genetic drift are minimal in the larger and more widespread population groups of Africa, Europe, and mainland Asia.

Reconstructing the history of human populations

Allele frequencies and DNA sequences in modern populations can be used as clues to their evolutionary origins. For example, American molecular biologist Rebecca Cann and her co-workers studied mitochondrial DNA sequences in samples from over 100 human populations. Mitochondria are organelles in the cytoplasm of eucaryotic cells (Chapter 6, p. 170) that produce much of the cell's energy and that also contain small strands of DNA independent of the DNA in the nucleus. Mitochondrial DNA is transmitted only maternally, from mother to both male and female offspring. Sperm from the father contain almost no cytoplasm and do not transmit mitochondrial DNA. Because mitochondrial DNA is smaller than chromosomal DNA in the nucleus, it is easier to sequence and is thus ideal for tracing evolutionary patterns. On the basis of these DNA sequences, Cann and her colleagues proposed a family tree of human populations using a maximum-parsimony computer model: of all possible family trees, the one shown in Figure 7.7 requires fewer mutational changes to have occurred than for any other tree. Another research team, headed by Luigi Cavalli-Sforza, used alleles of 120 genes to study the genetic similarities among 42 populations representing all the world's major population groups and many small ones as well. The findings of these two studies (and others) support the hypothesis of a divergence in the distant past between African and non-African populations, with the non-African populations later splitting into North Eurasian and Southeast Asian subgroups (see Figure 7.7). Australian Aborigines and Pacific Islanders are descended from the Southeast Asian subgroup, whereas Caucasians (Europeans, West Asians) and Native Americans (Amerind) are both descended from the North Eurasian group, which also includes Arctic peoples. The groups suggested by this study are geographically coherent and confirm certain well-documented patterns of migration. Existing linguistic evidence also matches these groupings, except for a few

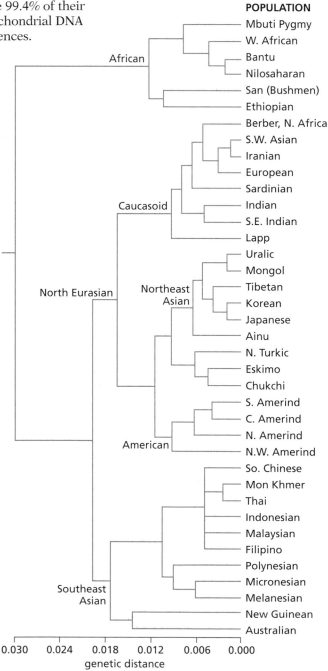

cases of cultural borrowing, which can be documented historically. Cavalli-Sforza's group estimates, largely on the basis of archaeological evidence, that the split between African and non-African populations took place 92,000 or more years ago. Other estimates have placed this split much earlier, back to the time of *Homo erectus*. The spread of human genes outward from Africa was either a very early event, or perhaps there were several such diffusions.

Studies such as those we have just described have sometimes been criticized for not being politically correct or for 'reinscribing racism.' A related criticism of the methodology is that geneticists with no training in anthropology are often tempted to lump together people who live close together even if there is good evidence that they have been historically and culturally separate. In other cases, people may maintain contact across considerable distances with other people who are culturally similar and speak the same language, and may consider themselves as belonging to the same group, even if population geneticists list them as separate because of the geographical distance between them. Although a good deal of interbreeding between groups always takes place, people more often choose their mates from what they consider as their own group. In order to assess what population groups actually exist (or existed historically), population geneticists need to cooperate with anthropologists familiar with the people being studied.

Paleontological and anthropological studies show that *Homo sapiens* has always been geographically widespread, with early populations spread across three continents, from Indonesia to Zambia and Western Europe. The earlier species *Homo erectus* was also geographically widespread. Despite this geographic spread, however, neighboring populations have always maintained genetic contact. Adaptation to local environments has caused populations to evolve geographic differences from one another, while matings between populations has maintained enough gene flow to prevent populations from becoming even more different. These two opposing tendencies form the basis for what American anthropologist Milford Wolpoff has called the **multiregional model** of the human species, which asserts that human populations have always maintained genetic contact with one another despite the differences resulting from local adaptation. The genetic contact maintains all human populations as one species, while the local adaptations have prevented geographic uniformity.

The study of allele frequencies has also been used to determine the origins of particular groups of people. One such study, for example, showed that Koreans are derived from a group that includes the Mongolians and Japanese but not the Chinese. Also, several studies have provided evidence for a Middle Eastern contribution (perhaps via Phoenecian sailors) to the populations of both Sicily and Sardinia. Studies of the Native Americans have shown that a minimum of three separate migrations were responsible for populating the Western Hemisphere, and more recent studies show that the situation is far more complex than this.

How did the adaptations come about that led to the various population differences in allele frequencies? The next two sections attempt to provide some of the answers to this question.

THOUGHT QUESTIONS

1 Random mating in a sexual species means that any two opposite-sex individuals have the same chance of mating as any other two. If there are a million individuals of the opposite sex, then each should have an identical chance (one in a million) of being chosen as a mate. Do you think human populations mate at random? Why or why not?

2 Is there ever a real population (of any species) in which the conditions specified by the Hardy–Weinberg equilibrium exist? How close do particular populations come?

3 If language has nothing to do with race, why do you suppose that researchers attempting to reconstruct the past history of human populations use linguistic evidence?

Malaria and Other Diseases Are Agents of Natural Selection

As any species evolves, biological differences among its populations arise largely through natural selection. Diseases are among the selective forces that can result in genetic differences among populations. In this section we consider some genetic traits that confer partial resistance to malaria. In malaria-ridden areas, natural selection acts to increase the frequency of alleles that confer partial resistance to malaria while decreasing the frequency of alleles that leave people susceptible to malaria. Many other selective forces have also operated over the course of human history, but resistance to malaria provides a series of well-studied examples.

New traits are produced by mutation (see Chapter 3, pp. 67–69) and are then subjected to natural selection, a process in which many traits die out in populations. The traits that survive natural selection are adaptive traits, or adaptations (Chapter 5), that is, traits that increase a population's ability to persist successfully in a particular environment. A good deal of human variation consists of adaptations that have resulted from natural selection operating over time, disease being a significant agent of that selective process.

CONNECTIONS
CHAPTERS 2, 3, 5

Malaria

On a worldwide basis, **malaria** causes over 110 million cases of illness each year and causes close to 2 million deaths, more than most other diseases. (Only malnutrition and tuberculosis cause more deaths each year, and measles causes about the same number.) Malaria also has a greater impact than most other diseases on the average human life expectancy because most of its victims are young, so that many more years of life are lost for each death that occurs. Malaria is more prevalent in tropical and subtropical regions than in temperate climates. The threat of malaria has largely been eliminated in the industrially developed countries through mosquito eradication programs and the draining of swamps, but as late as the first half of the twentieth century, malaria claimed many thousands of victims in Florida, Louisiana, Mississippi, and Virginia.

Historical and anthropological evidence confirms that malaria was rare (and therefore not a significant selective force) before the invention of agriculture. Even today, the disease is rare in undisturbed forests or in hunting-and-gathering societies. The clearing of forests for agricultural use opens up more swampy areas, and the building of irrigation canals or drainage ditches creates additional pools of stagnant water. The mosquitoes that carry malaria breed best in stagnant water open to direct sunlight. Agriculture therefore did much to change, in unintended directions, the agents of death (and thus the selective pressures) that act on human populations.

Life cycle of *Plasmodium*. Malaria is caused by one-celled protozoan parasites belonging to the genus *Plasmodium* (kingdom Protista, phylum Sporozoa), which live in human blood and liver cells. Of the four species of *Plasmodium* that cause malaria, *Plasmodium falciparum* is the most virulent. All species of *Plasmodium* have a complex life cycle, spending different parts of their life cycle in two different host species, mosquitoes and humans. The *Plasmodium* sexual stages (male and female gametocytes) are intracellular parasites that inhabit human red blood cells. When a female mosquito of the genus *Anopheles* is ready to lay her eggs, she first takes a blood meal from a person during which she ingests large numbers of red blood cells. (Mosquitoes rarely bite otherwise.) If the red blood cells contain *Plasmodium*, the male and female gametocytes combine in the mosquito's gut to form zygotes (fertilized eggs). The zygotes develop asexually through several stages within the mosquito, culminating in the infective forms (sporozoites), which migrate into the mosquito's salivary glands (Figure 7.8).

The mosquito's thin mouthparts function like a tiny soda straw or hypodermic needle. Shortly before consuming a blood meal, the female mosquito injects her saliva into her victim. The saliva contains anticoagulants that prevent the human blood from clotting inside the mosquito's mouthparts. When the mosquito injects saliva into a new human host, any sporozoites present in her salivary glands are injected along with it. These sporozoites enter the human bloodstream and are taken up by the liver. Each parasite then develops into thousands more, which may remain in the liver for years. Some parasites periodically escape from the liver into the bloodstream and invade the red blood cells. The parasites reproduce asexually within the red blood cells, producing the disease symptoms. The parasites digest the cell's oxygen-carrying hemoglobin molecules, and one stage also ruptures the red blood cells. Any impairment of the ability of the blood to carry oxygen to the body's tissues is called an anemia; all anemias leave their victims run-down and weakened. In malaria, the anemia is caused by destruction of both the hemoglobin and the red blood cells. Cell rupture also brings on fevers, headache, muscular pains, and liver and kidney damage. Within a given host, the asexual cycle of *Plasmodium* continues again and again until the patient either recovers or dies. In the red cells, the parasites can also develop into the sexually reproducing gametocyctes, which may be picked up by another mosquito in its next blood meal, spreading the disease.

Sickle-cell anemia and resistance to malaria

One of the symptoms of malaria is anemia. There are many other types of anemia. A very serious type was first discovered in 1910 by a Chicago physician named Charles Herrick. This strange and usually fatal disease also produced abnormally shaped red blood cells that sometimes resembled sickles. For this reason, Herrick called the disease **sickle-cell anemia**

Figure 7.8

Life cycle of the malaria parasite *Plasmodium*.

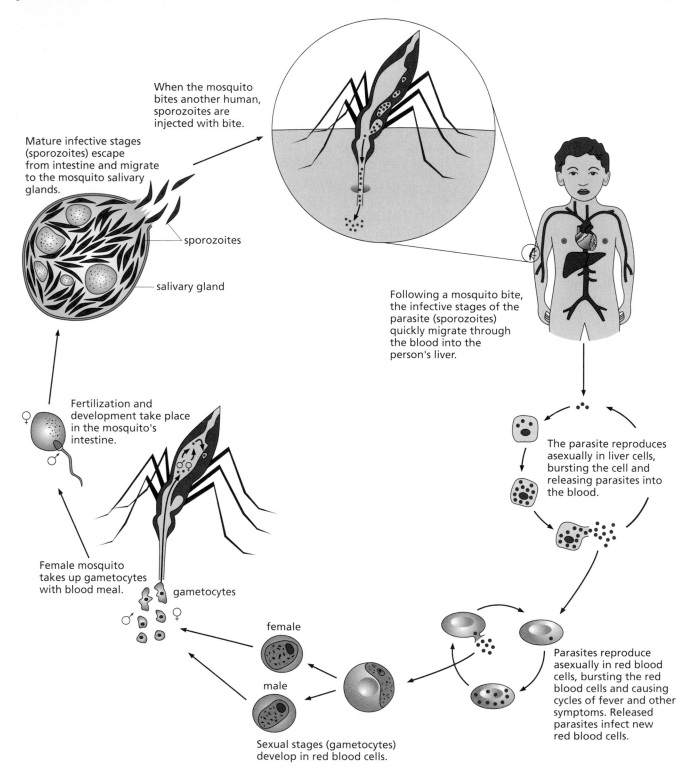

When the mosquito bites another human, sporozoites are injected with bite.

Mature infective stages (sporozoites) escape from intestine and migrate to the mosquito salivary glands.

sporozoites

salivary gland

Following a mosquito bite, the infective stages of the parasite (sporozoites) quickly migrate through the blood into the person's liver.

The parasite reproduces asexually in liver cells, bursting the cell and releasing parasites into the blood.

Fertilization and development take place in the mosquito's intestine.

Female mosquito takes up gametocytes with blood meal.

gametocytes

female

male

Sexual stages (gametocytes) develop in red blood cells.

Parasites reproduce asexually in red blood cells, bursting the red blood cells and causing cycles of fever and other symptoms. Released parasites infect new red blood cells.

A simple blood test was soon devised to test for the condition: a glass slide containing a bowl-shaped depression is used, and a drop of the patient's blood is placed inside the depression. A ring of petroleum jelly is placed around the margins of the depression and a cover glass is then applied, forming an airtight seal. As the red blood cells use up the available oxygen in the depression, the oxygen level decreases. Under these conditions, the red blood cells of a person with sickle-cell anemia assume their characteristic sickle-like shape, while normal red blood cells retain a circular biconcave shape (Figure 7.9). This blood test also allows the recognition of heterozygous carriers, a small percentage of whose blood cells sickle while the rest remain round.

Normal and abnormal hemoglobins. Sickle-cell anemia is caused by an abnormality in the molecules (called **hemoglobin**) that carry oxygen within the red blood cells. The hemoglobin molecule consists of four protein chains (two each of two different proteins) surrounding a ringlike 'heme' portion. Suspended in the middle of this ring is an iron atom that can bind one oxygen molecule (O_2), giving the hemoglobin its ability to transport oxygen and also its red color.

A change in a single amino acid, number 6 in one of the protein chains, is responsible for sickle-cell anemia. Normal adult hemoglobin (hemoglobin A) has glutamic acid in this position in the chain, while sickle-cell hemoglobin (hemoglobin S) has valine instead. This minute change alters the shape of the hemoglobin S molecules, straining the ringlike heme part of the molecule so that hemoglobin S does not carry oxygen as well as hemoglobin A. When they are not carrying oxygen, hemoglobin S molecules are stickier than normal hemoglobin. When the oxygen concentration in the blood is low, such as during physical exertion, hemoglobin S molecules adhere to one another and also to the inside of the red blood cell membrane, deforming the cells into the characteristic sickled shape. The difference in the proteins is hereditary and is caused by an altered codon in the hemoglobin gene on the DNA.

Figure 7.9

Normal red blood cells and red blood cells from a patient with sickle-cell anemia.

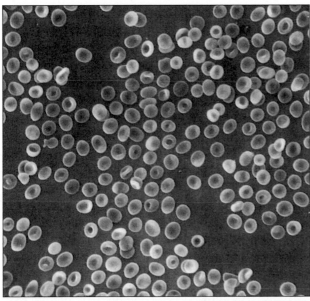

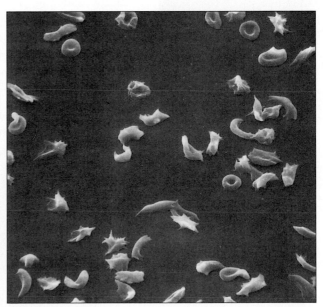

Normal cells   Sickle cells

The genetics of sickle-cell hemoglobin. Sickle-cell anemia is inherited as a simple Mendelian trait. People who die from sickle-cell anemia are always homozygous and their parents are almost always heterozygous, as are a certain number of siblings and other relatives. The gene for hemoglobin is designated Hb and the different alleles are designated by superscripts: Hb^A is the allele for normal hemoglobin and Hb^S is the allele for sickle-cell hemoglobin.

In U.S. and Caribbean populations, the vast majority of people carrying the Hb^S allele for sickle-cell hemoglobin are blacks of African ancestry. Tests of African populations also show high frequencies of the sickling allele, up to 25% in certain populations. In homozygous individuals (Hb^SHb^S), all the red blood cells are sickled at low oxygen concentrations, as commonly occurs during heavy exertion. Heterozygous individuals (Hb^AHb^S) have both types of hemoglobin and about one percent of their red blood cells can become sickled while the rest are normal in shape. Because both alleles produce a phenotypic result in heterozygotes, they are codominant, as we described earlier in connection with the AB blood type.

Symptoms of sickle-cell anemia. Most of the debilitating symptoms of the disease are consequences of the deformed, sickle-shaped cells brought on by exertion. The smallest blood vessels, capillaries, have a diameter only slightly larger than the diameter of blood cells. Because of their sickle shape and changed diameter, sickled cells cause flow resistance in the capillaries and thus impair microcirculation. In most of the body's organs, impaired microcirculation further reduces oxygen levels (hypoxia), which results immediately in a severely painful sickle-cell crisis. These crises begin in infancy. Damaged cells collect in the capillaries of the joints and result in painful swelling. The sickled cells are also more easily disrupted and destroyed than the normal-shaped round ones, resulting in a decreased oxygen-carrying capacity (anemia). The anemia and impaired circulation results in tissue damage to many organs, eventually resulting in death (Figure 7.10). In African populations, the death of homozygous Hb^SHb^S individuals often occurs before adulthood, but in the United States and the Caribbean, survival to reproductive age is now increasingly common. The reduction in red blood cell number and the sickle-cell crises also occur among heterozygotes, but not as severely.

Population genetics of sickle-cell anemia. When geneticists realized that sickle-cell anemia in the United States and Jamaica was largely confined to people of African descent, they began to investigate other populations. Using the blood test described earlier in this chapter, researchers investigated the frequency of the allele for hemoglobin S in many African and Eurasian populations. Over large parts of tropical Africa, researchers found remarkably high frequencies of the Hb^S allele, up to 25% or more. At first this appeared puzzling, because sickle-cell anemia was nearly always fatal before reproductive age. An allele whose effects are fatal in homozygous form should long ago have been eliminated by natural selection because people having sickle-cell children would have fewer children surviving to reproductive age.

Maps were made of the frequency of the sickle-cell allele. From these maps and from other evidence, it was noticed that the areas where the sickle-cell allele was frequent were also areas with a high incidence of malaria, particularly the variety caused by *Plasmodium falciparum* (Figure 7.11A and B).

Subsequent research confirmed the basic fact that the Hb^S allele, even in heterozygous form, confers important resistance to the most virulent form of malaria. Tests in which volunteers were exposed to *Anopheles* mosquitoes showed that the mosquitoes are far less likely to bite heterozygous Hb^A/Hb^S individuals than homozygous Hb^A/Hb^A individuals. Tests with the *Plasmodium falciparum* parasites showed that they thrive on the red blood cells of Hb^A/Hb^A individuals, who nearly always come down with a serious case of malaria after infection. However, when Hb^A/Hb^S heterozygotes or Hb^S/Hb^S individuals with sickle-cell anemia are infected with *Plasmodium falciparum*, their malaria symptoms are mild and they recover quickly because the parasite cannot complete its asexual cycle in their sickled blood cells. The protection that the Hb^S allele affords against malaria is sufficient to explain its persistence in those populations in which the incidence of malaria is high.

Hemoglobin S thus decreases the fitness of homozygotes by causing sickle-cell disease, but it increases the fitness of heterozygotes in areas where malaria occurs. In this way, malaria acts as an instrument of natural selection and has a dramatic influence on the allele frequencies of populations.

In addition to hemoglobin A and hemoglobin S, several other genetic variants of hemoglobin have been discovered. Some of these, such as Hb^C, also occur principally in areas where malaria is present and are thought to confer some resistance to malaria.

Figure 7.10

Development of the consequences of the Hb^S mutation in the hemoglobin gene. A small change in a gene can have many phenotypic consequences.

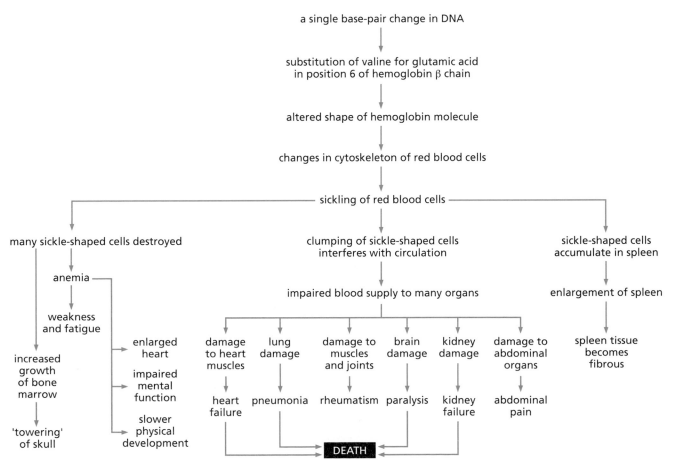

Other genetic traits that protect against malaria

Sickle-cell anemia is not the only heterozygous condition that protects against malaria. Two others are thalassemia and G6PD deficiency.

Thalassemia. In many countries bordering the Mediterranean Sea (including Spain, Italy, Greece, North Africa, Turkey, Lebanon, Israel, and Cyprus), many people have suffered from a different debilitating type of anemia known as **thalassemia**, (meaning 'sea blood' in Greek). The disease also occurs further east, especially in Southeast Asian countries such as Laos and Thailand (Figure 7.11C). Thalassemia is marked by a reduced amount of one or more of the protein chains in the hemoglobin molecule. The disease exists in a more serious, often fatal, homozygous

Figure 7.11

Distributions in the Eastern Hemisphere of *Plasmodium falciparum* malaria and several genetic conditions that protect against it.

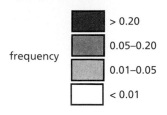

frequency
> 0.20
0.05–0.20
0.01–0.05
< 0.01

(A) Occurrence of *P. falciparum* malaria

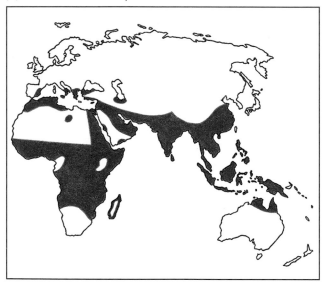

(B) Frequency of *Hb^s* allele

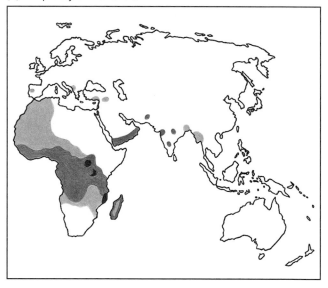

(C) Frequency of alleles for thalassemia
(actually the sum for alleles that lead to several different forms of thalassemia)

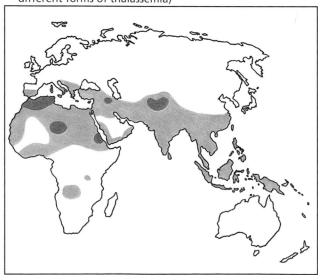

(D) Frequency of allele for G6PD deficiency

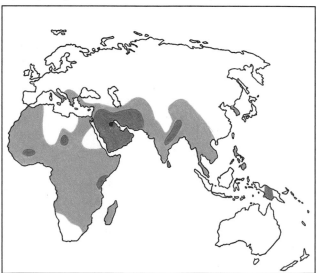

form called thalassemia major and a less severe heterozygous form called thalassemia minor. Red blood cells containing nonfunctional hemoglobin are destroyed in the spleen, producing anemia.

The symptoms of thalassemia vary, but all forms result in some decrease in oxygen transport in the blood. The bone marrow compensates by overproducing red blood cells, and this overproduction robs the body of much-needed protein and results in stunted growth and smaller stature.

Populations in which thalassemia occurs can now be screened for the genotypes that cause the disease, and genetic counseling can be provided to those found to carry the trait. Screening programs and newer methods of treatment have greatly reduced the problems caused by this disease in Italy, Greece, and elsewhere in the Mediterranean.

The geographical distribution of thalassemia follows closely the distribution of malaria in countries where sickle-cell anemia is infrequent or absent. For this reason, it has long been suspected that thalassemia confers a protective resistance to malaria, similar to that caused by sickle-cell anemia. The evidence is indirect: if heterozygous individuals (those with thalassemia minor) did not have some selective advantage such as malaria resistance, then the deaths caused by thalassemia major would have caused the genes for this trait to die out long ago.

G6PD deficiency. Blood sugar (glucose) is normally broken down within each cell in a series of reactions that begin with the formation of glucose 6-phosphate. Most of the glucose 6-phosphate is broken down into pyruvate (see Chapter 10, pp. 349–350) in a series of energy-producing reactions, but some is also used to make ribose (the sugar used in RNA) and to make reducing agents such as NADPH and glutathione. The removal of two hydrogen atoms from the glucose 6-phosphate molecule requires the enzyme glucose 6-phosphate dehydrogenase (G6PD). There are many people who have too little of this enzyme, a condition known as **G6PD deficiency**, or favism. G6PD deficiency results from a mutation in the gene that encodes the G6PD enzyme.

CONNECTIONS
CHAPTER 10

Under many or most conditions, people with G6PD deficiency remain perfectly healthy, but they occasionally suffer from an anemia in which the red blood cells rupture, spilling their hemoglobin, which then becomes physiologically useless but easy to detect by simple lab tests. This type of anemia, which is potentially fatal, can occur in G6PD-deficient people as a reaction to certain drugs (aspirin, quinine, quinidine, chloroquine, chloramphenicol, sulfanilamide, and others), in response to certain illnesses, or after eating fava beans (*Vicia faba*), a common legume of the Eastern Mediterranean and Middle East. The anemia may also exist chronically in a nonfatal form in people with G6PD deficiency.

G6PD deficiency has been shown to offer protection against *P. falciparum* malaria. It affects some 10 million people, and is thus the most common disorder offering protection against malaria. Most importantly, heterozygous carriers of the deficiency are also malaria-resistant, but the exact mechanism of the resistance has yet to be worked out.

G6PD deficiency occurs mostly in Mediterranean populations from Greece to Turkey and from Tunisia to the Middle East, and among Sephardic Jews. It also occurs south of this area into Africa and eastward across Iran and Pakistan to Southeast Asia and southern China (Figure 7.11D). The Greek mathematician Pythagoras may have suffered from this disorder, for his aversion to beans (one of the triggers of anemia in G6PD-deficient people) has become legendary. Pythagoras founded a religious cult in which the avoidance of beans was an important belief. Opponents of his cult once captured Pythagoras by chasing him toward a bean field, which they knew he would not cross.

Population genetics of malaria resistance

Polymorphism is the term used to describe a condition in which two or more alleles of the same gene are known in a given population at frequencies higher than the mutation rate. (This last restriction means that the alleles were inherited and could not all simply be the result of new mutation.) Polymorphism is a characteristic of the population, not of individuals; an individual may bear only one, or at most two, of the many alleles present in the population. Some alleles of polymorphic genes have harmful effects when homozygous, but they persist in populations because the same alleles also confer some important benefit (such as malaria resistance) when heterozygous. If the polymorphism persists for many generations, it is likely to be a **balanced polymorphism**. Balanced polymorphism arises when the homozygous genotypes suffer from some selective disadvantage or reduction in fitness, while the heterozygotes have the maximum fitness. For example, in a country in which malaria is present, $Hb^A Hb^A$ homozygotes have lower fitness because they are susceptible to malaria, and most $Hb^S Hb^S$ homozygotes die young from sickle-cell anemia. The $Hb^A Hb^S$ heterozygotes have maximal fitness because they are malaria-resistant and because they have enough normal red blood cells for them not to suffer from fatal sickle-cell anemia. Under conditions like these, natural selection brings about and perpetuates a situation in which both alleles persist.

The selection by malaria for genetic traits that offer resistance to it is at least as old as the open, swampy conditions (ideal for the breeding of mosquitoes) brought about by agriculture in warm climates. Evidence for this exists in the form of human bones found at a Neolithic archaeological site along the coast of Israel. Cultural remains found at this site show that it was an early farming community, one of the first in the area. Pollen analysis shows the presence of many plants characteristic of swampy areas. Some of the bones show characteristic increases in porosity (due to the increased production of red blood cells in the bone marrow) indicative of thalassemia.

Other diseases as selective factors

Hereditary diseases that confer some advantage in the heterozygous state are not confined to those that protect against malaria. In European populations of past centuries, tuberculosis, an infection caused by a bacterium called *Mycobacterium tuberculosis*, was an important force of selection, especially in crowded cities from the Middle Ages to the early twentieth century. One scientist has proposed that people heterozygous for the alleles that cause cystic fibrosis (an inherited lung disorder discussed in Chapter 3, p. 77) were protected against tuberculosis; they therefore survived tuberculosis epidemics in greater numbers than did people without cystic fibrosis alleles. As the heterozygotes

CONNECTIONS
CHAPTER 3

increased in number, some of them married one another, and, on average, one out of four of their children became afflicted with cystic fibrosis.

What about the geographic variation in blood groups and other genetic traits? There is evidence that at least some of this variation may also result from the natural selection brought about by various medical conditions. In a smallpox epidemic in Bihar province, India, researchers found that those who died were more often of blood group A, while survivors were more often of blood group B. In similar fashion, cholera selects against blood group O and favors blood group B. (Note that these studies demonstrated a difference in fitness, but did not explain the mechanism.) Other studies have shown statistical correlation of various blood types with other diseases: blood group O is correlated with an increased risk of duodenal ulcers and ovarian cancers, and blood group A with a slightly increased risk of stomach cancer. Associations of particular blood groups with cancers of the duodenum and the colon have also been postulated. Such statistical associations do not necessarily indicate a cause-and-effect relationship between the associated factors.

Fatal diseases are among the most striking agents of natural selection, but there are many other selective forces. We examine some of these other forces of natural selection in the next section.

THOUGHT QUESTIONS

1 How is an average life expectancy measured? Why is the average life expectancy of a population more affected by the deaths of children (e.g., from malaria) than by the deaths of elderly people?

2 All heterozygous carriers of the allele for G6PD deficiency are female. What does this tell you about the location of the G6PD gene? (You may need to review Chapter 3 to answer this question.)

Natural Selection by Physical Factors Causes More Population Variation

There are other agents of natural selection in addition to diseases. Among them are climatic factors such as temperature or sunlight, as well as climatic variation that makes food more scarce at some times of year or from one year to another. Like the genetically based traits that confer protection against disease, other genetic variation between populations has arisen in response to these other selective factors. In this section we look at some of these other factors and how they have selected in different geographic regions for differences in the genetically regulated aspects of physiology and of body shape and size.

Human variation in physiology and physique

During part of the Korean War (1950–1953), American soldiers were exposed to the fierce, frigid conditions of the Manchurian winter. Many soldiers were treated for frostbite. Most of the Euro-American (Caucasian) soldiers responded well to the medical treatment that was given, but a disproportionate number of African American soldiers did not and many of them lost fingers and toes as a result. Disturbed by these findings, the U.S. Army ordered tests on resistance to environmental extremes in soldiers of different racial backgrounds.

In one series of tests, army recruits were required to perform strenuous tasks (such as chopping wood) under a variety of climatic conditions. In a hot, humid climate, the African American soldiers were able to continue working the longest and performed the best as a group; Asian American and Native American soldiers performed nearly as well as the African Americans, and Euro-American recruits lost excessive fluids through sweating and became easily fatigued and dehydrated. Under dry, desert conditions, the Asian American and Native American soldiers did best, the African Americans were second best, and again the Euro-American soldiers became dehydrated. Under extremes of cold, it was the Euro-American soldiers who did best, followed closely by the Native American and Asian American soldiers; the African Americans shivered the most and some became too cold to continue. These tests demonstrated definite differences between groups in bodily resistance to physiological stress under a variety of environmental extremes. The significance of these differences was enhanced by the fact that, in other respects, the recruits represented a fairly homogeneous population: 18- to 25-year-old males who had all been screened by the army as being physically fit and free from disease and who had passed the same army physical and mental exams.

Other physiologists outside the Army conducted tests in which adult male volunteers immersed their arms in ice water almost to the shoulders. African Americans in general shivered the most and suffered the most rapid loss of body heat, as measured by a decline in body temperature. Euro-Americans and Asian Americans lasted longer without shivering, but they, too, eventually suffered loss of body heat. Only the Inuit (Eskimo) volunteers were able to keep their arms immersed indefinitely without any discomfort and without shivering. Subsequent studies that replicated these results made the additional finding that diet is also a factor: Inuit volunteers who ate high-protein, high-fat diets (traditional for the Inuit) did far better than other Inuit who had become acculturated to American dietary habits. It would be a mistake, however, to extrapolate findings from studies such as these beyond the groups used for the tests (adult males in good health) without further investigation. Many traits vary with age or sex or both.

Bergmann's rule. Genetically based differences in physiology that correlate with climate are the basis for a number of ecogeographic rules. Adaptations can also work indirectly, through variables such as body physique. Biologists have long noticed certain general patterns of geographic variation among mammals and birds. In one such pattern, called **Bergmann's rule**, body sizes tend to be larger in cold parts of the range and smaller in warm parts. This can be explained by the relationship of body size to mechanisms of heat generation and heat loss. For example, an animal twice as long in all directions as another animal has eight times the volume of muscle tissue generating heat ($2 \times 2 \times 2 = 8$) as the smaller animal but only four times the surface area over which heat is lost ($2 \times 2 = 4$). Thus, the larger animal is twice as efficient as the smaller ($8/4 = 2$) in conserving heat under cold conditions. A survey of human variation confirms that the largest average body masses are found among people living in cold places (like Siberia), while most tropical peoples within all racial groups are of small body mass, even when their limbs are long. These relationships of body size to climate result from natural selection acting on genetic variation within populations over long periods. It does not mean that a person of a certain genotype will grow larger if they move to a cold climate.

Allen's rule. Another broad, general phenotypic pattern in most geographically variable species of mammals and birds is **Allen's rule**: protruding parts like arms, legs, ears, and tails are longer and thinner in the warm parts of the range and shorter and thicker in cold regions. This rule is usually explained as an adaptation that conserves heat in cold places by reducing surface area and dissipates heat more effectively in warm places by increasing surface area. Human populations generally follow this rule: Inuit people have shorter, thicker limbs, while most tropical Africans have longer, thinner limbs (Figure 7.12). There are exceptions, however: a number of forest-dwelling populations along the Equator are much smaller than Allen's rule would predict, although they are usually thin-legged. Also, the tallest (Tutsi) and shortest (Mbuti) people on Earth live near one another in the Democratic Republic of Congo (formerly called Zaire), showing that climate is not the only instrument of natural selection influencing limb length or overall height within populations.

Diabetes and thrifty genes. Diabetes, a potentially life-threatening illness in many populations, may be an indirect result of one or more of the so-called 'thrifty genes' that protected certain people from starvation in past centuries. Ancestral Polynesians, for example, endured uncertain journeys over vast stretches of Pacific Ocean waters. Uncertain food supplies during such voyages selected for people who could withstand longer and longer periods of starvation and still remain active. The postulated 'thrifty genes' may have caused excess food, when it was available, to be converted into body fat that could be used for energy in times of famine. The result was a population that was stocky in build and resistant to starvation in periods when food supplies were low but that was also more susceptible to diabetes under modern conditions, when physical exhaustion is rare and food is always available. Diabetics fed on 'ordinary' diets have excess sugar in their blood, much of which is converted to fat and stored. Although diabetes is itself an unhealthy condition, the storage of fat may have been, under conditions like those described for the early Polynesians, an adaptive trait. Perhaps diabetes is an unfortunate modern consequence of having one or more alleles originally selected for their ability to convert sugar to body fat.

Figure 7.12

Bergmann's and Allen's rules illustrated by comparisons between arctic and tropical body forms.

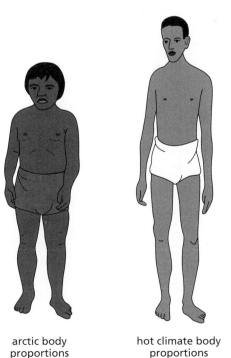

arctic body
proportions
(Inuit)

hot climate body
proportions
(Sudanese)

A similar history of selection for 'thrifty genes' (not necessarily the same ones) might also explain the late twentieth-century upsurge of diabetes in certain Native American populations, notably the Navajo and Pima of the southwestern United States. The risks that selected for 'thrifty genes' in the past were more significant in barren environments than in places in which the food supply was more assured. However, the commercial introduction of sugar-rich foods and a change from an active to a sedentary lifestyle have both raised the risks of diabetes, which are higher for sedentary people eating carbohydrate-rich diets. Because of these environmental changes, the genes that were once advantageous have in some cases turned into a liability, putting people of these genotypes at greater risk of diabetes. The Navajo and Pima have discovered that a return to frequent long-distance foot racing (a traditional activity they had nearly abandoned) has kept their populations healthier and has significantly lowered the incidence of diabetes in the runners. Not enough time has yet elapsed for the allele frequencies of the 'thrifty genes' to have again changed in this population, but the partial return to an earlier lifestyle has changed the environmental stresses and decreased the incidence of diabetes.

Natural selection, skin color, and disease resistance

The skin is the largest organ of the body and a major surface across which the body makes contact with the forces of natural selection in its environment. Human populations vary widely in skin color. Could these differences in skin color be adaptive?

Geographic variation in skin color. Skin color is one of the most visible human characteristics, and the one to which Americans have always paid the most attention when identifying race. Long-standing patterns of geographic variation are easier to understand if we ignore the population movements of the years since A.D. 1500 and consider only those populations still living where they did before that time.

Europe has for centuries been inhabited by light-skinned peoples, Africa and tropical southern Asia by dark-skinned peoples, and the drier, desert regions of Asia and the Americas by people with reddish or yellowish complexions. What is even more remarkable is that we find geographic variation along the same pattern within most continents, and in fact greater variation within the larger population groups than between such groups. For example, among the group of populations spread continuously from Europe across Western Asia to India, we find the lightest skin colors (also eye and hair colors) in Scandinavia and Scotland, progressively darker average colors (and darker hair) closer to the Mediterranean Sea, further darkening as we move through the Middle East and across Iran to Pakistan and India, and the darkest at the southern tip of India and on the island of Sri Lanka. A similar gradient (a cline) for skin color can be found among Asians, from northern Japan south through China into the Philippines and Indonesia.

Why would it be adaptive for people to be light-skinned in Europe but dark in Africa, Sri Lanka, and New Guinea? Notice that there are some very dark-skinned people outside Africa, and they generally have few other physical or genetic characteristics in common with Africans other than their dark skin colors. The natives of Sri Lanka, for example, have very straight hair and blood group frequencies totally different from those of Africa. One clue to this puzzle is that all very dark-skinned peoples have lived for millennia in tropical latitudes.

Sunlight as an agent of selection. Tropical regions receive on a year-round basis more direct sunlight than do temperate regions. In fact, the amount of sunlight received at ground level decreases with increases in latitude and corresponds more closely to belts of latitude than to variations in temperature. This is especially true for light in the ultraviolet region of the sun's spectrum.

If we exclude places where few people live, Europe receives the least sunlight of all the inhabited regions of the world. This is first and foremost a function of latitude. Europe includes populated regions of higher latitudes than on any other continent: London and fourteen other European capitals are located north of latitude 50°, while North America and Asia above this latitude contain few large cities and a great deal of sparsely inhabited land. Europe also has a frequent cloud cover that screens out even more of the Sun's rays. As a result of both high latitude and cloud cover, people in Europe receive much less exposure to ultraviolet light than most other people.

That sunlight levels select for body coloration is described by a third ecogeographic rule, **Gloger's rule**. While Bergmann's and Allen's rules, described earlier, take only temperature into account, Gloger's rule takes into account sunlight and humidity as well. Under Gloger's rule, most geographically variable species of birds and mammals have pale-colored or white populations in cold, moist regions, dark-colored or black populations in warm, moist regions, and reddish and yellowish colors in arid regions. We do not know all the reasons for this variation. Camouflage has been suggested as a cause, but vitamin D synthesis also has an important role.

Vitamin D is needed for the proper formation of bone. Children who do not receive adequate vitamin D during growth suffer from a condition called rickets, a disease of bone formation that may result in weakness and curvature of the bones (especially those of the legs) and in crippling bone deformities if left untreated. Sunlight is necessary for vitamin D synthesis. Many foods are rich in vitamin D, such as egg yolks and whole milk, but most vitamin D found in foods is in a biologically inactive form. The final step of vitamin D biosynthesis takes place just beneath the skin, with the aid of the ultraviolet rays of natural sunlight. This is why vitamin D is sometimes called the 'sunshine vitamin.' To get adequate amounts of vitamin D, a population must have both adequate intake of the vitamin in the diet and adequate exposure to sunlight. European populations have the lightest skin colors (and they get lighter the farther north you go) as an adaptation that allows maximum sunlight penetration into the skin. Europeans also have many cultural adaptations related to vitamin D intake, such as the eating of cheeses and other fat-rich milk products containing vitamin D. Northern Europeans place great value on outdoor activity at all times of the year, including such occasional extremes as nude dashes into the snow after the traditional sauna.

In northern Europe, people with dark skins could be at a very high risk of vitamin D deficiency because melanin pigment blocks out a large proportion of the Sun's ultraviolet rays. Very few dark-skinned people

lived in northern Europe even as immigrants. This has changed since World War II, when synthetic vitamin D became widely available. Because this prepared vitamin D is already in its active form, sunlight is no longer needed for its activation. Dark-skinned people can now live and remain healthy in northern latitudes without developing deficiency diseases.

At latitudes closer to the Equator, other problems exist. The same wavelengths of ultraviolet that are needed in the final step of vitamin D synthesis are also cancer-causing. Skin cancer (see Chapter 12) is generally a disease of those white-skinned people who are overexposed to the Sun's direct rays. Recently, it has also been found that ultraviolet radiation from overexposure to the Sun destroys up to half of the body's store of folate, an important vitamin that protects women against giving birth to children with spina bifida and other neural tube defects. Melanin pigment screens out ultraviolet radiation and thus protects against both cancer and folate deficiency. Populations of all racial groups living closer to the Equator have been selected over the millennia to have darker skins. Those individuals who had lighter skins were less fit in this Equatorial environment; that is, they more often got skin cancer and died earlier, or more often had deformed babies or miscarriages. Melanin pigment absorbs much of the ultraviolet light, protecting dark-skinned people against skin cancer and folate deficiency, while still allowing enough ultraviolet light through for adequate synthesis of Vitamin D.

There are no known genes that code specifically for skin color (except that the allele for albinism, present in all human populations, prevents synthesis of all melanin pigment and results in pale white skin). Apart from environmental effects (such as suntanning), there are probably dozens of genes that produce enzymes that influence the synthesis of melanin and other skin pigments—so many that it is difficult to study any one of them apart from the others. Nevertheless, natural selection has favored different levels of pigmentation in different geographic regions.

Nutritional sources of vitamin D in the far north. For the reasons given in the preceding sections, populations living in the high latitudes are generally light-skinned and populations that are adapted to living in tropical latitudes are generally dark-skinned. There is one very interesting exception: the Inuit populations of Arctic regions, sometimes known as Eskimos. (These people have always called themselves Inuit; the name 'Eskimo' was a pejorative name used by their enemies.) The Inuit are not very light-skinned, nor do they expose themselves much to sunlight. Most Inuit people live in places so cold that the exposure of bare skin poses a greater danger than any benefit of ultraviolet rays could overcome, and most Inuit are fully protected by clothing that offers hardly any exposure to the Sun. So how do they get enough vitamin D? The Inuit have discovered their own way of staying healthy. One of the world's richest sources of vitamin D is fish livers, especially those of cold-water fishes. (Cod liver oil is a very rich source of both A and D vitamins.) Moreover, the vitamin D in fish oils is fully synthesized and needs no sunlight to activate it. So, instead of having pale skins and traditions of exposing their skins to sunlight, the Inuit have traditions of catching cold-water fish (Figure 7.13) and eating them whole,

Figure 7.13

Traditional Inuit fishing. The Inuit get most of their vitamin D from eating whole fish, including the liver.

liver and all. These traditions have allowed them to stay healthy in a climate that is too cold and too sunless for most other populations.

In all of the above examples, a population that has lived in a particular geographic area for long periods has become adapted to the temperature, humidity, sunlight, and other conditions of their environment. The evidence presented in this chapter and in Chapter 5 suggests that natural selection is largely responsible for these adaptations.

CONNECTIONS CHAPTER 5

THOUGHT QUESTIONS

1 If people differ in their resistance to extreme cold or heat, does this mean that the difference is genetic? What would you need to know to answer this question? How could an experiment be arranged to test this?

2 Blood type O is statistically associated with duodenal ulcers, one of many such correlations between a blood type and a disease. Does a correlation demonstrate a cause? Does a correlation imply a mechanism of some kind? Does a correlation suggest new hypotheses? How can scientists learn more about whether there is a causal connection between the blood type and the disease?

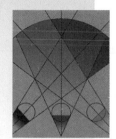

Concluding Remarks

Throughout the history of biology, scientists have developed various ways of describing groups of people. Some of these groupings have been known as races. Some concepts of race have attempted to find biological explanations for the racial groupings already established by various societies. Morphological concepts of race divided humans on the basis of their physical appearance. Biologists and anthropologists of the past gathered descriptive data about the physical characteristics of different populations and assumed that each group was distinct and unchanging. More recently, biologists have abandoned these concepts, in part because of the racism that has flowed from them, but also because these ideas no longer fit the data that we now have. The population genetics theory of human variation views human populations as varying continuously, with no group being uniquely different from any other. Biological differences among human populations are products of evolution. Like any other species, humans can evolve only when genetic variation is present in a population. When a population encounters some agent of natural selection, such as disease or climate, people with certain genotypes survive in greater numbers and leave more offspring than those with other genotypes. Over long periods, this process results in the adaptation of a population to its environment, with allele frequencies differing from one population to the next. This evolution continues today, although the increased mobility of people and technological alterations of the environment are slowly making populations less distinct than in past centuries. Populations vary in the frequencies of traits; they do not carry any unique traits. There is no biological phenotype, genotype, or DNA sequence that can assign an individual to a race or to a population. Although our biological concepts about race and other human variation have changed over time, racism will continue to exist if one group of people is held to be more valuable than another.

Chapter Summary

- Human **populations** vary geographically. Phenotypic and genotypic variation within populations usually exceeds variation between them.

- Phenotypic variation within a population can be either continuous or discontinuous. Continuously variable traits, in which all intermediate values are possible, can be described in terms of average values for each population. Discontinuously variable traits, such as those that are either present or absent, can be described in terms of phenotypic frequencies or **allele frequencies**.

- Differences among populations have historically been described in terms of culturally defined or morphological **races**.

- **Population genetics** allows us to describe groups of populations that differ from one another by certain characteristic allele frequencies.

- Most allele frequencies vary gradually and continuously among populations, without abrupt boundaries. Continuous variation from one population to another is best described in terms of geographic gradients, also called **clines**. Clines can be plotted on maps for average values of continuously variable traits such as height, or for population frequencies of discontinuously variable traits such as particular blood groups or alleles or DNA sequences.

- When more than one allele of a gene persists in a population this is called a **polymorphism**.

- The **Hardy–Weinberg equilibrium** describes the conditions under which allele frequencies remain constant in a population.

- Populations that were at one time small may have allele frequencies that have been shaped in part by **genetic drift**.

- Aside from genetic drift, most geographic variation among human populations has resulted from natural selection producing adaptation of the population to the environment.

- Disease is an important force of natural selection. Malaria, a widespread parasitic infection, has selected in different regions for high frequencies of alleles associated with sickle-cell anemia, thalassemia, and G6PD deficiency, all of which protect heterozygous individuals against malaria. Malaria and other diseases result in **balanced polymorphism** whenever the heterozygous genotype enjoys maximum fitness.

- Temperature selects for geographic variation in the alleles influencing body size and body shape.

- Ultraviolet light at different latitudes selects for geographic variation in the population frequencies of alleles influencing skin color. Alleles producing pale skin are selectively favored at high latitudes as an adaptation to absorb more ultraviolet light and prevent vitamin D deficiency. Alleles producing dark skin are favored near the Equator as a protection against skin cancer from too much ultraviolet exposure.

CONNECTIONS TO OTHER CHAPTERS

Chapter 1	Every study of human variation is conducted in a cultural context.
Chapter 1	Studies of human variation have ethical implications, including those arising from inappropriate use of the results.
Chapter 3	Many human variations have a genetic basis; such alleles arose ultimately from mutations.
Chapter 5	Human population variations reflect evolutionary processes, including mutation, natural selection, and genetic drift, all of which continue to work in modern populations.
Chapter 9	Nearly all human populations are growing, and some are growing much faster than others. Population growth and migrations change various allele frequencies.
Chapter 10	Different populations sometimes have different ways of meeting their nutritional requirements.
Chapter 12	Some types of cancer are more frequent in some human populations and less frequent in others.
Chapter 18	Human variation is an example of biodiversity at the population level.
Chapter 19	Because of damage to Earth's ozone layer, people are being exposed to increased ultraviolet radiation, which may select over time for a shift in allele frequencies leading to a darkening of skin pigmentation in human populations.

PRACTICE QUESTIONS

1. How many different genotypes can code for the blood group B phenotype? What are they? Are they heterozygous or homozygous?

2. How many different genotypes can code for the blood group AB phenotype? What are they? Are they heterozygous or homozygous?

3. How many different genotypes can code for the blood group O phenotype? What are they? Are they heterozygous or homozygous?

4. How many different genotypes can code for the Rh⁻ phenotype? Are they heterozygous or homozygous?

5. How many different genotypes can code for the Rh⁺ phenotype? Are they heterozygous or homozygous?

6. If the allele frequency of Hb^S in a population is 0.1, how many people in that population will be heterozygous $Hb^A Hb^S$? (Review the Hardy–Weinberg equation.)

7. How many different host species does the *Plasmodium* parasite need to complete its life cycle?

8. Why do people who are heterozygous for sickle-cell anemia have less severe anemia than people who are homozygous $Hb^S Hb^S$?

9. How does the bottleneck effect alter the allele frequencies of a population?

10. What is a balanced polymorphism? Give an example.

Issues

- What is sociobiology? How is it different from sociology?
- Why does sociobiology have so many critics? Who is objecting, and what are they objecting to?
- Is most behavior learned or inherited?
- What are the differences between instincts and other innate behaviors?
- How do learned behaviors relate to evolutionary change?
- To what extent can social behavior be modified?
- Why do the sexes behave differently in so many species?
- How different are humans from other species in social behavior? To what extent can findings in other species be extrapolated to humans?

Biological Concepts

- Evolutionary change (variation, natural selection, nonrandom mating, specialization and adaptation, human evolution)
- Population ecology (populations, regulation of population size)
- Learning and instinct (interaction of genotype and environment)
- Reproduction (asexual reproduction, sexual reproduction, mating systems, sexual dimorphism, reproductive strategies)
- Behavior (social behavior, communication, courtship and mating)

Chapter Outline

Sociobiology Deals With Social Behavior

Learned and inherited behavior

The paradigm of sociobiology

Research methods in sociobiology

Instincts

Social Organization Is Adaptive

Advantages and disadvantages of social groups

Simple forms of social organization

Altruism: an evolutionary puzzle

The evolution of eusociality

Reproductive Strategies Can Alter Fitness

Asexual versus sexual reproduction

Differences between the sexes

Mating systems

Primate Sociobiology Presents Added Complexities

Primate social behavior and its development

Reproductive strategies among primates

Some examples of human behaviors

8 Sociobiology

*B*ehavior that influences the behavior of other individuals of the same species is called **social behavior**. Examples of social behavior in animals include cooperative feeding, cooperative defense, aggression within the species, courtship, mating, and various forms of parental care. People also practice many forms of social behavior: nurturing their young, helping their neighbors, defending their possessions, and providing both material help and emotional support to their loved ones and to others. The population crisis discussed in Chapter 9 is a direct result of reproductive behavior. Some types of social behavior are often termed 'antisocial' behavior and result in problems for society. Examples include violence, crime, racist acts, sexist acts, and child abuse and neglect, all of which are social behaviors because they affect the behavior of other individuals. **Sociobiology** is the comparative study of social behaviors and social groupings among different species. The study of social behaviors in complex human societies is a separate discipline called sociology.

Can behaviors that cause problems be changed easily? Can beneficial behaviors (however defined or recognized) be substituted for destructive behaviors? Is most behavior rigid and unchangeable, or plastic and easily molded? Are we governed more strongly by our genetic background (nature) or by our upbringing (nurture)? The debate is very old. In Shakespeare's Tempest (4:1), Caliban is described as "a born devil on whose nature / Nurture can never stick." If human behavior were strongly determined by genes, then cultural influences, including education and training, would have only limited power to bring about changes in human behavior. Social reformers of all kinds usually support the opposite viewpoint, that human behavior can be modified almost at will, subject to few if any restrictions. Debates about alcoholism or homosexuality are often unproductive because some people assume that these are behaviors that could easily and voluntarily be changed, while others assume that these are permanent and deeply rooted in biological differences that may or may not be genetic. Differences in behavior between the sexes are likewise seen by some researchers as genetically constrained and by others as culturally controlled and easily changeable.

Although the most heated discussions arise from attempts to apply sociobiology to humans, sociobiology is a broad field of study and humans are but a single species. Most research in sociobiology focuses on nonhumans. Altruism, for example, poses a major research question in the sociobiology of all species. Among other broad-spectrum issues within sociobiology are the advantages of sociality itself, the kind of social organization found in each species, and the manner in which it evolved. Another issue is that of social relations between the sexes of each species, including the concept of reproductive strategies; in this chapter we show that parental care, infanticide, adultery, and altruism can all be viewed as components of reproductive strategies. The evolution of these strategies is an important field of investigation for sociobiology. In this chapter we examine the sociobiology paradigm and some of the major issues within the paradigm.

Sociobiology Deals With Social Behavior

Sociobiology means different things to different people. To scientists working in sociobiology, it is a field of study that deals with social behavior and its evolution. Sociobiologists usually explain behaviors in evolutionary terms. Although sociobiologists are more interested in the inherited components of behavior, they all acknowledge that much of behavior can also be modified by learning. They also acknowledge that natural selection can act only on those components of behavior that are inherited. One of the important research goals of sociobiology has therefore become the investigation of the relative importance of learned and inherited influences on particular behaviors. Sociobiology also has a number of critics who challenge the emphasis on inherited behavior patterns. These critics prefer to emphasize learning, including cultural learning in humans, as a strong influence on behavior. We will examine both viewpoints.

CONNECTIONS
CHAPTERS 5, 9

Learned and inherited behavior

Many behavioral patterns may be strongly influenced by experience in dealing with the environment, i.e., by **learning**. Nearly every behavior that has been carefully investigated also has some genetic component. Learned behavior may increase fitness, but only the genetic components (or predispositions) underlying the behavior can be influenced by natural selection (see Chapter 5). Natural selection can operate on the capacity for learning particular kinds of things, such as how to find one's way through the maze of one's surroundings. The character favored by selection in such cases is not the behavior itself, but rather the capacity to learn the behavior.

This is true of the ability to run through mazes, one of the most often studied types of learned behavior. Rats were tested for their ability to learn certain mazes, and the number of training sessions that it took the rats to learn the mazes was recorded. Their littermates, who were never tested themselves, were then selectively bred for several generations. Breeding the littermates of fast maze-learners resulted in a strain in which the average number of training sessions needed was low, while a strain of slow learners was bred from the littermates of individuals who needed more repetitions. The use of littermates in this experiment eliminated learning experience or other influences as determinants of the differences between the two selected strains. Notice that the behavior was not fully determined by inheritance; it still had to be learned. This behavioral trait is determined by many genes and environmental influences acting together. The *difference* between the two strains resulted from the buildup of gene combinations, which was only possible because some portion of the variation between groups was heritable.

Furthermore, to say that variation between groups is heritable *does not mean that the behavior is inherited as a fixed and unchangeable trait*. There are extremely few behaviors in any species (and none at all in humans) that are not subject to modification through learning. For example, nobody learns to play basketball like Michael Jordan or to play the cello like Yo Yo Ma without years of practice. Nobody can become even a mediocre basketball or cello player without lots of practice—a period of learning. However, some innate talent and ability are surely needed, or else any of us would be able to become a great basketball star or a world-renowned cellist simply by practicing enough.

Thus, it is important to emphasize that the oft-posed question of learned versus inherited behavior is a false dichotomy. Every learned behavior is based in part on some inherited capacity to learn, which may include the capacity to learn certain kinds of behaviors and not others, to respond to some stimuli and not others, to learn up to a certain level of complexity, and so on. Similarly, most behavior patterns with an inherited component can be modified to some extent by learning. These observations give rise to the testable hypothesis that *nearly every behavior pattern is at least partly learned and at least partly inherited*. Behaviors that do not require learning are called **innate**, and innate behaviors are assumed to have an inherited component. No behavior is 100% learned, and few are 100% inherited in any species (Figure 8.1). The methods used to distinguish between learned and innate components of behavior are described later in this chapter.

The paradigm of sociobiology

Sociobiology, the study of social behavior among different species, uses a scientific paradigm of the kind described in Chapter 1: one or more

Figure 8.1

Learned versus innate behavior patterns.

Nursing and suckling behaviors have strong innate components in most mammals (but this doesn't prevent bottle-feeding of many human infants).

Novel behaviors can be learned by many species.

Most forms of behavior show both learned and inherited components. Robins and many other birds innately peck at certain stimuli and thereby gain learning experiences about how to hunt more effectively and how to distinguish food objects from other objects.

theories, plus a set of value-laden assumptions, a vocabulary, and a methodological approach (Box 8.1). The formulation of sociobiology as a paradigm dates from the publication of the book *Sociobiology: the New Synthesis*, by the American evolutionary biologist Edward O. Wilson (1975). Many of the ideas of this paradigm can be traced to Charles Darwin's writings. What was new in 1975 was the way in which these ideas were put together to form the paradigm.

If people outside the paradigm had viewed sociobiology as no more than the study of social behavior, few objections would have been raised to it. However, sociobiology was frequently criticized for its focus on inherited behavior. As the many critics of sociobiology have emphasized, much of behavior is learned, and nearly all behavior can be modified by learning, particularly in mammals. Also, human behavior is strongly influenced by both language and culture, so many scientists who are otherwise sympathetic to sociobiology have cautioned against extrapolating sociobiological findings from other animals to humans.

Some nonscience critics of sociobiology are fearful of genetic determinism, the assertion that our individual characteristics are determined before birth and cannot be changed. As we discussed in Chapter 7, genetic determinism is feared because throughout history people in power have sought to control other people (other social classes, other races, and women) by teaching that existing inequalities were 'natural,' based on innate and unchangeable differences. Also, many people fear that the mere claim that a behavior is innate will discourage people from trying to change that behavior through education or similar means. The claim that behavior is innate can be particularly threatening to social reformers who pin their hopes for the future on the ability of people to modify their behavior.

Among biologists, those who believe in genetic determinism are decidedly in the minority. Most biologists, especially those who study animal behavior, are impressed with the degree to which behavior can change in response to environmental circumstances, including the behavior of other individuals. There are genetic constraints on what can and cannot be learned, but, within these limits, behavior is remarkably changeable (or 'plastic') in most animal species. No behavior is fully 'determined' either by genetics or by environment—almost every behavior is influenced by both of these factors throughout the lifetime of the individual.

We now examine the research methods used by sociobiologists.

Research methods in sociobiology

No behavior can be analyzed by any method until it has been adequately described. Sociobiology therefore includes a great many observational field studies of animals. How does one distinguish between the learned and innate components of a particular behavior? Sociobiologists use the following methods to investigate these components.

Rearing animals in isolation. A classic type of experiment is to raise an animal in isolation, in a soundproof room with bare walls and minimal opportunities for learning, including no opportunity to learn behavior from others. Behavior that the animal exhibits under these conditions is assumed to be largely innate. Experiments of this sort cannot ethically be done on humans.

BOX 8.1 The Sociobiology Paradigm

Research activity in science is often organized around paradigms (see Chapter 1). Here, in brief outline form, are some of the major points of the sociobiology paradigm:

1. Behavior is interesting to observe and to study. (This is a value judgment; people who do not share it will never be attracted to the paradigm.)

2. Much of the interesting behavior influences the behavior of other individuals, and is called 'social.' (This is a definition with an implied value judgment that people within the paradigm are expected to share.)

3. Social behavior has evolved and continues to evolve. (This is a central theory whose rejection would bring down the entire paradigm.)

4. The evolution of social behavior takes place by natural selection, along the lines outlined by Darwin: variations occur, and the variations that increase fitness persist more often than those that do not. (This is again a theory; it includes theoretical concepts such as 'fitness' and 'variation.')

5. Behavior is often modified by individual experience ('learning'). However, this learning takes place within limits set by the biology of the organism: the eyes limit what can be seen (likewise with other sense organs); the muscles and skeleton limit the possible responses; the structure of the brain limits the learning capacity, and so on. There are also many preexisting predispositions to respond to certain types of stimuli, to react in certain ways, and so on. These predispositions may have been learned at an earlier time, but *at least some of them* precede any learning and may be called 'innate.' (This is a central tenet of the paradigm, forming the basis for its further research.)

6. In the evolution of behavior, learned modifications are not directly inherited. Learned behaviors can contribute to fitness, but cannot be inherited. Only the innate predispositions and their biological underpinnings can be inherited, and only these inherited components can evolve. Natural selection can only work on the inherited aspects of behavior. (These ideas follow in part from the ways in which 'learned' and 'innate' are defined, and in part from the findings of evolutionary theory.)

7. It is therefore important to distinguish the learned and innate components of behavior, and to focus attention on the latter. This is a value judgment about the aims of research within the paradigm. It does not mean that learned behaviors are unimportant; it just means that sociobiologists would rather identify what is learned so that they can ignore it and spend the rest of their time studying the innate components. *It is this preference for studying the innate components of behavior that makes the sociobiology paradigm so controversial; most critics of sociobiology have the opposite preference.*

8. We can use modified Darwinian methods of investigation to study those components of behavior that evolve. One method is to measure variations in fitness by observing many individuals and studying the number of viable offspring successfully reared by each. Another method is to study the results of past evolution by comparing social behaviors among different populations or different species. (These are the basic research methods.)

9. Before comparisons can be made, however, there must first be an often lengthy period of observation and description. However, we realize that the presence of observers might modify the behavior that we wish to study. Because we are interested in behavior under 'natural conditions,' it follows that we should conduct most observations at a distance and interfere as little as possible. (These are more research methods.)

Rearing animals under different conditions. If the behavior is performed in the same way by animals or humans reared under strikingly different circumstances, then the behavior is largely innate. If, in contrast, the behavior varies according to the circumstances of rearing, then the variation can be attributed to environmental influences, although this does not rule out inherited influences, which may also be present. Cross-cultural studies are used to compare the behaviors of people raised in different societies or under different customs; innate behaviors are expected to be constant across various cultures, while learned behavior patterns are expected to vary.

Studying behavior in different genetic strains. If different strains or breeds of a species differ behaviorally in a consistent and characteristic way, then a strong inherited component exists. (This does not rule out learned components, which might also be present.)

Conducting adoption studies. If two populations differ in a particular behavior, it may be useful to study individuals from one group who are adopted early in life and raised by the other group. Under these conditions, behavior consistently resembling the population of birth demonstrates an inherited influence, while behavior resembling the population of rearing demonstrates a learned influence. Mixed or inconsistent results may indicate that both influences are present.

Conducting twin studies. If a trait is under strong genetic control, then identical twins should usually both exhibit the trait whenever either one does, while fraternal twins more often exhibit differences. Twin studies in humans are frequently criticized because the effects of learning cannot easily be separated from those of inheritance unless the twins are reared separately in families randomly chosen, conditions that are rarely even approximated. Some studies compare identical twins reared together to those reared apart, an experimental design that attempts to get around some of these difficulties.

Instincts

A subset of innate behaviors are called **instincts**. Instincts differ from other innate behaviors in being complex behavior patterns that are under strong genetic control. The classical test for whether a particular behavior is an instinct is whether the behavior appears at the appropriate time of life in an animal reared in isolation since birth or hatching. For example, if a songbird reared in a soundproof room sings the song of its species and sex upon reaching maturity, then the song is considered to be instinctive. By this test, many behaviors that have been studied in fishes, birds, and many invertebrates (including insects) have been shown to be largely instinctive. Behaviors related to courtship and mating usually have strong instinctive components in most species. Other behavior patterns that are frequently instinctive include automatic 'escape' behavior, nest-building behavior, orb-weaving in spiders, and various threat gestures. When instinctive behavior leaves a lasting product, such as a nest or a spider's web, these products are often so distinctive that they can be used to identify the species that created them.

Mammals generally rely more on learned behavior than on instinct. Among primates especially, many behaviors that are instinctive in other species have strong learned components. These behaviors may vary greatly among human societies.

Advantages of instincts. Short-lived animals rely heavily on instincts. For example, mayflies (insects of the order Ephemeroptera) have an adult life span of less than 24 hours. During this brief period they do not feed but have just enough time to find a mate, copulate, lay their eggs, and die. There is no time for learning to take place, nor is there any time for mistakes. The mayflies that accomplish their mission successfully are those that can perform their behavior correctly on the first try; they will probably never get a second chance. Selection over millions of years has therefore produced a series of adult behaviors that are instinctive and automatic, allowing no room for diversity or innovation. This is typical of instincts generally: behavior is instinctive in contexts in which uniformity and automatic response are adaptive and where innovation and diversity might be maladaptive. A greater complexity of behavior is possible with a simpler brain if the behavior is instinctive; learned behavior of equal complexity requires a more elaborate nervous system and also a long learning period during which many mistakes are made.

Mating behavior. Mating behavior includes both courtship (attracting a mate and becoming accepted as a mate) and the actual release or transfer of gametes. Mating behavior has a strong instinctive component in nearly all species, except in higher primates. Scientists can demonstrate the instinctive component of most forms of mating behavior by raising individuals in isolation until they are sexually mature, then testing them to see whether they can perform the behavior typical of their species.

Natural selection favors uniformity in mating behavior rather than diversity. Such unvarying behavior (called **stereotyped behavior**) is used for mate location and recognition in many species. The behavior that evolved in each species matches the type of signal that each is able to sense, so that visual mating signals are used by species with good vision, chemical signals by species with good chemical reception, and sounds by species with good sound discrimination. Many species of birds, frogs, and insects use sounds as mating signals, and the noncalling sex (usually female) responds only to mating calls of the proper pitch, duration, and pattern of repetition. Both the mating signals and the behavioral response to them are instinctive. Members of each sex know exactly what to listen for in the other sex and usually avoid nonconformers who deviate from the instinctive pattern. Sexual selection thus penalizes the nonconformers, who generally fail to mate and therefore leave no offspring. The flashing patterns of fireflies, though visual, are sexually selected in the same way. Because of sexual selection, mating calls or visual displays are precisely controlled within a narrow range for each species. Closely related species often differ in their mating calls and courtship patterns. Differences in mating calls and other courtship displays often serve as reproductive isolating mechanisms that prevent interbreeding between species (see Chapter 5).

Male birds of many species display conspicuously colored parts during courtship. Mating rituals that include beautiful, ornate displays

CONNECTIONS
CHAPTER 5

evolve as a consequence of sexual selection in those species where the discriminating sex (the one doing the choosing) consistently prefers the most conspicuous displays. Peacocks, lyre-birds, and birds-of-paradise are renowned for their beautiful and ornate male plumage (Figure 8.2). Male birds of species with less conspicuous plumage may concentrate instead on building an elaborate nest. The South Pacific bowerbirds build their nests within a large framework (a bower) that also serves as a place of mating. A few species even build an 'avenue' lined with colorful stones leading to the entrance of the bower. Generally, bowerbird species with ornate plumage do not build elaborate bowers, and the species that build impressive bowers do not have elaborate plumage.

Figure 8.2

An example of a conspicuous mating display in a peacock. Females of this species prefer males with the most conspicuous displays.

Territorial behavior. In many species, one or both sexes may show territorial behavior by defending a territory, either throughout the year or only during the mating season. The defense of a territory against intruders of the same species is common in many animal species. In some species, only males are territorial. Territorial behavior spaces individuals apart and encourages the losers to strike out in search of new territory, thus extending the range of the species wherever possible. Each territory must have sufficient food resources for a mating couple and their offspring, places for hiding and refuge, and at least one suitable nest site. Males without any territory are usually unable to attract mates and thus leave no offspring in that particular season. The specific boundaries of a territory and one's status as a territory holder or a trespasser are learned, but the general tendency to establish territories is instinctive in certain species (or sexes or seasons).

Territorial species may use gestures to threaten territorial rivals. Mammals who establish territories may mark their territory with their own scent. The intimidation of rivals by gestures or by the presence of odors serves to space individuals apart without causing injury or loss of life. Such ritualized forms of territorial defense are much more common than any form of fighting in which injuries are likely.

Nesting behavior. The choice of a nesting site may be an important part of territorial behavior. In some bird species, the male builds the nest and then offers it to the female as part of the mating ritual. In other species, male and female may cooperate in building the nest together as part of the mating ritual. Females may incubate the eggs alone, but males may provide other forms of assistance by bringing food or by defending the area against predators. In other species, the males and females take turns in guarding the nest and sitting on the eggs. Feeding the hatchlings may similarly be either a solitary or a shared task.

The behaviors just described are performed by individuals. Behaviors can also be performed by groups of organisms, a subject that we take up in the next section.

THOUGHT QUESTIONS

1 Does 'antisocial' behavior (such as assaulting others and causing them injury) fit the definition of social behavior? Do you think the definition should be modified? In what way?

2 Is sociobiology a subject area with room for many viewpoints, or is it a single viewpoint that enshrines both genetic determinism and sexism? Can sociobiology be studied without the assumptions of genetic determinism?

3 Can the methods used for gathering or analyzing data in sociobiology be the same for different species? To what extent do size (small versus large animals) or habitat (above ground, underground, underwater, in trees, etc.) require differences in field methods? What special problems in methodology arise when humans are being studied? Can the methods used for other species be applied to humans?

Social Organization Is Adaptive

Very few animal species consist of solitary individuals that spend all their time alone. Even in species whose members are solitary much of the time, individuals must come together for sexual reproduction. Most species, however, are far more social than this, and species that form social groups greatly outnumber those that consist primarily of solitary individuals. Social groups vary greatly in both size and cohesiveness. Simple pairs and family groups have only a few individuals. Larger social groupings include antelope herds, baboon troops, and fish schools, all of which may include up to a few hundred members. Still larger are the colonies of social insects, which may include many thousands or in some cases millions of individuals. Some social groups are loosely organized, with individuals staying together but seldom interacting, while others are organized into social hierarchies within which interactions are complex, as they are among humans and social insects.

Advantages and disadvantages of social groups

There are clear disadvantages to living in a social group. Chief among these is the competition for food and other resources (mates, hiding places, nesting sites), and this competition is made more intense by the fact that members of the same species generally have exactly the same ecological requirements (they seek the same foods, nesting sites, etc.). In those cases where food comes in portions small enough to be monopolized by a single individual, social groups are often small or non-existent. Social groupings also foster the spread of parasites and infectious diseases, and they may make the group easier for a predator to spot. Species that rely on camouflage avoid forming densely clustered social groups.

Despite these disadvantages, we find that species in which there are social groups far outnumber the species formed by solitary individuals. Clearly, there must be some great advantages for social groups.

Some advantages of social grouping are related to the obtaining of food. A large group of individuals searching for food together has a higher probability of finding it than a single individual. If food tends to be discovered in quantities much greater than a single individual needs, selection favors the formation of social groups.

Finding a mate is made easier if there are social groups. Indeed, many species that are solitary throughout most of the year come together on occasion, often during a particular season, and form social groups and mate. The risk of inbreeding increases if social groups are small and remain closed to the introduction of new genes; this risk is usually minimized by mechanisms for the exchange of genes or of individuals between populations.

Other advantages of social grouping relate to defense against predation. Social groups can often defend themselves more effectively than individuals can. Musk oxen, for example, respond to threats by standing close together with individuals facing outward in different directions (Figure 8.3). Even in species that do not practise group defense, members of a group may warn one another by giving alarm signals, or simply by fleeing as soon as a predator is spotted. Thus, belonging to a group gives all group members the advantage of greater (and earlier) alertness against predator attacks. For this reason, large but loosely organized

Figure 8.3

Musk oxen in a defensive formation. When musk oxen stand close together and face in different directions, no predator can surprise them.

flocks, schools, or herds are common among birds, fishes, and ungulates (hoofed mammals such as wildebeest and zebra) (Figure 8.4). Other advantages to group membership arise from the sharing of risks: a predator attacking the entire herd may capture one of its members at most, while the rest escape, so that each individual in a herd of 500 is exposed to only 1/500 of the risk of capture faced by a solitary individual. Actually, the risk may be even smaller because predators can more easily capture solitary individuals. Most herd animals taken by predators are individuals that have strayed from the herd.

Simple forms of social organization

Social organization refers to the ways in which social groupings are structured. The fact that social organization sometimes varies among closely related species suggests that social organization evolves. Studies on the inheritance of social status (dominance) within organized social groups point to a complex interplay of learned and inherited behavioral components in the establishment of social organization.

Groups without dominant individuals. Perhaps the simplest form of social organization is shown by brittle stars (see Figure 6.17, p. 192), marine organisms distantly related to sea stars. On encountering one another, brittle stars tend to stay together in clumps, even though there is no evidence of any more complex interaction.

The schooling behavior of fish is another very simple form of social organization. There are hydrodynamic advantages to schooling—swimming is made slightly easier by certain changes in water pressure caused by the swimming of the other fish—but these effects are small. The major advantage to schooling behavior may be that the fish hide behind one

CONNECTIONS
CHAPTER 6

another in such a way that most escape predators. Most fish school closer together when a predator is nearby (see Figure 8.4). When attacked, schools of fish or flocks of birds tend to scatter in every direction, a reaction that confuses many predators and that gives the individuals a chance to escape.

The size of social groups can vary greatly, often in response to ecological factors. For instance, the weaver birds live in many parts of Africa and Asia. In humid, forested regions, most weaver birds nest in pairs and feed on insects, while those species inhabiting grasslands and other drier habitats build large communal nests and eat a diet rich in seeds.

Groups with dominant individuals. One form of social organization is called a linear dominance hierarchy or 'pecking order.' Such hierarchies are found among domestic fowl and certain other captive animals (Figure 8.5). The top-ranked individual, usually a strong male, can successfully bully or threaten all the other individuals in the group, literally pecking at them in the case of birds. The second-ranked individual can intimidate all others except the top-ranked individual. The third-ranked individual can intimidate all except the first two, and so on. Occasionally, two closely ranked individuals may be tied for status, so that neither can dominate the other. For the most part, however, this type of organization results in the biggest bully getting whatever he wants, the second biggest

Figure 8.4

Social groups in various species.

Pelicans

Minnows schooling in the presence of predators

Wildebeest on the plains of Africa

Gannets on the coast of Quebec

getting whatever he wants as long as he steers clear of the top-ranked individual, and so on. Some feminist critics of sociobiology suggest that such male-dominated forms of social organization exist more in the minds of male sociobiologists than in the animals that they study. In at least some studies, pecking orders may reflect the artificial conditions of captivity and confinement.

Altruism: an evolutionary puzzle

Efforts to solve human social problems such as pollution often call for individuals or corporations to sacrifice their own interests for the common good, a practice called **altruism**. Altruistic behavior exists in many other species as well. As an example, consider the 'broken wing' display of certain female birds such as nighthawks. When guarding her nest against a predator, a female nighthawk may lead it away from the nest location, distracting its attention by limping or pretending to have a broken wing. Once she has drawn the predator sufficiently far from the nest, she flies away, leaving the predator confused. While protecting her young, she has increased her own danger. In evolutionary biology, altruism is defined as behavior that *decreases* the fitness of the performer while it increases the fitness of another individual. In this example, the female bird has decreased her own fitness by putting her life at risk for the sake of her offspring. Remember that fitness is defined as the relative number of fertile offspring produced by an individual (see Chapter 5). Only changes that increase fitness are perpetuated by natural selection.

Altruistic behavior poses a problem in evolutionary theory because natural selection might be expected to work against it. How could altruism evolve if it decreases fitness? Various hypotheses have been developed to explain this. In this section we examine several hypotheses that act at different levels of selection.

Selection at the species level. One early hypothesis for the evolution of altruism is that it benefits the species as a whole. However, careful examination of this hypothesis shows it to be unsatisfactory. If a species had both altruists and selfish individuals ('cheaters'), and if some part of this behavioral difference were controlled genetically, then selection would work against the altruists and in favor of the cheaters. Altruism may benefit all recipients of another individual's altruistic behavior, but the advantage is greater to selfish individuals than to other altruists. Under these conditions, natural selection should favor selfishness and eliminate altruism from the population.

Group selection. Another possible explanation for the evolution of altruism was proposed by the British ecologist V.C. Wynne-Edwards. If a

Figure 8.5

Domestic fowl showing a pecking order.

CONNECTIONS CHAPTER 5

species is subdivided into populations or social groups, then selection among these groups (group selection), may favor one group over another. In particular, a group containing altruists is favored *as a group* over other groups composed of selfish individuals only.

As we explained earlier, the defense of territory prevents excessive population density by spacing individuals apart and limiting population size. Wynne-Edwards describes the losers of territorial disputes as altruists who forgo mating for the benefit of the group as a whole. He argues that the mating of individuals without territories would lead to overpopulation, increased mortality, and a smaller resulting population size. Selection between groups would thus favor altruism. Other biologists who have examined this claim with mathematical models have shown that a loser who cheats (mates anyway) would greatly increase its fitness over one who does not mate, and that cheating behavior would thus be favored over altruism in every territorial species. Similar arguments have been advanced to show that other behaviors that achieve spacing or population control would also not be favored by group selection because cheaters would tend to leave more offspring than altruists.

Kin selection. Many biologists dissent from the group-selection hypothesis, seeking instead a simpler explanation based on individual selection. The currently favored explanation of altruism is based on the concept of **inclusive fitness**, defined as the total fitness of all copies of a particular genotype, including those that exist in relatives. Relatives are listed according to their degree of relationship, symbolized by *R*. For sexually reproducing organisms with the common types of mating systems, an individual shares half of its genotype (*R* = 1/2) with its parents, its children, and, on average, with its brothers or sisters (who share two parents). Also, an individual shares one-fourth of its genotype (*R* = 1/4) with grandchildren, half-siblings (who share one parent only), uncles, aunts, nieces, and nephews. The inclusive fitness of your genotype is the sum total of your individual fitness *plus* one-half the fitness of your parents, children, and full siblings who share half your genotype, *plus* one-fourth the fitness of those relatives who share one-fourth of your genotype, *plus* one-eighth of the fitness of your cousins who share one-eighth of your genotype, and so on. This concept allows us to define **kin selection** as the increased frequency of a genotype in the next generation on the basis of its inclusive fitness.

The conditions under which kin selection favors the evolution of altruism were specified by the British sociobiologist William D. Hamilton. Assume that altruistic behavior results in a certain reduction in fitness or 'cost' to the altruist, and a corresponding gain in fitness or 'benefit' to another individual who shares a fraction of the altruist's genotype. Hamilton reasoned that natural selection would favor altruism whenever the gain in inclusive fitness to the altruist's genotype exceeds the cost. If I perform an altruistic act that diminishes my individual fitness by a certain cost but raises my child's fitness or my sister's fitness (with whom I share half my genotype) by more than twice that cost, then the net effect on my inclusive fitness is positive. The probability that my genotype will be represented in future generations is increased because the benefit to my relatives (or to the fraction of my genotype that they share) exceeds the cost, so the net result is an increase in my inclusive fitness.

The above explanation, however, gives rise to an interesting prediction: kin selection favors altruism only if close relatives are more likely to benefit from altruistic acts than more distant relatives or nonrelatives. Studies of many species have confirmed this prediction: the beneficiaries of altruism are often close relatives of the altruist, and the frequency of altruistic acts varies in almost direct proportion to the degree of the relationship. In the Florida scrub jay, the offspring of the previous year are not mature enough to mate. Instead, they help the nestlings who are their own brothers and sisters (Figure 8.6). In doing so they contribute to the survival (and thus the fitness) of these near relatives, who share a portion of their own genotype. Ground squirrels emit an alarm call when a predator is spotted. The alarm call decreases the fitness of the caller, but increases inclusive fitness by warning the caller's kin (see Figure 8.6).

For kin selection to operate, it is not necessary that the altruist be able to distinguish relatives from nonrelatives; it is only necessary that close relatives are more likely to benefit from altruistic acts. Although kin selection does not *require* kin recognition, can animals assess the degree to which other organisms are related to themselves? In some social animals, individual recognition (based on growing up together) can be used. Other species, including mice, use odor cues. The odor of each animal is genetically influenced, and the diversity of genotypes results in a diversity of odors. Mice can detect by odor which individuals are the most closely related to themselves. Mice seem to use odor-based kin-recognition when they establish communal nests. Several females share a nest and nurse each other's offspring. A mother's inclusive fitness is maximized if she nurses only offspring that are closely related to her. Females who share a communal nest are usually related genetically.

Another explanation of altruism, based on game theory, is described on the book's Web site, under Resources: Reciprocal altruism.

Figure 8.6

Two examples of altruism favored by kin selection.

A one-year-old Florida scrub jay (right) assists in the care and feeding of its younger siblings.

A female ground squirrel (*Spermophilus beldingi*) stands guard against predators. If a coyote or hawk is spotted, the guard female emits an alarm call that attracts the predator and thus endangers the caller, but the alarm also warns the caller's next of kin and thus raises her inclusive fitness.

The evolution of eusociality

The highest degree of social cooperation is developed among the truly social, or **eusocial**, insects. Eusocial species are recognized by the possession of three characteristics: strictly defined subgroups called castes, cooperative care of the eggs and young larvae (cooperative brood care), and an overlap between generations. Eusociality occurs in the insect order Isoptera (termites) and particularly often in the order Hymenoptera (bees, wasps, and ants). A few bird species and one mammal (the naked mole rat, a burrowing type of rodent) approach eusociality in having 'helper' individuals who assist in caring for their siblings, but these helpers do not form a distinct caste.

Humans show some of the characteristics of eusocial behavior, but not to the extent shown by the eusocial insects. Humans have overlapping generations but often do not cooperatively care for their young and do not usually form castes. Assisting in the care of someone else's children (**alloparental behavior**) is, according to the American sociobiologist Sarah Hrdy, an important characteristic of our species (see Figure 1.4, p. 17), but the eusocial insects far surpass us in this behavior.

Eusociality in termites. Termites (order Isoptera) are a group of insects related to the cockroaches. Termite colonies are founded by a single reproductive pair called the 'king' and 'queen.' The queen grows many times larger than the other colony members, her offspring, who continually feed her and raise her additional offspring.

An important termite characteristic central to the understanding of their evolution is their chewing and digesting of wood. Termites can digest wood only with the help of symbiotic microorganisms (mostly flagellated unicellular organisms of the kingdom Protista) that live in their guts. The termites transmit these protists through regurgitated food passed to other members of the colony. This habit not only spreads the wood-digesting microorganisms throughout the colony, it also feeds those members of the colony, such as the queen, who do not feed themselves. In the evolution of eusociality among termites, the chewing of wood led to selection favoring the retention and transfer of the symbiotic microorganisms.

Along with food and microorganisms, termites also pass chemical secretions that communicate social information to other colony members. Chemicals that are used for communication are called **pheromones**. Some of these chemicals are similar to hormones, except that they are secreted by one individual and produce their effects in other individuals. One such chemical, secreted by the termite queen, inhibits most other individuals in the colony from becoming reproductively mature. Thus, the passing of food and symbiotic microorganisms throughout the colony was a precondition that probably led to the evolution of termite eusociality by providing each queen with the means to chemically control the reproduction of other individuals. These other individuals form several types of sterile castes, depending on the species. Many of the nonreproductive individuals are workers that feed the queen, tend her larvae, and enlarge the colony's living space. Other individuals serve as soldiers, fending off potential enemies that pose threats to the colony.

CONNECTIONS
CHAPTER 1

At seasonally timed intervals, winged reproductive individuals of both sexes are produced; these winged individuals emerge from the colony all at once and embark on nuptial flights during which mating takes place. Newly mated pairs become the founders of new colonies. Meanwhile, the original colony persists for the lifetime of the queen, a period of some 10–12 years.

Eusociality in the Hymenoptera. The insect order Hymenoptera (bees, wasps, and ants) has a much larger number of social species, of which an estimated 12,000 are species of ants. The American evolutionary biologist E.O. Wilson, considered to be the founder of modern sociobiology, is a specialist on ants. He has estimated that eusociality has evolved among the Hymenoptera as many as a dozen times and perhaps more. Why has eusociality evolved so many times in this one insect order, and so seldom in other animals? The clue seems to be found in hymenopteran sex determination and in its effects on inclusive fitness.

The social Hymenoptera have a unique form of sex determination (called **haplodiploidy**). Eggs that are unfertilized, and therefore haploid, nonetheless develop, but all develop into males. Eggs that are fertilized, and therefore diploid, all develop into females. Reproduction is sexual, but each reproductive female mates only once, for life, with a single male. All cells in the male are haploid, and he contributes the same haploid set of chromosomes to all his offspring. The daughters therefore share all the same alleles from their father. (By contrast, in the more usual form of sexual reproduction found in animals like ourselves, males are diploid and their haploid gametes do not all carry the same alleles; their children do not share all their father's alleles, but each gets a different assortment.)

In both forms of sexual reproduction, each female is diploid, and her gametes carry different alleles following meiosis. Her daughters each get half of her chromosomes, and half of her alleles, but each daughter gets a different sample. For each chromosome that a female receives from her mother, there is a 50:50 chance that her sister will receive the same maternal chromosome. Thus, on average, a female shares half of her maternal chromosomes, and half of her maternal alleles, with each sister (Figure 8.7). As a result, two sisters share all of the alleles from the half of their chromosomes obtained from their father, plus half of the alleles from the half of their chromosomes obtained from their mother. Sisters therefore share 1/2 + 1/4 = 3/4 or 75% of their alleles, on average (see Figure 8.7). A female, however, only shares half of her alleles with her mother or her daughters. By neglecting her own daughters (who share only one-half of her genotype) and raising her sisters instead (who share three-fourths of her genotype, on average), she is increasing her genetic fitness. For this reason, sex determination by haplodiploidy favors the evolution of eusociality in the Hymenoptera because most females can gain greater inclusive fitness by becoming sterile workers and by helping their mother (the queen) to raise her offspring (their sisters) than by raising offspring of their own. Ancestral Hymenoptera were solitary (and many solitary species still exist), but eusociality has evolved repeatedly and independently in this group of insects (Figure 8.8).

The queen bee or wasp usually secretes pheromones that inhibit the sexual development of other females in the colony. Other mechanisms determine which larvae develop into queens and which into sterile workers. For example, future queens are fed a nutritious 'royal jelly' that contains both nutrients and chemicals that stimulate their reproductive development. Also, whenever new queens emerge, one of them (usually the one emerging first) stings the others to death and thus emerges as the undisputed queen.

Figure 8.7

Haplodiploidy in a species with a haploid chromosome number of 2. Notice the four females shown with shaded borders in the first generation. Each of these females shares between 50% and 100% of her chromosomes with the others, who are all her sisters. On average, each of these females shares 3/4 of her chromosomes ($R = 3/4$) with her sisters but only 1/2 of her chromosomes ($R = 1/2$) with her own daughters (the females in the second generation).

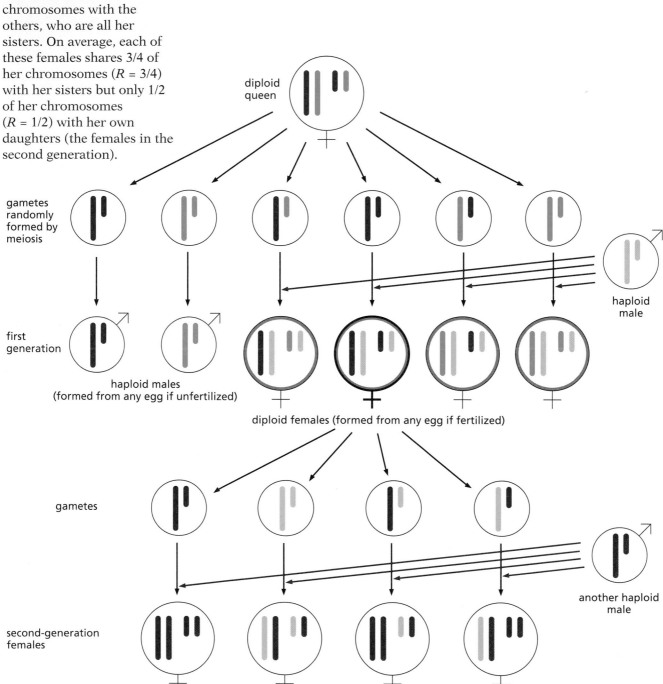

Most of the social behavior of eusocial insects is under instinctive control; in fact, the eusocial insects represent the highest complexity that instinctive behavior has ever reached. Antisocial behavior (meaning behavior that decreases the fitness of others) does not exist in these societies because antisocial individuals are quickly eliminated.

Figure 8.8

Eusocial insects.

A colony of ants

Honeybees swarming

THOUGHT QUESTIONS

1 Are the behaviors of individuals within a species more alike than the behaviors of individuals from different species?

2 As noted in this section, individuals share, on average, half of their genotype with

their siblings. Refer to the discussion of meiosis in Chapter 2 (pp. 42–44) and explain why this is so.

3 Why should humans be interested in the social behavior of birds, frogs, or insects?

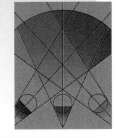

Reproductive Strategies Can Alter Fitness

Natural selection results in some genotypes leaving more copies of themselves in subsequent generations than other genotypes do. The manner in which these copies are produced can be called a reproductive strategy. **Reproductive strategies** include such features as the manner of reproduction (laying eggs or bearing live young), the litter size or number of eggs laid, the presence or absence of parental care, the presence or absence of sexual recombination, and, if there is a mating system, whether it is predominantly monogamous, polygamous, or promiscuous. Sexual behavior is an important part of reproductive strategy in many social species.

Asexual versus sexual reproduction

Reproduction of organisms can be either sexual or asexual. Many species (including all mammals and birds) are exclusively sexual, while bacteria are predominantly asexual, and certain other species (yeasts, aphids, and a variety of plants) can reproduce either way depending on the circumstances.

Asexual reproduction may be defined as reproduction without any genetic recombination. This type of reproduction has certain advantages over sexual reproduction. Within a group of organisms that includes species reproducing sexually and also species reproducing asexually, those reproducing asexually can generally do so faster and with lower energy costs. Asexual reproduction allows reproduction at an earlier age and a smaller body size, and it also avoids the costs associated with sexual reproduction. For an individual that discovers a large but finite or perishable supply of food or some other resource, asexual reproduction is an advantage because more offspring, and more generations of offspring, can be produced in a minimum of time, without any need of finding or courting a mate. Moreover, the numerous offspring are genetically identical to the original parent or founder, ensuring that favorable combinations of genes are perpetuated exactly. (The genetically identical asexual offspring of a single individual are referred to as a **clone**.)

In contrast, **sexual reproduction**, reproduction with genetic recombination, is more costly than asexual reproduction because of the time and energy expended in seeking, finding, and courting a mate, and in transferring or accepting sperm. Energy is also used in synthesizing structures that attract mates, and in the mating act itself. Mate attraction also makes a sexually reproducing individual more visible to predators, exposing that individual to increased risks. A major genetic cost is that of passing only 50% of one's alleles to each child, giving up the other half (during meiosis) to be replaced by those from one's mate. In view of these costs, it is amazing that sexual reproduction would be so widespread in both the animal and plant kingdoms. Sexual reproduction must have some great advantage.

The great advantage of sexual reproduction is genetic variety among the offspring. In the most common type of sexual reproduction, males produce sperm cells that contain the haploid number of chromosomes, and females produce eggs that are also haploid. Each sex cell (gamete) produced by an individual carries only half of that individual's genetic material, formed during meiosis by a random choice of one chromosome from each pair (see Chapter 2, pp. 42–44). Because each gamete-forming cell undergoes meiosis independently, the chromosomal choices are different each time, and the gametes thus vary among themselves. The combination of gametes with the gametes of the opposite sex is also random. The result is that *sexually produced offspring vary greatly in all genetically controlled traits*. This may be a disadvantage if tomorrow's (or next year's) conditions are identical to today's—and unchanging conditions do in fact favor asexual reproduction. However, if tomorrow's (or

next year's) environmental conditions are uncertain, then the best hedge against this type of uncertainty is to produce many *different kinds* of offspring, and sexual reproduction achieves this very efficiently. What we have just said pertains not only to the common forms of sexual reproduction, but also to other forms, such as the special kind found among the social insects: however much they differ in detail, all forms of sexual reproduction are characterized by greater variation among offspring than any form of asexual reproduction.

The hypothesis that sexual reproduction derives its adaptive advantage from the greater variation among the resultant offspring receives support from the study of certain insect species (such as aphids, also called plant lice) that are capable of producing either sexual or asexual generations. During the summer, when maturing crops offer dependable food supplies for several months in a row, these insects produce several asexual generations in quick succession. At the end of the season, however, these insects reproduce sexually, and the sexually produced eggs overwinter. When they emerge in the following spring, diverse genotypes of offspring find their way to the new stands of plants under new weather conditions, neither of which could have been predicted during the previous fall when the eggs were laid. Many genotypes perish, but a few survive and prosper by reproducing asexually during the new season. The important point is that the genotype that proves most fit in the spring is not necessarily the same one that produced successful offspring asexually during the preceding year. *Sexual reproduction is favored whenever future conditions are uncertain*, and experiments confirm that individuals laying eggs in the fall have more surviving offspring the next year if they reproduce sexually than if they reproduce asexually.

Differences between the sexes

In sexually reproducing species, the two sexes are not necessarily different. Some species, such as the green alga *Chlamydomonas* (kingdom Plantae, phylum Chlorophyta), have male and female haploid gametes that look identical, a condition called **isogamy** (meaning 'equal gametes'). But a pair of gametes may be at an advantage if at least one of them is capable of finding the other over greater distances, thus allowing more mating or mating from a wider choice of potential mates. In some cases there may also be an advantage for the resultant fertilized egg (zygote) if it possesses stored food or protective layers, each of which increases bulk. The advantages of motility and of large size can best be balanced if one of the gametes is large and the other is small and motile, a condition called **anisogamy** (meaning 'unequal gametes'). The larger, nonmotile gamete is called an **egg**, and the smaller, motile gamete is known as a **sperm**.

Males and females. Although it is possible that different-sized gametes could be produced by identical organisms, this does not usually happen. Instead, reproductive anatomy and behavior differ between the sexes in most species of animals and plants, and most of the familiar differences between males and females are explained within evolutionary theory as the consequences of anisogamy. Selection among sperm-producers, or **males**, favors the release of numerous gametes, each of which is of minimal size and maximum motility. The minimal size means that each individual sperm represents a trivial investment (in energy and materials) for the male that produces it. A male can easily produce thousands (or millions) of sperm, and he can compensate for a poor choice of mates by mating more often. Competition among males usually favors whichever one can produce the most gametes that combine successfully with the most eggs.

Selection among egg-producers, or **females**, generally favors a larger investment of parental resources, such as stored food, in each egg. Among numerous eggs, those with the most stored food or the strongest protective layers generally have the best chance of survival. This necessarily limits the number of eggs that a female can produce, and places a premium on egg *quality* rather than number.

Parental investment. There are further consequences of **parental investment**. If male parental investment is low in both energy and material costs, the price that a male pays for mating with a given female is very small. If their offspring are low in fitness (i.e., they have a small chance of survival), the male can simply mate again with other females. Low parental investment produces non-discriminating males.

Females, having fewer eggs, can produce more surviving offspring if they invest more care and protection in each one. This is especially true in mammalian females, which devote much time and energy to gestation, intrauterine feeding, and lactation. Because a female's parental investment is high, each of her offspring is more costly to her. If she mates with a low-fitness male, she greatly reduces her own fitness. She cannot simply make up for a poor choice by mating again because her capacity for repeated mating is generally limited by the large investment she must make in each of her offspring. Females thus have more at stake in each mating, and stand to gain more by choosing a mate who will father offspring who are more fit, or to lose more by choosing a mate who will father offspring who are less fit. In social species in which males vary in social status, a female can generally maximize her fitness by mating with a high-status male who can provide her and her offspring with a greater degree of protection. Selection thus favors females who are more discriminating in their mate choices, both as to social status and genetic fitness.

Mating systems

There are many types of **mating systems** known within sexual reproduction. In species in which care of the young requires the cooperation of both parents, parental investment tends to be high for both sexes. These conditions favor **monogamy**, or mating between one male and one female (Figure 8.9). If the rearing of their common offspring takes a long time, the formation of a permanent pair-bond (i.e., mating for life) is favored.

Another common situation is one in which only females care for the young, but males provide protection to both female and offspring. This situation generally favors the development of one form of **polygyny**, a mating system in which one male mates with several females. Many mammals form polygynous mating units; for example, male fur seals come ashore during the breeding season and establish territories, which they defend against other males (see Figure 8.9). The strongest male defends the best territory, an area where females can rear their pups within easy reach of the sea. Females are attracted to the territory (rather than to the male himself) and mate with males that hold territories. Males who lose territorial contests go off in search of other suitable territories. If they find none, they will not mate during that season.

Red deer, bighorn sheep, and certain other species of hoofed mammals (ungulates) form polygynous mating units in a different way. Adult males establish a dominance hierarchy, either through ritualized threat displays or through actual fighting. The females are most attracted to the dominant male. The dominant male gathers together as many females as he can, forming a 'harem.' Male social status in harem-forming species often correlates with fighting ability and with the size of horns, antlers, or other conspicuous features, so females can see at a glance which male is dominant. Females can ensure better protection against predators for themselves and their offspring by following and mating with the dominant male. Any genetic component of the characteristics correlated with social dominance is passed on to their offspring, who will thus inherit such characteristics as fighting ability, size, and the size of horns or other weapons. Nondominant (subordinate) males have fewer mating opportunities than dominant males. Many subordinates are simply young adults who will get their turn to become dominant the following year.

In addition to monogamy and polygyny, other types of mating systems include **polyandry**, an uncommon type in which one female mates with multiple males. The term 'polygamy' is sometimes used to include both polygyny and polyandry. Another mating system is **promiscuity**, in which members of both sexes mate with multiple partners and generally avoid forming permanent partnerships (see Figure 8.9).

We have seen that reproductive behaviors, as well as other social behaviors, vary greatly between species. We have primarily looked at examples from the animal kingdom, but even bacteria show some social behavior. When bacteria grow in groups, they make different proteins than when they grow singly. Bacteria in groups can, for example, influence one another in the timing of their cell cycles and metabolic events. This type of social behavior is called quorum sensing. Plants show other social behaviors, including the secretion of chemicals that inhibit the nearby growth of other individuals. Plants also show some ability for kin recognition in that some plants can assess the 'match' between proteins or other molecules derived from the male pollen and the female stigma. In the next section we look more closely at social behaviors and reproductive strategies in primates, including humans.

Figure 8.9

Examples of different mating systems.

Monogamy: a family of Canada geese

Polygyny: a large male fur seal surrounded by females

Promiscuity: baboons

THOUGHT QUESTIONS

1 What are the biological definitions of 'male' and 'female?' How do these compare with cultural definitions of the same words? Do 'male' and 'female' (or 'masculine' and 'feminine') mean different things in different cultures, or at different times in history?

2 In humans and other species, males tend to have greater muscle mass than females.

Under what conditions would you expect anatomical differences (in muscle mass, antlers, or size) to evolve? Is there a reason why such differences would be favored by natural selection?

3 Does the difference in gamete size in humans and other mammals tell us anything about our sexual behavior? Are human males 'destined' to be promiscuous?

Primate Sociobiology Presents Added Complexities

Primates are an order of mammals that includes monkeys, apes, lemurs, tarsiers, and humans. Primates are all extremely social animals. They are so interested in interacting with other members of their species that they sometimes go to great lengths to maintain an interaction or merely to look. We can devise an experiment to test this hypothesis. Set up a partition that completely obstructs the view through a window, and provide a lever that raises the partition for a predetermined length of time, affording a temporary view through the window. Most primates will then spend hours repeatedly pressing the lever and looking through the window, then going back to press the lever again for another view almost as soon as the partition falls. The rate of lever-pressing is higher if the window affords a view of moving objects (such as electric trains) rather than non-moving objects (such as furniture). The rate is higher still if an actively moving animal is visible through the window, and it is highest of all if the view includes other primates of the same species. Is it any wonder that people also spend hours looking through windows at the world around them, especially when other people's movements and interactions are visible? And, in addition to the live action visible through a real window, television provides a virtual window through which we can watch people interact even more.

Primate social behavior and its development

Social skills in both human and nonhuman primates depend strongly on learning that takes place early in life. All parents and future parents should be aware of the paramount importance of early childhood experiences in all later aspects of human life. The lasting importance of learning that takes place very early in life is one of many important areas in which humans and other primates are very similar. Because of the great similarity among all primates, findings based on experiments with nonhuman primates are often used to gain insights into human behavior, although findings in one species should not be uncritically applied from one species to another. However, when we compare the behaviors of different primate species, we usually find many more similarities than differences.

Early development of behavior. As we stated earlier, the standard test for an instinct requires that an animal be raised in isolation. Raising a primate in isolation, however, results in abnormal behavior resembling that of abused children. Sigmund Freud claimed that a baby's attachment to its mother is based initially on its need for nutrition. To test this hypothesis, Harry Harlow of the Wisconsin Primate Research Center raised infant rhesus monkeys (*Macaca mulatta*) with various forms of care but with no live mothers. Instead, dummy 'mothers' with colorful wooden 'heads' held baby bottles mounted in wire frames. Although the infant monkeys drank the milk, their behavior grew progressively more abnormal with time. The infants frequently cowered in the corner and were easily frightened. They formed no emotional attachments and seemed to ignore their 'mothers' except when they were hungry. Freud's hypothesis was falsified because the young monkeys failed to treat the wire model as a mother. Something more than the milk supply was needed for infants to form a bond with their mothers.

Harlow noticed that young monkeys liked the feel of terrycloth towels. He tried wrapping the wire mothers in a few layers of terrycloth to make them soft and clingy. The infant monkeys enjoyed clinging to these cloth-covered dummies, and the terrycloth retained the infant's own body heat during the periods of clinging. Harlow raised infant monkeys with two dummy 'mothers,' with and without terrycloth, one of them holding a baby bottle. Young monkeys spent countless hours clinging to the terrycloth 'mother,' regardless of which dummy held the bottle (Figure 8.10). When exposed to a novel or frightening stimulus, the infant monkeys would run to their terrycloth 'mothers' to cling for reassurance. After clinging for a while, the young monkeys were sufficiently reassured that they became brave enough to inspect the previously frightening stimulus. In many cases, their curiosity finally overcame their fear. Wire dummies, in contrast, never provided the behavior-changing reassurance.

Development of adult behavior. Rhesus monkeys raised with terrycloth 'mothers' seem to function normally until they become sexually mature, at which time behavioral deficits do appear. A normally raised male rhesus monkey 'mounts' a female during her reproductive cycle if she 'presents' to him (Figure 8.11), but the motherless males never mounted any females and seemed not to know how to behave in this situation. Motherless females did come into their reproductive cycles (their genitals swelled up and became bright pink), but they never

Figure 8.10

An infant rhesus monkey raised by two dummy 'mothers,' one made of bare wire and the other covered with soft terrycloth. Note that the infant maintains contact with the terrycloth 'mother,' even while nursing from the wire dummy.

'presented' to any test males, and they consistently rejected all sexual advances. A few such females were artificially inseminated under anaesthesia and became pregnant. When their babies were born, they showed no signs of maternal behavior, such as picking up their infants and holding them to the breast. Instead, they either ignored or rejected their infants, in some cases so forcibly that the infants had to be removed for their own safety. Sexual behavior and maternal behavior had never been learned in these monkeys, even though their behavior had seemed normal up to the time of sexual maturity. Adult social behavior has very strong learned components in rhesus monkeys and in other higher primate species, too.

Harlow continued his experiments, seeking to pinpoint what the motherless monkeys were failing to learn from the terrycloth dummies. Could the young monkeys receive a proper upbringing without a live mother? What conditions were minimally necessary? Remembering that wild juvenile primates associate with one another in play groups, Harlow let some of the young, motherless monkeys play together. He found that motherless monkeys who had opportunities both to cling to a terrycloth dummy and to play with one another developed normal adult social behaviors. By varying the length of the play period, Harlow was able to show that as little as half an hour of play per day was adequate to ensure that young monkeys would acquire normal adult behaviors. Harlow concluded that instincts were not sufficient to produce the proper sexual behavior or maternal behavior in these monkeys, but that a youthful period of social learning was also required.

Rough and tumble play. Most play in primates is "rough and tumble" play in which there is frequent and repeated body contact, including pushing, pulling, and climbing—just watch young children in a schoolyard to see examples. Primate play also includes a good deal of chasing and dodging, usually followed by more rough and tumble play. Although rough and tumble play is neither sexual nor maternal, it seems to teach many lessons, such as how to handle and perhaps restrain other individuals without hurting them. Hurting another individual, whether accidentally or not, brings

Figure 8.11

Normal mating behavior in rhesus monkeys.

an adult female 'presenting'   mounting and copulation

squeals of pain, generally causing adults to intervene and break up the activity. Play also teaches taking turns at different roles: pursuer and pursued, restrainer and restrained, climber and support, etc. In the context of play, the players learn how strong or weak other individuals are, and how much rough play each will tolerate. These lessons are later refined into dominance and submission relationships with other individuals and into sexual behaviors such as those in which male monkeys mount females. Mounting behavior arises during rough and tumble play, without regard to the sex of either individual; only after sexual maturity does it take on an explicitly sexual meaning. The defense and protection of smaller individuals, including picking them up and delicately cradling them, is also learned in play. In large, mixed-age social groups, there is usually an opportunity for subadult animals to practice the behaviors related to child care.

There are parallels in human behavior. Children learn many lessons in play, including cooperation, turn-taking, role-playing, counting and score-keeping, setting and following rules, and settling arguments and disputes. They also learn a good deal about each other's personalities: who plays fair, who cheats, who is a bully, who cries if they do not get their way, and so forth. Children often imitate adult roles in play, practicing many of the skills that they see adults using and that they may themselves use later in life: hunting, digging, child care, food preparation, useful and artistic crafts, and so on. Abused children and those deprived of the opportunities of exploratory and rough and tumble play with other children often fail to develop the proper adult social behaviors, including both marital behavior (which is much more than just sexual) and parental care.

Social organization. Most primates are extremely social, but the size and complexity of social groupings vary greatly. Closely related to the species that Harry Harlow studied are the baboons, monkeys of the genus *Papio*. *Papio hamadryas* is a harem-forming, polygynous species that lives in the rocky highlands of Ethiopia. Male hamadryas baboons are often aggressive fighters, and were revered by the ancient Egyptians for this trait. The other baboons, *Papio cynocephalus* and related species, live on the open, grassy savannas of Africa. The savanna baboons all share a form of social organization different from that of the hamadryas. In the wild, savanna baboons hardly ever fight. They express dominance largely through gestures such as staring at an opponent, showing their teeth, or slapping the ground (Figure 8.12). We can study dominance by observing pairwise encounters (between two individuals at a time) and noting which baboon more often gets what it wants. Dominance status generally follows size and fighting strength, although it is rarely contested and outright fighting is rare. A lengthier description of baboon social organization can be found on our Web site, under Resources: Baboons.

Grooming. Baboons, like other monkeys, are forever grooming one another—picking burrs and parasites from each other's fur (see Figure 8.12). As a gesture of friendliness, grooming is generally reciprocated, with groomer and recipient taking turns. Grooming is a pleasurable activity, and it helps form many social bonds. Infants and juveniles are often groomed by their mothers. Females who are not yet mothers themselves often practice grooming behavior and infant care. This 'mother-in-training' behavior, called 'allomothering' or 'aunt behavior,' is very important in many primate species, including humans. Through such experiences, older juvenile primates of both sexes learn the behavior patterns essential to parenting, while younger primates gain social experiences, learning experiences, and even substitute parents in the event of the parent's death or temporary removal.

Human examples of allomothering include holding and feeding other people's children, playing with children, and, of course, baby-sitting.

Figure 8.12

Examples of social behavior in monkeys.

Reproductive strategies among primates

Studies of primate behavior before the late 1960s were in most cases written by male scientists and tended to emphasize male behavior and dominance relations among males. Males were often described as making choices, while females were often depicted as either passive or 'coy.' Beginning with the early work of primatologists Jeanne Altmann (American), Phyllis Jay (American), and Jane Goodall (English), relationships among female primates began to receive equal or greater attention. The primatologists of the subsequent generation conducted many important new studies that focused on the social behavior of female primates.

One primatologist who has changed our views of primate sexual biology is Sarah Hrdy, whose sociobiology is influ-

Baboons grooming one another

Threat display of a male hamadryas baboon, showing his large canine teeth

Grooming behavior in rhesus monkeys

enced by a feminist outlook. Female primates, according to Hrdy, are much more sophisticated than previous researchers had imagined. Whereas the adult males use rather obvious means to maximize their inclusive fitness, Hrdy discovered that the means used by females were considerably more subtle and usually involved influencing the behavior of the males.

In her work on langur monkeys in India, Hrdy discovered the important ways in which female monkeys, although subordinate in power and strength to males, nevertheless managed to influence male reproductive choices and male social behavior to the female's own advantage. Male primates differ from one another in the number of offspring that they leave, and female primates frequently modify what males must do to achieve reproductive success. Female primates can often maximize their own reproductive success by the ways in which they influence male social behavior. Hrdy identified at least five ways in which female primates can maximize their reproductive fitness:

1. by choosing their mates,
2. by influencing males to support and protect them,
3. by competing with other females for resources,
4. by cooperating with other females (usually close relatives),
5. by increased efficiency in daily activities such as locomotion and obtaining food.

Through the work of these primatologists, we now know that females make important choices of their own and solicit male attention for a variety of reasons, showing that it often pays for them to be flirtatious rather than coy. For example, females of many species can mate with males who are not their usual partners, and they have often been observed to mate at times when they were already pregnant or otherwise unable to produce new offspring. Males can generally increase their reproductive fitness by mating with as many females as they can, indiscriminately. The optimal behavior for a female, however, depends on her own fitness and social status, as well as that of her possible mates. If a female is of high status herself, and is mated to a high-status male, then she has nothing to gain from mating with a lower-ranking male. In contrast, a female of low status, or one mated to a low-status male, could potentially increase her fitness by mating additionally with a high-status male. If he sires one of her offspring, then she has produced a higher-status offspring and raised her own fitness as a result. That is because the offspring of higher-status males have more opportunities to mate; therefore, females can maximize their fitness (leave more grandchildren) if they raise the offspring of high-ranking males. Moreover, even if their mating produces no offspring at all, the high-status male who has mated with her will maximize *his* fitness by protecting any female that he has mated with, as well as her subsequent offspring, because he would be operating under the assumption that these offspring might be his. Thus, females can gain important advantages from liaisons with high-ranking males, even at times when ovulation has not occurred and when subsequent pregnancy and childbirth are not possible.

Hrdy also discovered that male primates are sometimes infanticidal, and that female willingness to mate with powerful males was sometimes a strategy to discourage their infanticidal tendencies. Infanticide may occur among certain primate species whenever a new dominant male takes over a group. The new male can increase his fitness if he kills infants that are not his, especially if their mothers are lactating. Lactation inhibits the female reproductive cycle in most mammalian species; infanticide by the male causes lactation to end. Females enter estrus and the male gains access sooner to reproductive females. Once he has mated and produced offspring, however, the male will maximize his fitness if he defends all his mates and their offspring.

One of the many consequences of primate reproductive strategies is a difference between the sexes in how they pay attention to social rankings. Males in socially ranked species must pay attention to their own rank and status—they must remember who has ever threatened them or been intimidated by them. Females, however, must know much more, because each female must not only know her own status, but also that of every male in the group. In order to know whether one potential mate ranks higher than another, she must pay attention to *all* the social interactions among the males. In social species, females therefore generally take more interest than males in knowing about the social interactions of all other members of the group and in learning the status of all the males. Those who are better at paying attention to male–male interactions and correctly judging each male's social status and genetic fitness are at a selective advantage because they are better able to maximize their fitness by their behavior toward these males.

Both Hrdy and Jane Goodall have observed several instances in which competition between females produced outright hostility, even infanticide. Arguing from a sociobiological perspective, Hrdy explained that competition among unrelated females should be expected when their genetic self-interests are in conflict. A universal sisterhood, in which all females cooperate as a unit, would therefore never evolve. In evolutionary terms, such a sisterhood would not be a stable strategy, because an individual female would always be able to 'cheat' by refusing to cooperate, and by doing so she would raise her fitness and be favored by natural selection. Because evolution would never be expected to produce cooperative sisterhoods among unrelated females, Hrdy suggests that women who share her desire for such cooperative sisterhoods should strive to create them socially. Humans are not prisoners of biological destiny and are able to create social groupings and social behaviors that have not evolved.

Some examples of human behaviors

Much interest has focused on certain human behaviors and on the extent to which these behaviors are learned or inherited. Behaviors disapproved of by large segments of society have generally attracted the most attention. People who wish to change behavior that has a strong learned component generally seek to find *how* it is learned and how an alternative form of behavior can be learned instead. If the behavior has a strong inherited component, it will be more difficult to bring about change through education. Other forms of intervention that might be more appropriate in such cases include trying to identify any genes involved in the behavior. However, most of the behaviors of interest are complex and are probably influenced by many genes, making it harder to identify any of them or to modify them in any meaningful way.

We also hasten to add that political motives are often suspected of those who write about human behavior or who try to apply what we know of other species to the understanding of human behavior. History has taught us that various oppressors have claimed scientific support to justify slavery, political repression, and genocide. (The Nazis come to mind as the most obvious example, but there have been many others.) There is thus reason for caution, but sometimes the reactions have become uncivil and tempers have run high. Some biologists who intended no harm to anyone have been yelled at by disapproving crowds and have had eggs thrown at them. There are people, in other words, who fear that scientific study may again be used to justify unspeakable horror.

Aggression. Konrad Lorenz was a German scientist who studied animal behavior. He first won recognition (including a Nobel Prize) for his studies on imprinting, a form of learning that occurs early in life. In his later years, Lorenz wrote several controversial books in which he claimed that many human behaviors are instinctive. For example, in his book *On Aggression*, Lorenz claimed that aggression is largely instinctive in humans as well as other animals. As evidence, Lorenz argued first that aggression is widespread in many animal species and in various human societies. Second, he argued that the facial expressions and other gestures that accompany aggression and aggressive threats are similar in humans and animals and are also similar across many human societies.

Other scientists, however, have marshalled considerable evidence that aggression in humans has strong learned components.

1. Aggression takes many different forms in different societies, which use different weapons and different fighting traditions. If aggression were entirely instinctive, one would not expect it to be so variable.

2. Aggression is more prevalent in those societies that encourage it, and it generally takes the form that the society encourages. In the many societies that encourage aggression only in males or only in certain age groups, it occurs primarily in the groups in which it is encouraged. In societies that discourage aggression, it is much less common.

3. Within any society, some individuals are more aggressive than others. Individuals trained to be aggressive become aggressive, while most people raised to be less aggressive become less aggressive. We would not expect such large individual differences if aggression were inherited.

4. When aggressive behavior is desired, as in the military or in sports such as boxing and judo, it must be taught and practiced frequently.

Two special forms of aggression that have received a lot of attention are child abuse and rape. Studies examining criminal records in various countries all over the world have shown that child abuse among humans follows the same patterns as infanticide in other primate species. In particular, stepfathers (who are genetically unrelated to the children who live with them) are up to 100 times more likely to abuse or kill the children in their care than are genetic fathers.

Feminist writers such as Susan Brownmiller have portrayed rape, or forcible sexual intercourse, as an attempt by the rapist to dominate and control his victim, and thus as a crime of violence rather than of sex. Against this idea, sociobiologists Randy Thornhill and Craig Palmer argue that rape is very much about sex. They use statistical records from rape crisis centers to show that victims are most often in the prime reproductive age range and that married rape victims feel more heavily traumatized than unmarried ones. They claim that a predisposition to rape persists because rape does occasionally produce children who perpetuate the genes of the rapist. Therefore, these authors argue, rape is natural, though they hasten to add that it is still reprehensible behavior. Their hypothesis does not, however, explain why the overwhelming majority of men are *not* rapists.

Many studies show that most women prefer as mates men who are good-looking, healthy, strong, skillful, kind, respected by others, and high in social status and wealth. Thornhill and Palmer say that "men might resort to rape when they are socially disenfranchised, and thus unable to gain access to women through looks, wealth, or status." According to this hypothesis, the men who become rapists can leave more offspring if they rape than if they do not, because they are usually the men that women are unlikely to choose as mates.

Barbara Ehrenreich, a critic of the Thornhill–Palmer hypothesis, emphasizes that rapists make inferior husbands and fathers, and that the children of rape are thus far less fit than other children. The mothers of these children have been traumatized, and their fathers are in most cases gone, and when present they are neither good fathers nor good role models. Compared with the men that women would choose as mates, rapists are inferior in social standing, inferior in fitness, and inferior in their ability to raise fit children. This may explain why most men are not rapists: they can produce more children, and contribute as fathers to their children's fitness, by cultivating the behaviors that women value. The children of these men and the women who choose to marry them usually attain higher social status and are more socially and psychologically equipped to enter into normal and stable relationships themselves. They tend to leave more future children and are thus far more fit than the children born of rape.

Alcoholism. Alcoholism is a complex form of behavior that seems to have both learned and inherited components. To complicate matters, there are various degrees of alcoholism, and many individuals are classified as alcoholics by some criteria and not others. However, the greatest complication arises from the heterogeneity of the disorder: alcoholism manifests itself differently in men and in women, and it may also have different characteristics in different social classes.

Recent studies show that alcoholism may in fact exist in two or more separate forms. Type I alcoholism, also called late-onset or milieu-limited, typically arises after age 25 and is common in both sexes. It is characterized by psychological or emotional dependence (or loss of control), by guilt, and by fear of further dependence. This type of alcoholism frequently responds well to treatment. By contrast, type II alcoholism, also called male-limited, early-onset, or antisocial alcoholism, typically arises during the teenage years and is common in males only. It manifests itself in alcohol-seeking behavior, in novelty-seeking or risk-taking behavior generally, and in frequent impulsive and antisocial behavior including alcohol-related fighting and arrests. This type of alcoholism responds poorly to conventional forms of treatment.

Adoption studies in Denmark, Sweden, and the United States suggest that a predisposition for type II alcoholism may be inherited. The largest study, of 1775 adoptees in Sweden, found that the rate of alcoholism among the biological sons of type II alcoholic fathers raised in families without alcoholics was nine times the rate among other adoptees, including those adopted into type II alcoholic households. Type I alcoholism, however, shows a much smaller hereditary influence and may instead be subject to strong environmental influences. Some experts suggest that type I alcoholism is still heterogeneous and should be subdivided further.

Studies on alcoholism among twins show a higher rate of concordance in identical twins than in fraternal twins, meaning that, if one twin is an alcoholic, there is greater probability that the other is also an alcoholic if the twins are identical than if the twins are fraternal. (As described in Chapter 3 (pp. 71–72), a rate of concordance is the fraction of individuals who match in a certain trait.) The concordance is greater for type II alcoholism than for type I.

CONNECTIONS
CHAPTER 3

Sexual orientation. Some people regard variations in sexual orientation, including homosexuality, as innate and unchangeable, while others view them as learned behavior patterns that are subject to change. The available evidence, which is not very extensive, was summarized and reviewed in two books by the English-born neurobiologist Simon LeVay. Some small differences were observed between the brain structures of homosexual men and heterosexual men, but many of the homosexual men in the study had died from AIDS, so it is uncertain whether these differences resulted from AIDS or pre-dated the onset of that disease. If a difference in brain structure could be demonstrated between homosexual and heterosexual men, other questions would remain to be answered: did the structural difference precede the sexual orientation, or might the structural change have resulted from some aspect of a lifestyle difference? Scientists are only just beginning to examine such questions in homosexual men; studies examining lesbian women are even rarer.

Studies have been conducted on homosexual males who have twin brothers. The rate of concordance is higher for identical twins than for fraternal twins, meaning that, if one twin is homosexual, there is a much higher probability that the other twin is also homosexual if he is an identical twin than if he is a fraternal twin. Such a result is suggestive of at least some genetic influence, but the very real methodological problems of such twin studies makes it very difficult to rule out other possible influences. The biggest shortcoming of twin studies is that the environments in which the twins are raised are never chosen at random and are usually very similar, even in cases of adoption.

THOUGHT QUESTIONS

1 To what extent can sociobiological findings on animals be extrapolated to humans? Are animal studies relevant at all to the study of human behaviors such as alcoholism or homosexuality?

2 How important are fathers in early childhood development? What important *social* skills do children learn from interacting with their mothers? With their fathers? What do children learn from watching their parents interact with one another? What happens in families in which no father is present? What happens when no mother is present?

3 Think of the many ways in which humans learn (and subsequently practice) the social skill of evaluating the social status and motives of others. How much do we learn (or what skills do we exercise and practice) from play, from small-group discussions, from gossip, from novels, or from television? Do males and females participate in these activities in the same way? Why, or why not?

Concluding Remarks

Sociobiology, the comparative study of social behavior and social groups among organisms, is a subfield within evolutionary biology. Much of social behavior is learned, but only those aspects of social behavior that are inherited are subject to natural selection and therefore to evolutionary change. Sociobiology therefore focuses on inherited behaviors or capacities, although all sociobiologists agree that learning can modify those behaviors in many species. Often it is a predisposition for a behavior, or a capacity to learn a behavior, that is inherited, not the behavior itself. Sociobiology can predict when natural selection will favor altruism, social groups of differing sizes, behavior that is stereotyped as opposed to variable, or behavior that differs between the sexes. Many such hypotheses have already withstood repeated testing.

In humans, even though some components of behavior are inherited, every behavior can also be modified by learning. Twin studies, adoption studies, cross-cultural studies, and studies of other species can all provide important clues to the understanding of human behavior patterns. Many human behaviors vary across cultures; many are also strongly influenced by early childhood experiences. One of the most effective and cost-efficient ways in which we can improve human society is to provide each and every child with a safe childhood full of experiences from which to learn.

Chapter Summary

• **Sociobiology** is the evolutionary study of **social behavior** and **social organization** among all types of organisms.

- Organisms live in social groups because it affords such advantages as group defense, help in finding and exploiting food resources, and greater reproductive opportunities.

- **Altruistic** behavior is favored if it contributes to **inclusive fitness** through **kin selection**.

- Inclusive fitness has also favored the evolution of social cooperation and **eusocial** species.

- Among **reproductive strategies**, asexual reproduction is favored by natural selection in situations in which a quickly produced series of uniform offspring are advantageous, but **sexual reproduction** is favored whenever future conditions are uncertain and a greater variety of offspring is a greater advantage.

- Among sexually reproducing species, there are many different **mating systems**, including monogamy, polyandry, polygyny, and promiscuity.

- In many species, different levels of **parental investment** favor different reproductive strategies in **females** (**egg** producers) and in **males** (**sperm** producers).

- All behavioral characteristics that have been closely studied are influenced by both genetic and environmental influences to various degrees.

- Behaviors that can be performed without the opportunity for prior learning are considered **innate**, and innate behaviors that are complex are called **instincts**.

- In animal species, behaviors related to mating and courtship are more often **instincts**, while **learning** has a stronger influence on most locomotor behaviors.

- Learned behavior is highly important among primates, especially among humans.

CONNECTIONS TO OTHER CHAPTERS

Chapter 1	Sociobiology is a good example of a paradigm.
Chapter 2	Social behavior can differ among different genotypes.
Chapter 5	Social behavior can greatly affect the fitness of each genotype and can thus alter allele frequencies. Social behavior also evolves, and the evolution of behavior is of prime interest to sociobiologists.
Chapter 9	Mating is one of the most important kinds of social behavior. Population growth is a result of mating behavior on a large scale.
Chapter 10	Access to good nutrition is one important motivating force in social behavior.
Chapter 13	Social behavior in animals happens as the result of brain activity.
Chapter 14	Drugs can alter social behavior.
Chapter 15	Social support can promote healing; stress can interfere with healing.
Chapter 16	AIDS is spread by certain social behaviors.

PRACTICE QUESTIONS

1. For each of the following human behaviors, state at least one piece of evidence pointing to an important learned component for the behavior:

   a. eating with utensils

   b. speaking English

   c. hunting

2. For each of the following behaviors, present an argument for an important innate component of the behavior:

   a. tail wagging in dogs

   b. mooing in cows

   c. smiling in humans

3. Present at least one argument supporting each of the following assertions:

   a. that piano playing ability has important learned components

   b. that piano playing ability has important innate components

4. State at least six research methods used in sociobiology.

5. Which of the following behaviors is most likely to have strong instinctive components in a wide variety of species?

   a. attracting a mate

   b. climbing a tree

   c. finding and capturing food

   d. moving about one's habitat

   e. none of the above

6. Natural selection favors the instinctive control of behavior in all of the following situations *except*:

   a. in species with short life spans

   b. in outsmarting prey

   c. in courtship displays or mating calls

   d. in escaping from sudden danger

   e. in building a nest or weaving a web

7. Which of the following does NOT fit the definition of an altruistic act?

   a. a millionaire leaves money to charity in her will

   b. a taxi driver runs through red lights to get a pregnant woman to the hospital in time to deliver her baby

   c. a man runs in front of oncoming traffic to save a small child

   d. a firefighter runs into a burning building to rescue people who may be trapped inside

8. Which of the following belongs in a different group from the rest?

   a. ants

   b. bees

   c. termites

   d. wasps

9. Which of the following situations favors sexual reproduction?

   a. microorganisms reproducing inside a human host

   b. fungi growing in a fallen tree as it decays

   c. insects or worms exploiting a large and dependable food supply

   d. insects colonizing new food supplies by laying eggs in them

10. What are the three conditions that define eusociality?

11. How is kin selection defined?

Issues

- Is the Earth overpopulated?
- How fast are human populations growing?
- Why might a population explosion be detrimental to the social good?
- What methods are available to people who want to control their reproduction?
- Can we restrict reproduction without violating people's rights?
- Can we diminish population growth and its impact?

Biological Concepts

- Population ecology (populations, population density, growth rates, carrying capacity, population regulation)
- Reproductive biology (reproductive anatomy, reproductive physiology, reproductive cycles, hormonal controls)
- Biosphere (human influences, overpopulation and its effects, resource uses, habitat alteration)

Chapter Outline

Demography Helps to Predict Future Population Size

Population growth

Malthus' analysis of population growth

Growth within limits

Demographic transition

Human Reproductive Biology Helps Us to Understand Fertility and Infertility

Reproductive anatomy and physiology

Impaired fertility

Assisted reproduction

Can We Diminish Population Growth and Its Impact?

Birth control acting before fertilization

Birth control acting after fertilization

Cultural and ethical opposition to birth control

Population control movements

The education of women

Controlling population impact

9 The Population Explosion

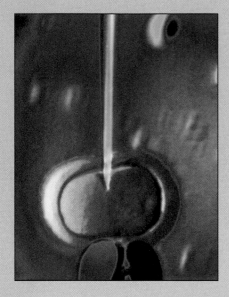

Imagine a world where people must share a room with 4 to 12 others. A room of one's own is a rare luxury. In fact, people who have any housing at all consider themselves fortunate, because so many people have none. Drinking water is in short supply each summer, and overworked sewer systems are breaking down all the time; many millions have no sewer system at all. Jobs are scarce, and well-paying jobs are almost unheard of. Beggars crowd every street, and each garbage can is searched through time after time by starving people looking for something to sustain them.

Some parts of the world already experience these conditions. Some experts predict that a future like this may be in store for all of us unless something is done soon, and on a massive scale, to control population growth.

The Earth is currently experiencing the most rapid population increase in all of human history. From 2.5 billion people in 1950, the total world population more than doubled to 6 billion in 1999 (Figure 9.1). At current rates of increase, the global population will double again in about 38 years. Each year, the world's population increases by some 94 million people.

In this chapter we consider the factors that control the size and the rate of growth of populations, including the biological controls on populations that operate independently of any conscious planning. The ecological principles that we discuss include models of population growth, limits to growth, and some of the consequences of excessive growth. Although we focus mainly on human populations, all these ecological principles apply equally well to populations of other organisms.

As with many of the other issues in this book, population growth cannot be looked at as a purely biological issue. There are political, religious, and ethical dimensions to population growth and its control, and so we consider some of these factors also. In addition, even in the face of a global population increase, there has also been increasing research on the biology of infertility, in part because an increasing number of couples who are not able to conceive their own children seek these reproductive therapies. Both global population growth and individual fertility and infertility raise ethical questions. We have seen earlier (Chapter 1) that the boundaries between the individual good and the social good are one of the subjects of ethics. Biology can inform ethical debate by assessing, for different scenarios, the biological risks to the individual and to society.

Demography Helps to Predict Future Population Size

The study of the biological factors that affect the sizes of populations is called **population ecology**; the study of human populations in numerical terms is known as **demography**. Recall that a **population** is defined as a set of potentially interbreeding individuals in a certain geographical location at a certain time (see also Chapter 5, p. 151). Our ability to understand population growth depends on our ability to make predictions, based on both population ecology and demography. Because populations are large aggregates, we need mathematical models to make these predictions and to study the factors that might influence and possibly curb population growth. These mathematical models were initially developed from the study of bacterial populations, but they pertain to the growth of all other species, including humans.

Mathematical models of population growth begin with the gathering of census data. A **census** is, at minimum, a head count of all the individuals living in a specified area, usually within recognized political boundaries. Early censuses of human populations were often inaccurate, and the opposition of the censused populations (who did not want to be taxed) only compounded the inaccuracy. Moreover, these early censuses were hardly ever repeated over the same stable boundaries at a later time, the only conditions under which population growth could be accurately assessed.

The United States implemented the first nationwide census of any modern nation in 1790, for the purpose of achieving proportional representation of the different states in Congress. (This remains one of the major functions of the U.S. census today.) In the first few decades of the nineteenth century, most European nations began to census their populations and have continued to do so at regular intervals.

Population growth

The rate of change of the size of a population depends on its birth rate, its death rate and the relation between the two. For a given time period, the **birth rate**, *B*, of a population is found by dividing the number

Figure 9.1

Graph showing the growth of the world's human population. The inset shows a street scene in Quito, Ecuador.

billions of people

6

5

4

3

2

1

0

500,000 years ago   8000 7000 6000 5000 4000 3000 2000 1000 B.C. 0 A.D. 1000 2000

← hunting and gathering phase →│← agricultural phase →│ industrial phase

of births during that period by the number of people already in the population, N:

$$B \text{ (each year)} = \frac{\text{number of births per year}}{N}$$

The birth rate, like any other rate, is always a fraction in which one number is divided by another. To illustrate with some actual numbers, if there are 10,000 people in a population ($N = 10{,}000$) and 1000 babies are born that year, then the birth rate is $B = 1000/10{,}000 = 1/10 = 0.1$ per year or 10% per year. The **death rate**, D (also called the mortality rate), is found in a similar manner to the birth rate, and represents the fraction of the population that dies within the time interval in question:

$$D = \frac{\text{number of deaths per year}}{N}$$

Notice that N, the population size, appears in the equations for both the birth rate and the death rate because both are expressed in terms of the size of the population, that is, both are fractions of N.

At the end of the year, the population will have changed because N will have increased by the number of births and decreased by the number of deaths. To express this mathematically, the above equations can be rearranged and then combined. By rearranging the equation for the birth rate, we get

$$\text{number of births per year} = B\,N$$

and by rearranging the equation for the death rate, we get

$$\text{number of deaths per year} = D\,N$$

Thus, to find the change in N per year, we add the births and subtract the deaths, to get

$$\text{change in } N \text{ per year} = B\,N - D\,N$$

We can rewrite this equation using the notation standard in mathematics for rates of change: if the change in N, called dN, is divided by the change in time, dT, we have

$$dN/dT = B\,N - D\,N$$

or

$$dN/dT = (B - D)\,N$$

The quantity $(B - D)$, is called the **growth rate**, and is symbolized as r. It is the difference between the birth rate and the death rate.

$$dN/dT = r\,N$$

A population increases if its birth rate, B, exceeds its death rate, D, and in this case r is positive. If $B = D$, then $r = 0$, and the population is stable, neither increasing nor decreasing. A population whose death rate exceeds its birth rate decreases and r is negative.

The growth rates of 130 different nations are shown in Figure 9.2A. Growth rate does not correlate with population density (numbers of people per square mile, Figure 9.2B). Some nations with high growth rates have low population densities (for example, Afghanistan and Mali), although density can be assumed to be increasing wherever the growth

rate is positive. Current data on population density and population growth rates are shown on our Web site (under Resources: Population data).

The preceding discussion assumed a 'closed' population, unaffected by people migrating into or out of it. This is a safe assumption for the world as a whole, but for a single country or region we must also add terms for both immigration (people entering) and emigration (people leaving). The United States, for example, currently has a birth rate not much higher than the death rate, but the population continues to grow because more people move to the United States each year from other countries than move away from the United States. In fact, one-fifth or more of the yearly population growth in the United States comes from immigration. To include migration rates in the calculation of r, we must write

$$r = B - D + i - m$$

where i is the rate of immigration (the number of immigrants in a year divided by the population size) and m is the similarly defined rate of emigration.

Figure 9.2

(A) Population growth rates (r) and (B) population densities around the world.

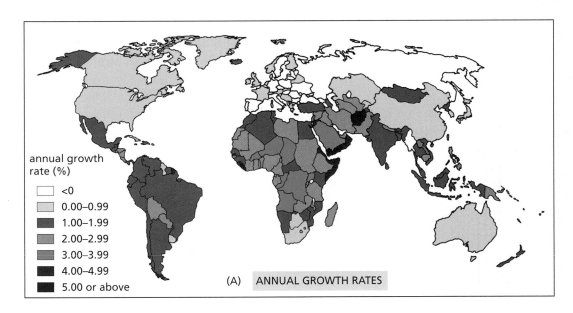

annual growth rate (%)

- <0
- $0.00–0.99$
- $1.00–1.99$
- $2.00–2.99$
- $3.00–3.99$
- $4.00–4.99$
- 5.00 or above

(A) ANNUAL GROWTH RATES

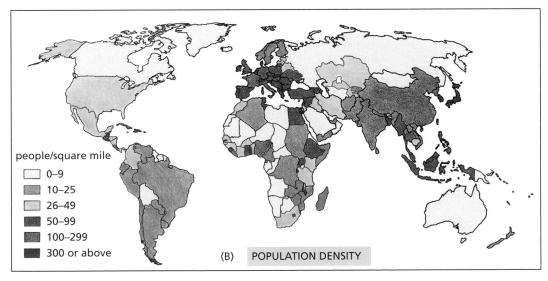

people/square mile

- $0–9$
- $10–25$
- $26–49$
- $50–99$
- $100–299$
- 300 or above

(B) POPULATION DENSITY

In most nations today, population growth results mostly from the excess of births over deaths rather than from the excess of immigration over emigration. In the 1990s the U.S. population seemed to have reached a balance between birth and death rates. More recently, however, the birth rate (about 1.8% annually, or one birth every 7 seconds) has increased to about twice the death rate (about 0.9% annually, or one death every 14 seconds), and there also continues to be an excess of immigration over emigration. The net change is a population increase of one person approximately every 11 seconds in the United States. Many factors, some of them perhaps temporary, have contributed to the recent changes. For example, many 'baby boomers' (people born between about 1945 and 1960) began having additional children many years after they first became parents, and there have also been increases in the number of people remarrying and starting second families.

Exponential (geometric) growth. Although the definition of the term r may include immigration and emigration, the overall equation for the growth rate has not changed, and remains

$$dN/dT = r\,N$$

The type of growth described by this equation is an example of geometric growth. A geometric series is one in which each number is multiplied by a constant to produce the next number in the series. For example, 1, 2, 4, 8, 16, ..., is a geometric series in which each number is multiplied by 2 to give the next number. The growth rate equation is a geometric series because each new value of N results from multiplying the previous value by $1 + r$. (Another familiar example of geometric increase is the growth of money by compound interest.) After an initial start, geometric growth always results in a rapid increase as the growth curve turns sharply upward.

Geometric growth is also called **exponential growth** because the expression for population growth can also be written as

$$N = N_0\,e^{rT}$$

where N is the population size at time T, N_0 is the initial population size (at time $T = 0$), e is the base of natural logarithms (approximately 2.71828), and r is the growth rate. In this form, the constant by which each number is multiplied (r) appears as an exponent, hence the term exponential growth. The same equation can be used to describe populations of bacteria, fish, or humans, although different time units (minutes, months, or years) may be used in each case. Graphing this equation gives a growth curve such as the blue line in Figure 9.3. (Logistic growth, also shown on Figure 9.3, is explained later.)

For exponentially (geometrically) growing populations, we can calculate the **doubling time**, the length of time it will take for the population to double. Mathematically, doubling time is expressed as

$$\text{doubling time} = 0.69315/r$$

where the number 0.69315 is the natural logarithm of 2.

In 1993, the United Nations calculated that the growth rate for the world's population was 1.8% per year, which will cause a doubling every

$$0.69315/0.018 = 38.5 \text{ years}$$

As we saw earlier, the value of *r* varies from place to place (see Figure 9.2A). The fastest-growing nations have growth rates at or above 4%, which will cause them to double their population every 17.3 years or more rapidly. Many more nations are growing at about 3% per year and will thus double their population every 23.1 years.

Malthus' analysis of population growth

The world's human population has long been on the increase, but the rate of change was slow before modern times (see Figure 9.1). During the seventeenth and eighteenth centuries, several European countries began to experience a great upswing in their populations, adding to the many motivations for sending forth expeditions to find and settle new lands. The need to clothe growing populations encouraged innovations that brought greater efficiency to textile production, marking the onset of the industrial revolution.

During the early part of the industrial revolution, philosophers and economists began to pay attention to the phenomenon of population growth. David Hume and Benjamin Franklin each described population increase as a blessing for civilization. The first person to emphasize the negative consequences of population growth was Thomas Robert Malthus. In his *Essay on the Principle of Population* (1798), Malthus explained the following dilemma:

1. A population tends to increase *geometrically* if its growth is unchecked. (As we saw above, a geometric series is one in which each number is *multiplied* by a constant to obtain the next number.)

2. The available food supply, in Malthus' view, increased only *arithmetically*. (An arithmetic series, like 3, 4, 5, 6, 7, …, is one in which a constant, in this case 1, is *added* to each number to obtain the next number.)

3. Because the population increases faster than the food supply, the increasing population compounds human misery and poverty, especially among the lower classes.

Malthus divided the factors controlling the population increase into two broad categories that he called preventive checks and positive checks. **Preventive checks** were those that could prevent births from occurring. These were usually voluntary measures, operating on an individual level. They included delayed marriage (also called 'moral restraint'), reduced family size, and several forms of 'vice' (practices that Malthus condemned, including birth control and homosexuality). The **positive checks** on population were those that would operate automatically after births had taken place, whenever the preventive checks were not sufficient. Malthus identified as positive checks overcrowding, poverty, epidemic diseases, rising crime rates, warfare, starvation, and famine. Malthus opposed the social welfare legislation of his time because he

Figure 9.3

Exponential growth compared with logistic growth. The steeper (more vertical) the slope, the more rapid the rate of growth. Graphing exponential growth always gives a characteristic J-shaped curve (blue).

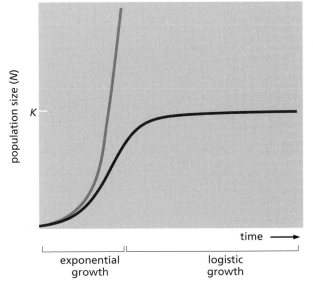

thought that any measure to improve the condition of the poor would only encourage them to reproduce faster, further outstripping their food supply and compounding their own misery. One outgrowth of this type of thought was the nineteenth-century theory attributing warfare to the economic needs confronting the population of each nation, the so-called economic theory of war.

Malthus noticed that, even in European countries with stable populations, the *rate* of population growth temporarily rose in the years following a famine or plague until the population was restored to its pre-disaster level. This phenomenon also takes place after many wars and economic hard times (witness the 'baby boom' in the years following World War II). Such events show that a population's potential for increase is much greater than is usually realized. The actual rate of population growth is kept lower than the potential rate of growth by positive and preventive checks.

Throughout the nineteenth and early twentieth centuries, technological progress in agriculture, especially in mechanized farming and in the use of chemical fertilizers, increased crop yields among the wealthy nations providing a plentiful food supply for the expanding population. Also, European population crises were in many cases relieved by large-scale emigration to other continents. Malthus and his gloomy predictions of starvation and misery were largely forgotten.

Following World War II, attempts to deal with poverty, disease, and food shortages around the world ran into the harsh reality that burgeoning populations were exacerbating all these problems. Public health improvements (in public sanitation, in mosquito control, in vaccination against infectious diseases, and in the delivery of medical care generally) were diminishing the death rates, especially among the young. The result was in many cases a staggering population growth.

The population growth affected different countries unequally. Many nations of the developing world (those countries which had not become industrialized) were trying to diversify their economies, modernize their industries, and improve their living standards. Improvements in health care and sanitation became more rapid after about 1950 and led to more births and fewer deaths; soon these nations found their financial and other resources stretched thin as their populations continued to increase. Economic development, in other words, was being slowed by population growth, and many leaders of the developing nations became increasingly concerned. Without great wealth, the housing and other needs of the expanding populations could not be met. The positive checks that Malthus had foreseen were operating. The developing world was coming face to face with a population crisis (see Figure 9.1).

The population crisis in the industrially developed countries of North America, Europe, and Japan was taking a different form: wealth was diverted into providing additional housing, roads, sewers, and needed services. However, this development may not be sustainable in the long run, because it depends on the use of nonrenewable resources such as fossil fuels and soil (see Chapters 11 and 18), and on resources and products imported from less developed countries, leaving less for them. The diversion of nonrenewable resources cannot indefinitely support stable or growing populations. By diverting resources, wealthy nations can postpone, but not avoid, the effects of global population growth.

CONNECTIONS CHAPTERS 11, 18

Growth within limits

Malthus' assumption of an arithmetical increase in the food supply has been questioned. Although the limited data available to Malthus were consistent with an arithmetical increase, there is no biological or other theory that explains why this should be so. However, most biologists do agree with Malthus' point that the growth in food supplies is slower than the rise in population. One reason this is so is because there is a loss of energy at each stage in which one type of plant or animal becomes food for another, a topic that we develop more fully in Chapter 11. Because food and other resources increase more slowly than population, any population growing exponentially will outstrip its food supply and other resources, including the available space in its habitat. Clearly, no population can continue growing exponentially. After growing exponentially for a while, a population usually follows a pattern called **logistic growth**.

Logistic growth. Logistic growth has been demonstrated in all types of experimental populations, and we know of no biological reason why this would not also apply to human populations.

Logistic growth can be modeled mathematically by the equation

$$dN/dT = r N (K - N)/K$$

In this equation, K is a new quantity called the **carrying capacity** of the environment. Like N, K is a number, not a rate. K refers to the maximum size of the population that can be sustained indefinitely by the environment. As population size approaches this carrying capacity, the population growth (symbolized by dN/dT) slows down, and when $N = K$ the population growth is zero (Figures 9.3 and 9.4). When the population

Figure 9.4

A variety of logistic growth curves.

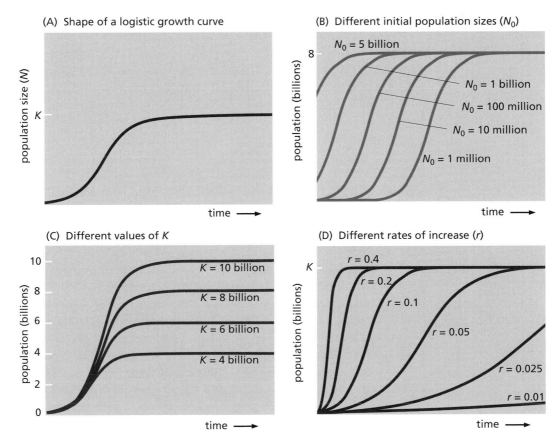

reaches whatever carrying capacity is imposed by the environment (see Figure 9.4C), the birth and death rates are balanced. If the population size overshoots K, then a population crash follows in which deaths outnumber births and the population size declines to the carrying capacity.

K is related to the amount of space in an environment and the other resources available, including the amount of energy that is in a form that can be used by living organisms. For animal or plant species living in an unchanging environment, K is generally constant. For our own species, K varies with changes in technology, especially technology that gets more usable energy from the same environment, for example by increases in the efficiency of energy use or of food production. A certain amount of land may support a particular population size of hunters and gatherers at a low carrying capacity (a low value of K). The development of agriculture generally results in an increased carrying capacity, and the industrial revolution (including the use of tractors and chemical fertilizers in agriculture) increases it still further.

Humans can avoid an environmentally imposed population crash by limiting the birth rate before the population reaches its carrying capacity; in addition, we need to consider the effects of human populations on the populations of other species. When two species interact, either one may modify the effective carrying capacity of the environment for the population size of the other. Population ecologists define competition as a type of interaction in which a species diminishes the population size of another. An increasing human population diminishes the sustainable population size of all species with which we compete, and is driving many of these other species toward extinction (see Chapter 18).

K-selection and r-selection. Biological species can have very different types of population structures and life cycles (see Chapter 8). Two basic types of population growth are those that are limited by ***K*-selection** and those that are limited by ***r*-selection** (Table 9.1). Each is favored by natural selection, but in different sets of circumstances. In environments where favorable conditions can change rapidly or disappear, high mortality rates are common, and natural selection favors prodigious reproduction of offspring (high r, therefore r-selection) to compensate for this devastating mortality. In more stable environments, population sizes stay close to carrying capacity (K), and natural selection favors efficient use of resources, especially energy. In K-selection, the advantage generally goes to whoever can most efficiently convert food resources into new adults of the next generation. Humans are an example of a K-selected species.

Demographic transition

Because humans are a K-selected species, human population increases in the past have coincided with major advances in technology that have allowed the carrying capacity (K) to reach new levels. The development of agriculture made it possible for human populations to increase well above the size permitted by hunting and gathering (see Figure 9.1).

Most of our understanding of the changes that accompany a population increase comes from studying the growth of a population with changes in economic development. Our current model of this process describes it as a **demographic transition**, characterized by an orderly succession of stages.

Stages of a demographic transition. As shown in Figure 9.5, the first stage of a demographic transition is a stable population in which a high death rate is balanced by a high birth rate. Population growth is, therefore, zero. Over the centuries, traditional societies with high death rates (especially from infant mortality and childhood infections) developed customs that encouraged high birth rates. In other words, high mortality rates encourage high birth rates; this is the pre-industrialized stage. Demographers estimate that overall mortality rates held steady or declined slowly in pre-industrial Europe, with occasional but temporary upsurges during wars and epidemics such as the great bubonic plague (the black death) that decimated European populations in the 1300s.

The second stage of the process is a period of exponential growth in which the population size climbs to a new high. This is brought about by technological changes (including industrialization) that result in falling mortality rates, or death control. In Malthusian terms, the improvements (in sanitation, health care, and nutrition, for example) result in the alleviation of positive checks and thus a lowering of D. However, it always takes at least a few generations for the cultural values and customs to change so as to permit a matching decline in birth rates. In the meantime, the memory of high death rates in the recent past continues to

Figure 9.5

Idealized stages of a demographic transition. Some authorities recognize only three stages by combining the middle two.

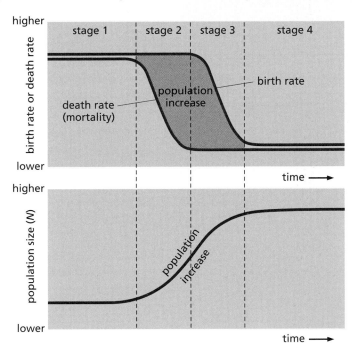

Table 9.1

Differences between K-selected and r-selected species.

	K-SELECTION	r-SELECTION
Ecological conditions	Favorable conditions dependable and change slowly (if at all) over time.	Unstable; very favorable and very unfavorable conditions appear and disappear erratically and unpredictably.
Population size	Stable, at or near the carrying capacity (K).	Fluctuates greatly over time, both above and below the carrying capacity; devastating mass mortality is frequent and often unpredictable.
Values of r	Low to moderate r; population growth is slower even when conditions are favorable.	High r leads to very rapid population growth under favorable conditions, which compensates for mass mortality under unfavorable conditions.
Reproduction	Always sexual.	Either asexual or sexual.
Body size at reproduction	Natural selection favors reproduction at a large body size.	Natural selection favors reproduction at a small body size and a young age.
Reproductive rate	Low; few offspring produced at once.	High; many eggs, seeds, or other reproductive stages produced at once.
Frequency of reproduction	Repeated many times throughout life.	Usually confined to a single occasion.
Offspring size	Offspring are individually large, and each represents a large proportion of its parents' reproductive output.	Released (as eggs or immature stages) at a small size, and widely scattered; each represents only a tiny fraction of its parents' reproductive output.
Parental investment	Generally high; may include provisioning of food for offspring and extensive parental care.	Little or no parental care or investment.
Dispersal	Dispersal of the population to new locations is generally slow. Newly founded populations grow more slowly.	High capacity for dispersal (spread of the population) to new habitats and locations. Most new locations are unsuitable, but the occasional favorable one permits a rapid explosion in numbers.
Mortality pattern	Most individuals live their full life span and die at an advanced age.	Mortality is extremely high among eggs, seeds, or larvae; most deaths occur very early in life.
Examples	Humans, cows, elephants	Carrion-feeding beetles, tapeworms, weeds

encourage high birth rates. In some cases, the birth rate may even rise as the result of better nutrition and the improved physiological condition and reproductive health of prospective mothers. The combination of high birth rates and lower death rates causes the population growth that characterizes the middle of the demographic transition.

The third stage of the transition is marked by a decline in birth rates as the population adjusts to new conditions and as the incentives for high birth rates are removed. Population growth follows the logistic growth equation during this stage. As the birth rate declines to match the new lower mortality rate, the population once again stabilizes, but at a larger population size (and a larger K) than before. The demographic transition is complete.

England's demographic transition began in the early 1700s and took some 250 to 300 years to complete. Other industrialized countries (United States, Canada, Japan, and some in Europe) took closer to 200 years to complete their demographic transitions. The remaining countries of the world began their demographic transitions only during the twentieth century, and did so much more suddenly, often going from high traditional mortality rates to low modern rates in a single generation. All countries in the world will eventually experience advances in sanitation and medical care, even if these are imported. The resulting reduction in death rates will bring about a demographic transition, even in the absence of economic development. The only exceptions may lie in countries where repeated famine and warfare (both positive checks) keep the death rates high.

The United States may have completed a demographic transition a few years ago, when birth and death rates became approximately equal, a condition known as **zero population growth**. However, as we noted earlier, the U.S. population is once again increasing, and it remains to be seen whether the long-range trend will more closely resemble the zero growth phase or the more recent increase.

Age structure of populations. The models we have examined so far treat all members of a population as being the same. However, we know that the probability of death and the probability of reproduction both vary with age. Thus, the true rate of population growth may vary depending upon the ages of individuals within the population. Grouping individuals by age gives the **age structure** of the population. The age structure can best be shown by a population pyramid, or **age pyramid**, such as those in Figure 9.6. Each horizontal layer on such a diagram represents the percentage of the population in a particular age group, with the youngest age groups on the bottom. Altogether, the age pyramid shows the distribution of individuals among the various age groups. To get a feeling for the scale of the age pyramids in Figure 9.6, notice that in Uganda the 0–4 age group constitutes about 20% of the population. Most age pyramids are divided by a vertical midline, with male age distribution shown on the left and female age distribution on the right. Notice in Figure 9.6 that the 0–4 age group has approximately equal numbers of males and females in all countries, but the 80+ age group has many more women than men in stable or declining populations. Among human populations, a pyramid with sloping sides and a wide base (many children) characterizes an expanding population. A shape maintaining more or less the same width throughout (except for the oldest few age classes) indicates a stable population. A stable age distribution is reached when the pyramid keeps the same shape as each age group grows older so that the numbers in each age group are replaced by an equal number advancing from the next younger group.

Sometimes there are age bulges, as with the post World War II baby boom, when the birth rate increased temporarily. The United States and some other countries experienced a second baby boom (or a baby boom 'echo') in the late twentieth century as members of the earlier baby boom reached their prime childbearing years. In these countries, schools must now cope with the largest generation of school-age children that has ever lived.

Predictions of future values of r can sometimes be made on the basis of age structure. Clearly, a population of 10,000 individuals with 4000 females of reproductive age has much more potential for increase than one with only 400 females of reproductive age. Calculations of the potential for future increase are often carried out by multiplying the number of females in each age group by the number of children that each of those females is likely to bear, then adding up these products for all age groups.

Life expectancy. In most of the industrialized nations, the control of many infectious diseases since the late 1800s and improvements in sanitation have resulted in decreased infant mortality. A few twentieth-century changes, such as increases in smoking, auto accidents, and hand guns, have increased death rates, but these have generally been offset by

Figure 9.6

Age pyramids for two rapidly growing populations (high r), a population with a moderate rate of growth, two stable populations (r near zero), and a declining population (negative r), based on United Nations data for the year 2000. Age groups from 20 to 39 are darker to emphasize that most reproduction occurs in these age groups.

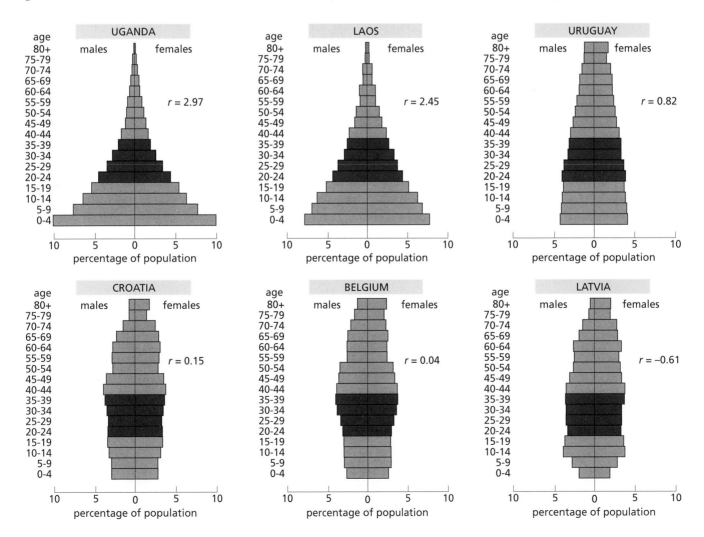

much greater declines in the death rates from famines and infectious diseases. One consequence of declining death rates, especially among the young, is a greater **life expectancy**, meaning the average maximum age that people attain in life, or the age to which a person born in a particular place and at a particular time can expect to live (see the graph on our Web site, under Resources: Life expectancy). Life expectancy is calculated on a statistical basis, taking into account the probability at any age of a person's proceeding on to the next age group. Therefore, both decreased mortality and increased longevity (the maximum age achieved by some individuals in the population) contribute to an increased life expectancy, with the largest increases resulting from reductions in childhood mortality. Life expectancy in the United States has risen from somewhere near 50 years in colonial times to over 75 years today, and the age group over 80 is the most rapidly growing segment of the U.S. population. This 'graying' of America and of many other countries results in a decrease in the proportion of the population in the most fertile age group and an increase in the proportion of older people, many of whom are dependent on the younger generation for their care. Numerous sociological changes follow the shift toward more people of advanced age and fewer children—more emphasis on medical care and less emphasis on schools, for example.

Demographic momentum. Even after the birth rate falls to the level of the mortality rate, the population may continue to increase for another generation or two because of a **demographic momentum**. The momentum is caused by an age structure opposite to the one just described for a graying population. A population in which a large fraction of individuals is not yet of reproductive age (see Figure 9.6), while only a small fraction is past the age of reproduction, has a future reproductive capacity larger than one in which a higher percentage of the population is beyond reproductive age. Because the younger age groups are more likely to reproduce in the next 20 years and less likely to die, a temporary increase in the birth rate (and a temporary decrease in the death rate) can easily be predicted. The population will continue to grow until the age distribution readjusts itself more evenly, that is, until a stable age distribution is reached. With a stable age distribution, the birth and death rates will no longer change, unless some external factor disrupts the stable situation.

World population estimates. Demographers estimate that the world's population was about 50 million in 7000 B.C. and increased to about 250 million by the time of Christ. The earliest date for which there are reliable population estimates is 1650, at which time the world's population stood at 500 million. By 1804, the world's population stood at an estimated one billion. Adding the next billion took 123 years (to 1927), but the third billion was added in only 33 years (by 1960). The world's population has increased even more rapidly since 1960, adding a fourth billion in only 14 years (by 1974), a fifth billion in 13 years (by 1987), and a sixth billion in just 12 years (by 1999) (see Figure 9.1).

The United Nations has published tables with detailed predictions for the further growth of the world's population under different sets of assumptions. According to the model that demographic experts consider most likely, the world's population will grow to about 9.4 billion in 2050 and will stabilize at around 11 billion by the year 2200. Other models, with different sets of assumptions, predict population values in the year 2050 as low as 7.9 billion or as high as 10.9 billion. Each of these assumes logistic growth, except for the model in which population growth continues exponentially at its present rate—that model predicts a population size of 296 billion in 2200, nearly 50 times the present value! Few, if any, biologists think that the latter is in any way sustainable; population growth will have to level off from its present rates.

Because people do not generally find raising the death rate to be an acceptable method of lowering population growth, decreasing the birth rate is the only other option. In the next section we study reproductive biology and the factors that contribute to fertility. In the final section of the chapter we see how fertility can be controlled to lower the birth rate.

THOUGHT QUESTIONS

1 What changes (biological, social or economic) could increase the carrying capacity (K) of the entire world or of one nation? Should we be more interested in controlling r or in modifying K?

2 Suppose that you count the number of people in a given town every year for 5 years and you discover that N has stayed about the same during that time period; say, at 10,000. You also know that no one has moved into the town or away from the town in that time. What can you deduce about B and D? Confirm this for yourself by solving the equations given earlier in the chapter. Can you tell from this information how many people were born in the town in any of the 5 years?

3 Study the age pyramids in Figure 9.6. Does the age distribution of the females or that of the males have a greater impact on future population size? Why?

4 What values of r are typical during the several stages of a demographic transition? See Figure 9.5, and review the way in which r is defined.

5 Which is likely to produce a greater increase in the number of people, a small population with a high r or a large population with a small r? Try some calculations with any values of N and r that you would like to examine. Use the simplest model that seems appropriate, then ask what changes the more complex models would bring.

6 From the data presented in this chapter, can we estimate the value of K (the carrying capacity) for the human population on planet Earth? Can we make a minimum or maximum estimate? Have we already reached the carrying capacity of the planet?

7 Which would you think would contribute more to an increase in life expectancy, decreased infant mortality or increased longevity?

8 Among the 'preventive checks,' Malthus listed several forms of 'vice,' including the following:
a. heterosexual intercourse outside marriage ("promiscuous intercourse"; "violations of the marriage bed");
b. sexual attraction to one's own sex (homosexuality), to animals (zoophilia), or to inanimate objects (fetishism);
c. "Improper arts to conceal the consequences of irregular connections" (i.e. birth control).

Malthus listed all these practices as preventive checks. Which of them would actually function to limit population growth? Which might be interpreted as preventive checks under certain assumptions that Malthus might have made? Do you think these assumptions are realistic? Do you think Malthus condemned these practices because of their effects on population or because of his own moral views?

9 What factors would need to be included in an equation to estimate what birth rate would produce zero population growth?

10 Which countries currently have the highest population growth rates? Are they wealthy or poor? Are they influential? Where do they stand in the current world order?

Human Reproductive Biology Helps Us to Understand Fertility and Infertility

In sexual reproduction, a male haploid gamete unites with a female haploid gamete. The two types of gametes are produced by individuals of different sexes. We begin this section by comparing the anatomy and physiology of the two sexes in humans.

Reproductive anatomy and physiology

Reproductive anatomy includes those structures that allow for hormonal secretion, gamete production, sexual intercourse, gestation of the fetus, and nourishment of the young infant. Males and females differ in appearance and in the type of gamete that they produce. These differences result largely from the actions of **hormones** that are present in both sexes, but in differing amounts and with different consequences.

Sex determination. The reproductive system, in the earliest stages of its development, is sexually indifferent, meaning that there are no indications as to the future sex of the embryo. The future gamete-producing reproductive organ (or gonad) is not yet male or female, but is just an indifferent gonad. As described in Chapter 2, several genes, including the *SRY* gene, begin to make products at this time. If the products of the genes *SRY*, *SOX9*, *DMRT1*, and *DMRT2* are all present and functional, then the developing gonad becomes a male gonad or **testis**. The embryonic testis then begins to secrete the hormone **testosterone** and another hormone called APH (anti-paramesonephric hormone). In adulthood, the testis will also produce sperm. A developing gonad that does not become a testis becomes an **ovary** instead. The ovary will secrete the hormone **estrogen** throughout life, and will produce eggs during adulthood. The developing testis and its hormonal secretions are essential to the development of male internal and external reproductive structures, but the corresponding female structures develop without requiring estrogen. Thus, a genetic male with any of genes *SOX9*, *DMRT1*, or *DMRT2* nonfunctional will develop a female phenotype, even in the presence of *SRY*. For this reason, embryologists say that the default sex of the embryo is female, and that male development can only occur when it is induced.

Maturation and puberty. Hormones are small molecules that are used for chemical communication throughout the body. Beginning during embryonic development, the ovaries secrete estradiol (the major estrogen) and the testes secrete testosterone and APH. The reproductive organs are formed during embryonic development under these hormonal influences, but they remain immature until **puberty**. At puberty, a group of hormones produce many changes in the body. At an age that varies greatly around an average of about 12 years, increasing levels of the pituitary hormone FSH (follicle-stimulating hormone) stimulate the final maturation of the ovaries in females to begin producing eggs and the testes in males to begin producing sperm. These, in turn, step up their secretion of other hormones (estrogen and testosterone), stimulating (among other changes) the development of **secondary sexual characteristics**. Such characteristics include the widening of the hips, growth of breasts, and redistribution of body fat in females; the growth of facial hair and the deepening of the voice in males; and the growth of long bones and pubic hair in both sexes. The term 'secondary sexual characteristics' means that these features, while characteristic of mature men and women, do not have a direct role in the production of gametes.

The age at which puberty occurs has declined steadily throughout the twentieth century in all countries in which such records have been kept. In Norway, which has kept records the longest, the average age at menarche (first menstruation) was just over 17 years in the 1840s but had declined to about 13.4 years by the 1950s. Data from other western European countries are consistent with this, and all countries show similar and steadily declining trends. One possible explanation is that puberty requires a critical weight (about 47 kg or 106 lb for females), and that improvements in childhood nutrition have allowed this critical weight to be achieved at earlier ages over successive decades. Consistent with this hypothesis is the trend in the United States over the past century, where a similar decline occurred, but where body weights have always tended to be above European averages. The average age at menarche has always been about 0.5–1.0 year younger in the United States than in western European countries, and is generally younger among African Americans than among U.S. whites. Beginning in the 1990s, this trend has begun to level off in many countries at an average age of between 12.0 and 12.5 years at menarche.

Male puberty is marked by the onset of sperm production (spermatogenesis). The pituitary hormone FSH initiates sperm production, and another pituitary hormone, luteinizing hormone (LH), induces other cells of the testis to secrete testosterone. Sperm production requires this testosterone for its completion, along with a small amount of estrogen. Female puberty is marked by the start of the menstrual cycle (see p. 299).

After puberty, FSH, estrogen, and testosterone continue to have other roles throughout the lifetime of the individual. In males, some of the testosterone is converted into another hormone called DHT (dihydrotestosterone), and small amounts of estrogen seem to be essential for spermatogenesis and other processes. In females, FSH, estrogen, and progesterone are all important in the menstrual cycle, and small amounts of testosterone are thought to be important in generating the sexual appetite or libido.

Male reproductive organs and sperm production. The sperm are male gametes, formed by the process of meiosis (see Chapter 2, pp. 42–44) in which each gamete receives half the adult number of chromosomes— one chromosome from each pair. Sperm have very little cytoplasm surrounding their nucleus, but they have a long tail that moves rapidly back and forth to propel the sperm along the reproductive tract of the female after intercourse (Figure 9.7).

CONNECTIONS
CHAPTER 2

Sperm are produced in the testes of the male (Figure 9.8). The hormone testosterone is secreted by the interstitial cells that are crowded into the spaces between the sperm-producing tubules in the testes. (Testosterone is also made in smaller amounts by the brain in both sexes and by the ovaries in females.) The sperm accumulate in a series of wrinkled ducts that form the epididymis. The epididymis also secretes a fluid (the seminal fluid) that carries the sperm through a long duct called the

Figure 9.7

(A) An egg surrounded by sperm. (B) Schematic diagram of a human sperm.

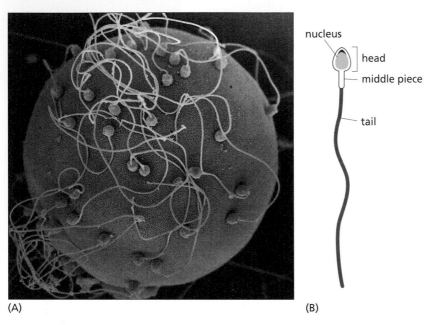

(A)

(B)

vas deferens through which the sperm leave the testes. The left vas deferens and right vas deferens merge within the prostate gland, where they join the urethra carrying urine from the urinary bladder. The secretions of the prostate gland and the seminal vesicles add certain nutrients (including the sugar fructose) that help the sperm to swim more vigorously. Near the base of the penis, the bulbourethral gland helps squirt the seminal fluid and sperm through the length of the penis during ejaculation. Between ejaculations, sperm and seminal fluid are held in the ejaculatory ducts.

Female reproductive organs and ovulation. The female reproductive organs (Figure 9.9) include the uterus and a pair of ovaries. The female gamete, or egg, is formed within the ovary by meiosis. The first meiotic division occurs before the egg is released from the ovary, but the second meiotic division, in which the chromosome number becomes haploid, is delayed until after the egg is fertilized. During the two cell divisions during meiosis in

Figure 9.8

The human male reproductive system.

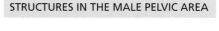

STRUCTURES IN THE MALE PELVIC AREA

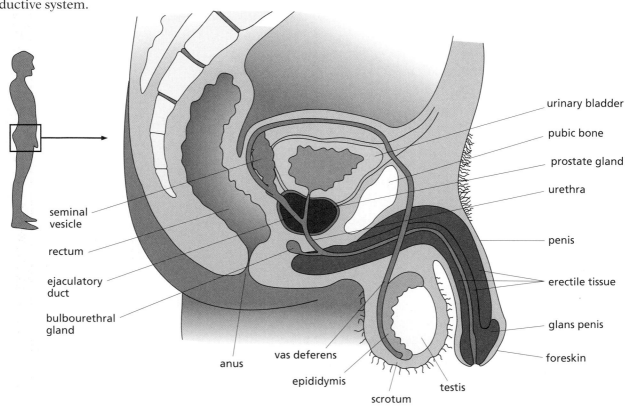

egg formation, the division of the cytoplasm between the offspring cells is extremely unequal. Only one of the four resultant cells becomes a mature egg with a large amount of cytoplasm to nourish the zygote after fertilization. The other three become polar bodies with very little cytoplasm, and these are usually lost. The nonreproductive cells surrounding the egg enlarge to form an ovarian follicle. The rupture of this follicle and the release of its egg are called **ovulation**.

The menstrual cycle. Egg production and ovulation in mammals is a hormonally regulated cycle. In humans, this cycle lasts about 28 days and is called the menstrual cycle.

The cycle of egg production and changes in the uterus is controlled by two ovarian hormones, estrogen and progesterone, and by two hormones (FSH and LH) secreted by the pituitary gland at the base of the brain (Figure 9.10). At the start of each cycle, which begins with menstruation, the pituitary secretes small amounts of FSH. The FSH stimulates two processes within the ovary: growth of an ovarian follicle, and production of the hormone estrogen, which reaches a peak concentration during the second week of the cycle. The estrogen stimulates the thickening of the lining of the uterus and the release of a second pituitary hormone, LH, that induces the release of the egg (ovulation), after which the tissue that surrounded the egg is left behind to form a scar tissue called the **corpus luteum**. The corpus luteum then grows. As it matures, it begins to secrete the hormone **progesterone**, which maintains the uterine lining in a thickened and receptive condition, ready for the implantation of an embryo should the egg be fertilized. If fertilization does not occur and no implantation takes place, the corpus luteum degenerates and the supply of progesterone drops sharply, causing the uterine lining to break down. The egg and the uterine lining are then sloughed off in the form of menstrual bleeding. The absence of progesterone also releases the pituitary to begin secreting FSH once again, initiating a new cycle.

Figure 9.9

The human female reproductive system.

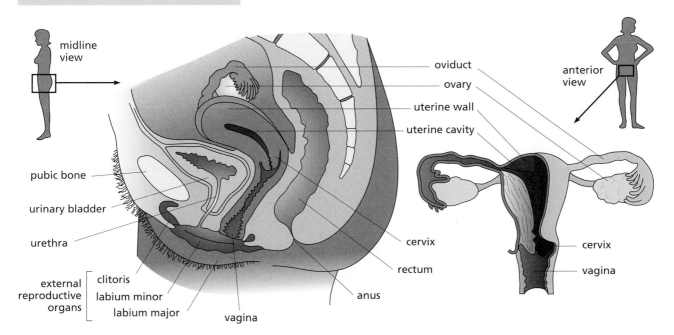

STRUCTURES IN THE FEMALE PELVIC AREA

midline view

oviduct
ovary
uterine wall
uterine cavity

anterior view

pubic bone

urinary bladder

urethra

cervix

rectum

external reproductive organs
clitoris
labium minor
labium major

anus

vagina

cervix

vagina

During the menstrual cycle, changes in the concentration of each hormone stimulate the production of the next hormone in the sequence. Hormones from the ovaries and hormones from the pituitary regulate each other. In several cases, the presence of a later hormone has an inhibiting effect on the secretion of the previous hormone. This regulation of a previous step of a cycle by a later step of the cycle is called a **feedback mechanism**. In this instance, feedback prevents the overproduction of any hormone and stops the production of a hormone once it has done its job.

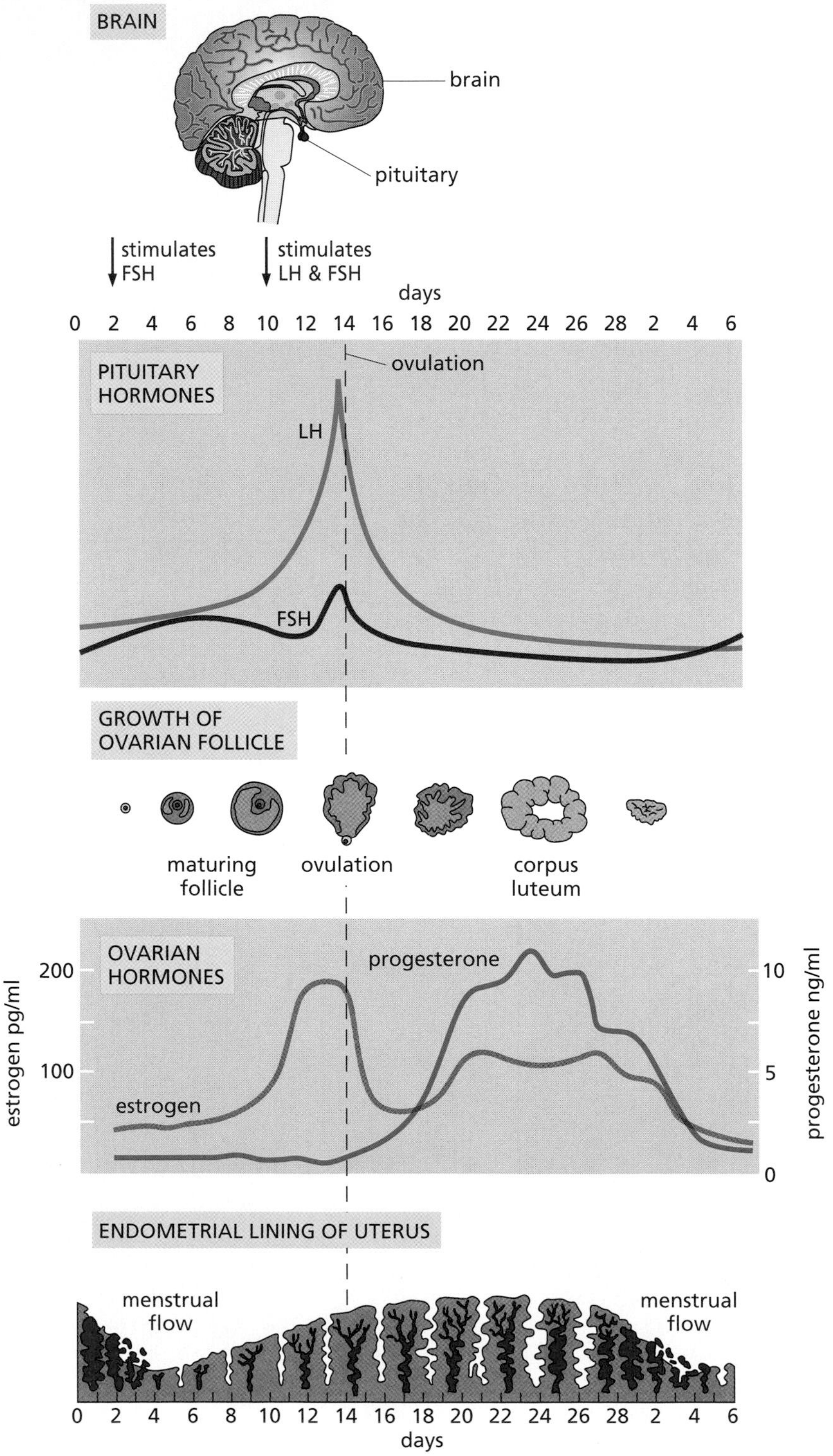

Fertilization and implantation. Neither the sperm nor the egg live very long. After its release from the follicle, the egg travels along the uterine (Fallopian) tubes or oviducts, and it is here that the egg is fertilized if sperm are present.

When the male ejaculates during sexual intercourse, approximately 300 million sperm are inserted into the vagina of the female. Of these, only about 3000 will successfully swim through the cervix and the uterus and into the oviducts, following hormonal signals released by the egg. Several sperm need to contact the egg to dissolve the coating that surrounds it. After the coating has dissolved, allowing a sperm to reach the plasma membrane of the egg, the sperm and egg plasma membranes fuse and the egg draws the sperm nucleus inside. The coating closes again, preventing the entry of any more sperm nuclei. The egg completes its second meiotic division and its haploid chromosomes join with the haploid chromosomes from the sperm, a process called **fertilization**.

If an egg is fertilized by a sperm, the resultant zygote continues travelling along the oviduct for 4–5 days, during which it undergoes several cell divisions and becomes an embryo. When it reaches the uterus, at a stage called the blastocyst, the embryo adheres to the inner (endometrial) lining of the uterine wall, a process called **implantation** (Figure 9.11).

Figure 9.10

Reproductive cycles in the human female.

If the embryo does not implant, it does not develop further and is shed from the uterus. Implantation triggers the growth of the **placenta**, consisting of intertwined tissues of the embryo and the endometrium of the mother, through which the developing embryo is nourished (see also Figure 14.9, p. 531). Most of the organs of the growing embryo form during the first month, after which the embryo is known as a **fetus**.

Impaired fertility

We have described the normal events of human reproduction, in which egg and sperm from a fertile female and a fertile male combine to make an embryo. Not all individuals are fertile, however. A couple is usually considered infertile if they have been unable to get pregnant after a year of trying. Infertility can result from lack of production of sperm or eggs, low motility of sperm, shortened viability of egg or sperm, or anatomical abnormalities preventing the sperm from contacting the egg. A number of diseases, most of them rare, can cause permanent sterility. These include: several chromosomal anomalies and other genetic disorders; developmental anomalies of the reproductive system, including failure of the testes to descend; and some cancers, including those of the reproductive organs. Many sexually transmitted diseases such as gonorrhea and syphilis cause a loss of fertility that can be permanent (see Chapter 17). Many other infectious diseases can lower fertility temporarily, but the fertility returns to previous levels or nearly so once the disease is cured.

A large number of other factors can also affect fertility in men and women. These are summarized in Table 9.2.

A further factor is age. The most fertile years for both men and women are between the ages of 20 and 24. Many people choose to delay having children until later than that. After age 35, female fertility (the percentage of women who become pregnant when attempting to become pregnant) is less than half what it was at its maximum. The rate of spontaneous miscarriage is around 20% for women in their thirties, increasing to over 30% for women aged 40–44. Male fertility also decreases with age, gradually from age 40–55, and then more steeply after age 55.

Since World War II, studies in various countries have documented a general decline in male sperm counts and a rise in the number of sperm abnormalities. From 1938 to 1990, the average sperm count for European men declined to about half its former value, from 113 million to 66 million sperm per milliliter. Although the reasons are unclear, many researchers suspect that the cause may be related to an environmental presence of industrial chemicals and pesticides that have estrogen-like effects.

In the next section we describe some of the methods that have been developed for overcoming infertility.

Figure 9.11

Fertilization in the oviduct and implantation in the uterus.

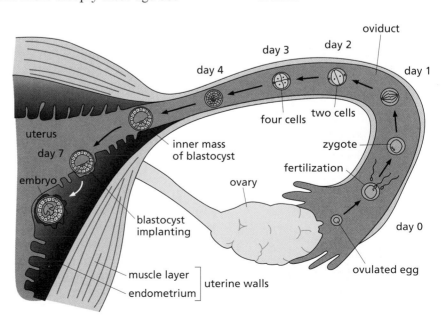

Table 9.2

Factors that can impair fertility.

FACTOR	DESCRIPTION OF EFFECTS
High blood pressure and other circulatory disorders	A major cause of erectile dysfunction in men (failure of the penis to enlarge and stiffen during sexual activity)
Opiate drugs (including morphine, heroin)	In women: loss of ovulation and of menstrual cycles
	In men: erectile dysfunction; loss of sexual interest (libido); low sperm counts
Alcohol	In women: suppresses ovulation and menstrual cycling (but the effect is inconsistent and affects different women to different degrees)
	In men: can lead to erectile dysfunction, loss of libido, sperm abnormalities, and a conversion of testosterone to estrogen that can lead to breast enlargement and sterility
Tobacco	Decreased fertility in both sexes; also earlier menopause in women
Marijuana	In both sexes: decline in sexual desire; lowering of fertility; hormonal changes
	Also in men: breast enlargement; decreased sperm count
Cocaine (chronic use)	In men: abnormal sperm and decreased sperm counts; erectile dysfunction
	In women: loss or disturbance of the menstrual cycles; also an 11-fold increase in uterine tube infertility
Drugs used to control high blood pressure; also many antipsychotic drugs and antidepressants	In men: erectile dysfunction; decreased libido; inability to achieve orgasm or to ejaculate (Inadequate study of women)
Many anticancer drugs	Inhibit both menstrual cycles and spermatogenesis; can lead to permanent sterility in both sexes
Stress (probably acting by stimulating the body's production of natural opiate-like compounds)	In women: suppresses ovulation; loss or reduction of menstrual cycles
	In men: lowered sperm counts; decreased sperm mobility; increased sperm malformation; decreased testosterone levels; erectile dysfunction; loss of sexual interest (libido)
Depression	Loss of sexual activity and interest
Anorexia; also physical overexertion, especially in female athletes and ballet dancers	In women: loss of menstruation, failure to ovulate
Heavy metals (mercury, lead, etc.)	Can cause permanent sterility in both sexes
Polychlorinated biphenyls (PCBs), dioxin, many organic pesticides (including DDT)	Estrogen-like effects impair male fertility; males exposed to these chemicals before puberty have very high rates of malformed internal and external reproductive structures including testicular disorders

Assisted reproduction

At the population level, there are many motivations for decreasing the birth rate, but on an individual level the urge to have children is very strong. Many couples who cannot conceive a child seek help in becoming pregnant. Usually the first step is a complete medical assessment of both partners to see if a medical cause can be found; in approximately 20–30% of cases no cause can be identified. In parallel with research on birth control technologies, much research has gone into the development of several new technologies for assisted reproduction. Much of the research on infertility has been done in countries such as India and China, which also have active population control programs, a reminder that individuals and societies may have conflicting goals.

If a couple remains infertile even after medical treatment, or if no medical treatment is appropriate, a variety of remedies can be suggested. In some cases, gametes from both partners are used to produce a child that is genetically theirs. In other cases, gametes are taken from one partner but not the other. Sperm donation is relatively easy and is more commonly practiced, but egg donation requires surgery and is thus less common. (Both sperm banks and egg banks have become profitable businesses in a number of countries.) A possibility always to be considered is adopting a child who is genetically unrelated to both parents.

In this section we examine a number of options for assisted reproduction.

Assisted gamete production. Various hormonal treatments are available to increase fertility in either men or women, for example by stimulating egg production and ovulation. Usually only one or a small number of eggs is released in a monthly cycle. Hormonal treatments for infertility may induce the release of many more eggs than normal and result in (fraternal, not identical) twins, triplets, and higher multiples if more than one egg is fertilized in the same ovulation cycle. The recent increase in the frequency of multiple births has been largely due to these treatments. (These treatments have no effect on the frequency of identical twins, which result from a fertilized embryo splitting.)

Often these methods are used together with other forms of assisted reproduction. In males with low sperm counts, sperm may be harvested, then concentrated by centrifugation and used in the process of artificial insemination described next.

Artificial insemination. Artificial insemination means introducing sperm into a female's reproductive tract other than through sexual intercourse. The sperm could be derived from a woman's husband or from another man. The procedure is fairly simple (and is routinely performed on cattle and certain other domesticated species, as are *in vitro* fertilization and surrogate pregnancy). However, the legal rights and responsibilities of a sperm donor other than the recipient's husband are unclear in a number of jurisdictions.

***In vitro* fertilization and embryo transfer.** An *in vitro* process is one that takes place in laboratory glassware rather than inside the body (*in vivo*). In the case of ***in vitro* fertilization**, eggs are harvested from a woman (usually after hormonal treatment to stimulate egg development) and fertilized in a glass or plastic dish, using sperm contributed either by her husband or by another man (Figure 9.12). Fertilized eggs are then allowed to develop to approximately the 64-cell stage, after which one or more of these embryos are implanted in a woman's uterus and allowed to develop to term. (The woman receiving the implanted embryo is usually the same woman who donated the eggs, but in some cases the recipient could be a surrogate.) Using sperm donated by some-

Figure 9.12

Outline of the procedure used for *in vitro* fertilization.

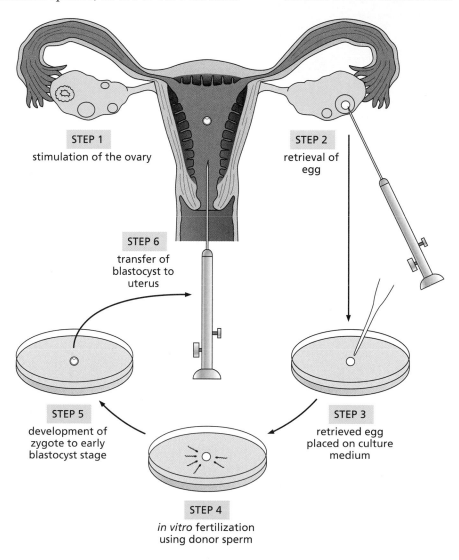

STEP 1
stimulation of the ovary

STEP 2
retrieval of egg

STEP 6
transfer of blastocyst to uterus

STEP 5
development of zygote to early blastocyst stage

STEP 3
retrieved egg placed on culture medium

STEP 4
in vitro fertilization using donor sperm

one other than the woman's husband raises the same kinds of legal issues (including custody and financial responsibility issues) as does artificial insemination.

In vitro fertilization was first successfully practiced on humans in 1978 and is now used in over 36,000 cases annually in the United States and Canada alone. Several newer modifications of this technique are also occasionally practiced. In one such technique, called ZIFT (zygote intrafallopian transfer), the zygote is transferred soon after fertilization rather than waiting until the blastocyst stage. In GIFT (gamete intrafallopian transfer), the eggs and sperm are inserted into the uterine tubes and fertilization takes place there rather than *in vitro*. In cases of low sperm count or low sperm motility, sperm heads or whole sperm may be injected into egg cells instead of allowing fertilization to take place by itself.

Persons who possess alleles that they do not wish to pass on to their offspring may seek *in vitro* fertilization and embryo testing before implantation, or artificial insemination using donated sperm. Genetic testing of the zygote or early embryo can be carried out before the use of any of the above techniques for implantation. Some researchers have been experimenting with techniques that test eight-cell embryos for genetic diseases (see Chapter 3). Because the testing process is usually destructive, only one of the eight cells is separated and tested, leaving the other seven cells available for implantation if desired; the organism resulting from a seven-cell embryo is just as normal as if all eight cells had been used. Embryos obtained by *in vitro* fertilization can thus be tested before they are implanted into a woman's uterus to complete the pregnancy. This technique, sometimes called Blastomere Analysis Before Implantation (BABI) or Preimplantation Genetic Diagnosis (PGD), has already been used to test human embryos *in vitro* for cystic fibrosis, allowing the selection of only those embryos that are free of the disease. Selected embryos can then be implanted, and the couple can be free of the fear that their child will be born with cystic fibrosis, a disease that is usually fatal. Once this technique becomes readily available for a wider variety of human conditions, it will become possible to avoid certain genetic diseases by this procedure, or to choose certain other characteristics, such as the child's sex.

Surrogate pregnancy. Surrogate pregnancy is the use of another woman's womb to carry a baby to term on behalf of a woman who cannot undergo the pregnancy herself, usually for medical reasons. In most cases, the baby is conceived by *in vitro* fertilization, if possible using egg and sperm cells donated by the couple who want the baby. The resulting embryo, which is the genetic offspring of the donor couple, is then implanted into another woman who agrees to act as a surrogate mother, usually for a fee. In addition, medical expenses are generally paid by the donor couple. The legal status and rights of the surrogate mother are subject to many ethical and legal questions. Surrogacy contracts have been outlawed or held invalid in a number of jurisdictions that view the birth mother (i.e. the surrogate) as the legal parent who is therefore 'selling' her baby if she receives any payment. Among the ethical issues raised are the exploitation of poor women by wealthy couples. Financial need is often a factor (one of many) in a woman's decision to become a surrogate. Other issues include the amount of compensation that can ethically or legally be given to the surrogate and the ways in which this situation is distinguished from 'baby-selling.' A final issue concerns the available alternative of adoption, which is generally less expensive and raises fewer legal and ethical objections.

The drive to reproduce is a strong force among all species, including humans. Not all humans want to have children, but for many it is something they desire greatly. When some people seek out assisted reproductive technologies to overcome infertility, it shows their motivation to have children that are genetically their own. In contrast, many people wish to limit the number of children that they have. In the next section we discuss the ways in which the timing and number of births can be controlled.

THOUGHT QUESTIONS

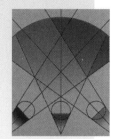

1 Are there hormones that are found only in males? Only in females?

2 Most forms of assisted reproduction are expensive and are likely to remain so, and success rates are generally low. They are considered elective procedures, and are therefore usually not covered by insurance.
a. Is it discriminatory to provide assisted reproduction to those who can afford to pay for it, when so many other people cannot?
b. Given the low success rate, should people be encouraged to give up after repeated failures? How many failures?
c. Should an upper age limit be imposed on assisted reproduction? Should different limits be put according to age on the number of attempts at assisted reproduction?

3 If an embryo is gestated in the uterus of a surrogate mother who did not contribute the egg, what rights should the surrogate have? What rights should the egg donor (and her partner) have? What legal rights do these parties have where you live? Should these parties have the right to negotiate and agree to a certain division of rights by contract? Why or why not?

4 In most cases of assisted reproduction, multiple eggs are fertilized, only some of which are then implanted and brought to term. What should be done with the rest? If they are stored (usually in a frozen condition), who has the right to decide their fate? Does that include the right to destroy them, to refuse to pay for their continued storage, or to decide where to implant them? Can one member of a couple exercise any of these rights if the other member objects? What if the couple divorces or if their relationship changes in another way?

5 Do reproductive health clinics offer assisted reproductive technologies to couples who are infertile because of reversible lifestyle decisions (such as cocaine or marijuana use)? Do you think they should?

Can We Diminish Population Growth and Its Impact?

As Malthus realized, many factors influence population growth. Improving nutrition and health care generally increases population by increasing fertility, decreasing infant mortality, and decreasing the death rate.

Deaths from accidents are being decreased by safety measures. To arrive at a stable population size without increasing the death rate, the main option available to people is a voluntary reduction of the birth rate.

We should make note of a distinction at this point: **population control** is usually understood to operate on the level of populations, while **birth control** methods generally operate by preventing births one at a time. A birth control method is not a successful population control method unless it is widely adopted.

Birth control is not new. It has always been practiced among human populations. Ancient texts in China, India, and Egypt mention abortifacients, i.e. drugs that induce abortions. In Egypt, the Ebers papyrus (1550 B.C.) describes a medicated tampon made with ground acacia seed. Fermentation of the seed in the female reproductive tract produces lactic acid, which is toxic to sperm. Many more birth control options are available today.

Birth control methods work by controlling fertility, so the term 'fertility control' is often used as a synonym for birth control. Because these methods allow the spacing and timing of the birth of children, they are also called family planning methods. Many of these methods prevent pregnancy by interfering with the reproductive anatomy or physiology of either the female or her male partner. The various methods of birth control form a spectrum of possibilities (Figure 9.13). By the timing of their action, they can be arranged into four distinct groups, which are highlighted in the gray boxes in the figure.

For each of the methods listed, a series of questions can be asked:

1. How does it work?
2. How effective is it in birth control?
3. What costs or risks are involved?
4. What kinds of objections have been raised against it?
5. Does it have any benefits apart from birth control, e.g. in the prevention of sexually transmitted disease?

Birth control acting before fertilization

Of all birth control methods, those that act to prevent pregnancy before fertilization (conception) are often called **contraceptive** measures.

Preventing gamete release by sterilization. Sterilization, the elimination of reproductive capacity, usually involves surgery and is usually permanent, although some methods are potentially reversible. One way to achieve male sterilization and allow hormone secretion to continue is by vasectomy, the surgical cutting and tying off of the sperm duct (the vas deferens; see Figure 9.8). Males with vasectomies continue to produce both testicular hormones and sperm for some time, but the sperm cannot reach the penis for release.

Among female sterilization methods, tubal ligation (tying off of the oviduct) is the only one done primarily as a birth control measure. Tubal ligation is analogous to male vasectomy; eggs and hormones are still produced, but the eggs are blocked from traveling to the uterus. Surgical removal of the uterus (hysterectomy) is performed for medical reasons other than birth control, but the removal of the uterus results in permanent sterility because the uterus is where the developing embryo grows.

Surgical removal of the ovaries, the organs that produce the female gametes, also results in sterility, but this is usually avoided for the same reason as male castration is avoided: the gonads produce many hormones, and their removal has widespread effects on the individual.

All these sterilization methods involve surgery, which makes them expensive to implement on a very large scale, especially in poor or medically underserved areas. As with all forms of surgery, there are risks, such as those of infection or from the use of anesthesia. However, both costs

Figure 9.13

Methods of birth control.

Method	effectiveness*		relative cost to provide
I. PREVENTING GAMETE RELEASE			
I.A. Sterilization Methods			
1. irreversible sterilization: (castration, hysterectomy, or ovariectomy)	100%		high
2. semipermanent sterilization:			
tubal ligation	99.6%		high
vasectomy	99.6%		medium
I.B. Hormonal Methods			
3. estrogen alone	general use 98%	experienced users 99.5%	medium
4. estrogen plus progesterone	general use 98%	experienced users 99.5%	medium
5. progestin only (injectible or 'minipill')	general use 97.5%	experienced users 99%	medium
6. male contraception	99%		medium
7. prolonged breast feeding	varies with nutritional status of mother		none
II. PREVENTING FERTILIZATION			
II.A. Abstinence Methods			
8. long-term sexual abstinence: (celibacy, delayed marriage)	100%		none
9. timed abstinence ('rhythm' methods)	general use 76%	experienced users 98%	none
10. withdrawal during intercourse (coitus interruptus)	general use 77%	experienced users 84%	none
II.B. Barrier Methods			
11. condom	general use 90%	experienced users 98%	low
12. vaginal diaphragm with spermicide	general use 81%	experienced users 98%	low
13. cervical cap with spermicide	general use 87%	experienced users 98%	medium
14. sponge with spermicide	general use 80%	experienced users 98%	low
II.C. Spermicidal Methods			
15. spermicidal creams, foams, and jellies alone	general use 82%	experienced users 97%	low
III. PREVENTING IMPLANTATION			
16. post-coital pills	not known		medium
17. intrauterine devices (IUDs)	general use 95%	experienced users 98.5%	medium
IV. INTERRUPTING GESTATION			
18. abortion	100%		high

* A 99% rate of effectiveness means that 1% of couples using that method will become pregnant in a year of use.

and risks are experienced on a one-time basis only and do not recur. All sterilization methods are completely effective as birth control methods without any further action on the part of the individual.

One of the greatest objections to all these methods is that they are permanent. Many people, even people who want birth control, do not want to become permanently sterile. In the United States, about 60% of men who undergo vasectomy later regret that they did so. Vasectomy and tubal ligation can in some cases be reversed, but success depends on microsurgical techniques, and reported success rates vary greatly. A new method for male sterilization has been developed in China, and is reversible under local anesthesia. In this method, a polyurethane elastomer is injected into the sperm duct, where it solidifies to form a plug that effectively blocks the passage of sperm.

Preventing gamete release by hormonal methods. Several birth control methods depend upon alterations of the female reproductive cycle. Hormonal birth control methods take advantage of the feedback mechanisms shown in Figure 9.10. For example, estrogen and progesterone both inhibit the secretion of FSH, so that supplying these hormones (or a similar compound) prevents ovarian follicles from reaching maturity and releasing their eggs. The hormones can be given as birth control pills, as injections, as implants just under the skin, or as patches on the skin. Regardless of the method of drug delivery, all hormonal methods work by preventing the egg from maturing and being released. Because hormones have many effects throughout the body, hormones used in birth control have many side effects, including the possibility of blood clots. For this reason, medical supervision is recommended and hormonal methods require a prescription in most countries.

Early birth control pills contained estrogen alone, but progesterone was later added (producing the combination birth control pills, also called 'combination oral contraceptive' or COC) to reduce the levels of estrogen and also its side effects. The continuous levels of these hormones prevent the usual hormonal cycling from taking place. The expense and the requirement of obtaining a prescription limit the use of birth control pills in many populations. Some developing countries have made birth control pills available *without* prescription, to encourage their more widespread use, but cost is still a problem. Birth control pills have become the most commonly used contraceptive method among the middle and upper classes in many countries. Most birth control pills sold today are of the combination type.

Newer types of birth control pills, developed in the 1980s, use progesterone-like compounds (progestins) only. These include the 'minipill', which suppresses ovulation only half the time, but instead works primarily by thickening the cervical mucus. Minipills have a relatively high failure rate, largely because they need to be taken on time (even a 3-hour delay can reduce their effectiveness). For this reason, subdermal implants of progestins have been used in many cases. Implants of a drug called levonorgestrel (trade name Norplant) can be inserted beneath the skin in the underarm region, where they slowly release steady dosages of the drug. The early version, called Norplant I, quite commonly resulted in irregular menstrual bleeding, but this side effect is observed much less commonly with the newer version, Norplant II.

Hormonal control of male fertility has been repeatedly suggested. One of the earliest such methods to be developed was a male contraceptive pill containing a drug called gossypol, derived from cottonseed. This drug was first developed around 1970 in China and is said to be about 99% effective. Although it does not affect the hormone testosterone, gossypol does somehow interfere with sperm production. Tests of gossypol have reported some toxic side effects, so an effort is now being made to develop a synthetic substitute.

Testosterone is known to inhibit LH and FSH production by the pituitary, which in turn can inhibit sperm production. However, complete suppression of sperm production by this method requires weekly testosterone injections, and these injections may have undesired side effects. It has also been discovered that the suppression of pituitary function is actually brought about by the small fraction of the testosterone that is converted into estrogen. As a consequence, several researchers have begun exploring the possibility that estrogen or progesterone-like drugs might be capable of functioning as a male contraceptive. The drug Depoprovera, similar to progesterone, has been used for this purpose in Europe for several years, and in California it has been used to suppress testicular function in convicted rapists. (In this use, the drug does inhibit spermatogenesis, but it has no effect on violent behavior.)

One of the newest proposed methods is the use of vaccines to achieve contraception by blocking reproductive hormones in males and thus inhibiting sperm production. In one approach being developed in India, men would be immunized against FSH, a hormone needed for sperm production; this method has been tested in animals and is now being tested in humans.

Preventing gamete release by extended breast feeding. An older hormonal method of birth control is the practice of extended breast feeding (delayed weaning), which is common in many traditional societies in African countries. Children in Africa are almost always breast-fed, and many are not weaned until they are 4–6 years old. While a woman is breast feeding, she is producing hormones that stimulate milk production. These same hormones also inhibit the rise and fall of the hormones produced by the ovary, thus interrupting the menstrual cycle and preventing egg maturation. Women who use this method of birth control do not wean their youngest child from the breast until they feel they are ready to have another child. This method of birth spacing is a widespread and seemingly effective practice in many parts of Africa, although studies have shown that it is unreliable among women living in the industrialized world at high caloric intake levels. For example, prolonged breast feeding has a contraceptive effect among the !Kung San (Bushmen) of South Africa and Namibia. These women have a low caloric intake and walk 4–6 miles a day, conditions that seldom occur among North American women. Because the ability of breast feeding to delay the return of the menstrual cycle is greatest among women who are physically active but who have a low caloric intake, improvements in maternal nutrition may actually decrease the effectiveness of delayed weaning as a method of birth control. Birth spacing by prolonged breast feeding is also most effective when babies suck vigorously and often. Any contribution to infant nutrition other than breast milk (e.g., by bottled milk or cereal) reduces the effect. In most third-world countries, the use of bottled milk reduces the effectiveness of birth spacing by delayed weaning. The closer spacing of births leads to an increase in birth rates at the population level, possibly offsetting the effects of birth control programs. Even so, one expert states; "In developing countries today, breast-feeding probably prevents more births than modern contraceptive methods."

Preventing fertilization by sexual abstinence. As a means of preventing fertilization, abstinence has the distinct advantage of being available to all people free of charge. However, the success of abstinence methods depends upon the determination of the people using them. Total voluntary abstinence (celibacy) has long been practiced as part of a regimen of religious devotion, but only by small numbers of people. Delayed marriage (with no other sexual activity) greatly reduces the birth rate, especially inasmuch as the years before age 30 are the most fertile period for a majority of women.

The so-called **rhythm method** is a form of partial abstinence, based upon the fact that a woman is fertile only for several days after ovulation, while the egg is in the oviduct or uterus; sexual intercourse performed at other times will generally not result in conception. Several versions of the rhythm method are practiced. The simplest version is a calendar method, in which a couple abstains from day 10 to day 17 (or, to be more certain, from day 7 to day 17) of a 28-day cycle in which the onset of menstrual bleeding is counted as day 1 (see Figure 9.10). Another version, with a higher success rate, is called the Billings method. In this method, the woman feels the mucus just inside her vagina with her fingers. Estrogen causes this mucus to become more slippery and elastic just before ovulation, when estrogen reaches a peak; after that, the secretion becomes scant and dry. A couple using this method abstains from intercourse from the onset of slippery mucus until 4 days after the peak day, but may have intercourse on the remaining dry days.

If practiced correctly, the rhythm method is highly effective, but its effectiveness depends on several conditions, including the regularity of a woman's menstrual cycles (this varies individually), the ability of the couple to keep a calendar and count the days without making a mistake, and the willingness of the couple to refrain from sex (or else to practice another method of birth control) during the woman's fertile period. The effectiveness of the rhythm method can be increased by monitoring the woman's vaginal temperature, since a rise in temperature indicates the time of ovulation more precisely.

Sexual intercourse is also called coitus. Coitus interruptus is the withdrawal of the penis before ejaculation occurs. Some couples have used this method effectively, but the majority find it unsatisfying or difficult to follow. Since some sperm can be released before ejaculation, coitus interruptus is not reliable. On a population-wide scale, it is generally not as effective as other methods.

Because of the high failure rates associated with abstinence methods, their advantage in low cost may be elusive. When the medical costs of an unwanted pregnancy are included, abstinence methods become among the most expensive methods available (Figure 9.14).

Preventing fertilization by barrier methods. **Barrier methods** are those that impose a barrier to the passage of sperm. Most **condoms** are designed for males and cover the penis, but a condom that is worn by women has also been developed. The male condom is the oldest of the barrier methods. First developed in England, traditional condoms were constructed of animal membranes (usually sheep intestine) and were therefore considered a luxury item. The development of rubber and then latex made condoms more widely available and also more reliable. Condoms have the added advantage of protecting against AIDS and other sexually transmitted diseases (see Chapters 16 and 17).

Other barrier methods include vaginal inserts (worn by women), such as cervical caps, vaginal diaphragms, and sponges. Vaginal diaphragms must initially be individually fitted by a physician or other trained medical worker, and must be inserted correctly into the vagina before intercourse and left in place for several hours thereafter. When properly placed, vaginal diaphragms block the movement of sperm from the vagina to the uterus, thereby preventing their joining with the egg. (One study in Brazil reported a higher-than-usual success rate if the diaphragm was left in place nearly all the time, but this result remains to be confirmed in other populations.) Cervical caps are made to fit over the narrow portion (cervix) of the uterus, where they also block sperm.

Spermicidal agents, chemicals that can kill sperm, in creams, foams, jellies, or suppositories, are often used together with a barrier method; the combination of barrier plus spermicide is much more effective than either method used alone. One of the newest methods is a sponge impregnated with spermicidal fluid. (Spermicides should not, however, be used with many types of condoms; the spermicide partly dissolves the condom, making it ineffective as a barrier to sperm or to sexually transmitted diseases.)

Barrier methods used with spermicides have extremely low failure rates when used by people familiar with their proper use; most pregnancies occurring with barrier methods are the result of improper use. Barrier methods are widely used in many countries.

Birth control acting after fertilization

Several birth control methods act after the egg has already been fertilized by a sperm.

Preventing implantation. Hormones or hormone analogues can be used to prevent implantation of a fertilized egg if taken within 72 hours after intercourse. These postcoital or 'morning-after' pills can thus be used for emergency contraception. Drugs of this kind contain either a high dose of estrogen or else a combination of estrogen and progesterone. In the United States, such drugs require a prescription, but emergency contraception kits containing these pills have been available in Europe for at least a decade, and school nurses in France can now dispense such drugs to teenage girls upon request.

CONNECTIONS
CHAPTERS 16, 17

Figure 9.14

Relative costs (over five years of use) of various birth control methods under the type of medical insurance plans common in the United States. For many methods, the largest costs are those associated with unwanted pregnancies resulting from the method's failure.

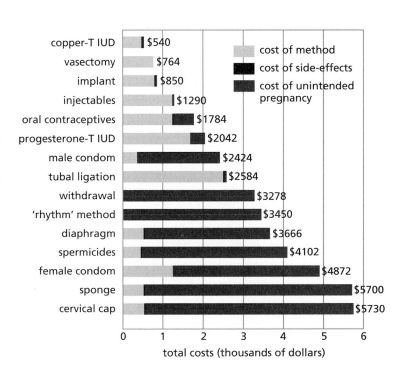

	cost of method
	cost of side-effects
	cost of unintended pregnancy

Method	Total cost
copper-T IUD	$540
vasectomy	$764
implant	$850
injectables	$1290
oral contraceptives	$1784
progesterone-T IUD	$2042
male condom	$2424
tubal ligation	$2584
withdrawal	$3278
'rhythm' method	$3450
diaphragm	$3666
spermicides	$4102
female condom	$4872
sponge	$5700
cervical cap	$5730

total costs (thousands of dollars)

One of the newest hormonal methods for preventing implantation is a drug called mifepristone (or RU-486), developed in France and now available elsewhere. In the United States, clinical testing found the drug safe and effective in preventing pregnancy. The drug was recommended for approval by the Food and Drug Administration (FDA), and is now legally available in the United States (with some restrictions) but is still not widely marketed, in part because of political pressure. Mifepristone blocks the action of the hormone progesterone, which is necessary for maintenance of a pregnancy. Mifepristone has its greatest potential use as a morning-after pill to prevent implantation from occurring during the first 5 days after intercourse.

An **intrauterine device** (IUD) is a small piece of plastic or wire, in one of several shapes (e.g., coiled or T-shaped), that is inserted by a physician into a woman's uterus, where it remains until removed by a physician. IUDs prevent pregnancy by preventing implantation, although the exact mechanism by which this occurs is not known. (Desert Bedouins have long practiced a similar method of birth control on their camels by inserting stones into the uteri of female camels to prevent pregnancy and removing the stone when breeding was again wanted.) A major advantage to this method is that, once inserted, the IUD works on its own with no need of further action on the part of the woman. On a worldwide basis, IUDs are used more than any other birth control method. This is largely due to the widespread use of IUDs in China, the world's most populous country, and the method is widely used throughout Asia and Europe. In the United States, the use of IUDs trails far behind the use of birth control pills and even sterilizations. The steroid-releasing Progestasert and a version of the copper-T are the most-used IUDs in the United States.

IUDs should not be used by women who have never been pregnant. However, women who use IUDs have a higher rate of satisfaction than with any other form of birth control, including pills. Today's IUDs are effective and their failure rate is low. Earlier IUDs were not as safe; one early model, the Dalkon Shield, caused infections and severe bleeding problems in many women, resulting in hysterectomies, large lawsuits, and the corporate bankruptcy of its manufacturer. Many people have shown renewed interest in IUDs in the last decade because of the development of newer, safer types.

Abortion. The termination of a pregnancy, including the cleaning out of the uterine lining and the expulsion of the embryo or fetus, is called **abortion**. The traditional method of dilation and curettage (enlarging the cervix, then scraping out the uterine interior with a spoonlike instrument) has now been supplemented by newer techniques such as vacuum aspiration (using a machine that uses suction to clean out the uterine contents). There are medical risks to the woman, including excessive bleeding, the chances of infection, and uterine injury, which can result in sterility. The sum total of risks to the life and health of the woman is less than the sum total of risks associated with completing the pregnancy and giving birth. The medical risks are especially low if the abortion is done early, during the first trimester, meaning the first 3-month portion of a 9-month pregnancy. Second-trimester abortions using saline injections can also be done safely in most cases. At whatever time an abortion is performed, the risks are much higher if it is done by an untrained person.

Worldwide, the highest abortion rates are in the countries of the former Soviet Union, where other methods of birth control are not readily available. In 1990, an estimated 11% of all Russian women aged 15–44 had undergone an abortion. In contrast, The Netherlands has an abortion rate of less than 1%, one of the lowest among the countries for which data are available. Surveys have also identified The Netherlands as the country with the highest rate of virginity among women entering into their first marriage. Both of these findings have been attributed to a societal attitude that encourages open discussions of sexuality and sexual matters.

The drug mifepristone, previously mentioned as a morning-after pill, can also be used to induce abortion after the embryo has become implanted in the uterus. Successful abortion has also been achieved in 96% of a group of 178 women given two drugs (methotrexate, followed days later by misoprostol) that are also legally available (by prescription) in the United States.

Abortion is seldom considered as a method of choice. Even among people who have no ethical objections to its use, abortion is typically thought of as a last resort, to be used if an earlier-acting method fails in a particular case. Abortions performed by untrained persons carry a high risk of injury, subsequent infertility, or death. Safe abortions, performed by trained personnel under antiseptic conditions, are expensive. Most population planners have advocated that safe abortions be made a widely available option as a backup when the other, less costly methods have failed.

Infanticide. Though not technically a means of birth control, **infanticide** has long been practiced as a means of population control in many parts of the world, especially in times of famine. There are records of infanticide from Medieval Europe, and the practice was still widespread in China, India, and other parts of Asia well into the twentieth century. In most cases, the infant is not directly killed, but is instead allowed to die through lack of care. In societies in which infanticide is practiced, female infanticide is more common than male infanticide. In China, infanticide is now officially outlawed, but boys outnumber girls in many areas. The government's strict population control policy allows only one child per couple (Figure 9.15), and social scientists suspect that female infanticide is still widely practiced by couples who want a boy but instead have a girl.

Cultural and ethical opposition to birth control

No single method of population control is best for all societies. Abortions and sterilizations, for example, require medically trained personnel. They are more expensive and more labor-intensive than other methods, and are therefore unlikely to become the methods most

Figure 9.15

A poster promoting birth control in China. The poster reads, "birth control benefits the nation and benefits the people."

widely used even among populations that have no objections to them. All methods need to be adapted to the customs of the people using them, and education in the use of certain methods may meet with resistance of various kinds. For instance, women in many Muslim societies are generally forbidden to discuss reproductive matters with anybody outside their families, including health care workers.

The attitudes of the Catholic church toward birth control have varied over the centuries; official opposition to most forms of birth control is historically recent. A considerable debate about birth control took place within the Catholic hierarchy in the 1960s, resulting in two Papal encyclicals, *Populorum progressio* (1967) and *Humanae vitae* (1968). The first of these acknowledges the population problem and the need for family planning in underdeveloped areas; the second denounces abortion, sterilization, and all forms of birth control except for the rhythm method. Surveys in many countries show that a large majority of Catholics use various forms of birth control despite the Church's official position. Attempts to spread birth control information have often been opposed by the Church, especially in Latin America, but Church teachings have not stopped Italy, a country over 98% Catholic, from achieving a stable (nongrowing) population, with one of the lowest birth rates in Europe.

Other religious groups have generally been more tolerant of contraceptive methods. Abortion, however, is opposed by Catholics, Protestant fundamentalists, Orthodox Jews, and Muslims. In its simplest terms, the principal argument voiced by these groups is that a fetus is a living human being and that killing it is an act of murder. People who wish to keep abortion legally available have argued several major points, including a woman's right to choose, a child's right to be wanted by his or her parents, and the need to control the world's population. This is a highly charged issue.

The abortion debate. Laws on abortion vary greatly from place to place and sometimes from one time period to another. In the United States, many states outlawed abortions until the Supreme Court ruling in *Roe v. Wade* (1973). As a result of this court opinion, women in the United States have a legally recognized right to an abortion under certain conditions. During the first trimester of pregnancy, this right can be exercised by the woman in consultation with her physician, without any interference from state laws. During the second trimester of pregnancy, state governments can impose waiting periods or certain other conditions, and during the third trimester they can limit abortions more strictly or outlaw them entirely. As a result, abortion practices vary from state to state. Practices also vary elsewhere: Ireland and most Moslem countries outlaw abortions, whereas most other European and Asian countries permit them.

Many of the questions usually raised in the course of the abortion debate revolve around matters of definition: When does 'life' begin? Is a fetus a 'person'? Is it a 'human being'?

Biological definitions of life. The usual definitions of 'life' (see Chapter 1, p. 11) mention properties such as cellular structure, motility, metabolism, homeostasis, the ability to respond to stimuli, and the presence of genetic material that is inherited. By these criteria, sperm cells, egg cells, and embryos are alive, and there is no age at which life begins. This, however, is a biological definition, and it still leaves for each society to decide the proper treatment, ethical status, and legal rights of human embryos. A "pro-life" advocate could use this definition to argue that abortion is a form of murder, whereas an advocate for reproductive choice could just as well argue that organs that are 'alive' by this definition are routinely removed surgically and discarded as 'medical waste,' and that an abortion is in this regard the moral equivalent of an appendectomy.

There are many other possible criteria for defining the start of human life, including the following.

- Anything that possesses human genetic material (DNA) could be considered human, so an embryo would be considered human throughout its development. Also human by this definition are human gametes (haploid sperm and egg cells) and the 'medical waste' such as human blood and surgically removed organs and tissues.

- A new individual could be defined to begin its life when it attains a unique individual genotype at the moment of fertilization.

- An embryo could be defined as a new and distinct individual once it loses the ability to split into separate individuals with separate personalities and experiences, at approximately 12 days after implantation.

- An embryo could be defined as a separate person only when it is capable of surviving outside the womb. This ability depends to a great extent on the ability of the lungs to remain open and function in aerating the blood. Only after the 25th week are the lungs sufficiently well endowed with a vascular blood supply to allow the baby's tissues to receive enough oxygen. Before the 25th week, the lungs tend to collapse when empty, and their internal mucous linings tend to stick together, preventing the lung from refilling. Many states in the United States use this criterion to define the legal status of a fetus as 'viable,' and therefore the killing of the fetus as an act of murder from this age forward.

- If life ceases when brain waves (EEGs) can no longer be detected, then the beginning of sentient life could be defined by the onset of these brain waves, at about 26–27 weeks of gestation. In their book, The Facts of Life (New York, Oxford University Press, 1992), Harold Morowitz and James Trefil argue for an 'acquisition of humanness' (and independent viability) at the time when most of the connections between nerve cells in the cerebral cortex are made. Electrical waves in the brain (EEGs) (see Chapter 13, pp. 494–495) begin at about this time, providing evidence that the fetus can now respond to certain stimuli and can be described as having 'experiences.'

- A baby could be defined as alive when its circulatory, respiratory, and digestive systems begin to function independently of its mother. These changes occur at birth, when the umbilical cord is cut and the baby takes its first breath.

Each of these definitions could be used to argue for a particular legal or ethical treatment of human embryos, but ultimately this is a social, ethical, and legal decision rather than a biological one. In fact, different societies (and sometimes different legal jurisdictions) use these criteria very differently.

Legal definitions of personhood. 'Personhood' is a social or legal concept, not a biological one, and it is differently defined in each culture or legal system. In the legal system followed in the English-speaking world, a 'person' is defined as a legal entity having certain legally recognized rights and duties. In this tradition, corporations and estates have the legal rights of 'persons.' The 'personhood' of a fetus is therefore a matter of legal definition and not of biology, and, because legislators can define personhood in various ways, the legal rights of a fetus can vary from one jurisdiction to another. Other cultures have their own ways of defining the rights or personhood of a fetus or newborn:

1.   in Japanese tradition, a baby is not considered a person until it utters its first cry, so killing it is not considered homicide;

2.   in parts of West Africa, a child is not considered human until it is a week old;

3.   the Ayatal aborigines of Formosa had no punishment for killing a child before it was given a name at age two or three years;

4.   natives of the Pacific island of Truk considered deformed infants to be ghosts and either burned or drowned them.

The point is that different cultures and different legal systems can reach remarkably different conclusions. Notice again that these are really not scientific questions, capable of being decided entirely by observational data.

Ethical considerations. Using the ethical principles discussed in Chapter 1, a deontologist who opposed abortions would simply argue that it is a wrongful act regardless of any type of medical or other evidence. Such a person would regard the matter as totally outside the bounds of science, because no possible observation or experimental evidence could change the wrongness of what they regard as a wrongful act. A utilitarian would weigh the possible consequences of an abortion (including the monetary costs and the medical risks) against the consequences of the birth if carried to term. Among the latter consequences are medical risks to the mother's health from the delivery, medical expenses, and the costs to the mother of raising the child (or the costs to society, if for some reason the mother does not raise the child properly). From a utilitarian viewpoint, the mother's wishes, abilities, and financial circumstances are all important in the evaluation; to the deontologist, they are all irrelevant. Much of the frustration of the whole abortion debate is that deontologists and utilitarians talk in such different terms and use such different arguments that neither has much hope of convincing the other of anything.

In the meantime, a small number of anti-abortion extremists have resorted to bombing clinics, and terrorizing or shooting doctors who perform abortions. Many hospitals have discontinued performing abortions to avoid being the targets of such intimidation. One result is the increasing concentration of abortions performed at clinics that do little else, making them easier targets for these extremists. Other results are that fewer doctors are being trained to perform abortions, while many other doctors who might otherwise perform abortions are refusing to do so for reasons of personal safety. These developments further reduce the pool of doctors and hospitals willing to help someone seeking an abortion. In some states it has become increasingly difficult for a woman seeking an abortion to find a doctor willing to perform one.

Population control movements

Organizations to promote birth control and control population growth were first formed in the nineteenth century. In England, several of these organizations called themselves 'Malthusian,' even though Malthus, a curate in the Anglican Church, was opposed to nearly all of the birth control methods available in his day. Knowledge about reproduction and birth control was not widely available before the mid-twentieth century. An 1832 book on the subject by an American physician, Charles Knowlton, was banned as immoral in the United States and elsewhere. Two British reformers, Charles Bradlaugh and Annie Besant, and an American, Margaret Sanger, worked through the late 1800s and early 1900s to make information and contraceptives more widely available. Besant strongly influenced Mahatma Gandhi and later migrated to India. Sanger always viewed birth control from the perspective of giving individual women more control over their own lives. The availability of birth control in the United States owes much to her tireless campaigns. She also traveled widely, spreading the message of birth control to India, China, and Japan.

The influence of Gandhi, Besant, and Sanger led India to become the first modern nation to institute a government-funded campaign to control its population. Beginning in 1951, India implemented population control measures that featured easy access to both contraception and abortion and an information campaign to encourage their widespread voluntary use. During the 1960s, nine other nations, including China, Egypt, and Pakistan, also implemented population control programs.

The most ambitious population control program in history was adopted in China in 1962, at a time when their population was around 700 million and was growing at just above a 2% annual rate. The campaign for "only one child for one family" (see Figure 9.15) was waged with special vigor. Parents who had only one child were given various benefits (such as pregnancy expenses paid for the first child only, free contraceptives, better housing, and educational benefits), while parents who bore more than one child were fined and sometimes imprisoned. The goal of this campaign was not just to limit population growth, but to reduce the population to its 1962 level of 700 million as quickly as possible. Because of a large demographic momentum (that is, an age structure with many children), China's officials realized that they would have to cut the birth rate to somewhat *below* the mortality rate for a time in order to achieve a stable population. China's population in 2000 was around 1.3 billion, but the annual growth rate has been cut to 0.71%.

The education of women

One of the most effective methods of reducing the number of children borne by each woman is to educate women. Studies in many parts of the world have shown that the population birth rate falls with each rise in the education level of women, even in the absence of any program aimed specifically at birth control (Figure 9.16). Around the world, the countries with the lowest rates of female literacy also have the highest population growth rates, whereas those with higher female literacy have lower growth rates. Countries with female literacy rates below 15% include Yemen, Niger, and Burkina Faso, all with population growth rates of 3.0% or more. Third-world countries with female literacy rates of 80% or higher include Jamaica, Sri Lanka, Lesotho, Botswana, and Tajikistan, all with population growth rates below 1.0%.

Third-world women with a seventh-grade education or higher tend to marry later than other women (4 years later, on average, and these are among the most fertile years). They also use voluntary means of birth control more often, have fewer (and healthier) children, and suffer far less often from either maternal or infant mortality in childbirth. The empowering of women (by giving them more education and more control over their reproductive lives) also raises the educational level of their children and results in more rapid economic development (at lower cost) than many other programs aimed more specifically at development.

In the United States, educational efforts are also an important part of most efforts to reduce pregnancy rates in teenagers. The rate of teenage pregnancy is lowest among women with the most years of schooling and highest among those with only a grade-school education or less.

Although most population control programs are carried out by national governments on their own populations, several programs are international in scope, including those run by the World Health Organization (a branch of the United Nations) and by the U.S. Agency for International Development (USAID). The United Nations has sponsored many conferences on population. Some past conferences emphasized the environmental impacts of population growth, but the 1994 conference in Cairo, Egypt, shifted the attention to the education of women. Women's rights advocates from a variety of countries stressed the need to improve both the education and legal status of women. Many people cautioned that overzealous government-sponsored programs aimed at population control could restrict the reproductive freedom of individual women as much as the earlier lack of birth control information. Although these

Figure 9.16

The education of women reduces the average number of children per family. Data for the graph are from the World Bank.

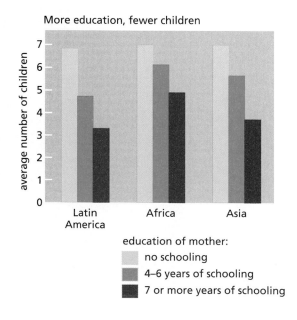

More education, fewer children

average number of children

Latin America   Africa   Asia

education of mother:
- no schooling
- 4–6 years of schooling
- 7 or more years of schooling

Teenage women in a Papua New Guinea classroom

people generally see the need to reduce the rate of population growth, they are deeply suspicious of programs that coerce individual women or restrict their freedom, and they are especially suspicious of programs urged upon third-world nations by male-dominated institutions in the industrial world. Instead, they favor programs to educate women and improve their legal status and reproductive choices. They point out that birth rates have diminished whenever the education, legal rights, and reproductive choices of women have improved.

Controlling population impact

Demographic transition brings about a marked increase in human population. In most parts of the world, the excess population tends to migrate to the cities, producing an overcrowding that strains the resources of those urban areas. In Europe and Japan, this process occurred gradually over a period of several hundred years (roughly, 1600 to 1900), giving cities a chance to adjust to their changing conditions. Many cities have accommodated high population densities (meaning large numbers of people per square mile) without widespread misery. Since World War II, urbanization in the third world has taken place much more rapidly than it did in Europe. Rapid, unplanned growth has strained most urban services to the point at which many of the newly arrived migrants have inadequate housing. Crowded slum areas or shanty towns often lack safe drinking water and may also suffer from chronic water shortage; sanitation and waste disposal are also frequent problems. It is often not crowding in itself that results in these problems, but crowding without sufficient facilities to support the population. Crime often increases and may become difficult to control, although many other factors beside population contribute to crime rates. Unemployment and economic hardship, when they occur, may compound these problems. The hardships of urban crowding usually fall disproportionately on the poor.

Pollution (see Chapter 19) tends to increase approximately in proportion to population, most obviously because of corresponding increases in household garbage and waste water. Densely populated areas are dependent on food, water and fuel coming in from a much wider radius; consequently their environmental impact is felt far beyond their political borders. As population increases, more forests are cleared for agricultural use and more trees are cut down to build houses. The destruction of habitat for other organisms, particularly of forests, is one result (see Chapter 18). The loss of arable land (through topsoil erosion, desertification, and other processes) is accelerated by population growth, as is the depletion of nonrenewable resources such as minerals or fossil fuels.

CONNECTIONS CHAPTERS 11, 18, 19

Effects of consumption patterns. The impact on the environment is not, however, solely a function of the number of people. The amount of the world's resources that each person consumes is not equal around the globe. On average, the amount of resources consumed by a person in the United States is 54 times that consumed by a person in a developing country. The impact of this consumption is magnified still further by the fact that much of this consumption is of nonrenewable resources. In addition, resources that might be renewable are often consumed or discarded in ways that make them nonrenewable. The enormous size of municipal solid waste disposal sites in industrialized countries is testament to these consumption patterns. These landfills are among the largest structures ever built by humans, and the materials within them are unavailable for reuse or biodegradation.

If the rate of energy use does not exceed the rate at which energy is captured from the sun in photosynthesis (see Chapter 11), then the energy use is considered **sustainable**. In many industrialized countries, however, present patterns of energy consumption are already unsustainable because they remove far more energy from global ecosystems than they produce.

Discussions of the world's population crisis frequently become linked to discussions of the environmental crisis. Many people, especially in the third world, believe that the population crisis is only a small part of a greater environmental crisis.

This environmental crisis, they say, is made worse by the industrial world's overconsumption more than by the third world's population increase. Frances Moore Lappé is one of several American writers holding such views. Some analysts even question whether the industrial world's concerns over population are misdirected (and possibly racist). Third-world countries, they say, could well support far larger populations than they do now if it were not for the export of so many of their resources to support the patterns of overconsumption that have become so typical of the industrial world. If the industrial countries, they say, were to give up their lavish patterns of consumption, then the third world could well support a larger human population (at a larger carrying capacity) than it does in current circumstances.

Others take what may be called a neo-Malthusian position. Paul Ehrlich, for example, views most other problems as consequences of overpopulation. If the population were smaller, he argues, most environmental problems would diminish or even disappear. Because some countries (mostly in Europe) have already limited their population growth, the greatest efforts should be directed at those nations (mostly in the third world; see Figure 9.2) that have the highest population growth rates.

Overpopulation and overconsumption need not be viewed as opposing viewpoints. Population growth and profligate consumption are both widely recognized as problems, and each makes the other worse. Some people see one of these as the bigger problem; some people see the other. Efforts directed at addressing either problem can only help to ameliorate both.

Limits on carrying capacity. Many scientists tell us that we will soon reach or even exceed the carrying capacity of the planet. In fact, this is one point on which people concerned with overpopulation and those concerned with overconsumption agree, although they postulate different causes for this condition. One of the few dissenters, economist Julian Simon, observes that the technological revolutions of past centuries have repeatedly brought about demographic shifts, each of which has increased the carrying capacity. He predicts that future technological revolutions will continue to enlarge the planet's carrying capacity indefinitely. Nearly all other scientists and writers who have contemplated the subject of population believe instead that the planet's carrying capacity has a limit.

Can the global carrying capacity be increased further? The answer is not known with certainty, but it depends in part on whether we assume the Earth's natural resources to be renewable and unlimited (Julian Simon's view) or limited and nonrenewable (the majority viewpoint). Those who accept the limits imposed by nonrenewable resources will be driven to the conclusion that carrying capacity cannot be increased very much. In fact, if we maintain our present patterns of consumption, we may not even be able to sustain the present population levels forever.

Human populations, like all other populations, are biological entities, requiring energy flow to survive. Populations are therefore subject to the laws of physics (energy is neither created nor destroyed), and populations cannot exceed the limits imposed by the availability of energy. When they approach K, populations are controlled by biological factors, such as starvation and disease. So the question "Should populations be controlled?" is academic because populations will be controlled by the forces of biology and physics, regardless of our answer. The relevant questions are, "Should we exercise preventive efforts at population control?" and, if we should, "How should we do so?"

THOUGHT QUESTIONS

1 Is an increase in population the only factor that puts resources in limited supply? Does population growth affect the availability of housing or medical care in the same way that it affects the availability of drinking water, sanitation controls, and food?

2 Find out whether your college makes birth control information and birth control itself available to the student population. Are certain methods favored over others? Why?

3 Why are most forms of birth control aimed at women rather than men? Why are most population control campaigns aimed at women? In societies in which women have little or no control over their lives, are they likely to be able to carry out family planning? Will family planning give them greater control?

4 Do you think it is proper to view birth control information as a freedom-of-speech issue? Do you think birth control methods should be taught in the public schools? Why or why not?

5 What social benefits are likely if population growth is controlled? What social and ethical problems need to be considered? What individual rights are at risk? In your opinion, what is the best way for a government to control its country's population growth without restricting the reproductive freedom of its women?

6 In what countries is abortion legal? Where is abortion illegal? Can you suggest reasons for these differences?

7 Can biology have any useful role in the abortion debate? What role? Would any biological data be persuasive to a person who opposed abortion on the basis of deontological principles? Would data be persuasive to someone using utilitarian ethics?

8 Most stem cells (useful in cancer research or cancer therapies) come from discarded embryos and aborted fetuses. Fertilized eggs are the ultimate stem cells. However, U.S. law prohibits the National Institutes of Health (NIH) from funding any experiments that deliberately create or destroy a human embryo. How does this restriction on research relate to the abortion debate in the United States? Do you agree with the restriction? If you do not, how would you change the law?

9 In Practice Question 15 on p. 324, what considerations have been ignored? Do you think they may safely be ignored?

Concluding Remarks

Individual decision making in family planning is sometimes at odds with government decisions aimed at population control and also sometimes with various religious teachings. There are also many other reasons why people might resent strangers urging them to modify their most personal behaviors in one way or another. All of these factors, moreover, vary from place to place, and the lessons learned in one country or population cannot necessarily be applied uncritically to other populations elsewhere. However, any attempts to implement change based on biological data will necessarily take place in a context of many, often competing, social values. Many scientists think that moving to sustainable consumption levels may partly alleviate the problems of population growth, but only temporarily. In addition, motivating people to decrease their consumption may be just as difficult as motivating them to have fewer children. Even well-planned efforts to address social, economic, or environmental problems may prove to be inadequate as resources are stretched to the breaking point in the face of increasing population pressure. "Whatever your cause," says one slogan, "it's a lost cause unless we can control population."

Chapter Summary

- The principles of population ecology are applicable to all species.

- A new population shows rapid **exponential growth** at first, but its **growth rate (r)** levels off when it approaches the **carrying capacity (K)** of its environment, a phenomenon called **logistic growth**.

- The effects of population growth beyond the carrying capacity are numerous: consumption of resources is increased, pollution is increased, and any inefficiency in the utilization of resources results in starvation and death.

- Humans and other **K-selected** species are characterized by stable populations at or near carrying capacity (K) and by most individuals living a long time. In contrast, **r-selected** species show more rapid growth, higher mortality early in life, and unstable population sizes.

- Human populations have grown markedly after each major advance in technology. Each major population increase has taken the form of a **demographic transition**, beginning with declining mortality and ending when the **birth rate (B)** declines to match the **death rate (D)**.

- Many factors promote infertility, including those that interfere with the menstrual cycle in females or with sperm quality in males.

- **Ovulation** and many other aspects of sexual development and of fertility are controlled by **hormones**.

- Understanding reproductive biology can help us to treat infertility and also to control the birth rate.

- Many methods of **birth control** are available. They differ in their biological mechanisms, their costs, their medical risks, and their acceptance by different groups of people.

- Many studies have found that improvements in the education and legal status of women lowers the birth rate in a cost-effective manner and brings other benefits besides.

CONNECTIONS TO OTHER CHAPTERS

Chapter 1	Abortion and other forms of birth control raise important ethical issues.
Chapter 3	New reproductive technologies can be used to enhance fertility.
Chapter 5	Population size responds to evolutionary forces, such as competition from other species.
Chapter 8	Different species have different reproductive strategies that control birth rates.
Chapter 10	Undernutrition and malnutrition increase the death rate and also lower rates of fertility.
Chapter 11	Plants are the main biological energy producers on which the life of humans and other consumer organisms depends. Populations cannot be sustained above levels that can be supported by producer organisms.
Chapter 15	Overcrowded conditions cause stress in many species.
Chapter 16	AIDS has greatly increased the death rate in many countries.
Chapter 17	Emerging infectious diseases increase infant mortality and decrease life expectancy.
Chapter 19	Increasing population size usually makes pollution problems worse. Needs for water and sewage treatment increase with population size.

PRACTICE QUESTIONS

1. What is the birth rate B in a nation with 4 million men and 3.9 million women, in which 200,000 children are born in one year? What is the birth rate B in a nation with 3 million men and 4.9 million women, in which 200,000 children are born in one year?

2. What is the death rate D in a nation of 100 million people in which 500,000 people died in a year and 200,000 of those who died were children below the age of 2 years? What is the death rate D in a nation of 100 million people in which 500,000 people died in a year and 50,000 of those who died were children below the age of 2 years?

3. What is the growth rate r in a nation of 500 million people if the birth rate is 2% and the death rate is 1% (assuming no immigration or emigration)? What is the growth rate r in a nation of 50 million people if the birth rate is 2% and the death rate is 1% (assuming no immigration or emigration)?

4. How many people will be added in one year to a population of 500 million people with a growth rate of 2%? How many will be added if the growth rate is 4%? How many will be added at growth rates of 2% or 4% if the initial population was 5 billion?

5. What is the doubling time of a population of 500 million people if the growth rate is 2%? What is the doubling time of a population of 5 billion people if the growth rate is 2%?

6. If a population is at its carrying capacity K of 500 million and its birth rate is 3%, what is its death rate (assuming no immigration or emigration)? If a population is at its carrying capacity K of 5 billion and its birth rate is 3%, what is its death rate (again assuming no immigration or emigration)?

7. A population of 4.5 million has a birth rate of 0.067 (or 6.7%) and a death rate of 0.024 (2.4%). Find the

growth rate (r) and the number of years that it will take for the population to double.

8. For the population in the previous question, find:

   a. the population increase this year;

   b. the size of the population after a year of increase;

   c. the population increase in the second year;

   d. the population size after 2 years of increase.

9. Which will increase more rapidly this year: a population of 3.3 million with a growth rate of $r = 2.0\%$, or a population of 5 million with a growth rate of 1.1%?

10. Where in the body is testosterone produced? Where in the body is estrogen produced?

11. During the menstrual cycle, what hormones are secreted by the ovaries? Other hormones are involved, in addition to the ovarian hormones. Where in the body are these other hormones produced?

12. How do hormonal contraceptive methods induce infertility?

13. How do barrier contraceptives prevent fertilization?

14. A particular birth control method is 98% effective. In a nation of 8 million people, if 400,000 women use this method, how many of them will become pregnant this year?

15. Suppose you are working for a family-planning program in a nation whose population is growing rapidly. Among several available methods of birth control are the following.

    Method A costs $1600 per woman, lasts an average of 20 years, and is 99% effective.

    Method B costs $20 per month for each woman and is 78% effective.

    Method C costs $200 per year for each woman and is 88% effective.

    Which method is most cost-effective (reduces pregnancies by the largest amount per $1000 spent)? (See also Thought Question 9 on p. 321.)

Issues

- Do all humans have the same dietary requirements?
- How are human diets related to good health? How are they related to chronic diseases?
- What is malnutrition? What are its causes and consequences?
- What social factors contribute to obesity or heart disease?
- Why do some people deliberately starve themselves?

Biological Concepts

- Organ systems (digestive system, circulatory system)
- Cell membranes (diffusion, active transport)
- Acidity
- Molecular structure (chemical and physical basis of biology: water, carbohydrates, lipids, proteins, enzymes, polar and nonpolar molecules)
- Energy and metabolism (energy conversion and storage, chemical bond energy, ATP, calories, glycolysis, Krebs cycle, oxidation–reduction reactions)
- Evolution (lactose intolerance)
- Homeostasis
- Health and disease (macronutrient malnutrition, micronutrient malnutrition, fiber, eating disorders, ecological factors)
- Species interactions (mutualism)

Chapter Outline

10 Nutrition and Health

*W*hen she weighed 65 kg (140 pounds), Melanie thought of herself as fat and ugly. Her menstrual periods stopped when her weight dropped to 45 kg (100 pounds). Now that she weighs 40 kg (90 pounds), all her friends tell her she is too skinny, but she is sure they are wrong because she still thinks of herself as chubby. She wants to lose even more weight. Melanie has an eating disorder known as anorexia nervosa. Her body is not getting the nutrition it needs. She could die if the situation remains untreated.

Melanie's father went in last week for a routine checkup. Although he was feeling fine, the doctor told him he had high blood pressure and needed to control his fat intake. Unless he lowers the fat content of his diet, he faces an increased risk of having a heart attack. He must now learn to eat a low-fat, high-fiber diet, which will lower his chances for getting heart disease, the number one cause of death in most industrialized countries.

All of us need food, but our dietary requirements vary according to our body size, age, sex, level of activity, and previous state of health. In addition, there are variations caused by hereditary differences in body constitution, metabolic rates, and other factors. The world's populations have found many different ways of meeting these nutritional needs. Different diets arose in different parts of the world because different kinds of plants grew best in each climate and in each type of soil, and each culture has its own preferences and prohibitions that limit their uses of the available foods in their environment—no culture makes use of all foodstuffs available to them.

In this chapter we examine the body's use of food, human dietary requirements, correlations of diet with the incidence of chronic diseases, and the effects of malnutrition that can result from eating too little or from eating the wrong foods. Malnutrition is one of the major health problems of the world, particularly among the poor and in areas of turmoil. Malnutrition can also result from eating disorders among people with sufficient access to good foods.

All Humans Have Dietary Requirements for Good Health

Foremost among our dietary needs is a need for energy, as measured in kilocalories (kcal). A **kilocalorie** is the amount of energy required to raise the temperature of a kilogram of water by one degree Celsius. The 'calories' that dieters count are actually kilocalories. Your body's need for caloric energy depends on many factors, such as body weight, level of activity, and gender (Table 10.1). A person who is completely inactive (e.g., in a hospital bed) requires a minimal number of calories, and this amount can be converted into a **basal metabolic rate**, the rate at which an inactive person uses energy.

Caloric intake is the most important measure of dietary sufficiency. In most industrialized countries, most people are adequately nourished or overnourished. In the industrialized countries, particularly in the

Table 10.1

Calculating your body's caloric needs.

A. BASAL METABOLIC RATE

First, find the number of kilocalories required to maintain basal metabolism, that is, to keep you alive when you are inactive and lying down:

Average adult woman	21.6 kcal/kg body weight per day
Average adult man	24.0 kcal/kg body weight per day

This works out to about 1225 kcal daily for a 125-pound woman or 1850 kcal for a 170-pound man. (1 pound = 0.454 kg)

B. LEVEL OF ACTIVITY

Multiply the figure obtained above by a factor depending on your normal level of activity:

1.35 for sedentary activity (e.g., telephone sales, TV viewing)
1.45 for light activity (e.g., college studies, office work with occasional errands, light housekeeping)
1.55 for moderate activity (e.g., nursing, vigorous housekeeping, waiting on tables, light carpentry)
1.65 for heavy activity (e.g., pick-and-shovel work, bricklaying, full-time competitive athletics)

C. OTHER FACTORS

Figures obtained above need to be increased by as much as 10% for any of the following conditions:
Growth (children 15 years old and younger)
Pregnancy
Recovery from a major illness or injury

D. INDIVIDUAL DIFFERENCES

Figures calculated above are only guidelines or averages. Your individual need may either be greater or smaller. If you maintain a steady caloric intake on a day-to-day basis and you gain weight, your caloric intake is greater than your caloric needs. Conversely, if you lose weight, your intake is less than your caloric needs.

United States, many people are overweight. A large number of overweight people (and quite a few who are *not* overweight) have tried to modify their food intake by dieting, either for weight loss or because of a greater interest in health. In a general sense, a diet is the sum of all foods eaten by a person or by a population. In popular usage, being 'on a diet' means something more restrictive: a conscious choice of foods to bring about some desired outcome. Diets that bring about weight loss do so by reducing caloric intake. Other diets, such as 'heart-healthy' diets, aim to promote long-term health by exclusion or inclusion of particular types of foods. Dieting can be carried to an extreme, however, and diets that are nutritionally very unbalanced can be harmful.

On a worldwide basis, inadequate caloric intake is the most widespread nutritional problem. Starvation kills millions of people each year, most of them children. Starvation and malnutrition are most noticeable in the nonindustrialized, or third-world, countries, but are also present in impoverished areas, both rural and urban, within many industrial nations. Many other nutritional problems, such as vitamin deficiencies, exist in undernourished people; most of these other problems are hard to treat if the caloric intake remains inadequate.

Most of what we call food can be classified chemically into three types of major constituents and several minor constituents. The major constituents, called **macronutrients**, include carbohydrates, proteins, and lipids (fats); the minor constituents, called **micronutrients**, include vitamins and minerals. Food is used to fuel all the activities of life, and macronutrients are the major sources for energy. In addition, people require fiber and micronutrients, which are not energy sources but have other vital functions. Next we examine the biology of these components of foods to help us understand why a balanced diet is necessary to maintain good health; we also see the consequences of increased or decreased intake of specific foods.

Carbohydrates

Most people around the world derive the majority of their calories from **carbohydrates**, which include starches and sugars. Plants store energy as carbohydrates, so they are a good dietary source for carbohydrates. Cereal grains such as wheat, rice, oats, and corn are the most nutritious source of carbohydrates because they also contain important vitamins, protein, and fiber. Breads, pastas, and other foods made from cereal grains retain all their nutritional value as long as the whole grain is used. Fruits and fruit products (including juices) generally contain sugars such as fructose or sucrose, together with important vitamins, minerals, and fibers. However, refined sucrose (table sugar) lacks these other nutrients and can also contribute to tooth decay (Box 10.1). Most vegetables contain carbohydrates but are even more important as sources of vitamins, minerals, and fiber.

Carbohydrates are molecules formed principally of three types of atoms: carbon, hydrogen and oxygen (Figure 10.1). A single carbohydrate unit is called a monosaccharide or simple sugar. Simple sugars differ from each other by their number of carbon atoms and the placement of their chemical bonds. More complex carbohydrates are built by hooking these monosaccharides together in pairs (disaccharides) or into larger structures (polysaccharides). Starch is a common polysaccharide composed of repeated units of the sugar glucose (see Figure 10.1).

Carbohydrates are largely soluble in water because of an important similarity in the types of bonds in carbohydrates and in water. In a chemical bond two atoms are held together by the sharing of electrons between the two atoms. These bonds can be either **polar** or **nonpolar** (Figure 10.2). Polar bonds have an unequal distribution of electrons and thus of electrical charge, whereas nonpolar bonds have a much more equal distribution of electrical charge. Water (H_2O) is one of the most polar liquids, with electrons unequally shared between hydrogen and oxygen atoms. Carbohydrates have many polar bonds between hydrogen and oxygen atoms and also between carbon and oxygen atoms. Because of the high proportion of oxygen atoms and of polar bonds (see Figure 10.1), carbohydrates tend to be soluble in water.

The human body's daily need for carbohydrates is measured in terms of total caloric intake, as indicated in Table 10.1. In terms of caloric content, all carbohydrates, both sugars and starches, are the same, providing

BOX 10.1 How Does Sugar Contribute to Tooth Decay?

The sugar that we add to coffee or cereal is known chemically as sucrose. There are many other sugars: fructose (fruit sugar), lactose (milk sugar), and dextrose (a synonym for glucose). Many bacteria live in our mouths and use these dietary sugars for their metabolic energy. One type of mouth bacteria make a gluelike substance that attaches them to the tooth surface, and to make this substance they require sucrose. Once the bacteria are glued to the tooth they can use other sugars (including sorbitol, the sugar in 'sugarless gum') as energy sources. When bacteria extract energy from sugars, acids are produced, and these acids dissolve tooth enamel, resulting in cavities. Without sucrose, the bacteria cannot make the glue and the acids are not trapped so closely against the enamel surface.

4 kilocalories per gram (kcal/g). From an energy standpoint, it makes little difference if the carbohydrates are eaten in the form of sugar or starch, or whether the sugar is fructose or sucrose. There is, however, a difference in the rate of absorption: starches generally take a few hours to be digested into absorbable sugars, while dietary sugars are capable of being absorbed within minutes. A meal containing both sugars and starches therefore maintains the body's energy level (or blood glucose) more evenly over a longer period.

In most populations, an increase in the consumption of carbohydrate-rich foods (especially whole grains) is desirable. Many diets in the nonindustrialized countries supply inadequate calories, and carbohydrates provide the most efficient and most economical means of improving

Figure 10.1

Chemical structure of selected carbohydrates.

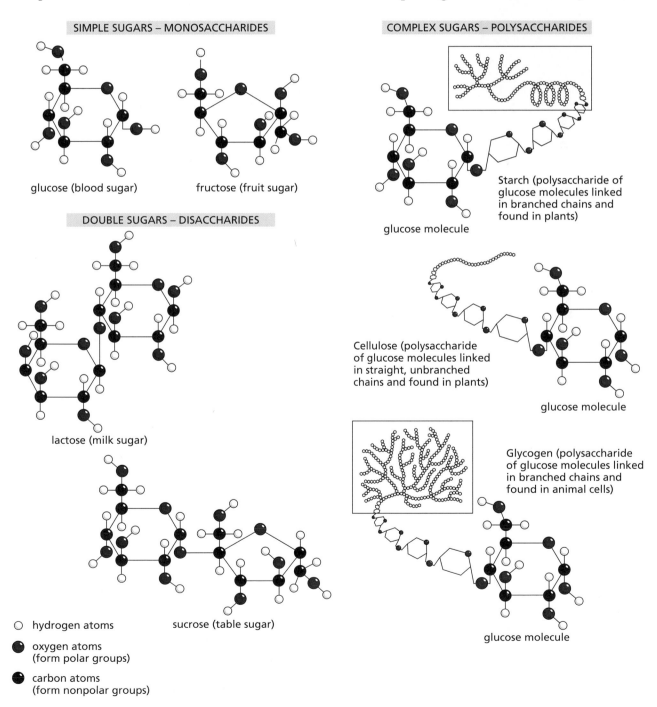

SIMPLE SUGARS – MONOSACCHARIDES

glucose (blood sugar)   fructose (fruit sugar)

DOUBLE SUGARS – DISACCHARIDES

lactose (milk sugar)

sucrose (table sugar)

○ hydrogen atoms

● oxygen atoms (form polar groups)

● carbon atoms (form nonpolar groups)

COMPLEX SUGARS – POLYSACCHARIDES

Starch (polysaccharide of glucose molecules linked in branched chains and found in plants)

glucose molecule

Cellulose (polysaccharide of glucose molecules linked in straight, unbranched chains and found in plants)

glucose molecule

Glycogen (polysaccharide of glucose molecules linked in branched chains and found in animal cells)

glucose molecule

these diets. Fewer kilocalories of labor, or fewer dollars, are needed to produce a kilocalorie of carbohydrate food than a kilocalorie of most fat-rich or protein-rich foods. In the United States, the replacement of dietary fats by complex carbohydrates, especially from whole-grain sources, would have many indirect health benefits, including a reduction in risks for heart attacks and certain forms of cancer.

Lipids

Lipids are organic compounds that do not dissolve in water because they are made mostly of hydrogen and carbon atoms organized into nonpolar hydrocarbon chains. Common lipids are fats, oils, waxes, phospholipids, and steroids. Dietary lipids are mostly **triglycerides**, molecules in which glycerol (a three-carbon molecule) is linked to three long chains of carbons and hydrogens called fatty acids (Figure 10.3). Triglycerides that are solid at room temperature are commonly called fats; those that are liquid at room temperature are commonly called oils. As sources of caloric energy, fats and oils contain almost 9 kcal/g, which is over twice as much as carbohydrates. A small amount of lipid is a dietary necessity, in part because the fat-soluble vitamins (especially A and D) cannot be absorbed without it. Lipids are also a source of fatty acids, which are the nonpolar portion of the phospholipid molecules that form cell membranes. Two particular fatty acids (linoleic and arachidonic acids) are required from dietary sources because they cannot be made by the body, but they are required only in very small amounts (about 3 g or one tablespoonful per person per day). Most nonstarving people have an adequate intake of lipids. In the United States, many people consume too much lipid. The body tends to store excess lipid (and some excess carbohydrate) as fatty deposits within numerous adipose (fat-storing) cells.

Fatty acids, cholesterol, and cell membranes. **Saturated fats** are fats whose fatty acids have only single bonds (see Figure 10.3). Most saturated fats are derived from animal sources (or from a few tropical plants such as palm and coconut), and most are solid at room temperature. **Unsaturated fats**, often derived from plant sources, have double bonds as well as single bonds in their fatty acid chains, causing the molecule to bend (see Figure 10.3). Those containing only one double bond are sometimes called monounsaturated; those with multiple double bonds are polyunsaturated. Both types are usually liquid at room temperature because the bends made by the double bonds prevent the molecules from packing too tightly together and solidifying.

Fats are important in cell membranes. Fatty acids from dietary fats become incorporated into cell membranes as part of molecules called **phospholipids**. The fatty acid portion is nonpolar, as we have seen, but the other end of a phospholipid is polar. In water, phospholipids orient

Figure 10.2

Polar molecules and nonpolar molecules.

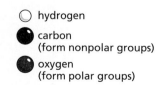

○ hydrogen

● carbon (form nonpolar groups)

◕ oxygen (form polar groups)

water (H_2O): a polar molecule with unequally shared electrons

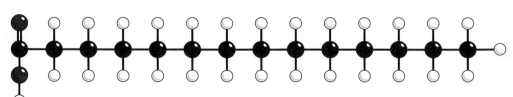

a fatty acid: a nonpolar molecule held together by nonpolar bonds

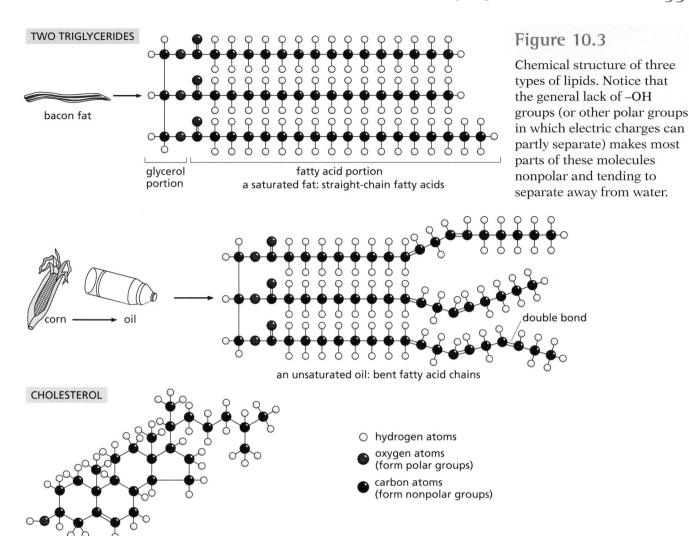

Figure 10.3

Chemical structure of three types of lipids. Notice that the general lack of –OH groups (or other polar groups in which electric charges can partly separate) makes most parts of these molecules nonpolar and tending to separate away from water.

spontaneously to form bilayer membranes in which the phospholipid polar heads face the watery surfaces and the nonpolar fatty acid tails are protected from the water by forming the interior of the bilayer (Figure 10.4). Membrane proteins are embedded in this phospholipid bilayer.

Cells remove the fatty acid chains from dietary triglycerides and incorporate the chains into membrane phospholipids. When unsaturated fatty acids are incorporated into the phospholipid cell membrane, the bends prevent their tight packing in the membrane, keeping the membrane more fluid. The phospholipid molecules need to be fluid to allow the embedded proteins to function. Conversely, if the diet is high in saturated fats, the membrane is less fluid, which reduces the functioning of membrane proteins such as those involved in nutrient absorption into cells. It is hypothesized that when dietary lipids cannot be properly absorbed into cells they may tend to build up on blood vessel walls, contributing to heart disease.

Another important dietary lipid is **cholesterol**, a fat-soluble molecule that is an important constituent of animal cell membranes (see Figure 10.3, and notice the absence of chains in the chemical structure). Along with unsaturated fatty acids, cholesterol helps to keep the membranes fluid, thereby keeping the cell and the organism functioning properly. Cholesterol is also the precursor of several important hormones.

We need cholesterol in small quantities, but our bodies can usually synthesize this amount, so little or none is needed from food. Plant cell membranes do not contain cholesterol, so plant products are always cholesterol-free, although some (like coconut oil) contain saturated fatty acids that are easily converted into cholesterol by the body. All dietary fatty acids are broken down into one of the major starting materials of cholesterol synthesis; cholesterol synthesis is thus increased by nearly all fatty foods, even if they are advertised as 'cholesterol free.'

Because the body makes about 75–80% of its own cholesterol, and makes it from dietary fats, most of the cholesterol circulating in the bloodstream comes from dietary fats (especially saturated fats), not from dietary cholesterol. Excess cholesterol, like excess amounts of other lipids, can build up on blood vessel walls and increase your risk of disease. Most foods that contain cholesterol are also high in saturated fats, so avoiding either also helps you to avoid the other. Eggs are exceptional in having a lot of cholesterol with few other fats.

Figure 10.4

The structure of phospholipids and cell membranes. The cell membrane shown is an animal cell membrane, and therefore contains cholesterol.

Structure of a phospholipid molecule

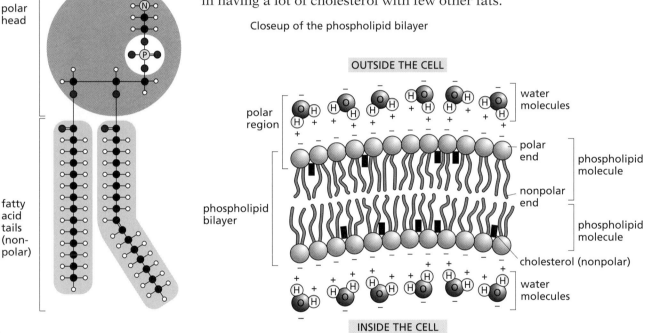

Closeup of the phospholipid bilayer

General structure of the plasma membrane

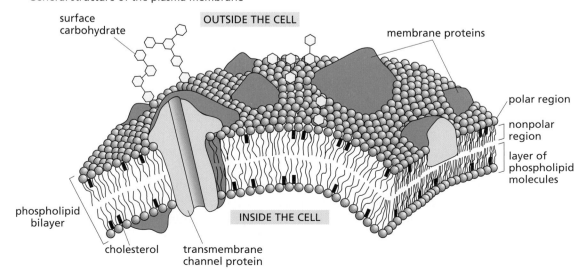

Proteins

The body uses **proteins** for tissue growth and repair, including the healing of wounds, replacement of skin and mucous membranes, and manufacture of antibodies (see Chapter 15). Proteins are important components of all cell membranes and can function to transport other molecules across cell membranes or as receptor molecules. Many proteins of the cellular interior provide structure, motility, and contractility to muscles and other cells. Other proteins such as collagen and elastin are located outside cells and give connective tissues their strength and thus help to support the entire body. Keratin, another protein found outside cells, is essential for healthy skin and is the main constituent of hair and fingernails. A much larger assortment of proteins function as **enzymes**, as described later in this chapter. Some enzymes (such as those used in digestion) function outside cells (extracellularly); many others function inside cells (intracellularly).

Some body proteins are needed only in small quantities, but our muscles, blood, skin, and connective tissues need proteins in large amounts. Tendons and certain other body parts are made of proteins that are relatively stable once they have been formed, but blood, skin, bone tissue, bone marrow, and many internal membrane surfaces all undergo constant reworking, repair, and replacement, requiring new protein supplies throughout life. Protein requirements are even higher, per unit of body weight, in growing infants and children, in pregnant or lactating women, and in persons recovering from a major illness or injury.

Dietary amino acids. Proteins are built from chains of smaller chemicals called amino acids. The digestive system breaks down the proteins in food into individual amino acids. After they have been absorbed by the body, these amino acids can then be used to build the body's own proteins. How a protein functions depends to a large extent on its three-dimensional shape after the linear sequence of amino acids has folded. The way in which a protein folds, and whether it is stable in the watery cytoplasm of the cell or in the nonpolar cell membrane, is determined by the arrangement of the polar and nonpolar side groups of its amino acids (Figure 10.5).

Because proteins are synthesized by adding one amino acid at a time to the end of a growing chain (see Chapter 3, pp. 66–67), if one type of amino acid is missing from the cell, the synthesis of any protein needing that amino acid stops. An amino acid that is present in small quantities and is used up before other amino acids is called a **limiting amino acid**.

Proteins are necessary in the diet. The daily requirement is 0.8 g per kilogram of body weight, for example, about 45 g for a 125-pound (57 kg) woman, or 64 g for a 175-pound (80 kg) man. Each species has its own capacities for making certain of the amino acids and therefore has its own dietary requirements for those it cannot make. Of the 20 standard amino acids, 8 cannot be synthesized by the human body and are therefore considered essential in the human diet; a ninth amino acid is essential in human infants. The human body can make the remaining amino acids from these 'essential' amino acids.

Complete and incomplete proteins. Most animal proteins are **complete proteins** in that they contain all the amino acids essential in the human diet. Soy protein is also complete, but most plant proteins lack at least one essential amino acid needed by humans. When an incomplete protein is eaten, the body uses all the amino acids until one of them, the limiting amino acid, becomes depleted. After the limiting amino acid is used up, the body uses the remaining amino acids to produce energy instead of making proteins, because dietary protein cannot be stored for later use in the way that carbohydrates and lipids can.

To get around the problem of incomplete plant proteins in the human diet, we can eat them in combinations in which one plant protein supplies an essential amino acid missing in another one. The Iroquois and many other Native Americans commonly obtained complete protein by combining beans, squash, and corn in their diets. Most bean proteins, for example, are deficient in the amino acids valine, cysteine, and methionine, while corn proteins are deficient in the amino acids lysine and tryptophan. Alone, neither one of these proteins is nutritionally complete for humans, but in combination (as in corn tortillas with a bean filling, or succotash, a mixture of beans and corn cooked together) the two plant sources provide a nutritionally complete assortment of amino acids because each has the essential amino acids that the other lacks.

Vegetarian diets. The amino acid inadequacy of plant proteins poses special problems for vegetarian (meat-avoiding) diets. Vegetarian diets are generally rich in carbohydrates, fiber, vitamins, and minerals, but they may be deficient in certain amino acids unless care is taken to combine several plant proteins at once. Some vegetarians avoid meat but consume fish or milk or eggs; these 'ovolacto vegetarians' can usually meet their protein needs without much difficulty, especially if they combine proteins from both plant and animal sources in the same meal. Strict vegetarians, also called vegans, who do not eat food from any animal source (including milk or eggs), need to carefully combine plant proteins sources so as to supply their bodies with nutritionally complete protein. For example, legumes (beans, peas, and peanuts) can be combined with whole grains (such as rice, corn, or wheat). Nuts and seeds contain protein and can be used to supplement amino acids missing from plant proteins from

Figure 10.5

Chemical structure of part of a protein.

- ◖ nitrogen atoms
- ○ hydrogen atoms
- ● oxygen atoms
- ● carbon atoms

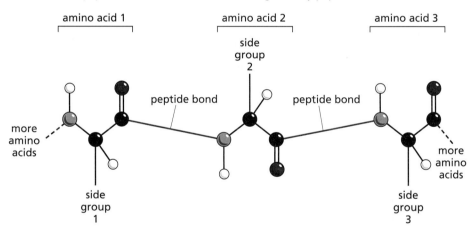

A tripeptide of three amino acids held together by peptide bonds (red lines)

amino acid 1   amino acid 2   amino acid 3

side group 2

peptide bond   peptide bond

more amino acids

side group 1

more amino acids

side group 3

the breaking of the peptide bonds releases the individual amino acids

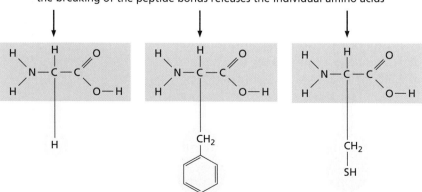

Glycine has the simplest side group, a hydrogen atom.

Seven amino acids have side groups that are insoluble in water. Phenylalanine is such an amino acid.

Nine amino acids have side groups that are soluble in water. Cysteine is such an amino acid.

other sources. Some vegetables also contain individual amino acids that can serve the same function. Our Web site contains vegetarian recipes (under Resources: Vegetarian) and additional references.

Because animal cells store energy principally as fat, proteins obtained from animals are accompanied by fat. Plant cells, in contrast, store energy in the form of complex carbohydrates such as starch, and plant proteins are thus accompanied by very little fat.

Plant-rich diets have other advantages. A given amount of arable land can support a larger human population if that land is used for raising crops for human consumption, including sources of plant proteins, than if the same land is used to raise food for animals that humans can eat. It takes 5–16 pounds of grain protein to produce one pound of meat protein. In well-fed countries with plenty of land, such as Australia or the United States, large tracts can be used for grazing or for the raising of crops primarily for animal consumption. However, poor countries of high population density can ill afford to feed crops to animals. Most of the world's poor eat little meat and get most of their protein from vegetable sources, or in some cases from fish. All whole-grain cereals contain some protein. If this protein is eaten with beans or other legumes, a high-quality protein source is created that is much less expensive than meat and contains far less saturated fat.

The consequences of inadequate protein intake are discussed in a later section.

Fiber

Not all nutrients are required as sources of calories. One example is fiber, material that the body cannot digest and absorb. Human diets should include both soluble fiber (pectin, gums, mucilages) and insoluble fiber (mostly cellulose), and both types are present in most fruits, vegetables, legumes, and whole grains. Many of these fibers are complex carbohydrate molecules, of which cellulose (see Figure 10.1) is an example. An increase in dietary fiber reduces the incidence of several cancers, especially those of the intestine, but scientists are not sure about the exact mechanism of this effect. One intriguing possibility is that protection against these cancers depends on the rate of movement of food through the intestine, and that fiber maintains the optimal rate of food movement. Higher rates cleanse the intestine of potentially toxic chemicals, while lower rates allow these chemicals to remain in one place long enough to undergo fermentation by bacteria into cancer-causing substances (carcinogens; see Chapter 12). Another possibility is that harmful carcinogens are frequently present inside the intestine for whatever reason, but a mucous secretion protects the intestinal lining from them; the insoluble fiber rubbing against the intestinal lining stimulates the lining to secrete more of this protective mucus.

Soluble fiber such as oat bran may reduce the level of serum cholesterol and the risk of heart disease. The mechanism for this effect is not known with certainty, but one hypothesis is that certain soluble fibers bind strongly to bile, synthesized from cholesterol and secreted into the intestinal tract to aid in fat digestion. Without the soluble fiber, the bile would be reabsorbed by the intestinal lining and reused, but the soluble fiber prevents this reabsorption and ensures that the bile is eliminated with the stools. Without recycled bile, new bile must be synthesized from cholesterol, which the body withdraws from the blood, lowering blood cholesterol levels. Diets that are high in fiber are statistically associated with lower rates of coronary heart disease and stroke.

Vitamins

Plants are good sources of micronutrients—vitamins and minerals—because plants need these substances and use them for their own metabolism.

Vitamins are complex nutrients needed only in very small quantities. Most vitamins are coenzymes, the nonprotein portions of enzymes needed for the enzymes to function as catalysts (see p. 343). Enzymes (and their coenzymes) are needed only in very small quantities because they are used and reused in the chemical reactions that they regulate. There are over a dozen vitamins categorized into two groups—water-soluble and fat-soluble (Table 10.2). They may be obtained either from pills or from food. The reasons for preferring vitamins in food are as follows.

- They are much less expensive this way.
- Foods rich in vitamins are also rich in other important substances, including minerals, fiber, and protein, nutrients that have other important health benefits. We do not know the complete nutritional requirements of any organism more complex than bacteria, and undoubtedly our food contains many unknown but needed nutrients. These other nutrients, known and unknown, are not obtained from vitamin pills.
- Some vitamins are more easily absorbed by the body in the combinations with other ingredients that exist in food than they are in the combinations that exist in vitamin pills.
- Purified vitamins can be toxic if taken in excessive amounts, an unlikely danger with vitamins contained in foods.

Vitamin overdoses and deficiencies. The amounts of vitamins recommended for maintaining good health are called recommended dietary allowances (RDAs). Most of these amounts are the same for most healthy adults, but menstruation, pregnancy, and lactation can alter some values in women. Nutritional requirements also differ for growing children and for people recovering from a major illness or injury.

Either too much or too little of a vitamin can result in disease. Vitamin overdoses are possible, but are more likely with fat-soluble vitamins. Water-soluble vitamins, including vitamin C and the B group of vitamins, do not accumulate in the body. When you eat more than you need, the excess is simply excreted in the urine. They must therefore be taken in regularly. Because they are not stored, these vitamins cannot easily build up to toxic overdoses, especially if you get them from foods. It is, however, possible to overdose on water-soluble B vitamins taken in pill form or as concentrated liquids, particularly vitamin B_6. Vitamin B_6 (pyridoxine) is a coenzyme for many of the enzymes of amino acid synthesis; it therefore helps to build proteins and is sometimes used by body-builders. Daily doses of 500 mg or more can be dangerously toxic to the nervous system and liver. Fat-soluble vitamins, namely A, D, E, and K (see Table 10.2), accumulate in the body's fat tissues and can build up over time. Overdoses of these vitamins, especially vitamins A and D, can be toxic.

More often (on a worldwide scale, but not as often in the industrialized countries), people can have vitamin deficiencies. Disorders of fat absorption often cause deficiencies in fat-soluble vitamins because these vitamins are transported and absorbed along with dietary fats. People with such disorders may have plenty of the vitamins in their blood, but their cells are unable to absorb them. Diseases result from deficiencies of each of the vitamins; in fact, research on the cause of these diseases led to the discovery of vitamins.

Vitamin B_1. Vitamin B_1 (thiamine) was the first vitamin to be chemically characterized. While stationed on the island of Java in the 1890s, the Dutch physician Christiaan Eijkman noticed that polyneuritis, a neurological disease in chickens, had many symptoms similar to those of a human disease called beri-beri. Both diseases caused muscle weakness and leg paralysis, resulting in an inability to stand up; both diseases were

Table 10.2

Vitamins and minerals in human health.

	IMPORTANCE FOR GOOD HEALTH	GOOD FOOD SOURCE
WATER-SOLUBLE VITAMINS		
Vitamin B_1 (thiamine)	Helps to break down pyruvate; maintains healthy nerves, muscles, and blood vessels; prevents beri-beri	Meat, whole grains, legumes
Vitamin B_2 (riboflavin)	Important in wound healing and in metabolism of carbohydrates; prevents dryness of skin, nose, mouth, and tongue	Yeast, liver, kidney
Vitamin B_3 (niacin)	Maintains healthy nerves and skin; prevents pellagra	Legumes, fish, whole grains
Vitamin B_6 (pyridoxine)	Coenzyme used in amino acid metabolism; prevents microcytic anemia	Whole grains (except rice), yeast, liver, mackerel, avocado, banana, meat, vegetables, eggs
Vitamin B_{12} (cyanocobalamin)	Required for DNA synthesis and cell division; prevents pernicious anemia (incomplete red blood cell development)	Meat, liver, eggs, dairy products, whole grains
Folic acid	Used in synthesis of hemoglobin, DNA, and RNA; prevents megaloblastic anemia and spina bifida	Asparagus, liver, kidney, fresh greens, vegetables, yeast
Pantothenic acid	Needed to make coenzyme A for carbohydrate and lipid metabolism	Liver, eggs, legumes, dairy products, whole grains
Biotin	Used in fatty acid synthesis and other reactions using CO_2	Eggs, liver, tomatoes, yeast
Vitamin C (ascorbic acid)	Antioxidant; used in synthesis of collagen (in connective tissues) and epinephrine (in nerve cells); promotes wound healing; protects mucous membranes; prevents scurvy	Fresh fruit (especially citrus and strawberries), fresh vegetables, liver, raw meat
FAT-SOLUBLE VITAMINS		
Vitamin A (retinol)	Antioxidant; precursor of visual pigments; prevents night blindness and xerophthalmia	Yellow and dark green vegetables, some fruits, fish oils, creamy dairy products
Vitamin D (calciferol)	Promotes calcium absorption and bone formation; prevents rickets and osteomalacia	Eggs, liver, fish, cheese, butter
Vitamin E (tocopherol)	Antioxidant; protects cell membranes against organic peroxides; maintains health of reproductive system	Whole grains, nuts, legumes, vegetable oils
Vitamin K	Essential for blood clotting; prevents hemorrhage	Green leafy vegetables
MINERALS		
Electrolytes (Na^+, K^+, Cl^-)	Maintain balance of fluids in body; maintain cell membrane potentials	Raisins, prunes; K^+ also in dates
Calcium	Part of crystal structure of bones and teeth; maintains muscle and nerve membranes	Dairy products, peas, canned fish with bones (sardines, salmon), vegetables
Phosphorus	Part of crystal structure of bones and teeth	Dairy products, corn, broccoli, peas, potatoes, prunes
Magnesium	Maintains muscle and nerve membranes	Meat, milk, fish, green vegetables
Iron	Part of hemoglobin; used in energy-producing reactions	Meat, egg yolks, whole grains, beans, vegetables
Iodine	Maintains thyroid gland; prevents goiter	Fish and other seafood products
Fluorine	Strengthens crystal structure of tooth enamel	Drinking water, tea
Zinc	Promotes bone growth and wound healing	Seafood, meat, dairy products, whole grains, eggs
Copper	Cofactor for enzymes used to build proteins, including collagen, elastin, and hair	Nuts, raisins, shellfish, liver
Selenium	Statistically associated with lower death rates from heart disease, stroke, and cancer	Vegetables, meat, grains, seafood

Figure 10.6

Structure of a cereal grain seed.

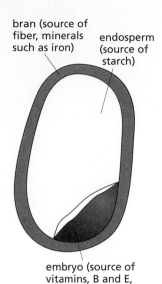

bran (source of fiber, minerals such as iron)

endosperm (source of starch)

embryo (source of vitamins, B and E, minerals, and protein)

A grain is highly nutritious, as can be seen in the schematic diagram. Grains turned into flour lose the bran and embryo. In addition, flour is often treated to make it white, losing further nutritional value. Enriched white flour is a modern phenomenon whereby millers, by law, have to artificially add some of the nutrients back into flour.

CONNECTIONS   CHAPTER 7

fatal if they persisted. Eijkman noticed that the chickens got polyneuritis only when they were fed on polished white rice, but the disease cleared up when rice bran was added to their feed or when whole brown rice was used. From the bran or skins of the unpolished rice grains (Figure 10.6), thiamine was later isolated and was found effective in both treating and preventing beri-beri in humans. Because thiamine is a vital (necessary) substance and is also an amine (a chemical containing an –NH$_2$ group), it was called a 'vital amine.' This term was later shortened to 'vitamine' and then 'vitamin', and the name was applied to the entire class of substances needed only in small quantities. Eijkman's discovery therefore led to the concept of vitamins, and he earned a Nobel Prize for his discovery in 1929. Beri-beri occurs primarily in people whose dietary carbohydrates come from a single, highly refined source such as white rice or white (unenriched) flour. Many countries now have laws requiring the addition of thiamine (and other B vitamins) to refined flour. For this reason, beri-beri is now rare in the industrialized world, although it does occur in severe alcoholics whose dietary intake is inadequate.

Other B vitamins. Other water-soluble vitamins are described in Table 10.2. Many vitamins owe their discovery to research on diseases such as beri-beri (called vitamin deficiency diseases). Vitamin B$_6$ deficiency (microcytic anemia) is seen frequently in people whose diets consist mostly of rice. A deficiency of niacin (Vitamin B$_3$) causes pellagra, a disease of the skin and nervous system. The body can make its own niacin from the amino acid tryptophan, which occurs in many proteins. But populations in which corn is the only protein source are often subject to pellagra because corn is particularly deficient in both niacin and available tryptophan. Hominy grits are made by a process that makes tryptophan (and thus niacin) available from corn protein.

Vitamin C. A deficiency of vitamin C causes scurvy, a disease once common in sailors at sea and among prisoners. A British naval surgeon discovered in the 1600s that limes and other fresh fruits would both prevent and cure scurvy. British ships then began carrying limes and became so well known for this practice that British sailors came to be called 'limeys.' An inflammation of the mucous membranes, as in a cold, increases the body's need for vitamin C. Vitamin C may therefore decrease the severity of the symptoms of such an infection, but it cannot on this account be considered a 'cure' or a prevention for the common cold, as has sometimes been claimed.

People who take large doses of vitamin C can suffer the symptoms of scurvy when they stop taking the vitamin. In addition, megadoses can produce hemolytic anemia (red blood cell deficiency caused by the rupture of red blood cells) in people with the G6PD metabolic deficiency (see Chapter 7, p. 233) found in African American, Asian, and Sephardic Jewish populations. In addition, individuals who are genetically predisposed to gout find that high doses of vitamin C can sometimes bring on the condition by raising blood levels of uric acid. Vitamin C megadoses can produce deficiencies of another vitamin, B$_{12}$, in people who are iron deficient. Even in healthy people, megadoses of vitamin C can irritate the bowel sufficiently to result in diarrhea.

Antioxidant vitamins. Vitamin A (retinol) is essential in the synthesis of the light-sensitive chemicals (retinal) used in vision. Vitamin A is also an

antioxidant, meaning that it protects body tissues from chemicals that would rob those tissues of electrons. The removal of electrons from any molecule is a process that chemists call **oxidation**. Chemicals that bring about oxidation by taking up electrons are called oxidizing agents; among the most highly reactive oxidizing agents are a group of chemicals, called **free radicals**, that have one or more unpaired electrons and a strong tendency to remove electrons from other molecules. Free radicals can thus damage many cellular molecules and are hypothesized to play a role in initiating some cancers (see Chapter 12). Free radicals are present in many things including cigarette smoke, car exhaust fumes, and meat that has been roasted over an open flame. Vitamin A and other antioxidants protect the body by destroying free radicals.

Vitamin A can be obtained from animal sources such as dairy products or fish. Many vegetables also contain a vitamin A precursor, the orange–yellowish pigment beta-carotene, which is split after ingestion to produce two molecules of vitamin A. High consumption rates of whole foods rich in beta-carotene are statistically associated with lower rates of lung cancer, but the causal link between the two is unclear. Laboratory studies on beta-carotene, mostly in rodents, have shown that it suppresses or retards the growth of chemically induced cancers of the skin, breast, bladder, esophagus, pancreas, and colon.

Vitamin E (tocopherol) is another antioxidant vitamin that is especially important in breaking down a group of strong oxidizing agents called peroxides. Vitamin E also helps to prevent spontaneous abortions and stillbirths in pregnant rats, and for this reason it has acquired a reputation as an antisterility vitamin. However, health claims related to the effects of this vitamin on sexual function remain unproved. Vitamin E overdose results in low blood sugar and headache, fatigue, blurred vision, muscle weakness, intestinal upset, and higher rates of lung diseases such as emphysema. Vitamin E occurs in several forms, of which alpha-tocopherol is the most potent. It is destroyed by freezing and also by cooking food.

CONNECTIONS
CHAPTERS 7, 12

Other fat-soluble vitamins. Vitamin D (calciferol) is discussed at greater length in Chapter 7 (pp. 239–241). It is essential to the body's use of calcium in bone formation. Vitamin K is essential to blood clotting because it serves as a cofactor in reactions that produce blood-clotting factors from their inactive precursors. Most people get adequate amounts of vitamin K from the bacteria that live in their intestines. However, newborn infants, whose intestines have not yet been colonized by bacteria, and persons whose intestinal bacteria have been killed off by antibiotics, need more dietary vitamin K until gut bacteria have become established or reestablished.

Minerals

Minerals are inorganic (non-carbon-containing) ions and atoms necessary for proper physiological functioning. The ions of sodium (Na^+), potassium (K^+), and chloride (Cl^-) are the principal electrolytes (charged particles) of the body. Differences in the concentration of ions on opposite sides of a cell membrane are both a type of concentration gradient and a type of electrical gradient, which together are called a membrane potential. Like chemical bonds, membrane potentials are a means by which cells store energy in a usable form. One example of a membrane potential is the electrical potential of nerve cell membranes (see Chapter 13, pp. 467–468) created by the distribution of sodium and potassium ions. Because the body's electrically excitable cells (nerve and muscle cells) respond to changes in these membrane potentials (see Chapter 13, pp. 468–471), maintenance of these electrolytes within a very narrow concentration range is very important. If the number of ions, particularly of sodium ions, gets too high, the body compensates by retaining water that would otherwise be excreted in the urine. Because blood pressure is related to the volume of fluid in the circulatory system, too high a concentration of sodium in the body tissues results in high blood pressure (hypertension). This is an otherwise symptomless condition that increases the risks for vascular (blood vessel) diseases such as stroke and coronary artery disease. Overuse of salt (sodium chloride) makes this condition worse, but is seldom the original cause. Many people in the United States consume too much sodium and not enough potassium. Potassium must be present in the proper amounts; either too much or too little can lead to heart failure and death. Raisins, prunes, dates, and bananas are good sources of potassium.

Other minerals important for human health are iron, calcium, fluoride, and a group of trace minerals.

Iron. Soluble iron is needed for the formation of blood hemoglobin and as a cofactor for many enzymes. A deficiency of iron in cells causes an anemia that is more common in older people with poor dietary habits and in menstruating women. Iron deficiency anemia is in fact the single most common nutritional deficiency in most industrialized countries, including the United States. Menstruating women need almost twice as much iron as men, and pregnant women need even more for the proper synthesis of hemoglobin in the fetus's blood. Vitamin C increases the cellular absorption of iron; so if vitamin C supplies are inadequate, cellular uptake of iron is inadequate too. Some people may have iron deficiencies resulting from low levels of the proteins that transport iron in and out of cells.

Calcium. Calcium (Ca^{2+}) is needed as an intracellular messenger for many processes, including muscle contraction (see Chapter 13, pp. 485–486). In addition, the crystal structure of bones and teeth is composed of calcium combined with phosphate and other minerals. Vitamin D promotes calcium absorption, so most people suffering from vitamin D deficiency have symptoms of calcium deficiency as well. High dietary levels of protein can sometimes contribute to an increase in the rate of excretion of calcium by the kidneys.

Many older women suffer from low bone density and bone brittleness (osteoporosis). Although bone and teeth seem to be very unchanging because of their solidity, they are actually living tissues that are constantly exchanging molecules with the surrounding fluids. There is a balance between the calcium in bone and the calcium in blood; in osteoporosis the balance shifts and calcium dissolves out of bone. Although low blood calcium levels are involved, the problem is not so simple that it can be solved by increasing the dietary intake of calcium later in life. Estrogenic hormones are important, and so is vitamin D, which promotes calcium absorption, but the exact processes are poorly understood. Supplementary doses of both calcium and vitamin D are recommended for postmenopausal women, although most bone loss within the first five years after menopause is caused by estrogen withdrawal, not by any nutritional deficiency. Higher levels of exercise in women aged 18 to 25 can increase their bone density and forestall the development of osteoporosis later in life. Vegetarians have an increased likelihood of osteoporosis at an early age is if they are not careful to eat a complete diet.

Fluoride. Fluoride (the ion F^-) is important in the growth of strong teeth during the childhood years. Insufficient fluoride results in a greater incidence of tooth decay. Drinking water is the most important dietary source of fluoride. In some areas the natural sources of drinking water contain high levels of fluoride. The observation that people in these areas had lower incidences of cavities led to a search for possible factors. Epidemiological research showed that nearby areas had similar diets and climate but higher rates of cavities. Analysis of the water showed higher fluoride levels in the low-cavity areas than in areas with a higher incidence of cavities. Many municipalities now add fluoride (in carefully measured amounts) to the drinking water supply as a preventive measure against tooth decay. Fluoride is also available as drops for breast-fed infants and others who do not have access to fluoridated water. Fluoride is toxic in very high doses. If high doses are accidentally ingested, milk can neutralize the fluoride.

Trace minerals. Most of the remaining mineral nutrients are sometimes called trace minerals because they are needed by the body only in very small quantities. Deficiencies of these trace minerals were more common in the past, when vegetables were grown locally in soils deficient in one or another trace minerals, and when domestic animals grazing on plants growing in the same soil were the main supply of animal food. In the industrial world, such nutritional deficiencies are much less likely because our food supply comes from numerous sources grown in a variety of different soils and climates.

In addition to the trace minerals listed in Table 10.2, chromium and manganese are needed in carbohydrate metabolism; cobalt is an important part of the vitamin B_{12} molecule; molybdenum and nickel are required in the metabolism of nucleic acids; and silicon, tin, and vanadium are needed in trace amounts for proper growth including the development of bone and connective tissue. Diets adequate in other nutrients usually supply sufficient amounts of these trace minerals.

Because of regional variations in the mineral content of soils, the mineral content of foods that grow in those soils also varies. Therefore, mineral deficiencies often vary geographically. Zinc deficiency, for example, is common in the Middle East. Iodine deficiency is found primarily in certain inland locations, such as the high Andes, the Himalayas, and parts of central Africa.

Newly recognized micronutrients

In addition to the micronutrients associated with various deficiency diseases, various other micronutrients are associated with improved human health and reduced cancer risks. Like vitamins, these organic micronutrients are used in very small quantities, and many of them function as antioxidants. The health benefits of the additional micronutrients have been discovered in just the past few decades. The name 'phytochemical' is sometimes applied to a diverse group of plant-derived chemicals found in foods that are associated with good health. Phytochemicals include lycopenes found in tomatoes, alkyl sulfides found in onions and garlic, certain flavonoids in foods such as red wine and green tea, curcumin in turmeric (a spice used in many curries), and many other compounds. Instead of being associated with the prevention of a deficiency disease like scurvy or beri-beri, these micronutrients are associated with lower long-term risks of heart disease and cancer. Maintaining good health involves much more than the mere avoidance of deficiency diseases. Much of the evidence for the health benefits of these nutrients comes from epidemiological studies. For this reason, the mechanisms by which any of these nutrients act is in most cases unknown.

For several reasons, it is much more difficult to establish recommended dietary levels for any of these new micronutrients. For traditional vitamins, a recommended dietary allowance has generally been based on the amount needed by experimental animals to avoid developing a deficiency disease, but these new micronutrients are not associated with any known deficiency disease, so it is difficult to assess whether an experimental animal has had an adequate or an inadequate amount. Another problem is that these micronutrents may work in groups or in combination with one another. For those that function as antioxidants, it is unclear whether an increase in one can compensate for a decrease in another, which would make it almost impossible to set a recommended daily amount for any one of them in isolation from the others. Also, some of these nutrients may work best in combinations, making it difficult to study any one of them in isolation. Given these uncertainties, it makes more sense to seek these nutrients in the natural combinations that occur in foods, rather than one at a time in pill form.

Most micronutrient needs can be met economically, even in poor countries, from grain and vegetable sources. Grains contain most B vitamins and also vitamin E and several important minerals including zinc. Fresh vegetables contain additional vitamins (including A and C), and several important minerals including calcium and iron. Fresh fruits provide additional vitamin C. Legumes provide calcium and iron in addition to proteins. Fruits, vegetables, grains, legumes, and even some spices are sources of a variety of phytochemicals. In general, all micronutrients can be supplied from plant sources, except for vitamin B_{12}, the one important vitamin that cannot be supplied from plant sources alone.

THOUGHT QUESTIONS

1 Why are sugars absorbed from the intestine faster than starches?

2 How is it possible for different species (such as rats and people) to have different vitamin requirements? What does this mean on a biochemical level?

3 How is it possible for some studies to show that calcium supplements forestall osteoporosis whereas other studies do not?

4 In Mexico before European contact, the diet consisted mostly of corn, beans, chili peppers, and squash; these same foods are still the major elements of most Mexican diets, especially in rural areas. Can you think of any biological reasons why this diet has proved so stable?

Digestion Processes Food into Chemical Substances That the Body Can Absorb and Use for Energy

All organisms need energy to carry out life processes. Plants get this energy from sunlight through photosynthesis (see Chapter 11, pp. 368–372). Most other organisms, including humans, get their energy from the foods that they eat. However, most of the foods we eat cannot be used in the forms in which they are eaten. To be useful to our bodies, food must first be converted into substances that the body can absorb. Digestion is the process that breaks down food into these absorbable, energy-yielding products. Digestion also functions to eliminate undigestible waste.

Chemical and mechanical processes in digestion

Digestion has two aspects: chemical digestion and mechanical digestion. Chemical digestion breaks foods down chemically using enzymes, which are substances that promote or speed up a chemical reaction without themselves being used up in the reaction (Figure 10.7). This speeding up of reactions is known in chemistry as catalysis, and enzymes are therefore biological catalysts. Nearly all enzymes are proteins. Some enzymes, such as DNA polymerase (see Chapter 2, p. 59) help to synthesize molecules. The enzymes of digestion, in contrast, help to break down molecules. Chemical digestion works on the surfaces of food fragments. Mechanical digestion exposes new surface areas to chemical digestion by breaking fragments into smaller fragments and by removing partly digested surface material.

The digestive system

The digestive system is one of the organ systems in the body. An organ is a group of tissues that are integrated structurally and functionally. An organ system is a group of organs that perform different parts of the same process. Thus, the digestive system is a group of organs that together digest food. The body plan of the human digestive system is a common one: from roundworms (phylum Nematoda) to humans, most animals have digestive systems that are a continuous hollow tube, called a gut, with an entrance at one end and an exit at the other. As we go through the next sections, locate the human digestive organs on Figure 10.8.

The mouth. Food is taken in through the mouth, and mechanical digestion begins in the mouth when the food is chewed. Chemical digestion of starches (carbohydrates) begins in the mouth with the enzyme called

CONNECTIONS
CHAPTERS 2, 11

Figure 10.7

Enzymes: biological catalysts.

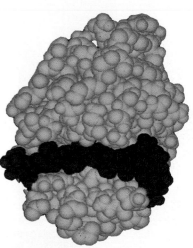

space-filling model of the enzyme lysozyme (blue) with a large sugar molecule (red) bound to it

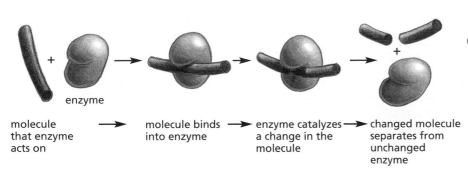

molecule that enzyme acts on → molecule binds into enzyme → enzyme catalyzes a change in the molecule → changed molecule separates from unchanged enzyme

enzyme

salivary amylase. This enzyme, present in saliva, breaks down starches into smaller units (sugars). Starches are usually not in the mouth for long enough to be completely digested, however, and their digestion is completed later. Another salivary enzyme called lysozyme catalyzes the breakdown of large sugar molecules (polysaccharides) into smaller units (see Figure 10.7).

The stomach. Once food is swallowed, it passes quickly through the esophagus and into the stomach. The stomach performs mechanical digestion through rhythmic contractions that knead the food back and forth, mixing it thoroughly, rubbing food particles against one another, and exposing new surface areas. The main activity in the stomach is the digestion of protein, accomplished with the aid of the enzyme pepsin, which breaks large protein molecules up into smaller fragments called **peptides**. Like many other protein-digesting enzymes, pepsin is secreted in an inactive form, which protects the glands that secrete the enzyme from digesting themselves. The inactive form is converted into active

Figure 10.8

The human digestive system.

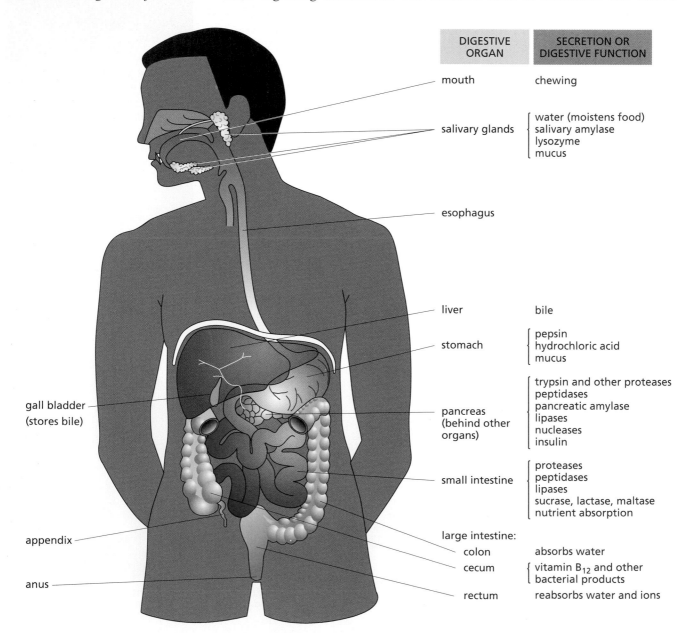

DIGESTIVE ORGAN	SECRETION OR DIGESTIVE FUNCTION
mouth	chewing
salivary glands	water (moistens food) salivary amylase lysozyme mucus
esophagus	
liver	bile
stomach	pepsin hydrochloric acid mucus
pancreas (behind other organs)	trypsin and other proteases peptidases pancreatic amylase lipases nucleases insulin
small intestine	proteases peptidases lipases sucrase, lactase, maltase nutrient absorption
large intestine:	
colon	absorbs water
cecum	vitamin B_{12} and other bacterial products
rectum	reabsorbs water and ions

gall bladder (stores bile)

appendix

anus

pepsin by digestive enzymes. Pepsin works best in an acidic solution, which the stomach provides by secreting hydrochloric acid (HCl). Acidity is measured on a scale called the pH scale; the lower the pH, the more acidic the solution (Figure 10.9). Fluids in the stomach are among the most acidic in biological systems, with a typical pH of 2. The stomach also secretes a mucus that protects the stomach lining (which is partly protein) from the pepsin and acid.

The small intestine: processing of fat. The lower end of the stomach empties into the small intestine, where the pH is no longer acidic. The term 'small' refers to the diameter, which is about 3 cm; the small intestine is actually very long (20 feet, or 6 m). Here, the food receives the secretions of the liver, called bile. In the watery environment of the intestine, fats tend to come out of solution and form large globules that coalesce to form even larger globules whenever they collide. Bile breaks up fat globules into smaller droplets and keeps these small droplets separate. Fats are insoluble in water because their chemical bonds are nonpolar, while those of water are polar (see Figure 10.2).

Globule formation is prevented by bile. One portion of each bile molecule is polar and consequently stable in water; another portion is nonpolar and is thus unstable in water but stable in fat. The nonpolar portions of the bile molecules dissolve in the fat droplets, leaving the polar portions of these molecules exposed on the surface, in contact with the watery intestinal fluids. The polar coating helps the fat droplets to mix with the water and also prevents the small droplets formed by mechanical action from coming back together to form large globules (Figure 10.10). This maintains the larger surface area of many small fat droplets and this increases the efficiency of digestion because a much greater surface is accessible for digestion and absorption.

Bile is secreted by the liver in a steady dribble, but is used in large amounts when fats or oils are present in the intestine. Bile from the liver accumulates in the gall bladder until it is needed, and is then released all at once under the stimulus of a digestive hormone. A hormone is a chemical messenger that causes a specific

Figure 10.9

The pH scale. The pH of a solution tells us the concentration of hydrogen ions (H⁺) in the solution. The H⁺ concentration can be expressed as moles of H⁺ per liter of solution, as shown to the left of the bar. It is more common, however, to express the concentration as pH, shown to the right of the bar. The pH scale is a reciprocal scale: the lower the pH, the higher the concentration of H⁺ (and the more acidic the solution) and the higher the pH, the lower the concentration of H⁺ (and the more basic the solution). The pH scale is also a logarithmic scale; that is, each number differs by a factor of 10 from the next number. A solution with a pH of 2 thus has 10 times more H⁺ ions than a solution with a pH of 3.

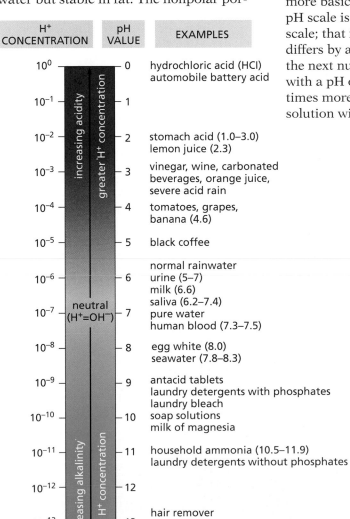

physiological change in one or more target organs. There are many kinds of hormones with very different functions; in Chapter 9 we saw the functions of some of the reproductive hormones. The intestinal lining secretes a digestive hormone whenever fats are present in the intestine; it acts on the gall bladder—its target organ—stimulating the release of bile into the intestine.

The small intestine: digestive enzymes. Farther down the small intestine, chemical digestion is completed by enzymes secreted by the pancreas and by the intestine's own lining. Enzymes are often named by combining the suffix '-ase' with the name of the molecule on which the enzyme works: proteases break down proteins, lipases break down lipids, and so forth. Among the intestinal enzymes are the following:

- Proteases, protein-digesting enzymes such as trypsin and chymotrypsin, secreted by the pancreas. Other proteases are secreted by the small intestine. Like the pepsin in the stomach, these enzymes break the chemical bonds between certain amino acids, thus breaking the proteins into smaller chains of amino acids called **peptides**. Each protease is specific and breaks only the bonds between certain specific amino acids.

- Peptidases, enzymes that complete the final stages of protein digestion by breaking peptides down into individual amino acids. Both the pancreas and intestinal lining secrete peptidases.

- Pancreatic amylase, an enzyme secreted by the pancreas. This enzyme continues the job, begun in the mouth, of breaking starches down into sugars.

- Lipases, fat-digesting enzymes, secreted by both the pancreas and the intestinal lining. These enzymes break down fats and oils into glycerol and fatty acids, molecules small enough to be absorbed.

- Sugar-digesting enzymes such as sucrase and lactase, secreted by the small intestine, which break down larger sugars (sucrose or lactose) into simple sugars such as glucose and fructose.

The presence of digestive enzymes may vary in human populations. We saw in Chapter 7 that northern Europe receives less ultraviolet radiation than other regions of the world, and populations living in northern Europe therefore have less sunlight to help them synthesize vitamin D. To supply this vitamin, most Europeans consume dairy products rich in vitamin D, and these dairy products also contain significant amounts of lactose, the sugar in milk. Thus, natural selection acted on European populations to favor those individuals who possessed the enzyme lactase, needed to digest lactose. Outside Europe, ultraviolet radiation is usually sufficient for the synthesis of large quantities of vitamin D, so dairy products are not needed in the adult diet. Because they do not need to digest lactose, people in these populations often do not have the enzyme lactase. When a person without lactase

CONNECTIONS
CHAPTERS 7, 9

Figure 10.10

The action of bile salts in breaking up fats.

fat droplets in water

without bile salts

droplets merge to form larger fat droplets with less surface in contact with water

with bile salts

polar portion of bile salt molecules

nonpolar portion

droplets remain separate because of polar surfaces formed by bile molecules, so surface area remains large

consumes most dairy products, the unused lactose is fermented by gut bacteria, producing large amounts of carbon dioxide gas that results in painful cramps, diarrhea, and sometimes vomiting. This condition is called lactose intolerance.

The small intestine: nutrient absorption. The absorptive part of the intestine (the ileum) is lined on the inside with thousands of tiny finger-like tufts (villi), which greatly increase the surface area through which the products of digestion are absorbed. Absorbable products include simple sugars, glycerol, fatty acids, and amino acids. Also absorbed are water and mineral salts, including dissolved ions (charged atomic particles) of sodium, calcium, and chloride, which do not require digestion to make them absorbable. The process of absorption is further described later.

The large intestine: our mutualistic relationship with intestinal bacteria. The material that has not been absorbed by the ileum passes into the large intestine, most of which is also called the colon. In comparison with the small intestine, the large intestine has a larger diameter (2.5 inches or 6–7 cm), but is much shorter (4 feet, or 1.2 m). This part of the intestine is inhabited by many bacteria, and certain nutrients produced by the bacteria are absorbed here. Mammals cannot make the enzymes that digest cellulose, the major constituent of plant cell walls. Bacteria that live in the intestine, and especially in a small dead-end portion (called the cecum or caecum), must do it for them. The cecum is especially important (and also much larger) in plant-eating mammals such as horses and rabbits, which consume large amounts of cellulose. Because we are mammals, humans cannot make the enzymes that degrade cellulose, and we also do not have the right species of intestinal bacteria to digest cellulose for us, so we cannot digest cellulose at all. We do, however, get some necessary nutrients from our intestinal bacteria. Humans cannot make vitamin K and biotin, needed for the synthesis of blood-clotting factors and for fatty acid synthesis, but our gut bacteria can synthesize them and we can then absorb these micronutrients.

Gut bacteria live in a form of symbiosis with vertebrate organisms. **Symbiosis** means simply that two organisms live together; **mutualism** is the form of symbiosis in which the two species are beneficial to each other. Mutualistic gut bacteria derive nutrients from the food taken in by their human host. In exchange, they synthesize vitamins that humans need and break many complex molecules into simpler components that are more easily absorbed from the intestine. The symbiosis may be disrupted by factors such as antibiotics, which kill the bacteria.

The large intestine: water absorption. The remainder of the large intestine consists of a straight portion called the rectum, which leads to a final opening called the anus. In the colon and rectum water is absorbed, mostly by diffusion, from the material passing through the gut, making this material a firmer consistency. Much of this material, which is called feces, is undigested food, but more than half is intestinal bacteria, which are rapidly replaced by bacterial cell division in the intestine. It is partly these bacteria and partly the bile pigments that give feces their characteristic brownish color.

Conversion of macronutrients into cellular energy

Carbohydrates, fats, and proteins are all macronutrients, the principal sources of calories for the body. After large macromolecules are broken

Figure 10.11

Membrane transport mechanisms.

ATP	adenosine triphosphate
ADP	adenosine diphosphate
P$_i$	inorganic phosphate
LDL	low-density lipoprotein

down into simple subunits in the digestive system, the subunits are absorbed into the cells. They are absorbed first by the cells lining the digestive tract, and are then transported via the blood to all of the other cells of the body. Once inside a cell, these simple subunits can begin a series of reactions, which result in the storing of energy in the form of a molecule called ATP (adenosine triphosphate). ATP is one of the principal molecules in which chemical energy is stored for later use by the cell. Food therefore provides the ability to make the ATP our cells use for all their work.

Absorption. Food cannot be converted into energy until it has been absorbed. Absorption takes place through the cell membrane of the cells lining the intestine. The polar chemical structure of most products of digestion means that they cannot enter the cell directly, because the interior of a cell membrane is nonpolar. The membrane thus acts as a controller for what enters (or leaves) the cell. Chemicals are absorbed by one of four mechanisms (Figure 10.11); these mechanisms also bring chemicals into cells elsewhere throughout the body.

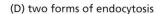

(A) passive diffusion

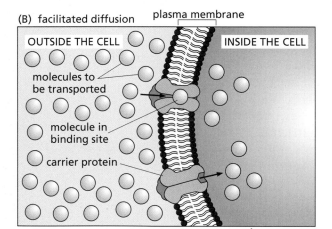

(D) two forms of endocytosis

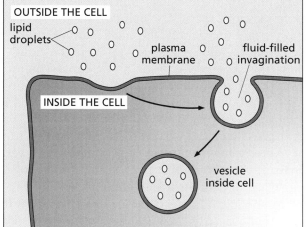

uptake without receptors

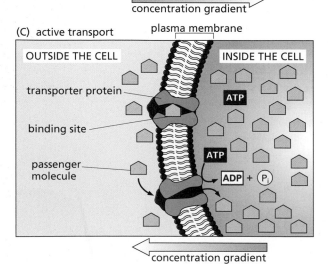

receptor-mediated endocytosis

- Some small molecules enter the cells by **diffusion**, a process that requires no added energy and is therefore sometimes called 'passive diffusion' (see Figure 10.11A). Diffusion works only if a **concentration gradient** exists; each substance diffuses from a place where it is more concentrated to a place where it is less concentrated. Small, uncharged molecules such as water (H_2O) or oxygen (O_2) diffuse directly through the cell membrane. Charged molecules cannot cross the membrane but may diffuse through **channels**, protein-lined 'holes' in the membrane.

- Small, polar molecules are often moved into cells with the help of protein molecules (called carrier proteins) that extend through the cell membrane. When this transport proceeds with a concentration gradient, it is called facilitated diffusion (see Figure 10.11B).

- Other molecules are absorbed *against* the concentration gradient, that is, from an area of lower concentration to an area of higher concentration of that type of molecule. This process requires an input of energy (usually from the breakdown of ATP) and is called **active transport** (see Figure 10.11C). Membrane proteins bind to the molecule being transported and use energy to carry it across the membrane.

- Large particles can be taken into the cell by a process called **endocytosis** (see Figure 10.11D), in which the plasma membrane is pulled in toward the interior of the cell. The plasma membrane forms a pit that may contain large particles; then the margins of the pit draw closed, and the pit pinches off to form a vesicle inside the cell. This bulk process transports many molecules at once, either suspended in liquid or attached to membrane proteins called **receptors**.

Energy-releasing pathways. The small molecules that are absorbed are then broken down into even smaller molecules, as can be followed on Figure 10.12.

- The long carbon chains of fatty acids are broken down in stages. In each stage, two carbons at a time are broken off from the long carbon chains to make an acetyl group. These acetyl groups are each put onto a carrier molecule called coenzyme A, so that the complex is called acetyl coenzyme A (acetyl CoA).

- The various amino acids are degraded into either pyruvate or acetyl CoA or one of the molecules in the Krebs cycle (see below).

- Sugars such as glucose are converted into a three-carbon molecule called pyruvate by a process called **glycolysis**. Pyruvate is then converted into acetyl CoA. During glycolysis, limited amounts of the energy-rich molecule ATP are synthesized. Glycolysis is one of the oldest known metabolic pathways and is found in most living organisms.

Pyruvate and acetyl CoA produced from the breakdown of proteins, carbohydrates or fats are then transported from the cytoplasm into organelles called mitochondria (see Chapter 6, pp. 169–170), where further energy is extracted from them in a cycle of reactions called the **Krebs cycle**. In each cycle, two ATP molecules are synthesized. The

CONNECTIONS
CHAPTER 6

enzymes for this process are proteins in the interior of the mitochondria, including some that are attached to the inner mitochondrial membrane. One molecule of a compound called oxaloacetate combines with the two carbons carried by acetyl CoA, and a series of oxidation reactions removes the electrons from hydrogen atoms, thereby extracting energy at each of several biochemical steps (see Figure 10.12).

The hydrogens from these oxidation reactions are taken up by a molecule called NAD$^+$ (forming NADH) and then donated to a series of mitochondrial membrane protein complexes, called the electron transport chain, that pass the electrons along from one protein to the next. These protein complexes are oriented in the membrane so that each time an electron passes from one electron carrier to the next electron carrier in the chain, a proton (H$^+$) is pumped from one side of the mitochondrial membrane to the other. An unequal distribution of protons on opposite sides of the membrane (called a **proton gradient**) is thus formed. This proton

Figure 10.12

How the major products of digestion are broken down in a series of energy-yielding reactions.

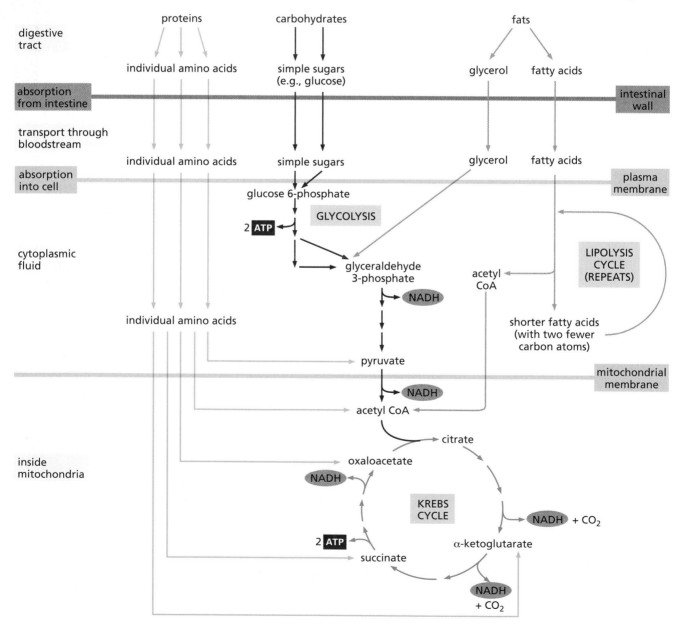

gradient is both a chemical gradient and a gradient of electrical charge, two important forms in which energy can be stored for later use by the cell. Some of this stored energy is used by an enzyme called ATP synthetase, producing additional ATP. After their energy has been extracted in steps, the electrons finally end up combining with oxygen, which is why we need to breathe in oxygen from the atmosphere. The energy-extracting processes in the mitochondria are shown in Figure 10.13.

Several steps in the Krebs cycle and in electron transport require vitamins as coenzymes. As an example of how a vitamin functions as a coenzyme, consider the role of the vitamin thiamine in the breakdown of the molecule pyruvate. An enzyme that contains thiamine as one of its constituent parts combines with the pyruvate molecule, releases CO_2, then emerges from a later reaction in its original form. Because the thiamine is not used up, it can participate in the reaction again and again. For this reason, only minute amounts of thiamine are needed to facilitate the breakdown of large quantities of pyruvate formed in carbohydrate metabolism. Vitamins B_2 and B_3 form parts of the larger molecules that carry electrons from the Krebs cycle to the electron transport chain (see Figure 10.13).

Digestion thus culminates with chemical changes at the cellular level. The process of turning foods into ATP begins when we eat and ends with electrons being taken up by oxygen. Each cell breaks down sugars and other organic molecules, converting their chemical bond energy to ATP. This process is called cellular respiration. Ultimately many of the carbon atoms from the food molecules are combined with oxygen as carbon dioxide (CO_2), which the body exhales. The hydrogen atoms from the food are split: their protons form gradients and their electrons are transported, recombining in the end with oxygen to form water (H_2O). At each intermediate step some energy is extracted and eventually converted into ATP. Other kinds of organisms extract energy by transporting carbon and hydrogen to molecules other than oxygen, enabling them to live in the absence of oxygen, but much more energy can be extracted from food when oxygen is used as the final acceptor. Eating causes the cellular ATP-production process to swing into high gear, producing ATP for immediate use and for storage against future needs.

Figure 10.13

Energy-producing processes that take place inside the mitochondria. Electrons are shown as e^-, and protons (hydrogen ions) as H^+.

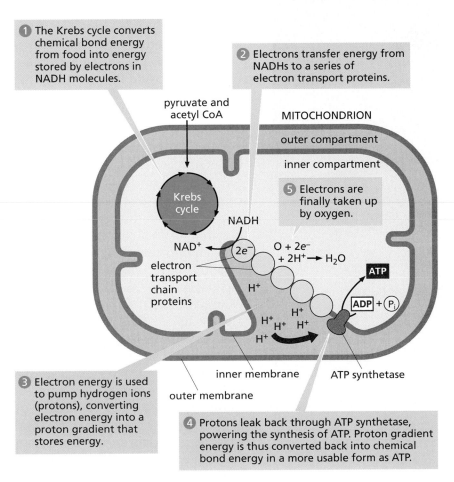

1 The Krebs cycle converts chemical bond energy from food into energy stored by electrons in NADH molecules.

2 Electrons transfer energy from NADHs to a series of electron transport proteins.

pyruvate and acetyl CoA

MITOCHONDRION

outer compartment

inner compartment

5 Electrons are finally taken up by oxygen.

Krebs cycle

NADH

NAD$^+$

$2e^-$

electron transport chain proteins

$O + 2e^- + 2H^+ \rightarrow H_2O$

H^+

ATP

ADP + P$_i$

H^+ H^+ H^+ H^+ H^+

inner membrane

ATP synthetase

3 Electron energy is used to pump hydrogen ions (protons), converting electron energy into a proton gradient that stores energy.

outer membrane

4 Protons leak back through ATP synthetase, powering the synthesis of ATP. Proton gradient energy is thus converted back into chemical bond energy in a more usable form as ATP.

THOUGHT QUESTIONS

1 Why is it important for food to spend the proper length of time in the stomach? What would happen if food left the stomach too soon?

2 The contents of the digestive tract are pushed along rather slowly by rhythmic muscular action (peristalsis). How does this relate to the body's need for a long intestinal tract?

3 What would be the consequences of a mutation that prevented a person's body from making a membrane protein necessary for the active transport of sugar from the intestine?

4 Why do food substances need to be digested into smaller molecules before they can be absorbed from the intestine?

Absorbed Nutrients Circulate Throughout the Body

After nutrient molecules have been absorbed in the small intestine, they circulate throughout the body. Other materials, including dissolved ions, oxygen, and cells of the immune system, also circulate throughout the body.

Circulatory system

The circulation of materials is carried out by the circulatory system. In all vertebrates, this system consists of blood, contained within a series of blood vessels, and the heart, a muscular pump that keeps the blood circulating.

CONNECTIONS
CHAPTERS 7, 15

The blood consists of a fluid material (the plasma), containing a series of cells and platelets. The cells include the red blood cells (erythrocytes), which contain the oxygen-carrying molecule hemoglobin (see Chapter 7, pp. 229–231), and several types of white blood cells, which form the immune system (see Chapter 15, pp. 542–543). The platelets are cell fragments that release materials important in blood clotting. Also important in clotting is a soluble protein, fibrinogen, one of several soluble proteins that circulate within the plasma. When a wound brings fibrinogen into contact with the air, it can turn into an insoluble tangle in which cells become trapped. This tangle and its trapped cells form the blood clot that blocks further blood loss and initiates the process of wound healing.

The blood circulates through a series of larger and smaller blood vessels. The vessels leading away from the heart are called arteries. The vessels leading back toward the heart are called veins. The arteries branch into finer and finer vessels. The veins are arranged as a series of tributaries that flow into larger vessels and eventually back to the heart. The thinnest vessels, called capillaries, carry blood from the smallest arteries to the smallest veins (Figure 10.14). The capillary walls are a single layer of cells. Materials diffuse from capillaries into tissues and from tissues into capillaries across the cells of the capillary walls. The very large capillary surface area permits diffusion on a large scale. Throughout the body, no cell of any tissue is very far from a capillary.

Nutrients absorbed in the gut enter the capillaries of the gut lining and flow to the liver via the hepatic portal vein. If the blood contains more glucose than the body needs immediately, the excess is converted into the storage molecule **glycogen**. Glycogen storage takes place in most body cells, but the largest amount is stored in the liver. As the body uses up blood glucose, the liver cells convert glycogen back into glucose and release it into the bloodstream as needed, a mechanism that ensures a dependable but moderately low concentration of glucose in the blood. The storage of glycogen and the efficient use of glucose both require the hormone **insulin**, secreted by special clumps of cells within the pancreas. Persons in whom these cells have degenerated cannot produce enough insulin; their condition is known as insulin-dependent diabetes mellitus (IDDM, or type I diabetes). The symptoms of diabetes can be controlled by supplying insulin or by controlling weight and diet, but there is no known cure for the disease itself.

Blood from the liver and the body's other organs flows through veins to the heart. Most veins have thin, flexible walls, and the blood within them flows at a relatively low fluid pressure. (Arteries, in contrast, have thick walls that can withstand the high, pulsating fluid pressure of the blood within them.) The blood within veins is propelled by the massaging action of nearby muscles and other organs. Valves within the veins keep the blood flowing in one direction and prevent it from flowing backwards.

In addition to distributing nutrients, the circulatory system also distributes oxygen to the body and transports carbon dioxide (a waste product of cellular respiration) from body cells to the lungs (which then exhale it). The way in which blood circulates through the heart keeps the oxygen and carbon dioxide from mixing.

The heart

The heart is a muscular organ whose rhythmic contractions keep the blood circulating throughout the body. In all mammals, the heart contains four chambers (Figure 10.15). Oxygen-poor blood from the body's various organs enters the right atrium and is pumped into the right ventricle. Contraction of the right ventricle propels the blood out through the pulmonary arteries and into the lungs. Here, the oxygen in the lung's tiny pockets, or alveoli (see Chapter 14, pp. 503–504), diffuses into the blood, while carbon dioxide diffuses out and is exhaled. Oxygen-rich blood from the lungs returns to the left side of the heart, where it enters the left atrium. Contraction of

Figure 10.14

The human circulatory system.

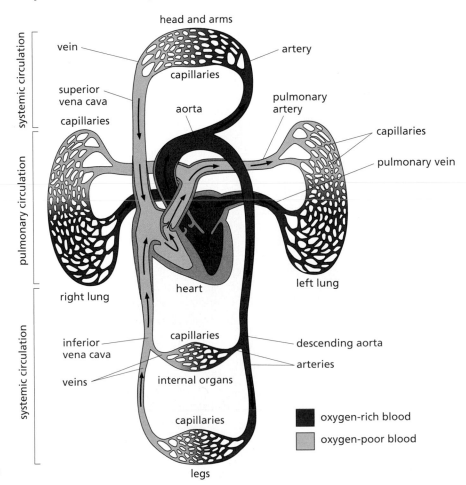

the left atrium pushes the blood into the left ventricle, the heart's largest chamber. Contraction of the left ventricle propels the blood throughout the arteries and into the body's various organs; for this reason, it has the thickest walls. Oxygen diffuses from the blood into the body's many cells across the thin capillary walls, and cellular wastes, including carbon dioxide, diffuse from the cells into the blood. The veins collect this oxygen-poor blood from the body's various organs and carry it back to the heart, where the pattern of circulation repeats (see Figure 10.14).

The heart maintains its own rhythmic pattern of contractions. A heart cut from a living animal and placed in a salt solution will continue to beat for many hours. The rhythm is maintained even in the absence of any nerve input, showing that the heart's rhythm originates in the heart itself.

Cardiovascular disease

Cardiovascular disease includes both heart disease and diseases of the blood vessels. In the United States each year 500,000 people die of heart disease, making it the number one cause of death. Another 1.5 million have nonfatal heart attacks. Each year almost another 500,000 die of strokes, a blood vessel disease. Men have more cardiovascular disease earlier than women (2:1 male:female ratio overall), but cardiovascular disease is nevertheless the major killer of both men and women. Risk factors for cardiovascular disease include smoking, obesity, high-fat diets, lack of exercise, hypertension, atherosclerosis, high cholesterol levels, stress, and genetic predisposition. Proper diet, exercise, weight loss, and stress reduction are the major ways of reducing the risks.

Too much dietary fat. The study of disease factors in large populations is called **epidemiology**. Epidemiological studies point to a connection between certain types of dietary fats and cardiovascular disease (heart disease and strokes). The United States, Australia, and New Zealand—all meat-producing countries—have a high consumption rate of meat products per person and also high incidences of cardiovascular diseases. Most meats are high in saturated fats. People in Mediterranean countries consume much of their lipid in the form of olive oil, an unsaturated fat, and their cardiovascular disease rates are lower. Heart attacks are very rare among the Inuit (Eskimos), whose diet contains large amounts of cold-water fish, a good source of a type of fatty acid called omega-3 fatty acid that has been shown to guard against the production of chemicals that damage cell

Figure 10.15

The human heart.

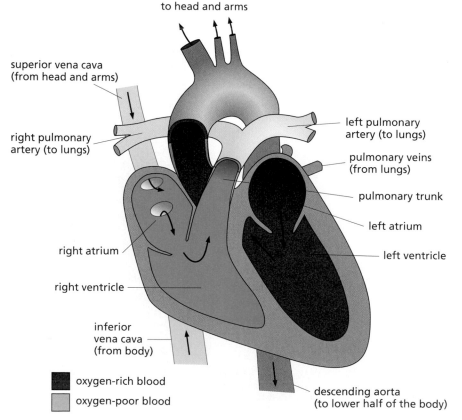

to head and arms

superior vena cava
(from head and arms)

right pulmonary
artery (to lungs)

left pulmonary
artery (to lungs)

pulmonary veins
(from lungs)

pulmonary trunk

left atrium

right atrium

left ventricle

right ventricle

inferior
vena cava
(from body)

oxygen-rich blood

oxygen-poor blood

descending aorta
(to lower half of the body)

membranes. The Japanese also tend to have low consumption rates of saturated fats and low rates of cardiovascular disease.

Epidemiology also provides clues about whether the association between dietary fats and cardiovascular disease is more closely related to diet or to genetically inherited traits: Japanese people in Japan have much lower rates of heart disease or stroke than do Japanese living in Hawaii or California, whose rates are similar to those of their non-Japanese neighbors. These findings (and similar ones on other immigrant groups) all point to diet, not heredity, as the major difference responsible for the different disease rates between populations.

Because saturated fats have been linked to a greater risk of cardiovascular disease, many experts recommend that saturated fats be replaced with unsaturated fats in most diets. Advertising has convinced many people that unsaturated fats—especially the polyunsaturated kind—are desirable, but this is true only if those fats replace saturated fats. Most experts recommend that the quantities of all dietary fats be reduced to lower the risk of cardiovascular disease.

Atherosclerosis. Dietary fat can increase the risks for cardiovascular disease in several ways. One way is that excess dietary fat can result in fat deposits that build up in the arteries, causing atherosclerosis (Figure 10.16), a type of cardiovascular disease that can lead to heart attacks. The fat deposits obstruct the blood vessels, making the passages narrower; eventually these deposits may calcify and make the vessels more rigid. Atherosclerosis contributes to hypertension (high blood pressure), although a person can have hypertension without having atherosclerosis.

LDLs and HDLs: lipid transport particles. In contrast with carbohydrate molecules, lipid molecules contain few oxygen and nitrogen atoms and have mostly nonpolar bonds in which electrons are shared equally around carbon and hydrogen atoms (see Figure 10.2). Because the bonds in lipids are nonpolar, lipids are not water soluble. Blood plasma is mainly water, so lipids must be transported through the blood from one part of the body to another by transport particles such as LDLs (low-density lipoproteins) and HDLs (high-density lipoproteins). These transport particles are proteins that bind lipids in such a way that they can move through body fluids.

People eating identical diets may not have the same serum cholesterol level. This difference seems to have a genetic component and relates in part to each person's ability to make LDLs and HDLs. No one has very much free (unbound) cholesterol because cholesterol is very nonpolar, so what is called serum cholesterol is actually the total of all the cholesterol bound to HDLs and LDLs. HDLs transport cholesterol, phospholipids, and triglycerides out of tissues to the liver, where they are used in the synthesis of bile acids, while LDLs transport cholesterol and lipids into tissues and cells. The HDL:LDL

Figure 10.16

Atherosclerotic plaque reducing the effective diameter of an artery. Although the outer diameter of the vessel has not changed, the inner diameter (lumen) through which the blood flows has become smaller by the formation of lipid deposits on the inside of the vessel, making the blood pressure higher. These deposits may also become calcified, further increasing blood pressure.

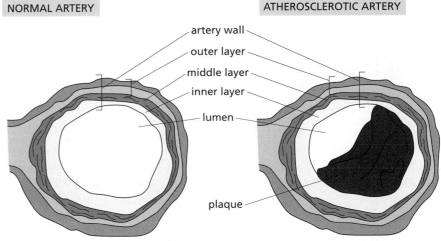

NORMAL ARTERY   ATHEROSCLEROTIC ARTERY

artery wall
outer layer
middle layer
inner layer
lumen
plaque

ratio is the ratio of outbound to inbound lipids. A high HDL:LDL ratio thus indicates that the proportion of 'good cholesterol' (lipids on their way out) is higher than the proportion of 'bad cholesterol' (lipids on their way into cells). A very low ratio, meaning a preponderance of inbound lipids, correlates strongly with an increased risk of atherosclerosis in the arteries of the heart (coronary arteries), a condition that can precipitate a heart attack.

Some people are genetically prone to high cholesterol levels because their cells lack LDL receptors on their surfaces. When cells need cholesterol, they get the cholesterol from the LDLs in the bloodstream by binding these particles to LDL receptors and internalizing the receptors and LDLs by endocytosis (see Figure 10.11D). If the LDL receptors are missing or nonfunctional, the cells manufacture their own cholesterol even when LDL levels are already high, because they cannot take up LDL from the blood. Thus, diet is not the only factor leading to atherosclerosis; problems with lipid transport and lipid uptake are also factors.

THOUGHT QUESTIONS

1 Cancer is not as frequent a cause of death as is heart disease, yet cancer research seems to get more publicity than research on heart disease. Examine some newspapers and magazines to see whether this is a correct impression. If it is, what factors might contribute to this difference?

2 The human circulatory system is a continuous, closed system; in other words, none of the blood vessels are open-ended. Why must this be so?

Malnutrition Contributes to Poor Health

The term **malnutrition** literally means 'bad nutrition.' Malnutrition can result from eating too little or from eating the wrong foods, and is one of the major health problems of the world, particularly among the poor and in areas of turmoil. Malnutrition also exists in people with access to sufficient foods, sometimes as a result of eating disorders.

Eating disorders predominate in the industrialized nations

Obesity. In the United States, the Center for Disease Control and Prevention (CDC) estimates that 40% of adults are obese. **Obesity** is defined as a body weight 20% or more above the ideal for the particular subject's sex and height, or a Body Mass Index (BMI) of 30 or more (Box 10.2). Obesity increases the risk for high blood pressure, heart disease, stroke, diabetes, and several cancers (including those of the colon, breast, and prostate). The CDC has declared that obesity is one of the greatest threats to public health in the United States.

Numerous factors contribute to obesity. There are some genetic contributions to the relative proportion of body fat or carbohydrate used for

BOX 10.2 Obesity and the Body Mass Index (BMI)

The new definitions of overweight and obesity are based on the Body Mass Index (BMI), calculated as described below. A person with a BMI of 25 or higher is considered overweight, and a person with a BMI of 30 or above is considered obese. On the basis of these criteria, the CDC estimates that over 60% of adults in the United States are overweight or obese, and they have declared obesity to be a "public health epidemic." The CDC (at www.cdc.gov/nccdphp/dnpa/obesity) lists a total of 20 disease conditions to which overweight or obesity contributes.

TO CALCULATE YOUR BODY MASS INDEX:

Divide your weight by your height, then divide by your height again. (The formula is BMI = w/h^2.)

- If you measured your weight in kilograms and your height in meters, the result is your BMI.
- If you used kilograms and centimeters, multiply the result by 10,000 to get your BMI.
- If you used pounds and inches, multiply the result by 703.7 to get your BMI.

According to the CDC, the incidence of overweight and obesity has been increasing in the United States.

Percent of U.S. adults	Overweight but not obese (BMI 25–29.9)	Obese (BMI 30 or above)	Total overweight (BMI 25 or above)
1980	33%	15%	48%
1999–2000	34%	30%	64%

energy production and to the likelihood of energy surplus being stored as fat versus lean tissue. However, environmental factors are much stronger for most people than are genetic ones. Factors that encourage obesity include decreased exercise, increased calorie intake, and increased consumption of fatty foods. In many cases the amount of excess weight is proportional to the number of meals eaten away from home, where portions are large and fat content is high.

The marketing of food (including its advertising and packaging) encourages unhealthy behavior and overeating. High-calorie or high-fat foods are advertised heavily. Many foods, especially snack foods, are packaged in individual serving units that are convenient for a person on the go (often in containers that fit into the cup holders of automobiles), and these individual serving units can be much larger than any serving recommended by nutritionists, sometimes as much as two or three times as large. Fast food is often marketed in "super size" portions, and customers are encouraged to have side orders of fried foods and beverages consisting of calorie-rich and vitamin-poor sodas and shakes. Eating in the car or while working or watching TV are forms of 'unconscious' eating that can prevent a person from being aware of how much they have eaten.

Many people attempt to reduce their weight by dieting. Some major facts about dieting are widely agreed upon by nutritionists. For example, many different diets can be effective in helping to reduce weight. The diets that work do so by reducing caloric intake. Some diets that are nutritionally very unbalanced can be harmful. Of course, no person can achieve results on a diet that they cannot stay on, and different people find different diets easier or harder to stick to. Unfortunately, most people who lose weight by dieting gain back what they lost (often more than what they lost) within a few months or so after stopping their diet.

Anorexia and bulimia. In the middle and upper classes of the industrialized nations, some people suffer from a condition called **anorexia nervosa**. The ratio of anorexic women to anorexic men is about 9:1. Anorexic individuals suffer from a mistaken perception of their body size. They imagine themselves to be heavier than they really are, and desire to be thinner as a result, a feature that clearly distinguishes anorexia from all other forms of undernourishment (Figure 10.17). Anorexics also respond poorly to body cues of hunger and satiety. The misperception of hunger, satiety, and body size are early symptoms that precede the most noticeable feature of the disease, what American (German-born) physician Dr. Hilde Bruch has called a "relentless pursuit of thinness," a self-imposed undernourishment that borders on starvation. At the same time, there is usually an absorbing or obsessive interest in food, which may include talking or reading about food, preparing food, collecting recipes, or serving food to others, while all the time avoiding eating.

One of the surest signs of the disease in anorexic women is that they usually stop menstruating because of a lack of the cholesterol needed for synthesis of the hormones that regulate the menstrual cycle (see Chapter 9, pp. 299–300). Other symptoms include changes in brain activity. The brain must be constantly supplied with glucose to provide cellular energy for nerve cell function; when it is not, many mental functions may be impaired. These impairments may manifest themselves in anorexic persons as deception (hiding things, keeping secrets) and a distrust of others, but only late in the process, after starvation has already set in. Untreated anorexia is usually fatal.

Anorexia is most common among Caucasian women between the ages of 15 and 30 years with an average or above-average level of education. The disease also occurs in Japan and in parts of Southeast Asia, but only in the uppermost social strata. Anorexia is virtually unknown among people living in poverty or in undernourished populations anywhere. It never seems to occur where food is scarce, or in times of famine. Even in countries where it occurs, it seems to vanish during economic hard times, such as the Depression of the 1930s.

Anorexia was once extremely rare. Its marked increase in the United States since World War II has been attributed by several experts to a general standard of beauty that has increasingly glorified thinness, as measured by such criteria as waist and hip measurements of Miss America

Figure 10.17

One of the earliest signs of anorexia is misperception of one's own body.

contestants, *Playboy* centerfolds, and models and ballet dancers more generally. In fact, women in professions such as modeling and ballet dancing are particularly likely to develop anorexia. Also, female athletes in sports such as rowing (where competition is organized by weight classes) are at high risk for developing a 'female athlete triad,' consisting of eating disorders such as anorexia, combined with loss of menstruation and a loss of bone mass that may cause osteoporosis later in life. The loss of menstruation resulting from overexercise has been attributed to the depletion of the body's estrogen.

Many anorexics also suffer from a related condition called **bulimia**, although bulimia can also occur independently of anorexia. Bulimia is characterized by occasional binge eating of everything in sight, usually including large quantities of high-calorie 'forbidden' foods, in total disregard of any concept of a balanced diet. Immediately after a binge, bulimics typically force themselves to vomit or else purge themselves with an overdose of a laxative. Both the binge and the purge are usually done in secret; bulimics (like anorexics) become extremely skillful at hiding their condition from others. Persistent bulimia can lead to ulcers and other problems of the digestive tract, and also to chemical erosion of the teeth from the frequent contact of the teeth with the acidic secretions of the stomach. Bulimia occurs in both sexes, but more often in women. Bulimia is especially common among educated women who have easy access to unrestricted amounts of food, a situation common on many college campuses.

Starvation

Around the world, more people die each year of starvation than of any single disease or other preventable cause. Death by starvation occurs primarily in poor countries, but it also occurs in pockets of poverty within wealthier nations. The immediate cause of death by starvation is usually inadequate caloric intake. In populations with very low caloric intake, even a small or temporary interruption in the food supply can lead to mass starvation: just one bad crop, or political disruption that prevents the planting or harvesting of the crop, can have severe consequences.

Inadequate caloric intake can also result in protein deficiencies that take several forms, including both low total protein and low levels of particular amino acids. If protein intake is inadequate for either reason, a protein deficiency called kwashiorkor develops. Kwashiorkor occurs when carbohydrate intake is adequate but protein intake is not.

In all organisms, cells and proteins are constantly being broken down, and in a healthy, adequately nourished body, they are constantly being replenished. When protein intake is insufficient for this replenishment to occur, there is considerable loss of muscle tissue, and the death of many cells releases numerous dissolved ions into the surrounding tissue. These ions retain water and contribute to tissue swelling (edema) that makes the loss of muscle tissue harder to see. Children suffering from kwashiorkor have swollen abdomens, a fact often noticed in photographs from protein-deficient areas. Their large bellies conceal the fact that these children are actually starving to death.

If protein intake is inadequate and carbohydrate intake is also inadequate, the combined deficiency produces a condition called marasmus, in which the body slowly digests its own tissues and wastes away. When carbohydrate is not available as an energy source, amino acids are used to produce energy. Because amino acids are not stored except in the form of the body's structural and functional proteins, the use of amino acids for metabolic energy degrades these proteins. Once the body's muscle mass falls below a certain minimum, marasmus is always fatal.

Ecological factors contributing to poor diets

Malnutrition is recognized as a worldwide disease with regional differences in its cause but with similar outcomes everywhere. Here we take Africa as an example, although many other examples exist. Many populations in Africa experience either chronic or periodic protein deficiency; Africa has therefore been described as a protein-starved continent. Fresh vegetables are widely available, so diets are high in fiber and most vitamins. Vegetables (including legumes) can fill nearly all of a population's nutritional needs as long as supplies are adequate and as long as some form of nutritionally complete protein is obtainable from animal sources (including fish) or from a combination of plant proteins. There are lakes and rivers where fishing provides adequate protein resources, and there are cattle-herding groups, such as the Masai and Fulani, who can meet their protein needs from their animals. Across most of the continent, however, animal protein sources are not readily available, and the supplies of grains that could offer protein by amino acid balancing are not always adequate.

Grain supplies are inadequate owing to a combination of ecological and social factors. Tsetse flies, which spread blood-borne diseases, have made much of the land uninhabitable to most domestic animals. This not only limits the availability of meat proteins, but also limits the supply of draft animals that can pull plows and till the soil. Tractors and fuel are too expensive for most farmers in Africa, and many places are either too dry or too wet to support agriculture. The world's largest desert, the Sahara, occupies most of Africa's northern half; other deserts exist in Somalia and Namibia. Most of these deserts, including the vast Sahara, are growing larger each year as animals such as sheep and goats overgraze and destroy the plants on the desert fringes, a process known as desertification (see Chapter 18, pp. 669–672). In other parts, high rainfall leaches important minerals from the soil, leaving soils deficient in the minerals necessary for plant growth (see Chapter 11, p. 373 and pp. 384–386). Rice, wheat, and most other grains grow rather poorly in many African soils. Some success has been achieved with millet and corn (maize), but raising a sufficient quantity and variety of grains to provide complete protein is difficult. In most places in Africa, populations can maintain adequate nutrition in years when there are ample harvests and efficient distribution. Unfortunately, these two conditions are not always met. Protein deficiencies are made worse by political and military upheavals that drive people from their farms or that prevent the planting, harvesting and distribution of crops. Protein starvation (marasmus and kwashiorkor) is all too common, particularly among children.

CONNECTIONS CHAPTERS 11, 18

Effects of poverty and war on health

Climatic, economic, and political factors all contribute to the unequal production and unequal distribution of food across the planet. Poverty exists in all nations of the world. Many populations are marginally nourished and are therefore more vulnerable to a year of drought or a bad harvest, but malnutrition exists in every part of the globe, especially among the poor. Even in regions in which nourishment is usually adequate, people can become undernourished because of the disruptions of life associated with war.

Wartime starvation. The effects of starvation on humans have been studied retrospectively in people who were, at an earlier time, subjected to starvation by war. Such a study on the short-term and long-term effects of starvation was conducted by using the birth records of infants born in the Netherlands in the winter of 1944–1945, when many people in certain cities experienced starvation. The food shortages varied from city to city, but the other effects of war were equal across the country. It was found that when food rations dropped below a threshold level, fertility in the population decreased and the decrease was greatest among people in the lower classes. Fetuses carried by women who experienced starvation in the first trimester had a higher rate of abnormal development of the central nervous system. Maternal undernourishment in that period also carried forward to premature births, very low birth weight, and an increase in infant death rate immediately after birth. Maternal undernourishment in the third trimester produced the greatest increase in the infant death rate in the three months after birth. Brain cells were depleted in infants who died. Among survivors, however, undernourishment in infancy was not correlated with long-term effects on mental development when males were tested at age 18. Other retrospective studies have found similar results.

Long-term effects of childhood undernutrition. There are several indications that undernutrition during childhood decreases brain development and capacity. Infants who died of kwashiorkor or marasmus had less DNA, protein, and lipid in their brain cells compared with infants who died at similar ages of causes not related to nutrition. Evidence suggests that both the severity and the length of the period of malnutrition affect intellectual development. Iron deficiency, particularly in the early years of life, can also impair mental functions.

Malnutrition can also result from children's failure to eat properly because of a depressed mental state. Malnutrition of this type is called 'failure to thrive.' Infants must learn that their needs will be met, and this learned capacity is termed 'basic trust.' When circumstances are such that a child is not regularly fed and nurtured, it fails to develop trust. Loss of a caregiver or a diminution or loss of nurturing care can create depression in infants, who then lose trust and become malnourished. Medical intervention cannot rescue the child unless a trusting relationship is established. Adults recover their physical and mental capacities if they are rescued from starvation. For children, however, nutritional replacement will not lead to full recovery unless emotional and psychological support is also provided. The recovery of children from famine requires much more than food.

Micronutrient malnutrition

Malnutrition can still exist when total caloric intake is sufficient or even when it is excessive. Because the roles of micronutrients include the proper functioning of enzymes, micronutrient deficiencies prevent the proper utilization of other foods. As we saw earlier, mineral deficiencies can result when diets are restricted to those foods grown in mineral deficient soils. Diets that include only a few types of foods, particularly processed 'junk foods,' may be lacking in necessary micronutrients. It is for this reason that Japanese children are encouraged to "eat at least 100 different foods each week." Limited incomes, limited mobility, and limited availability of fresh foods all make this goal difficult to achieve and generally make nutritional problems worse. Micronutrient malnutrition may be a danger that is overlooked among people who lose interest in eating, including elderly people, people with chronic diseases including cancer, and people who have mental illnesses such as depression. Some elderly people may also forget what they have or have not eaten. Thus, many people need more than just food in order to be properly nourished.

THOUGHT QUESTIONS

1 Think about the definition of obesity as a weight 20% or more above the ideal weight for a person's sex and height. What is meant by 'ideal?' Who sets these 'ideals?' Are there cultural differences in what is considered 'ideal?' How do cultural definitions of 'ideal' relate to biological definitions?

2 Why do you suppose anorexia occurs only among well-fed populations, and never among the poor or in times of famine? What does this imply about the possible causes of the condition?

3 Will efforts aimed at improving nutrition in people with various forms of malnourishment be readily accepted by the people they are meant to help? What factors determine acceptance?

4 If sufficient food can be produced, will every person have good nutrition? Why or why not?

5 Why might starvation have greater long-term effects in children than in adults?

Concluding Remarks

During the first half of the twentieth century, vitamin deficiencies were major public health concerns. Vitamin and mineral deficiencies can lead to various deficiency diseases, birth defects, or neurological damage. Nutritional deficiencies and deficiency diseases have declined since then in the industrial world, the results of better eating habits, more varied diets, and vitamin-fortified foods. Greater interest now centers on whether certain foods can promote good health and reduce the risks of chronic diseases. Since the 1960s, public health officials and nutritionists have increasingly turned their attention to heart disease, stroke, cancer, and chronic health problems such as obesity and high blood pressure that can influence the risks for these fatal diseases. Heart disease is the number one cause of death in many industrialized countries and is significantly associated with a diet that is too high in fat, particularly saturated fats.

Malnutrition still exists, however, and it can take many forms. Inadequate caloric intake can result from crop failures, from poverty, or from eating disorders. Poverty and war usually worsen nutrition and contribute to stress among civilians. Adequate nutrition depends on getting all the necessary nutrients without taking in too much of any one of them. Much of the world's population gets inadequate food and inadequate protein. In many industrialized countries, problems are caused instead by high-fat diets with inadequate fiber. The best way to avoid both types of problems is to eat a varied diet while keeping fat intake low. Sensible eating habits such as these are important for good health. If we want to promote health for ourselves, our families, and the rest of humankind, then a very important step is ensuring that each person has an adequate and balanced diet.

Chapter Summary

- Digestion is a process in which materials consumed as food are broken down into substances that the body can absorb and then use. Digestion consists of **chemical digestion** with the aid of **enzymes**, and **mechanical digestion**.

- The **macronutrients** are the major sources for energy, measured in **kilocalories**, and including **carbohydrates**, **lipids** (fats and oils), and **proteins**. Lipids are also needed for the formation of cell membranes, and proteins are needed as **enzymes**, receptors, and membrane transporters. **Complete proteins** have all the amino acids needed in human nutrition. Proteins are digested and used until one of the needed amino acids, called the **limiting amino acid**, is depleted, after which the remaining protein is used as an energy source in the manner of carbohydrates.

- Molecules or parts of molecules may be either **polar** and thus stable in contact with water or else **nonpolar** and thus water-avoiding.

- Digested food enters cells by passive **diffusion**, facilitated diffusion, **active transport**, or **endocytosis**. Diffusion always acts along a **concentration gradient**, from a place of higher concentration to a place of lower concentration; active transport always operates against such a concentration gradient. Concentration gradients can serve as a form of energy storage.

- Cellular energy is derived in the mitochondria by **oxidation** reactions in the **Krebs cycle** and electron transport to form ATP.

- Material that the body cannot absorb constitutes **fiber**.

- **Micronutrients** (**vitamins** and **minerals**, including electrolytes) are needed for good health.

- **Epidemiology** is the statistical study of diseases and risks in large populations. Data from epidemiology can help to provide evidence for certain health risks and benefits, even when the mechanisms of action are not known.

- Good nutrition reduces such chronic conditions as high blood pressure and **obesity**, and lowers the risks for **cardiovascular disease** and certain cancers.

- **Malnutrition** can result from inadequate food, inadequate protein, or vitamin and mineral deficiencies. Eating disorders such as **anorexia** and **bulimia** can lead to malnutrition.

CONNECTIONS TO OTHER CHAPTERS

Chapter 3	Some genetic differences exist in the body's ability to digest certain substances (as in phenylketonuria) or to use certain nutrients after their absorption (as in diabetes).
Chapter 7	Nutritional requirements may differ among human populations for various inherited and environmental reasons.
Chapter 9	Unchecked population growth puts more people at risk for malnutrition.
Chapter 11	Crop improvements may help to alleviate starvation in many populations.
Chapter 12	Certain cancers have frequencies that vary according to diet: high-fat diets promote certain cancers, whereas high-fiber diets lower many cancer risks.
Chapter 15	Poor nutritional status impairs the immune system.
Chapter 16	People with HIV infection stay less sick if they have good nutrition, but appetite is often suppressed (and nutrition suffers) as AIDS progresses.
Chapter 18	Biodiversity is threatened by the need to clear more land for farming to feed the world's population. Developing and conserving better and more varied crop plants will feed more people and support greater biodiversity at the same time.

PRACTICE QUESTIONS

1. Explain the differences between chemical digestion and mechanical digestion.

2. How many times greater is the hydrogen ion concentration in stomach acid than in blood?

3. What makes a molecule polar? What makes a molecule nonpolar? Can some molecules be both polar and nonpolar?

4. What kinds of food molecules can supply precursors to the Krebs cycle?

5. What type of transport is used to bring glucose into a cell?

6. What hormone regulates the uptake of glucose by cells?

7. What type of transport brings lipids into cells? How are nonpolar lipids able to travel through the blood to get to cells?

8. How does atherosclerosis result in an increase in blood pressure?

9. What is the HDL:LDL ratio and why is it important?

10. What are the functions of cholesterol in the human body?

11. Which food molecules produce the most ATP?

12. Where in the cell is ATP produced?

13. What are some of the functions of the water-soluble vitamins in the body?

14. How is an electrolyte different from other mineral micronutrients? What are some of the functions of each in the body?

15. Which can lead to health problems: vitamin deficiencies or vitamin megadoses?

Plants to Feed the World

Issues

- What is plant science?
- How has plant science changed the world?
- Can plant science feed the world?
- Why is the 'law of unintended consequences' so important?
- Are genetically engineered plants different from other plants?

Biological Concepts

- Photosynthesis (autotrophs, light-dependent reactions, dark reactions)
- Osmosis
- Nitrogen cycle
- Plant structure and function (tissues, water transport, gas exchange)
- Ecosystems (trophic levels: producer, consumer, decomposer; soil; sustainable agriculture; pest control; integrated pest management)
- Limiting factors (fertilizers, irrigation)
- Plant genetic engineering

11 Plants to Feed the World

Hunger, starvation, and malnutrition are endemic in many parts of the world (see Chapter 10). Rapid increases in the world's population (see Chapter 9) have intensified these problems. Because all the food that we eat comes either directly or indirectly from plants, any attempt to produce adequate food for the world's growing human population must focus on the production of food by plants. However, the amount of land under cultivation has its geographic limits, and extending these limits carries a very high biological cost in terms of the destruction of natural ecosystems (see Chapter 18). Another way to increase crop production is to increase the yields or the nutrient content of crop species through such techniques as soil improvement, the use of fertilizers and advanced irrigation techniques, pest control, and the use of different plants or new genetic strains of plants. In this chapter we concentrate on food production by plants because that is where most of our food comes from.

Plants are essential to all of life on Earth. In addition to food, plants also provide the oxygen that humans and other organisms breathe. Without plants, most other forms of life would soon die out. Plants are the world's richest energy source. On a worldwide basis, the amount of energy produced by plants is about 6×10^{17} kilocalories per year (abbreviated kcal/year), or the equivalent of a sugar cube 5 km (or 3 miles) on each side, and we are rapidly outgrowing this energy resource. This chapter discusses how plants make energy available to other organisms and how humans, through application of this knowledge, might learn to grow more crop plants and grow them more efficiently.

Plants Capture the Sun's Energy and Make Many Useful Products

As we saw in Chapter 6, plants are a kingdom of living organisms that have eucaryotic cells containing specialized structures for capturing energy from the sun's light. Plants are essential to life on Earth as they do not simply consume all of the light energy that they capture; they convert much of this light energy into a form that can be used by organisms unable to use the light energy directly. And plants do even more: in the course of building and maintaining their own bodies, plants make products that serve many needs of other organisms. How exactly plants trap and convert light energy, the ultimate source of the plant products that we depend on, is discussed in this section.

Plant products of use to humans

Humans and other animals find many ways of benefiting from the use of plants. Most importantly, plant products are eaten as foods, that is, as sources of energy and nutrients (see Chapter 10). Many of these foods are the carbohydrates (and some are the oils) that plants have stored for their own use as energy sources for a later time. Agriculture, meaning the

cultivation of plants as food for humans, is often regarded as the one crucial achievement that led also to the establishment of human civilization.
The cultivated plants most important in the development of civilization
were the cereal grains (wheat, rice, corn, oats, and others). Today, wheat,
rice, corn, and potatoes provide more of the world's food than all other
crops combined. Of the almost 250,000 species of plants known, some
80,000 are edible by humans. Of these, however, only about 30 form the
major crop plants.

People also use plants as sources of beverages, flavorings, fragrances,
dyes, poisons, decorations, building materials, and medicines. Beverages
made from plants include coffee, tea, cola, beer, wine, spirits, and many
juices. Many plant parts are used as spices, fragrances, and flavorings,
from barks such as cinnamon, seeds such as black pepper, and roots such
as ginger to flowers and flower parts such as cloves, saffron, and vanilla
(Table 11.1). Often, an essential oil or other ingredient is squeezed or
extracted from the appropriate plant part and used in concentrated form.
Some of these extracts are used as fragrances or food ingredients; others
are used as animal poisons. For example, roots containing rotenone are
used by native South Americans to help in capturing fish by temporarily
paralyzing them. Rotenone is also used as a pesticide because it paralyzes and kills insects. Manioc, a tropical root, is the source of both an
arrow poison and tapioca, a food that also serves as a thickening agent.

Wood is a commercially important plant product used the world over
as a fuel and as a building material in the form of lumber. The history of
human civilization would have been very different without spears, axes,
hoes, boats, and many other objects made mostly of wood. Also, paper
and paper products are made largely from wood.

Our modern arsenal of prescription medicines is derived largely from
plant products, many of them tropical (Table 11.2). Other drugs such as
aspirin were originally derived from plants (the bark of willow trees,

CONNECTIONS ▶ CHAPTERS 5, 9, 10, 18

Table 11.1

A few of the many vascular flowering plants used as spices and fragrances.

PLANT TAXON	PLANT NAME	PLANT USE	PLANT PART USED
CLASS DICOTYLEDONAE			
Mint family	*Mentha* spp.	Mint	Leaves or essential oil
	Ocimum basilicum	Basil	Leaves
	Origanum vulgare	Oregano	Leaves
Pepper family	*Piper nigrum*	Pepper	Seeds
Nightshade family	*Capsicum frutescens*	Chili peppers, paprika	Fruit
Laurel family	*Cinnamomum* spp.	Cinnamon	Inner bark
Myrtle family	*Eugenia aromatica*	Cloves	Whole, unopened flower buds
Ginger family	*Zingiber officinalis*	Ginger	Roots
Mustard family	*Brassica nigra*	Mustard	Seeds
Rose family	*Rosa* spp.	Rose	Essential oil
CLASS MONOCOTYLEDONAE			
Lily family	*Allium sativum*	Garlic	Bulbs
Iris family	*Crocus sativus*	Saffron	Petals
Orchid family	*Vanilla* spp.	Vanilla	Immature seed capsule

NOTES: In many cases, the plant part used is dried and ground into a powder. The abbreviation spp. means that various species in the same genus are
used.

Salix spp.), but now are manufactured synthetically. Willow bark, or tea-like infusions made from willow bark, were used to treat aches and pains for centuries by the Greeks and by Native Americans, among others. Many herbal medicines continue to be used in many parts of the world.

Photosynthesis

Plants use the energy that they capture from the sun to make energy-rich carbohydrates by a process called **photosynthesis**. They store the carbohydrates for their own use as energy sources for a later time. When humans and other animals eat plants, they harvest some of this energy.

Energy producers and energy harvesters. Organisms, such as plants, that can use light or other inorganic energy sources to make all of their own organic (carbon-containing) molecules from simpler molecules are called **autotrophs**. Most other organisms, such as animals, are **heterotrophs**; they cannot make their own organic compounds from inorganic materials and are therefore absolutely dependent on the organic compounds made by plants and other autotrophs. Rather than obtaining energy from light, most heterotrophs use the energy stored by other organisms, including plants, in the form of chemical bonds.

Most autotrophs are also called producers because they produce compounds usable by other organisms, including ourselves. Most heterotrophs are consumers that must obtain their energy by eating other organisms. Plants are the ultimate source of food energy not only for the primary consumers that eat plants directly, but also for the secondary and higher-order consumers that eat other consumers. When organisms die, their complex organic molecules are broken down by other heterotrophs called decomposer organisms; the breakdown products can then be recycled and used by other living things.

Energy enters the biological world as sunlight and flows through producer, consumer, and decomposer organisms in turn (Figure 11.1). At each step, some energy is lost by being converted into forms that are not usable by organisms. Most of this unusable energy escapes as heat. Because of the loss of heat, the process would run down and stop altogether unless new energy were continually supplied. The new energy for the living world is sunlight. Plants are essential to the global energy flow

Table 11.2

A few of the many medicinal plants.

PLANT		DRUG	
LATIN NAME	COMMON NAME	NAME	USE
Digitalis purpurea	Purple foxglove	digitalis	strengthens heart contractions
Rauwolfia serpentia	India snakeroot	reserpine	lowers blood pressure
Atropa belladonna	Deadly nightshade	atropine	blocks neurotransmitters, antispasmodic
		belladonna	blocks neurotransmitters, antispasmodic
Datura spp.	Jimson weed (thorn apple)	scopalamine	sedative, controls nausea
Papaver somniferum	Opium poppies	codeine	cough suppressant, pain killer
		morphine	pain killer
Cinchon ledgeriana	Cinchona tree bark	quinine	malaria prevention
Catharanthus roseus	Madagascar rose periwinkle	vinblastine	cancer chemotherapy
		vincristine	cancer chemotherapy
Taxus brevifolia	Pacific yew	taxol	cancer chemotherapy

because they are the principal means by which light energy is captured and changed into forms that other organisms can use.

Energy and pigments. By the process of photosynthesis, plants gather energy from sunlight and use this energy to make carbohydrates (sugars and starches; see Chapter 10, pp. 328–329) from atmospheric carbon dioxide and water. The overall process can be summarized by the following equation:

$$6\,CO_2 \;+\; 12\,H_2O \;\xrightarrow{\text{light energy}}\; C_6H_{12}O_6O \;+\; 6\,O_2 \;+\; 6\,H_2O$$

carbon   water   glucose   oxygen   water
dioxide   (a sugar)

Plants thus sustain the composition of the atmosphere by using up carbon dioxide and releasing oxygen in exchange. They thus sustain animal (including human) life by supplying the very oxygen that we breathe.

The capture of light energy for photosynthesis takes place in certain light-sensitive molecules called pigments, of which **chlorophyll** is the most important. Each of these molecules absorbs some wavelengths of light and not others. Chlorophyll absorbs blue and red light but not green light. The colors that we see are those that are reflected rather than absorbed, which is why so much of the living world looks green (Figure 11.2). In addition to chlorophyll, plants and other photosynthetic organisms possess various other pigments, such as carotenes and xanthophylls, which absorb light of other colors and pass the energy on to chlorophyll. These pigments are useful to the organism because they enable it to use light energy of different wavelengths. Because particular light-absorbing pigments are found only in certain groups of photosynthetic organisms, pigments are used to identify these groups and reconstruct their evolution.

Figure 11.1

Energy flow through a biological system. Energy enters as sunlight. Producers convert the sunlight to chemical bond energy usable by other organisms, the consumers. Energy locked in the chemical bonds of dead producers and consumers is released by decomposer organisms. Energy is also lost as heat at each step, so more energy must continuously enter the system for the process to continue.

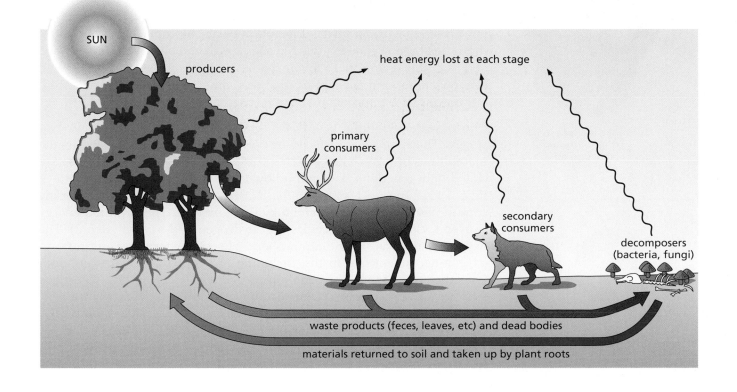

Figure 11.2

Perception of color.

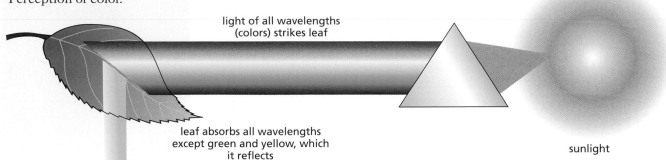

light of all wavelengths
(colors) strikes leaf

leaf absorbs all wavelengths
except green and yellow, which
it reflects

sunlight

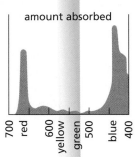

amount absorbed

700 red 600 yellow green 500 blue 400

absorption spectrum
of chlorophyll *a*
*(numbers show wavelengths
in nanometers)*

we see the reflected
light, so the leaf
looks green

CONNECTIONS   CHAPTERS 1, 6

For example, similarity between the pigments of green algae, bryophytes, and vascular plants has been used to argue that bryophytes and vascular plants probably evolved from green algae (see Chapter 6).

Many of the broad-leaved trees of temperate regions stop making chlorophyll in the autumn. In the absence of chlorophyll, the other pigments in the leaves become apparent. These pigments absorb green and blue and reflect many other colors of light, resulting in the fantastic rainbow of leaf colors in the fall foliage (see Figure 1.1, p. 5).

In all photosynthesizing organisms that have a eucaryotic cell structure (plants, including algae; see Chapter 6), the photosynthetic pigments are contained in organelles known as **chloroplasts**, which are located within the cells of the green parts, especially leaves. The fluid interior of these chloroplasts contains stacks of flattened membrane vesicles (Figure 11.3). Photosynthesis takes place along the membranes of these stacked vesicles (called thylakoids).

Although most photosynthesizing organisms are plants, there are also several photosynthesizing species from the kingdom Eubacteria, including a few bacteria and all cyanobacteria. These simpler photosynthesizing organisms have a procaryotic cell structure (see Box 6.2, pp. 170–171) with no internal membranes or chloroplasts, so photosynthesis in these organisms takes place along the cell plasma membrane and in the cytoplasm. Photosynthetic procaryotes (bacteria and cyanobacteria) and eucaryotes (diatoms, dinoflagellates, and others) fill our oceans with nutrients that can be used by other organisms.

Photosynthesis: light reactions. Photosynthesis takes place in two stages. The first stage consists of the reactions that take place in the membranes of the stacked vesicles; these require light energy and so are called light reactions. You can follow the steps of the light reactions in Figure 11.4. In step 1, photosynthetic pigments in the vesicle membrane capture light energy, represented by the gold rays. The captured energy is used to split water molecules into hydrogen and oxygen (step 2). The oxygen is released to the atmosphere (step 3), where it is essential to oxygen-dependent organisms, including humans. Each hydrogen atom is further split into a hydrogen ion (H^+) and an electron (e^-). Hydrogen ions are also called protons and are shown as gray circles in Figure 11.4. The hydrogen ions accumulate inside the stacked vesicle membranes (step 4). The electrons are shown as the blue circles and blue arrows in Figure 11.4. The electrons move down a chain of electron transport proteins that are part of the vesicle membrane (step 5), and are ultimately

Figure 11.3

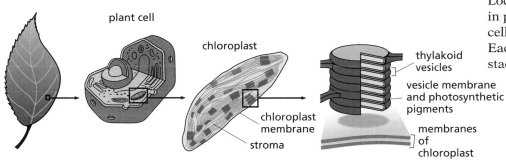

Location of photosynthesis in plant cells. Within plant cells there are chloroplasts. Each chloroplast contains stacks of thylakoids.

delivered to a molecule called $NADP^+$, forming NADPH (step 6). $NADP^+$ is an energy-carrying molecule that is made in part of niacin, a vitamin described in Chapter 10 (pp. 337–338).

Some of the electron transport proteins are ion pumps. These use some of the electrons' energy to pump hydrogen ions into the interior space of the stacked membranes (step 7), thus forming an ion gradient. As with other ion gradients, such as those across the membranes of mitochondria (see Chapter 10, pp. 350–351), the hydrogen ion gradient stores energy. The stored energy is then used to power the synthesis of the energy-rich molecule ATP (step 8). The plant cell uses ATP for most biological activities that require an energy source, just as we saw for animal cells in Chapter 10.

In summary, in the light reactions of photosynthesis, captured light energy is converted into the high-energy chemical bonds in NADPH and

Figure 11.4

The light reactions of photosynthesis. The flows of energy, electrons, and hydrogen ions are shown, as are the syntheses of ATP and NADPH, both of which contain high-energy chemical bonds.

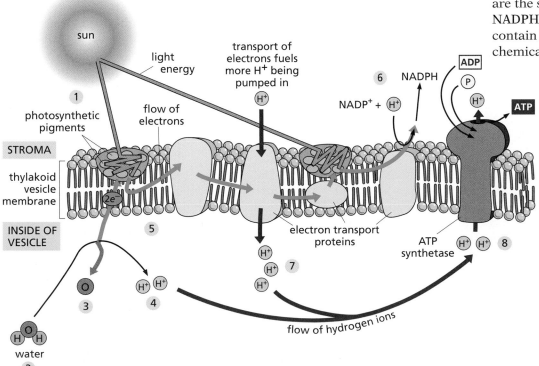

Figure 11.5

Summary of the reactions of photosynthesis.

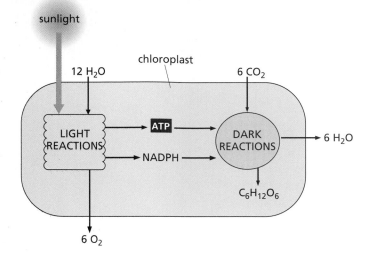

$$12 H_2O + 6 CO_2 \xrightarrow{\text{light energy}} 6 O_2 + C_6H_{12}O_6 + 6 H_2O$$

sunlight

chloroplast

12 H₂O

6 CO₂

LIGHT REACTIONS

ATP

DARK REACTIONS

6 H₂O

NADPH

C₆H₁₂O₆

6 O₂

CONNECTIONS

CHAPTER 10

ATP. This stored, light-derived energy is later used in the dark reactions that form the remainder of photosynthesis. Some light energy is also unavoidably transformed into heat.

Photosynthesis: dark reactions. The ATP and NADPH created in the stacked membrane vesicles are not released from the chloroplasts; they move from these vesicles into the surrounding fluid of the chloroplasts (see Figure 11.3). In this fluid, the ATP and NADPH participate in the second of the two stages of photosynthesis. The ATP supplies energy and the NADPH provides hydrogen for the synthesis of glucose ($C_6H_{12}O_6$) from carbon dioxide (CO_2), the source of the carbon and oxygen atoms. Because the reactions that use ATP and NADPH do not directly require light, they are called the dark reactions. The net outcome of the dark reactions is that atmospheric carbon dioxide is 'fixed,' or incorporated, into plant organic material as the sugar glucose. Figure 11.5 summarizes the overall process of photosynthesis.

Under most conditions, glucose is immediately used as an energy source in metabolism or converted into sucrose, fructose, starch, or other carbohydrates (see Chapter 10, pp. 328–329) for long-term energy storage. If the energy is later needed by the plant at a time when photosynthesis is not possible or in a nonphotosynthetic part of the plant, storage compounds in plants can be converted back into glucose and broken down to supply energy. Table sugar is sucrose, the storage product found in sugar cane and sugar beets. Starch is the storage product in potatoes and cereal grains. In addition to these carbohydrate storage molecules, many plants (including corn, palm, and most nuts) use oils (see Chapter 10, pp. 330–331) as storage products in their seeds; the energy stored in seeds is used when the seed germinates. Growth of the new plant depends on energy from the seeds until enough new leaves have been produced to carry on photosynthesis for themselves.

THOUGHT QUESTIONS

1 Consider the value judgments that may be hidden in the term *useful* (as in 'useful products'). Are plants that are useful to humans more valuable than plants that are not? Is utility the only criterion for making something valuable? Is economic value the only way of measuring value? What might some of the other ways be? How would you measure them?

2 What are some ways in which large plants (e.g., trees) are used by other plants? What are some ways in which plants are used by animals? List as many possible uses as you can.

3 Do animals have any molecules that can absorb light energy? Where might you expect to find them?

4 Some producer organisms are autotrophs, but are all producers autotrophs? Can a consumer organism also be considered to be a producer organism if it is eaten by another consumer organism?

5 The sun is a renewable, free energy source. Humans cannot photosynthesize, but in what other ways have they made use of the sun's energy? What other possible uses can be developed in the future?

Nitrogen Cycles Through the World's Ecosystems

An **ecosystem** includes all the species that live together and interact, plus all the physical resources (including water, soil, and atmosphere) with which they interact. Materials are naturally recycled throughout ecosystems, and each chemical element has its own cyclical pattern. Some of these patterns are simple. Oxygen, for example, combines with organic compounds in all organisms (e.g., during the cellular respiration described in Chapter 10), and is released to the atmosphere during photosynthesis. Carbon is incorporated into organic compounds during photosynthesis, and it is released to the atmosphere as CO_2 as an end product of both respiration and decomposition. Thus, of the several chemical elements that plants need to make their essential biological molecules, carbon can be obtained from atmospheric carbon dioxide during photosynthesis, and water can serve as the source for hydrogen and oxygen as well as dissolved ions.

Nitrogen, however, follows a more complex cycle.

Nitrogen for plant products

In addition to molecules produced as a result of photosynthesis, plants must also synthesize nucleotides for DNA and RNA and amino acids for proteins. Synthesis of these and other molecules requires nitrogen. Amino acids can be made from simpler compounds by the addition of an amino group ($-NH_2$). The amino group can be supplied from soluble ammonium compounds (containing the NH_4^+ ion). Amino groups can also be transferred from one amino acid to another. Once nitrogen-containing amino acids have been synthesized, they can be used to build proteins. Amino acids are also used as the starting materials for all the other nitrogen-containing compounds that the plant needs, including DNA and RNA, vitamins, plant hormones, and pigments.

Nitrate: a limiting nutrient. Although plants need a source of amino groups ($-NH_2$) to help make amino acids and proteins, most plants cannot absorb NH_4^+ ions directly. Instead they get their nitrogen from the soil as dissolved nitrates (NO_3^- ions) or nitrites (NO_2^- ions), which they then convert into ammonia (NH_3). Ammonia reacts with water to form ammonium hydroxide (NH_4OH), the source of the NH_4^+ ions needed for amino acid synthesis. However, a plant does not accumulate any more ammonia than it immediately uses because excess, unused NH_4OH is very alkaline and damaging to most organic tissues.

Many plant species are limited in where they can live by the availability of dissolved nitrates. A nutrient whose absence makes a species unable to grow in a particular place or on a particular food source is called a **limiting nutrient**. If the amount of the limiting nutrient were increased (by the use of fertilizers, for example), more growth would take place, assuming that other nutrients were present in adequate amounts. For many plants in many places, nitrates are a limiting nutrient.

Where does the nitrogen in nitrates come from? Is Earth's supply of nitrogen adequate? To find answers to these questions we need to take a global view.

The nitrogen cycle. Nitrogen moves through the world's ecosystems, changing from one molecular form to another, forming a loop called the **nitrogen cycle**. Plants and other living organisms are an important part of this cycle: chemical end products released by one type of organism are used in biochemical reactions by other organisms. Some of these organisms can fix (incorporate) atmospheric nitrogen into molecular forms that other organisms can use, while other organisms release nitrogen as a gas into the atmosphere. In each ecosystem, living organisms are thus united with each other and with their physical surroundings (including the atmosphere) by the nitrogen cycle. The nitrogen cycle is a good example of a nutrient cycle; all such cycles are loops in which materials such as nitrogen are exchanged among producers, consumers, decomposers, and their surroundings, including the atmosphere.

The key stages of the nitrogen cycle are shown in Figure 11.6. Follow first around the outer circle, starting with free nitrogen (N_2) in the atmosphere. Earth has an abundant source of nitrogen in nitrogen gas (N_2), which makes up about 78% of our atmosphere. Plants need nitrogen, but, like other eucaryotic organisms, they are unable to make the enzymes necessary to use nitrogen from the atmosphere. They are therefore dependent upon certain procaryotic microorganisms that have the ability to convert atmospheric nitrogen into ammonia (NH_3). Incorporation of atmospheric nitrogen into ammonia is called nitrogen fixation. There are several kinds of nitrogen-fixing organisms in the soil, including many cyanobacteria (see Chapter 6, pp. 168–169) such as

Figure 11.6

The nitrogen cycle.

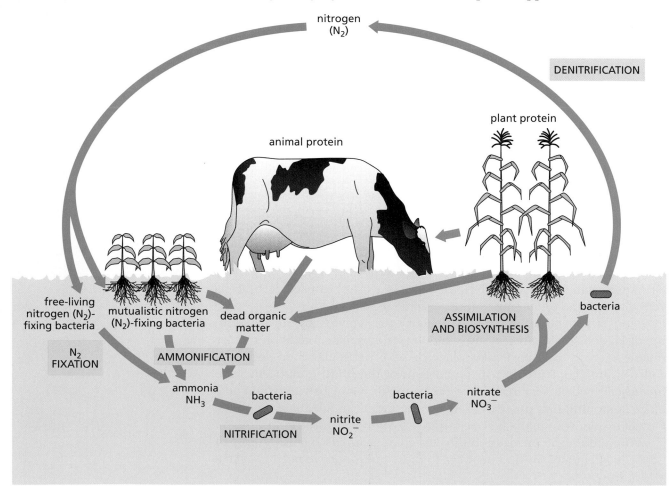

Nostoc and many nonphotosynthetic bacteria such as *Rhizobium, Azotobacter, Klebsiella,* and *Clostridium.*

Most plants, however, cannot take up ammonia; they can only take up nitrogen as nitrite (NO_2^-) ions or as nitrate (NO_3^-) ions. Ammonium compounds produced by nitrogen-fixing bacteria in the soil can be converted into nitrites and then into nitrates by still other bacteria (a process called nitrification). Bacteria such as *Nitrosomonas* convert ammonia to nitrites and obtain their energy from this conversion process. *Nitrobacter*, another bacterium, then converts the nitrites into nitrates. Plants can absorb both nitrites and nitrates. After the nitrites and nitrates have been absorbed, plants convert NO_2^- or NO_3^- back into NH_3 to be used in the biological synthesis of amino acids, proteins, and other nitrogen compounds.

The nitrogen cycle is completed by still other soil bacteria. Excess nitrates that are not absorbed by plants are converted by these bacteria into nitrogen (N_2). The N_2 is then released into the atmosphere as a gas. In a balanced nitrogen cycle, the amount of nitrogen released as a gas equals the amount converted into nitrogen compounds.

Thus, in the nitrogen cycle as a whole, N_2 is taken from the atmosphere, cycled through many organisms, and returned to the atmosphere. Most of these processes are carried out by bacteria. Nitrogen also cycles through more complex organisms, as when plants take up nitrites and nitrates and use the nitrogen to make amino acids, proteins, and nucleic acids. Animals get their nitrogen by eating plants or other animals. When plants and animals die or give off waste products, their nitrogen is returned to the soil by still other species of bacteria that convert these waste products into ammonia. This process (called ammonification) is a second source of the ammonia needed for making nitrates. The cycling of nitrogen from ammonia to nitrites, nitrates, plant proteins, animal proteins, and back to ammonia forms a subcycle within the nitrogen cycle, as shown by the inner circle in Figure 11.6.

Mutualistic relationships

As we have seen, ammonia can enter the soil via free-living nitrogen-fixing bacteria or by the breakdown of dead organic material. A third source of soil ammonia is by bacterial nitrogen fixation within the roots of some plants. Some species of plants ensure the availability of nitrogen-fixing microorganisms by growing root nodules not far below the soil surface (Figure 11.7). These root nodules actively attract the growth of nearby nitrogen-fixing soil bacteria, chiefly *Rhizobium*. The root nodule serves as a culture chamber for these microorganisms, which then carry out nitrogen fixation inside the plant roots.

The relationship between the two organisms is called mutualism, an interaction from which both species benefit. Bacteria produce NH_3, which, because it is already inside the plant, is taken up directly without the need for conversion to NO_2^- or NO_3^-. The plants thus benefit from the mutualism because they have a built-in supply of nitrogen for biosynthesis. Bacteria have to spend a lot of energy to fix nitrogen. The reaction that fixes the nitrogen

Figure 11.7

Underground nitrogen fixation in root nodules.

root of a legume   *Rhizobium*-containing nodules

uses a great deal of energy, as does the synthesis of the enzyme (nitrogenase) needed to carry out the reaction. The bacteria benefit from the mutualism because they get the energy for both enzyme synthesis and nitrogen fixation from the plant in the form of organic compounds.

Plants of the family Leguminosae, including beans, peanuts, peas, locusts, and alfalfa, are all capable of adding nitrogen compounds to the soil because of their mutualistic association with nitrogen-fixing *Rhizobium* bacteria. This is the basis for the use of 'green manure,' the planting and subsequent plowing under of a legume crop in a field that has been depleted of nitrates by the earlier growing of plants that absorb large quantities of nitrates from the soil. When the legume is plowed under, the nitrogen compounds that it has produced with the help of its mutualistic nitrogen-fixing bacteria are returned to the soil in a form that other bacteria can convert to nitrates. Green manure thus adds nitrogen compounds to the soil, reducing or eliminating the need for the addition of fertilizers containing nitrate or ammonia.

Most plants do not have mutualistic nitrogen-fixing bacteria living in their roots. They therefore need to absorb their nitrogen compounds from the soil, principally as nitrates. A few plant species, however, have evolved other means of obtaining nitrogen.

Plants living in nitrogen-poor soils

In many soils, the nitrogen cycle is balanced: the amount of nitrogen released to the atmosphere is equal to the amount taken up from the atmosphere. However, in waterlogged situations, where conditions may be anaerobic (lacking in oxygen), the balance may be lost. The bacteria that produce nitrates require oxygen to live and so die off. The bacteria that convert nitrates into N_2 do not require oxygen to live, so this conversion continues and nitrogen is lost. Thus, the available nitrates get converted entirely to N_2 and are not replaced, leaving none to be taken up by plants. Without available nitrates, plants cannot absorb adequate amounts of nitrogen.

Plants that have evolved to live in nitrogen-poor soils have several different ways of coping with the scarcity. Some plants can absorb the ammonium ions (NH_4^+) that form in wet, decaying leaf litter, but they do so by exchanging the NH_4^+ ions for hydrogen ions (H^+), making the soil more acidic. Microbial decay of the leaves of the plants releases even more acid, eventually killing all those plant and microbial species that are not acid-tolerant. The bacteria that convert ammonium ions to nitrites and then to nitrates are not acid-tolerant and thus are inhibited in this environment. The habitat that results is an acid bog with high organic content but few nitrates.

One rather unusual solution to the problem of obtaining nitrogen in an acid bog is to digest animal proteins. Few plants, however, can be so fortunate as to have an animal die and leave its carcass in the soil just within reach of their root system. How, then, are they to obtain animal protein? Carnivorous plants have evolved adaptations to trap and kill small animals (mostly insects) and derive nitrogen from the digested proteins. Most carnivorous plants live in such nitrogen-poor habitats as acid bogs, where moisture and insects are both usually abundant. A variety of mechanisms have evolved by which these plants trap their prey, digest their proteins, and absorb the resulting amino acids. The traps can either be passive or active (Figure 11.8). Passive mechanisms can include pitfalls (as in pitcher plants), sticky materials that form a passive flypaper, and traps shaped like lobster pots. Active mechanisms can include trapdoors, active flypapers (such as the sundew, whose hairs bend to enclose and further hold their victim), or pads that swing shut when an object touches them (such as the Venus flytrap; see also Figure 11.12).

It is clear that some plants have evolved some intriguing structures to obtain nitrogen, while others have developed complex mutualisms and niches within the nitrogen cycle. Most nitrogen, as well as water, comes in through the roots of the plants; most carbohydrates are made by photosynthesis in the leaves. Yet all of these compounds are needed throughout the plant. Vascular plants have specialized tissues that carry out different functions in different parts of the plant, including specialized vascular tissues that transport fluids from one location to another throughout the plant.

Figure 11.8

Plants that obtain nitrogen from animals.

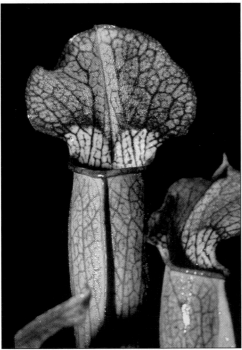

Pitcher plant (*Sarracenia oreophila*), showing a passive trap (pitfall type), with slippery inside surfaces and fluid pool at bottom containing digestive enzymes

external view

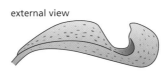

cutaway view

Sarracenia psittacina, a pitcher plant with two passive traps. The entrance works like a lobster trap because insects that have entered the chamber have difficulty finding the opening by which they came in. When they crawl into the long tube, the hairs inside form a second trap that allows them to crawl deeper in only one direction.

sensitive hairs closed over fly

The sundew plant, *Drosera*, whose sticky hairs close over an insect, forming an active trap (flypaper type).

THOUGHT QUESTIONS

1 Why do you think plants take up NO_3^- or NO_2^- ions but not NH_4^+ ions?

2 Some plants have mutualistic associations with bacteria in root nodules. Do humans also have mutualistic associations with bacteria?

3 What is the definition of acidity? How does the uptake of NH_4^+ ions by some plants make the soil around them more acidic?

Plants Use Specialized Tissues and Transport Mechanisms

As we have seen, photosynthesis in most plants is carried out principally in the leaves. Other plant parts are also specialized for particular functions. Roots are specialized for water absorption and fruits for reproduction and dispersal. The division of labor among different parts of the plant would not be possible without the specialization of plant tissues, nor would it be possible without efficient mechanisms for the transport of materials from one part of the plant to another. These adaptations are described in Chapter 6, but a brief review is presented here.

Most plants are vascular plants, possessing separate organs such as roots, stems, and leaves. The roots absorb water and nutrients from the soil. The leaves carry out photosynthesis and produce sugars and other carbohydrates. Gas exchange in the leaves brings in CO_2 and releases both oxygen and water to the atmosphere. The vascular tissues, especially in the stems, conduct water upwards from the roots and distribute photosynthetic products to all parts of the plant. The vascular tissues are strong enough to support the weight of the plant and allow it to grow very tall—up to 80 m (250 feet) in some tropical trees.

Tissue specialization in plants

In many algae and other simple plants, each cell carries out all essential functions, including photosynthesis, absorption of nutrients, and excretion of waste products. Most plants, like animals, have groups of similar cells organized into tissues, and groups of tissues organized into organs such as leaves and roots. For example, each leaf is an organ, while each cell layer within a leaf is a tissue. The simplest plants containing separate types of tissues are the mosses, liverworts, and hornworts, often grouped as Bryophyta and also known as nonvascular plants.

The most familiar and ecologically dominant group of plants are the **vascular plants**, including all plants that have vascular (conducting) tissues that transport fluids, generally through tubular cells surrounded by rigid cell walls. Vascular plants contain two types of conducting tissue:

xylem, which usually conducts water and minerals upward, and phloem, which conducts a water solution of photosynthetic products (mostly sugars) in both directions but more often downward. The existence of these vascular tissues allows the parts of the plant to specialize (Figure 11.9). The roots grow underground, anchor the plant in the soil, and absorb water and dissolved nutrients. The vascular tissues conduct the water and nutrients from the roots through the xylem of the stem to the above-ground parts, where the water is needed for both photosynthesis and support of the upper parts of the plant. The roots can receive photosynthetic products through the phloem from the chlorophyll-containing tissues above. Roots thus need not contain chlorophyll or carry out photosynthesis themselves, so they are not green and do not require light.

Vascular tissues have rigid cell walls, and, as we see in the next section, water pressure makes plant tissues even stronger as well as helping plants to stand upright. Thus, in addition to transport, these adaptations in stems allow many vascular plants to grow tall without having the skeletons that support vertebrate animals.

Water transport in plants

Water moves over great distances within plant bodies. Plants would not be able to absorb nutrients, conduct photosynthesis, synthesize chemical compounds, or get rid of waste products if water were not able to move in and out of plant cells and through the plant as a whole.

Osmosis. Like all cell membranes, plant cell membranes stay intact only because they are surrounded by water. Lipid bilayer membranes are impermeable to ions yet are permeable to water. This allows water to move by a process called **osmosis** whenever opposite sides of the membrane contain solutions at different concentrations. We usually think of the concentration of a solution in terms of the number of dissolved ions or other particles in a given volume of water. We have seen that a membrane can store chemical potential energy in the form of differences in ionic concentration on the membrane's two sides. But we can also think of the solution in terms of the concentration of water, that is, the numbers of molecules of water per volume of solution: the higher the concentration of ions and other dissolved materials, the lower the concentration of water. All substances tend, according to the laws of physics, to diffuse from a region of higher concentration to a region of lower concentration. Osmosis is a special type of diffusion in which water passes through a membrane from a region of high water concentration (and low ion concentration) to a region of low water concentration (and high ion concentration), but the ions are prevented by their charge from moving through the membrane. There is no

Figure 11.9

Different functions carried out by specialized parts of vascular plants. Detail at the top shows a cross-section through a leaf.

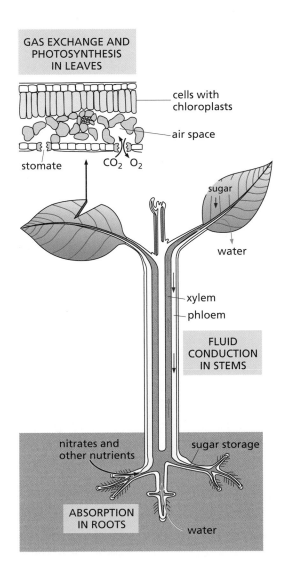

Figure 11.10

Osmosis in plant cells. Water moves across cell membranes toward the side with the lower water concentration and higher ion concentration.

attraction or repulsion in the process of osmosis, just a flow of water molecules from an area where their number is higher to an area where their number is lower (Figure 11.10).

Plant cells are surrounded by a rigid cell wall made of cellulose, which lies outside the plant cell membrane. When water flows into plant cells by osmosis, the cell membrane eventually pushes against this cell wall, producing a form of fluid pressure (called turgor) that makes plant tissues stiff (see Figure 11.10). It is thus water pressure, rather than a

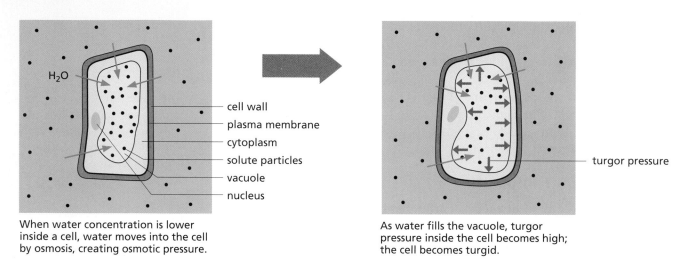

cell wall
plasma membrane
cytoplasm
solute particles
vacuole
nucleus

H_2O

When water concentration is lower inside a cell, water moves into the cell by osmosis, creating osmotic pressure.

turgor pressure

As water fills the vacuole, turgor pressure inside the cell becomes high; the cell becomes turgid.

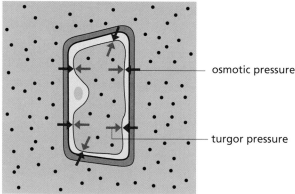

osmotic pressure

turgor pressure

When concentration is the same inside and outside, there is no osmotic flow; osmotic pressure equals turgor pressure.

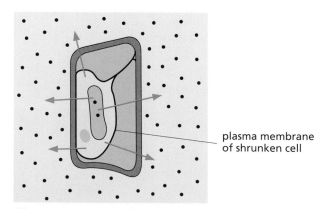

plasma membrane of shrunken cell

When water concentration is lower outside, a cell shrinks as water flows out.

skeleton, that keeps nonwoody plants upright. In woody plants, cell walls are stiffened with a rigid material (called lignin) that provides additional support.

Water transport. A plant moves water from its roots to all of its cells. The rate of water movement is greatest when the plant is in sunlight, a time when the water loss by evaporation from the leaves (called transpiration) is greatest. Transpiration lowers water pressure in the leaves, and the higher pressure from below moves the water upward.

Transpiration is controlled by openings called stomates in the undersides of leaves. The opening and closing of the stomates is accomplished by large 'guard cells' on either side, which change shape in response to increasing and decreasing water pressure (Figure 11.11). When water pressure is low (which happens when water supplies are low), the guard cells are closed. When water pressure increases in adjoining cells, ion channels in the guard cell membranes open, allowing K^+ and Cl^- ions to enter the guard cells. Water molecules flow by osmosis, increasing pressure in the guard cells and resulting in a change in their shape. In the new shape, a space between the two guard cells opens and allows the exchange of CO_2, O_2, and water vapor. When the stomates are open, gases flow in and out: carbon dioxide (CO_2), needed for photosynthesis, is taken up, and oxygen (O_2), a product of photosynthesis, is given off (see Figure 11.9). Water vapor is also given off because the water concentration of the air is lower than that of the xylem.

Transpiration also serves to cool the plant. Some of the light energy absorbed by plants is transformed into heat and cannot be used in photosynthesis. Plants avoid overheating by evaporating water from their leaves through transpiration. Most of the water absorbed by any plant is eventually lost through the leaves in this way.

Rapid movement in plants. Whereas animals can use nerve cells to stimulate movement through muscle contraction, plants have neither muscles nor nerves and thus have only limited powers of movement. Some plants, however, can move their parts quite rapidly; their ability to regulate turgor pressure is the key to how plants such as the Venus fly trap (*Dionaea muscipula*) and the sensitive plant *Mimosa*

Figure 11.11

Opening of a stomate.

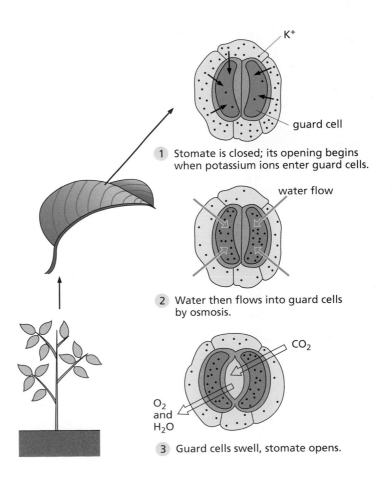

1 Stomate is closed; its opening begins when potassium ions enter guard cells.

2 Water then flows into guard cells by osmosis.

3 Guard cells swell, stomate opens.

pudica are capable of rapid movements (Figure 11.12). Touching one of these plants causes ion channels in the tissue touched to open, and ions to flow out from the cells. Water flows out in response, and turgor drops quickly. The cytoplasm of a plant cell is connected to the cytoplasm of adjoining cells through small openings. The changes of ion concentration are thus passed along to the next cell, where ion channels are stimulated, passing along the change in turgor. When turgor drops in the cells where a *Mimosa* leaflet joins the stem, the leaflets collapse towards each other (see Figure 11.12). When changes in turgor have propagated to the base of the leaf stem, which may take more than one touch, the whole leaf droops. The plant will return to its former position when water flow reverses, reestablishing turgor pressure. Thus, in plants, movement can be accomplished without contractile muscle fibers.

Now that we have learned some things about how plants obtain their nutrients and energy, we can return to the major question posed by this chapter: can plant science feed the world?

Figure 11.12

Rapid and selective movement in plants.

When this sensitive plant (*Mimosa pudica*) is touched, the leaflets fold together within a fraction of a second.

When an insect touches the sensitive hairs of this Venus fly trap, the leaf halves snap together in less than half a second, trapping the insect.

THOUGHT QUESTIONS

1 Do plants have sense organs?

2 How can plants move without nerves and muscles?

3 How are plants able to stand upright?

4 Why can plants grow without soil but not without water?

Crop Yields Can Be Increased by Overcoming Various Limiting Factors

To feed the world's growing human population, we must increase crop production. The amount of land under cultivation is, however, limited, and most efforts to increase yields of crop species have therefore focused on such techniques as soil improvement, fertilizers, irrigation, and pest control. Each of these methods seeks to increase yield in a different way: by supplying a limiting nutrient, by supplying water (which is often limiting in the same way), or by controlling natural enemies of the crop species. Moreover, these improvements must be carried out sustainably, that is, without lasting harm to the ecosystem that supports the growth of crop plants. (In a later section we also examine the development of new strains of crop plants.)

Fertilizers

In those locations where a nutrient is a limiting factor for the growth of plants, crop yields can often be dramatically increased by adding the appropriate fertilizer to the soil. A **fertilizer** is a substance that supplies organic or inorganic nutrients needed by plants. In the largest number of locations, nitrogen is the limiting nutrient for many plant species, and nitrates are among the most important fertilizers used in agriculture. In many places, crop yields can be dramatically increased by adding nitrates to the soil. Phosphorus (P), in the form of phosphates (PO_4^{3-}), can also be an important limiting nutrient, especially for ornamental flowers or crops whose edible portion is a seed, a flower, or a fruit. Phosphorus often signals the plant to begin setting flowers, seeds, and fruit. Potassium (K^+) is often important as well. It is used by plants to control the opening and closing of stomates (see Figure 11.11), and it may be scarce (and therefore limiting) in certain soils. The hazards as well as the benefits of applying fertilizers need to be considered.

Fertilizers can be either organic, including various manures and composts, or inorganic. The nutrients in organic fertilizers are complex molecules that break down slowly. The slow breakdown gives these fertilizers the advantage of slow release, a process that provides a steady supply of soluble nutrients that is more likely to be absorbed by growing plants than to be washed away as are many inorganic fertilizers. One disadvantage of organic fertilizers is their high bulk and cost of transportation, features that often limit their use to the immediate locality of their production.

Manures. Waste products from domestic animals are used all over the world as fertilizers, and these animal manures are often readily available near where crops are grown. In China, human and animal wastes have been used as fertilizers for centuries. This practice provides needed nutrients to crops, but it also has the undesirable effect of spreading parasites (especially flukes of the animal phylum Platyhelminthes) and other infectious diseases. In some countries, animal wastes are treated to kill parasites or to reduce water content (and therefore transportation costs).

We have seen that plants can accumulate nitrogen compounds. Green manure consists of the remains of plants that are plowed into the soil at the end of the season, providing organic fertilizers without need of transportation. For example, grasses in a field previously used as pasture may be plowed under. Some plants are grown only for this purpose. More often, some part of the plant is harvested and the remainder is plowed under; lima beans are an example.

Many farmers plant a certain fraction of their fields with plants that accumulate nitrogen; the location of the various plants is then rotated from year to year (a practice called 'crop rotation'), giving each field a turn to accumulate nitrogen in the soil. The alternation of cereal grains with soybeans or alfalfa (both legumes) is a form of crop rotation that contributes to high agricultural productivity in many large regions of North America and Asia. Organic farmers use green manuring, sometimes in combination with the use of various animal and plant wastes as fertilizer. The benefit of crop rotation for maintaining the fertility of the soil was recognized in ancient Rome and several other ancient civilizations.

Composting. Organic material (leaves, grass clippings, food waste) can be layered with manure or soil. The mixture provides a place for microorganisms to live and they aid the decomposition of the plant and manure layers. Many United States municipalities encourage composting, either by teaching people how to compost in their own backyards or by having centralized composting facilities to which people can bring their yard wastes. Even so, less than 1% of organic wastes are composted in the United States, while much more is done in Europe. It is estimated that solid waste in United States landfills could be reduced by 20% if organic material were composted.

Chemical fertilizers. Inorganic chemical fertilizers are usually nitrate or phosphate salts sold as powders or granules to be applied to fields, usually by mechanical equipment. When they were first introduced, chemical fertilizers were cheap and readily available. The benefits in increased crop production were immediately evident. Bountiful harvests created surplus crops and the use of chemical fertilizers increased greatly in countries that could afford them.

Ecological problems associated with the prolonged use of inorganic chemical fertilizers became evident only after several decades of general use, in accordance with what some call the 'law of unintended consequences'. This has also been stated as the first law of ecology: any change imposed upon nature has a number of effects, many of which are unpredictable—"you can never change just one thing." Excessive use of inorganic fertilizers can kill off the soil organisms required for maintaining soil fertility, thus requiring the use of even more fertilizer for subsequent crops.

Compared with organic fertilizers, inorganic minerals are relatively expensive to produce or extract, and most of them must be transported over great distances. Cheap local supplies have in most cases been depleted, and many are now mined far from the places where they are used. For example, large quantities of nitrate minerals are mined in Chile for use as fertilizers in the Northern Hemisphere. Fossil fuels are used in the mining and transportation of fertilizers and by the tractors used to spread these fertilizers on fields. Most chemical fertilizers are costly to transport to where they are needed, and in most of the developing nations of the world they are prohibitively expensive. Organic fertilizers are often available more locally, and therefore need not be transported as far.

Sources of phosphorus. Fertilizers are often needed to supply phosphorus. Important nutrients, including both calcium and phosphates, are slowly released from fish meals and bone meals, which are powdered, dried bone. The phosphates provided by these fertilizers are insoluble; they are converted to soluble form only slowly, often by the very plants that use them. If the practice is begun in soil that is nutritionally deficient, the slow solubility is a disadvantage at first; chemical fertilizers may therefore be necessary during the initial phase-in period. Fish or fish meals also contain additional phosphorus in the form of nucleic acids from the nuclei and organic phospholipids from the membranes of the cells of the dead animal. Native Americans throughout the eastern woodlands traditionally grew their corn in mounds, under which they customarily buried a fish that provided both phosphate and nitrogen as fertilizer for the corn.

The presence of phosphorus in fertilizers is often associated with problems of runoff and stream pollution. Fertilizers are concentrated nutrient sources for all plants, not just for the crop plants for which they are intended. When fertilizers run off from agricultural fields into bodies of water, they supply nutrients to algae in the water. The growth of the algae may have previously been held in check by the lack of some limiting nutrient, such as phosphorus. Once that nutrient is present, the growth of these algae can be so rapid that they use up all the oxygen in the water, causing the death of many fish. The decomposition of dead fish releases more nutrients that further accelerate the growth of the algae. This process of nutrient enrichment, algal growth, and oxygen depletion is called **eutrophication**. Eutrophication led to the deaths of most fish, and the commercial fishing industry, in Lake Erie in the 1960s. International regulations that cut down on runoff pollution have allowed the lake to recover to a great extent. Eutrophication remains a problem in more than half of the lakes in the United States (Figure 11.13).

The problems of runoff and eutrophication can originate from animal wastes (e.g., from dairy or hog farms or from human sewage) or from inorganic chemical fertilizers, especially soluble nitrates. Laundry detergents containing phosphates were also a major source of nutrients, but public awareness pressured manufacturers into changing their formulations, so that many detergents are now made phosphate-free. The amount of rainfall and its

Figure 11.13

Eutrophication: algal blooms accelerated by fertilizer runoff. The two portions of this lake in Ontario were separated by a plastic curtain, and phosphates were added to one side of the curtain only. An algal bloom occurred only on the far side, to which the phosphates had been added, making the water look more cloudy.

phosphates added   plastic sheet dividing   no phosphates added
                              the lake

seasonal distribution can greatly influence runoff, as can human factors such as pavement near the edges of bodies of water. Runoff can sometimes be minimized by applying fertilizers at the proper time of year, just before the nutrients are most needed by the plants and are most likely to be absorbed.

Soil improvement and conservation

Soil is the loose material derived from weathered rocks and supplemented with organic material from decaying organisms. This organic material supports the growth of plants and other species that, in turn, add to the soil-building process. The type of soil formed depends on the parent rock type, the climate, the removal of soluble materials by percolating groundwater, and the mechanical and chemical activities of living organisms such as bacteria and protists, earthworms, and plant roots. Soil is thus the product of both biological and geological processes (Figure 11.14).

Humus. Organic material in the soil is broken down by bacteria and fungi, which function as decomposers. The partially decomposed organic matter is called humus and it serves several important functions in the top layer of the soil. It binds the particles of weathered rock together to form small aggregates, which give the soil its structure. The small holes between particles help to hold water and oxygen and provide space for the growth of minute root hairs, extensions of single cells on the roots of plants. Although the larger roots provide the anchorage for the plant and in some plants can reach deeply into the ground in search of water, it is the root hairs that provide the enormous surface area for absorption of nutrients. Because root hairs are so delicate, they cannot push into soil, but must grow into already existing spaces in the soil structure.

Humus has an overall negative charge, which helps to hold positively charged nutrient ions such as potassium (K^+), calcium (Ca^{2+}), and ammonium (NH_4^+) in the topsoil, where they are available to plants' roots. Inorganic fertilizers can provide chemical nutrients, but they do not contribute to humus and soil texture as organic fertilizers can. The plowing under of organic matter helps to build humus. Rather than being left bare, fields can be planted with grasses during the season when cash crops do not grow, or with legumes that will be plowed under when it is time to sow the crops. This both prevents erosion and adds to humus. Humus can also be produced by composting; the decomposer bacteria from the soil turn the organic matter into humus over a period of weeks or months.

Soil as a nonrenewable resource. Topsoil is eroding faster than it is being formed on approximately one-third of the world's croplands. Soil that is lost can be regenerated by biological processes, but only very slowly—one inch of topsoil takes between 200 and 1000 years to renew, depending on the climate and other factors. Groundwater pollution, excess

Figure 11.14

Soil formation from the weathering of rock (in this case, limestone) and from biological processes.

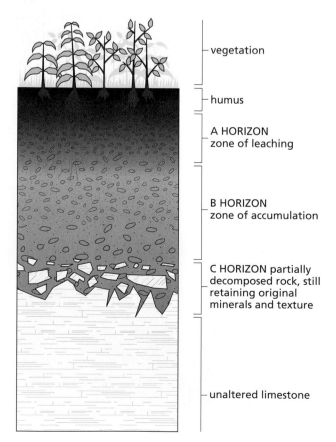

vegetation

humus

A HORIZON
zone of leaching

B HORIZON
zone of accumulation

C HORIZON partially
decomposed rock, still
retaining original
minerals and texture

unaltered limestone

fertilizer, and runoff from highway salt can kill soil microorganisms and stop the regeneration of topsoil. Because soil cannot be regenerated within a few years, it must be treated in the same way as a nonrenewable resource. Treating soil as a nonrenewable resource, farming and other land uses are best done in ways that are sustainable (see Chapter 18, pp. 674–676). Sustainable practices are those that lead to no net loss of a resource. Important practices in sustainable agriculture include crop rotation, use of organic, humus-building fertilizers, and the prevention of erosion.

CONNECTIONS
CHAPTER 18

Irrigation

Plants need water to supply the hydrogen ions used in photosynthesis. They also need water for transport of their nutrients and the products of their synthesis reactions. About 90% of the water absorbed by a plant is evaporated through its leaves during transpiration, dissipating excess heat. Also, the rigidity and strength of plants are based partly on the properties of water, as explained earlier.

Irrigation is the process of supplying added water, a vital substance for the growth of plants. In much of the world, crop yields are limited by a shortage of water. Water is thus a limiting resource for agriculture in most places where crops could be grown.

Irrigation is expensive in regions where water is scarce and where irrigation is therefore most needed. Traditional methods of spray irrigation lose much of the water to evaporation. Newer methods include drip irrigation, in which irrigation tubes with tiny holes are laid at or below ground level. The tiny holes, spaced every few inches, are designed to leak or drip water slowly into the soil, providing moisture but minimizing evaporation. All irrigation systems, whether drip or traditional, require a large initial investment in pipes or ditches, pumping stations, and the like, plus a supply of fresh water or desalinized water (seawater from which the salts have been removed). Freshwater sources can become the subject of political disputes between neighboring governments, such as those between Israel and Jordan or between California and Arizona.

Hydroponics

Although plants cannot live without water, they can live without soil. Some plants grow naturally without soil. The growing of crop plants without soil is called hydroponics (Figure 11.15). In places where there is very little soil, or where the soil is unsuitable, hydroponics offers a possible alternative agricultural method. In a hydroponic system, the plants are grown with their roots immersed in tanks through which water carrying dissolved nutrients is allowed to flow. The water is recirculated

Figure 11.15

Hydroponics: growing plants in water tanks, without soil. (The water surface is often covered over to retard evaporation.)

(and therefore conserved), and its dissolved materials are frequently monitored and adjusted.

Advantages of hydroponics. A well-managed hydroponic system can produce greater yields than traditional soil-based systems. Because hydroponic systems are generally inside greenhouses, where they are protected from insects and other plant pests, the produce that results is free from the blemishes often caused by these pests. Much of the labor traditionally concentrated on soil care, including tilling, planting, fumigation, and irrigation, is eliminated. Water is provided to the plants directly and more efficiently, and is recycled so as to minimize its use. This is extremely important in arid climates, where sunshine may abound but where water is scarce. Instead of traditional application of fertilizers that can diffuse beyond the reach of plant roots, mineral nutrients are added directly to the water, where the excess not taken up by the plants remains available and can be recycled. In this way, the use of nutrient supplements is minimized, and runoff problems are also minimized. Disease and pest control can be handled easily by adding the necessary chemicals to the recycled hydroponic water, with far less danger that these substances will spread to local water supplies or to domestic animals and humans.

Disadvantages of hydroponics. Many of the disadvantages of hydroponic systems have to do with cost: the initial construction costs; the costs of maintaining equipment and greenhouse facilities; the costs of nutrients; and the salaries of trained personnel for constant monitoring for nutritional problems (such as nutrient depletion) and waterborne diseases. Hydroponic systems may be in locations distant from the Equator, where the provision of artificial light and heat can add to the cost.

Plants are subject to disease, as are all living organisms; waterborne diseases are those in which the pathogen (disease-causing organism) can live in water. Even if a pathogen does not grow well in water, its ability to survive in water might enable it to be transmitted from plant to plant in a hydroponic system. Most hydroponic farms concentrate on a single plant species, a practice known as **monoculture** that brings a risk of the rapid spread of disease, whether the monoculture is hydroponic or land-based. The risk relates to the fact that many pathogens are able to infect only one or a few host species. When a pathogen encounters a monoculture, it can spread very quickly because of the proximity of other individuals of the same host species.

For these reasons, hydroponic systems are easier to justify economically for plants that are economically profitable in small quantities (e.g., pharmaceutical plants or certain vegetables) rather than for staple crops such as cereal grains. Hydroponics is potentially useful where either growing space or soil are limited, as in proposed space stations. On Earth, hydroponics works best in warm climates such as Israel or southern Italy, where light energy is abundant and artificial heating is unnecessary, or in countries such as Japan, where people are willing to pay extra for produce that is 'picture perfect.'

Chemical pest control

Each year, 30% or more of many crops are destroyed by insects and other crop pests. In addition to insects and their larvae, which damage the plants themselves, other crop pests, including various rodents, damage many crops in the field and also consume stored grains. Fungi also destroy up to 25% of stored crops. Nutrients that are consumed by pests are not available to the plant or to humans for food.

Monoculture and pests. Some of the farming practices that have increased crop yields over the last few centuries have also increased the susceptibility of crops to damage by pests. For example, large, mechanized farms (farms that are heavily dependent on the use of machinery) may plant monocultures that are thousands of hectares in size. (The hectare, or metric unit of land area, is an area 100 m by 100 m, equivalent to about 2.47 acres.) Monocultures are especially suitable for mechanized agriculture (Figure 11.16) and large-scale operations tend to be more economical than small-scale ones; as a result, monoculture is extremely profitable for a while. However, as we have seen, monocultures make it easier for a pest species to spread rapidly, especially if neighboring farms are also planted with the same crop. If neighboring fields were planted with different crops, the spread of pest species would be interrupted or made more difficult. Also, planting the same crop year after year enables pest species to survive from year to year in the form of eggs or larval stages. In contrast, planting different crops in rotation interrupts the life cycle of a pest that depends entirely on one particular crop species for its propagation.

Pesticides. Any substance that kills organisms that we consider undesirable is called a **pesticide**. The ideal pesticide would (1) kill only the target species, (2) have no effect on nontarget species (including humans), (3) avoid the development of resistant genotypes in the target species, and (4) be capable of breaking down into harmless substances in the soil by natural processes in a reasonably short period. It is hard to develop a pesticide that has all these characteristics. The ideal pesticide unfortunately does not exist.

Chemical pesticides have been used since ancient times. For example, the Romans dusted sulfur, a fungicide, on their grapes. The use of chemical pesticides, including arsenic and copper compounds, increased enormously during the nineteenth and twentieth centuries. These compounds have largely been replaced by pesticides derived from petroleum.

Figure 11.16

Monoculture: the planting of a single species only.

A wheat field in Nebraska.

A lettuce field in Ontario.

Chemical pesticides can greatly reduce the amount of crop damage due to pests, and they continue to be used in many countries because they increase crop yields.

For much of the twentieth century, economic pressure encouraged the use of pesticides on crops and post-harvest treatment with fungicides. The post-harvest treatments have given many farm products longer shelf lives, allowing transportation across longer distances and a globalization of food production. As a result, people in the industrialized world have come to expect perfect, blemish-free produce at almost any time of the year, even for crops that do not grow at all in their local area.

A classic example of a chemical pesticide is DDT. First developed in the late 1930s, DDT was found to kill large numbers of different insect species. During World War II, DDT was sprayed on soldiers to control body lice and similar pests. Its effectiveness in insect control was followed by extensive spraying campaigns on crops after the war. Because it killed so many species of insects, DDT reduced crop damage and greatly increased the food supply in many countries. DDT was also sprayed onto the surface of bodies of water to control mosquitoes and other disease-carrying insects that have aquatic stages in their life cycles, thereby reducing human disease. Disease reduction and increased food supplies both contributed to population increases in many countries (see Chapter 9). The use of DDT marked the time of greatest optimism in the use of chemical pesticides. This era was brought to a close by the discovery of DDT's toxic effects on nontarget species, findings that were convincingly made public by Rachel Carson's book, *Silent Spring* (1962). The term 'nontarget species' is an apt one for a phenomenon that is an example of the operation of the law of unintended consequences: the target species is the one you intended to kill, and the nontarget species are those that are killed unintentionally. Often, unintentional effects are not detected immediately, not because of ill-will, but because it can be difficult to predict where, when, and how the unintended effects will show up.

CONNECTIONS   CHAPTER 9

Negative consequences of pesticide use. There are many problems associated with the use of chemical pesticides such as DDT. The pesticides themselves are generally expensive; most of them are petroleum derivatives, and a great deal of energy is used in their extraction and further synthesis. Attempts to control pests with chemical pesticides have in several cases brought about increased levels of pest-related devastation several decades later. Pesticides like DDT are toxic to a wide variety of harmful and beneficial species alike. They may kill so many of the target species' natural enemies that the population size of the target species subsequently increases (after a time delay) above its earlier levels. Another problem with frequent pesticide use is that the target species develop pesticide-resistant mutations and are no longer killed by the spraying. Over 400 insect species, for example, are now DDT-resistant. Widespread use of the same pesticide year after year favors the evolution of pest populations with mutations that make them pesticide-resistant. Once they originate, these resistant strains of pest species spread rapidly because of selection by the pesticide itself.

As with fertilizers, there is also a runoff problem. A pesticide can find its way into groundwater supplies and from there into streams and lakes.

Pesticides may contaminate drinking water supplies in this way, and they may also poison the fish in our lakes and streams.

Biomagnification. DDT and many other long-lasting insecticides that are applied to crops become concentrated in the bodies of the pests that eat those crops, and then further concentrated (and thus more toxic) in the bodies of animals that eat the pests, a principle known as **biomagnification**. In addition to pesticides, a wide range of other pollutants, including heavy metals such as mercury, can become concentrated through biomagnification. To understand biomagnification, recall that producer organisms such as plants supply food energy to the primary consumers that eat them. The primary consumers, in their turn, supply food energy to secondary consumers, and so on. In each conversion, much of the food energy is lost in the form of heat, so that 10,000 kilocalories (kcal) of sunlight provided to the grass produces only 1000 kcal of grass energy to the cows that eat the grass; the cows in turn provide only 100 kcal of energy to the people that eat beef steak. These relations can be represented in a food pyramid (also called a trophic pyramid), as shown in Figure 11.17. Similar food pyramids could be drawn in proportion to biomass (quantity of biological tissues) or numbers of individual organisms rather than energy; most such diagrams would have basically the same pyramidal shape. The energy relations in a food pyramid are one reason why it is inefficient and wasteful to have intermediate steps between the producer plants and the top-level consumers, and thus why a certain supply of crops will support more people if the people eat the crops directly instead of eating animals that eat the crops.

When pesticides are used, an additional problem arises. Many pesticides are not broken down by decomposer organisms and thus persist in the environment. Also, many pesticides accumulate in biological tissues without being excreted. If a long-lasting pesticide is taken up by plants at the base of a food pyramid, the same amount of the pesticide is concentrated into a smaller and smaller amount of biological tissue with each

Figure 11.17

A simple food pyramid. The width of each column is proportional to the amount of energy available as food for the next higher step up the pyramid. At each energy conversion, 90% or more of the energy is lost, leaving only 10% available as food, as expressed by the width of each step.

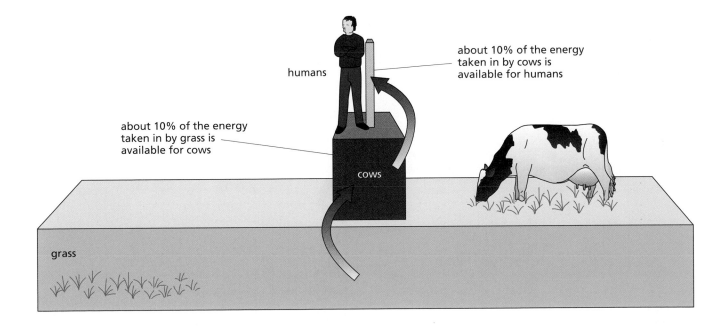

humans

about 10% of the energy taken in by cows is available for humans

about 10% of the energy taken in by grass is available for cows

COWS

grass

successive conversion. The concentration of the pesticide increases with each successive step in the food pyramid as the result of biomagnification.

As a case in point, Figure 11.18 shows the biomagnification of the long-lasting pesticide DDT. In Long Island Sound, New York, the DDT concentration was measured as 0.000003 parts per million (ppm) in the water, but was 25 ppm in the tissues of fish-eating birds such as ospreys and eagles, an increase in concentration of more than 8 million times.

DDT and other chemical pesticides that concentrate in fish and fish-eating birds interfere with calcium metabolism, causing a thinning of the shells of the eggs laid by the birds. The thin shells break before the chicks can develop, and the bird populations, unable to reproduce, decline. For birds that reproduce only once a year, laying only one or a few eggs at a time, it can take many decades after the pollution is removed for the species to recover its former numbers. For this and other reasons, DDT is now banned in most industrialized countries, though it is still used in some parts of the world.

Pesticides and neurotransmitters. Many pesticides kill insect pests by interfering with the chemicals (neurotransmitters) that conduct impulses from one nerve cell to the next (see Chapter 13). Blocking the enzymes that break down these chemicals induces continuous muscle contraction (see Chapter 13, p. 487). Because some of the same chemicals are used in the nervous systems of vertebrates (including humans) as in insects,

Figure 11.18

Biomagnification of DDT in Long Island Sound, New York, around 1970. The width of each column is proportional to the biomass.

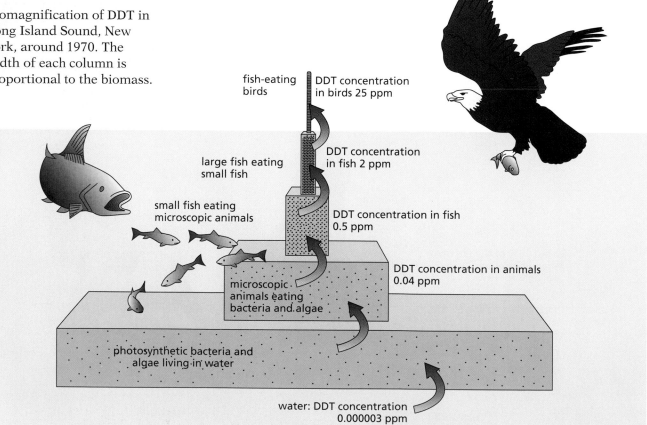

these pesticides can also impair nerve function in humans. Although the recommended concentrations properly used for pest control would not kill a person, they can cause permanent injury. The concentrated forms in which the pesticides are manufactured, transported and stored are more dangerous and may be lethal. About a thousand cases of pesticide poisoning are reported every year in the United States, primarily among agricultural workers, and the actual incidence is probably much higher. Containers with warning labels are sometimes not in the language of the people using the pesticide. Safety standards, where they exist, are often not enforced. High concentrations can also result from biomagnification. Unfortunately, the banning of DDT use in many countries has resulted in the development and use of other chemicals that are even more toxic to nontarget species, including humans.

Integrated pest management

Integrated pest management is a newer approach to crop pest management, one that uses a combination of techniques (Figure 11.19). The term 'management' means that pest populations are kept under control, so that they stay below the levels at which they cause economic harm. Total pest eradication is in most cases viewed as a goal that can be achieved only at an unacceptably high cost (including the cost to the environment or to society as a whole) or that cannot be achieved at any cost. The term 'integrated' means that all available tools are used in a mix of strategies that includes chemical controls (such as pesticides), biological controls (such as maintaining a population of the pest's natural enemies), cultural control (such as public education), and regulatory control (such as public policy legislation). Integrated pest management avoids or reduces most of the risks of chemical pesticides.

Figure 11.19

Various techniques of integrated pest management.

(A) Several small traps set to monitor the levels of pest populations.

(B) Closeup view of a trap that uses pheromones to attract target species for monitoring.

(C) Corn earworm, an insect whose presence in corn ears can be monitored visually.

(D) A plant pest, the saddleback caterpillar, parasitized by a wasp. Larval wasps eat the caterpillar and kill it; new wasps emerge from the pupae (cocoons) seen here and lay their eggs in another saddleback caterpillar.

Integrated pest management requires the monitoring of pest populations to assess the possible damage that they may do (see Figure 11.19A–C). This allows the use of just enough pesticide to reduce pest populations to acceptable levels, saving expense and reducing runoff and possible harm to nontarget species. Because integrated pest management relies more on biological controls than previous techniques, it requires a good working knowledge of the ecology of the pest species, including knowledge of its natural enemies (see Figure 11.19D).

Economic impact level. An important feature of any integrated pest management program is the concept of an economically acceptable level of the pest population, which is generally termed an **economic impact level**. The economic impact level is the threshold level above which corrective action must be taken. Pest populations are constantly monitored, and as long as the populations stay below the economic impact level, they are left alone and the cost of countermeasures is saved. The cost of corrective measures includes the cost of expendable materials such as pesticides, the cost of using and maintaining necessary equipment, the cost of labor, and the costs to the environment (including cleanup). For integrated pest management to become widely adopted, it must result in a net saving most of the time. As compared with 'calendar' spraying (spraying at a particular time of year, without any regard to the level of the pest population or the need to spray), integrated pest management saves costs in chemicals and equipment, but there are costs in monitoring pest populations and in using biological controls.

Introduction of predator species. Planting crops in smaller, separated patches instead of larger, single-species blocks is one way in which the spread of pests can be controlled without the use of chemical pesticides. Planting seasons can sometimes be modified so as to interrupt the life cycles of the pests. The most important techniques in integrated pest management, however, are those that take advantage of the natural enemies that keep the pest species in check. If a predator that preys upon the pest can be identified, then measures that encourage the growth, development, and proliferation of the predator may be able to keep the pests in check. For example, the bacterium *Bacillus thuringiensis* (Bt) can attack the larvae of insect pests and prevent them from destructively feeding on crops; therefore, many plant growers spray these bacteria on their crops. There are many subtypes of *B. thuringiensis*, each producing a toxin specific for only certain types of insects, so the *B. thuringiensis* type needs to be matched with the pest. Integrated pest management can often cost less than the application of chemical pesticides. For example, a predator species need not be applied repeatedly because it will reproduce naturally on its own, particularly if conditions that support all life stages of the predator species are maintained. As another advantage, because only predators with a very specific and limited range of prey species should be selected as biological controls, there should be less damage to nontarget species.

For example, cotton has long been a crop of commercial importance in the southern United States, India, Egypt, and elsewhere. Cotton pests include the boll weevil and the pink bollworm. Spraying with chemical pesticides initially reduced the levels of these pests, but, by the 1960s, pesticide resistance had developed in both pest species. Despite increased spraying, pest populations continued to increase. Worse yet, the chemical sprays destroyed many of the pests' natural enemies, such as the spined soldier bug, and the destruction of the natural predators allowed other pest species, such as the tobacco budworm (previously unimportant as a pest of cotton), to become significant pests—in some cases more devastating than the original ones.

In both Texas and Peru, integrated pest management techniques have been used successfully to control cotton pests. Soldier bugs and other natural predators are collected, reared, and released on the cotton fields, while chemical spraying has been greatly reduced and, although not eliminated entirely, is used only selectively. The planting season is timed so as to disrupt the life cycle of the pink bollworm moth; when the moths emerge, they can find no cotton plants on which to lay their eggs. Stalks and other unused parts of the plants are shredded and plowed under soon after each harvest, denying the pest insects places to hide and over-winter until the next growing season. In some places, corn and wheat are interplanted with the cotton to help the growth of populations of natural predators and to reduce the ability of the cotton pests to spread from one field to the next.

Alfalfa is another important crop in which integrated pest management techniques have been used successfully. As a nitrogen-rich legume, alfalfa is useful in crop rotation, and its high-quality protein is valued as an animal feed for many domestic animal species. In the United States, alfalfa ranks fourth (behind corn, cotton, and soybeans, and ahead of wheat) in area under cultivation. The principal pests of alfalfa are two related species of alfalfa weevils (genus *Hypera*). At least nine natural enemies of these weevils have been identified, most of them wasps that parasitize either the weevil larvae or other stages of the weevil life cycle. The weevil's life cycle can also be disrupted by harvesting alfalfa early. Alfalfa pests were formerly controlled with chemical pesticides, but this practice was sharply curtailed when one such pesticide, heptachlor, showed up in the milk produced by cows that had eaten the treated alfalfa.

Use of pheromones. Also part of integrated pest management is the spraying of pheromones, hormones that function in animal communication (see Chapter 8, p. 260). The pheromones are targeted at disrupting reproductive activities, thus preventing the production of new larvae—it is usually the larvae that destroy crops by their feeding activities. An example of such a pheromone is glossyplure, used by female pink bollworm moths to attract their mates. Spraying this pheromone on cotton fields confuses the male moths and interferes with their ability to locate the females, resulting in a natural birth control that is very specific to the pink bollworm and that has no effect on other species. It is, moreover, a chemical to which the bollworm can never develop a natural resistance without impairing its own ability to mate. Pest control via pheromones depends on decreasing the pest population size. Pheromones do not directly affect the juvenile stages of the insect, which in many species is the stage that causes the actual crop damage.

CONNECTIONS
CHAPTER 8

THOUGHT QUESTIONS

1 The runoff of fertilizers from agricultural fields often produces algal blooms. Can you explain why this would be so? (What limits algal growth under normal conditions?) Would the problem be greater with organic fertilizers or with inorganic ones? Why do you think so?

2 Compare the volume occupied by the same weight of commercially available potting soil and sand. Compare the amount of water that can be held by equal weights of soil and sand. Now compare those results with the amount of water that can be held by soil from your area. Does the soil in your area contain a lot of humus? What ways can you think of to improve the quality of your soil?

3 Can you see any evidence of erosion in the area where you live? What natural processes or human activities might contribute to soil erosion?

4 Once the toxic effects of DDT on nontarget species received widespread publicity, agricultural use of the chemical was banned in the United States and in many European countries. The United Nations considered imposing a worldwide ban, but this effort was stopped by the insistence of many third-world nations that they needed the pesticide to help control both crop pests and mosquitoes. Do you think DDT should have been banned in countries like the United States? Do you think the ban should have been extended worldwide?

5 What do you think would happen if DDT were allowed to be used in limited amounts? How might the limits be enforced? What would be done if a farmer found that he or she could kill more pests (and thus increase crop yields) by using more DDT and causing potential future harm to the environment? How intrusive would enforcement agencies need to be? Is a total ban more practical than a limited ban? Would rationing work? Would an 'agricultural prescription' system (similar to medical prescriptions for drugs) work?

6 Once pest resistance to a pesticide arises, it can spread quickly through any pest populations treated with the pesticide. Explain this fact using your knowledge of genetic mutations and natural selection.

Crop Yields Can Be Increased Further by Altering Plant Genomes

Many plant characteristics, including the size, texture, and sweetness of the edible portion, are at least partially genetically determined. Also under genetic influence are many factors that determine the hardiness of crop plants, their drought resistance, their rate of growth under different soil conditions, their dependence on artificial fertilizers, and their resistance to various pests and plant diseases. Therefore, the yield, both in terms of the amount of crop per hectare (or per acre) and the amount of nutrition per unit of crop, can be increased by improving plant genomes. Other possible goals of genetic improvement include the development of new crop plants such as drought-resistant strains, pest-resistant strains, or strains with diminished nutrient requirements, capable of growth on marginal or poor soils.

The kinds of changes to a species that can be accomplished by traditional methods of selective breeding are limited by the genetic variation that exists within the species or its close relatives. Genetic engineering offers a newer method for customizing food crops by giving them genetic traits that they normally lack. We discuss traditional methods first and then genetic engineering afterwards.

Altering plant genomes is not new

Selective breeding is also called **artificial selection**. As carried out by both animal and plant breeders, the practice was already well known in Charles Darwin's time, and served as a model for his theory of natural selection (see Chapter 5, pp. 128–130). Darwin realized that great changes in agriculturally important plants and animals had been made within his own lifetime by British animal and plant breeders. These breeders chose the individuals of the species that best exemplified the trait they desired. They allowed these individuals to mate, while preventing mating between individuals that did not have the desired trait.

Artificial selection can be used to change almost any trait of a crop species in one direction or the other. A closely related wild species may offer a desired trait, such as a nutritionally more complete protein, in which case the wild species may be crossed with the crop plant as a first step toward the production of a nutritionally superior strain. If this makes you wonder how the concept of crossing members of different species can be reconciled with the biological species definition given in Chapter 5 (p. 151), remember that the definition refers to populations (not individuals) that do not *naturally* interbreed. It is often possible to get individuals under domesticated conditions to do what is not natural for entire populations, for example, by dusting pollen artificially from a cultivated plant species onto a wild relative.

Figure 11.20 shows the results of 50 years of selection to produce corn plants with high or low oil content, or high or low protein content. However, attempts to change only one trait at a time can often result in the production of an inferior strain. For example, it does no good to select for corn plants with larger kernels or larger ears unless the stalks and root systems are capable of supporting them and unless the plants are sufficiently drought resistant and disease resistant to survive under field conditions. Modern breeding practices include the selection for several traits at once, resulting in harmonious combinations of traits that are well adapted to function together as a whole.

By selectively planting strains with desirable traits and by avoiding the use of genetic strains with less desirable traits, agricultural scientists in many nations have dramatically increased crop yields in the past few hundred years. The seeds of high-yielding strains command a high price. Around the world, nations that have achieved the most efficient

Figure 11.20

The results of 50 years of selection on the oil content and protein content of corn (*Zea mays*). By selectively breeding only those plants that had the highest or lowest protein or oil content in each generation, plant scientists have changed the inherited characteristics of each strain.

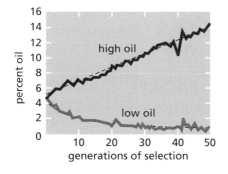

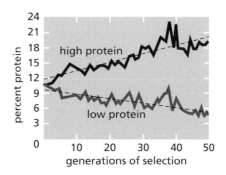

agricultural production (high cash value yields of major crops per worker-day or per cash unit invested) have generally become wealthy, while those countries with the least efficient agricultural production are generally among the poorest. Thus there is a high correlation between the affluence of a nation as a whole and the efficiency of its agricultural production. The development of new strains of crops, each suited to a particular climate and soil type, is among the most important components of agricultural efficiency, rivaling even the mechanization of agricultural work. Since about 1920 these strains have contributed to the increase in crop yields in industrialized countries. In the United States, for example, crop production doubled between 1940 and 1990 even though some land was taken out of cultivation.

The green revolution. In the 1960s and 1970s, an effort was made to export many new and improved genetic strains of plants from North America and Europe to other parts of the world. This effort, loosely termed the 'green revolution,' was aimed at improving both the agricultural yields and the nutritional content of crops in the recipient countries. For example, some agricultural scientists developed a more nutritious variety of corn, high in lysine, an amino acid in which corn is usually deficient. High-lysine corn provides more complete protein for human nutrition (see Chapter 10, pp. 333–334). Yield was increased by the development of wheat and rice strains with short stems. They produce more grain on less stem and mature earlier so that more than one crop can be planted in a year. Many of these strains were developed in third-world countries under the auspices of international plant breeding institutes established there.

Greatly improved crop strains are, however, subject to the law of unintended consequences. For example, many of these new strains grew well with mechanized agriculture, but getting comparably increased yields in the third world meant the adoption not just of the new plant strains but of irrigation and fertilizer use. Although production on farms was increased by 50–100%, it never increased to the extent that it had on research stations. The results of the promised 'green revolution' have been mixed, as is summarized in this quote:

> Forty years after the first adoption cycles of the Green Revolution began in Mexico, and 15 years after they came to completion in Asia, we see that the world is not much better off. A similar percentage (10–15%) of the world population that was undernourished in the 1950s and 1960s is undernourished in the 1990s. The increases in agricultural production, while impressive, have kept just ahead of population growth, and not led to a more even distribution of food to all people. In addition, many problems plague sustainability in the high-input system, in the Third World just as well as in the developed countries (M.J. Chrispeels and D. Sadava, *Plants, Genes and Agriculture*. Boston, Jones and Bartlett, 1994).

Altering plant strains through genetic engineering

As we noted earlier, artificial selection is limited by the genetic variation that exists within the species or its close relatives. Genetic engineering offers a newer method for customizing crops by giving them genetic traits that they normally lack. These may include better nutritional qualities, pest or herbicide resistance, the ability to live in nitrogen-poor soils and other marginal habitats, or the ability to fix atmospheric nitrogen and make their own nitrates. In this section we examine some techniques of genetic engineering in plants and the uses to which plant genetic engineering is put. A later section provides insight into the controversy that surrounds this area of biology.

Genetic engineering is not simple. The genetic traits to be changed must first be identified. As in animals, many plant traits are not controlled by single genes and thus are not easily altered by genetic engineering. Most of the traits related to hardiness (drought resistance or cold resistance) are multigene traits. Other traits, for example pest or herbicide resistance, have been successfully introduced by transferring single genes into plants, often a gene from another species.

Whether or not a genetically altered plant actually makes a desired protein depends on whether the desired gene has been inserted into a portion of the genome that is transcribed into mRNA. This means that the gene must be located 'downstream' from the signals that control gene expression (see Chapter 12, pp. 419–421). We also need to know how the altered plant will function at the ecological level, knowledge that we most often do not have until after the genetic engineering has been performed.

Plant genetic engineering follows the general concepts of genetic engineering that were described in Chapter 4. First, the gene of interest must be isolated and the gene must be coupled with (or relocated near) an easily identified marker such as a gene for antibiotic resistance. Second, the new gene must somehow be inserted into the plant genome. Third, because gene uptake is always a chance event with only a low to moderate chance of success, the organisms that successfully took up the new gene must be screened and separated from those that did not. Last of all, the genetically altered plant needs to be cloned so that many such plants can be produced.

Organizing the genes and markers to be inserted. Before being inserted into a plant cell, the gene or genes of interest must be assembled into a continuous piece of DNA. This DNA contains the nucleotide sequences of the gene associated with the trait being modified. In addition, it must contain regulatory DNA sequences that will enable this gene to be transcribed and translated into protein. The regulatory DNA may also control the tissues of the plant in which the protein gene product will be made. Finally the DNA assembly must carry what is called a 'marker gene'. As we see shortly, the product of this marker gene allows the separation of those cells that took up the assembled gene complex from those cells that did not. The parts of the DNA assembly are shown in Figure 11.21.

Methods of inserting new genes into plants. Once the necessary DNA segments have been put together into one piece of DNA, the whole piece can be inserted into the plant cell. Scientists can use at least three different methods to insert a new or altered gene into a plant genome: viruses, bacterial plasmids, and mechanical insertion.

Figure 11.21

Structure of a DNA assembly for use in genetic engineering.

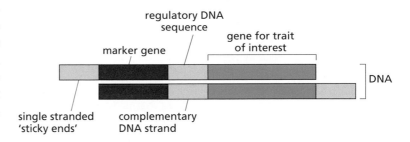

One method of gene insertion is to use viruses. Restriction enzymes are first used to make 'sticky ends' on the assembled DNA piece (see Figure 11.21) and on the viral DNA (see Chapter 4, pp. 97–98). The sticky ends bind the gene to the viral DNA, and the scientists then make use of the ability of viruses to incorporate themselves into the host genome (see Chapter 4, pp. 100–101, and Chapter 16, pp. 584–585). The virus enters the plant cells and adds its DNA and the new gene to the plant's DNA.

One virus used in genetic engineering experiments in plants is the tobacco mosaic virus. Like most viruses, the tobacco mosaic virus is restricted in its choice of hosts. Because the virus reproduces only in tobacco plants, it very unlikely that it could accidentally spread new traits to other species. Tobacco mosaic virus has the advantage that both its biology and that of its host, the tobacco plant (*Nicotiana tabaccum*), have been intensively studied for decades. In cases in which it does not matter what plant is used in producing a particular compound, the tobacco plant is a logical choice, because methods for its cultivation and for the growth and insertion of tobacco mosaic virus are well known.

Because tobacco mosaic virus does not enter cells of species other than the tobacco plant, this virus is unsuitable for changing most crop species. For these species, genetic engineers have experimented principally with *Agrobacterium tumefaciens*, a bacterial species that causes tumors in many plant species. *Agrobacterium*, like many bacteria, can carry DNA fragments known as plasmids. Plasmids can be used to carry new genes into host species, as was described in Chapter 4 (pp. 98–99). In the normal life cycle of *Agrobacterium*, plasmids can introduce bacterial genes into the cells of the host plants, and these genes cause tumors to form. The plasmids used in genetic engineering are modified so that they are still capable of introducing bacterial genes into the host plants, but they lack the gene that causes the tumor formation, so these plasmids no longer induce tumors.

As was described for viruses, restriction enzymes can be used to engineer the plasmid to carry the gene assembly described earlier (see Figure 11.21). The engineered plasmid is then induced to enter the *Agrobacterium* cells. The bacteria are incubated with pieces of the plant to be transformed. The bacteria then insert the plasmid carrying both the desired genes and regulatory DNA sequences into the genome of the recipient plant cells (Figure 11.22). Many different plant species can serve as hosts to *Agrobacterium* and can incorporate the plasmid. Many other important crop species, such as corn, cannot incorporate *Agrobacterium*. Several other plasmids are useful in only one or two host species.

In any kind of genetic engineering, one of the stumbling blocks is getting the inserted DNA past the cell membrane and nuclear membrane of the recipient cell. In addition to host-specific virus or plasmid transfer methods, mechanical methods of gene transfer have been developed. These methods are less specific and can therefore be used on nearly any plant species.

In the particle-gun method, plant pieces are literally shot with tiny pellets whose surfaces carry the DNA for the gene of interest and its regulatory DNA sequences (see Figure 11.22). The force of the bombardment rams some of the pellets through the plant cell walls and membranes. The 'naked DNA' carried by the pellets will sometimes incorporate into the DNA of the plant cells.

In another nonspecific method called electroporation, temporary holes are made in plant cell membranes by disrupting them with electrical current. While these holes are open, DNA can enter the cytoplasm and then the nucleus, before the hole closes again.

Figure 11.22

Methods for inserting new genes into plant cells.

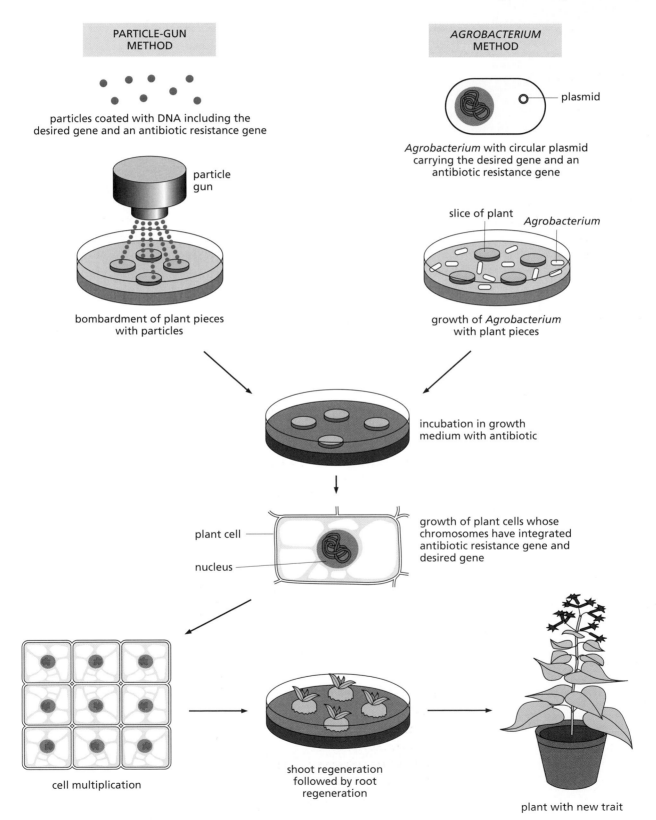

PARTICLE-GUN METHOD

particles coated with DNA including the desired gene and an antibiotic resistance gene

particle gun

bombardment of plant pieces with particles

AGROBACTERIUM METHOD

plasmid

Agrobacterium with circular plasmid carrying the desired gene and an antibiotic resistance gene

slice of plant   *Agrobacterium*

growth of *Agrobacterium* with plant pieces

incubation in growth medium with antibiotic

plant cell

nucleus

growth of plant cells whose chromosomes have integrated antibiotic resistance gene and desired gene

cell multiplication

shoot regeneration followed by root regeneration

plant with new trait

Screening and cloning. Regardless of the method of entry, the uptake and incorporation of new DNA do not happen in most plant cells that are treated. Scientists therefore need a method of screening to find out which cells actually have the new gene. As mentioned earlier, this is done in most cases by incorporating another gene along with the gene for the trait being modified. This additional 'marker' gene is usually one that confers resistance to some antibiotic, as shown in Figure 11.22. After treatment of the plant pieces by some insertion method, they are incubated in growth medium containing the antibiotic to which the marker gene confers resistance. Those cells that have not taken up the transferred genes are killed by the antibiotic, while those cells in which the new genes have been successfully incorporated will survive and grow.

Once it has been determined which cells have incorporated the gene of interest, those cells can be used to grow complete plants. Cloning (and therefore genetic engineering) is much easier to accomplish in plants than in animals because many plant species can regenerate a whole individual from just one or a few adult cells. Many genetically identical plants can thus be grown and produced asexually from a single cell containing the integrated DNA (see Figure 11.22). Cloning the plants in this way is usually done by growing the cells in small dishes in growth medium containing plant growth hormones, from which complete plants will develop without sexual reproduction or seed formation. Plants are particularly amenable to cloning, and methods have been developed for the cloning of many species of plants. Cloning techniques are not just restricted to growing genetically engineered plants: plants with desirable traits that have been developed by traditional plant breeding (artificial selection) can also be grown by such techniques. In fact, the cloning of traditionally developed plant strains is in far more general use and has had a far greater impact on agriculture around the world than any form of genetic engineering.

Uses of transgenic plants

Genetic engineering using *Agrobacterium* was first achieved in 1983. Since then, many plant species have received genetically engineered genes by using the techniques just described. In all, **transgenic plants** (meaning those with genes derived from another species) have been produced in over 20 species, including tomatoes, potatoes, carrots, alfalfa, corn, soybeans, peas, cotton, rice, and sugar beets. An estimated 70% of all soybeans and alfalfa grown in the United States is now genetically modified, along with a smaller percentage (about 35%) of corn.

Transgenic plants with altered nutritional content. Although genetic engineering holds the promise of producing plant strains with higher or more complete nutritional content, very few of the strains that have been developed have actually been changed in this way. One that has is the potato, in which the starch content has been increased by insertion of a bacterial gene. Potatoes are already 21–22% starch, but an even higher starch content makes the potatoes better for processing into potato chips and frozen french fries, the major ways in which potatoes are consumed in the United States.

The canola plant (*Brassica napus*), whose seeds are the source of canola oil, has been engineered to produce lauric acid, a saturated fatty acid used in the food industry (for food additives such as coffee creamers and cake and candy coatings) and in the detergent industry. Lauric acid is not ordinarily made by *B. napus*. It is naturally synthesized by tropical plants such as palm and coconut and laurels, but by inserting a gene from the California bay laurel tree, scientists have modified a canola strain to produce lauric acid.

Pest-resistant transgenic plants. Sometimes genetic engineering is used to modify some trait that makes growing agricultural crops more efficient, thus increasing yield. Currently, the most widespread transgenic plants are those in which the gene for a toxin from the bacterium *Bacillus thuringiensis* (Bt) has been inserted. *B. thuringiensis* toxin is a natural insecticide and the bacteria are widely used for spraying onto plants as part of integrated pest management. Transgenic plants containing the toxin gene make the toxin themselves, doing away with the need for repeated spraying of the bacteria onto crops. The toxin protects the crop plants against damaging pests but does no harm to the natural enemies of those pests. Cotton, potatoes, and corn with Bt toxin genes are all commercially available.

Resistance to nematodes has also been engineered into potatoes, tomatoes, and sugar beets. Nematodes are roundworms (kingdom Animalia, phylum Nematoda), which do a tremendous amount of damage to the roots of plants, in many parts of the world but particularly in tropical climates. NemaGene™ is a patented gene for a protease inhibitor enzyme that kills nematodes as they feed on the plant roots. This gene has been engineered with regulatory DNA sequences that ensure that it is transcribed and translated only in the roots of the plant, not in the parts of the plant that humans eat.

Herbicide-resistant transgenic plants. Another use for genetic engineering in plants is the introduction of genes for herbicide tolerance. If a crop plant is given a gene that allows it to resist (or tolerate) a particular herbicide, then the herbicide can be used as a weedkiller to control weed species that would otherwise compete with the crop for water and other limited resources. A bacterial gene conferring resistance to Roundup™, a popular wide-spectrum herbicide, has been introduced into the most commonly planted strains of corn and soybeans, making them resistant to the herbicide. Fields can now be sprayed with the herbicide, killing the weeds but sparing the crops that carry the resistance gene. Over half of all soybeans grown in the United States now have this genetically engineered gene.

Plants with longer shelf lives. In 1994, the Flavr-Savr™ tomato was introduced into markets in the United States as the world's first commercially available genetically engineered fresh produce. These tomatoes were modified by another method of genetic engineering. Rather than inserting a gene from another species, here a tomato gene was turned off. A gene had been identified that codes for a tomato enzyme that softens the tomatoes as they ripen. In production of the Flavr-Savr™, knowledge of the DNA sequence of this gene was used to synthesize its complementary DNA. The complementary DNA along with regulatory DNA was inserted into a strain of tomato plants using a plasmid. When the plant cells transcribe the complementary DNA into mRNA, this mRNA is complementary to the normal mRNA and binds to the normal mRNA, decreasing its translation into protein. The engineered plants thus secrete less of a softening enzyme, which allows the tomatoes to be vine ripened, thus becoming more flavorful, and still last through transportation and storage. The Flavr-Savr™ tomato also has less water in proportion to the amount of pulp, making it advantageous for the manufacture of ketchup, sauces, and soups by reducing the amount of energy needed to boil off the water. It is in far greater use in these cooked commercial products than it is as fresh produce (indeed, the Campbell's Soup Company commissioned its development by the Calgene Corporation).

Molecular farming. One type of genetic engineering that uses tobacco plants is called molecular farming. The goal here is not to make a better tobacco plant but to use the tobacco plant as a biological factory to produce, say, a medicine even small quantities of which would be valuable. The genetically engineered medicine would need to be purified to remove the nicotine and other tobacco plant molecules, but this might not be any more expensive than purifying the medicine from its original plant source. Further, the tobacco plant may grow in places where the original plant source will not, or it may grow faster. Success in such endeavors might encourage tobacco farmers to grow more plants for pharmaceutical uses and fewer for cigarettes. Especially if the pharmaceutical plants became more profitable, tobacco farmers would have reason to switch away from growing the crop that is currently the leading cause of lung cancer and other diseases such as emphysema.

A recent example of such molecular farming is the insertion into tobacco plants of a gene from cows that codes for lysozyme. Lysozyme is found naturally in the saliva of many animal species, where it has an important antibacterial function because of its ability to digest the cell walls of many bacterial species. Cow lysozyme produced by tobacco plants makes the tobacco leaves resistant to those bacteria. The lysozyme so produced can be used to treat seeds from many plant species. Disease-transmitting bacteria on seeds are a major agricultural problem. Treating seeds with dilute lysozyme produced by genetically engineered tobacco plants clears harmful bacteria from the seeds.

Future transgenic plants. The above-mentioned genetically modified species are all now commercially available. Many more strains are in development but are not yet marketed. Plant strains have been engineered to make both soybeans and canola oil more nutritionally complete by the inclusion of methionine, an amino acid not normally produced in these plants. Recently, a strain of rice has been produced that contains genes for beta-carotene. Beta-carotene is metabolized in the body to produce vitamin A (see Chapter 10). Vitamin A deficiency is a significant cause of blindness: 50,000 children a month are affected worldwide, by some estimates. The beta-carotene gives the rice a yellow color; hence, its name Golden Rice. The hope is that such enhanced rice could be a factor in reducing worldwide vitamin A deficiency. Currently the level of beta-carotene achieved in the rice is not sufficient to meet minimum daily requirements for the vitamin, but researchers hope to increase the amounts.

Plants resistant to viral and fungal infections are being developed, as well as plants more tolerant of high-salinity soils or drought conditions, or with increased ability to fix nitrogen. Many edible vaccines are being

CONNECTIONS
CHAPTER 10

developed; for example, bananas have been modified to make a viral molecule that, when eaten by people, immunizes them against the viral disease hepatitis B. Plants have also been engineered to produce nonfood products, such as polyhydroxybutyrate, which is used in the manufacture of plastics that can be broken down by decomposer organisms. The Monsanto Corporation has a research project that it calls the Blue Gene project, in which a transgene for a blue pigment inserted into cotton plants makes them produce blue cotton, decreasing the need for chemical dyeing to produce denim for blue jeans.

Risks and concerns

Whether these genetically modified plants will become commercialized, and whether those commercially available will gain widespread acceptance, will depend in part on overcoming opposition to genetic engineering. Part of the opposition is biological and part of it is ethical.

Ethical arguments. Ethical questions include both deontological and utilitarian concerns. A few critics, notably Jeremy Rifkin, have consistently opposed all biotechnology, especially transgenic research. We are, these critics say, attempting to alter nature by going considerably beyond the bounds that nature intended. Rifkin goes so far as to question whether *any* transgenic research can be ethical. In the terminology explained in Chapter 1, Rifkin might be described as a deontologist who believes that any transplantation of a gene from one species to another is inherently unethical, a stance that prevents the experimental measurement of certain risks (among other things). One possible answer to such criticisms is to point out that genomes are being rearranged all the time in nature. Gene transfer (introgression) between related plant species happens fairly often in plants, and cross-breeding has been transferring genes between domestic plant strains of the same species for centuries.

CONNECTIONS
CHAPTER 1

Other deontological objections include those from some vegetarians who are opposed to products that might contain a gene from an animal. Religious groups have other objections, such as Muslim fears that genetically engineered foods might contain genes transferred from pigs, which may not be eaten under Muslin dietary laws.

Various Christian and Jewish theologians view genetically modified plants as equivalent to plants modified by more traditional agricultural methods. This is the view generally taken by scientists also, although they are more likely to consider this to be a scientific, rather than an ethical, issue. For example, the Scientific Committee on Problems of the Environment (SCOPE), a committee of the International Council of Scientific Unions, has stated that "there are no convincing scientific grounds for distinguishing engineered organisms from natural ones" (H.A. Mooney and G. Bernardi, *Introduction of Genetically Modified Organisms Into the Environment*. New York, Wiley, 1990). Also, "because organisms of either type could pose unforeseen hazards, some safety testing is desirable before large-scale propagation."

Utilitarian ethical arguments center more on the premise that the risks of genetic engineering are poorly understood and quantitatively uncertain. Risks probably vary from one plant species to another, making it important for the questions to be raised (and the research conducted to answer them) again and again for each new application. For example,

the use of the *Agrobacterium* plasmid has been criticized because the original form of this plasmid stimulates plant tumor formation. Suppose that a genetically altered strain of a plant containing an *Agrobacterium* plasmid managed to reacquire the gene for plant tumor formation, either by mutation or (more likely) by genetic recombination with wild strains of *Agrobacterium*; plants carrying the plasmid might then grow tumors. The probability of such an event must be carefully estimated if reliable risk–benefit ratios (see Chapter 1) are to be obtained and if defensive measures against such mutant strains are to be planned in advance. Potential economic benefits from genetically engineered crops are generally easy to estimate; risks often are much more uncertain. Many utilitarians are not opposed to genetic engineering in principle, but argue that the risks are great and should be evaluated before we proceed.

Concerns about damage to the environment. The risks (or potential dangers) of biotechnology do exist, and some of them have been alluded to earlier. Many biologists, particularly ecologists, argue that there are biological questions that need to be answered and that will not be addressed if we think only at the molecular level. When molecules operate in plants and when those plants may be introduced into the environment, we need to consider the functioning of the ecosystem as a whole. The Union of Concerned Scientists has recently published a summary of their concerns, which are in two areas: the possible escape of genetically altered strains as 'superweeds'; and the spread of plant viruses. They argue that because agricultural crops cannot be isolated from their surrounding ecosystems, transgenic plants pose risks that other genetically engineered species, such as bacteria grown in factory vats for the production of medicines, do not. They caution that transgenic plants could escape from cultivation and become weeds. They give as examples various plants, such as kudzu and purple loosestrife in the United States, that have been introduced into new locales and have overrun the environment, altering the habitats of other plants and of animals. (Scientists in SCOPE do not see this as an equivalent example; genetically engineered crops, they say, would most probably be introduced into areas where the unengineered form of the crop had already been grown, so natural ecological balances should still apply.) A larger risk, though, is whether engineered plants would cross-pollinate or otherwise transfer their new genes to weeds. Plant species frequently do cross-pollinate with weeds of related species. This might be particularly worrisome if herbicide-resistant plants passed their transgenes to weeds, making the weeds herbicide resistant. The Union of Concerned Scientists cautions that small field trials are not necessarily good predictors of full-scale agricultural conditions, particularly if the field trials have not been designed to examine ecological effects. They propose expanded testing protocols that should be done before crop species are approved.

The issue of the spread of plant viruses, or of the creation of new viruses from recombinations of the virus used to insert the transgenes with normal viruses already present in the plant, is also of concern. Although viruses are known to recombine, the risk is considered to be low, simply because viruses are not now frequently used for gene insertion into plants.

A further ecological concern had to do with the *Bt* bacterial pesticide genes that have now been engineered into several crop plant species.

Opposition to *Bt*-containing crops was enhanced by reports that pollen from such plants killed the larvae of monarch butterflies. The initial reports merely established the fact that, under laboratory conditions, enough Bt protein is produced by these plants to kill some butterflies. The report did not contain controls, such as a comparison with the number of butterflies killed by Bt or other pesticides sprayed onto plants. Although the initial report continues to receive much attention, many further studies have found that under field conditions the effect on monarchs is negligible.

Concerns about human health. Another issue raised by biologists is a food safety issue, but again it has to do with thinking in a more integrated way about how systems function. At issue here are the antibiotic resistance genes that are engineered into plants along with the gene of interest (see Figure 11.22). The unanswered question is whether these genes will contribute to certain bacteria becoming resistant to the antibiotics. Many disease-causing bacteria are already becoming resistant to antibiotics owing to their overuse in animal feeds and their improper use medically (for example, in the treatment of colds, which are viral diseases against which antibiotics have no effect). If transgenes in food add to this growing problem, it could help undo the control of infectious diseases that has been achieved in the past 50 years.

Much of the opposition of member nations within the European Union (EU) to genetically engineered foods centers on this last point. The EU banned Bt transgenic corn in 1996 because it also contains an inserted gene for ampicillin resistance. While this would not pose any direct threat to human health, the possibility exists that the resistance gene could be transferred to gut bacteria. Laboratory tests show that most, but not absolutely all, of the DNA ingested in food is destroyed by strong acids present during digestion. The possibility of resistance transfer is further diminished in any food that is cooked, because the DNA and its proteins would be destroyed in the cooking. Because ampicillin is used to treat many human diseases, and as resistance to it often confers resistance to many other penicillin-type antibiotics, the development of resistant bacteria in the gut, which could possibly transfer resistance to pathogenic bacteria, is a risk that many scientists feel should be more thoroughly investigated. The EU reversed its outright ban, but now requires labeling on any food that contains live genetically modified organisms, or has modified ingredients that are not equivalent to, or materials that are not present in, the original, or has substances that might be objected to on ethical grounds (such as animal genes that might be opposed by vegetarians).

In the United States, opposition to corn containing Bt has had a different basis. Much of the publicity has been about a particular corn variety called StarLink™. This corn contains both the *Bt* transgene and another gene that makes it tolerant to a commonly used herbicide. Because Bt is a pesticide, products containing it are regulated in the United States by the Environmental Protection Agency (EPA), not by the Food and Drug Agency (FDA). In tests done as part of the approval process, it was found that the protein product of the *Bt* gene (called Cry9C) could partially survive cooking and digestion. Initially there were also concerns about the possible capacity of the protein to induce allergies. Consequently, the corn was approved only for use in animal feeds. In

September of 2000, however, traces of the Bt protein were found in human food products. The Aventis company voluntarily agreed to recall the food products, cancel its EPA registration for the corn, and pay the farmers to remove their corn from the market. Although further testing has cast some doubt on both the protein's persistence and on its allergic potential, the corn is currently off the market and mired in lawsuits.

The potential of other transgenic proteins to cause problems for people with allergies remains a concern. This is one reason for the desire to have genetically modified foods accurately labeled. The StarLink™ corn experience has cast doubt on the feasibility of ever separating foods for human use from other uses, leading some to call instead for truth in labeling. Others see practical barriers to such labeling. Products refined from transgenic crops include oil and corn syrup from corn, and protein, oil, lecithin, and several vitamins from soybeans, all of which are now a common part of the food supply in the United States. The transgenes have to do with pest or herbicide resistance, not with the character of the food produced by the plant. Once a refined product like corn syrup has been extracted, there is no way of determining the genetic background of the plant source.

Social and agricultural concerns. Another line of opposition to genetically engineered crops is that they do not contribute to efforts to develop sustainable agricultural practices. The Union of Concerned Scientists states that we should be developing sustainable practices that prevent environmental problems in the first place, rather than focussing on solving problems after they are created. The proponents of genetic engineering point to crops modified to be pesticide resistant as an example of ways in which plant engineering could cut down on pesticide use. On the other hand, herbicide-resistant crops have no benefit if they are not used in conjunction with the matching herbicide. (In fact, farmers who use Monsanto Corporation's RoundupReady™ soybean seeds are required to sign an agreement with the company to use only Monsanto's Roundup™ herbicide, and are faced with heavy fines imposed by the company if they do not.)

Some see genetically engineered plants, in part because they are developed and marketed by large multinational corporations, as inherently contributing to monoculture practices, thereby also contributing to the loss of biological diversity, a topic that we examine in greater detail in Chapter 18.

The other half of a risk–benefit equation is the benefit side. Many see the benefits of engineered plants as potentially immense. Others see the profits as potentially immense, but the benefits to society as very small. The countries in which farmers will be able to afford these seeds are countries that are already awash in excess food. Examples are raised such as the engineering of lauric acid into canola, which will benefit North American farmers that grow canola, but this will be at the expense of tropical farmers who grow palm and coconut, the natural sources of lauric acid. Because engineering is seldom done on the primary food crops used in the third world it is unlikely to be of any direct benefit in these countries. Moreover, the use of patented or trademarked plant strains, or the increasing use of plant strains that require mechanized agricultural methods, will probably make third-world farmers more dependent on imported seed supplies and on foreign debt. Many groups of scientists also caution that genetic engineering will not solve the world's food problems; they see these problems as being due largely to the unequal distribution of food, not to a lack of food production. They point to the example of the green revolution, which has tremendously increased production but has not done away with hunger.

The possible benefits of the genetic engineering of crops are very large. The monetary costs may also be large, but they are very difficult to estimate because of the great uncertainties involved. The biological dangers are not fully known. Given the law of unintended consequences, difficulties should not be underestimated. The risks may prove to be minimal, or they may prove to be significant, and the dangers may prove to be easily controlled even in worst-case scenarios—we will never know unless we undertake the relevant investigations for each species. Only if we have investigated the possible dangers will we be able to assess the possible risks. That is one reason why it may be desirable to proceed with testing, why all applications should be closely monitored, and why many people await the evaluation of risks as well as benefits.

THOUGHT QUESTIONS

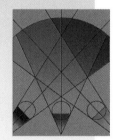

1 Artificial selection and genetic engineering are both ways of modifying plants. In what ways are the two methods similar? In what ways are they different?

2 Experimenters attempting to alter plant strains by selecting for one trait sometimes end up changing not only that trait but some other trait along with it. Use your knowledge of genetics to explain why this is so.

3 Do genetically engineered foods increase nutritional quality for consumers or are they of more benefit to mechanized farming and the food processing industries? Develop arguments on both sides.

4 Jeremy Rifkin and other critics of genetic engineering have argued that the escape of a genetically engineered strain from cultivation would be a chaotic event whose consequences are inherently unknowable. Do you agree or disagree? Is there any way of planning for such events? If a genetically altered strain of plants (say, tomatoes that stay on the vine longer to develop better flavor or color) were found growing outside cultivated areas, what should our response be?

Concluding Remarks

As the world's population continues to increase, methods are being developed to make agricultural production more efficient. Overcoming various limiting factors, such as nutrients or water, increases food production. Sustainable agricultural practices aim at ensuring that production will remain high far into the future. Plants need light energy for photosynthesis, and light energy is more abundant in countries close to the Equator, including many poor countries. Wherever light energy is naturally abundant, the need for fertilizers can be diminished by planting crops that harbor symbiotic nitrogen-fixing organisms in their roots. Alternatively, the genes for nitrogen fixation or for other desirable traits can be genetically engineered into plants that do not naturally possess them. Genetic engineering can also be used to make plants more nutritious and more resistant to drought and to pests.

Although there is considerable opposition to genetic engineering, most groups of scientists conclude that the risk–benefit ratio still argues in favor of continuing research on genetic engineering of food crops. Genetically engineered crop plants may increase food yields and nutritional value. The possible benefits of the genetic engineering of crops are thus immense, although the dollar costs may also be immense. Many scientists caution that increased food production by itself will not solve the world's food problems. They see these problems as being also due to the social, economic and political forces that result in unequal distribution of food. Plant science will be an important part of any solution, but world food problems will not be solved by science alone.

Chapter Summary

- Plants make carbohydrates by the process of **photosynthesis**, using water and atmospheric carbon dioxide as raw materials and sunlight as an energy source. Sunlight is absorbed by plant pigments such as **chlorophyll**, located in plant cell organelles called **chloroplasts**.

- Because plants can make all of their own organic compounds they are **autotrophs**; organisms that cannot, including all animals, are **heterotrophs**, dependent on autotrophs for their food.

- Proteins, nucleic acids, vitamins, and other plant products require nitrogen for their synthesis. Most plants get their nitrogen from dissolved nitrates, which limits the distribution of many plant species to soils that contain adequate nitrogen. Some plants form mutualisms with microorganisms that can fix atmospheric nitrogen and convert it into a form that the plants can use. The cycling of nitrogen through the biosphere and atmosphere is called the **nitrogen cycle**.

- **Vascular plants** absorb water through their roots and evaporate water through the stomates in their leaves. **Osmosis** generates water pressure, and this pressure contributes strength to most plant tissues; a lack of water pressure causes wilting.

- Crop yields can be increased by supplying **limiting nutrients** through **fertilizers** or soil improvement, by supplying water, by controlling pests that compete with the plant or with humans for the energy produced by plants, and by altering the traits of the plants either by **artificial selection** or by genetic engineering.

- Pesticides and chemicals can become concentrated in biological tissues by **biomagnification**.

- **Monocultures** allow the rapid expansion of pest species.

- **Integrated pest management (IPM)** can reduce our dependence on chemically produced pesticides and herbicides. IPM techniques begin with a strategy of setting an **economic impact level** for each pest, below which countermeasures will not be taken and money and labor will be saved.

- **Transgenic plants** may also be used to introduce desirable traits into crop plants, such as making them hardier in the field or making them more nutritious.

CONNECTIONS TO OTHER CHAPTERS

Chapter 1	Genetic engineering of crops raises ethical issues. Pest control and fertilizer uses have both costs and benefits.
Chapter 4	Genetic engineering techniques can be used on crop species as well as other species.
Chapter 5	Artificial selection can be compared to natural selection.
Chapter 6	Plants have adapted to their environments in the course of evolution.
Chapter 7	Some human populations have evolved that cannot digest certain plant crops that are useful nutrition for most humans.
Chapter 9	Feeding the world's growing population will be aided by increased crop yields and more nutritious crops.
Chapter 10	Undernutrition and malnutrition affect human health. Plants are the source of most nutrients that we need.
Chapter 12	Several anti-cancer drugs are plant products. The plasmid used in plant genetic engineering originally induced tumors in host plants.
Chapter 13	Plants do not possess nervous systems or contractile muscle fibers, in contrast with animals.
Chapter 14	Most drugs are plant products, and several are psychoactive in humans.
Chapter 18	Clearing more land for agriculture threatens biodiversity and destroys ecosystems.

PRACTICE QUESTIONS

1. What is the difference between an autotroph and a heterotroph?

2. Name at least:
   a. one microscopic heterotroph
   b. two heterotrophs larger than your thumb
   c. two autotrophs

3. The major chemical process in the light reactions of photosynthesis involves the splitting of _____ and the release of _____.

4. In the dark reactions of photosynthesis, _____ from the atmosphere is incorporated into organic molecules such as _____.

5. What form of energy enters photosynthesis? In what form is that energy stored at the end of photosynthesis? In what other forms is energy stored during photosynthesis?

6. What carbohydrates can be obtained by eating plants? What function(s) do these carbohydrates serve in the plant?

7. Name three types of compounds containing nitrogen that plants need to make. Do humans need nitrogen for these same compounds?

8. How do plants obtain nitrogen? How do humans obtain nitrogen? Can either plants or humans obtain nitrogen from the air?

9. What is a limiting nutrient? What are some examples of nutrients that can be limiting for the growth of plants? How can farmers supply limiting nutrients to plants?

10. Which part of a vascular plant is responsible for each of the following?

   a. uptake of water

   b. the bulk of photosynthesis

   c. the major portion of fluid transport

   d. holding the plant up

   e. anchoring the plant in place

11. What molecules move through membranes during osmosis?

12. What are the functions of humus?

13. Under integrated pest management, name:

   a. two items that cost more time or money than in traditional forms of pest management

   b. two items that cost less time or money than in traditional forms of pest management

14. Name four ways of introducing a gene into a strain of plants in which it is not already present.

Issues

- Why is cancer so important?
- What are stem cells and why are they so important?
- What is cloning? Why has it already been banned in some countries?
- What are the causes of cancer? Are they more genetic or more environmental?
- What are the treatments for cancer?
- Can we test people for predisposition to disease? Should we?
- Can we prevent cancer?

Biological Concepts

- The cell cycle and its regulation (receptors, growth factors)
- Levels of organization (cells, tissues)
- Gene expression and regulation (cell differentiation in embryos and adult animals, potentiality, stem cells)
- Embryonic development
- Interaction of genotype and environment (carcinogens, mutagens, diet)
- Scaling (ratio of surface area to volume)
- Health and disease (homeostasis, risk)
- Molecular biology of cancer (oncogenes, tumor suppressor genes)

Chapter Outline

Multicellular Organisms Are Organized Groups of Cells and Tissues

Compartmentalization
Specialization
Cooperation and homeostasis

Cell Division Is Closely Regulated in Normal Cells

The cell cycle
Regulation of cell division
Regulation of gene expression
Limits to cell division

Development Begins with Undifferentiated Cells Called Embryonic Stem Cells

Cellular differentiation and tissue formation
Stem cells
Cloning
Ethical and scientific questions

Cancer Results When Cell Division Is Uncontrolled

Properties of cancer cells
The genetic basis of cancer
Accumulation of many mutations
Progression to cancer

Cancers Have Complex Causes and Multiple Risk Factors

Inherited predispositions for cancers
Increasing age
Viruses
Physical and chemical carcinogens
Dietary factors
Internal resistance to cancer
Social and economic factors

We Can Treat Many Cancers and Lower our Risks for Many More

Surgery, radiation, and chemotherapy
New cancer treatments
Cancer detection and predisposition
Cancer management
Cancer prevention

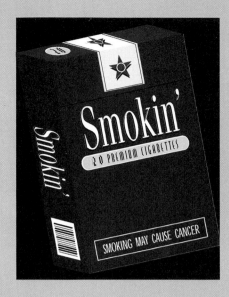

12 Stem Cells, Cell Division, and Cancer

*C*ancer is now the second leading cause of death in most industrialized countries, second only to heart disease. In the Netherlands and some other countries, cancer ranks first. Cancer is also one of the most dreaded illnesses, and people who learn that they have cancer often suffer additionally from fear of the disease.

Because there are many forms of cancer, it is more accurate to speak of cancers in the plural. **Cancers** of all types result from the same problem: cell division that is out of control. The cancerous cells no longer respond to the signals that normally limit the frequency of cell division. These signals keep normal cells functioning in an integrated manner, a necessity in all multicellular organisms whose cells are organized into specialized tissues. The activity of normal cells in tissues is sometimes compared to the behavior of animals in social groups (see Chapter 8): like the behavior of one animal, the behavior of one cell influences that of others. In contrast with normal cells, cancer cells do not behave in an integrated, social way. Instead, they grow chaotically, under their own direction, gradually pushing normal cells aside or growing right over them. Research on cancer has taught us a lot about the biology of normal cells, including stem cells, cells that have the ability to develop into many different kinds of tissue. The basic rules of normal cellular behavior are common to all multicellular species. In this chapter we consider first how cell growth and behavior are regulated in normal cells. Then we see what goes wrong in cancer and what brings it about. We also consider how we can most effectively reduce our risks for various cancers.

Multicellular Organisms Are Organized Groups of Cells and Tissues

All living organisms are composed of one or more compartments called **cells**. Most organisms larger than a certain microscopic size are subdivided into many cells. The first organisms consisted of only single cells performing all life functions in that single compartment, with little or no spatial separation among their functions. Bacteria continue to live very successfully as single-celled organisms. So why should multicellular life forms have evolved at all? Multicellularity evolved because it offers living things several advantages over unicellularity.

Compartmentalization

Compartmentalization into cells permits organisms to become much larger than they could be as single cells. Physical restrictions are imposed on living things by the ratio of their surface area to their volume. The requirements for energy and the production of wastes both increase in proportion to the volume of an organism, while the organism's ability to absorb nutrients and to release wastes varies with its surface area. However, as an organism enlarges, its volume grows faster than its surface area. Look at the cubes shown in Figure 12.1. What is the volume of cube A and of cube B? What is the surface area of cube A and of cube B? When you calculated the volume of cube A you multiplied three numbers: height times width times depth. (Another way to say this is that volume is proportional to length cubed.) When you calculated surface area you multiplied only two numbers—height times width—and then added up the number of surfaces. (Another way to say this is that area is proportional to length squared.) As you found out, the volume of cube B is 27 times the volume of cube A, yet the surface area of cube B is only 9 times the surface area of cube A. By subdividing a cube, we make more surface area. Subdivided cube C has the same volume as cube B, but three times the surface area.

Just as with the cubes in Figure 12.1, compartmentalization of the interior volume of an organism into cells keeps the volume the same, but increases the effective surface area. This maintains an efficient ratio of surface area to volume so that nutrient intake stays balanced with metabolism. Again in Figure 12.1, the ratio of surface area to volume of cube A is 6, but for the larger cube B the ratio is only 2. Subdividing cube C returns the ratio of surface area to volume to 6, the same as it was in the small cube A. Subdividing an organism into cells achieves the same effect.

Specialization

An advantage that multicellular aggregates and organisms have is that not every cell needs to perform every function. This allows specialization. In sponges (kingdom Animalia, phylum Porifera), cells are specialized but are not organized into tissues (see Figure 6.13, p. 187). The cells of all other animals form **tissues**. A tissue consists of similar cells and their products located together (structurally integrated) and functioning together (functionally integrated). The inner and outer cell layers of animals in the phylum Cnidaria, also called coelenterates (see Figure 6.13, p. 187), are separate tissues with different functions. Although the cells in the two layers are different, each cell is still changeable, so that the Cnidaria are able to regenerate an entire organism from a small piece. The cellular flexibility of Cnidaria (and other organisms capable of regeneration from parts) contrasts with the situation in more complex multicellular organisms. In these complex organisms, the fate of a cell becomes more and more restricted as it divides, through a process called **differentiation**. Initially, a cell is capable of performing a variety of functions, but as the cell differentiates (literally, 'becomes different'), it progressively loses some of its abilities and becomes specialized at doing only a few things very well. Some cells, such as human muscle or nerve cells, lose so many important abilities during differentiation that they may become incapable of further cell division. Other cells, such as those of human bone marrow and adult stem cells, retain throughout life a good deal of the flexibility characteristic of cells in developing organisms. How cells 'know' what type of tissue to become has long been one of the major questions in biology. Much of what we currently know about normal cell division and differentiation has been aided by the comparison of normal cells and cancer cells.

Figure 12.1

Subdividing a volume into cells increases the effective surface area. Cube A is one unit on each side; cube B is three units on each side. Cube C is also three units on each side but is subdivided into smaller cubes, each 1 unit per side. What is the total volume of each cube? What is the total surface area? What is the ratio of surface area to volume?

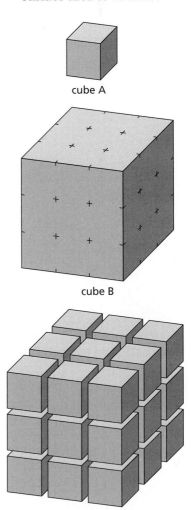

cube A

cube B

subdivided cube C

Cooperation and homeostasis

The organization of cells into tissues allows multicellular organisms to specialize their functions while maintaining an efficient ratio of surface area to volume. For specialization to be beneficial, the behavior of one type of cell must be integrated with the behavior of other cells. Tissues are further integrated into organs and organ systems, in which two or more types of tissues coordinate to perform more complex functions, such as reproduction (see Chapter 9), digestion (see Chapter 10), external sensing (see Chapter 13), respiration, circulation, or excretion (see Chapter 14). A multicellular organism can thus be considered a complex ecosystem: a human organism, for example, is an ecosystem of some ten trillion individual cells. In the course of evolution, specialization based on cooperation worked well and was favored by natural selection. All multicellular organisms—fungi, plants, and animals—have continued this basic plan.

The proper functioning of the whole organism depends on the continued integration and cooperation of all the cells. When this integration is functioning properly, that is, when the ecosystem of cells is stable, we may consider that organism to be in a state of health. According to French physiologist Claude Bernard (1813–1878), cells are responsible for maintaining a 'milieu intérieur' within each cell and within the body as a whole. Good health is defined as the maintenance of more or less constant conditions within this internal environment, a process that Bernard named **homeostasis**. This does not mean that during homeostasis there are no changes within the organism. Just the opposite is true: molecules and cells are constantly being made and being broken down, but these changes occur around a balance point. Homeostasis is the ability to return to that balance point. Disruption of homeostasis produces illness. As we will see, cancer is a disruption of the cellular homeostasis in which cell division is no longer in balance with cell death.

THOUGHT QUESTIONS

1 When the weight of some material in solution remains the same, how do changes in cell volume influence the concentration of the material?

2 What would an organism be like if all of its cells were the same?

3 Many animals, including insects, fishes, amphibians, and reptiles, do not maintain a constant internal temperature but allow their internal temperature to change with the external temperature. Are these animals in homeostasis? Why or why not? (Many of these animals can regulate their temperature behaviorally by moving to different locations.)

4 How could multicellularity in animals or plants have evolved through natural selection?

Cell Division Is Closely Regulated in Normal Cells

When biologists say that a process is regulated, it means that there are natural mechanisms that cause either more of the process or less of the process to occur in a given time period. The proper functioning of multicellular organisms depends on the regulation and integration of the processes of all their cells, particularly the process of cell division. We learned about mitosis in Chapter 2. Mitosis is one step of a larger process of cell division called the cell cycle. Scientists are finding that the molecular signals that regulate cell division and the cell cycle are remarkably similar in highly diverse organisms, from fungi to humans.

The cell cycle

Normal cells grow only a small fraction of the time. They continually make new proteins and other cellular chemicals to replace ones that have been used or damaged, but most of the time they do not increase in size. When cells do grow, they soon reach the size at which their ratio of surface area to volume makes them inefficient. Instead of becoming increasingly inefficient, the cells divide. When we talk about how fast cells grow, we really mean how frequently they divide, not how fast they enlarge.

The complex process of eucaryotic cell division recurs whenever the cell divides; it is called the **cell cycle**. You can follow the steps of the cell cycle in Figure 12.2. The cell cycle begins with a phase called G_1 in which protein synthesis is increased. Then, if the cell receives signals, it enters the synthesis, or S, phase, marked by DNA synthesis and the replication of both DNA strands (see Figure 2.22, p. 58). When DNA synthesis is complete, the cell enters the G_2 phase, in which preparations are made for mitosis. Mitosis itself (see Chapter 2, p. 42) constitutes the M phase, at the end of the cell cycle. One or both of the two offspring cells can then re-enter the cell cycle and divide again.

Most of the time, however, both offspring cells spend most of their time in a "resting stage," or G_0, between cell divisions. During the resting stage, other cellular metabolic processes proceed but the cell does not re-enter the cell cycle to divide again unless it is signaled to do so. The duration of the cell cycle (G_1 through M) is fairly constant with a species, but G_0 varies greatly. For single-celled organisms, the

Figure 12.2

The cell cycle. Hours shown are the approximate lengths of time for each phase in a cell with a 24-hour cell cycle, typical of many eucaryotic cells.

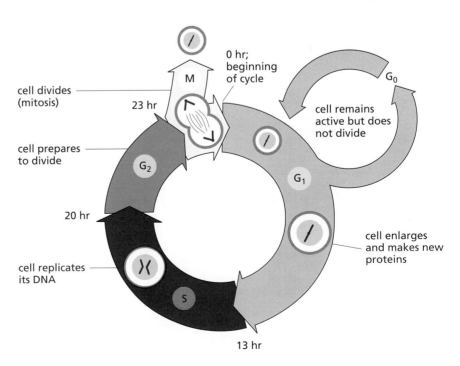

length of time in the resting stage depends on the availability of nutrients. The length of G_0 in multicellular organisms varies with the developmental stage. When an individual animal or plant is developing, the rate of increase in the number of cells can be very rapid and cells spend little or no time in G_0. In most species, most types of cells spend more time in G_0 once adulthood is reached. The neighboring cells only signal a cell to enter the G_1 and then the S phase when some cell has died and needs to be replaced. Replacements are obviously needed if there has been an injury; but even in uninjured tissues some cells die and others divide to replace them. How often this occurs depends on the tissue.

Regulation of cell division

The cell cycle (and thus cell division) is a tightly regulated process in all types of organisms, both single-celled and multicellular. Body tissues have some way of determining how big they should be, and they stop growing when they reach that size. We will soon see that the major difference between normal tissues and cancerous tissues is that cancers grow without any such limits.

For cell division to take place in normal tissues, a number of conditions must be optimal: (1) there must be available space for the new cell; (2) signals must be properly communicated; and (3) the dividing cell must be attached to a surface. In multicellular organisms, the size of most cells is restricted not only by surface area but also by the cell's being confined to a space of a certain size within a tissue. In adult organisms, cells do not divide unless a previous cell has died or been damaged, opening a space for a new cell. Contact with neighboring cells suppresses cell division in normal cells, a condition called **contact inhibition**.

Normal cells receive signals of various kinds from their external environment and do not divide unless they receive signals that send them out of the G_0 resting phase and into the G_1 phase of the cell cycle (see Figure 12.2). These signals are usually small molecules, called **growth factors**, secreted by other cells into the spaces between cells. Like most molecules, growth factors cannot cross the cell membrane. So how do they tell cells to divide?

The multistep process that tells cells to divide is shown in Figure 12.3. The first messengers, cytokines, bind to specific receptors, molecules that extend through the membrane of the cell. The term *specific* means that a given receptor can bind to only one particular type of molecule. Binding of a growth factor to its specific receptor on the exterior of the membrane

Figure 12.3

How growth factors signal cell division in normal cells.

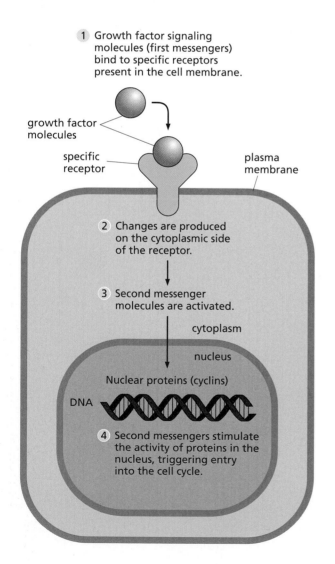

1 Growth factor signaling molecules (first messengers) bind to specific receptors present in the cell membrane.

growth factor molecules

specific receptor

plasma membrane

2 Changes are produced on the cytoplasmic side of the receptor.

3 Second messenger molecules are activated.

cytoplasm

nucleus

Nuclear proteins (cyclins)

DNA

4 Second messengers stimulate the activity of proteins in the nucleus, triggering entry into the cell cycle.

changes the portion of the receptor molecule that is on the interior of the membrane. The changes are then passed along through the cytoplasm to the nucleus of the cell by a network of molecules called **second messengers**. In response to second messengers, the concentrations of proteins (called cyclins) in the nucleus change. When the concentration of cyclin D is high, the cell enters the S phase, committing it to division. The growth factor itself remains outside the cell, but *information* has been transmitted across the cellular membrane and to the cell nucleus, triggering cell division.

Most cells have an additional requirement: they divide only when they are attached to a surface. When cells are isolated and grown in the laboratory (tissue culture) they attach to the plastic dish or to a coating on the dish (Figure 12.4). In multicellular organisms, cells attach to complex organic molecules outside the cells (the extracellular matrix in some tissues or the basement membrane in others). A normal cell may be prevented from dividing if it loses its ability to adhere to such an external structure, or if the structure changes in a way that prevents adherence. Even in the presence of growth factors, tissue cells do not divide unless they are attached.

The response of a cell to divide thus depends on the presence and normal functioning of signal molecules, receptors for these signals, second messengers and cyclin nuclear proteins, and attachment to an external support.

Regulation of gene expression

Cell division, like other cellular processes, depends on the presence of the right proteins (growth factors, receptors, and cyclins for example) at the right time and in the right amounts. These proteins are made from the DNA of a gene in a process called **gene expression**. This is a complex process with two major steps called transcription (using DNA to make

Figure 12.4

Most normal cells need to be attached to divide; cancer cells do not.

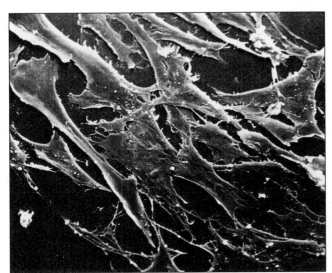

Normal cells in tissue culture growing attached to a culture dish; these cells lose their ability to divide when they become detached.

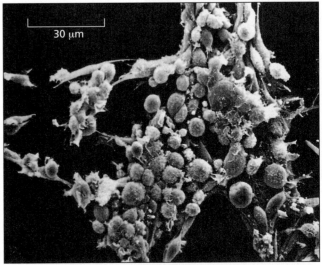

Cancer-forming cells have rounded up and lost their attachment, but, unlike most normal cells, they continue to divide when unattached.

RNA) and translation (using RNA to make protein), as described in Chapter 2. Gene expression is regulated to control whether a protein is produced and how much of the protein is made.

The regulation of transcription is summarized in Figure 12.5. Transcription begins when an enzyme called RNA polymerase binds to a special DNA sequence known as a promoter sequence (see Figure 12.5A). Each gene has its own **promoter**. Proteins needed in only small amounts are expressed by genes present as single copies in the genome and controlled by promoters that only weakly bind RNA polymerase. When the polymerase enzyme falls off the promoter, transcription stops. A protein needed in very large amounts is often coded for by multiple copies of the same gene, controlled by a promoter that strongly binds the polymerase enzyme. When the polymerase stays attached to the promoter longer, more copies of messenger RNA (mRNA) are transcribed. Cell division proteins are transcribed when they are turned on by the binding of RNA polymerase to their promoters, triggered by the second messengers resulting from growth factor signals.

On the DNA near a promoter there are regulatory gene sequences that can either enhance or repress transcription by changing how RNA polymerase binds to the promoter. If enhancers bind, RNA polymerase binds more strongly and more copies of mRNA are transcribed from that gene (see Figure 12.5B). If repressors bind to the regulatory sequences, RNA polymerase is blocked from the promoter and transcription is halted (see Figure 12.5C). Transcription repressors include the tumor suppressor proteins that slow or prevent cell division. These proteins inhibit RNA polymerase from binding to the promoters of the genes for proteins that signal the cell to enter the cell cycle. Thus, these signal proteins are not transcribed and translated, and the cell does not divide.

Repressors themselves are also regulated. A repressor can be prevented from binding either by an inhibitor blocking its DNA-binding site (see Figure 12.5D), or by mutations that change the shape of the repressor

Figure 12.5

Regulation of transcription.

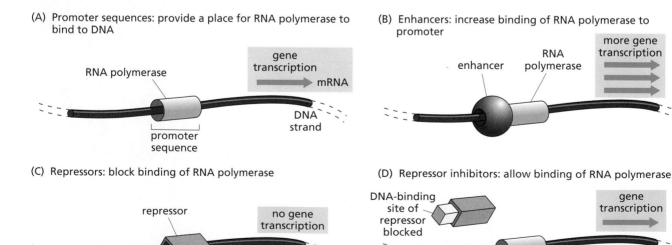

(A) Promoter sequences: provide a place for RNA polymerase to bind to DNA

(B) Enhancers: increase binding of RNA polymerase to promoter

(C) Repressors: block binding of RNA polymerase

(D) Repressor inhibitors: allow binding of RNA polymerase

protein. In either case, transcription is once again allowed. If the gene product was a cell division signal, the cell divides, possibly dividing repeatedly until a tumor results.

We have just seen how transcription is regulated (Figure 12.5, and Figure 12.6, step 1); gene expression can be regulated at several later steps as well (Figure 12.6, steps 2 through 5). In eucaryotic cells the mRNA synthesized during transcription needs to leave the nucleus before it can be translated into proteins, because the ribosomes are in the cytoplasm. Many mRNAs must be chemically modified before they can leave the nucleus (Figure 12.6, step 2); if they are not modified, the mRNA is not translated, and gene expression is halted. An example of mRNA modification is the removal of the non-coding regions (exons) as we saw in Chapter 4. Once the mRNA is in the cytoplasm, translation can take place and the rate of translation is regulated. Rapid translation produces more copies of a protein, while slow translation produces less (Figure 12.6, step 3). Further regulation occurs after translation. The amino acid sequence first produced in translation is often not the amino acid sequence of the final protein. Some amino acids may need to be removed, or other chemical groups added, before the protein can fold properly into its functional shape. Without this processing, functional proteins are not produced (Figure 12.6, step 4). Finally, protein activity can be regulated by the binding of other molecules, called effector molecules, which change the protein shape to either slow down or speed up the activity of the protein (Figure 12.6, step 5).

Cell division is thus regulated by all of the mechanisms shown in Figures 12.5 and 12.6. These mechanisms control the concentration and activity of regulatory proteins, including growth factors, nuclear cyclins, and some second messengers. Many of the genes that control and coordinate cell division were identified by research on cancer cells.

Figure 12.6

Five steps in the process of gene expression at which regulation can take place.

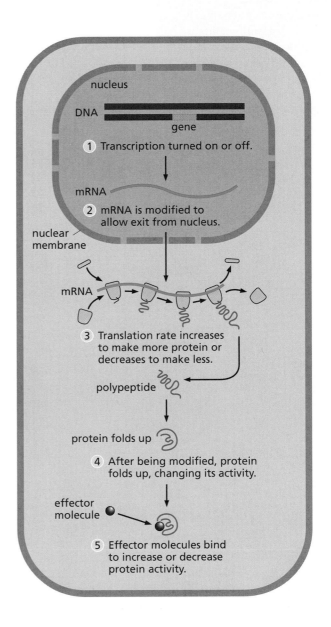

Limits to cell division

Normal cells of most tissues seem to have a limit to the number of times that they can divide. After a certain number of divisions the cells die rather than divide, even when optimal conditions exist. There seems to be a biological clock of some sort that keeps track of the number of cell divisions.

One candidate for this 'clock' is the end portion of each chromosome, called a **telomere**. Telomeres are thought to function in maintaining the

Figure 12.7

Loss of telomeres and preservation of telomeres by the enzyme telomerase. In normal somatic cells, the loss of part of the telomeres from the ends of chromosomes at each cell division gradually leads to cell aging and loss of the ability to divide. In cancer cells, the enzyme telomerase is active and restores the telomeres, allowing the cells to continue dividing without limit.

integrity of chromosomes. Each time a cell divides, a few dozen base pairs are lost from the telomere. The chromosome becomes progressively shorter on a molecular scale, although it does not lose enough length for this to be visible microscopically. When the telomere has shortened to a certain length, the cell can no longer divide (Figure 12.7).

In bacteria and in the cells that produce gametes in eucaryotes, an enzyme called telomerase restores the bases lost from telomeres, thus maintaining the length of the chromosome. These cells can thus continue to divide indefinitely. As we will see, telomerase is active in cancer cells, maintaining the telomeres no matter how many times the cell divides. Cancer cells thus have no limit to the number of times that they can divide and are therefore considered **immortal**.

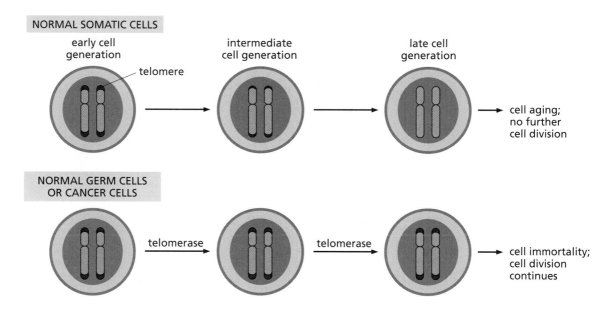

THOUGHT QUESTIONS

1 Why is it an adaptive advantage for an organism to have certain proteins such as insulin produced by just one type of cell rather than produced by all cells throughout the organism?

2 Does a cell's DNA determine what type of cell it becomes? What other factors, if any, are involved?

3 In what way might research on cancer also lead to a better understanding of the aging process? How might it further our understanding of birth defects?

Development Begins with Undifferentiated Cells Called Embryonic Stem Cells

A fertilized egg (zygote) is a single cell whose cellular descendants are capable of forming all the different cell types within the body. The long list of possibilities includes skin cells, muscle cells, glandular cells, bone cells, liver cells, and so forth. Within a developing multicellular organism, cells that are dividing also become different. Differentiation takes place in steps. At each successive cell division and differentiation, the range of possible future identities for that cell lineage is narrowed, until it narrows to a single cell type. Once a cell lineage has differentiated as muscle cells, for example, all progeny cells are committed to being muscle cells.

The undifferentiated cells are called **stem cells**. Like cell division, differentiation is tightly regulated by the control of gene expression. Much of what we know about cell differentiation has come from embryology, the study of the development of an organism from a zygote, and from the study of stem cells. Studies of normal differentiation have taught us much about the abnormal conditions that exist in cancer.

Cellular differentiation and tissue formation

The list of possible types that a cell may become is called its **potentiality**. The zygote has maximum potentiality because it gives rise to all cell types. The potentiality of cells has been investigated by transplanting cells from the embryos of experimental animals. Up until the eight-cell stage in a mammalian embryo, each of the cells—called **embryonic stem cells**—could develop into a complete organism. As the cells continue to divide, they first form a hollow ball called a blastula (see Figure 6.14, p. 188). Cells in a blastula begin to differentiate and form tissue layers (ectoderm, mesoderm, and endoderm), a process that begins at a landmark called the dorsal lip. Cells in each layer are restricted to becoming certain types of tissues. At the stage where the embryo begins to form differentiated cell layers, it is called a gastrula. These steps in embryonic development are shown in Figure 12.8A.

A group of cells removed from the ectodermal layer of the embryo at the gastrula stage and transplanted elsewhere on the same embryo can form various tissue types, but only types that are ectodermal. Their potentiality is still quite broad, but not as broad as that of the zygote. As shown in Figure 12.8B, each of the gastrula cell layers is destined to become certain types of cells. Cells transplanted at a later time have a further narrowed potentiality. An ectodermal cell is restricted to one of two groups, epidermal cells or cells of the nervous system (see Figure 12.8B). Finally, at a still later stage, the fate of these cells is completely **determined**, so that eye lens cells, for example, can form only eye lens tissue (see Figure 12.8B).

We have seen that, as a cell becomes differentiated, its potentiality becomes restricted and that the cell somehow seems to 'know' what these restrictions are. Do these restricted potentialities result from a loss of genes as a cell differentiates? To find out, a British cell biologist named

Figure 12.8

Differentiation of various cell types.

J.B. Gurdon exposed some frog eggs (phylum Chordata, class Amphibia) to ultraviolet radiation. Because ultraviolet radiation is absorbed by DNA, a sufficient dose of ultraviolet can be used to destroy the egg nucleus without damaging its cytoplasm. Gurdon then carefully inserted into each of these eggs the nucleus of a differentiated cell type, such as a skin cell. The resulting cell thus had cytoplasm from an egg but a nucleus

(A) Cell layers form as a blastula develops into a gastrula.

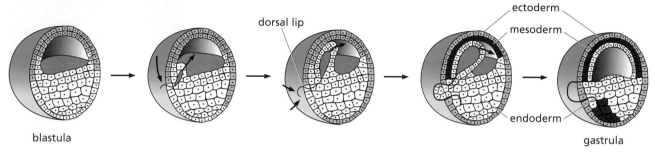

(B) As cells in the ectoderm, mesoderm, and endoderm divide, they differentiate, eventually becoming specialized cells.

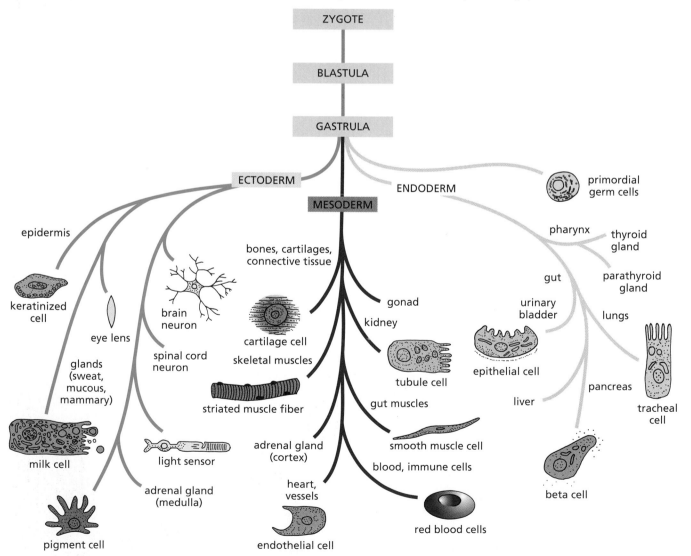

from a differentiated cell. Gurdon was able to show that this cell, like a zygote, produces an entire tadpole (Figure 12.9). The various types of cells of the body do not differ in the genes that they contain. Each nucleus usually keeps its full genome, and it is thus not the loss of genes that restricts potentiality.

Tissue differentiation. If every cell contains all the genes, how do cells become specialized in tissues? Differentiation—the process of becoming different—is, like cell division, coordinated by the regulation of gene expression. To a great extent, differentiation is a process of controlling what proteins (including enzymes) are made by a particular cell at a particular time. Although the DNA of all cells continues to carry the instructions for building all of the different cell types that make up that organism, each cell is somehow restricted to expressing only certain genes. Gene expression varies during the lifetime of the individual; some proteins are needed only by the developing organism and are not made in later stages of development. Other proteins are made only by the cells of specific tissues (Figure 12.10).

Tissue induction by organizer cytokines. At each cell division, a developing cell receives chemical signals that determine whether it will differentiate and what type of cell it will form. These chemical signals that carry information from one cell to another are called **cytokines**. (The growth factors we saw earlier are another type of cytokine.) Any region of an embryo that produces cytokine signals that cause cells to differentiate is called an organizer. The effect is local, meaning that most of the influences on cell differentiation come from neighboring cells. The concept of an organizer was first developed by the German embryologists Hans Spemann and Hilde Mangold, work for which Spemann later won the Nobel Prize. In their early experiments, Spemann and Mangold transplanted various parts of frog gastrulas into different positions on other frog gastrulas and observed developmental changes. Through such experiments they were able to show that a part of the embryo called the dorsal lip acts as an organizer that induces (causes) the overlying tissue to form a neural plate, the earliest part of the nervous system to be formed. As Figure 12.11 shows, if a dorsal lip is transplanted, a second neural plate forms above the transplanted dorsal lip cells, in addition to the normal neural plate that forms above the dorsal lip of the host.

Organizers do not actually contribute cells to the tissues that they stimulate. Spemann demonstrated this by transplanting cells from a chick embryo into the embryo of a duck. Ducks and chickens are closely enough related for a transplant of embryo cells from one to grow in the other. Because these species have different chromosome numbers, the origin of cells in a structure developing after a transplant can be ascertained. In all vertebrate embryos, the neural tube forms from ectodermal cells above a structure called the notochord. Spemann transplanted notochord tissue from a chick to a duck and it induced the formation of a second neural tube in the duck embryo. He was able to show that the second neural tube was made of duck cells, even though the transplanted tissue consisted of chick cells (Figure 12.12). The transplanted chick cells, in other words, formed no part of the neural tube that they induced, thus falsifying the hypothesis that neural tube cells arise from notochord

Figure 12.9

Gurdon's experiment demonstrating that a differentiated cell contains all the genes needed for the development of a complete organism. The nucleus of a frog egg was destroyed by ultraviolet irradiation and was replaced by the nucleus from the fully differentiated skin cell of another frog. The egg with its transplanted nucleus was allowed to grow and it developed into a normal tadpole.

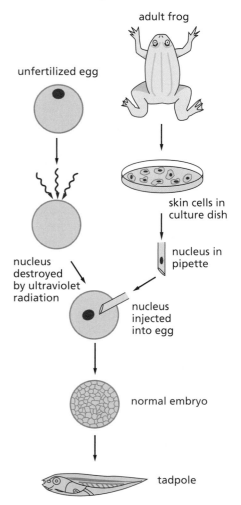

Figure 12.10

Virtually all somatic cells contain the complete genome, but different genes are expressed by the cells of different tissues.

	genetic material	example of genes switched on		
		rhodopsin	smooth muscle actin	insulin
rod light sensor	complete genome (all genes)	✓		
smooth muscle cell	complete genome (all genes)		✓	
pancreas beta cell	complete genome (all genes)			✓

cells. Duck cells had divided to form the tube in response to a signal from the organizer region transplanted from the chick. In later experiments, a chemical extract of the notochord was combined with egg white; this extract was able to induce a neural tube in precisely the way that the notochord had, suggesting that the signal from the organizer was a chemical substance rather than a group of cells.

In the development of normal tissues, cells need not arise in their end location; tissue formation relies on many migrating cells that travel through the organism until they find their 'proper' location, where they adhere and join the tissue. They are generally partly differentiated at the time of their migration and become fully differentiated when exposed to growth factors in their new microenvironment. Such molecular 'addresses' can be in the form of membrane receptors that bind specifically to molecules expressed only on certain types of tissues. Abnormalities in cellular adhesion and cell migration are pertinent to the spread of cancer to other tissues.

Stem cells

Stem cells are cells that are: (1) in an undifferentiated state; (2) able to differentiate into more committed cell types; and (3) able to renew themselves by cell division. They can be derived either from embryonic cells or from adult tissue.

The discovery that stem cells can be induced to grow into many different cell types in tissue culture has led to the idea that they may some

Figure 12.11

Spemann and Mangold's experiments on tissue formation. By manipulating cells from the gastrula stage of frog embryos, these scientists showed that neural plates form at the locations of both host and transplanted dorsal lip cells. Both neural plates developed to form brains and other head structures.

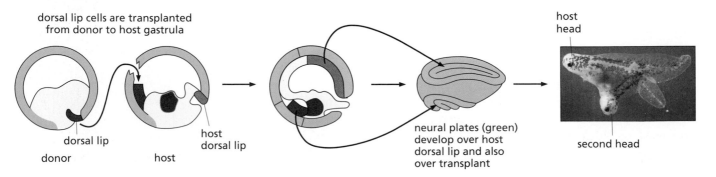

dorsal lip cells are transplanted from donor to host gastrula

dorsal lip

donor

host dorsal lip

host

neural plates (green) develop over host dorsal lip and also over transplant

host head

second head

day be used therapeutically. A person's own stem cells could be removed, isolated, grown in tissue culture with cytokines, and directed to differentiate into replacement cells as a cure for people suffering from such diseases as diabetes (insulin-producing beta cells) or Parkinson's disease (neurons—Chapter 13). Stem cells are being hailed as the dawning of regenerative medicine.

Embryonic stem cells. We have just seen how embryonic stem cells give rise to all of the tissues and organs of a new organism. Embryonic stem cells were isolated in the 1970s from early mouse embryos called blastocysts. In 1998, this process was copied in the laboratory using human blastocysts. A blastocyst is a hollow, ball-shaped cluster of about 60–200 mammalian cells from an early stage in development—the stage when implantation to the uterine wall takes place. Blastocysts maintained in culture in the laboratory for 6 months or more and without showing any signs of differentiating are referred to as embryonic stem cell lines. They can be stored frozen in liquid nitrogen for use at a later date. Under the influence of a growth medium containing different cytokines, they can be induced to differentiate into various types of cells.

Research is under way to see if embryonic stem cells transplanted into adult organisms can successfully replace damaged or degenerated

Figure 12.12

Chick–duck transplants to distinguish between cell donation and cell induction. Transplanting chick notochord tissue induces the formation of a second neural tube in a duck embryo. Do the transplanted chick notochord cells become the secondary neural tube or do they induce neural tube formation by duck cells? Chicks and ducks differ in chromosome number, so Spemann was able to determine that the secondary neural tube is made of duck cells that were induced to form the structure by the chick cells.

duck
ectoderm

normal (duck)
notochord

transplanted
(chick) notochord
tissue

neural fold

primary
(duck)
neural tube

secondary
(induced)
neural tube

original
notochord

transplanted
notochord

duck cells

chick cells

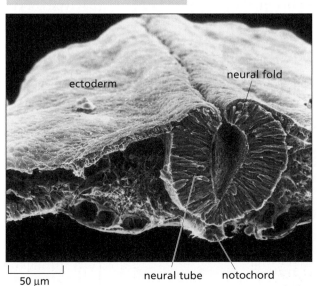

DEVELOPING NEURAL TUBE IN A
TWO-DAY-OLD CHICK EMBRYO

ectoderm

neural fold

neural tube   notochord

50 μm

tissues. Success has been demonstrated in mice, where mouse embryonic stem cells were directed to differentiate into the neurons normally defective in Parkinson's disease. When transferred to the mouse brain, the stem-cell-derived neurons improved motor function. Human embryonic stem cells have been induced to form neurons in the laboratory, and human fetal tissue that presumably contained embryonic stem cells has already been successfully transplanted. These discoveries lend great hope to the human sufferers of several diseases, including Parkinson's.

Human embryonic stem cell research is a highly controversial area because it involves the destruction of a human embryo. Virtually all of the existing human embryonic stem cell lines were derived from excess embryos from *in vitro* fertilization clinics. As we saw in Chapter 9, infertile couples can sometimes successfully conceive a child when their eggs and sperm are mixed in a dish, incubated in the laboratory, and the embryo is implanted into the woman's uterus. In all cases, more embryos are created than are ever implanted. The extra embryos sometimes remain frozen and, with the couple's informed consent, can be used for research; sometimes they are discarded. In 1996, prior to the development of methods for maintaining human embryonic stem cells, Britain mandated the destruction of 3600 embryos that had been in long-term storage in infertility clinics. These embryos could have become sources of embryonic stem cells. On August 9, 2001, very early in his presidency, George W. Bush authorized the use of 64 already existing human embryonic stem cell lines for further research, and declared that no U.S. federal funding could go toward developing any new lines. Because it is early in the development of laboratory techniques for stem cell research, it is not yet known whether these 64 cell lines have the same potentialities for differentiation as one another. Nor is it known if they all have the same capacity to regenerate cells in ways that could help cure disease. Of course, these U.S. guidelines do not restrict what researchers in other countries, or U.S. researchers in private companies, may do.

Adult stem cells. Stem cells do not exist just in embryos but are found in some adult tissues and are important for normal tissue function. Very few types of cells are permanent: some cells die and must be replaced by cell division and differentiation throughout the lifetime of the organism (Table 12.1). **Adult stem cells** are partly differentiated cells present in some tissues of adult organisms whose normal function is to divide and replace cells that are lost through routine physiological processes. These stem cells are located in several areas where cells are continually being lost: skin, gut lining, uterine cervix, bone marrow, and many glands. When a stem cell divides normally, one daughter cell usually remains undifferentiated as a stem cell and the other differentiates and is therefore less likely to continue dividing (Figure 12.13). These processes are tightly coordinated in adult organisms, so that cells lost from specific tissues are replaced by the correct number and type of cells.

One of the first locations in the body in which adult stem cells were identified was the bone marrow, the porous interior of the major bones. Both white and red blood cells are produced from bone marrow stem cells throughout a person's life. New blood cells are produced only to replace blood cells that have been lost through injury or that have reached the end of their lifespan. Both the types of cells and the numbers

CONNECTIONS ▶ CHAPTER 9

Table 12.1

Average life spans of human differentiated cell types.

CELL TYPE	LIFE SPAN (DAYS)
Intestinal lining	1.3
Stomach lining	2.9
Tongue surface	3.5
Uterine cervix	5.7
Stomach mucus	6.4
Cornea	7
Epidermis: abdomen	7
Epidermis: cheek	10
Lung alveolus	21
Lung bronchus	167
Kidney	170
Bladder lining	333
Liver	450
Adrenal cortex	750
Brain nerve	27,375 + (75+ years)

of each type that are produced are tightly regulated. Thus, the bone marrow maintains a supply of cells with the ability to replenish the blood cells throughout a person's lifetime. Bone marrow transplants can reestablish blood cells and an immune system in individuals lacking them. Transplantation of bone marrow from one person to another requires the exact matching of a set of inherited cell-surface proteins, otherwise the transplant will not be successful. In some cases a person's own bone marrow cells can be removed and later transplanted back, for example after the person's own immune cells have been killed by cancer therapy. This works very well because there is no fear of rejection of the cells by the person's immune system.

Although adult stem cells are probably not as undifferentiated as embryonic stem cells, they are still capable of forming many types of cells in addition to the type found in the tissue from which they were derived. In tissue culture, bone marrow stem cells can develop into other types of cells, including muscle or nerves, given the right set of cytokine signals. Transplants of these adult stem cells are being investigated for their future potential to regenerate other types of tissue in addition to blood cells. It appears that, in tissue culture, it may be possible to bring adult stem cells back to full potentiality, that is, the ability to differentiate into any and all kinds of cells. This is not yet known for certain, however.

Cloning

The term **cloning** means the asexual production of a group of genetically identical cells or organisms. We saw the term in Chapter 4, where the cloning of asexually reproducing bacterial cells was described in the context of genetic engineering. Here the usage of the term is somewhat different because it is cells or individuals of species that normally reproduce sexually that are being produced asexually. In cloning, there is no genetic recombination, so the genotype of the progeny is identical to the starting genotype.

There are two basic purposes for which cloning may be used. **Therapeutic cloning** is asexual cell growth whose outcome is the production of cells or tissues that might be used in treating illness, injury or disability. **Reproductive cloning** is any asexual cell growth whose purpose is the making of a complete individual. The two types of cloning use the same method to initiate the process, but in therapeutic cloning the end point is reached when stem cells are produced, and in reproductive cloning the end point is the birth of a new individual.

Therapeutic cloning. One barrier to the use of stem cells for regenerative medicine is that they themselves express the cell-surface molecules that are the barriers to all types of tissue and organ transplantation. Thus they could only be transplanted into a recipient who has the same set of these molecules, or else they would be rejected by the recipient's immune system (see Chapter 15). This is one reason why many scientists disagree with the U.S. ban on establishing more embryonic stem cell lines. The 60 or so that exist represent a very limited sample of these cell-surface molecules that form the barriers to transplantation.

Therapeutic cloning would get around the problem of transplantation barriers. In the method used currently, the nucleus of a cell from a patient is transferred into an enucleated egg from a donor. Under certain

Figure 12.13

When a stem cell divides, each daughter cell can remain as an undifferentiated stem cell or can differentiate into a tissue cell.

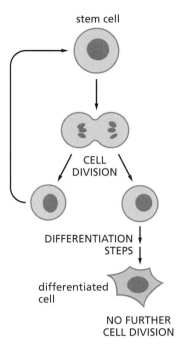

conditions in the laboratory, the egg will begin to divide, as though it were a fertilized embryo, up to the blastocyst stage. The inner cell mass is removed and these embryonic stem cells are grown in tissue culture. The cell-surface molecules will be of the same type as that coded for by the genome that was donated by the patient. These embryonic stem cells have full potentiality for regeneration of cells or tissues for the patient, and because they have the patient's cell-surface molecules, the cells or tissues derived from them will not be rejected when they are transplanted into the patient.

This procedure is a form of cloning because the embryonic cells produced are genetically identical to the patient. Some researchers prefer the term *nuclear transfer for tissue replacement*, both because it is a more accurate description of the method and because it may engender less negative reaction than the word 'cloning.'

There is another method being investigated for asexually producing embryos. It is called parthenogenesis. This process occurs naturally in some diploid species, such as certain salamanders. Unfertilized eggs contain two haploid genomes, one that normally merges with the sperm and the other that is expelled from the egg at the time of fertilization. In laboratory cloning by parthenogenesis, donor eggs are put into culture, and treatment with chemical signals causes both haploid genomes to be retained, giving the egg a full genome. The diploid egg is then allowed to divide and form a blastocyst, as in the nuclear transfer method of cloning. This technique has not yet been successful in mammals, and at best would only be able to produce cells genetically identical to the woman who donated the egg.

Reproductive cloning. Reproductive cloning uses essentially the same kind of nuclear transfer as is used for therapeutic cloning. Gurdon's experiment, in which a nucleus from a differentiated frog cell replaced the nucleus of a frog egg, was the first example of reproductive cloning. This procedure results in a complete new individual having the complete genome of another individual, the individual that donated the nucleus.

Although it was many decades ago that Gurdon successfully cloned frogs, it was only recently (1997) that the technical difficulties of producing a mammal in this way were overcome and a sheep was cloned. 'Dolly' became famous (and infamous) overnight. Differentiated cells from a sheep's udder were incubated in conditions that relaxed the differentiation signals on their DNA. (What these signals are is not yet fully known.) The nucleus from one of these cells was then put into an egg cell whose own nucleus had been destroyed. The resulting egg cell was allowed to divide in tissue culture to the blastocyst stage and then was implanted in a surrogate mother sheep who subsequently gave birth to Dolly, who was genetically identical to the animal that donated the nucleus.

Although Dolly appeared normal at birth and developed normally, it was not known whether Dolly would age prematurely, and there are now some indications that she did. She died in February of 2003 at the age of six, although the normal life expectancy for a sheep is as much as 16 years. One reason scientists thought that Dolly might age prematurely is that the donor nucleus probably had chromosomes with shortened telomeres, typical of the donor's age, so that Dolly's cells had already used up some of their limit of cell divisions.

Still more recently (1999), calves were cloned from nuclei taken from mammary cells that had been shed into the cow's milk. Mice, a cat, and a few other mammals have also been successfully cloned. Since mice develop more quickly than sheep or cows, scientists have been able to establish that the cloned mice are capable of reaching sexual maturity and reproducing.

Ethical and scientific questions

The ethical questions raised by stem cell research center mostly on the derivation of embryonic stem cells. Among the opponents of stem cell research are religious conservatives who argue that embryos should be given the moral status of a human being and not be destroyed even in the interest of scientific research. Other people hold that a new genome is not yet a unique individual. Biological evidence in support of this view includes the fact that an embryo may split to become twins, and thus become two individuals. Also, within the first 2 weeks in the uterus, two embryos may fuse and result ultimately in one individual. Such uncommon fusions are called chimeras, and the resultant person will have some cells with one genome and some with the other. Human chimeras can have cells that are genetically two different skin colors or two different sexes. In this view, humanness develops later, making the use of a blastocyst ethically acceptable. (See Chapter 9, pp. 315–316, where we discuss the concept of what qualifies as humanness.) Proponents of stem-cell research share this view, including many scientists, patients and people involved in the biotechnology industry.

Interestingly, healing and the promotion of health are part of all religious traditions. So most people see the use of adult stem cells for therapy as a potential good. We do not yet know whether adult stem cells will have the full potentiality of embryonic stem cells, so many people are reluctant to restrict the research only to adult stem cells. Some, though certainly not all, people of all religions have accepted *in vitro* fertilization as a method of infertility treatment and people go into such treatments knowing that more embryos will be created than are used. Ethicists from most major religions have accepted the use of the excess embryos for stem cell research. Very few ethicists, however, support the creation of new embryos "as a means only" for the production of stem cells or of cloned individuals.

Another set of ethical questions have to do with 'ownership' or patent rights. Can a human embryo be owned? Advanced Cellular Research in Massachusetts has received a patent on nuclear transfer techniques for cloning other animals, not humans. Britain, Australia, Israel, Japan, Portugal, Singapore, Belgium and Luxemborg have approved therapeutic cloning. Britain specifically forbids implantation of a cloned blastocyst, thus forbidding reproductive cloning. Frozen embryos produced for *in vitro* fertilization for infertile couples can be used for research in Canada, but no embryos can be created by cloning or by *in vitro* fertilization specifically to be used for research. The United States is the only industrially developed nation where efforts are under way to make all human cloning, therapeutic or reproductive, illegal.

Other people urge caution because there are so many still unanswered scientific questions. Most scientists oppose human reproductive cloning on this basis (and many do on moral grounds as well). With regard to stem cells, many scientists wonder whether transplanted stem cells have the possibility of becoming cancerous cells in the recipient. In the next sections we examine the characteristics of cancer cells and their similarities to stem cells that make some scientists cautious.

THOUGHT QUESTIONS

1 Should cloning of humans be allowed? Why or why not?

2 Should a human clone and his or her DNA donor be identical just because they have identical DNA? Identical twins share the same genome. Are they identical people?

3 What effect does the cytoplasm have on gene expression? If two identical nuclei were transferred into denucleated eggs from different individuals, would the clones develop differently?

Cancer Results When Cell Division Is Uncontrolled

Now that we have seen how cell division and differentiation are controlled in normal cells, we can examine these processes in cancer cells. Cancer is more than new growth of cells; it is the growth of cells that have escaped from the controls that operate in normal cells. Cancers can arise in any tissue whose cells are dividing. All multicellular organisms can develop cancer. In this section we discuss primarily human cancers, although much of what follows also applies to cancer in other species.

Properties of cancer cells

In cancer cells, control of cell division has been lost. The process that a cell undergoes in changing from a normal cell to an unregulated, less differentiated, 'immortal' cell is called **transformation**. The transformed state is traceable to changes in the DNA, and is therefore passed on to all progeny cells. Cancer can result from the transformation of just a single cell.

After cells have been transformed, they exhibit many characteristics that differ from those of normal cells (Table 12.2). Cancer cells continue to divide indefinitely. Some cancer cells have been maintained in tissue culture for decades. In many cases, cancer cells are less differentiated than the cells from which they arose. The membrane transport systems of transformed cells carry nutrient molecules into the cell at a higher rate. In the body this gives transformed cells a competitive advantage

Table 12.2

Characteristics of normal cells and transformed (cancer) cells.

CELLULAR BEHAVIOR	NORMAL	TRANSFORMED
Limit to the number of cell divisions	Finite	No limit (immortal)
Differentiation	Present	Inhibited
Transport of nutrients across cell membrane	Slower	Faster
Nutrient requirement	Higher	Lower
Contact inhibition	Present	None
Requirement for attachment	Present	None
Adhesiveness	High	Low
Secretion of protein-degrading enzymes	Low	High
Genetic material	Stable	Unstable

over normal cells. Transformed cells are not inhibited by contact with other cells. In tissue culture, their growth does not stop when they have formed one-cell-thick monolayers, but instead continues, forming piles of cells growing over and on top of each other (Figure 12.14). Cancer cells grow this way inside organisms, and the growing piles of cells are called **tumors**.

Transformed cells grow without the need to be attached (see Figure 12.4B); in fact, this is the characteristic that best predicts whether a cell growing in culture will form a tumor if put into an animal. Changes back and forth from attached growth to unattached growth are believed to spread some tumors to new locations.

Cells become transformed when they are dividing; therefore, cells that are terminally differentiated and will never again divide cannot become transformed, for example nerve cells in the brain and muscle cells of the heart. In contrast, many types of cancers arise from the transformation of stem cells. Stem cells divide often, and respond more easily to cell division signals than do highly differentiated cells. When stem cells divide normally, one daughter cell remains undifferentiated as a stem cell and the other differentiates (see Figure 12.13) and is therefore less likely to continue dividing. Unlike a normal stem cell, a transformed stem cell divides into two daughter cells that both remain undifferentiated and each can continue to proliferate. One concept of experimental therapy for stem cell cancers is to give drugs that promote cellular differentiation and thus slow down the progression of the cancer. Conversely, one of the possible problems resulting from stem cell transplants or therapeutic cloning is that the cells may not stop dividing and thus will act like cancerous cells.

The genetic basis of cancer

A transformed cell is a less-differentiated 'immortal' cell that no longer responds to the signals that normally regulate cell division and cell differentiation. These signals have not been completely identified, but this much seems cer-

Figure 12.14

Cell division of transformed cells is no longer inhibited by contact with neighboring cells.

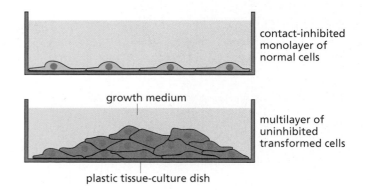

contact-inhibited monolayer of normal cells

growth medium

multilayer of uninhibited transformed cells

plastic tissue-culture dish

tain: cancerous growth signals are aberrant forms of the normal growth signals, and the aberration is located in the cell's DNA. The normal growth regulatory genes fall into two categories: genes encoding proteins that promote cell division (**proto-oncogenes**) and genes whose protein products normally inhibit cell division (**tumor suppressor genes**). Mutations in either type of regulatory gene increase the probability that cancer will arise.

Oncogenes and proto-oncogenes. The signaling of cell division in normal cells was discussed earlier in this chapter and is summarized in Figure 12.3. The genes whose products are the growth factors, receptors, second messengers, and cyclins shown in Figure 12.3 are the proto-oncogenes. Proto-oncogenes are thus genes whose products signal and regulate normal cell division. American cell biologists J. Michael Bishop and Harold Varmus received the 1989 Nobel Prize for their discovery of proto-oncogenes, leading to a new era in our knowledge of cell division. The abnormal, mutated forms of these proto-oncogenes that lead to cell transformation and cancer are called **oncogenes**. Oncogenes, the cancer-causing mutants, were actually discovered before the normal proto-oncogenes. In this way, research on cancer has led to a better understanding of the ways in which normal cell division is controlled.

Oncogenes differ from proto-oncogenes in any of three basic ways: the timing and quantity of their expression, the structure of their protein products, and the degree to which their protein products are regulated by cellular signals. The expression of a proto-oncogene responds to cellular controls (see Figure 12.3), but the expression of an oncogene does not. The protein product of an oncogene may differ by as little as a single amino acid from the protein product of a corresponding proto-oncogene, but this small change in structure can be enough to remove the protein from control by the cell's regulatory mechanism.

The mutation of a proto-oncogene to an oncogene can alter the cell division signals at any of five steps and trigger uncontrolled cell division. One type of oncogene codes for a modified growth factor receptor that, unlike its normal proto-oncogene counterpart, continuously activates second messengers (thus triggering cell division) without having bound its usual growth factor cytokines (Figure 12.15, type 1). Another type of oncogene causes a cell to secrete growth factors for which it has receptors, allowing the cell to stimulate itself to divide rather than needing signals from its neighbors (Figure 12.15, type 2). A third type of oncogene codes for altered cellular second

Figure 12.15

Continuous signaling of cell division in transformed cells by the protein products of five types of oncogenes that lead to the cell's escape from the regulation of cell division.

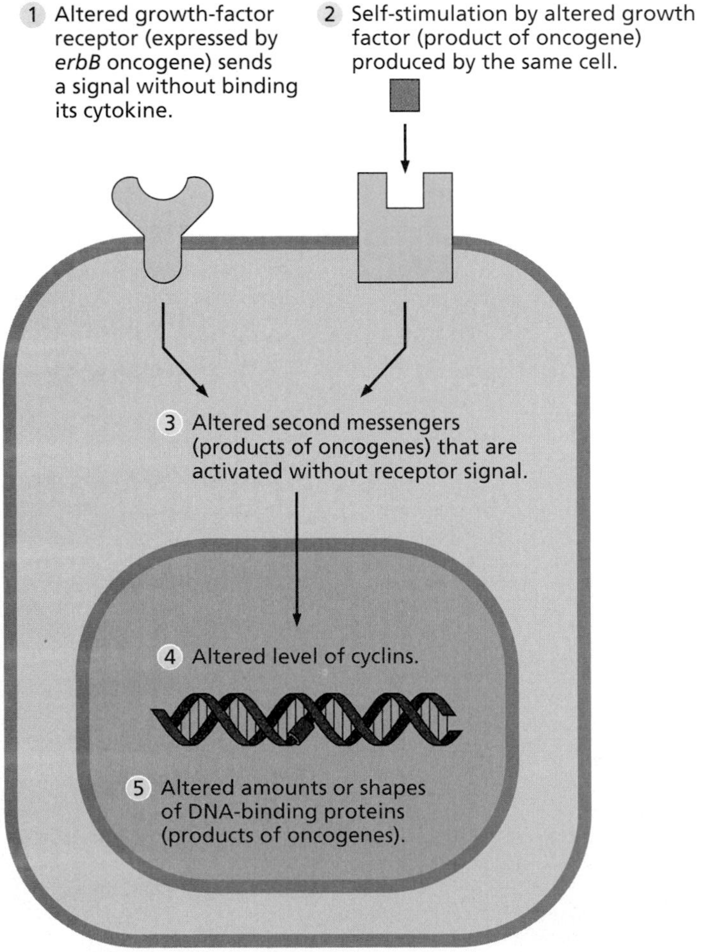

1. Altered growth-factor receptor (expressed by *erbB* oncogene) sends a signal without binding its cytokine.

2. Self-stimulation by altered growth factor (product of oncogene) produced by the same cell.

3. Altered second messengers (products of oncogenes) that are activated without receptor signal.

4. Altered level of cyclins.

5. Altered amounts or shapes of DNA-binding proteins (products of oncogenes).

messenger molecules that carry 'activate cell division' commands across the cytoplasm in the absence of any growth factor signal from outside the cell (Figure 12.15, type 3). Still other oncogenes alter the regulatory steps inside the nucleus, affecting the concentrations of cyclins (Figure 12.15, type 4) or of certain DNA-binding proteins (Figure 12.15, type 5).

Tumor suppressor genes. The protein products of tumor suppressor genes normally repress cell division. If these genes are altered, their repressor activity may be removed. Inactivation of a cell-division repressor leads to cell division (see Figure 12.5C and D). A tumor suppressor gene, *p53*, is mutated in as many as 55% of noninherited cancers. When something goes wrong inside a cell, the normal p53 protein halts cell division and causes the abnormal cells to die. When the *p53* gene is mutated, the altered p53 protein does not halt cell division, and cells with damaged DNA continue to live and divide, passing on their accumulated mutations to their progeny cells.

Accumulation of many mutations

The transformation of cells may require a combination of changes in several proto-oncogenes and tumor suppressor genes rather than a change in just one. If single mutations were the primary cause of cancer, the rate of incidence for new cancers would be the same for individuals of every age. That is clearly not the case for most cancers; cancer rates increase with age, particularly in advanced age. It is estimated that five or six such mutations must occur *in a single cell* before it becomes transformed to a cancer cell. Most cancers arise in somatic cells (body cells) rather than in gametes. Somatic mutations are passed along to the progeny cells in that individual but are not passed on to the individual's offspring. In the rarer inherited cancers, some mutation in gamete DNA is passed on to sons or daughters. Because these individuals have inherited one or more of the mutations needed for transformation, fewer somatic mutations need to accumulate for some cells to become fully transformed and to progress to cancer. Inheriting such mutations thus predisposes the individual to get cancer more readily than someone else living in the same environment.

In a human lifetime, there are on the order of 10^{16} cell divisions. Because of the limitations on the accuracy of DNA replication, at every cell division each gene has about a 1 in 10^6 chance of being copied wrongly. This gives, for the approximately 10^{16} cell divisions, a mutation rate of 10^{16} divided by 10^6, or about 10^{10} mutations per gene in a human lifetime just from mistakes in replication.

Several mechanisms prevent the overwhelming majority of these mutations from initiating a cancer. Most mutations are corrected as they occur by 'spell-checking' proteins (Figure 12.16). Cells with mutations that are not successfully corrected are usually induced to die by the p53 protein. To contribute to transformation, uncorrected mutations must be in a proto-oncogene or tumor suppressor gene. Because these genes are a tiny fraction of the whole genome, the probability that one of them will mutate at random and that the mutation will be uncorrected is low. Additionally, for cancer to arise, mutations in many growth regulatory genes must all accumulate in the same cell.

Mutations within genes are not the only way that proto-oncogenes are altered to become oncogenes. In other cases there may be an increase in the number of copies of the unmutated proto-oncogene in the DNA (gene amplification), causing it to stimulate uncontrolled cell division by producing abnormally large amounts of the normal protein. Genes moved to another chromosome or another part of a chromosome as a result of abnormal crossovers can become cancer causing. In their new location they may no longer be under the control of the regulatory elements that operated at their previous location in the genome.

Progression to cancer

We see the characteristics of transformed cells when we study them at the cellular level. However, cancer occurs in whole, multicellular organisms, not in isolated cells. As mentioned earlier, an organism can be considered an ecological system of many billions of cells. The interacting growth and differentiation signals keep the cellular ecosystem stable. After cells are transformed, progression to a tumor depends on many ecological factors. Mutated progeny cells may be killed or they may not outgrow the normal cells and thus never progress to a tumor. Alternatively, the transformed cell and its progeny may continue to divide, taking up space and nutrients required by their neighbors, passing on the mutation to each new progeny cell (Figure 12.17). Normal cells begin to die off, not because they are killed outright by cancer cells, but because they are deprived of space and nutrients. With the decline in the number of normal cells comes a reduction of their normal function, and the organism begins to show signs of illness. The particular symptoms depend on the type of cancer and the type of normal cells that are lost.

Transformed cells within organs may form solid tumors within those organs (see Figure 12.17). For a tumor to be visible on X-ray, the original transformed cell must divide repeatedly until there are about 10^8 cells in the tumor. For a tumor to be large enough to be felt (about 1 cm in diameter), approximately 10^9 cells are needed. By the time that tumors are this size, they have begun to influence their environment. The tumor cells may secrete growth factors (called angiogenic growth factors) that

Figure 12.16

DNA mismatch mutations and their repair.
(A) A mismatch leads to a permanent mutation on one DNA strand if not corrected.
(B) A mismatch repair protein (spell-checking protein) acts on a mutated strand, cutting out that DNA and allowing the correct base to be added.

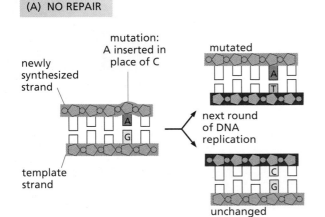

induce nearby blood vessels to develop new branches that grow into the tumor. These are normal blood vessel cells whose growth is induced by the tumor cells. A tumor may thus contain normal cells as well as transformed cells.

A tumor is said to be benign if it is contained in one location and has not broken through the basement membrane to which normal cells are attached. Benign tumors, as their name suggests, often cause no health problems for the individual. Benign tumors can become large enough to interrupt the functioning of normal tissues, but their removal by surgery is generally successful because they have not intermingled with normal tissue. Tumor cells that invade normal tissues, rather than just pushing them out of the way, are said to be **malignant** (see Figure 12.17). For cells to be invasive, they must produce protein-degrading enzymes such as collagenase, an enzyme that dissolves the collagen connective tissue that holds groups of cells together (see Table 12.2). The term 'cancer' is generally reserved for malignant tumors.

Because malignant tumors produce enzymes that allow them to invade other tissue, they often spread to new locations, a process known as **metastasis**. In this process, one or more of the transformed cells lose their attachment to the other cells of the tumor, break through the basement membrane, and spread via the circulation to other areas of the body (see Figure 12.17). In the new location they regain attachment and continue to divide, forming new tumors. The new tumors are of the same type as the original tumor and thus when viewed with a microscope are seen to be different from the cells around them. Cancers that have begun to metastasize are far more serious and more resistant to treatment than those that have not, because no amount of surgery can eliminate all the cancerous cells that have spread.

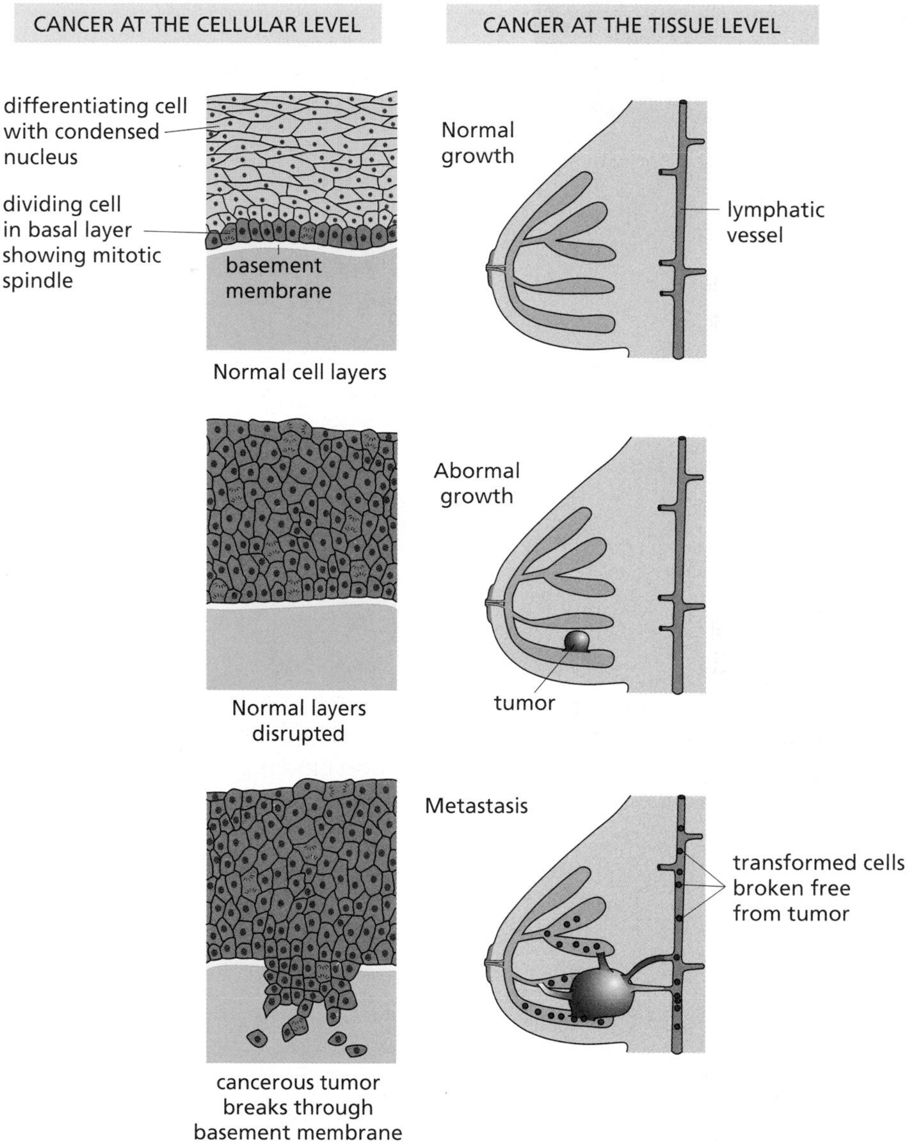

Figure 12.17

The growth of a tumor in the human breast.

CANCER AT THE CELLULAR LEVEL

differentiating cell with condensed nucleus

dividing cell in basal layer showing mitotic spindle

basement membrane

Normal cell layers

Normal layers disrupted

cancerous tumor breaks through basement membrane

CANCER AT THE TISSUE LEVEL

Normal growth

lymphatic vessel

Abormal growth

tumor

Metastasis

transformed cells broken free from tumor

THOUGHT QUESTIONS

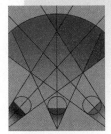

1 How do stem cells differ from transformed cells? How do stem cells differ from muscle cells or blood cells?

2 In what ways are benign and malignant tumors the same and in what ways are they different?

3 What does the statement "Cancer is a disease of the genes, but it is not a genetic disease" mean?

Cancers Have Complex Causes and Multiple Risk Factors

The study of disease at the population level constitutes the science of epidemiology. The basic epidemiological data for various forms of cancer have been compiled for the United States since 1950. The incidence for various cancers in the United States is given in Table 12.3.

Epidemiology uses descriptive statistics to find patterns in the incidence of diseases. Those patterns indicate possible risk factors that can suggest hypotheses that can be further tested in other ways. In general, and with a number of exceptions, the causes of adult cancers seem to be mainly environmental, not genetic. Evidence to support this conclusion comes from epidemiological data for the United States and several Euro-

Table 12.3

Cancer Cases, Deaths and 5-Year Survival Rates in the United States, arranged in order of the number of deaths (totals estimated by the National Cancer Institute from incidence rates found in the Surveillance, Epidemiology and End Results [SEER] Program).

TYPE OF CANCER	NEW CASES*	DEATHS*	5-YEAR SURVIVAL RATE (%)‡
Lung	171,600	158,900	13.4
Colon	94,700	47,900	61.0
Breast	175,000†	43,700†	83.2
Prostate	179,300	37,000	85.5
Pancreas	28,600	28,600	3.6
Lymphoma	64,000	27,000	51.0
Leukemia	30,200	22,100	68.6
Urinary bladder	54,200	12,100	80.7
Skin, all	1,000,000	9200	–
Rectum	34,700	8700	61.0
Oral cavity and pharynx	29,800	8100	–
Skin, melanoma	44,200	7300	86.6
Uterus, cervix	12,800	4800	68.3
Uterus, endometrium	37,400	6400	83.2

* 1999 American Cancer Society Facts and Figures.
† Includes 1300 cases and 400 deaths in men.
‡ Percentage of people treated for some cancers who are still alive 5 years later, compared with percentage in a comparable, cancer-free population; Scientific American, September 1996.

pean countries, showing in each case a marked increase in cancer rates throughout the twentieth century. Most of this increase in cancer pertains to a single type, cancer of the lung (Figure 12.18A). The increase coincided with advancing industrialization and other changes in the environment, but very little change in the gene pool. In Germany, cancer caused only 3.3% of all deaths in 1900, but 20.9% of all deaths in 1967, a more than sixfold increase. From 1950 to 1979, lung cancer death rates more than tripled for both U.S. males and U.S. females, with higher rates of increase among nonwhites than among whites. In each case, the death rate increases paralleled an increase in the consumption of cigarettes, with a lag of about 15–20 years (Figure 12.18B). Since 1960, cigarette consumption by males has decreased and in 1990 the lung cancer death rate among males began to decrease. Cancers of the pancreas and large intestine increased more slowly over the same period, while stomach and rectal cancers declined. The rapid change and irregular pattern of change both fit much better the hypothesis of environmental causes than the alternative hypothesis of a genetic cause or causes.

Human cancers are named according to the type of cell from which the cancer is derived. A cancer that arises in epithelial tissue (sheet-like tissue or glandular tissue) is called a carcinoma; a cancer that arises in connective tissue is called a sarcoma. There are also subtypes of tumors: a mesothelioma, for example, is a sarcoma of the lining of the abdominal cavity. A cancer that arises among white blood cells (leucocytes) is called a leukemia if the cells are circulating throughout the body via the bloodstream, but it is called a lymphoma if it is a solid tumor in lymphoid (leukocyte-containing) tissue.

Figure 12.18

Deaths from cancers in the United States.

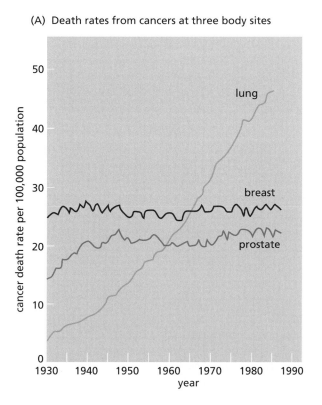

(A) Death rates from cancers at three body sites

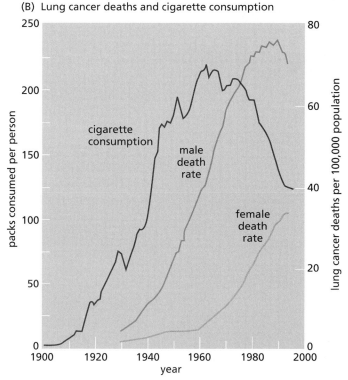

(B) Lung cancer deaths and cigarette consumption

Some blood cell cancers (leukemias) are more common among children. Approximately 85% of childhood cancers are acute lymphocytic leukemias that arise from stem cells in the bone marrow. Although these are very aggressive, they have a good cure rate because children still have many normal cells to take over after therapy.

About 85% of adult cancers are carcinomas, including cancers of the lungs, breast, colon, rectum, pancreas, skin, prostate, and uterus (see Table 12.3). The incidence of these cancers (and many others) increases with age, so that cancers become more and more significant as causes of death with advancing age. Environmental or lifestyle factors are believed to affect most of these adult cancers. The following factors have been suspected, on the basis of epidemiological evidence, of causing at least one type of cancer or of increasing the rate at which at least some cancers occur: genes, increasing age, viruses, ionizing radiation, ultraviolet radiation, diet, stress, mental state, weak immune systems, unsafe sexual behavior, hormones, alcohol, tobacco, and some chemical substances.

In this section we examine the evidence from epidemiological studies and animal studies that have suggested the many possible causes of cancer. We will see how these seemingly disparate causes may be working by very similar pathways at the cellular and molecular levels. Keep in mind that when we speak of 'causes' of cancer we often mean factors that are associated in epidemiological studies with increased incidence in populations. As such, these factors are more properly called 'risk factors,' not causes. A multitude of factors contributes to whether any particular person gets cancer. We generally cannot say that one thing 'caused' a particular cancer. As Clark Heath of the American Cancer Society has said, "Cancer cases are clinically nonspecific—you can't look at a leukemia case clinically and say, 'Ah, this is a radiation-caused leukemia.' " (*Scientific American*, September 1996, p. 86.)

Inherited predispositions for cancers

Cancer is not inherited, but a predisposition for some cancers can be. For example, retinoblastoma is a rare cancer of the eye, caused by defective alleles of the gene for a protein called pRB, which helps control the cell cycle. Of the people who carry the defective alleles, 80–90% develop retinoblastoma. Another rare cancer is xeroderma pigmentosum, a skin cancer that results from a defect in the mechanism of DNA repair. Almost all persons with mutant alleles for this DNA repair protein develop cancer because of their extreme sensitivity to ultraviolet radiation. (Most cancer-related genes have much lower rates of associated cancers than do retinoblastoma or xeroderma pigmentosum.) Recently, a genetic mutation associated with a rare form of colon cancer was located and identified as being in a gene coding for a DNA-checking protein. When DNA is replicated, the wrong bases can be put into the growing strands. As we saw in Figure 12.16B, specific proteins called checking proteins check for mistakes, like a spell-checking program on a computer. Enzymes then repair the mistake. People with this inherited predisposition for colon cancer inherit a mutated gene and produce defective checking proteins. Mutations throughout the genome are thus likely to be passed along to progeny cells (see Figure 12.16A). Other cancers have been associated with this same defective checking protein; the defect is not confined to colon cells. If a mistake is uncorrected in a growth control gene, the cell may become transformed. The cancer cannot be said to be caused by the inherited allele, but by increasing the mutation rate throughout the genome the defective checking protein increases the probability that a mutation will occur in a growth control gene.

Five to ten percent of breast cancers are associated with a genetic predisposition. At least two genes have been identified in families in which multiple members have early-onset (premenopausal) breast cancer. These genes are called *BRCA1* and *BRCA2*, for breast cancer 1 and 2. These have often been referred to as 'breast cancer genes.' However, although their presence increases the probability that a person (male or female) will get breast cancer, it does not guarantee it, so they should more properly be called 'breast cancer predisposition mutations.' The incidence of *BRCA1* mutations in the general U.S. population is estimated to be one person in 1000. The sequences of the normal genes is now known and they are among the largest human genes known. Their protein products bear little similarity to any other known human proteins. Their normal protein products act as tumor suppressors, inhibiting cell division when DNA is damaged. *BRCA* gene mutations that make the suppressor proteins nonfunctional allow cells with damaged DNA to divide when they normally would not (see Figure 12.5D).

Increasing age

Far more cancers seem to be environmentally caused than genetically caused, even after taking into account genetic predispositions for some cancers. The more common type of breast cancer, a late-onset disease of postmenopausal women, is not linked to inheritance. The strongest risk factor for these breast cancers is age. The incidence of all cancers increases with age, presumably because there has been more time for environmental exposures to produce accumulated mutations. In fact, one of the reasons for the present-day higher incidence of cancers (and chronic diseases such as heart disease) is that people are living much longer because mortality from infectious diseases is lower.

From data on the incidence of cancers, the probability of acquiring cancer at different ages can be calculated. Table 12.4 gives data for breast cancer. As can be seen, the probability increases with age. Because age is such a strong factor in all health studies, epidemiologists must 'control

Table 12.4

Age-specific probabilities of developing breast cancer.

| AT AGE | PROBABILITY OF DEVELOPING BREAST CANCER IN THE NEXT 10 YEARS | |
	%	1 IN:
20	0.04	2500
30	0.40	250
40	1.49	67
50	2.54	39
60	3.43	29

Based on 1997 U.S. data from the American Cancer Society.

for age;' that is, they must either compare groups of the same ages, or use mathematical formulas to 'age-adjust' the data.

Cancer data are often shown as the probability of acquiring cancer by age 75. Examples are shown in Figure 12.19. Here data of the type shown in Table 12.4 have been added up to give the 'lifetime probability' of acquiring cancer. Note that these numbers do not mean, for example, that a white woman's chances of acquiring breast cancer are 10%. The chances for each age group are the numbers shown in Table 12.4, which are much lower.

Data of the type shown in Figure 12.19 are useful in comparing the probabilities for different types of cancers. They are also useful in comparing the probabilities for different segments of the population. In the United States, health statistics are summarized by sex and by race. These are data for populations and do not mean that any individual's probability is the number shown. Individual risk is increased or decreased by all of the factors mentioned in this section.

Viruses

Several cancers are known to be associated with viruses and other infectious agents. In 1911, American pathologist Peyton Rous showed that a tumor of connective tissues (a sarcoma) in chickens was caused by a virus that was later named Rous sarcoma virus. (Rous received a Nobel Prize for this work, but not until 1966.) Chickens infected with this virus develop sarcomas. Viruses seem to be associated with cancer in at least two different ways. First, a virus infection may cause a decrease in the activity of the immune system. Decreased immunity increases the likelihood that a transformed cell will progress to cancer. An example of this type is Kaposi's sarcoma, which occurs in people with AIDS (see Chapter 16, pp. 576–577). Second, some viruses carry genes that, when inserted

CONNECTIONS CHAPTER 16

Figure 12.19

Lifetime probabilities of acquiring various types of cancers.

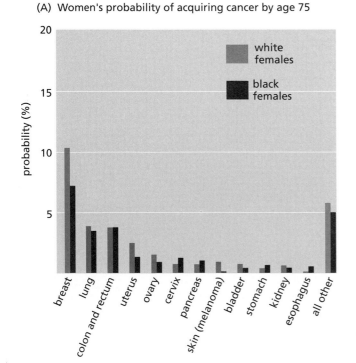

(A) Women's probability of acquiring cancer by age 75

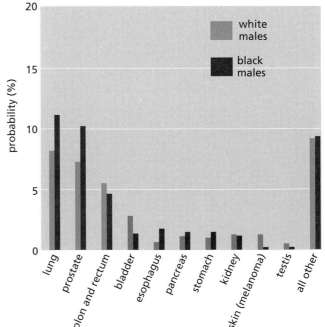

(B) Men's probability of acquiring cancer by age 75

into the host DNA, cause the host cell to become transformed into a cancerous cell. These genes are therefore oncogenes. The viruses do not cause the mutation; rather, they carry an entire mutated gene into the cell they infect.

The incidence of liver cancer is high in third-world countries. A very high proportion of the people who develop liver cancer have previously had hepatitis B, a viral infection of the liver. Some people, after recovering from the acute symptoms of hepatitis, remain infected carriers of the virus. Over 250 million people worldwide are carriers of hepatitis B virus; carriers have a 100-fold higher risk of developing liver cancer, although this may be as long as 40 years after having hepatitis. Hepatitis C virus is also associated with liver cancer, particularly in Japan. Worldwide, as many as 80% of liver cancers are caused by viral infections.

Cancers of the reproductive organs, especially cancer of the uterine cervix, are statistically related to both male and female sexual behavior. Incidence rates for cervical cancer are higher among women who were younger at the time of their first intercourse or who have had multiple sexual partners. The rates of cervical cancers are also high in those countries in which women tend to have few sexual partners and to marry as virgins but where a tradition of *machismo* often encourages men to seek multiple sexual partners. This epidemiological evidence argues that male promiscuity is an important risk factor for cervical cancer even though it is women who develop the disease. Sexually transmitted human papilloma viruses often cause genital warts, but some types also cause cancer of the cervix, which is thus a sexually transmitted form of cancer. Papilloma viruses account for more than 80% of cancers of the genitals and anus.

For the association of each of these viruses with cancer, the epidemiological evidence has been verified by animal and tissue-culture experimentation. Although the incidence of virally induced tumors is low in the United States, such tumors account for 20% of all cancers worldwide.

Physical and chemical carcinogens

A large and growing number of external agents are known to cause cancer; such agents are called **carcinogens**. Evidence that carcinogens cause cancer comes from studies in which animals are exposed to them. Evidence also comes from epidemiological studies of occupationally exposed persons, such as industrial workers who handle the agents. Still other carcinogens are discovered by recognizing epidemiological clusters of persons with unusually high incidence rates for particular cancers living in the area surrounding an industrial plant or waste disposal site. There are two types of carcinogens, those that induce DNA mutations (**mutagens**) and those that promote the progression of transformed cells into cancer. Some mutagens are physical agents, energy sources with high enough power to damage DNA. Other mutagens are chemical agents; these too work by damaging DNA.

Physical carcinogens (radiation). Some carcinogens are physical agents, particularly certain types of energy sources. Ultraviolet (UV) radiation, for example from sunlight, can cause susceptible people to develop skin cancers, including malignant melanoma, a cancer of the pigment cells (melanocytes). Melanomas are dangerous because they readily spread around the body and thus are difficult to treat once they have metastasized. Melanomas kill more women in their twenties than does breast cancer. Light-skinned people are more susceptible to these cancers. Melanoma affects some 44,000 Americans annually and a total of 1 million people get some form of skin cancer in the United States each year. Most of these are squamous or basal cell carcinomas, which are less aggressive and more treatable than melanoma. Ninety percent of skin cancers are attributable to UV radiation, particularly UV B rays, which cause DNA damage. The adult incidence of skin cancer is correlated with the number of sunburns that a person received as a child. Epidemiological evidence shows that even exposures to natural levels of UV radiation increase the incidence rates for skin cancers, which are much higher in the southern half of the United States than in the northern half.

Ionizing radiation, such as that produced by radioactive substances, was clearly shown to be carcinogenic by studies on the Japanese survivors of the 1945 bombings of Hiroshima and Nagasaki. The French (Polish-born) chemist Marie Curie (1867–1934), a two-time winner of the Nobel Prize and pioneer in the study of radioactive elements, died of a leukemia induced by her frequent handling of these elements. X-ray machines also produce ionizing radiation, so the level of exposure is carefully controlled to minimize the exposure. Medical and dental diagnostic X-rays do not increase cancer incidence. The only medical uses of radiation associated with increased cancer risk are the very high radiation doses used in cancer therapy itself. Although these treatments are necessary to save a patient, they do induce DNA damage from which new cancers may arise a decade or more later. Radiation treatments, like all medical treatments, are evaluated by estimating risk–benefit ratios. When someone is very sick and would die without treatment, the probable benefit from the treatment may make a higher level of risk acceptable to the patient and her or his physician.

Ionizing radiation associated with cancer is also caused by radon. Radon is a radioactive element that occurs naturally in certain types of rocks. When such rock is uncovered, radon may be given off as a gas. A person with long-term exposure to radon gas may develop lung cancer; however, radon accounts for less than 10% of lung cancers, whereas smoking accounts for more than 85%. Proper ventilation removes virtually all of the risk from radon.

Low-frequency electric and magnetic fields from power lines or household appliances and radio-frequency electromagnetic radiation from cell phones or microwaves are too low in energy to cause ionizing damage to DNA. Any time that charged particles move, whether they are electrons in a power line or ions across cell membranes, electric and magnetic energy fields are created. All living things, because of ions moving through them, are sources of electrical and magnetic energy. The ambient level of energy from a cell phone is less than 1% of the electromagnetic radiation given off by the person holding the phone.

Chemical carcinogens. Other carcinogens are chemicals. Exposure to the chemicals in tobacco smoke, including second-hand smoke, is the largest single risk factor for cancer in the industrialized world. People who begin to smoke when they are teenagers or in college are more than 10 times more likely to develop lung cancer than people who have never smoked (Figure 12.20A). Although a person's risk of cancer decreases after he or she stops smoking, the risk never drops as low as that for people who have never smoked (Figure 12.20B). The danger of second-hand smoke is evident from the fact that nonsmoking women whose husbands smoke have higher cancer rates than nonsmoking women with nonsmoking husbands. Exposure to tobacco smoke is also a risk factor for heart disease and emphysema.

Tobacco smoke contains dozens of known carcinogens including nitrosamines, formaldehyde and other aldehydes, arsenic, nickel, cadmium, and benzo[a]pyrene. These are also present in smokeless tobacco (chewing tobacco and snuff). Both smoked and smokeless tobacco also greatly increase the risk for cancers of the mouth and throat (oral cavity and pharynx).

A large number of industrial chemicals have been shown to be carcinogenic, including vinyl chloride (used in the making of many plastics), formaldehyde, asbestos, nickel, arsenic, benzene, chromium, cadmium, and polychlorinated biphenyls (PCBs) (Table 12.5).

There is a clear correlation between the ability of an agent to act as a mutagen and induce the mutation of DNA (Box 12.1) and its ability to act as a carcinogen and induce cancer. The terms mutagen and carcinogen are not synonymous, however. Transformation is initiated by a mutation in the DNA; the mutation or mutations can be caused by a virus, a chemical, radiation, or by spontaneous mistakes in replication. The presence of mutagens greatly increases the mutation rate throughout the genome, increasing the likelihood that some mutation will be carcinogenic and turn a proto-oncogene into an oncogene.

Tumor initiators and tumor promoters. Tumor initiators are agents that begin the process of transformation by causing permanent damage in the DNA. Mutagens, including tobacco smoke, are tumor initiators. In a cell whose DNA is damaged in this way, transformation can be completed by exposure at a later time to a tumor promoter.

Tumor promoters by themselves do not cause mutation detected by the Ames test described in Box 12.1. They can induce cell division, and if a dividing cell contains a mutation from an earlier exposure to an initiator,

Figure 12.20

The effects of smoking on the incidence of cancer.

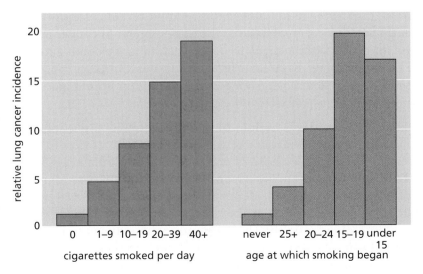

(A) Lung cancer risk for smokers and nonsmokers

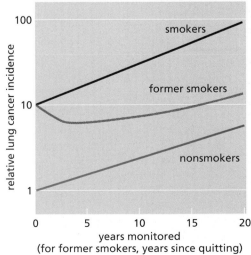

(B) Lung cancer risk in former smokers compared with smokers and people who never smoked

Table 12.5

Carcinogens in the workplace.

CARCINOGEN	CANCER TYPE	EXPOSURE OF GENERAL POPULATION	EXAMPLES OF WORKERS FREQUENTLY EXPOSED OR EXPOSURE SOURCES
Chemical Agent			
Arsenic	Lung, skin	Rare	Insecticide and herbicide sprayers; oil refinery workers
Asbestos	Lung, other sites	Uncommon	Brake-lining; shipyard; insulation and demolition workers
Benzene	Bone marrow	Common	Painters; distillers and petrochemical workers; dye users; furniture finishers; rubber workers
Diesel exhaust	Lung	Common	Railroad and bus-garage workers; truck operators; miners
Formaldehyde	Nose, pharynx	Rare	Hospital laboratory workers; manufacture of wood products, paper, textiles, garments and metal products
Heavy metals (cadmium, uranium, nickel)	Prostate	Rare	Metal workers
Man-made mineral fibers	Lung	Uncommon	Wall and pipe insulation; duct wrapping
Hair dyes	Bladder	Uncommon	Hairdressers and barbers (inadequate evidence for customers)
Mineral oils	Skin	Common	Metal machining
Nonarsenical pesticides	Lung	Common	Sprayers; agricultural workers
Painting materials	Lung	Uncommon	Professional painters
Polychlorinated biphenyls (PCBs)	Liver, skin	Uncommon	Heat transfer and hydraulic fluids and lubricants; inks; adhesives; insecticides
Soot	Skin	Uncommon	Chimney sweeps and cleaners; bricklayers; insulators; firefighters; heating-unit service workers
Vinyl chloride	Liver	Uncommon	Plastic workers
Physical Agent			
Ionizing radiation	Bone marrow, several others	Common	Sunlight; nuclear materials; medicinal products and procedures
Radon	Lung	Uncommon	Mines; underground structures

Data modified from Scientific American, *September 1996.*

its chance of acquiring additional mutations that lead to complete transformation are increased. Because the DNA damage from an initiator is permanent, a tumor promoter can have its effect years after exposure to the initiator. Tumor promoters include alcohol, phenobarbital, dioxin, saccharin, asbestos, and tobacco.

Dietary factors

The American Cancer Society states that "for the majority of Americans who do not use tobacco products, dietary choices and physical activity are the most important modifiable determinants of cancer risk." Evidence that dietary factors contribute to the development of cancer is best established for cancers of the digestive tract, including the colon and rectum. The evidence comes from laboratory studies of animals exposed to experimentally controlled diets, from clinical studies on human patients, and from epidemiological studies of large populations.

Dietary fiber and fats. Diets high in fiber and low in fats are associated with a lower incidence of cancers of the intestinal tract (including the colon and rectum) and also those of the pancreas and breast. In countries where fiber consumption is high and fat consumption is very low, as in most of equatorial Africa, incidence rates of colon and rectal cancer are only a small fraction of what they are in the industrialized world. Australia, New Zealand, and the United States, where diets are lower in fiber and higher in fats, have high rates of colon and rectal cancers. In fact, diet is more strongly correlated with cancer incidence than is industrial pollution. Studies comparing the cancer rates in Iceland and New

Zealand, where diets are similar to those in the United States but where there is far less industrialization, have shown that the cancer incidence is the same in these countries as it is in the United States. In contrast, cancer rates overall, and for many specific types of cancer, are much lower in Japan, which, like the United States, is an industrialized nation, but one in which dietary fat intake is very low. The incidence of cancers among Seventh-Day Adventists who are vegetarian and do not smoke or drink is much lower than the incidence in their neighbors, despite both groups' living in the same conditions and being exposed to the same environmental pollutants. Several studies have also shown that eating fresh vegetables, particularly those rich in vitamins A, C, and E, and beta-carotene (a vitamin A precursor), reduces the incidence of many cancers.

Salty and pickled foods. Cancer of the stomach follows a different epidemiological pattern correlated with a different set of dietary factors. This cancer is most frequent in Japan and in certain Latin American countries, where it seems to be correlated with the eating of very salty foods and pickled vegetables. The incidence of this cancer in Japanese immigrants to Hawaii and California decreases after a generation or two, while that for cancers of the colon, rectum, and breast increases. Among Japanese Americans in Hawaii, the incidence of stomach cancers correlates closely with the retention of other aspects of Japanese culture: that segment of the Japanese American population who maintain more of their traditional culture have higher rates of stomach cancer than those who adopt more Western cultural practices. This evidence suggests that diet has a larger role than genetics in the incidence of stomach cancer.

Alcohol. Ethyl alcohol has been identified as a risk factor for cancer by a number of researchers, but the increased cancer risk is much greater in people who also smoke. The risk of developing a cancer of the mouth or throat, for example, is much higher in people who both smoke and drink (Figure 12.21). This is an example of a **synergistic effect**, meaning that the increased risk due to two causes is much more than the

Figure 12.21

Synergism between alcohol and cigarettes in producing cancers of the mouth and throat.

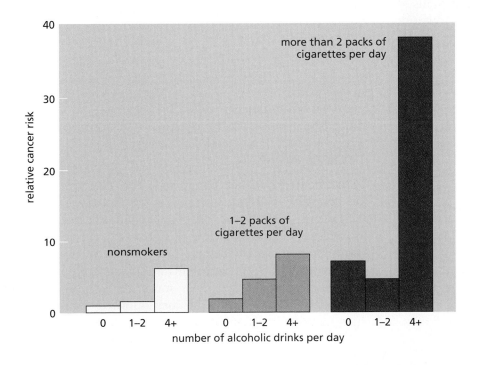

BOX 12.1 The Ames Test

Tens of thousands of known chemical substances have never been tested as possible carcinogens in animals. Animal testing is expensive and slow; it would take many, many decades (and many billions of research dollars) to test all these substances. Clearly, we need a quick screening method that tells us which substances are more likely to be carcinogenic; these substances can be tested first, while the testing of less likely carcinogens can wait.

The Ames test, devised by cell biologist Bruce Ames of Cornell University, is a screening method that detects mutagens capable of causing a particular type of mutation in a culture of *Salmonella* bacteria, as shown in the diagram below. The bacteria used are from a strain called *his⁻*, which are unable to synthesize histidine, an amino acid required for the manufacture of bacterial proteins and hence for bacterial growth. Most bacteria are *his⁺*, meaning that they can make their own histidine from other materials. In the Ames test, *his⁻* bacteria are grown in a medium containing just a small amount of histidine, which allows just enough growth for mutations to have a chance to occur. Soon, however, the histidine is used up, and the bacteria die unless they have mutated from *his⁻* to *his⁺* and thus have become able to make their own histidine. The rate of spontaneous mutation is very low. If a

chemical is added to the culture medium and many more bacterial colonies grow than in a culture without this addition, the chemical can be assumed to have caused the mutations—i.e., to be a mutagen. Counts of the numbers of colonies also identify stronger and weaker mutagens.

Remember that the Ames test was designed as a *screening method* for carcinogens. The basis of the Ames test is the observation that many known carcinogens are also mutagenic. This is simply a statement of *correlation*; it does not necessarily indicate a causal relation between mutagenesis and carcinogenesis. Not every mutagen is a carcinogen, so the Ames test is only preliminary. It focuses our attention on chemicals that are mutagenic in bacteria, and thus more likely to be carcinogenic in animals. We can then proceed with the animal testing of these substances.

Bruce Ames, the originator of the Ames test, has also pointed out that nearly *any* substance is mutagenic in a sufficiently high dose. We are surrounded with thousands of naturally occurring carcinogens (mostly weak ones), and yet we do not all get cancer from them. Perhaps, he argues, we should study our mechanisms of defense against these natural carcinogens rather than concentrating on merely identifying one carcinogen after another.

---

additive combination of their effects taken separately. Alcohol and tobacco are also synergistic in producing other forms of cancer.

Internal resistance to cancer

A good deal of evidence shows that people vary in their resistance to cancer. People with the same exposure to all known risks do not get cancer at the same rate. People also vary in their recovery rates once they get cancer. Individual variation in hormones, stress, mental outlook, and immune function may be involved.

Hormones have been implicated in some types of cancers, especially uterine cancer, which occurs more often in women who have been exposed to certain estrogens, including the synthetic hormone diethyl-

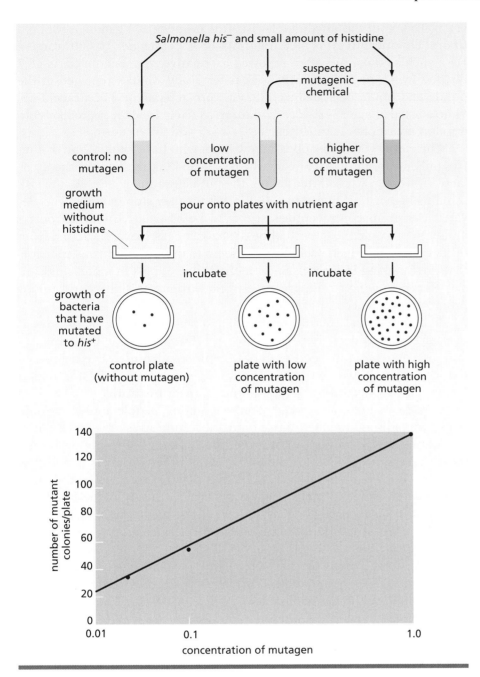

stilbestrol (DES). Hormones also influence breast cancer rates, although the process is unclear. The risk for some breast cancers can be reduced by ovariectomy (removal of the ovaries) or by taking the estrogen-inhibiting drug tamoxifen.

Oral contraceptives contain two hormones, estrogen and progesterone (see Chapter 9). A comprehensive report that reanalyzed data from more than 50 studies found a slight elevation in risk of breast cancer in women who took oral contraceptives (the 'pill'). For every 10,000 women who are currently using the pill and who started using it between the ages of 25 and 29, 48.7 are expected to develop breast cancer in the next 10 years. Among women of the same age who have never used the pill, 44 out of 10,000 are expected to develop breast cancer in the next 10

CONNECTIONS

CHAPTER 9

years. Therefore the 'attributable risk' (the number that can be attributed to oral contraceptives) is 48.7 minus 44 or 4.7 cases per 10,000 women. Data such as these are often reported as 'relative risk' calculated as 48.7 divided by 44, or 1.16. Relative risk is generally what is reported to the public and would be stated as follows: women using oral contraceptives (who started at age 25–29) are 1.16 times as likely (or 16% more likely) to develop breast cancer in 10 years.

The elevated risk disappears in women who have been off birth control pills for more than 10 years, if they started taking them at age 20 or later. In women who started taking the pill before the age of 20, the risk remains higher for longer than 10 years after they stop using it. The elevated risk from contraceptives seems to be associated with estrogen's ability to promote the growth of breast cancer cells that have arisen from other causes. Estrogen itself does not seem to initiate cancerous transformation. Smoking is synergistic with birth control pills in increasing cancer, possibly owing to the presence of both tumor initiators and tumor promoters in tobacco.

In contrast, oral contraceptives are associated with *decreased* rates of ovarian cancer. A decrease by as many as 1700 cases in the United States per year may be associated with oral contraceptive use. The decreased risk continues in women for as long as 15 years after they stop taking contraceptives, according to the National Cancer Institute.

Certain cancers are more common among people under chronic stress. Stress is a difficult variable to measure, and stress levels are usually reported simply as 'high' or 'normal.' Many studies of stress are flawed in that researchers failed to take into account other variables in addition to stress. For example, night workers and daytime workers in the same industry may differ in many other ways besides stress levels. Psychologists who have focused their attention on the means by which people deal with their stress have found lower cancer rates among people with better coping skills.

Evidence is increasing that people with weakened immune systems develop cancers more frequently. This factor is striking in conditions that severely damage the immune system, such as AIDS (see Chapter 16), but is also present in people whose immune systems are weakened less drastically from other causes, including chronic stress, sleep disorders, and so forth. Far more cells are mutated and transformed than ever develop into cancers. A healthy and active immune system eliminates most of these cells as they arise. Any weakening of the immune system increases the number of transformed cells that grow and proliferate, and a higher rate of cancer is one of the results. People's immune systems also weaken with age, which is consistent with the finding that the incidence of new cancers increases with age.

Social and economic factors

Social and economic factors have also been shown to be correlated with incidence rates and especially with survival rates for various cancers. In the United States, some studies have shown that whites and blacks (African Americans) have comparable incidence rates for certain cancers, but the mortality rates for blacks are higher because they receive far fewer routine medical exams that would detect the common cancers in their earliest and most treatable stages.

More recent studies have found higher incidence and mortality rates for most types of cancers for urban black populations than for urban white populations. Much of the excess cancer rate among blacks was in those with low income levels and low educational attainment. In other words, a large amount of the difference in cancer incidence rates could be explained by differences in income and factors related to income. When blacks and whites of comparable socioeconomic status were compared, many of these differences disappeared or were reversed. Colon cancer showed no difference in incidence by race, and rectal cancer was more common in whites than in blacks. Blacks had higher rates for cancers of the stomach, prostate, and uterine cervix. Female breast cancer showed a higher rate in white women than in black women, and among white women it showed a higher incidence in the higher income brackets than in the lower ones.

As we have seen, many factors may contribute to cancer incidence rates, including genetic predisposition, lifestyle, and exposure to environmental carcinogens. A summary of the relative contributions of various biological causes is given in Table 12.6. These are rough averages for global cancer incidence. The percentage of cancers attributable to viruses is 20% worldwide, but is much lower in the United States. The percentage attributable to diet also varies from one part of the world to another, as does the percentage attributable to chemical carcinogens in the environment. With the exception of the low percentage of cancers that may be attributable to genetic predisposition, most cancers are preventable by changes that can be made by individuals or by societies, a topic we explore further in the next section.

Table 12.6

Summary of various causes of cancer worldwide.

CAUSE	RELATIVE PERCENTAGE OF CANCER DEATHS*
Smoking	30
Diet	30
Alcohol	3
Food additives (salt)	1
Sedentary lifestyle	3
Radiation	2†
Pollutants (air, water)	2
Viruses	20
Chemical carcinogens	Variable‡
Genetic susceptibility	<10

* Derived from statistical analysis of epidemiological data. A figure of 30 means that 30% of all cancer deaths worldwide are attributable to that particular cause.
† Over 90% of skin cancers are caused by UV radiation.
‡ The percentage of deaths in the general population attributable to carcinogen exposure is low, but can be locally very high, for example in industries with high or prolonged exposure.

THOUGHT QUESTIONS

1 Not everyone who smokes gets cancer. Does this mean that smoking is not a risk factor for cancer?

2 What is the difference between a risk factor and a cause? Can we say what caused cancer in a given individual?

3 Why do different individuals respond differently to cancer risk factors?

4 Does an increase in the percentage of deaths due to cancer necessarily mean that cancer rates have increased? What else could explain such findings? How could you go about determining which of the possible explanations best fits the data?

5 Recently the genetic defect associated with an inherited form of colon cancer was identified as a defect in a DNA-checking protein. This finding was reported in the lay press as the discovery of 'the colon cancer gene.' Is this name misleading? To what extent can a defective repair mechanism be considered to be the same thing as a cause of a cancer?

6 Tobacco smoke contains chemicals that are tumor initiators and other chemicals that are tumor promoters. How does this combination contribute to the carcinogenicity of tobacco smoke?

7 Tobacco and alcohol act synergistically in increasing cancer risks. Can you explain this in terms of what is happening inside cells?

We Can Treat Many Cancers and Lower Our Risks for Many More

An understanding of the mechanisms that produce cancer has greatly increased our understanding of how to prevent it. Treatments have improved and continue to improve so that survival rates for most cancers have increased. In this section we examine medicine's current strategies for treatment and for prevention.

Surgery, radiation, and chemotherapy

Most of the present-day treatments of cancer use one or more of three types of treatments: surgery, radiation, and chemotherapy. Surgery is limited to those cancers that produce visible tumors. Because single cancer cells that metastasize can lead to later cancer recurrence, surgery is often combined with radiation or chemotherapy. In either radiation therapy or chemotherapy, the strategy is the same: cancer cells are dividing cells, so agents that interfere with cell division should stop cancer cells. Radiation causes breaks in the DNA of dividing cells that are so large that the cell cannot repair them and the cell cannot live with the damage. Chemotherapeutic drugs prevent DNA synthesis at several steps. Some of the drugs inhibit the synthesis of the nucleotides needed to build DNA; some substitute for certain nucleotides in newly synthesized DNA, preventing its further replication; and some inhibit an enzyme needed to unwind and rewind the double helix during its replication. Other chemotherapeutic drugs, some of which are natural plant products, prevent RNA synthesis or block mitosis. Still others act by damaging the DNA strands, thus preventing cell division and killing the cell.

An important drawback to radiation and chemotherapy treatments is that, because they damage DNA, they increase the risk for the development of secondary cancers from damaged cells that survive the treatments. Another drawback is that both radiation and chemotherapy are nonspecific: both kill any type of dividing cell. Two examples are hair follicle cells and immune cells. A high proportion of cells in hair follicles are dividing, so hair loss frequently accompanies these treatments. A large percentage of cells of the immune system are also dividing and so are killed. Not all hair follicle cells and immune cells are killed because not all were dividing at the time of treatment and so they repopulate. Hair grows back and people regain their immune cells. During the time when people's immune systems are compromised, they need to avoid exposure to infectious diseases. Both radiation and chemotherapy may destroy a particular type of immune cell, memory cells, which 'remember' which diseases the person has been exposed to or vaccinated against (see Chapter 15). If these memory cells are killed, even a person who has regained the ability to form new immune responses has lost previous immunities and therefore may need to be revaccinated.

CONNECTIONS
CHAPTER 15

A further risk from chemotherapeutic drugs is that they put a selective pressure on the population of transformed cells. As a result, any cells that become resistant to the drug quickly overgrow the drug-susceptible cells. Several drugs, each of which works by a different mechanism of action, are often used in combination to minimize the development of drug resistance.

Despite the drawbacks and risks, surgery, radiation, and chemotherapy have been very effective, increasing the survival rates of many types of cancers. Moreover, in most patients, leukemia, Hodgkin's lymphoma, and testicular cancer can now be cured by these treatments.

New cancer treatments

Research on cancer treatments continues. New chemotherapeutic agents, both natural and artificial, continue to be sought and developed. The ultimate goal of new anticancer therapies is to exploit some property that distinguishes cancer cells from normal cells.

The SERMs (selective estrogen receptor modulators) are new chemotherapeutic drugs with more specific action. The hormone estrogen binds to receptors inside cells and causes some types of cells to divide, including some types of breast cancer cells. However, estrogen has many normal functions, including maintenance of bone density and cholesterol regulation. The SERMs block estrogen's ability to promote breast cancer cell division but do not block its desirable effects. Tamoxifen is one such drug and has proved effective in reducing mortality from breast cancer and in decreasing its onset in women at high risk due to genetic predisposition or age (over 60).

Another new type of chemotherapeutic drug starves tumors. As mentioned earlier, many tumors have the ability to induce the body to grow new blood vessels, which then bring needed blood to the tumors. Drugs called angiogenesis inhibitors block this process, thereby cutting off the nutritional supply of the tumors. Treating tumors with these drugs shows great promise and has received much publicity. At present several angiogenesis inhibitors are in clinical trials.

In addition to treatments with direct effects on tumors, new treatments are being developed to boost the immune system's ability to fight off tumors. If the immune system gets rid of many transformed cells, why can we not vaccinate people against cancer as we do against many infectious diseases? One reason is that the immune system can act only against cells it perceives to be nonself (see Chapter 15). The surface molecules of cancer cells are often the same as those on normal cells; they are just expressed in the wrong amounts or at the wrong times. The immune system cannot distinguish the last two possibilities as differing from the normal: it can detect only new or different cell-surface molecules. Some cancers, especially cancers initiated by chemical carcinogens, do have new or altered molecules at their surface. Unfortunately these tumor-associated antigens are different in each person and even in two different chemically induced tumors in the same person. What new molecules will be present cannot be predicted, so vaccines cannot be developed against them. Some tumors induced by viruses do have common tumor-associated antigens, and vaccines can be developed. Feline leukemia is the most successful example so far of a cancer that can be prevented by vaccination.

Antibodies can block growth factor receptors. About 25–30% of breast cancers have cells with abnormally high levels of the growth factor receptor molecule called HER2/neu. More receptors mean that the cell can be triggered to divide by lower concentrations of growth factor. A new drug called Herceptin is being considered for approval by the FDA in the United States for use along with conventional chemotherapy for the treatment of metastatic breast cancer. This drug is an antibody that binds specifically to the receptor, blocking binding of the growth factor and thus inhibiting cell division. Initial studies have found that it shrinks tumors in some women and delays tumor progression in others.

Another entirely new type of treatment, called photodynamic therapy, has been approved for esophageal cancer in the United States, and some other countries have approved its use for other cancers as well. The patient is given an intravenous injection of a photosensitizing dye. Over the next 24–48 hours most of the dye accumulates in tumor cells (normal tissues remove the drug). Then the patient's tumor is illuminated by using fiber optics for 30–60 minutes. Light reacts with the dye to kill many tumor cells and sensitize others to being killed by the immune system.

A good deal of cancer research is aimed at assessing the efficacy of new therapies. After extensive laboratory research, new therapies are tested on patients who have given their informed consent (see Chapter 1) to being part of the studies. Only by gathering data in properly designed and controlled clinical trials can the risks and benefits of new treatments be demonstrated.

Because cancer is greatly feared and not always curable, some people put their hopes in unproven remedies. Various unconventional cancer therapies and treatments have been publicized in the past few decades.

Some of these, such as laetrile, achieved a large and devoted following. The supporters of laetrile finally became so influential that the National Cancer Institute conducted careful clinical trials and announced in 1981 that laetrile had proved to be worthless as a cancer treatment.

Cancer detection and predisposition

Early detection greatly increases the probability of successful cancer treatment. Breast self-examination is very effective at finding tumors while they are treatable. Diagnostic breast X-rays (mammograms) detect smaller tumors than can be felt, but are less effective in younger women than in postmenopausal women whose breast tissue is less dense. Microscopic examination of tissue from the cervix taken during a medical examination is effective at early detection of cervical cancer. The test is called a Pap smear. Testicular cancer, although rare, is the most frequent cancer in men between the ages of 15 and 34; monthly self-examinations to detect lumps in the testes are an important method of early detection.

An emphasis on early detection has contributed to the increased survival rate from some cancers. For other cancers, it is not so clear what is meant by an increase in survival rate and how this relates to a 'cure.' When we try to evaluate the meaning of statistical statements like 'survival rates,' we need to know a lot about how the numbers were gathered and what definitions are being used for certain terms. What is often reported as survival rate refers to the proportion of persons with cancer still living after 5 years compared with the proportion of surviving persons without cancer. For example, it is estimated that it takes 9 years for a breast cancer to develop, spread, and kill a person. If past methods of detection led to discovery of the cancer 7 years after its inception, very few people would have been alive 5 years after its discovery. Now, with better detection methods and better public education for breast self-examination, the 5-year survival rate looks much improved. Does this mean that therapies have improved, or does it simply mean that people are now finding the cancers at 2 years into their development rather than at 7 years? It is not always easy to distinguish advances made through better or earlier diagnosis from advances made in the treatment of cancers once they have reached comparable stages of development.

In addition to the tests mentioned above, two types of laboratory cancer tests currently exist. One type is for the early detection of existing cancers. The other type is genetic testing for cancer predisposition.

An example of the first type is the PSA test for prostate cancer. This test measures the level of PSA (prostate-specific antigen) in a person's blood. This protein is elevated when prostate cancer begins to develop. Thirty percent of people with elevated PSA are found to have cancer when the test is followed by a biopsy (tissue sample examined by microscope). The PSA test can detect cancer up to 5 years before there are other symptoms. The clinical question then becomes what to do about it. Cancers detected by PSA tests are generally still localized and can therefore be successfully removed surgically. However, prostate cancer is very slow-growing, and it has been said that most men die with prostate cancer, not because of it. One-third of men have some form of prostate cancer by the time they are over the age of 50, but only 3% die from it. For many individuals, after a biopsy has confirmed the presence of prostate cancer the recommended treatment was therefore to do nothing. However, a recent study has shown that men with newly diagnosed prostate cancer who have their prostate surgically removed have a 48% reduction in their risk of dying from this disease (4.6% mortality) compared with a similar group of men who underwent 'watchful waiting' (8.9% mortality).

The second type of laboratory test does not detect cancer, but instead identifies DNA sequences that are statistically correlated with an increased probability of some day acquiring the disease. Tests of this type are available for *BRCA1* and *BRCA2* (breast cancer predisposition mutations), DNA mismatch repair genes (predisposing to some kinds of colon and uterine cancers), *p53* (predisposing to brain and other tumors) and a few other tumor suppressor genes and oncogenes.

These tests for genetic predispositions for cancer are controversial for many reasons. They are very expensive and do not give much more useful information than is gained from knowing your family medical history. A negative test does not mean that a person will not get cancer; cancers arise spontaneously in the same way as they do in people with no family history of the disease, so the recommended preventive measures discussed below should still be followed. A positive test is not a guarantee that a person will get cancer. The probability is increased but we cannot tell by how much. The increase in risk is different for different mutations, but none increases the probability to 100%. The genetic mutation cannot be repaired, so there is little that a person with increased risk can or should do beyond what is recommended for everyone (regular check-ups and the lifestyle choices summarized below). Still, because some of these cancers are difficult to detect early, being aware of a predisposition for them can ensure that physical examinations are done even more thoroughly than usual, and perhaps more often.

Some women with increased breast cancer risk due to mutations in *BRCA1* (or family history) have opted for 'prophylactic mastectomies,' that is, removal of their breasts before there is any evidence of disease. A recent study was widely publicized as showing that women who had their breasts removed reduced their risk of dying by 90%. While this statement is not untrue, it is only part of the story and is an example of reporting 'relative risk' instead of 'absolute risk.' In the study, 639 women had their breasts removed. On the basis of calculations made from the number of deaths among their sisters who faced the same increased susceptibility but who had not had their breasts removed, it was estimated that 20 of the 639 women would have died. Only 2 actually did die, so the relative risk was decreased by 90% [$100 \times (20 - 2)/20 = 90\%$]. When absolute risk is considered, it can also be correctly stated that 97% of the women had their breasts removed unnecessarily [$100 \times (639 - 20)/639 = 97\%$]. The difficulty for any woman faced with such a choice is that there is no way to predict whether she will be one of the 18 saved by the procedure or one of the 619 who did not need it.

Cancer management

Some people feel that more research dollars should be spent on cancer management rather than cancer treatment. Cancer management includes the development of drugs or strategies to minimize the side-effects of cancer treatments. Examples include cold capping, a procedure in which a cold pack is applied to the scalp during chemotherapy so as to slow cell division in hair follicle cells, thus decreasing hair loss. Another possibility is the development of anti-nausea drugs. The possible medicinal use of marijuana to overcome nausea and restore appetite in chemotherapy patients is being studied (see Chapter 14).

Cancer management also includes support groups and grief therapy. The aim of these approaches is to improve the quality of life—to treat the person, not the disease. Women who were in support groups after recurrent breast cancer lived longer than those who were not. Survival rates have been found in some cases to be influenced by mental attitude (see Chapter 15). Patients who were optimistic, who were aggressive, or who were 'determined fighters' had statistically longer survival times and higher cure rates than those who were pessimistic or who resigned themselves early to their fate. There is a large body of psychological literature on 'learned helplessness,' a phenomenon in which a person or an animal experiences repeated stresses from which there is no escape and for which no remedy is available. Such individuals 'learn' that there is nothing they can do to change anything, a lesson that they then apply to other areas of their lives. When such people get cancer, their learned helplessness results in a much lower survival rate and a shorter life span.

CONNECTIONS
CHAPTERS 10, 14, 15

Cancer prevention

As we learn more about the causes of cancer, it seems that one of the more successful strategies may be cancer prevention. Preventing cancer may be far easier than curing it. Smoking remains a major cause of cancer. The Centers for Disease Control and Prevention state that cigarette smoke causes 30 times more lung cancer deaths than all regulated air pollutants combined. Exposure to secondhand smoke, for example as a result of living in a home with someone who smokes, causes the deaths from lung cancer of 3000 nonsmokers a year in the United States. Tobacco smoke contains both tumor initiators and tumor promoters, and it suppresses the immune system. In addition to its cancer risks, exposure to secondhand smoke is also responsible for nearly 300,000 infections per year in infants younger than 18 months, and has been implicated as an important contributing factor to Sudden Infant Death Syndrome (SIDS).

Some dietary regimens have been associated with a decreased risk of cancer: lower total calories and low fat, for example. Food containing antioxidants may help; the antioxidants include beta-carotene and vitamins A, C, and E (see Chapter 10). The American Cancer Society has issued guidelines promoting fresh foods high in vitamins A and C, especially fresh vegetables of the plant family Cruciferae. Cruciferous vegetables include cabbage, radish, turnip, broccoli, cauliflower, kale, kohlrabi, mustard greens, and brussels sprouts. Clinical studies have shown that some vitamin supplements are not as effective in preventing cancer as the same vitamins obtained from foods, probably because other food ingredients are also at work.

High-fat diets may be an important risk factor in cancers of the colon and rectum, and in postmenopausal cancer of the breast. There is some evidence that high-fiber diets lower the risks of colon and rectal cancers.

Given the available evidence, the most important actions you can take to lower your cancer risks are the following.

1. *Don't smoke!* Also, avoid secondhand smoke from poorly ventilated rooms where others smoke. *These are the single greatest steps you can take to reduce your cancer risk, far outweighing all other possible measures.*

2. Follow a diet low in fats, high in fiber, and high in antioxidants such as beta-carotene and vitamins A and C.

3. Avoid occupational exposures to potential carcinogens; minimize exposure through the appropriate use of safety equipment.

4. Avoid exposure to radioactive substances and X-rays above necessary minimum levels; avoid needless exposure to ultraviolet radiation from the sun or tanning booths.

5. As you age, be sure to get checkups at regular intervals, including screening that detects the common cancers in their earliest and most easily treated stages. If you are a woman, learn to practice breast self-examination and, if you are a man, testicular self-examination.

THOUGHT QUESTIONS

1 Explain how the theory of evolution accounts for the development of cancer cells that are resistant to chemotherapy.

2 What characteristics would you look for in an ideal chemotherapeutic drug for the treatment of cancer?

3 Secondhand smoke is a cancer risk. Does this biological reality change the ethical debate about smoking in restaurants or smoking in the workplace? What rights are in conflict on either side?

4 Do you agree with the following statements? When someone's chances for survival are predicted to be very low, any and all treatments are justified. In other words, any treatment is good as long as it is not harmful. Try to apply this thinking to such unproven remedies as laetrile.

5 Is it ever ethically permissible to give up on treatment? Are there things other than treatments that can be done for a dying person? Do you think people who have terminal cancer should be told of their condition? Try to justify your answer. What ethical assumptions underlie your argument?

Concluding Remarks

Research directed toward understanding the basic biology of cancer continues. Some people feel that treatment cannot be rationally designed unless the underlying biology is known, while others feel that such basic understanding is not important. The former group point to the development of new treatments such as tamoxifen and Herceptin. The latter group use arguments such as the following: we still do not know the basic biology underlying the disease polio, but development of a vaccine for its prevention has eliminated our need to know. The real dilemma is a problem in the allocation of resources. How much money should we spend on treating cancer patients, how much on improving methods of treatment, how much on laboratory research to discover more information on the causes of cancer, and how much on cancer prevention activities? How much funding should be aimed at particular types of cancers, such as breast cancer as compared with colon cancer? There are no clear answers here because we cannot accurately predict how well or how soon funds spent on certain activities (especially research) will translate into a reduction of cancer incidence rates or cancer deaths. Cancer prevention is clearly very cost-effective, but the cost-effectiveness of the other alternatives may be very difficult to assess.

On an individual level, we can reduce our exposure to cancer risk factors by many choices that we make in our lives. However, not all types of exposure among those listed above are matters of personal choice. Dumping of carcinogens on the land and water of poor people with little or no political power has become a global environmental issue. Air pollution affects people at great distances from the source. Therefore, as part of any effort to prevent cancer, people need to work together to prevent or remove environmental hazards from work places and communities.

Chapter Summary

- All organisms are built of **cells** that maintain an efficient ratio of surface area to volume for the organism.

- Normal cells occasionally enter the **cell cycle** and divide. They always stop dividing when enough cells are present because of such phenomena as contact inhibition and other processes that maintain **homeostasis** of cell number.

- Cells also **differentiate** as they divide, becoming more and more fully **determined**, that is, restricted in their **potentiality** to form different kinds of cells that in most species are organized into **tissues**.

- **Stem cells** retain the ability to differentiate into many different kinds of cells.

- **Cloning** can begin with stem cells, increasing them in number for therapeutic purposes. Cloning for reproductive purposes begins with the insertion of a diploid nucleus into an enucleated egg to reproduce a new individual asexually.

- Both cell division and cell differentiation result from differential **gene expression**, and both are influenced by molecular signals (**cytokines**) secreted by other cells. Cytokines are bound by receptors and this information is then transferred to the cell nucleus by **second messengers**. Cytokines that trigger cell division are called **growth factors**.

- Cancer cells are cells that have undergone **transformation** and differ from normal cells in having abnormal responses to growth control signals—they do not stop dividing (they are 'immortal'), and they remain less differentiated.

- A few **cancers** have genetic predispositions, but most are caused by environmental factors. These factors include exposures to certain viruses and infective agents, dietary and other behavioral factors, and exposure to a long list of **carcinogens** including ionizing radiation (radioactivity), ultraviolet radiation, tobacco smoke, and a variety of chemicals.

- Mutagens damage DNA, and other chemicals can then increase the risk that these mutated cells will develop into cancer. Mutagens and tumor promoters have synergistic effects; that is, the increased risk of exposure to both is greater than the sum of the risks of the two separately.

- Cancerous **tumors** can be removed surgically, but other forms of cancer therapy target any cells that are dividing, destroying many healthy cells along with the cancer. New, more specific chemotherapies are being developed, including therapies based on boosting the body's immune system.

- We can best reduce our risks for cancer by avoiding tobacco smoke and other carcinogens (including ultraviolet and ionizing radiations) and by eating a diet low in fats, high in fiber, and rich in beta-carotene and vitamins A and C.

CONNECTIONS TO OTHER CHAPTERS

Chapter 2	Cancers are caused by mutated growth-control genes or by DNA damage during chromosome crossovers.
Chapter 3	Cancers can result from changes that bring normal genes to abnormal locations in the genome.
Chapter 7	Several cancers have different incidence rates in different human populations.
Chapter 10	High-fat diets with low fiber content increase the risks for several cancers.
Chapter 11	Many cancer-fighting drugs are plant products.
Chapter 15	Good mental outlook and immunological health can improve cancer survival rates.
Chapter 16	Certain otherwise rare cancers occur more frequently in AIDS patients.
Chapter 18	Rainforest plants have been sources for new cancer-fighting drugs.
Chapter 19	Pollution increases the incidence rates of several forms of cancer.

PRACTICE QUESTIONS

1. What are the four phases of the cell cycle and what happens in the cell during each phase? Is G_0 part of the cell cycle?

2. In which cells is cell division inhibited by contact with other cells: normal cells or cancer cells?

3. Compartmentalizing an organism by dividing it into cells increases which of the following: the volume or the surface area?

4. Which of the following develop by differentiation from the zygote: tracheal cells in the lungs, muscle cells, or cells of the eye? Which develop from the gastrula? Which develop from the endoderm?

5. How long does an average human tongue cell live? How long do human liver cells live on average? Human nerve cells?

6. When the telomere region of the chromosomes becomes too short, which of the following happens?

    a. The cell can no longer divide.

    b. The cell becomes cancerous.

    c. The cell can no longer differentiate.

7. Do the protein products of proto-oncogenes induce cells to divide or do they prevent cells from dividing? What about the protein products of tumor suppressor genes?

8. How many cells need to be transformed for cancer to develop? Does every transformed cell result in cancer? Why or why not?

9. Which of the following causes more cases of cancer: heredity, smoking, or viruses?

10. What processes are induced in a cell by binding of a growth factor to its receptor? Does a growth factor induce these processes in every cell?

11. Are the same genes transcribed and translated in every cell of the body?

12. Does cellular differentiation take place in adult animals or only in embryos?

The Nervous System and Senses

Issues

- What happens when the nervous system malfunctions?
- Can mental illness be understood and treated biochemically?
- How can changes in brain chemistry interfere with message transmission?
- How can many dissimilar malfunctions have similar causes?

Biological Concepts

- Evolution (comparative anatomy, specialization and adaptation, form and function)
- Hierarchy of organization (ions, neurons, brain, behavior)
- Matter and energy (membranes, ion potentials, action potentials)
- Detection of environmental stimuli (receptors, neurotransmitters, sense organs)
- Movement (muscular system)
- Homeostasis (feedback mechanisms, nervous system)
- Health and disease
- Behavior (sleep, learning)

Chapter Outline

The Nervous System Carries Messages Throughout the Body

The nervous system and neurons

Nerve impulses: how messages travel along neurons

Neurotransmitters: how messages travel between neurons

Dopamine pathways in the brain: Parkinsonism and Huntington's disease

Messages Are Routed to and from the Brain

Message input: sense organs

Message processing in the brain

Message output: muscle contraction

The Brain Stores and Rehearses Messages

Learning: storing brain activity

Memory formation and consolidation

Alzheimer's disease: a lack of acetylcholine

Biological rhythms: time-of-day messages

Dreams: practice in sending messages

Mental illness and neurotransmitters in the brain

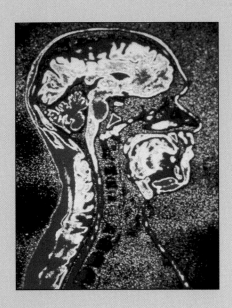

13 The Nervous System and Senses

A man in his sixties finds it difficult to walk and shuffles his feet along the floor, just an inch or two at a time. Last week he tried to walk by lifting his feet, but he stumbled after only two quick but awkward steps. A woman in her seventies used to work crossword puzzles and solve math problems with ease, but she has forgotten how to do these tasks. Her friends, whom she no longer recognizes, say that she is not the same person that they knew a few years ago. A younger woman has lost all motivation to work, to keep herself clean, or just to go on living. "What for?" she asks, "none of it matters." These three people have three very different diseases: Parkinsonism, Alzheimer's disease, and depression. Could it be that all their assorted symptoms are caused by chemical imbalances in the brain?

The brain coordinates the activities of the rest of the body, including moving, sleeping, eating, and breathing. Changes in the brain can thus produce diseases such as Parkinsonism, whose symptoms include uncoordinated movements.

Does the brain also produce activities that we associate with the mind? Does biochemical activity in the brain produce perceptions, emotions, moods, and personality? Does it produce thoughts, dreams, and hopes? A branch of biology called neurobiology is the scientific study of the brain and nervous system. A central theory of neurobiology is that the mind and the brain are one and the same. As a framework that guides (and limits) research, the assumption that mind equals brain is a good example of a research paradigm (see Chapter 1). There is probably no theory in biology that is more controversial, both among biologists and between biologists and the public.

There is now considerable evidence in support of this theory. Electrical and biochemical activities have been measured in the brain during dreams and thought. Some diseases, such as Alzheimer's disease, in which there is brain degeneration, are accompanied by changes in personality. Mental illnesses such as depression are associated with changes in brain chemistry and can in many cases be treated with drugs.

In this chapter we examine the workings of the brain and the nervous system and consider the extent to which the mind is another name for the brain. The nervous system is adapted for sending and receiving messages. You will learn how these messages travel electrically along cell membranes and are carried from cell to cell by chemicals called neurotransmitters. You will also learn how the brain is organized, how sense organs handle incoming messages, how muscles react by contracting, how the brain processes and stores messages, and how mental processes in both health and disease derive from activity at the chemical and cellular levels.

The Nervous System Carries Messages Throughout the Body

Central to the science of neurobiology is the theory that all functions and dysfunctions of the nervous system, including mental activities and mental illnesses, result from the actions and interactions of cells and chemicals of the nervous system. The diseases described in this chapter all involve malfunctions related to chemicals in the brain; several are also characterized by the degeneration of particular groups of cells.

In 1817, Dr. James Parkinson first described a disease marked by muscle tremors, including a distinctive 'pill-rolling' movement of the thumb and forefinger. The disease is also characterized by a walking gait in which the body above the waist leans forward while the feet shuffle slowly in small steps and are barely lifted from the ground. People with Parkinsonism have difficulty in initiating voluntary movements. This difficulty is not a true paralysis because, by stopping other activities and concentrating on their voluntary movements, Parkinson patients can temporarily improve their performance.

For voluntary movements of any kind to take place, the brain must initiate and send messages to the muscles. Parkinsonism presents some of the clearest evidence of a malfunction in this message-sending system. However, before we can understand this malfunction in greater detail, we must first describe the cells of the nervous system and how the sending of messages usually works.

CONNECTIONS CHAPTER 1

The nervous system and neurons

Nearly all animals (and only animals) have nerve cells arranged in some type of nervous system. The simplest nervous systems are netlike arrangements of interconnected cells with no control center; hydras and other cnidarians have such nerve nets. A bit more organized are the nervous systems of flatworms, which have a concentration of sense organs (eyes, hearing organs) at the front end of the animal. Coordinating the input from these sense organs are a pair of enlarged control centers called 'cerebral ganglia,' the beginnings of a simple brain. Many worms, especially parasitic worms, have greatly simplified nervous systems; that of the roundworm *Caenorhabditis elegans* has fewer than two dozen individual nerve cells. More complex animals need more of a nervous system to coordinate the movements of legs and other appendages, and those with more elaborate sense organs need a larger central nervous system to receive and interpret messages from these sense organs. Especially among mollusks and arthropods, animals with better sensory capabilities and locomotor skills have larger brains to act as centers of coordination and control.

The largest brains are found among the vertebrates or backboned animals of the phylum Chordata. In these animals, the brain continues rearward into a long spinal cord, and brain and spinal cord together make up the central nervous system or CNS. The rest of the nervous system is called the peripheral nervous system, and it includes the remainder of the nerves (called 'peripheral nerves') and the special sense organs such as the eyes and ears (Figure 13.1).

Figure 13.1

Major divisions of the nervous system.

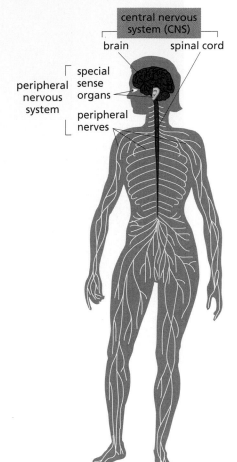

The functions of the nervous system, including those of the brain, are carried out largely by nerve cells, also called **neurons**, cells that carry nerve impulses along their membrane surfaces. Each neuron contains a cell body that includes a nucleus and surrounding cytoplasm. Extending out from this cell body are the branches called dendrites, which conduct nerve impulses toward the cell body. Each neuron also has another extension, called an **axon**, that conducts impulses away from the cell body. Some neurons are oriented so that their axons conduct impulses from a sense organ inward toward the central nervous system. Other neurons conduct impulses away from the central nervous system to end either at a muscle or a gland (Figure 13.2A). Nerve impulses pass from neuron to neuron (or from neuron to muscle) across a gap known as a **synapse**.

Many, but not all, axons are surrounded by a series of special cells whose rolled-up plasma membranes form a structure called the **myelin sheath** (Figure 13.2B and C). The myelinated axon looks a bit like a string of sausages because the sheath is thinner where the adjacent cells of myelin sheath meet one another. The myelin sheath acts as an insulator that prevents nerve impulses from spreading sideways from one axon to another, and it keeps the impulses traveling along the length of the axon. When myelin is disrupted by disease, transmission of nerve impulses becomes disordered. In the disease called multiple sclerosis (MS), the myelin sheath is destroyed by cells of the body's own immune system. The resulting disturbance of nerve conduction results in weakness or trembling of arms or legs and in hazy or double vision, among other symptoms.

Aggregations of neurons or their parts have distinctive names in the nervous system. Bundles of axons are called **nerves** throughout the peripheral nervous system and tracts within the brain and spinal cord. Clumps of cell bodies are called ganglia throughout the peripheral nervous system and nuclei within the central nervous system, except for one group of brain structures called the basal ganglia.

In addition to neurons, the nervous system contains other types of cells (called neuroglia). In the brain, these cells have cellular extensions that wrap around the neurons in the brain and other cellular extensions that wrap around small blood vessels on the surface of the brain. Through these extensions, these cells carry nourishment from the small blood vessels to the neurons. In addition to nutritive functions, the neuroglia also provide structural support and electrical insulation for the brain tissue.

Nerve impulses: how messages travel along neurons

Much of what we know about nerve impulses was learned from studying the giant axons of the squid (Figure 13.3A), a member of a group called the Cephalopoda (which includes the octopus and chambered nautilus), which is a class of the phylum Mollusca (see Figure 5.7, p. 135). The animals of this class need to react both quickly and forcefully to stimuli that could signal danger; they all have giant axons as part of their quick-response system. These axons are many times the diameter of typical axons in humans and so are ideal for studying nerve impulses, as we will see below.

Electrical potentials. One way to study nerve impulses within a single neuron is by using a very sensitive voltmeter attached to needlelike probes (electrodes) that conduct electricity. Voltmeters read differences in the amount of electric charge in contact with their two electrodes. A giant axon has a large enough diameter that one of the electrodes can be placed inside the axon and the other electrode outside (Figure 13.3B). When placed this way, voltmeter readings show a difference in charge between the inside and outside of the axon, a difference called an electrical potential, measured in units called millivolts (mV). Unlike electrical potentials in nonliving systems (such as electrical wiring), which are due to electrons, electrical potentials in living systems are due to differences in the concentration of charged particles called ions. Electrical potentials

Figure 13.2

The neuron and its myelin sheath.

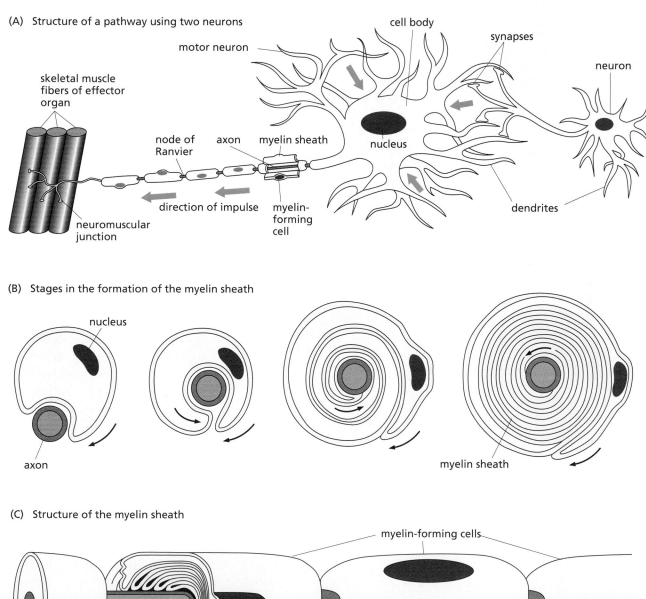

(A) Structure of a pathway using two neurons

(B) Stages in the formation of the myelin sheath

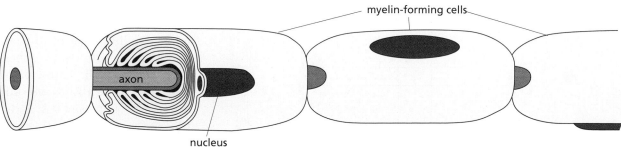

(C) Structure of the myelin sheath

in the nervous system result from differences in concentration of sodium ions (Na^+) and potassium ions (K^+) inside and outside the neuron. There are two chemical gradients: more sodium ions outside the cell than inside and more potassium ions inside than outside (Figure 13.3C). There is also an electrical potential because there are more positive charges outside the cell than inside. The excess of positive charge outside the cell can be measured as an electrical potential of about –70 mV across the cell membrane of the neuron, meaning that the inside of the cell is negatively charged in comparison with the outside. (The magnitude of this potential is different in different types of neurons.) When a neuron is not conducting an impulse, the electrical potential across its cell membrane remains fairly constant and is called a **resting potential** (Figure 13.3B and C).

Measuring nerve impulses. We say that a membrane is polarized when there is an electrical potential across it. When the electrical potential decreases, we say it has depolarized. A nerve impulse is a wave of depolarization traveling along the cell membrane. The easiest way to demonstrate this is by using a voltmeter with electrodes applied to two points along the outside of the axon. The voltmeter compares the charge at the second site outside the axon with the charge at the first site. When the neuron is not conducting an impulse, the concentration of positive ions is the same at the two sites so the two electrodes detect the same charge, and the voltmeter needle reads zero. When the first site depolarizes because some positive ions move inside the axon, the concentration of positive ions, and thus the charge, is less at the first electrode than at the second and the voltmeter reads negative. Later, successive other sites down the axon depolarize, while the concentration of positive ions at the first site is restored. When the concentration of positive ions, and thus the charge, is higher at the first electrode than at the second, the voltmeter reads positive. As a nerve impulse passes, the needle deflects first one way and then the other, as shown in Figure 13.4.

Along the length of an axon, the nerve impulse travels at a relatively rapid rate that increases with the diameter of the axon and is faster with the presence of a myelin sheath. Unmyelinated neurons generally have small diameters, between 0.3 and 1.3 micrometers (1 μm = 10^{-6} m) and conduction velocities of 0.5–2.3

Figure 13.3

Resting potentials in the giant axons of the squid.

(A) Location of the giant axons in the squid (*Loligo*)

(B) Measurement of the resting potential in an axon

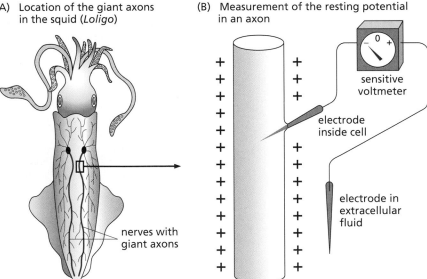

sensitive voltmeter

electrode inside cell

electrode in extracellular fluid

nerves with giant axons

(C) The distribution of Na^+ and K^+ ions responsible for the resting potential

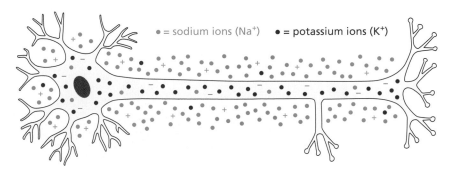

● = sodium ions (Na^+)   ● = potassium ions (K^+)

meters per second (abbreviated m/sec). Myelinated neurons have larger diameters (3–20 μm) and faster conduction velocities, varying from 3 m/sec in the smallest fibers to 120 m/sec in the largest. The giant squid axons, so named because of their large diameters, have a very high conduction velocity.

Neurons are stimulated to depolarize in a small portion of their cell membranes; however, not every localized depolarization leads to a sustained impulse. Depolarizations that are strong enough trigger a nerve impulse that travels the length of the axon. We now look at the propagation of a sustained nerve impulse in more detail.

Action potentials. Recall that at its resting potential a neuron has an excess of sodium ions outside the cell and an excess of potassium ions inside (see Figure 13.3). Such neurons are electrically excitable; that is, they can be stimulated by changes in electrical charge. When a nerve cell membrane is sufficiently stimulated, channels through the membrane open and let sodium ions flow in, locally depolarizing the membrane for a short time. This can be measured as a change in voltage by placing electrodes as in Figure 13.3, and voltage changes over time can be recorded as graphs. A slight depolarization of the membrane is transmitted a very short distance to adjacent portions of the cell membrane; these effects decrease rapidly with distance and may thus disappear quickly. However, a greater depolarization that reaches or exceeds some **threshold** triggers nearby membrane channels to open. The rapid inflow of sodium ions results in a characteristic type of electrical discharge known as an **action potential**. The action potential appears as a 'spike' on a graph of

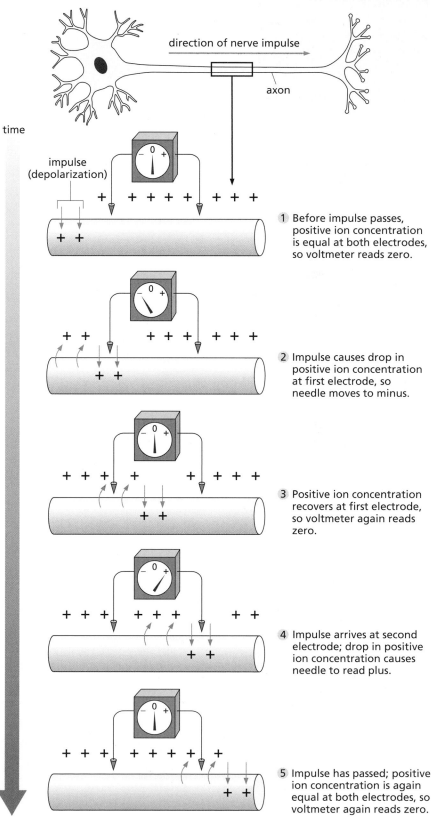

Figure 13.4

Detection of a nerve impulse by using a sensitive voltmeter.

direction of nerve impulse

axon

time

impulse (depolarization)

1 Before impulse passes, positive ion concentration is equal at both electrodes, so voltmeter reads zero.

2 Impulse causes drop in positive ion concentration at first electrode, so needle moves to minus.

3 Positive ion concentration recovers at first electrode, so voltmeter again reads zero.

4 Impulse arrives at second electrode; drop in positive ion concentration causes needle to read plus.

5 Impulse has passed; positive ion concentration is again equal at both electrodes, so voltmeter again reads zero.

Figure 13.5

Generation of an action potential in a nerve cell. The action potential corresponds to steps 2 through 4 and extends from one resting potential to the next.

potential versus time, as shown in Figure 13.5. From a resting potential of −70 mV (step 1), a portion of the axon membrane depolarizes owing to the opening of a small number of sodium channels (step 2). If the

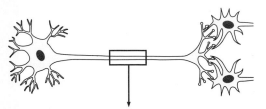

time

① Resting potential when no impulse is present

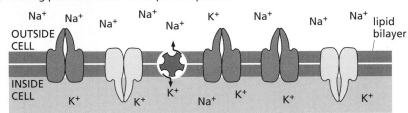

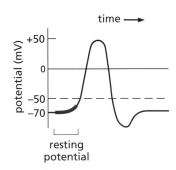

② Stimulus causes a few channels to open; sodium ions leak in, depolarizing the membrane

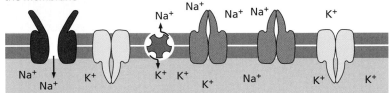

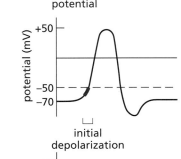

③ At −50 mV, all sodium channels open (*bright red*); sodium rushes in, causing charge to reverse and form an action potential

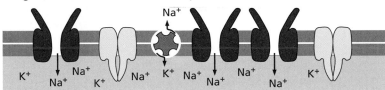

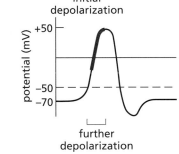

④ Sodium channels close as potassium channels open (*solid green*); potassium ions rush out, repolarizing the membrane

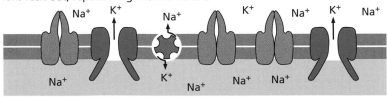

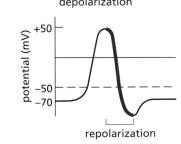

⑤ Closing of potassium channels restores resting potential; sodium–potassium pump (*purple*) redistributes ions

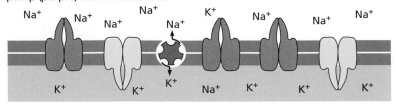

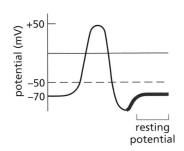

depolarization passes the threshold value of –50 mV (for this neuron), more sodium channels open, inducing a depolarization to zero, then a reversal of charge to +50 mV (step 3). Depolarization ends as the sodium channels close; the membrane becomes repolarized by the opening of other channels, the potassium channels, and by the outward flow of potassium ions (step 4). The electrical potential becomes negative once again, and returns to resting potential as the potassium channels close. Because there are now more sodium ions inside the axon than is normal and more potassium ions outside, the original ion distribution must be restored. This is accomplished by **sodium–potassium pumps**, a series of membrane proteins engaged in the active transport of sodium ions and potassium ions in opposite directions across the cell membrane. (Active transport and ion gradients are described in Chapter 10.) For every two potassium ions transported into the neuron, three sodium ions are transported across the membrane to the outside. Because each sodium ion and each potassium ion carry a single positive charge, this also moves three positive charges out of the cell for every two that are moved in. Active transport thus restores the sodium and potassium concentration gradients and additionally contributes a few millivolts to the resting potential (step 5).

The action potential spike in one location causes the adjacent portion of the cell membrane to depolarize, repeating step 2 and producing another action potential equal in strength to the first. In this way, an action potential rapidly spreads along the entire neuron membrane with no reduction in its size or intensity, with each spike being the trigger for the next depolarization. It is this traveling of successive action potentials along the neuron that we recognize as a nerve impulse. Action potentials maintain their direction of travel because, once sodium channels have opened and then shut, they are prevented from reopening for a short period.

Neurotransmitters: how messages travel between neurons

Even though neurons are electrically excitable, the transfer of information from cell to cell (across a synapse) is generally chemical, not electrical. Any chemical substance that can stimulate or inhibit an action potential is called a **neurotransmitter**. The vast majority of synapses between nerve cells are chemical synapses, in which a chemical neurotransmitter crosses the narrow space separating two adjacent cells. In a chemical synapse, depolarization of the membrane in the presynaptic or transmitting neuron (Figure 13.6A) causes it to release neurotransmitter molecules (Figure 13.6B). The neurotransmitter released into the synapse binds to receptors on the next cell (the postsynaptic cell), opening channels in its membrane. Ions flow through these channels, thus altering the electrical potential of the postsynaptic or receiving cell (Figure 13.6C). The neurotransmitter thus carries the impulse across the synapse but never enters the postsynaptic cell (Figure 13.6D).

Experiments demonstrating neurotransmitters. The concept of a chemical neurotransmitter was first suggested by the British physiologist Henry H. Dale but was first demonstrated by his German-American colleague Otto Loewi, with whom Dale shared the Nobel Prize in 1936. Dale and Loewi already knew that the vagus nerve, running down from the brain into the abdominal cavity, sends branches to the heart that slow down the heartbeat. They also knew that the heart of a frog separated from the rest of the animal and maintained in a physiological salt solution (a solution containing various ions at concentrations close to those in the intact organism) would keep beating for hours. Loewi dissected out the beating hearts of two frogs and placed them in salt solutions. One of his preparations contained nothing but the heart, but the other contained a carefully preserved vagus nerve. When Loewi stimulated the vagus nerve, the heart connected to this nerve slowed down. Loewi then used a dropper to take some of the fluid surrounding this heart and transfer it to the other heart, which no longer had any nerves leading to it. The second heart also slowed down, showing that some chemical had been present in the first preparation that could be transferred by dropper to the second, for this was the only connection between the two. The chemical was then isolated and identified as acetylcholine, the first neurotransmitter to be studied experimentally. Loewi's preparation also allowed the testing of various drugs or other substances that could block or otherwise modify the effect of acetylcholine. These drugs, in turn, could often be used to study whether a particular synapse used acetylcholine as a neurotransmitter.

Types of neurotransmitters. A large number of chemicals are now known to be able to function as neurotransmitters; some of these are listed in Table 13.1, grouped by their chemical structure. Most neurons in the peripheral nervous system use the neurotransmitter acetylcholine; a few use norepinephrine. But neurons in the brain use many different neurotransmitters, including all the ones in Table 13.1. Each particular neuron secretes one type of neurotransmitter primarily or exclusively. Each cell that responds to a neurotransmitter has a specific receptor for that neurotransmitter.

Removal of neurotransmitters from the synapse. After a neurotransmitter evokes its response, there are mechanisms that stop further

Figure 13.6

How a chemical synapse works.

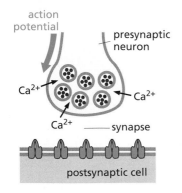

(A) Depolarization by an action potential triggers entry of calcium ions into the cytoplasm of the presynaptic neuron.

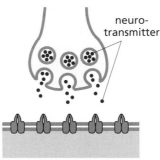

(B) The calcium triggers the cytoplasmic vesicles containing neurotransmitters to release the neurotransmitters into the synapse.

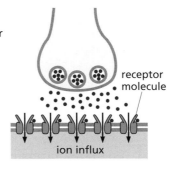

(C) The neurotransmitter binds to receptor molecules on the postsynaptic cell, opening channels that allow positive ions to enter that cell's cytoplasm, depolarizing its membrane.

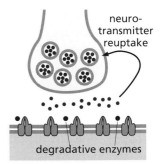

(D) After a brief time, the neurotransmitter is either taken back into the cell that secreted it, or else it is degraded by an enzyme.

transmission by removing loose (unbound) neurotransmitter molecules from the synapse (see Figure 13.6D). Many neurotransmitters are reabsorbed by the cell that secreted them, permitting the same molecules to be recycled and reused. This process, called reuptake, is typical of many synapses in the brain. Also, several neurotransmitters can be chemically degraded by enzymes. The amine neurotransmitters can be degraded by the enzyme MAO (monoamine oxidase). Another enzyme, cholinesterase, breaks down acetylcholine molecules in synapses outside the brain. Any process interfering with the chemical breakdown or reuptake of neurotransmitters lets the neurotransmitter stay in the synapse and thus excessively stimulate the postsynaptic cell; likewise, any enhancement of chemical breakdown or reuptake may decrease neurotransmission. As we shall see, changes in neurotransmission can result in disease.

Dopamine pathways in the brain: Parkinsonism and Huntington's disease

One of the amine neurotransmitters (see Table 13.1) is dopamine. Dopamine transmits messages across synapses between neurons in the brain. Some of the neurons that are stimulated by dopamine act, in turn, on peripheral neurons that stimulate voluntary muscle cells. Dopamine does not act directly on muscle cells; rather it acts within the brain to smooth and coordinate signals to the muscles. Two diseases have helped us to understand what can happen when dopamine secretion is out of balance. In Parkinsonism there is too little dopamine and in Huntington's disease there is too much.

Parkinsonism: too little dopamine. When autopsies are performed on the brains of Parkinson patients, a consistent finding is the degeneration of a bundle of darkly pigmented neurons (called the 'substantia nigra'). Experimental staining shows that these neurons secrete the neurotransmitter dopamine. Among the cells stimulated by this dopamine are those of the basal ganglia. The degeneration of these neurons in Parkinson patients thus deprives the basal ganglia of dopamine. The neurons of the basal ganglia stimulate and coordinate muscle movements by acting on the acetylcholine-secreting neurons that trigger muscle contraction. If sufficient dopamine is not present, the cells of the basal ganglia do not function normally, and the person suffers from muscle tremors and has difficulty walking.

Evidence for the hypothesis that dopamine underproduction has a large role in Parkinsonism comes from the effectiveness of the drug L-DOPA (levodopa) in temporarily alleviating many of the symptoms of Parkinsonism. Because DOPA is a precursor of dopamine, it is hypothesized that supplying L-DOPA (a synthetic form of DOPA) can increase dopamine production and thus relieve the symptoms of dopamine deficiency. Unfortunately, this form of treatment increases dopamine production everywhere, not just to the portion of the brain that needs it. The excess of dopamine in other places may cause serious side effects, such as schizophrenialike symptoms.

One promising form of therapy that has already been tested is the implantation of fetal tissue into the brains of Parkinson patients. The

Table 13.1

Neurotransmitters.

AMINE NEUROTRANSMITTERS
Epinephrine (also called adrenalin)
Norepinephrine (also called noradrenaline)
Dopamine
Serotonin
Histamine

AMINO ACID NEUROTRANSMITTERS
Aspartic acid (aspartate)
Glutamic acid (glutamate)
Gamma-aminobutyric acid (GABA)
Glycine

PROTEIN OR PEPTIDE NEUROTRANSMITTERS (NEUROPEPTIDES)
Somatostatin
α-endorphin
β-endorphin
Leu-enkephalin
Met-enkephalin
Substance P

GAS NEUROTRANSMITTER
Nitric oxide (NO)

ESTER NEUROTRANSMITTER
Acetylcholine

hypothesis is that the fetal tissue will grow and replace the missing or damaged dopamine-secreting cells. Adult tissue is unsuitable for this purpose because the brain loses much of its capacity to regenerate new neurons at an early age; even tissue from newborn babies or infants has much less regenerative capacity than fetal tissue. This is one reason why many people in the medical community welcomed the lifting in 1993 of an earlier ban on the use of fetal tissues in medical therapy and in medical research in the United States. However, the use of fetal tissue raises objections from opponents of abortion because the tissue is obtained in most cases from aborted fetuses (see Chapters 9 and 12).

Huntington's disease: too much dopamine. If Parkinsonism is a disease in which voluntary movements are made difficult by a lack of dopamine, then too much dopamine might cause excessive movements. This is precisely what happens in Huntington's disease, a degenerative neurological disease whose genetic basis was discussed in Chapter 3 (p. 78). Huntington's disease is marked by uncontrollable spasms or twitches of many muscles, usually beginning between ages 40 and 50. As the disease progresses, the spasms become more pronounced, and the patient gradually loses control of all motor functions and of mental processes. A slow death occurs within a few years of the disease's onset. Autopsies of Huntington patients reveal that some of the brain cells in the basal ganglia have been destroyed. Because these cells normally inhibit the production and release of dopamine, one major effect of their destruction (and one sign of the disease) is an overproduction of dopamine. Drugs known to inhibit the action of dopamine can temporarily reduce the symptoms of Huntington's disease.

Feedback systems. How does the destruction of some of the cells in the basal ganglia result in an overproduction of dopamine? One answer may lie in another neurotransmitter, gamma-aminobutyric acid (GABA). GABA functions in many neuronal pathways by inhibiting a neuron that has fired from firing again unless another stimulus, larger than the first, is received. Because the GABA-secreting neuron inhibits a neuron at an earlier point in the pathway, the information is 'feeding back' and is thus described as a **feedback system**. Feedback occurs in many biological systems whenever a later step regulates an earlier step in any process. One common type of feedback is a feedback inhibition (or 'negative' feedback) in which a later step inhibits an earlier step. Feedback systems of this kind function to keep certain variables such as body temperature within narrow limits, turning metabolic processes higher when the body is cold or taking measures (such as sweating) to dissipate heat when the body is warm. We saw in Chapter 9 (p. 300) that levels of certain hormones are regulated by feedback from other hormones. In the nervous system, feedback can often be seen literally as neurons that feed information back to the earlier neurons that stimulated them.

GABA normally has this type of feedback effect on the dopamine-secreting neurons mentioned earlier. Because it acts to inhibit dopamine, it is an example of feedback inhibition. The destruction of the GABA-secreting neurons in Huntington's disease removes this inhibition. Because the dopamine-secreting neurons are then no longer

CONNECTIONS
CHAPTERS 3, 9, 12

inhibited, they begin to overproduce dopamine (Figure 13.7). Too much dopamine leads to overstimulation of other GABA-secreting neurons, ones that trigger the acetylcholine-secreting neurons, overproducing muscle contractions.

Figure 13.7

Feedback inhibition usually prevents dopamine overproduction, but impairment of the feedback pathway in Huntington's disease allows dopamine overproduction.

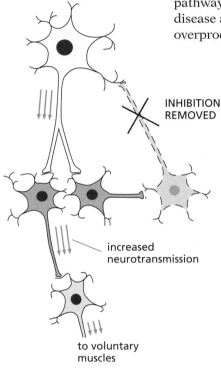

SUBSTANTIA NIGRA

BASAL GANGLIA

NORMAL PATHWAY WITH FEEDBACK

dopamine-secreting neurons

direction of nerve impulse

FEEDBACK INHIBITION

GABA-secreting neurons

acetylcholine-secreting neurons

to voluntary muscles

PATHWAY MODIFIED BY HUNTINGTON'S DISEASE

INHIBITION REMOVED

increased neurotransmission

to voluntary muscles

THOUGHT QUESTIONS

1 To test the clinical benefits of a technique such as the surgical implantation of fetal cells into the brains of Parkinson patients, a double-blind comparison needs to be made with a group of control subjects. The control subjects in such a study are generally given sham surgery, meaning that a hole is drilled in their skull and a probe inserted into their brains, but no fetal cells are introduced. Because of a strong placebo effect, patients receiving this treatment often improve at least temporarily. Do you think it is ethical to conduct tests in this way? Do you think it is ethical to adopt such a procedure without this kind of clinical testing? Construct both utilitarian and deontological arguments for your position.

2 Suppose a scientist proposes the hypothesis that, in a select group of hard-to-reach cells, a particular amino acid functions as a neurotransmitter. How would you test this hypothesis? What kinds of drugs or poisons would you look for to help you study the properties of the hypothesized neurotransmitter?

3 Explain how Loewi's experiment demonstrated that neurotransmitters are chemicals, not electric current.

4 What are the arguments for and against the use of fetal tissue for the treatment of Parkinsonism?

5 If a neuron is stimulated halfway along its axon, will an impulse travel in both directions on the axon?

Messages Are Routed to and from the Brain

Some types of neuron messages are internal. As seen in the feedback system illustrated in Figure 13.7, neurons can carry information about the internal workings of the organism. They can also regulate internal processes, both within the brain (see pp. 481–485) and in other parts of the body (see Chapter 15). Many other messages processed by the brain originate outside the organism. The nervous system is one of the major body systems through which the individual receives information about its environment. External stimuli cause specialized cells to trigger depolarizations (changes in membrane potentials) that travel to other neurons by means of neurotransmitters, and eventually to the brain, where huge numbers of incoming messages are processed. Some, but not all, of these messages become conscious perceptions.

Message input: sense organs

The brain receives sensory input from the outside world through a variety of sense organs, all containing many specialized sensory cells. Different types of sensory cells respond to different types of stimuli. Although sensory reception is a function of these sense organs, sensory perception, which involves interpretation, is largely a function of the brain.

The skin: reception of touch, temperature and pain. The main functions of skin, our body's largest organ, are protective: it protects us from temperature variations, harmful chemicals, water loss, and pathogens that might cause disease. Our skin also protects us by sending messages to the rest of the body from its various sensory nerve endings. Five types of specialized sensory endings are known, corresponding to five different skin senses: light touch, deep pressure, warmth, cold, and pain (Figure 13.8). When these are stimulated, their axons depolarize. The altered

Figure 13.8

Nerve endings in the skin.

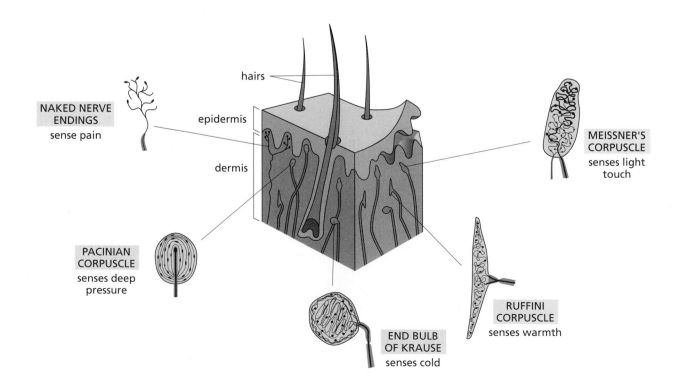

membrane potential produced in sensory cells is called a generator potential. Unlike an action potential, which operates on an all-or-none principle, the magnitude of a generator potential depends on the strength of the stimulus.

The different sensory receptors can be distinguished using a microscope but are much too small to be stimulated individually in an experiment. So how do we know which type of ending serves which function? Our knowledge is largely based on the spatial distribution of the endings over the body's surface. For example, the places most sensitive to touch (such as the lips and the fingertips) have the highest densities of the endings thought to sense light touch, while the places most sensitive to cold (such as the cornea of the eye or the tip of the penis) have the highest densities of the nerve endings thought to sense cold.

Several of these sensory receptors can report anomalous sensations when they are overstimulated. For example, temperatures a few degrees lower than the surroundings are perceived as cold, but much colder temperatures may be perceived as heat or burning, and temperatures still lower are simply perceived as pain. In fact, a sufficiently large stimulus of any sort, even a sound much too loud or a light much too bright, is perceived by the brain as pain. The nerve endings that are specifically pain receptors are naked nerve endings, which are easily overstimulated to produce sensations of pain. In contrast, the deep pressure receptors are surrounded by onionlike layers, ensuring that the nerve ending within can only be stimulated to produce changes in membrane potential if there is enough pressure to deform the shape of these layered capsules deep inside the skin.

Our sense of touch requires direct pressure on the skin. Fishes have an additional sense organ for touch (called the 'lateral line system') that is sensitive to very small changes in water pressure caused by nearby obstacles or by the swimming movements of other fish (Figure 13.9). Some fishes also have sense organs that can detect electric currents. At least some of these species also produce electric current and use these currents as a means of communicating with each other.

The eye: reception of light. The organ of vision in all vertebrates is the eye (Figure 13.10A), which is surrounded by two outer protective layers, the choroid and the sclera. Light reaches the interior of the eye through a transparent front layer, the cornea. The light is concentrated by a transparent lens, which changes shape to bring into focus objects at varying distances from the eye. A suspensory ligament holds the lens in place, while a series of smooth muscles change the shape of the

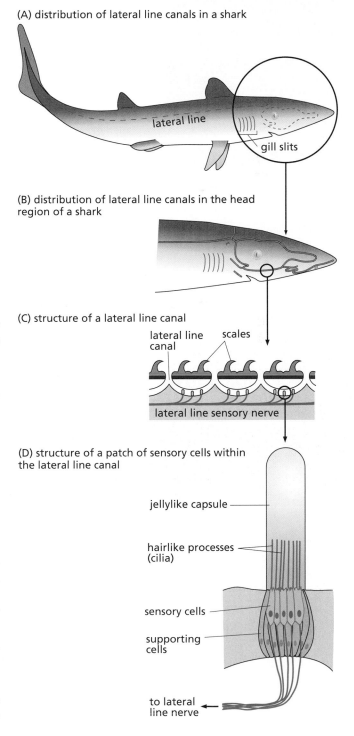

Figure 13.9

The lateral line system of fishes.

(A) distribution of lateral line canals in a shark

lateral line

gill slits

(B) distribution of lateral line canals in the head region of a shark

(C) structure of a lateral line canal

lateral line canal   scales

lateral line sensory nerve

(D) structure of a patch of sensory cells within the lateral line canal

jellylike capsule

hairlike processes (cilia)

sensory cells

supporting cells

to lateral line nerve

Figure 13.10

The human eye.

(A) Anatomy of the eye

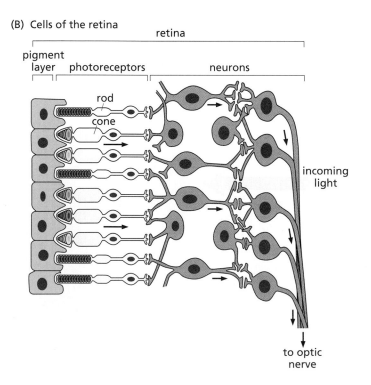

(B) Cells of the retina

lens. The iris diaphragm, which gives each of us our individual eye color, controls the amount of light that enters the lens through the pupil, an opening that enlarges in response to dim light and becomes smaller in response to bright light.

The lens focuses light on the retina, the sensory sheet along the rear of the eye that responds to light. Because the retina is too far from the lens in a nearsighted person, the lens of a nearsighted person is not able to focus light on the retina. In a farsighted person, the opposite is true—the lens is too close to the retina. For both nearsighted and farsighted people, glasses or contact lenses bring the focal point of the light to the retina. While a person is young the lens itself can change shape to accommodate for the vision of either near or distant objects. However, as a person ages, the lens becomes less flexible and the range of accommodation diminishes. For this reason, many people who had good vision earlier in life may need to wear glasses as they age.

The retina contains photoreceptor (light-receiving) cells called rods and cones, and light that falls on the retina causes changes in the chemical structure of pigment molecules in these cells. These pigment changes alter the membrane potential in the rods and cones, inducing impulses. The impulses travel along the retina to the optic nerve, which carries the information to the areas of the brain responsible for vision (Figure 13.10B). Rod cells respond in dim light. Cone cells require brighter light to trigger impulses, and different ones respond to different wavelengths of light. The brain receives messages from cones sensitive to different wavelengths and processes these messages, creating our perception of color.

As we mentioned in Chapter 5, many details of eye structure differ across the animal kingdom (Figure 13.11). Vertebrate eyes have a retina that conducts a nerve impulse from back to front, while various mollusks have retinas that conduct impulses from front to back. Insect eyes are constructed on a totally different optical principle, and they are sensitive to a wider range of wavelengths than are vertebrate eyes, especially in the ultraviolet range.

The ear: reception of sound and balance. The ear is a highly elaborate structure. In humans and other mammals, it is arranged in outer, middle, and inner parts (Figure 13.12A). The outer ear, in which sound waves travel as vibrations in air, consists of an external flap, the pinna, and a tube called the ear canal. The pinna acts as a funnel to focus the sound waves into the tube leading to the eardrum (tympanic membrane), which marks the boundary between the outer ear and middle ear. The middle ear is connected to the back of the throat by the Eustachian tube, which allows pressure to equalize on the two sides of the eardrum. If the Eustachian tube swells shut, as can happen in middle-ear infections or changes in altitude, pressure builds up inside the middle ear preventing the vibrations of the ear bones. This can lead to partial hearing loss, which is usually a temporary loss that is alleviated when the tube opens.

Sound waves in the middle ear travel to the inner ear as vibrations through a series of small bones—the hammer, anvil and stirrup bones. The inner ear is divided into a cochlea, responsible for sensing sound, and the semicircular canals, responsible for sensing gravity and balance. Both portions of the inner ear are filled with fluid, and the sensory nerve endings respond to the movement of fluids. The fluid in the coiled cochlea picks up the vibrations from the bones of the middle ear. The cochlea contains a sensory membrane whose hairlike cells respond to these vibrations (Figure 13.12B and C). Very loud noises (as from industrial sources or rock concerts) can damage these hair cells and result in a permanent loss of hearing (Figure 13.12D). Hair cells at different positions along the cochlea trigger action potentials in response to different pitches or frequencies of sound.

The range of pitch detectable by different species varies greatly from that detectable by humans. Whales detect extremely low-pitched sound waves, which travel far and are used by them for communication across great distances. Dogs detect much higher-pitched sounds than we can, and bats higher still. The high-pitched squeaks of bats are reflected off objects, and the bats use these echoes for echolocation, permitting them to avoid obstacles at night and to catch insects in mid-flight.

The other portion of the inner ear includes the three semicircular canals, oriented at right angles to one another. Accelerations due to gravity and to body movements result in the movement of the fluid within these canals, and this information is sensed by a patch of sensory cells near the end of each canal. People with damage to the nerve serving these patches have difficulty in standing up and maintaining balance.

Figure 13.11

Differences in the structure of eyes in different groups of animals.

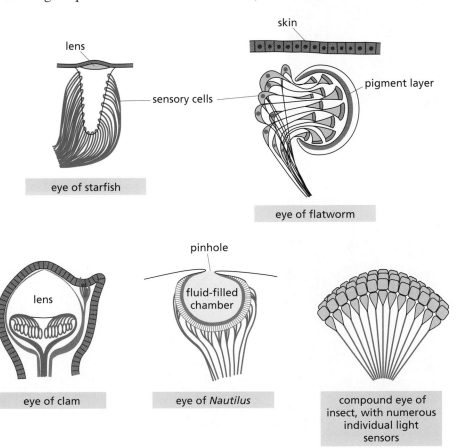

Figure 13.12

The human ear.

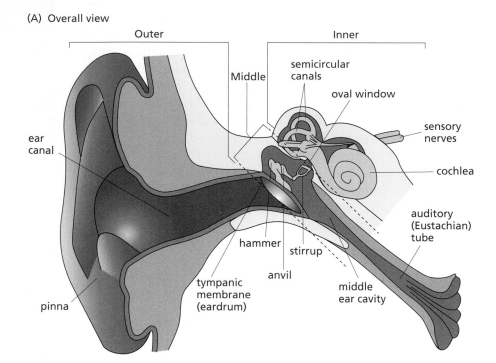

(A) Overall view

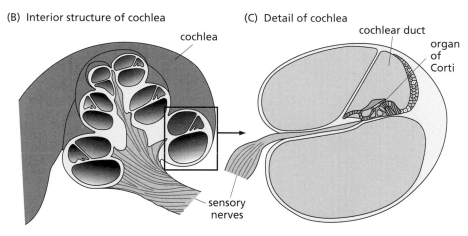

(B) Interior structure of cochlea

(C) Detail of cochlea

(D) Normal sensory hair cells from organ of Corti (left) and hair cells damaged by exposure to toxic chemicals or loud sound (right)

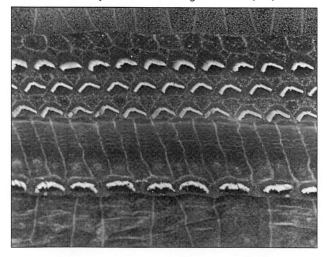
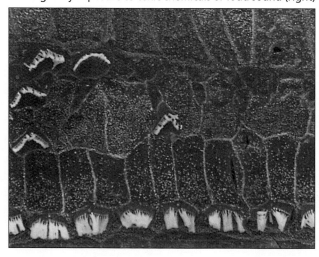

The tongue: reception of chemical signals as taste. Humans sense taste through a series of taste buds (Figure 13.13A) located along the tongue (Figure 13.13B) and the roof of the mouth. Four different basic tastes in humans were for many years described as sweet, salty, sour, and bitter. The taste buds for these different tastes all look the same, but the receptors are sensitive to different chemicals: the sweetness receptors trigger action potentials in response to sugars, as do the salt receptors in response to sodium ions (Na$^+$), the sour receptors in response to the hydrogen ions (H$^+$) present in acids, and the bitter receptors in response to various other chemicals, including bitter-tasting plant compounds called alkaloids. A fifth basic taste has been identified by Japanese researchers. The receptors, called umami receptors, respond to glutamic acid and to MSG (monosodium glutamate).

The nose: reception of chemical signals as smell. Although we are sensitive to thousands of different odors, the mechanism of smell is poorly understood. There is even disagreement as to the number of basic odors, with some experts naming as few as 7 while others list 20 or 30. Nerve endings sensitive to smell are most abundant in the lining of the nose (Figure 13.13C), although many vertebrates also have a similar sense organ opening into the roof of the mouth. These nerve endings trigger action potentials in response to chemicals just as the taste buds do. We perceive these chemical messages as smells, not tastes, because the nerves from the nasal lining go to a different area of the brain from those from the tongue. However, taste and smell perception interact to a large extent.

Many vertebrates also have openings from the roof of the mouth into an odor-sensitive organ (called the vomeronasal organ). Snakes will smell by darting their tongue in and out very quickly, picking up odorous particles in the air and touching them to the entrance of this sense organ.

Virtually all organisms, including single-celled procaryotes and eucaryotes, can respond to chemical signals by means of receptors. Only animals have nervous systems, however, so only animals respond to chemical signals by producing nerve impulses. The sense of smell is well developed among the vertebrates, most of which are more acutely sensitive to odors than humans are.

Message processing in the brain

Central to the function of the entire nervous system is the brain itself, which receives information from various parts of the body through peripheral nerves. The brain is the center where most activity of the nervous system takes place, where most decisions are reached, and where most bodily activities are both directed and coordinated. Nearly all activities that the body performs are carried out in response to signals sent from the brain to the rest of the body through the peripheral nerves. We are aware of many of these signals; such signals are often said to represent

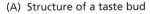

Figure 13.13

Organs of taste and smell.

(A) Structure of a taste bud

sensory nerve endings

(B) Distribution of greatest sensitivity to different tastes on the tongue surface

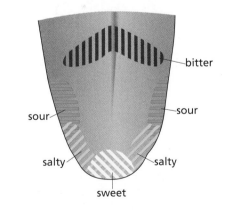

bitter

sour

sour

salty

salty

sweet

(C) Cells lining the nose

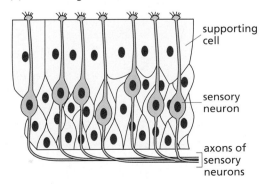

supporting cell

sensory neuron

axons of sensory neurons

'conscious' activity, although it is difficult to define consciousness in a manner that permits good experimental investigation.

Studying the brain. Our knowledge of the brain and its functions derives from many types of study. Many studies are done using experimental animals. Brain function can be studied in experimental animals by implanting a recording electrode in the brain to measure electrical activity during various brain activities. Another method uses an electrode through which brain activity can be stimulated in experimental animals for observation of their behavioral responses. Another technique is to destroy a portion of the brain (usually with an electric current) and to study the resulting changes in behavior.

Invasive techniques cannot ethically be used for studying brain function in humans. Electroencephalograms (EEG), in which electrodes are pasted to the scalp, measure electrical activity within the brain (see pp. 494–495). A related technique, positron emission tomography (PET), creates a computer-integrated picture of various brain activities, such as glucose metabolism and blood flow, within the living brain.

The brain can also be studied anatomically by dissections of the brain and its parts, and also by microscopic examination of the brain and its cells. What are some of the major anatomical features of the brain?

Brain anatomy. The brain is an immense network of neurons, yet it weighs less than three pounds (about 1.3 kg) in adult humans.

The folds and constrictions in the partly developed brains of embryos allow us to recognize several major brain divisions. These same divisions are also recognizable in the brains of adult animals of various species. The major regions of the brain are the forebrain, midbrain, and hindbrain (Figure 13.14), each composed of millions of neurons and each containing a central cavity. These divisions are similar in various vertebrate species, but their proportions differ (see Figure 13.14). Comparison of the brains of various species reveals a great deal about evolution and adaptation. The midbrain and hindbrain make up a larger proportion of the total in primitive vertebrates, while the forebrain, especially the cerebral hemispheres, is larger in mammals, particularly primates. In general, animals with more complex behavior patterns have larger brains with larger and more highly folded forebrain areas. The increasing size and complexity of the forebrain are especially apparent in humans and other species that rely primarily on learned behavior. Other changes in proportions are related to the senses that each species uses: brain regions concerned with vision are larger in those species that rely upon vision, whereas species that rely on smell or hearing have larger brain regions devoted to those functions.

Next we look in more detail at the anatomy of the human brain (Figure 13.15).

Forebrain. The forebrain includes the cerebrum, olfactory bulbs, hippocampus, and diencephalon. In early vertebrates, the forebrain was concerned with the sense of smell, and much of it still is through the olfactory bulbs, which process nerve impulses from sensory cells in the nose. Smells can influence both hormonal secretions and emotional responses such as anger, fear, and sexual response, which are therefore also processed in the forebrain, specifically in the diencephalon. In many cases, these responses can take place on an unconscious level, meaning that we may not even be aware that the changes are taking place. The hippocampus, concerned with certain types of learning, is located off center in the forebrain.

The cerebrum is divided into two halves, the right and left cerebral hemispheres. Thoughts and actions originate in the cerebral hemispheres, as do most of our 'higher' functions of intellectual thought and reasoning. The cerebral hemispheres of mammals are much larger than those of other vertebrates. This is particularly true in humans and closely related primates, in which the cerebral hemispheres make up the largest part of the brain. Conscious activity and higher thought originate in the highly folded surface layer of the cerebral hemispheres known as the cerebral cortex. Below the cortex lie many series of neuron interconnections that constitute the cerebral white matter. Deeper still lie several clumps of neuron cell bodies known as basal ganglia. Several of these structures are involved in the disorders described elsewhere in this chapter.

Figure 13.14

Correspondences among the brains of several vertebrate species (not to scale). In the chimpanzee and human brains, the olfactory bulbs and midbrain are concealed beneath the expanded cerebrum.

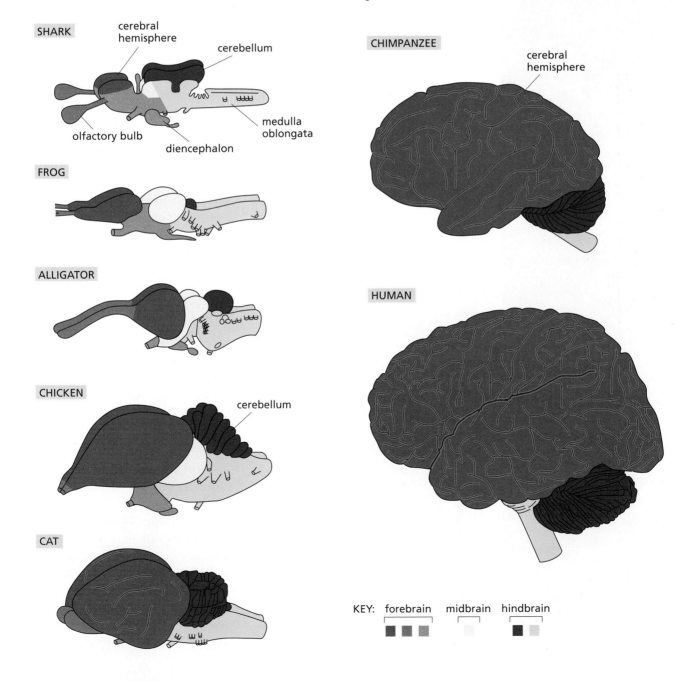

SHARK

cerebral hemisphere

cerebellum

olfactory bulb

diencephalon

medulla oblongata

FROG

ALLIGATOR

CHICKEN

cerebellum

CAT

CHIMPANZEE

cerebral hemisphere

HUMAN

KEY: forebrain   midbrain   hindbrain

Figure 13.15

Section through the human brain. Only structures located in the midline plane are visible in this view; off-center structures (such as the olfactory bulbs) cannot be seen.

Midbrain. The midbrain includes a ventral tegmental area, containing the brain's positive reward centers (see Chapter 14). Also located in the midbrain, but extending into part of the hindbrain, is the reticular formation, important in keeping us awake and alert.

Hindbrain. The hindbrain includes the cerebellum and the medulla oblongata. Functionally, the cerebellum is concerned primarily with balance, processing neuron impulses from the semicircular canals of the inner ear. The cerebellum also coordinates complex muscle movements. The medulla is concerned with such involuntary functions as breathing, functions that must continue even in sleep.

Blood–brain barrier. The brain has few internal blood vessels; most of the arteries and veins that supply the brain run along the brain surface only. The cells deep in the interior must therefore receive most of their nutrition through the cerebrospinal fluid, which fills the brain's interior cavities. There is no direct flow of fluid from the blood to the cerebrospinal fluid, hence the name blood–brain barrier. This cerebrospinal fluid communicates with the blood supply across a thin membrane (the tela choroidea). Nutrients and other small molecules cross this membrane or move through the neuroglia (nutritive cells) mentioned earlier, but many types of molecules, particularly larger molecules, cannot cross the blood–brain barrier. This barrier thus functions to prevent many molecules that have access to the rest of the body from entering the brain (see Chapter 14).

Epilepsy: abnormal message processing. Messages come to the brain from the peripheral nervous system and the sense organs. These signals trigger action potentials in a series of interconnected neurons in different parts of the brain. It is this message processing that produces our perceptions of the world.

In some disorders, neurons in the brain trigger action potentials spontaneously, without an outside signal. Epilepsy

(A) Major regions of the brain

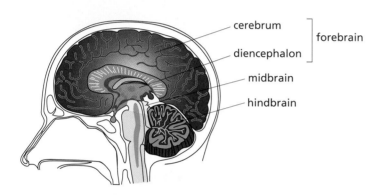

cerebrum
diencephalon
forebrain
midbrain
hindbrain

(B) Forebrain and hindbrain structures

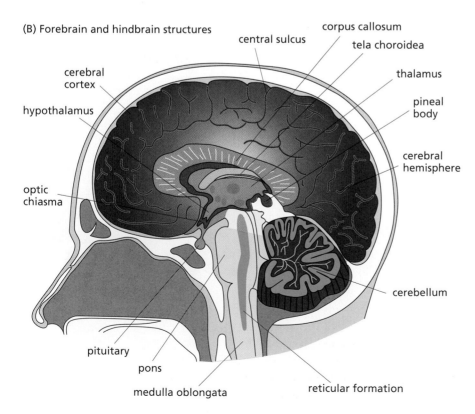

central sulcus
corpus callosum
tela choroidea
thalamus
pineal body
cerebral hemisphere
cerebellum
reticular formation
cerebral cortex
hypothalamus
optic chiasma
pituitary
pons
medulla oblongata

is a disorder marked by brain seizures, usually mild, characterized by uncontrolled electrical activity in the cerebral cortex. Both the cerebral cortex and the hippocampus contain many feedback pathways involving GABA-secreting neurons. As you probably recall, GABA inhibits a neuron that has fired from firing again (see Figure 13.7). An impairment of one or more of these GABA feedback pathways is hypothesized to allow a stimulated neuron to keep firing, perhaps causing an epileptic seizure. Some of the drugs that block GABA synthesis or GABA receptors can bring on epileptic seizures, thus lending support to the hypothesis. What is not fully explained by this hypothesis is why the seizures are temporary, why long periods intervene between them, and why certain events precipitate the onset of seizures.

In epilepsy the processing of messages by the brain is abnormal. Some people with epilepsy experience 'auras'. These take many forms depending on the part of the brain in which the spontaneous neuron signals occur. The person might have a visual hallucination or smell a bad smell when no actual source of such a sight or smell is present. A specific thought may be triggered, or a vague feeling of a place being familiar, even if it is not. Another of the symptoms of uncontrolled brain activity in epilepsy is muscle seizures in which muscles are continually stimulated to contract. To understand how changes in brain activity could affect muscle activity we next examine the pathways of normal muscle contraction.

Message output: muscle contraction

In addition to message processing by neurons forming synapses to other neurons within the brain and nervous system, the brain also coordinates the functioning of other structures within the body. Neurons, as well as forming synapses with other neurons, can also form synapses with cells of other types, such as muscle cells and gland cells. Glandular secretions are regulated by the autonomic nervous system, a part of the peripheral nervous system that is described in Chapter 15.

Voluntary movements are produced when the brain sends nerve impulses to the body's skeletal muscles. Because the muscles are attached to the skeleton, contraction of the muscles brings about movement of the skeleton and of the body as a whole. Most movements would be jerky and uncontrolled if only a single muscle were involved; controlled, steady movements usually require the simultaneous contraction of several muscles that pull in different directions to smooth and steady the movement. Coordination of messages to thousands of cells is a function of the brain. Earlier we saw that diseases (Parkinsonism and Huntington's disease) can result when the brain does not function properly. Now we examine in more detail how neurons induce muscle contraction in health.

Muscle contraction at the molecular level. Neurons whose axons form synapses with muscle cells rather than with other neurons are called motor neurons (Figure 13.16A). Muscle cell membranes, like the cell membranes of neurons, carry an electrical potential and are electrically excitable. The muscle cell's electrical potential can be depolarized after the cell receives a neurotransmitter signal from the motor neuron. Motor neurons generally secrete the neurotransmitter acetylcholine, and muscle cell membranes have specific receptors for this neurotransmitter.

When a motor neuron releases acetylcholine into the synapse with a muscle cell, the acetylcholine binds to receptors on the muscle cell membrane, causing the membrane to depolarize. This depolarization of the muscle cell membrane causes a release of calcium ions from vesicles within the cell. These calcium ions trigger reactions that allow cross-bridges to form between the two major muscle proteins, actin and myosin. The cross-bridges pull the actin filaments, increasing their overlap with the myosin filaments and producing the muscle contraction (Figure 13.16B and C). Many filaments of actin are pulled over many filaments of myosin to produce a forcible contraction of the muscle fiber as a whole.

Actin and myosin may be arranged in orderly bands (see Figure 13.16A), giving certain types of muscle fiber a striated (cross-banded) appearance common to both skeletal and heart muscle tissue. Another type, smooth muscle tissue, found in blood vessels and internal organs, lacks this cross-banding because the actin and myosin fibers are arranged at random intervals.

Muscle relaxation. Muscle cell contraction is a neuron-stimulated process; muscle cell relaxation is not. When the muscle cell membrane is no longer receiving depolarization signals, calcium ions are removed from the cell cytoplasm. Removal of calcium breaks the cross-bridges between actin and myosin, allowing them to return automatically to their original positions relative to each other (see Figure 13.16B). They do not receive another neuron signal that pulls them back apart. Removal of calcium ions does require energy, however. This energy is supplied as ATP. When an organism dies, no more ATP is made, so calcium cannot be removed and the muscle cells stay contracted. This produces rigor mortis.

Within the whole organism, controlled muscle contraction may bend an arm. Relaxation of those same muscles does not return the arm to its original position; for that movement another set of muscles must contract. Skeletal muscles are therefore arranged in opposing sets, with flexor muscles bending limbs and extensor muscles straightening them (Figure 13.17).

As we saw earlier (see Figure 13.6), neurotransmitters released into a synapse are normally taken back up by the neuron that secreted them or are

Figure 13.16

Structure of a portion of skeletal muscle.

(A) Overall structure of skeletal muscle, showing cross-striations

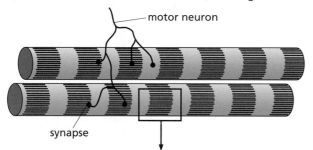

motor neuron

synapse

(B) Enlargement showing arrangement of actin and myosin filaments in a relaxed portion of skeletal muscle

myosin   actin

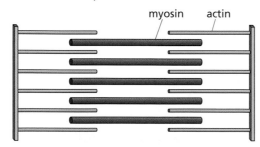

(C) The same portion of muscle, contracted

degraded chemically. Motor neurons release acetylcholine into the synapse, where it is normally broken down by the enzyme cholinesterase, causing contraction to cease. Some insecticides work by blocking the action of cholinesterase at these synapses. Then any synapse stimulated by acetylcholine remains continually stimulated (because the acetylcholine is never broken down), and the insect receiving the insecticide dies with most of its muscles in a state of rigid contraction. Because most animal nervous systems use acetylcholine as a neurotransmitter, such pesticides are toxic to all animals, great and small, including insects, pets, and humans. The toxic dose, however, depends on body size, so an amount that might kill most insects would only make your pet sick and might not noticeably affect you at all.

Many messages to and from the brain pass through neuron synapses connecting the peripheral and central nervous systems. We have seen that sense organs in the peripheral nervous system send messages to the brain. The brain also sends messages to the peripheral nervous system, for example through motor neurons that induce muscle contraction. Many other messages, and message storage, pass through synapses within the brain itself, a subject we take up in the next section.

Figure 13.17

The major muscles that flex and extend the arm. Most muscles act together with other muscles that steady their action by pulling in the opposite direction.

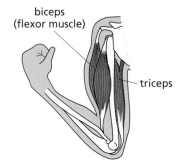

FLEXION
Biceps contracts forcefully; triceps contracts only slightly, to steady the biceps

biceps (flexor muscle)

triceps

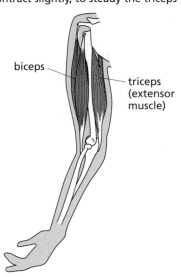

EXTENSION
Triceps contracts forcefully; biceps may contract slightly, to steady the triceps

biceps

triceps (extensor muscle)

THOUGHT QUESTIONS

1 The function of a portion of the brain of an experimental animal can be investigated by destroying it and studying the defect produced. What are this technique's limitations? Could you study how a piano or an automobile works by inserting a probe, destroying some local region, and studying the resulting defects? Could you study the function of a radio in this way, or a computer? Which of these is more comparable to the brain in complexity?

2 What does it mean to say that there are a particular number of tastes, such as four or five, and not six? How would we

evaluate the claim that "umami" is a fifth taste but "banana" is not? Compare this with the several different types of receptors in the skin.

3 Why would your being on a roller coaster or in a stunt airplane confuse your inner ear with misleading stimuli?

4 What causes muscle contraction to stop? What happens to an organism if muscle contraction does not stop?

5 Why does the body have so many different kinds of neurotransmitters?

The Brain Stores and Rehearses Messages

Neuron activity within the brain produces many of the functions that we associate with the mind. Some messages are stored as memory for future retrieval. Some of these modify our future behavior, a process that constitutes learning. At times, but especially when we sleep, the brain may rehearse sending and interpreting certain messages without remembering them. Brain activity produces personality, emotions, thoughts, and dreams; and, as we will see, mental illnesses can result from abnormal brain activity.

Learning: storing brain activity

When we change in response to changes in the world around us, this is known as **adaptation** (a separate meaning, unrelated to the adaptations that arise from the operation of natural selection). Adaptation can take place through many physiological, immunological, and neurological mechanisms and can be either conscious or unconscious. **Learning**, which consists of lasting changes in behavior or knowledge in response to experience, is an important form of adaptation brought about by the nervous system. Different types of learning are distinguished by the types of information that are learned and by ways in which the nervous system processes and stores the information. **Declarative learning** is mostly conscious remembrance of persons, places, things, and concepts, requiring the actions of neurons in the hippocampus and certain parts of the cerebral cortex. Memory of how to do things, **procedural learning**, does not require the hippocampus or the temporal lobe of the cerebral cortex, and is not necessarily conscious. People with hippocampal damage can still learn how to do new things, although they will not consciously recall that they can.

Procedural learning. Procedural learning can be very simple—in fact, simpler animals such as mollusks and insects are capable of procedural learning because it does not require the forebrain structures that evolved in vertebrates. In human development, procedural learning becomes possible earlier than declarative learning. Infants at first learn procedurally: how to eat, how to move, how to respond to gravity.

The three simplest kinds of procedural learning are habituation, sensitization, and classical conditioning. These simple kinds of learning are distinguished from declarative learning in that they can be involuntary: the learner changes his or her behavior without showing any awareness of the learning process. Most animals, no matter how minimal their nervous systems, can learn in these simple ways.

When a stimulus is presented repeatedly, an animal may learn no longer to respond to it. This is known as **habituation**. Habituation occurs both behaviorally and at the level of the neuron. Single neurons can stop making action potentials owing to changes in their receptors or to an increase in the threshold for generating action potentials. When a stimulus does not result in harm, an organism may learn to change its behavior and not expend energy in responding to the stimulus (Figure 13.18A). Humans habituate to all kinds of signals: if you move to a new location, you see and hear many things that people who have lived there for a while have learned not to notice any more. After a time, you no longer see or hear them either. If something unusual happens, the habituation can be overcome. Habituation is thus context-specific to some extent.

Sensitization (Figure 13.18B) is the opposite of habituation. An organism becomes sensitized when an intense and aversive stimulus, such as a loud gunshot, increases subsequent responses to other stimuli. On the cellular level, for a prolonged time, neurons become more capable of generating an action potential, a change known as long-term potentiation. Several hypotheses have been suggested to account for long-term potentiation. One recent hypothesis involves a type of receptor known as an NMDA receptor (*N*-methyl-D-aspartate receptor). These receptors respond to the secretion of glutamate (a neurotransmitter) by producing nitric oxide. The nitric oxide then diffuses back to the glutamate-secreting (presynaptic) cell, where it enhances the future release of glutamate. This type of feedback is called a positive feedback loop because a condition stimulates more of itself and often leads to an extreme condition. In this case, glutamate secretion enhances future glutamate secretion. The extreme overstimulation can lead to glutamate poisoning (characterized by convulsions), a condition made worse by the food additive MSG (monosodium glutamate).

A third type of simple procedural learning is called **classical conditioning** (Figure 13.18C), a change in which an organism learns to associate a stimulus with a particular response. Classical conditioning is also called Pavlovian conditioning because it was first demonstrated by the Russian physiologist Ivan Pavlov in about 1900. Dogs salivate when they see or smell food. If a bell is rung each time food is presented, a dog learns after a very few repetitions to salivate when it hears the bell ring, even if no food is present.

Humans can learn through classical conditioning, very often without being consciously aware of it. Fears sometimes become associated with various objects, colors, or smells when we are young because those stimuli were present when we were hurt in some way, or because they remind us of other unpleasant stimuli. One person

Figure 13.18

Three kinds of procedural learning.

(A) A person becomes habituated to a ticking clock

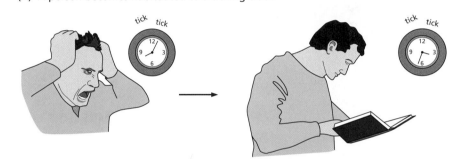

(B) A deer becomes sensitized by a loud noise

deer is startled by loud noise   deer is now more sensitive to other sounds

(C) A dog undergoes Pavlovian conditioning

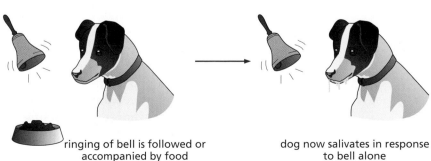

ringing of bell is followed or accompanied by food   dog now salivates in response to bell alone

we know grew up with a decades-long aversion to gelatin desserts. It seems that at the kindergarten he attended, such desserts were brought in on a tray, piled in cubes whose wiggling movements reminded him of certain caterpillars. Caterpillars are something that most children would avoid eating, so the shaking movements in this case conditioned an aversion to the desserts. Adults too can become conditioned. In one case, several people became nauseated every time they saw the carpeting in a hospital. It turned out that the carpet was the same color as the chemotherapeutic drugs that these people had taken for treatment of their cancers, drugs that had made them sick to their stomachs. The people had become conditioned, associating the color of the carpet with the cause of their nausea.

Declarative learning. In contrast to procedural learning, more complex types of learning require the activity of the hippocampus and cerebral cortex. There is evidence that complexity of experience actually contributes to the size of the cortex. Rats raised in 'enriched environments'—in large cages with other rats and with 'toys'—develop a thicker and more elaborated cortex. In humans, at about the age of two, the age at which the number of brain cells reaches its maximum and a critical level of complexity of connections is achieved, declarative learning begins. Although the exact age varies from child to child, the capacity for declarative learning always develops later than the capacity for procedural learning.

Memory formation and consolidation

To become a **memory**, a piece of information must be acquired, stored, and retrieved. Many acquired pieces of information can be retrieved for only a short period. For example, you may be able to recall having heard a particular sound if someone asks you about the sound within a few minutes of the time you heard it, but not after that. Information that is quickly forgotten is said to be part of short-term memory. Short-term memory is mostly chemical, having to do with temporary changes in neurotransmitters and their receptors. Long-term storage of information requires actual structural changes (the formation of new synapses) within many parts of the brain.

Long-term memory. One part of the brain that is essential to change a sensory input from a short-term into a long-term memory is the hippocampus (Figure 13.19). People who have suffered damage to the hippocampus are unable to form new long-term memories. They can recall things from before the time of the damage, indicating that the storage sites themselves are not in the hippocampus. Their recall of newly acquired information or experiences is limited to the short term, after which it is forgotten; thus, normal long-term retention of a memory may be said to require a 'lack of forgetting.' This is especially true of memories that can be consciously known.

The hippocampus turns short-term memories into long-term ones by making new cellular connections with other parts of the brain. Neurons form new synapses connecting several neurons in loops, with each cell synapsing on the next cell in the loop. The stimulation of any one of these neurons results in information transfer around the whole loop.

Stimulation of these assemblies of cells may need to continue for years before a memory is permanently stored.

As time passes after the acquisition, memory consolidation occurs. While long-term memories are forming, and even after they have been stored, they are organized and restructured on the basis of even more recent experiences. Existing knowledge is constantly being reordered in the light of new knowledge. Memory consolidation relies on information processing, one innate aspect of which is the capacity to make generalizations from specific experiences. If we live in a city and have walked in a forest only once, we mentally picture all forests as being like the one we walked in. Further experience, either 'in person' or acquired through seeing pictures or reading stories, enables us to reformulate the initial generalization that we made, replacing it with another generalization. In addition, we can still recall some very specific aspects of the particular forest that we first walked in.

Emotional states can sometimes modify the process of memory consolidation: we are more apt to remember something that we would normally not consider worth remembering if we associate it with an event that had great emotional meaning for us (either positive or negative). People who were old enough at the time of the assassination of John F. Kennedy, the tearing down of the Berlin Wall, or the September 11, 2001 attacks on the World Trade Center and Pentagon can remember vivid details of where they were and what they were doing. The same is generally true of events with great personal meaning, such as weddings, deaths, or natural disasters such as earthquakes or floods.

Figure 13.19

Horizontal section through the brain, showing the hippocampus, an important structure in the formation of certain types of memory. (Only one hippocampus is shown; the other is symmetrically located.)

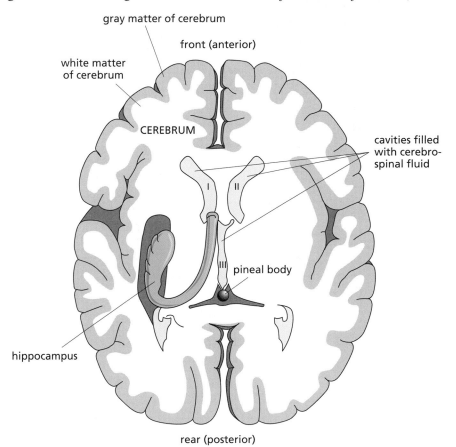

Abstraction and generalization. Memory consolidation also relies on the capacity to conceptualize, an extension of the ability to generalize. Researchers can demonstrate the capacity for generalization by using what are called 'oddity problems'. Monkeys are shown three or four objects, all alike except one, and they are rewarded for picking the different one. If a set of objects consists of three toy trucks and a car, they must learn to pick the car. Presented with a totally different set, two oranges and an apple, they must learn to pick the apple. After many such sets, each set different, have been presented, the monkeys develop a concept of 'oddness' and pick the single object immediately. This type of declarative learning requires activity in the temporal lobe of the cerebral cortex. Much of human learning and memory is processed by the cerebral cortex, consolidating memories and forming concepts, although other parts of the brain are also involved.

Memory formation requires the hippocampus, but memory consolidation requires more parts of the brain, particularly the cerebral cortex. The cortex is also important for memory retrieval. In a series of experiments in the 1940s, a neurosurgeon named Wilder Penfield electrically stimulated the cerebral cortex of conscious patients who were undergoing brain surgery for various neurological diseases. Such stimuli cause a person to recall a memory so vividly that they feel they are reliving the experience.

Alzheimer's disease: a lack of acetylcholine

Alzheimer's disease is a form of progressive mental deterioration in which there is memory loss and a loss of control of body functions that ends in complete dependence and death.

Clinical evidence suggests that nerve transmission across certain synapses that use acetylcholine is impaired in Alzheimer patients. Acetylcholine, in addition to being secreted by motor neurons, is also secreted by some neurons in the brain in synapses to other neurons, especially in the hippocampus and cerebral cortex. Drugs that inhibit cholinesterase can temporarily improve memory and other brain functions in Alzheimer patients, although these drugs cannot arrest the course of the disease. Inhibiting cholinesterase amplifies the effect of any existing acetylcholine. This fact, together with the gradual and progressive nature of the mental decline, has caused many workers to hypothesize that the disease symptoms are caused by a gradual loss of acetylcholine receptors in postsynaptic neurons; others believe that the primary defect is in the synthesis of acetylcholine itself. Recently, receptors for another neurotransmitter, glutamate, have also been implicated. Drugs that enhance glutamate reception also improve long-term memory formation in elderly patients.

In addition to changes in neurotransmitters, Alzheimer patients have abnormal deposits of a protein called amyloid in their brains. In 1993, a gene called *apo-e4* was identified as being present in a majority of patients with the most common form of Alzheimer's disease. The product of this gene is a lipoprotein related to the formation of amyloid. Autopsies reveal that amyloid deposits occur throughout the brains of people who had Alzheimer's disease, but are especially prevalent in the hippocampus and the parts of the cortex that are involved in memory.

Biological rhythms: time-of-day messages

In addition to the conscious processes of declarative learning, the brain sends itself many unconscious messages. Our bodies respond all the time to messages that tell us what time of day it is. In response to these messages, we establish a biological rhythm that governs our pattern of sleep and wakefulness.

Circadian rhythms and their control. There are many kinds of biological rhythms, biological processes that repeat at somewhat predictable intervals. Those of approximately 24 hours' duration are called circadian rhythms (Latin *circa*, 'about,' and *die*, 'day'). Various biological functions can be monitored on a 24-hour basis, and most of them show some recurrent circadian rhythm.

Where do circadian rhythms originate? If the rhythms originate internally, how do they keep tuned to the 24-hour cycle of the world around us, and how can they adjust to different time zones, seasons, and work shifts? If, in contrast, the rhythms originate externally, how do external cues regulate the body's cycles, and can these external cues be easily manipulated? The first serious attempts to answer these questions began with isolation experiments, such as those conducted in Mammoth Cave in Kentucky. In these experiments, volunteers who were kept from all sources of natural light were allowed to set their own daily routines. Nearly all individuals maintained fairly constant circadian rhythms in their sleep–wake cycles and also in body temperature, activity, and a variety of physiological measurements. Circadian rhythms were all maintained despite the constant environment and were rather uniform for each individual. Most significantly, the circadian rhythms maintained in these isolation experiments were generally slightly more than 24 hours, between 25 and 26 hours for most individuals. The maintenance of such a rhythm after its drift from synchrony with the day–night cycle of the outside world showed that internal rhythm-keeping mechanisms existed.

How, then, does the external world exert its influence, causing most of us to maintain a 24-hour daily rhythm? Our current understanding is that the external environment provides us with certain time-related clues. The most important of these clues is the natural rise and fall of light intensity throughout the day; other important clues include our own activity and our bombardment with external stimuli (including social stimuli) during daylight hours.

Important to circadian rhythms are a group of neurons located in the hypothalamus (see Figure 13.15). Destruction of these neurons abolishes all circadian rhythms. Under experimental conditions of continuous darkness or continuously dim lighting, these neurons maintain an internal circadian rhythm with a cycle of slightly more than 24 hours in length. The rhythm is also regulated by the pineal body, a structure about the size of a pencil eraser, located on the roof of the forebrain (see Figures 13.15 and 13.19). Under natural conditions, the light received by the pineal body adjusts the rhythm to follow a 24-hour cycle, with a peak of activity in the late morning and with greatly reduced activity throughout the hours of darkness. During times of darkness, the pineal body secretes a hormone called melatonin, which is not secreted during times of illumination. By illuminating various parts of the body separately from the rest, biologists have experimentally determined that the pineal body is sensitive to the light that it receives right through the skull and brain! If a laser (a highly focused beam of light) is focused on the pineal body through the head, and is turned on and off in a 24-hour rhythm, all parts of the body follow the established circadian rhythms, even when the rest of the body is in darkness. If, instead, the head is kept in darkness while other parts of the body are illuminated, the effect is the same as if the body and head were both in total darkness.

Gradual or slight disturbances in a person's 24-hour rhythm can result from short-distance travel, seasonal changes, and the semiannual change of clocks at the beginning and end of daylight savings time. The effects of these changes are minimal in most cases. More drastic effects are felt in the phenomenon known as jet lag, the disturbance of our 24-hour rhythms as a result of travel through several successive time zones. The major symptom of jet lag is fatigue, plus a desire to sleep or remain awake at inappropriate times for a few days until the body readjusts to the new cycle. The rigors of travel add to the fatigue, but travel north to south within a time zone is much less fatiguing than travel east to west across time zones. People who have traveled long distances by air are also statistically more susceptible to infection. While some of this may be due to other conditions of air travel, the increase in susceptibility is greater among people who have flown east–west than among those who have flown south–north, suggesting that interrupted circadian rhythms are involved. People whose bodies are not able to adjust to any particular rhythm because of irregular work schedules also show an increase in fatigue-related events such as the number of accidents.

Sleep. One mental state that shows a strong circadian rhythm is sleep. At the end of each day, we usually have a strong urge to sleep; we can postpone this urge, but only to a limited extent. People deprived of sleep do not function well when awake (they make more mistakes, for example); and people who awake from a 'good night's sleep' feel refreshed and alert.

As we have seen, the electrical activities of single cells take place across their membranes. The cumulative effect of ions flowing across the membranes of many cells can be seen in an electroencephalogram (EEG), a graph of electric activity obtained from electrodes pasted onto the scalp (Figure 13.20A). By using an EEG, we can detect different patterns of ion flow in the brain. EEGs show several levels or stages of sleep. Stage 1 is characterized by regular respiration, slowed heart rate, drifting mental imagery (similar to daydreaming), and low-voltage EEG wave patterns of about 7–10 Hz. (Hz stands for Hertz, a measure of frequency equivalent to cycles, or waves, per second.) Subjects aroused from stage 1 sleep will often say, "I wasn't sleeping." Sleep stages 2, 3, and 4 also have characteristic EEG patterns (Figure 13.20B). A further sleep stage is characterized by rapid eye movements (REMs) noticeable as movements of the eyeball beneath the closed eyelids of sleeping subjects, including cats and dogs as well as humans. About 80% of human subjects awakened during REM sleep tell of some dream that they were having, often in vivid detail, while subjects awakened during other sleep phases do not remember any dreams.

During a typical night's sleep, an adult goes through about four or five sleep cycles. As shown by the steps in Figure 13.20C, each cycle goes through stages 1, 2, 3, and 4, in that order, then in reverse order through stages 3, 2 and 1, followed by a REM period, after which the next cycle begins. The proportion of REM sleep increases with each successive cycle, as shown by the increasing width of the REM bars in the later sleep cycles in Figure 13.20C. The later sleep cycles may also skip some of the stages, for example proceeding only to stage 3 before reversing back to 2, 1 and REM, and at a still later cycle, proceeding only to stage 2 before returning to 1 and REM. If subjects are allowed to awaken by themselves (with no alarm clock or other external stimulus), they usually awaken near the end of a cycle—that is, following a REM stage or during stage 1 (see Figure 13.20C). Although this pattern is fairly typical, sleep cycle patterns vary widely with the person and the circumstances.

Evidence continues to mount that REM sleep is extremely important. Volunteers who are deprived of REM sleep for a day or two begin to daydream and have their thoughts wander. They also make more mistakes and have more accidents, even if their total amount of sleep and their amounts of all the other sleep stages are normal or above normal. If finally allowed to sleep as long as they wish, REM-deprived subjects sleep longer than usual, and a higher than usual proportion of their sleep is REM sleep. People deprived of any of the non-REM stages do not show any abnormal symptoms.

Some drugs can alter the natural occurrence of sleep. Caffeine, amphetamine, and other stimulants may interfere with the onset of sleep, although sensitivity to this effect varies with the person. Alcohol and barbiturate drugs bring on sleep more readily, especially in persons already tired, but this drug-induced sleep is less restful because it has longer stage 3 and stage 4 intervals and shorter REM episodes. Muscle relaxants have no effect on sleep except to overcome muscle tension, which may sometimes inhibit the onset of sleep. (Most drugs taken as aids to sleep are either barbiturates or muscle relaxants.) There is no totally safe sleeping pill, and all drugs that alter sleep patterns usually have other effects as well.

The role of neurotransmitters in the control of sleep and wakefulness is unclear. The rate at which neurotransmitters or certain other chemicals are produced or degraded during sleep can be studied by labeling these

Figure 13.20

Electrical activity in the brain during different stages of sleep.

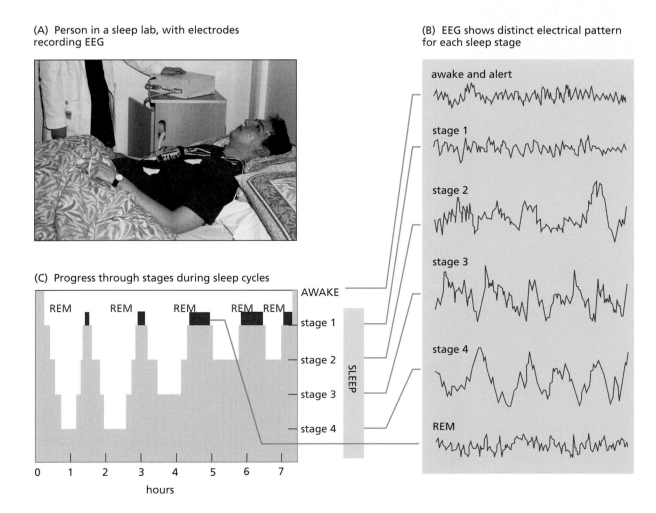

(A) Person in a sleep lab, with electrodes recording EEG

(B) EEG shows distinct electrical pattern for each sleep stage

awake and alert

stage 1

stage 2

stage 3

stage 4

REM

(C) Progress through stages during sleep cycles

REM   REM   REM   REM   REM

AWAKE

stage 1

stage 2

stage 3

stage 4

SLEEP

0   1   2   3   4   5   6   7

hours

chemicals radioactively. For example, studies in which animals are fed radioactive tryptophan (a chemical precursor from which the neurotransmitter serotonin is synthesized) show that the rate at which tryptophan is converted into serotonin increases during sleep. Likewise, studies have found that serotonin breaks down more rapidly during sleep. Such experiments have enabled us to locate areas of neurotransmitter activity during sleep and wakefulness. In these studies, as in other biological studies with radioactively labeled substances, the radioactive dosage is kept low to minimize the risks to the experimental subjects and experimenter.

An important part of the brain governing sleep and wakefulness is a group of neurons called the reticular activating system, which radiates outward and upward from the reticular formation of the brain stem (Figure 13.21). These neurons send 'alertness' signals through the diencephalon to widely scattered parts of the cerebral hemispheres. These signals seem to accompany most types of sensory input but do not seem to vary depending on the type of stimulus. Their message seems to be simply "pay attention!" The reticular activating system is usually more active by day and quiescent at night. Low-level activity allows sleep, but higher activity maintains wakefulness. Fortunately, it is also possible for the reticular activating system to awaken us to an emergency in the middle of the night or to deviate in other ways from its usual 24-hour rhythm, as conditions demand.

Dreams: practice in sending messages

Dreams are an important phenomenon of sleep. Philosophers and poets, fortune-tellers and psychiatrists have each had their ways of interpreting dreams. Modern-day dream researchers have used several techniques to study dreams. One method is to have the subject keep a 'dream diary' in which they record as much of a dream as they can remember upon awakening. Among the findings using this technique are the following: most dreams are visual, including those of people who became blind after age five or six, but excluding those of people who were blind from birth or infancy. Other senses (hearing, touch, smell, taste) also are mentioned in dream reports, but less often than vision. (These other senses predominate, however, in the dreams of people blind since birth or infancy.) Familiar persons, places, and types of events appear in most dreams, but not always congruously: the dreamer sometimes experiences familiar activities in the wrong setting, and places that are supposedly far away or unfamiliar often look very familiar in the dream. There are frequent changes of scene, of mood, or of persons present. Stimuli in the dreamer's environment are often incorporated in the dream

Figure 13.21

The reticular activating system.

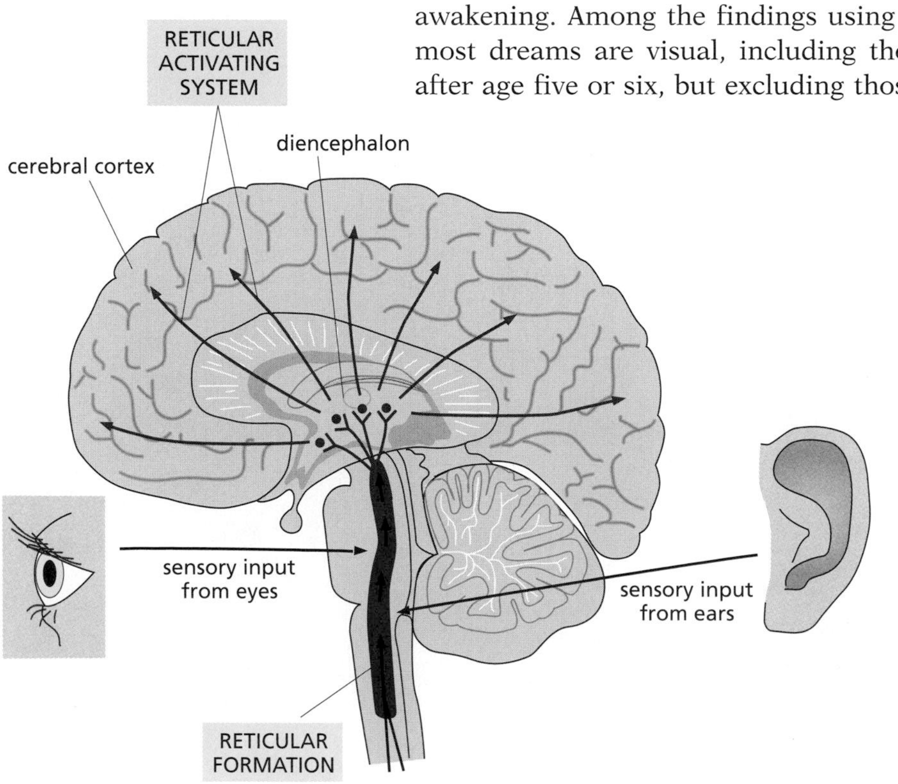

imagery itself. Actual sounds, smells, flashes of light, and other sensations from the sleeper's environment may be worked into the dream.

Perhaps one of the most important and most constant feature of our dreams is that we remember so little about them during our waking hours. There seems to be a good reason for this. The EEGs recorded during REM sleep show spontaneous electrical activity in the brain. This is our brain's way of rehearsing its message-sending functions, and it might be related to the strengthening (through practice) of existing synaptic connections or the establishing of new ones.

Our brains are usually programmed to interpret any such electrical activity as a coherent picture of the world around us. When we are awake, our brains usually record these coherent pictures so that we can remember them at some later time. In dreams, the integrating mechanisms are still at work, so a coherent or semicoherent picture of the world is drawn, but the mechanism whereby these pictures are remembered for later use is suppressed, at least most of the time. When people have disturbing nightmares, what is usually most disturbing about them is that they are remembered as if they had been real. Of all people, schizophrenics have the greatest difficulty in distinguishing reality from dream activity.

Mental illness and neurotransmitters in the brain

We have described brain functioning during normal mental states. Several mental illnesses are associated with chemical imbalances in the brain, such as abnormalities of neurotransmitters.

Depression and serotonin. Depression is a disorder marked by feelings of total helplessness, despair, and frequent thoughts of suicide. Many people suffering from depression attempt suicide, and some succeed. The smallest task, such as getting out of bed in the morning, can seem overwhelming. Everyone has unhappy or pessimistic feelings from time to time, but a person suffering from depression has these feelings nearly all the time and to a severe degree. Depression is about twice as prevalent among women as among men.

Patients suffering from depression have smaller amounts of several neurotransmitters than do other people. In particular, the brains of depressed patients who commit suicide are found at autopsy to contain low concentrations of serotonin.

A number of drugs are effective in treating the symptoms of depression. These drugs act in either of two ways. Some drugs act by blocking the action of the enzyme MAO, which degrades many neurotransmitters, including serotonin, mostly in synapses. Other drugs, including Prozac, Elavil, and Tofranil, inhibit the reuptake of serotonin and several other neurotransmitters by the presynaptic neurons that secreted them. In either case, the effect is the same: the serotonin remains active for a longer time after it is secreted and results in a greater (or a more lasting) stimulus being passed on to the next neuron. Drugs that inhibit the body's production of serotonin can reverse the effects of antidepressant drugs, but drugs that inhibit the production of several other neurotransmitters do not have this effect.

How might decreased amounts of a neurotransmitter bring about the symptoms of depression? One possible explanation is that, when something good happens, most people receive a pleasurable stimulus in the form of a stimulation to the brain's positive reward centers, and this stimulation makes them more likely to repeat whatever behavior led to the reinforcement (see Chapter 14, pp. 522–523). However, the reinforcement mechanism is not working in patients who are depressed, so they are never rewarded, nor do they learn to repeat whatever behavior led them to the sensation.

Schizophrenia and dopamine. Schizophrenia is a disorder characterized by frequent delusions and auditory or visual hallucinations. Schizophrenia seems to result from an excess of the neurotransmitter dopamine; drugs that stimulate or mimic dopamine (such as amphetamines or L-DOPA) make schizophrenic symptoms worse, and drugs that block dopamine (such as chlorpromazine and haloperidol) generally lessen symptoms. Further evidence comes from schizophrenic side effects observed when drugs are used to treat Parkinsonism, and Parkinsonian side effects observed when drugs are used to treat schizophrenia. Recently, it has been found that, in addition to dopamine, serotonin may also be involved in schizophrenia.

THOUGHT QUESTIONS

1 Suppose that you were investigating why depression occurs more often in women than in men. How might you test whether differences in upbringing and other cultural influences were at work? Would it be useful to study the prevalence of depression in different cultures? What methodological, ethical, or social problems would such a study face? Would you examine many people or few? What additional difficulties would you face if depression were defined differently in each culture? Could you investigate depression by studying an outcome such as suicide? What new ethical or social problems might arise from the results of such a study?

2 Is it ethical to give a drug that changes someone's personality? If a drug is prescribed to treat depression, and the patient kills someone, can the doctor be held responsible?

3 If a person's social situation may be contributing to a problem such as depression, is it proper for a doctor to treat the condition with drugs without also addressing the social situation?

Concluding Remarks

Many of the functions that people have historically attributed to the mind have been shown by neurobiology to originate in activity in the brain. Neurons transmitting action potentials and stimulating other neurons across synapses can account for mental activities such as sensations, dreams, learning, and memories. Mental illness can also be seen to have a neurochemical basis, at least in part. However, whether these brain and nervous system activities actually *are* the mind, or whether they are the mechanisms by which the mind becomes material, is a question that philosophers will continue to debate. Certainly a person is not an isolated collection of biochemicals but exists in a social relationship to other people and is necessarily linked to the rest of the world. So, although much progress has been made in drug treatments of mental illness by thinking of the mind and the brain as synonymous, it is a concept that could be carried too far if it leads us to ignore the importance of the whole person and her/his relation to others and to the environment.

Chapter Summary

- The work of the brain is carried out by nerve cells (**neurons**). Nerves are bundles of neurons.

- Neurons maintain differences in ion concentration across their cell membrane and thus a difference in electrical charge known as the **resting potential**.

- If a **threshold** of depolarization is exceeded, neurons carry nerve impulses in the form of successive **action potentials** down the length of their **axons**. **Sodium–potassium pumps** then restore the ion distribution, which reestablishes the resting potential.

- Neurons communicate with one another across spaces called **synapses** by means of chemical **neurotransmitters** such as acetylcholine, norepinephrine, serotonin, dopamine, and GABA. Decreased dopamine neurotransmission may lead to Parkinsonism, while excessive dopamine neurotransmission may produce the uncontrolled movements that characterize Huntington's disease.

- Neurons control the activity of other neurons by **feedback systems**. Feedback can be either inhibitory (feedback inhibition, negative feedback) or stimulatory (positive feedback).

- Sense organs in the peripheral nervous system are responsible for receiving external stimuli such as visual images, sounds, tastes, smells, touch, pressure, heat, cold, and pain.

- Sensory messages are carried to the central nervous system and processed in the brain; such processing produces our perceptions of the stimuli.

- The brain also sends out messages such as those that stimulate muscle contraction.

- **Learning** requires activity in the brain but not necessarily consciousness of the activity.

- **Procedural learning** includes **habituation**, **sensitization**, and **classical conditioning**, none of which require conscious awareness or activity in the hippocampus or the temporal lobe of the cerebral cortex.

- **Declarative learning** is conscious learning, requiring the hippocampus and cerebral cortex for stimulus processing, **memory** consolidation, and the formation of generalizations and abstractions.

- Many biological functions follow a **circadian rhythm** that is set internally and fine-tuned by light acting on the pineal body within the brain.

- Sleep follows definite stages, each with a distinctive EEG pattern. Dreams coincide with periods of rapid eye movements (REMs) and result from the brain's own practice at sending messages.

- Abnormal neurotransmitter concentrations are associated with some mental illnesses.

CONNECTIONS TO OTHER CHAPTERS

Chapter 1	Equating the mind and the brain is an example of a research paradigm.
Chapter 1	Ethical issues are raised by the use of fetal tissue for research and for therapy.
Chapter 3	Susceptibility to some brain disorders has a genetic basis.
Chapter 5	Similarities in the brains of different species have resulted from evolution.
Chapter 8	Social behavior is, in part, learned.
Chapter 10	Neurotransmitters are made from dietary amino acids.
Chapter 14	Psychoactive drugs alter the functions of the brain, in some cases by affecting neurotransmission.
Chapter 15	Brain activity can influence our general health. Sleep deprivation, for example, can impair the immune system.
Chapter 18	Many drugs that influence brain activity are obtained from tropical rainforest plants.
Chapter 19	Brain disorders may arise from environmental pollution.

PRACTICE QUESTIONS

1. In a neuron that is not conducting an impulse, there is an excess of _____ ions inside the cell and an excess of _____ ions just outside the cell membrane.

2. An action potential is caused by a large number of _____ ions moving across the cell membrane toward the _____.

3. Which neurotransmitter(s):

   a. is a gas?

   b. is broken down by cholinesterase?

   c. is overproduced in Huntington's disease?

   d. was the first to be experimentally investigated?

   e. do depressed patients often have in insufficient amounts?

   f. are amines?

4. _____ is a neurotransmitter but not an amine; the enzyme _____ breaks it down in the synapse.

5. Name three parts of the forebrain.

6. For each of the following, name the sense organ that contains it:

   a. anvil bone

   b. iris diaphragm

   c. cones

   d. naked nerve endings

   e. tympanic membrane

   f. cochlea

   g. an area sensitive to acidic compounds

7. Each of the following is an example of what form of learning? Please be as specific as you can.

   a. Learning to associate horses with the word 'horse'.

   b. Being more alert to other stimuli when the fire alarm rings.

   c. Learning not to pay attention to traffic noises in one's neighborhood.

   d. Learning to recognize a TV show by its theme music.

8. How does the removal of GABA produce more dopamine?

9. What is the effect of monoamine oxidase inhibitors on neurotransmission, and why?

10. Name the types of sensory input that induce generator potentials in neurons of each of the following sense organs: skin, eye, ear, tongue, nose.

11. What are the functions of the blood–brain barrier?

12. What are the major ways in which brain anatomy has changed during evolution?

13. What are the major functions of each of the three divisions of the brain: forebrain, midbrain, and hindbrain?

Issues

- What is a drug?
- How do drugs interact with one another?
- What is addiction?
- Why are some drugs addictive?
- Is drug abuse the same as addiction?
- Does drug abuse lead to addiction?
- What are the social effects of drug abuse?
- Are all forms of drug addiction treated as crimes?
- Why do some argue that marijuana should be legalized, while others think that it should not?

Biological Concepts

- Molecules (structure, diffusion)
- Membranes and cell-surface receptors (agonists, antagonists)
- Energy and metabolism
- Organ systems (respiratory system, excretory system, placental circulation)
- Homeostasis (drug tolerance, drug metabolism, excretion)
- Central nervous system (brain)
- Behavior (reinforcement, addiction, behavior modification)
- Health and disease (public health issues)

Chapter Outline

Drugs Are Chemicals That Alter Biological Processes

Drugs and their activity

Routes of drug entry into the body

Distribution of drugs throughout the body

Elimination of drugs from the body

Drug receptors and drug action on cells

Side-effects and drug interactions

Drug safety

Psychoactive Drugs Affect the Mind

Opiates and opiate receptors

Marijuana and THC receptors

Nicotine and nicotinic receptors

Amphetamines: agonists of norepinephrine

LSD: an agonist of serotonin

Caffeine: a general cellular stimulant

Alcohol: a CNS depressant

Most Psychoactive Drugs Are Addictive

Dependence and withdrawal

Brain reward centers and drug-seeking behaviors

Drug tolerance

Drug Abuse Impairs Health

Drug effects on the health of drug users

Drug effects on embryonic and fetal development

Drug abuse: public health and social attitudes

14 Drugs and Addiction

Most of us are accustomed to the idea of taking drugs when we are sick. Many of these drugs are prescribed to fight bacterial infections, to fight cancer, or to regulate the body's physiological processes. Many other drugs are taken without medical supervision and for a wide variety of purposes. Many of these drugs are legal; some are not. The United States is the number one drug-producing and the number one drug-using country in the world. There is said to be a 'drug problem' that has led to a 'war on drugs;' yet there does not seem to be societal agreement on the answers to some very basic questions: What is a drug? What are the various legitimate uses of drugs? What is addiction and what makes a drug addictive? In this chapter we examine some recent research in biology and in related fields relevant to these questions. We also examine the basic biological principles that underlie our understanding of how drugs work. An understanding of the respiratory, circulatory, and excretory systems is needed to see how drugs enter and become distributed around the body, how long they stay in the body, and how they are eventually removed.

Drugs Are Chemicals That Alter Biological Processes

The term **drug** can have many meanings, depending on the context. To biologists, a drug is any chemical substance that alters the function of a living organism other than by supplying energy or needed nutrients. In a medical context, a drug may be thought of as any agent used to treat or prevent disease. Those who work in the field of drug addiction define a **psychoactive drug** as any chemical substance that alters consciousness, mood, or perception. As with all definitions in science, these are open-ended; no one definition can cover every possible situation. Each of these definitions expresses a slightly different concept, and each is correct within its contextual field.

This chapter emphasizes psychoactive drugs, that is, those that alter consciousness, mood, or perception. All psychoactive drugs alter biological functions, and are thus considered to be drugs on the basis of the first definition. Many, but not all, are also drugs by the medical definition.

Drugs and their activity

To understand how psychoactive drugs work, we need to know some general principles that apply all types of drugs. The study of drugs, their properties, and their effects is called pharmacology. Pharmacological principles explain how drugs can be both effective and dangerous, how both concentration and time affect the activity of drugs, and how interactions between drugs can change drug activity.

The activity of any drug varies with its dose, meaning the amount given at one time. There is an effective dose, the amount 'effective' in producing the desired change; in medicine, the effective dose is also called a therapeutic dose. For almost all drugs there is also a toxic dose, the amount at which the drug produces harmful effects, and a lethal dose, the amount that kills the organism. The more commonly used term, overdose, includes both toxic and lethal doses. Some drugs have a wide 'margin of safety'; that is, the toxic dose is many hundreds or thousands of times as much as the therapeutic dose. For other drugs, the toxic dose or even the lethal dose may be very close to the effective dose.

The exact amount that is effective, toxic, or lethal differs with the chemical structure of the drug. It also differs from one individual to the next depending on body size and many other physiological variables. We discuss some of these variables later in the chapter.

The drug concentration achieved at any given location in the body depends on several factors: how the drug enters the body, how it is carried around the body, how it is transformed by the body's cells, and how it is removed from the body.

Routes of drug entry into the body

Most drugs enter the body through the digestive system or the respiratory system; a few are injected directly into the bloodstream. Drugs that are taken orally and swallowed must be able to withstand the acidic environment of the stomach and then be absorbable by the cells of the intestinal lining in the same ways in which food substances are taken up (see Chapter 10, pp. 348–349). Drugs that enter through the respiratory system face different obstacles.

The respiratory system. The lungs (Figure 14.1) are the body organ by which vertebrate land animals take up oxygen and give off carbon dioxide, a waste product of their metabolism. This process is controlled by the respiratory system and has two parts: the mechanics of breathing and the exchange of gases.

Breathing is accomplished, not by the lungs themselves, but by the diaphragm, a muscular layer below the lungs. When the muscles of the diaphragm contract, they pull the diaphragm downward, expanding the chest cavity. The laws of physics tell us that the pressure of a gas depends on the number of gas molecules in a given volume; therefore, when the volume of the chest cavity becomes larger, the pressure inside it decreases. The air pressure inside the lungs is then less than the air pressure outside the body. Physics also tells us that gases flow from areas of higher pressure to areas of lower pressure, so the enlargement of the chest cavity causes air to enter the lungs (inhalation). (Hiccups result when the downward movement of the diaphragm is more rapid than normal.) When the diaphragm relaxes, it returns to its higher position, decreasing the volume of the chest cavity and pushing air back out (exhalation) (see Figure 14.1A).

Air passes into the lungs from the trachea, or windpipe, through branching airways called bronchi into thin-walled sacs called alveoli. Oxygen gas and carbon dioxide are exchanged between alveoli and the blood vessels (capillaries) of very small diameter that surround them (Figure 14.1B). The gases move by passive diffusion (see Chapter 10, pp. 348–349) across the cell membranes of the cells of the alveoli and the cells of the capillaries. Oxygen is in higher concentration in the inhaled air in the alveoli than it is in the capillaries, so oxygen diffuses *into* the

Figure 14.1

The human respiratory system. The blood receives oxygen in the lungs, and the veins therefore carry oxygen-rich blood back to the heart while the arteries carry deoxygenated blood, the reverse of the usual situation.

blood. Carbon dioxide is in higher concentration in the blood than it is in the inhaled air, so it diffuses *out of* the blood into the alveoli, to be exhaled. The rate of diffusion depends on the difference in concentration and on the amount of surface area over which the diffusion can occur. The huge number of capillaries and the compartmentalized, saclike structure of the alveoli provide an enormous surface area (Figure 14.1C), making diffusion very rapid (the effects of compartmentalization on surface area are discussed in Chapter 12, pp. 414–415).

Gaseous drugs, including many anesthetics, can enter the lungs, diffuse across cell membranes, and enter the blood by the same rapid mechanism as does oxygen. Toxic inhalants, including various solvents and propellants, can enter in the same way. Particulate matter, such as the

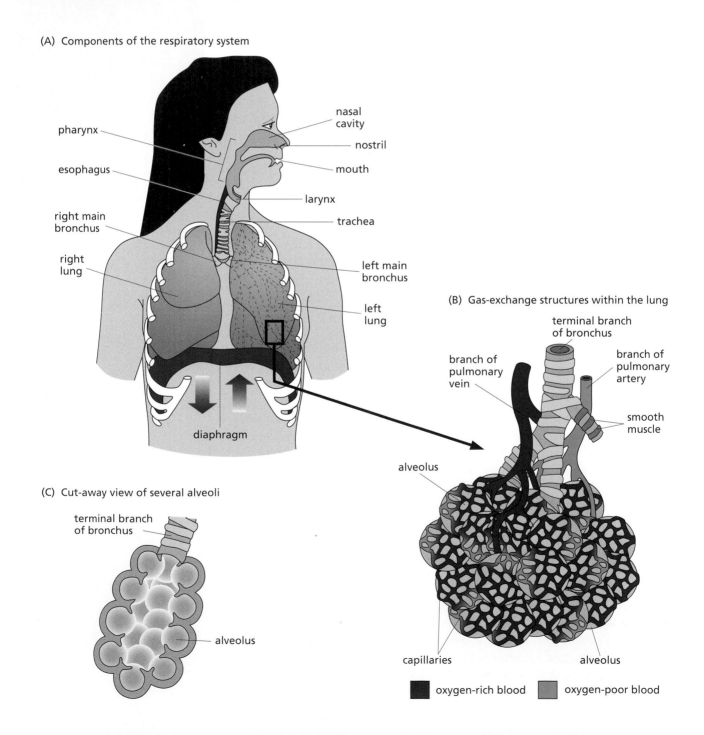

(A) Components of the respiratory system

pharynx
esophagus
right main bronchus
right lung

nasal cavity
nostril
mouth
larynx
trachea
left main bronchus
left lung

diaphragm

(B) Gas-exchange structures within the lung

terminal branch of bronchus
branch of pulmonary vein
branch of pulmonary artery
smooth muscle
alveolus

(C) Cut-away view of several alveoli

terminal branch of bronchus
alveolus

capillaries   alveolus

■ oxygen-rich blood   ■ oxygen-poor blood

chemicals in smoke, can adhere to the inner surfaces of the alveoli, causing damage to those surfaces while delivering drugs to the bloodstream (see below).

Route of entry and effective drug dose. The way in which a drug enters the body often affects its resulting concentration in body tissues. As an example, consider cocaine, a product of the coca plant, *Erythroxylon coca*, native to the high Andes. Cocaine exists in many forms that differ in both the concentration of the drug and in its molecular form. Purification of the drug makes it easier for the user to receive a toxic or lethal dose. The addictive potential (see below), while always great, also increases with each purification.

Coca leaves were and are chewed by South American Indians, especially those living at high altitudes. The concentrations absorbed from the gut when coca leaves are chewed are quite low, because the cocaine is present in an ionized form. Ionized molecules carry a charge and thus do not readily diffuse across the nonpolar portion of the plasma membranes of the cells lining the gut (see Chapter 10, pp. 348–349).

In contrast, when cocaine is purified into a powder, although the cocaine itself is still in an ionized form, sniffing the powder increases both the rate of absorption and the amount absorbed into the capillaries under the lining of the nasal passages. Cocaine can be further purified, resulting in the un-ionized form called free-base cocaine and the still more highly purified form known as crack. Un-ionized molecules are less polar than ionized molecules and are therefore more rapidly absorbed across cell membranes. In addition, free-base and crack cocaine are no longer powders. They are smoked rather than sniffed, moving the drug beyond the nasal passages and into the lungs. Smoking these drugs further increases the rate of absorption, because the surface area of the lungs over which the drug is absorbed is much larger than the surface area of the nasal passages.

Whether cocaine is sniffed or smoked, it is inhaled not as a gas but as small particles. All forms of smoke, including smoke from crack cocaine, cigarettes, marijuana, and the burning of fossil fuels (industrial smoke and automobile exhaust), are particulate. For a chemical to be absorbed from these particles, the particle must first adhere to the lung tissue. Various chemicals can be highly concentrated within a particle, even if the concentration is low when calculated as amount of chemical per volume of air. Thus an adherent particle may cause substantial damage to the lung tissue, inhibiting the normal functioning of the tissue and damaging the body's health. Similarly, sniffing particles, such as cocaine powder, can cause local damage to the cells that line the nasal passages.

Other drugs may be administered by injection, that is, with a needle. In a drug injection, the needle can be placed in many locations, including into a muscle (intramuscular), under the skin (subcutaneous), or directly into a vein (intravenous, or IV). Each of these routes may be appropriate for different therapeutic drugs. For example, insulin, a drug given to control a sugar transport disease called diabetes, is usually given intramuscularly so that it will be released into the blood over a period of time rather than all at once; insulin cannot be given orally because it is a protein that would be digested in the gut.

Uptake from an intravenous injection is faster than it is from other routes, because entry is not dependent on absorption. When a drug must be absorbed from the muscles, the gut, the nasal passages, or the lungs, only a portion of the drug is taken up; intravenous injection puts the drug directly into the bloodstream. Accidental overdoses can more easily arise from injection than from other routes of administration. The biological consequences of drug abuse became more severe after the invention of the hypodermic needle and syringe in 1853. All forms of drug injection can be dangerous if contaminated needles are shared because these practices can transmit infections such as hepatitis and AIDS from one user to another (see Chapter 16, p. 596).

Distribution of drugs throughout the body

Although the concentration of drug is initially highest at the local site of entry, the blood vessels (veins and arteries; see Chapter 10, p. 352) quickly distribute substances throughout the body. Transport through veins and arteries is one aspect of drug distribution; transport into the tissues is another. Drugs and other molecules can leave the bloodstream and enter the body's tissues via the smallest blood vessels, the capillaries. In most areas of the body, almost every tissue cell is within a few cells of a capillary. Areas of the body that have many blood capillaries, such as the lungs, liver, kidneys, and heart, receive a higher dose of most drugs and receive it faster than those areas with fewer capillaries (e.g., skin, muscle, and fat).

Various parts of the circulatory system let different molecules through. To act on the nerve cells of the central nervous system, psychoactive drugs must pass across the blood–brain barrier (see Chapter 13, p. 484). Drugs that do not easily cross this blood–brain barrier remain in the bloodstream while they are in the brain, even though they can leave the bloodstream in other locations of the body. Most antibiotics, for example, do not cross the blood–brain barrier, making it difficult to treat bacterial infections in the brain. In contrast, ethyl alcohol is particularly effective at crossing the blood–brain barrier and in causing both short-term and long-term damage to brain cells. (Chemists recognize many types of alcohol, but the only one that we consider here is ethyl alcohol, commonly used in alcoholic beverages.)

CONNECTIONS
CHAPTERS 10, 13, 16

Elimination of drugs from the body

In contrast to food molecules, which may be stored, most drugs begin to be broken down or removed from the body as soon as they enter it. One way in which drugs are eliminated from the body is in the urine produced by the excretory system. Another way is for the active form of a drug to be chemically altered, in a process called drug metabolism, so that it is no longer active. Drug metabolism can occur in a central location such as the liver, or in scattered locations, as in the inactivation of neurotransmitters within synapses between nerve cells (see Chapter 13, pp. 472–473). In this section we look at these two mechanisms by which drugs can leave the body.

Metabolic elimination. Many drugs are chemically altered by the body into substances that no longer produce the drug's effects, although some of these breakdown products may have effects of their own. One example of a drug that is eliminated by drug metabolism is ethyl alcohol (ethanol), which is metabolized by the cells of the liver. The metabolic breakdown of ethyl alcohol involves several steps, but the overall rate is limited by the amount of the proton carrier NAD^+ available to be reduced to NADH (see Chapter 10, p. 350) and by the levels of the enzyme alcohol dehydrogenase. The liver can metabolize 7–10 ml (about 0.25–0.33 fluid ounces) of alcohol per hour; if the rate of intake exceeds the rate of metabolism, intoxication results. As Table 14.1 shows, the intake of most forms of alcoholic beverages can easily exceed this limit.

One step in the metabolic breakdown of alcohol involves the production of acetaldehyde, which is immediately broken down by another enzyme, acetaldehyde dehydrogenase. Some people become sick from small amounts of alcohol because they lack this second enzyme, leading to a buildup of acetaldehyde, a toxic chemical. In some populations, the proportion of people who lack acetaldehyde dehydrogenase is quite high; for example, as many as 50% of all people in Japan and China lack this enzyme. Disulfiram (Antabuse), a therapeutic drug used in the treatment of alcoholism, works by inhibiting acetaldehyde dehydrogenase in people with normal levels of this enzyme. People who take this drug and then drink alcohol become very sick.

CONNECTIONS
CHAPTERS 10, 13

The excretory system. In addition to drug metabolism, drugs are also eliminated by the same system that rids the body of the normal waste products of metabolism, such as the nitrogen compounds derived from the breakdown of proteins. The process of removing these waste materials from the body is called **excretion**. The major route for the excretion of drugs and other substances is through the urine. The excretion of urine is vital to maintaining the blood and tissues at the proper concentrations of many types of ions. For example, calcium (Ca^{2+}) is an important second messenger in muscle contraction, and potassium (K^+), sodium (Na^+), and chloride (Cl^-) each function in maintaining charge gradients across membranes and in the propagation of nerve action potentials (see Chapter 13, pp. 469–471). Levels of these ions in the blood are monitored and controlled by feedback loops in the kidneys.

Urine is produced in three steps—filtration, reabsorption, and secretion—all in the nephrons of the kidneys. The first step, filtration, takes place in the many thousand glomeruli within the nephrons. Each glomerulus consists of a mesh of capillaries surrounded by the Bowman's capsule (Figure 14.2). The circulatory system brings blood in via the renal (kidney) artery, and low-molecular-weight substances, including ions, amino acids, glucose, urea, and water, are filtered out of the blood and

Table 14.1

Beverages containing equivalent amounts of ethyl alcohol (ethanol).

BEVERAGE	ALCOHOL CONTENT (%) (VARIES SOMEWHAT)	AMOUNT OF BEVERAGE EQUIVALENT TO 'ONE DRINK'	QUANTITY OF ETHANOL
Beer and ale	4	12 oz can (350 ml)	1/2 oz (15 ml)
Table wine	12–15	4 oz glass (100 ml)	1/2 oz (15 ml)
Dessert wine (sherry or fortified wine)	20	2/3 glass (2.5 oz or 70 ml)	1/2 oz (15 ml)
Distilled liquor (whiskey, vodka, etc.)	40	1.25 oz (35 ml)	1/2 oz (15 ml)
Liqueur	40	1.25 oz (35 ml)	1/2 oz (15 ml)

Figure 14.2

The major organs of the human excretory system, responsible for the production of urine.

into the Bowman's capsule, which empties into the proximal tubule of the kidney. Larger molecules such as proteins are retained in the blood.

The second step, reabsorption, takes place in the proximal tubules and in the loops of Henle, both of which are surrounded by capillaries. Here the ions, amino acids, and glucose that the body can reuse are reabsorbed, meaning that they move back from the tubules into the blood. Reabsorption is by active transport (see Chapter 10, pp. 348–349) via transporter proteins in the membranes of the nephron cells. Much of the water is also reabsorbed.

The third step, secretion, takes place farther along, in the distal tubules. Higher-molecular-weight substances including drugs and toxins are secreted from the blood into the tubules. After these three stages, the liquid in the tubules is called urine. Urine goes to the collecting ducts and then from the kidneys to the bladder, where it is stored and from

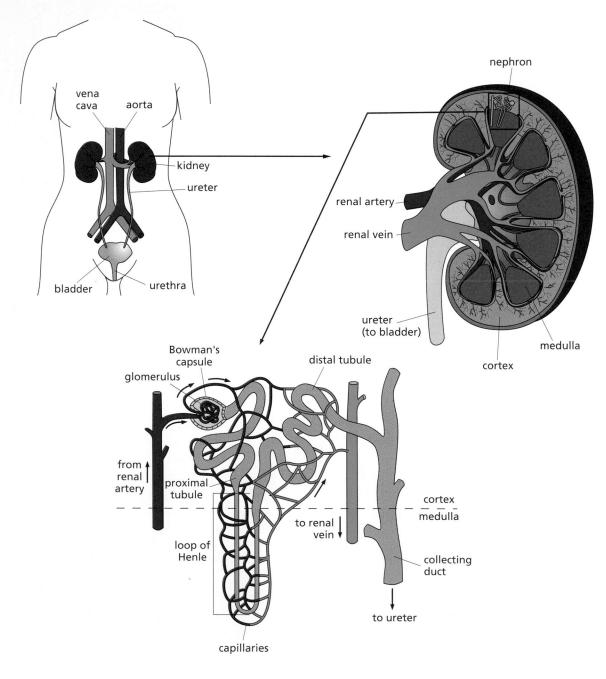

which it is excreted. Many drugs, or their breakdown products (metabolites), enter the nephrons at either the filtration or secretion steps and are excreted in the urine.

Drugs are also excreted in smaller amounts via saliva and sweat, and in air exhaled from the lungs. In nursing mothers, drugs may also be excreted into breast milk, with obvious consequences for the infant if the excreted drugs are still active, which they very often are.

Drug half-lives. The concentration of a drug in its active form can be measured, usually in the blood. The length of time required for the drug concentration to be decreased by half is called the **half-life** of the drug (Figure 14.3A). Cocaine, for example, has a half-life of 5–15 minutes, meaning that, regardless of the mode of intake of the drug or the initial starting dose, half of it will be gone in 5–15 minutes. A drug with a half-life of 10 minutes remains active for a shorter time than a drug with a half-life of 10 hours (Figure 14.3B), even if they start out at the same initial concentration in the blood. Drugs that are inactivated in a simple one-step process often have a shorter half-life than those whose inactivation pathways are more complex. For example, the half-life of alcohol is very short in most people, while marijuana remains in the blood for up to a week. Taking more of a drug before the original dose is completely eliminated increases the concentration of the drug to a level higher than the amount actually taken in the second dose (Figure 14.3C).

Drug receptors and drug action on cells

The activity of a drug depends on its dosage and its concentration in tissue. In many cases, drug activity also depends on the existence of specific receptors for the drug.

Specific receptors for drugs. After a drug has reached the site of its action, it acts on individual cells at that site. The ability of a drug to act on a particular cell most often depends on the presence of a cellular **receptor** for that drug. Picture a receptor as being like a glove. For a drug to bind to that receptor, the drug must have the shape of a hand. A small hand fits into a large glove, but, for any glove, there is some hand size that fits best. A large hand does not fit into a small glove. Those substances that have the best fit to a receptor have the greatest activity. Those that do not fit have no activity.

Many chemical molecules can have right-handed or left-handed shapes depending on the directions in which their bonds are arranged. That is, two molecules composed of identical atoms can differ in their shape. Right-handed molecules are said to be mirror images of left-handed molecules. As an aid in visualizing this, hold your hands flat. Place them in front of you, with the fingers of both hands pointing up and with the thumbs of both hands pointing toward the left (so that the palm of your left hand is towards you and the palm of your right hand is

Figure 14.3

Half-lives of various drugs. The half-life of a drug ($t_{1/2}$) is the time required for its active concentration to be reduced by half.

(A) Regardless of initial dose, a drug's concentration is reduced by half in one half-life.

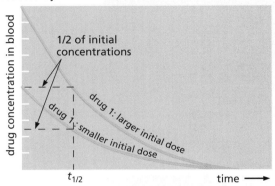

(B) Drugs with longer half-lives stay active for longer.

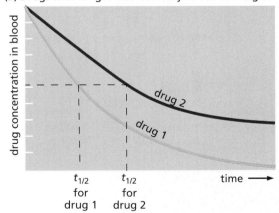

(C) Taking a second dose of drug before the first dose is completely eliminated raises the concentration, even if the two doses taken are equal.

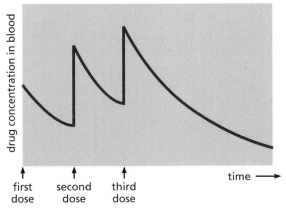

away from you). Look at your right hand in a mirror. Does the image of your right hand in the mirror look like your left hand without the mirror? Now put your right hand into the right glove of a fitted pair, then try to put it into the left glove. Your right hand fits into only one glove of the pair. Similarly, biological receptors are specific for right-handed or left-handed molecules, not both.

Tissue locations of receptors and locations of drug actions. Receptors can be located on the cell surface (plasma membrane receptors) or within the cytoplasm of the cell. Not all types of cells have all types of receptors, and the number of copies of a particular receptor molecule on a cell may change over time. The only cells that can respond to a particular drug are those with receptors for that drug. Thus, whether a drug is active in a particular tissue depends both on the ability of the drug to get to the tissue and on the presence of receptors for that drug on the cells of that tissue.

The binding of neurotransmitters to receptors on postsynaptic cells (see Chapter 13, pp. 471–472) is equivalent to drug–receptor binding. In fact, many psychoactive drugs act as either agonists or antagonists of neurotransmitters. An **agonist** is a substance that elicits a particular response or stimulates a receptor. For example, an agonist of a neurotransmitter is any other substance that produces the same response as that neurotransmitter (Figure 14.4A and B). An **antagonist** is a substance that inhibits a response (Figure 14.4C). The effects induced in all of the receptor-bearing cells throughout the body interact and produce functional effects in the body as a whole.

Side-effects and drug interactions

Because drugs are not taken by isolated cells, many effects are likely to be produced, especially for drugs that circulate throughout the body. Although a drug is usually intended to produce a specific effect, drugs almost always have effects other than the intended ones; these other effects are called **side-effects**. Side-effects may be weak or strong and can vary from person to person. Side-effects can also vary from beneficial or harmless to harmful or even lethal. Calling something a side-effect simply means that it is not the main effect or the reason for which the drug was used.

Types of drug–drug interactions. When two or more drugs are taken, they may interact in various ways. The simplest interaction is called an **additive effect**. When two drugs have an additive effect, the response to taking them together equals the sum of the responses produced by each drug individually (Figure 14.5A). In a **synergistic interaction**, one drug affects the action of the other so that the total response is *greater* than the sum of the responses to each drug separately (Figure 14.5B). An

Figure 14.4

Drug agonists and antagonists.

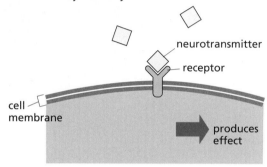

(A) Natural response to a chemical produced by the body

neurotransmitter

receptor

cell membrane

produces effect

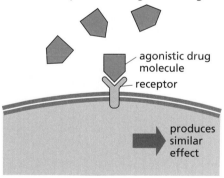

(B) Similar response to an agonistic drug

agonistic drug molecule

receptor

produces similar effect

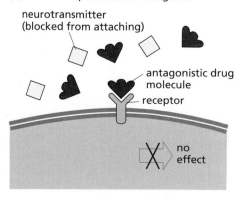

(C) Lack of response to an antagonist

neurotransmitter (blocked from attaching)

antagonistic drug molecule

receptor

no effect

antagonistic interaction is one in which one drug inhibits the action of another so that the total response is *less* than the response to the two drugs individually (Figure 14.5C).

Mechanisms of interactions. One mechanism for drug interactions is for one drug to change the threshold for response to the other drug. The **threshold** is a value (in this case, a drug dose) below which no effect is detectable. One drug may lower the threshold for response to the second drug, making the body responsive to a lower concentration of the second drug. This could happen, for example, if one drug increased the number of receptors or the affinity of the cell receptors for the other drug. In contrast, one drug may antagonize the action of a second drug by raising the threshold for response to the second drug or decreasing the number or affinity of receptors for it. Again, this is similar to the actions of neurotransmitters (see Chapter 13, pp. 471–472).

Differences in drug half-lives have implications for drug dosage and interactions between drugs. Because drugs with long half-lives remain in the body for a long time, a second dose adds to the remaining fraction of the first dose to produce a concentration higher than would result from either dose separately (see Figure 14.3C). Drugs with long half-lives also have the potential to interact with other drugs taken a long time after the first drug. Thus, it is not necessary that two drugs be taken at the same time for them to interact.

Another way in which drugs can interact is by altering the body's ability to metabolize other drugs. Barbiturates are eliminated from the body by enzymes contained in liver cells, and the use of barbiturates causes more of these enzymes to be made. Because these same enzymes eliminate other drugs, barbiturate use lowers the body concentrations of other medications, including steroid hormones. Body concentrations of estradiol, a steroid in birth control pills, is decreased by barbiturates, so taking birth control pills along with using barbiturates may result in an unwanted pregnancy.

Frequency of drug–drug interactions. Drug–drug interactions are an increasingly important problem, because many people take medication

Figure 14.5

How drugs interact.

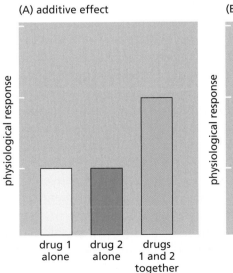

(A) additive effect

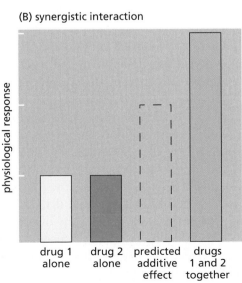

(B) synergistic interaction

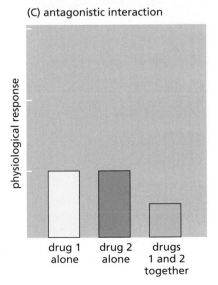

(C) antagonistic interaction

on a lifetime basis for the control of chronic conditions such as high blood pressure. A recent study on an elderly population demonstrated that the probability of having some adverse interactions from medications was 75% if they were taking five different medicines and 100% if they were taking eight or more medications on a regular basis. However, potentially harmful drug interactions are not solely a problem for elderly people. There can also be harmful interactions between various street drugs, between street drugs and medications, or between any of these and alcohol.

Drug safety

People obtain drugs in various ways. Some are sold by prescription only, a process that ensures at least some medical supervision. Other drugs, called 'over the counter' (OTC) remedies, are sold freely in stores. In many countries, a government agency supervises the formulation, but not the sale, of these OTC drugs. Plant products containing drugs are less widely regulated. In the United States, 'herbal remedies' and 'food supplements' can be advertised and sold with essentially no regulation, and the same is true of beverages containing caffeine. Alcohol and tobacco products are largely unregulated for adults, although various laws restrict their sale to minors. In addition, there are various 'street drugs,' sold illegally but not legally available. The definition of drugs covers a very wide spectrum.

Many variables can influence the safety of drug use. Drugs that are prescribed by a physician are usually safe if taken as prescribed, but harmful side-effects and drug interactions can still occur, even if directions are followed. The fact that a medicine is available only by prescription usually means that there is potential harm from unsupervised use. Most jurisdictions regulate the content of prescription medications to ensure that the stated ingredients are formulated to benefit most users with only minimal risk, if taken as directed, and that the formulation is appropriately labeled.

Over-the-counter drugs. Over-the-counter (OTC) drugs are those that can legally be sold without a prescription. Although people tend to view OTC drugs as 'safe' because a prescription is not required, there are actually thousands of deaths from OTC drugs each year and over a million cases of drug poisoning, generally from overdose. Aspirin is second only to barbiturates as the drug most frequently used in suicides and suicide attempts.

Several OTC preparations contain combinations of many drugs. Most of the OTC sleeping pills are combinations of aspirin and antihistamines. Cold remedies often have many ingredients and the liquid ones contain high concentrations (up to 25%) of alcohol. These combination drugs are marketed by urging people to 'cover the bases' by treating all possible symptoms. There is the unstated assumption in such advertising that the taking of unneeded drugs is harmless; there is no mention of possible negative effects of any of the ingredients.

While OTC drugs, when used as intended, are generally safe, their use in the wrong dose or for purposes other than those designated on the package can be harmful. People in the United States often have cultural expectations that problems can be fixed rapidly and with little effort on the part of the patient. This expectation, encouraged by advertising, is certainly a factor in the tremendous use of OTC drugs. Many of the OTC drugs, particularly those with psychoactivity, have a potential for abuse.

There are an estimated 350,000 pharmaceutical products (brand names) in the United States alone, with annual sales of $5 billion. New drugs seeking entry into the U.S. market must now undergo an extensive review process by the Food and Drug Administration (FDA). The drug must be proved to be effective in its intended use and must be demonstrated to be free of adverse side-effects. Many of the OTC drugs on the market have not undergone this process because they were 'grandfathered,' that is, exempted from testing because of their many years of prior use. Although official terminology lists these drugs as GRAS (Generally Regarded As Safe), many studies have shown that common OTC drugs, including aspirin, have side-effects that cause harm in so many cases that they would not be able to meet the more stringent standards applied to new drugs.

Drug contaminants and additives. Although prescription drugs and OTC drugs are not always 'safe' unless taken as directed (and even then are not without their side-effects), consumers can at least be assured that the product contains the drug it claims to contain and that the quantities of it are standardized from one batch to the next. Although no medication is sold as an unmixed, pure compound, the purchaser can be assured that harmful impurities are not present. None of these assurances exist, however, for herbal remedies, food additives, or street drugs. Street drugs can contain many impurities left over from chemical synthesis, and other substances are sometimes added deliberately. Strychnine is sometimes added to LSD ('white acid'), supposedly to sensitize the nerves to the LSD. It is also sometimes added to marijuana without the knowledge of the purchaser. By acting on the nerve cells, strychnine causes abnormal muscle contractions (convulsions) and is therefore sometimes used as rat poison. There is no specific antidote for strychnine, so it cannot be counteracted once it is taken, and it can cause permanent nerve damage in the brain and elsewhere.

Herbal medicines. Herbal medicines have a long history of use in many cultures, and a majority of the medicines that we now use were originally derived from plant sources. Many people have been returning to the use of herbal medications, sometimes favoring them as an alternative to the medicines sold by drug companies. At least some of this trend stems from a belief that such 'natural' remedies are better or safer than drugs obtained in pill form. Although many of these herbs are very active as medications, it is not accurate to think of them as always being 'safe.' Unprocessed herbs have unpredictable variations in the quantities of both the desired ingredients and the potentially harmful ones. Moreover, many of the most potent psychoactive drugs are plant products. Valerian is a plant containing a chemical called chatinine, which is a tranquilizer. Scotch broom contains a strong sedative, cytisine. Other herbal products, flowers and some spices (nutmeg, sassafras, mace, saffron, crocus and parsley) are strong stimulants of the central nervous system.

Chemically, the active ingredient in a herbal medication is identical to the same active ingredient if it is in a medicine produced by a drug company. Herbal medications are present in an unpurified form; that is, the active ingredient is present along with many other chemicals whose effects are not known and many of which are unidentified. A potential problem with herbal medications is the inconsistency of dose, both in the amount present in a particular plant and in the amount present in a particular extract. For example, many herbal remedies are taken as teas, that is, extracted in hot water. The length of brewing time and the temperature of the water greatly affect the amount of drug extracted.

Unlike OTC drugs, which are regulated by law to contain predictable doses of labeled medications and to contain no harmful ingredients, herbal remedies are unregulated in most places. A customer buying an herbal remedy cannot be assured that harmful substances are absent, or that the product in question has been tested and shown to be effective, or even that the product contains any active ingredient at all. Also, knowing that a substance is 'natural' offers no assurance that it is harmless: opium, strychnine, snake venom, and botulism toxin are all deadly poisons and all are perfectly natural.

In most countries, prescription medications and OTC medications are generally regulated by a government agency such as the FDA. A regulatory loophole, however, permits substances to escape regulatory oversight in the United States if they are marketed as 'food supplements.' Unfortunately, the term 'food supplement' is so ill-defined that nearly any substance can be so designated by the company that sells it, even if it fits the biological definition of a drug. Many herbal formulations are marketed this way in the United States, including echinacea, *Ginkgo biloba*, saw palmetto, and St. John's wort. Ephedra, the dried leaves of the plant *Ephedra sinica*, contains the stimulant ephedrine, and one company selling this product is currently under investigation in the United States because several deaths have resulted from its use. Tests on *Ginkgo biloba* show that it has no therapeutic effect on normal, healthy people. Without regulation, companies can market these drugs without testing them for dosage, efficacy, harmful side-effects, or interactions with other drugs, and they can also make health claims that have never been substantiated by research (although they are prohibited from making therapeutic medicinal claims). For these reasons, people taking such substances should realize that they are dealing with untested risks.

Studies on the safety or efficacy of various herbal remedies have shown that they are in many cases untrustworthy. Many herbal remedies are sold or advertised with claims that are not substantiated by any research. Herbal formulations also vary greatly in the level of their therapeutic ingredients; some formulations contain no detectable levels at all. Even when herbal remedies do contain active ingredients, their therapeutic effects are frequently inferior to those of prescription medications. Patients who self-medicate with herbal remedies are often denying themselves the benefits of prescription medicines known to be more effective. Patients who take herbal remedies together with prescription or OTC medications should keep their physician informed because of the increased risks of overdose or of drug interactions.

THOUGHT QUESTIONS

1 The physiological effects of a drug usually depend on its concentration in body tissues. If two people, one of whom weighs 60 kg (about 130 pounds) and the other 90 kg (about 200 pounds), take the same *amount* of a drug (one beer each, for example, or two aspirins), will the concentration reached in their body tissues be the same? If other factors were equal, which person do you think would be more strongly affected by the drug?

2 How would you set up an experiment to monitor the rate at which a drug is delivered to various body tissues? Consider both animal and human test subjects.

3 Why do left-handed and right-handed forms of the same molecule frequently differ in their ability to act as drugs? When a difference of this kind affects the activity of a drug molecule, what does it indicate?

4 How long will it take for the blood concentration of a drug whose half-life is 20 hours to be reduced to one-eighth of its peak concentration?

5 Devise a model of the mechanism of action of two drugs with their receptors that would account for additivity between the two drugs. How would you modify this model to account for antagonism? How would you modify it to account for synergism?

Psychoactive Drugs Affect the Mind

The brain coordinates the activities of the rest of the body (moving, sleeping, eating, breathing), but does it also produce activities that we associate with the mind? As we saw in Chapter 13, there is now considerable evidence in support of this theory. Electrical and biochemical activities have been measured in the brain during dreams and thought. Some diseases in which there is brain degeneration can be accompanied by changes in personality. Mental illnesses such as depression are associated with changes in brain neurotransmitters and can be treated with drugs (see Chapter 13). Additional evidence for the 'mind equals brain' theory comes from research on psychoactive drugs. Psychoactive drugs act directly on the nerve cells of the brain or central nervous system (CNS) to produce changes in consciousness, mood, or perception in highly drug-specific ways. Both sensation and perception can be altered, sometimes permanently, by drugs. People may become paranoid due to permanent damage from drugs; that is, they perceive danger that is not present in their incoming sensations. They may also experience hallucinations, chemically induced changes in sensation.

Some psychoactive drugs (opiates, marijuana, and nicotine) work through specific receptors, and some (like amphetamines and hallucinogens) work because they are structurally similar to a neurotransmitter and bind to receptors for that neurotransmitter. Drugs that do not work via receptors (e.g., caffeine and alcohol) have a more generalized effect because they act on many types of cells.

Opiates and opiate receptors

Opiates are narcotics, that is, drugs that cause either a drowsy stupor or sleep. Most narcotics also cause some degree of euphoria (literally, 'good feeling'), and most are highly addictive. In low doses, most narcotics can be used as painkillers (analgesics). High doses of narcotics can produce coma and death. Most of our common narcotics, including heroin, morphine and codeine, are derived from the opium poppy (*Papaver somniferum*) and thus are known as opiates. The painkilling and euphoria-inducing properties of opiates have been known for thousands of years in Asia, as has their ability to cause addiction. Pain reduction results from changes in the release of the neurotransmitters acetylcholine, norepinephrine, dopamine, and a pain-related substance called substance P. In people who are not in pain, opiates produce euphoria by action on neurons in certain locations in the brain.

Opiates and similar chemicals (opioids) act on cells via one or more specific opiate receptors to produce their various effects. Synthetic antagonist drugs have been made that block the binding of opiates to their receptors. These narcotic antagonists, of which naloxone (Narcan) and naltrexone are examples, are useful in the treatment of opiate overdoses. Some other synthetic drugs act as both agonists and antagonists: they bind to opiate receptors, thus blocking the effects of narcotics like morphine (antagonistic action), but their own binding produces some euphoric effect (agonistic action), sometimes leading to their abuse.

Opiate receptors are found on the membranes of many other types of cells in addition to neurons. Opiates consequently relax various muscles, including those of the colon. Consequently, morphine is an ingredient in some prescription anti-diarrheal medications (Paregoric), and severe constipation is an effect of long-term opiate use.

Several neurobiologists hypothesized that opiate receptors would never have evolved unless they served some adaptive function other than just allowing addictive behaviors to be learned. This type of thinking led to the hypothesis that opiate receptors must have some normal physiological molecules that could bind to them. The search for such molecules led to the discovery of endorphins and enkephalins, sometimes called endogenous opiates because they are produced within the body (endogenous), rather than being taken in from the outside (exogenous). These are peptides (short chains of amino acids) that have a molecular shape very similar to that of a portion of certain opiate molecules. The current hypothesis is that the endogenous opiates act as inhibitory neurotransmitters, decreasing the activity of the neurons that normally signal pain and stress. These same endogenous opiates have other effects throughout the body, including actions on the immune system, which are discussed in Chapter 15.

Marijuana and THC receptors

A number of psychoactive drugs are contained in marijuana smoke, the most active of which is Δ_9-tetrahydrocannabinol (THC). There are receptors for THC in the parts of the brain that influence mood. Binding of THC to these receptors produces an altered sense of time, an enhanced feeling of closeness to other people, and an intensity of sensory stimuli. In higher doses, marijuana can also cause hallucinations. Unlike the opiate receptors, for which endogenous brain chemicals have been found, endogenous substances that bind to THC receptors have not yet been found. However, researchers have found that certain compounds found in chocolate may bind to THC receptors. People who refer to their love of chocolate as an addiction may thus be close to the truth.

Stimulation of the THC receptors causes a release of norepinephrine by the nerves in the median forebrain bundle, producing euphoric effects. There are also THC receptors on the cells of the hypothalamus (see Figure 13.15, p. 484), a secretory part of the brain that regulates the steroid sex hormones. Long-term marijuana use decreases testosterone levels in males and alters the menstrual cycle in females.

Marijuana has long been recognized as a potentially dangerous and addictive drug, and its use is illegal in most countries. Later in this chapter we examine the possible medical uses of marijuana and the efforts by some people to relax legal restrictions on such medical use.

Nicotine and nicotinic receptors

Cigarette smoke contains over 1000 drugs, a large number of which are carcinogens (see Chapter 12, p. 443–445). The primary psychoactive drug among them is nicotine. In the brain, nicotine acts to stimulate the cerebral cortex, possibly by a direct effect on the cortical neurons, which have a series of nicotinic receptors. Nicotine also acts by stimulating nicotinic receptors on the neurons in the sympathetic ganglia, releasing the neurotransmitters acetylcholine, epinephrine and norepinephrine. In the brain, norepinephrine produces increased awareness. In the rest of the body, these neurotransmitters produce a variety of physiological effects: increased heart rate and blood pressure, constriction of blood vessels, and changes in carbohydrate and fat metabolism.

Amphetamines: agonists of norepinephrine

CONNECTIONS CHAPTERS 12, 13

Amphetamines are an example of a type of drug called central nervous system (CNS) stimulants. All amphetamines are derivatives of ephedrine, a drug originally obtained from the mah huang plant (*Ephedra sinica*). CNS stimulants increase behavioral activity by increasing the activity of the reticular activating system. The reticular activating system is the portion of the brain that normally maintains a baseline level of neuronal activity in the brain as a whole, thus keeping the body at a baseline level of wakefulness and awareness (see Chapter 13, p. 496). CNS stimulants have side-effects on organs outside the brain. Because of their effects on judgment and their effects on other organ systems, such as the heart and diaphragm (see Figure 14.1), they are dangerous drugs, accounting for 40% of all drug-related trips to the emergency room and 50% of all sudden deaths due to drugs.

Amphetamines mimic the effects of the neurotransmitter norepinephrine by binding to norepinephrine receptors. They can also indirectly increase norepinephrine activity by blocking its reuptake from the synapse and by inhibiting monoamine oxidase (MAO), the enzyme that normally breaks down norepinephrine. Either mechanism results in more norepinephrine remaining in the synapse to act on the receiving (postsynaptic) neuron (see Figure 13.6, p. 472). Prolonged use of high doses of amphetamines can induce a form of psychosis that includes aggressiveness, delusions, and hallucinations, possibly because of an oversupply of an enzyme involved in norepinephrine synthesis.

A controversial drug of this group is methylphenidate (Ritalin). In most human subjects, it has mild amphetaminelike effects similar to those described above. However, the drug has quite the opposite effect (called a 'paradoxical effect') in children who have attention deficit hyperactive disorder (ADHD, formerly called ADD)—it reduces their hyperactivity. The reason for this paradoxical effect seems to be related to the fact that hyperactive children actually have a lower than normal function in the reticular activating system, the area of the brain that keeps the rest of the brain alert. Such children may need to be constantly moving around to arouse their reticular activating system. Ritalin's chemical stimulation of this area obviates the children's need for movement.

One currently abused drug of this group is 'ecstasy' (technically methylenedioxymethamphetamine or MDMA), a drug associated with many all-night 'rave' dance parties because it suppresses feelings of fatigue. Ecstasy triggers the release of all stores of the neurotransmitter serotonin from brain neurons. It also blocks the reuptake of serotonin, thus producing prolonged alterations of sensory perceptions. In addition, one of the normal targets of serotonin is the hypothalamus, the part of the brain that regulates body temperature and thirst. Prolonged triggering of the hypothalamus can dehydrate the body and raise body temperature to damaging, even fatal, levels.

LSD: an agonist of serotonin

Lysergic acid diethylamide (LSD) is derived from the fungus *Claviceps purpurea*, which grows on rye. In contrast to ecstasy, which triggers release of serotonin itself, LSD and the related compound psilocybin (from mushrooms found in Central and South America) are structurally similar to serotonin. These drugs therefore act as serotonin agonists and activate the nerve cells that normally respond to serotonin, leading to altered sensory perception and hallucinations. 'Altered perception' means a change in the awareness of stimuli that actually exist. Colors may become brighter, or sounds more clear. Perceptions of the sizes of objects and of speed or time may be altered. Hallucination, in contrast, is the perception of things for which no outside physical stimuli have been received. Hallucinations can be visual, auditory, olfactory, or cognitive. Heavy use of these serotonin agonists leads to permanent brain damage, with symptoms ranging from impairments of memory, attention span, and abstract thinking to severe, long-lasting psychotic reactions.

Caffeine: a general cellular stimulant

Not all psychoactive drugs produce their effects via action on CNS neurotransmitters and their receptors. Some produce their effects within cells. Caffeine, for example, works inside cells to increase their rate of metabolism. Caffeine thus has a general stimulatory effect on cells throughout the body, including neurons in the brain. Like other CNS stimulants, caffeine increases the general level of awareness via action on the neurons of the reticular formation of the brain. In higher doses, or in more susceptible individuals, it can also produce insomnia (inability to sleep), anxiety, and irritability. It increases the heart rate, the respiratory rate, and the rate of excretion of urine by the kidney. It dilates peripheral blood vessels, but constricts the blood vessels of the CNS, which produces headaches in some people at high concentrations.

Caffeine is derived from several types of plants, including the beans of the coffee plant (*Coffea arabica*), the leaves of the tea plant (*Thea sinensis* or *Camellia sinensis*), the seeds of the cocoa plant (*Theobroma cacao*, from which we get chocolate), and nuts from the kola tree (*Cola acuminata*, an African tree, the source of cola beverages). Table 14.2 shows the amounts of caffeine in different beverages and nonprescription medications.

Alcohol: a CNS depressant

Ethyl alcohol belongs to a category of drugs called 'CNS depressants' because they depress the functioning of the CNS by inhibiting the transmission of signals in the reticular activating system. Also included in this category are barbiturates and tranquilizers. Because these drugs lower the general level of awareness, they are also called sedatives or hypnotics. Because they all affect the reticular activating system, when two or more CNS depressants are taken together, the effect is stronger (either additively or synergistically) than when either is used alone. The actions of barbiturates and tranquilizers are mediated through receptors on neurons in the brain, whereas alcohol produces a more generalized effect because it acts by making all cell membranes more fluid.

Alcohol is soluble in both water and fat and is thus readily able to pass through the plasma membrane of the cells forming the blood–brain barrier. The portions of the brain affected depend on the dose: the higher the dose, the deeper into the brain the alcohol penetrates. Even low doses of alcohol can impair a person's response time, with devastating (often fatal) consequences if that person is driving a motor vehicle or boat. In the United States, alcohol is involved in over 60% of all motor vehicle fatalities and in over half of all drownings. The likelihood of sexual abuse is also greatly increased if alcohol has been consumed first.

Higher doses of alcohol suppress the reticular activating system's stimulation of those portions of the brain involved in involuntary processes, such as the brainstem and the medulla oblongata. Depression of the medulla can result in the cessation of breathing (respiratory arrest)

Table 14.2

Caffeine content of beverages and nonprescription medications.

FORM OF INTAKE	APPROXIMATE CAFFEINE DOSE (MG)
HOT BEVERAGES, PER 6 OZ CUP (170 ML):	
Coffee, brewed	100–180
Coffee, instant	100–120
Tea, brewed from bag or leaves	35–90
Cocoa	5–50
Coffee, decaffeinated	2–4
CARBONATED BEVERAGES, PER 12 OZ CAN (350 ML):	
Cola drinks; also several others	35–60
NONPRESCRIPTION (OTC) DRUGS, PER TABLET:	
Stimulants:	
Vivarin, Caffedrine	200
NoDoz	100
Analgesics (pain relievers):	
Excedrin extra strength	65
Midol maximum strength	60
Anacin, Bromo Seltzer, Cope, Emprin	32

and death. Each year, thousands of college students engage in 'binge drinking'—consuming large quantities of alcohol in a short time. Binge drinking often results in coma ('passing out') and can result in death, including death from the alcohol directly as well as death from falls or other accidents while intoxicated.

The effects of alcohol on behavior can be predicted by the blood alcohol level (Table 14.3). Blood alcohol level is measured as the number of grams of alcohol in each 100 ml of blood. Thus 1 g of alcohol per 100 ml equals a 1% blood alcohol content and 100 mg equals 0.1%. Since alcohol is evenly distributed throughout the body, the actual blood alcohol content that results from drinking a given amount of alcohol varies with the blood volume of the person, which is approximately proportional to the muscle weight of the person. A blood alcohol level of 0.1% is the level at which a person can be charged with Driving While Intoxicated (DWI) or Operating [a motor vehicle] Under the Influence (OUI) of alcohol in most states of the United States, and several states have amended their laws to make the limit even lower. At a blood alcohol content of 0.05%, the probability of being involved in an automobile accident is 2–3 times higher than for someone who has not been drinking.

Table 14.3

Brain and behavioral effects of alcohol consumption. The greater the amounts of alcohol consumed, the deeper within the brain the alcohol penetrates after it is absorbed across the blood–brain barrier. Blood alcohol content is given for a 150 pound (68 kg) person; in general, persons weighing less will experience higher blood alcohol levels after consuming the same amounts.

BLOOD ALCOHOL CONTENT (%)	NUMBER OF DRINKS*	EFFECTS
0.01–0.04	1–2	Slightly impaired judgment
		Lessening of inhibitions and restraints
		Alteration of mood
0.05–0.06	3–4	Disrupted judgment
		Impaired muscle coordination
		Lessening of mental function
0.07–0.10	5–6	Deeper areas of cortex affected
		Slower reaction time
		Exaggerated emotions
		Talkativeness or social withdrawal
		Mental impairment
		Visual impairment
0.11–0.16	7–8	Cerebellum affected
		Staggering
		Slurred speech
		Blurred vision
		Greater impairment of judgment, coordination, and mental function
0.17–0.20	9–10	Midbrain affected
		Inability to walk or do simple tasks
		Double vision
		Outbursts of emotion
0.21–0.39	11–15	Lower brain affected
		Stupor and confusion
		Increased potential for violence
		Noncomprehension of events
0.40–0.50	16–25	Activity of lower brain centers severely depressed
		Loss of consciousness; shock
0.51 or more	26 or more	Failure of brain to regulate heart and breathing
		Coma and death

Based on equivalent amounts of alcohol: 12 oz of beer, 4 oz of wine, 1.25 oz of liquor (see Table 14.1).

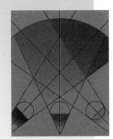

THOUGHT QUESTIONS

1 Does the finding that many drugs act directly on the cells of the brain to alter perception, mood, and consciousness necessarily mean that there is no 'mind' or 'spiritual essence' apart from the physical entity of the brain?

2 Can a person accept the scientific findings in neuroscience without accepting the central theory that the mind and the brain are one? Can the central theory of neurobiology ever be proved beyond doubt?

3 Alcohol concentrations in the body are measured as 'blood alcohol' levels. After alcohol is consumed, and before any of it is eliminated from the body, does all of it remain in the blood? Where else does it go?

4 In many countries the drinking age is the same as the driving age. In the United States, the driving age is lower than the drinking age. What kinds of impacts has this had in various societies?

5 How many drinks is someone likely to consume on a binge? Look at Table 14.3. What effects are likely to result from this number of drinks?

Most Psychoactive Drugs Are Addictive

Some drugs that have uses as medicines, and many others that do not, are also used socially. All cultures have used at least some drugs, particularly psychoactive drugs, for nonmedical purposes that can be described as social. For example, in many cultures, wine is a common accompaniment to food, and wine is also used in many religious ceremonies.

Excessive or harmful social use of a drug is considered **drug abuse**, or **substance abuse**. Drug abuse is a major problem that affects many thousands of people, their families, and many other people with whom they interact. Most drugs that are abused socially are psychoactive drugs. These drugs have both direct effects on the user and indirect effects on other people owing to the user's altered behavior.

If psychoactive drugs have been used by all cultures, why not approve their use? In most cultures in which the social use of psychoactive drugs has been endorsed by tradition, the uses have been highly ritualized or ceremonial, and the decision of how much drug to use is not left up to the individual. Psychoactive drugs such as alcohol impair a number of higher-order mental functions (see Table 14.3). One of these is judgment, the very mental function needed for a person to be able to distinguish between use and abuse. In addition, most, but not all, drugs that are abused are addictive.

Addiction has been defined as a compulsive 'physiological and psychological' need for a substance, implying that there is both a biological basis and a mental basis for addiction. However, as more psychologists have accepted the neurobiology paradigm that 'mind equals brain' and that all brain functions are biochemically based, the distinction between physiological and psychological addiction has become increasingly blurred. The above definition of addiction carries with it the assumption that all addictive drugs are psychoactive. The numbers of people in the United States using addictive drugs are given in Figure 14.6.

Dependence and withdrawal

Addictive drugs cause a physiological **dependence**, meaning that the person can no longer function normally without the drug. Once a person has become dependent on a drug, cessation of drug taking produces the biological symptoms of **withdrawal**. The length of time required for the development of dependence varies with the drug, as does the severity of withdrawal. Dependence on morphine and related drugs develops very quickly; withdrawal begins within 48 hours of the last dose and lasts for about 10 days. People are very ill during this time, but withdrawal is rarely fatal. Withdrawal from alcohol dependence is physically much more severe and can sometimes be fatal.

One aspect of addiction is psychological dependence. In this form of addiction, the physiology of the brain has become dependent on the drug in such a way that the person can no longer function normally without the drug. The sensations felt during withdrawal tend to be the opposite of the sensations produced by the original drug taking. For example, a person dependent on depressant drugs may become anxious and agitated during withdrawal. People who have become dependent on painkillers feel pain when the drug is withdrawn, long after the original, biological source of the pain has been eliminated, and these pain sensations often lead to resumption of the drug.

Cocaine, caffeine, nicotine, and marijuana all produce psychological dependence. Withdrawal from these drugs is typified by symptoms that affect the physiology of the brain rather than the physiology of the entire body, and because of this they were originally thought to be nonaddictive. In light of the neurobiology paradigm, it is now known that all four are addictive. In fact, if we define the level of addiction as the degree of difficulty of getting through withdrawal without returning to the drug, nicotine must be considered as one of the most highly addictive drugs known.

One aspect of psychological drug dependence in humans involves the context—the places in which the drugs are taken and the people sharing the experience. 'Conditioned withdrawal syndrome' refers to the fact that visual cues associated with drug taking (for example, seeing one of these people or places) can bring on the physiological symptoms of withdrawal in drug-dependent people even when the body is not actually in withdrawal, possibly leading them to seek the drug again. For this reason, many drug recovery programs recommend that recovering addicts stay away from specific people and locations associated with drug use.

Brain reward centers and drug-seeking behaviors

Neuroscientists have discovered both negative and positive reward, or reinforcement, systems in the brain. Certain things make us feel good (positive reward, or positive reinforcement). Stimulation of the nerves in the positive reinforcement system leads to a repetition of the behavior. Basic biological functions like eating and sexual activity are repeated because they activate these nerves. Throughout our

Figure 14.6

Users of addictive drugs in the United States in 1991. People are listed as users if they use the drug at least once a week (or, for heroin, at least once in the past year).

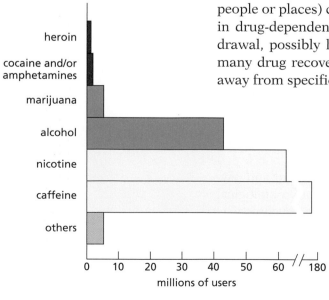

lives we learn other experiences that stimulate these centers. Our ability to derive pleasure from certain experiences, such as the feeling of 'a job well done' or the feelings evoked by a beautiful painting or a sunset, is 'learned' to a large extent; our families, religions, cultures, and other influences operate from birth to teach us to view certain experiences as positive and certain experiences as negative.

Two hypotheses of drug addiction. One hypothesis of drug addiction is that people use drugs to escape from some kind of pain. This hypothesis suggests that drug-taking behavior results from the attempt to inhibit or avoid the negative reward and to avoid withdrawal. The avoidance of a negative consequence, or removal of an unpleasant stimulus, is called 'negative reinforcement'.

A newer hypothesis suggests that drug addiction results from stimulation of the positive reinforcement centers. Of the psychoactive drugs, the ones that cause addiction are those that stimulate the positive reward system in the brain. They directly stimulate these centers, neurochemically producing the positive sensation. This hypothesis is supported by the findings that people may become addicted very rapidly, well before physiological dependence has begun. The two hypotheses may be valid for different drugs or different individuals, meaning that there may be more than one mechanism for addiction.

Evidence from behavioral experiments with rats. Much of the research on drug-seeking behavior has been done with rats. Rats have been fitted with tubes so that they can give themselves drugs, either into their bodies or directly into specific parts of their brains. These rats quickly learn to self-administer addictive drugs, and they increase the frequency of self-administration if allowed. They do not, however, self-administer nonaddictive drugs. Although physiological dependence on the addictive drugs does develop, the experiments suggest that dependence is the result of addiction rather than its cause.

A part of the brainstem known as the ventral tegmental area (VTA) (Figure 14.7) is thought to be the positive reinforcement center, or 'pleasure center.' This has been demonstrated in several types of experiments. If electrodes are placed into the ventral tegmental area, rats will activate the electrodes, electrically stimulating that area of the brain. They quickly learn to repeat the behavior, giving themselves repeated electrical stimulation, preferring it even over food. The same type of experiment can be done on rhesus monkeys, with the same results: they refuse food if their choice is between eating (even after starvation) and stimulating their ventral tegmental areas electrically. Electrodes placed in other areas do not produce repeated self-stimulation, and destruction of the ventral tegmental area of the brain stops the behavior.

In another type of experiment on rats, tubes are placed into an area of the brain so that by pushing a lever the rat can administer drugs directly to the area. By their action on neurons or their receptors, psychoactive drugs stimulate nerve impulses in the brain, producing the same effects as elicited by electrical stimulation of those same neurons. The negative reinforcement centers are in the periventricular areas of the brain, while the positive reinforcement center, as previously stated, is in the ventral tegmental area. The nerves of the ventral tegmental area make synaptic

Figure 14.7

Positive reinforcement areas of the brain, and the locations of action of addictive drugs on synapses in the cells of these areas.

connections to the nerves of another brain area, the nucleus accumbens (NA), which is involved in the processing or interpretation of the signal (see Figure 14.7). Self-administration of drug to the negative reward center is not reinforcing and is therefore not repeated, while self-administration to the ventral tegmental area or the nucleus accumbens is.

When rats are able to take drugs by a more normal route, rather than directly into their brains, the results are similar: they repeat the drug taking if the drug is addictive, and do not when the drug is nonaddictive.

Brain pathways for effects of addictive drugs

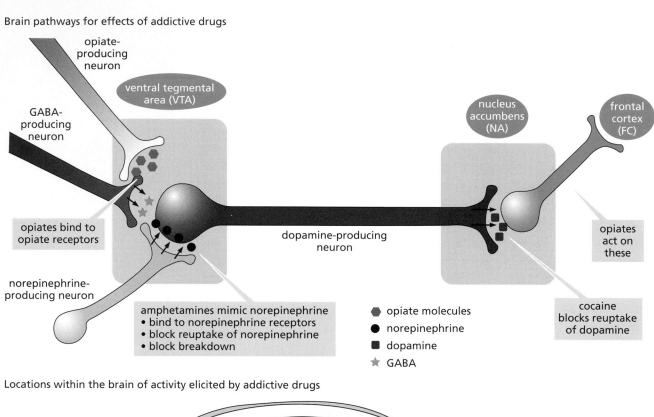

Locations within the brain of activity elicited by addictive drugs

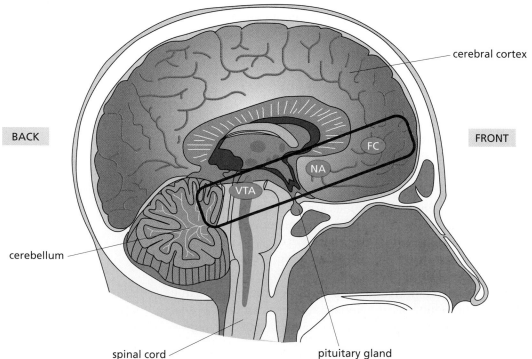

Measurements of changes in electrical activity in different parts of the CNS have shown that activity increases in the ventral tegmental area (and generally nowhere else) after the administration of an addictive drug.

The ventral tegmental area of the brain includes the reticular formation, the central part of the reticular activating system described above. Many of the drugs that act on the reticular activating system also act on the positive reinforcement area of the ventral tegmental area. Amphetamines indirectly stimulate the neurons of the ventral tegmental area, elevating mood. For this reason they have been successful in the treatment of depression. Cocaine acts on brain cells of the ventral tegmental area that secrete dopamine. Most researchers think that the euphoria produced by cocaine is due to its effects on the dopamine-secreting cells because the euphoria can be stopped with drugs that block dopamine receptors (see Figure 14.7). Opiates, marijuana, caffeine, and alcohol all produce ventral tegmental self-reinforcing effects.

Not all psychoactive drugs are addictive. Hallucinogens, for example, do not produce repeated self-administration by rats and therefore are not considered addictive under this hypothesis. Nicotine does not initially produce self-administration in rats; in fact, initially it is strongly aversive. However, after a rat has been exposed to nicotine several times, it begins to self-administer the drug and this becomes a strongly persistent behavior, that is, a behavior that is hard to break.

Conditioned learning in drug addiction. Addiction is both a biological response and a learned behavioral response in which the behavior being learned is the drug-seeking and drug-taking behavior. In this type of learning (called operant conditioning), behavior is learned as a result of its consequences. Drug seeking and drug taking can be learned this way if taking the drug is usually associated with stimulation of the positive reinforcement centers in the brain. Contextual cues are important. As is true in classical conditioning, one stimulus can become associated with another (see Chapter 13, pp. 488–489). Being in certain places or being with people with whom a person has taken drugs provide strong learned cues that bring on a physical sensation of craving for the drug. Thinking of those places or longing for those people brings on a craving for the drug and can also bring on the sensations produced by the drug itself. Seeing or thinking about aspects of the drug taking itself is often sufficient to bring on these feelings. A person dependent on cocaine reported that the sight of someone wearing a gold watch would bring on the sensations of a cocaine high because gold watches were one of the items he had often stolen to purchase his cocaine. The contribution of the 'drug culture' to the reinforcement of drug-taking behavior is enormous: many of the symbols and rituals associated with the culture become contextual cues. Attempts to block and reverse drug dependence must take into account these learned associations, which are called 'conditioned place preferences.' The rehabilitation of drug addicts is usually very difficult because of the strength of these learned associations. However, the chances of successful rehabilitation are increased if the addict can be helped to develop an aversion to the old behaviors while learning to substitute new behaviors.

Effects of long-term use. One of the effects of long-term use of psychoactive drugs is that they erase the ability of the ventral tegmental

CONNECTIONS
CHAPTER 13

nerves to respond to the normal positive signals: appreciation of a good meal, enjoyment of the company of friends, and happiness from helping others may all disappear. We tend to interpret positive experiences as 'pleasure,' so the positive reinforcement centers have sometimes been called the 'pleasure centers.' Drugs compete with the normal neurotransmitters in these brain centers. Long-term use of addictive drugs decreases the number of receptors on the nerve cells (see below) so that these centers are only triggered in the drug-abusing person by taking drugs, and no longer by the experiences that used to be pleasurable.

Drug tolerance

One of the biological effects of addictive drugs is that they produce **tolerance**, which is also called 'homeostatic compensation' (i.e., the body adjusting to new conditions). This means that the same dose of drug exerts a decreased effect when administered repeatedly; greater amounts of the drug must be taken to produce the same effect. Some well-meaning people have suggested that addicted people be given 'all the drugs they want' to keep them off the street. The simple biological reason why such an approach would not work is that drug tolerance develops. A person needs more and more drug to produce the original psychoactive effect. Higher doses affect other body systems and begin to produce negative mental states such as hostility and paranoia. The increasing doses may often become toxic or lethal.

Drug tolerance can be shown graphically as a shift of the drug's dose–response curve to the right (Figure 14.8). There are two broad categories of mechanisms that produce tolerance: metabolic and cellular. Metabolic tolerance is the body's production of an increased amount of the enzymes for the breakdown of the drug. Tolerance to barbiturates develops at least in part because the drug stimulates the synthesis of the liver enzymes responsible for its elimination, so the rate of elimination increases with repeated use.

Cellular tolerance results from changes in the receptors for the drug, principally the receptors on the nerve cells in the brain. Tolerance results from either a drug-induced decrease in the number of receptors or an increase in the threshold needed to trigger a response. Heroin produces these changes in the brain within a week or two of daily use. If use is more frequent than once a day, a higher level of tolerance develops; that is, even fewer receptors are present on the nerve cells so an even higher dose is required to produce an effect.

Some drugs can cause permanent damage to receptors, and therefore tolerance to these drugs becomes permanent. For most drugs, however, tolerance is not permanent but disappears gradually with time. The time period varies greatly from one drug to another.

So far we have emphasized the effects of psychoactive drugs on the brain. However, these drugs also have effects throughout the body, as we see in the next section.

Figure 14.8

Effect of drug tolerance on the dose–response curve. A higher dose (X_2) is required to produce the same response after drug tolerance develops.

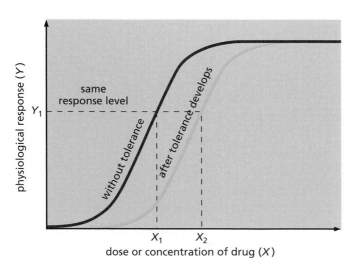

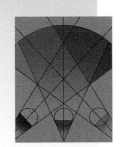
Drug Abuse Impairs Health

In addition to addiction, the use of psychoactive drugs can impair the health of individual drug users in other ways. It can also affect their gametes, and thus the development of their children.

Drug effects on the health of drug users

Drug abuse negatively affects the health of individuals and society. In the United States, deaths from drug overdoses increased from 6500 in 1979 to 10,000 in 1988. Most of these deaths were from heroin and cocaine, with the remainder being mostly the result of the misuse of legal drugs such as alcohol. Many more deaths also resulted from chronic use of drugs.

Alcohol. Among the most harmful of drugs, particularly in view of the frequency of its use, is alcohol. While each of the other types of CNS depressants is designated as a controlled substance, subject to fines and/or imprisonment for possession or sale in the United States and in many other countries, alcohol is not. The reasons for the difference are certainly not biological, because ethyl alcohol is a powerful depressant comparable to the others in terms of both short-term and long-term risks to health.

Most of the biological effects of alcohol on the body are **acute effects**, reversible in a matter of hours or days. Over time, however, permanent damage results from the **chronic effects**. Alcohol-induced biochemical imbalances permanently damage tissues such as the brain, liver, and muscle tissue (including the heart), resulting in dementia, cirrhosis and other liver diseases, and cardiovascular disease, respectively. Tissue damage may in turn contribute to altered uptake of and decreased sensitivity to medically necessary drugs, including antibiotics, antidiabetic drugs, and other medications. Alcohol in the gut also destroys certain vitamins and interferes with the absorption of others. This is why vitamin deficiencies rarely seen in industrial countries occur in those countries among alcoholics.

Alcohol depresses the immune system, leaving alcoholics very susceptible to infectious disease, including tuberculosis. Because alcoholics frequently substitute alcohol for food, they are often malnourished either in total calories or in micronutrients, which can further depress the immune system (see Chapter 15, p. 554). Alcohol consumption is also associated with an increased risk for cancer (see Chapter 12, pp. 447–448).

Caffeine. Caffeine, which is even more widely used than alcohol, can have significant negative effects on health. The consumption of more than 10 cups of coffee a day increases chromosome damage (which can lead to birth defects), respiratory difficulties, and heart and circulatory problems.

Tobacco. Tobacco is carcinogenic in any form. Smoking tobacco is correlated with lung cancer, whereas chewing tobacco is correlated with cancer in the mouth, a form of cancer that often metastasizes to other parts of the body. Tobacco and alcohol together produce a synergistic increase in the number of cancer deaths, beyond what either would produce without the other (see Figure 12.21, p. 447). Nicotine suppresses the immune system; consequently smokers have a high incidence of other lung diseases, including emphysema and respiratory tract infections such as pneumonia and bronchitis. Passive smoke (smoke produced by someone else's smoking activity) has been shown to increase death rates from cancer in adults and from sudden infant death syndrome, and it also increases the incidence of pneumonia and bronchitis in the first year of an infant's life. Because an infant's immune system is only partly developed, passive smoke can be much more damaging to infants than to adults.

In addition to direct effects on immune cells, tobacco smoke (including passive smoke) paralyzes the cilia on the lining of the respiratory tract. When we breathe, dust, bacteria, and other particles enter our respiratory tract along with the air. Many of these particles are trapped on a sticky fluid (mucus) that coats the respiratory lining. The cells lining the upper respiratory tract have many hairlike projections (cilia) on their surfaces. These cilia beat rhythmically in a coordinated way; the beating of the cilia moves the mucus and its trapped material upward, out of the respiratory tract. Tobacco smoke stops the beating of the cilia; inhaled bacteria and viruses consequently work their way down into the lungs instead of being eliminated. People who smoke therefore have a much higher incidence of respiratory tract infections and other infectious diseases of the lungs than do nonsmokers. The particulate matter from smoke, which can include cancer-causing chemicals in highly concentrated form, also works its way into the lungs and begins the process of transformation of normal cells into cancer cells (see Chapter 12, pp. 432–433). Smokers therefore have higher rates of lung cancer than nonsmokers.

Marijuana. The present-day controversy over the use and abuse of marijuana has led to a large amount of research on its effects. Drug tolerance does develop, as does physiological dependence in some heavy users. Although marijuana has a few possible medicinal uses, it is also known to have several biologically adverse effects. Possible medicinal uses include the reduction of eyeball pressure in glaucoma patients, the stimulation of appetite and suppression of nausea in cancer patients undergoing chemotherapy, and the stimulation of appetite in AIDS patients. In the United States, over a dozen states, from Maine to California, have legalized the medical use of marijuana under certain conditions, and similar measures have been proposed elsewhere. Nevertheless, federal laws still prohibit the sale or use of marijuana, and federal officials have vowed to enforce these laws. Germany, Australia, and the Netherlands now allow medical marijuana to be sold by prescription to certain patients, and the government of Canada has pledged to make medical marijuana available from government sources to patients who need it.

The harmful effects of marijuana are much better understood than the benefits. When smoked in cigarette form, much of the particulate matter in marijuana smoke stays in the lungs and builds up to form tar. Marijuana smoke produces more tar per weight of plant material than does tobacco smoke, and the tar is equally carcinogenic. It also inhibits the immune cells that clear debris from the lungs and protect against airborne infectious bacteria and viruses. All forms of marijuana alter the production of reproductive hormones, decreasing the production of sperm in men and ovulation in women. Men who use marijuana over long periods of time often develop fatty enlargement of the breasts (gynecomastia).

In 1999 the Institute of Medicine of the National Academy of Sciences released a report evaluating the health consequences of marijuana smoking and of the use of purified THC and other chemically related compounds (cannabinoids). Among their conclusions are the following:

- THC and other cannabinoids have some potential medical uses that deserve further study.

- "The accumulated data indicate a potential therapeutic value for cannabinoid drugs, particularly for symptoms such as pain relief, control of nausea and vomiting, and appetite stimulation."

- Cannabinoids can be addictive. The brain develops tolerance to them, dependence develops in some users, and withdrawal symptoms may occur when cannabinoid use is discontinued. These effects, however, are not as severe as with certain other drugs that are currently legal, such as diazepam (Valium) and other benzodiazepines, and should not preclude the development of useful medications.

- Smoked marijuana is "a crude THC delivery system that also delivers harmful substances" such as the tars and other compounds that impair the health of the respiratory and immune systems. Further and more carefully controlled studies are needed on the adverse health effects of smoked marijuana.

- Delivery of cannabinoid drugs through the smoking of marijuana is problematical at best. For any possible therapeutic uses of cannabinoids, a safer method of drug delivery needs to be developed, possibly by use of an inhaler.

- "Marijuana is not the most common, and is rarely the first, 'gateway' to illicit drug use;" tobacco and alcohol are both more common in this context. Despite some people's concern that medical users of marijuana-based drugs would become addicts or would proceed to abuse other drugs, "at this point there are no convincing data to support this concern."

In July of 2003, the government of Canada announced that it would supply medical marijuana directly to patients with certain legitimate medical needs.

Designer drugs. The term 'designer drugs' refers to those drugs that are slight structural alterations of existing drugs. Designer drugs are often

made to get around laws that ban particular drugs by name. New designer drugs may be legal until laws are rewritten to cover them, but they can be dangerous nevertheless.

MPTP and MPPP are two designer derivatives of Demerol (meperidine), itself a derivative of opium. These two have a psychoactivity similar to other opiates, but are also potent neurotoxins (nerve cell poisons). They particularly destroy the nerve cells in the area of the brain that controls movement, causing movement defects similar to those found in Parkinson's disease, a condition that otherwise mostly affects people over age 50 (see Chapter 13, pp. 473–474). In 1985, 400 cases of Parkinsonism in young people were found to be due to MPTP. After these cases, MPTP and MPPP were made illegal in the United States.

Prescription painkillers. The medical management of pain has been influenced by social attitudes regarding drug addiction. Because of the dangers of addiction, there has long been a reluctance to prescribe painkillers such as morphine. Now, however, more doctors are willing to prescribe such medicines to alleviate the pain experienced by patients, including those with terminal illnesses. Studies have shown that addiction seldom results when painkillers are used for pain management in persons who have never abused drugs.

One such drug is Oxycontin, an opiate now often prescribed as a painkiller. Unfortunately, this drug has also become more widely used by drug addicts, including those who rob pharmacies to obtain it. Oxycontin is designed to release its active ingredient slowly over a period of time. Some addicts have crushed these pills in order to defeat the time-release mechanism and release the drug all at once.

Steroids. Although most commonly abused drugs are psychoactive drugs, not all drugs that are abused are psychoactive. Anabolic steroids are abused instead because of their hormonal effects on physical development. A number of athletes including both female and male body builders, weight lifters, swimmers, runners, football players, and others, have used these drugs because they cause an increase in muscle mass, resulting in a bulkier and more powerful physique. These drugs are not addictive, but their many dangerous side-effects include damage to the reproductive organs and the circulatory system, especially the heart, as well as increased hair in some places and premature baldness in others. Deaths have occurred among amateur and professional athletes from the abuse of steroids.

Drug effects on embryonic and fetal development

Aside from the biological effects on the person taking a drug, there are many effects on developing embryos. Many drugs can affect fetuses *in utero*. Additional harm can result from damaged gametes from either a mother or a father who has used drugs.

The placenta in pregnant women is very rich in capillaries. The mother's blood does not circulate directly into the fetus, but maternal blood vessels in the placenta are close to the fetal blood supply (Figure 14.9). This close association of maternal and fetal blood systems effectively delivers nutrients and other molecules from the mother's blood into the fetal capillaries of the chorionic villi. Thus, in a pregnant woman, drugs transported throughout her body are also distributed to the fetus. In many cases, doses that are toxic or lethal for the fetus are much lower than the doses harmful to adult tissues. Drugs with only a small effect on the mother may therefore profoundly and permanently damage the fetus.

Caffeine. Some studies on rats show that caffeine intake comparable to 12–24 cups of coffee per day resulted in offspring with missing toes, while a dose comparable to as little as 2 cups per day delayed skeletal development.

The effects of caffeine intake on pregnancy have been investigated using an experimental design called a retrospective study. In this type of study, after an outcome has occurred, data are collected on the prior activities of the people in the study. Of 16 pregnant women whose estimated daily intake was 600 mg or more of caffeine (eight cups), 15 had miscarriages, stillbirths or premature births. In this study, the caffeine intake of men was also examined. In a subgroup of 13 fathers with a daily

Figure 14.9

Maternal and fetal circulation in the placenta. Fetal blood flows through vessels in the umbilical cord to the chorionic villi, which are in close contact with maternal blood. The fetus's blood receives oxygen in the placenta; the umbilical vein carries oxygen-rich blood back to the heart and the umbilical arteries carry an oxygen-poor mix of blood from the entire fetus.

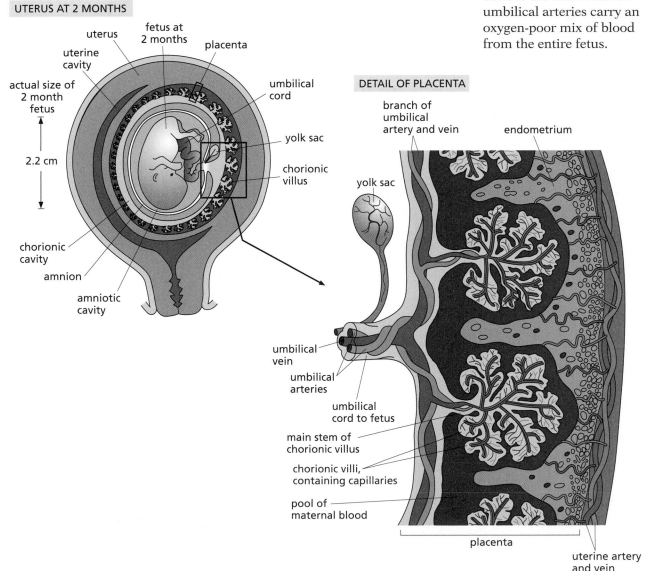

intake of 600 mg, although the mother's intake was less than 400 mg, only five of the births were normal. All of the births in which both the parents consumed less than 300–450 mg (four to six cups) were normal. Any paternal effects were presumably caused by chromosome damage before conception. Another retrospective study on pregnant women showed that an amount of caffeine equivalent to as little as half a cup a day increased the frequency of miscarriages; this study also demonstrated that some of the effects of caffeine might occur *before* pregnancy. However, other studies of caffeine have reached inconsistent and inconclusive results for a number of reasons. Among these reasons are the difficulty of measuring caffeine intake when people use different brewing methods and cup sizes, the lack of control for noncaffeine ingredients in caffeine-containing beverages, and a variety of other methodological differences among the studies such as inconsistencies in the number of people and the types of beverages studied. Another difficulty with retrospective studies is that it is often impossible for a person to reliably recall their drug intake, particularly so for the so-called 'soft' drugs such as caffeine.

Nicotine. Nicotine damages the placenta, increasing the likelihood of miscarriages, premature births, and damage to the fetus. Nicotine crosses the placenta very quickly and remains in the fetal circulation longer than it does in the mother's bloodstream (see Figure 14.9). Nicotine causes oxygen deprivation in the fetus, as do carbon monoxide and cyanide from cigarette smoke. Because oxygen is required in the production of ATP, cells deprived of oxygen have less ATP and are less able to perform the cell synthesis functions necessary to produce new cells in the growing fetus, especially in the brain. Oxygen deprivation is made worse by nicotine-induced damage to the blood vessels, including those of the placenta.

Alcohol. Alcohol that is consumed by a pregnant woman is quickly distributed into the blood of the fetus at the same concentration as is present in the mother's blood, causing severe and permanent mental and physical birth defects called 'fetal alcohol syndrome'. The prevalence in the United States in 1983 was 1–3 affected children per 1000 total births and 23–29 per 1000 births to alcohol-abusing mothers. The period during which the fetus is most sensitive to damage from alcohol is the first month, often before pregnancy is recognized. There is also evidence that alcohol abuse by women before conception correlates with decreased fetal growth, even when the mother abstains during pregnancy itself. There are few data on the fetal effects of heavy alcohol consumption on the part of the father. Alcohol is toxic to sperm and five or more drinks daily decrease the number of sperm produced.

Drug combinations. Alcohol, caffeine, and nicotine all increase the blood levels of the neurotransmitter acetylcholine, lowering placental blood flow. The effects increase with the dose of the drug and also with its duration in the body. Because the fetus lacks the enzymes for breaking down either alcohol or caffeine, the concentrations of these drugs stay higher longer in the fetus than in the maternal circulation. Because there is a higher incidence of smoking in people who abuse alcohol, the interactions of these drugs are also significant. Marijuana also crosses the placenta and is correlated with low birth weight and prematurity. Barbiturates readily cross the placenta, and the use of barbiturates by pregnant women can cause birth defects. The combination of marijuana or barbiturates with any of the other drugs mentioned increase the risks to the fetus.

Drug persistence after birth. Drugs passed on to the fetus *in utero* may remain in the child for a long time, particularly if the enzymes needed to metabolize these drugs are not present. Phencyclidine (PCP), known as 'angel dust,' can still be present in the blood of a five-year-old child of a PCP-using mother. Because children's brains and immune systems continue to develop after they are born, toxic drugs may continue to interfere with the development of the brain and the immune system in young children long after birth.

Drug abuse: public health and social attitudes

Recall that drug abuse has been defined as drug use that negatively affects the health of individuals or society. We have examined the effects of drugs on cells and on individuals, but an ecological perspective on biology teaches us that actions in cells and organisms generally have consequences for populations.

There are different concepts of how to protect society from the consequences of drug use by some members of society. One concept views drug addiction as a crime and drug abuse as a law enforcement problem. U.S. efforts to halt the trade in dangerous drugs have relied almost exclusively on this approach, which emphasizes arrests and interdictions instead of prevention or rehabilitation. Billions of government dollars have been spent on the 'war on drugs,' yet drug use continues to flourish and those who profit from the drug trade grow richer with each passing year.

The U.S. Comprehensive Drug Abuse Prevention and Control Act of 1970, commonly called the 'Controlled Substances Act,' established five categories of controlled substances according to their potential for abuse and whether they had a recognized medicinal use in 1970. Although the classification is not perfect, it represents an attempt to classify potentially dangerous drugs in a consistent manner. Many of the substances regulated by this law are listed on our Web site (under Resources: Controlled substances). The distinctions made by this classification are legal rather than medical. For example, drugs in Schedule I are illegal to possess or to prescribe. Drugs in Schedules II and III can be prescribed only by certain physicians registered with the U.S. Drug Enforcement Agency (DEA), and certain records must be kept and reported to the DEA periodically. Each schedule also specifies what kinds of researchers or medical professionals may possess or handle the drug, what kinds of records they must keep, and what penalties can be imposed on people who possess these substances illegally.

Another concept, referred to as 'harm reduction', is based on a view of addiction as a disease and drug abuse as a public health problem. Many European countries have followed harm reduction strategies, particularly in dealing with such drugs as heroin. The drug abuse rates (and crime rates) of these countries are much lower than those of the United States, which has followed the crime concept for the most part. In Great Britain, for example, heroin can legally be prescribed to those who are addicted. The approach seems to work in several ways: harm to the addicts from overdose or impure formulation is minimized. There is no incentive for the addict to commit crimes to pay for drugs or to recruit others into becoming addicts. (Many drug dealers are addicts who recruit others so that they will have a steady supply of customers and profits to support their own habit.) British rates of heroin use have dropped under the harm reduction approach, while rates of heroin use in the United States have risen. The illegal heroin trade has withered in Great Britain because it is no longer profitable, and new cases of addiction are rare because the drug is available only to persons registered as already addicted. Most promising of all is the fact that about 25% of British heroin addicts spontaneously give up the habit on their own.

The approach in several other European countries, including Germany, Switzerland, and The Netherlands, allows drug users freedom from arrest if they follow a few simple rules (staying in certain locations, for example). Instead of spending money mostly on enforcing drug laws (as in the United States), more government funds are spent in these countries on public health campaigns aimed at education, prevention, and rehabilitation. Although the unlicensed selling of drugs is illegal in these countries, the criminal justice system is used very little in attempts to minimize the harm done by addictive drugs, either to the addicts themselves or to society as a whole.

Education is at the heart of most measures aimed at preventing drug abuse, including most harm reduction strategies. Where prevention of addiction has been tried, it is cheaper and more successful than rehabilitation. Education about the risks of drug use does decrease drug use (Figure 14.10). European countries that emphasize education and prevention have much lower drug abuse rates than the rest of the industrialized world.

Occasional critics on both the political left and right have suggested legalizing various dangerous drugs, taxing them, and treating them as public health problems in the way that we treat alcohol and tobacco, with heavy reliance on education and prevention programs. Few people who work in the field of drug addiction favor this approach because these drugs are truly dangerous, and because addiction is easy to establish and very difficult to break. Legalization without the structures that are part of harm reduction policies would, they feel, result in increases in health risks. In 1993 U.S. Surgeon General Joycelyn Elders suggested that legalization approaches be studied because she didn't know all their possible consequences, but her modest suggestion was immediately quashed by President Clinton and others.

The harm done by alcohol shows that a drug need not be illegal to cause considerable social harm. In addition to its negative effects on the user's health, alcohol abuse also causes harm to others. Alcohol use currently causes deaths in motor-vehicle accidents, boating accidents, drownings, and many other causes

Figure 14.10

Influence of perception of risk on the use of marijuana. The availability of marijuana in the United States remained uniform over the years covered by this graph, but use decreased in inverse proportion to the perception of risk.

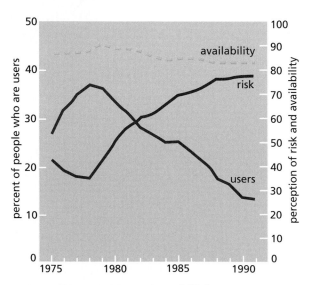

users = using once or more in past 30 days
risk = percent saying there is great risk of harm in regular use
availability = percent saying fairly easy or very easy to get

of accidental injury including industrial accidents. It also is responsible for much employee absenteeism, job loss, and school failure. Alcohol is frequently a factor in acquaintance rape (also called 'date rape'), child neglect, child abuse, spouse abuse, divorce, and suicide. Alcohol and other psychoactive drugs can also do great harm in safety-sensitive occupations such as commercial transportation (airline pilots, air traffic controllers, railroad engineers), power plant operations, the nuclear industry, and much of the military. We tend to hear about the increases in criminal activity and gangster influence when alcohol use was illegal during the Prohibition on alcoholic beverages in the period from 1920 to 1933 in the United States, but it is also true that the incidence of alcohol-related accidents and disease was greatly *decreased* during Prohibition.

THOUGHT QUESTIONS

1 Divide a piece of paper into three columns. In one column list all the characteristics you can think of to describe addictive drugs; in the second, list the characteristics of psychoactive drugs; in the third column list the characteristics of drugs of abuse. Also list specific drugs in each column. Are all addictive drugs drugs of abuse? Are all drugs of abuse addictive? Are all psychoactive drugs addictive? Are all psychoactive drugs drugs of abuse?

2 Many factors contribute to making a drug dangerous. In what ways are street drugs more dangerous than a chemically similar drug obtained from a licensed manufacturer?

3 Does the American cultural expectation of a quick fix for life's pains contribute to our tremendous use of legal drugs? Does the widespread and somewhat casual use of legal drugs contribute to drug abuse?

4 Read the package inserts for some drugs that you have purchased over the counter. Do these inserts indicate that there are any potential negative effects? What kind of warnings do the package labels contain? Study the advertisements for these medications. Do the advertisements and the labels give you the same impression of the products?

5 Use the information given in this chapter to argue either for or against the legalization of marijuana for medical uses. What type of ethical reasoning are you using? What facts about marijuana are you using in your argument? What do we not yet know about marijuana that might cause you to change your position?

6 Select a substance sold as a 'food supplement' and find out what research (if any) has ever been conducted to test:
a. Whether it has any of the effects claimed for it,
b. Whether it is safe (specify at what dose and under what conditions),
c. Whether it has harmful side-effects, and
d Whether it interacts with other drugs.

If any research studies have been conducted, did they use adequately large samples? Was there a control group? Was the supplement compared with a placebo in a double-blind study?

7 Is information about the biological effects of drugs an effective prevention or a deterrent against drug use? See whether you can find any published information on the effectiveness of various educational programs.

8 What criteria could be used to distinguish between the use and abuse of caffeine? Can these same criteria be applied to other drugs?

9 What ethical considerations govern the use of animals in testing drugs for safety and effectiveness? What about the use of human volunteers?

Concluding Remarks

Where is the boundary between the rights of the individual and the rights of the group in relation to drug use? Many people are inclined to leave this matter up to the individual if there is no 'harm to society.' However, in many if not all forms of drug abuse, there is clearly a harm to society. Two obvious effects of drugs on populations are the bodily harm done to others by persons under the influence of a psychoactive drug and the commission of crimes to pay for the drugs. In some instances, once the societal effects of individual drug abuse have been documented, laws have been passed to limit these effects. The U.S. public has accepted laws designed to protect the nondrug-using citizen from the harmful effects of others' use (or abuse) but has not accepted laws that are perceived as infringement on individual rights. For example, laws designed to protect others from exposure to secondhand smoke have been successful at limiting where or when people may smoke, while attempts to pass laws prohibiting individuals from smoking at all have not been successful. A similar approach is being tried on college campuses in educational efforts to raise awareness about the secondhand effects of binge drinking, which include higher risks for rape and for sexually transmitted diseases including AIDS.

In Chapter 1 we outlined steps for arriving at policy decisions on societal issues that are influenced by science. Have these methods been followed on the issues presented by drug use and abuse? Nicotine addiction causes far more deaths (from lung cancer) than all other drugs combined, yet it is legal! Alcohol abuse ruins more families and careers than do illegal drugs, and causes more fatal accidents, yet it is legal in most places. Marijuana users, on the other hand, cause far less harm to others, yet the substance is illegal in most places. Caffeine is not regarded by most people as a drug and is readily available even to children, yet it is certainly addictive. The inconsistencies go on and on. Clearly, decisions as to which drugs should be legal and which should be illegal are not always made on scientific criteria.

Chapter Summary

- A **drug** is a chemical substance that produces one or more biological effects, and usually several.

- The effects of any drug depend on the molecular structure of the drug, on its dose or concentration, and in many cases on the bodily location of specific receptors for the drug.

- All drugs, whether used medically or nonmedically, must enter the body by some route, usually orally or via the respiratory system. The drug must then be distributed around the body, usually by the circulatory system, and eventually eliminated either by metabolic breakdown or by **excretion**. The length of time that a drug remains metabolically active in the body is measured by its half-life.

- Drugs have both **acute effects** and **chronic effects**.

- Drugs have their effects on cells either by direct action on the cell membranes or by stimulating **receptor** molecules to alter one or more cellular functions. Actions on cells in different tissues produce different physiological effects.

- There are always other effects in addition to that for which the drug was taken, and these **side-effects** are every bit as real as the intended effect.

- Drugs can interact either **additively**, **synergistically**, or **antagonistically**.

- **Psychoactive drugs** cross the blood–brain barrier and act directly on the nerve cells of the central nervous system. Those that produce **addiction** do so by stimulating activity in the positive reinforcement center of the brain. Addictive drugs induce **dependence** and **tolerance** to the drug, as well as **withdrawal** symptoms that appear when the drug is stopped. The possible side-effects of psychoactive drugs include permanent damage to the brain cells and interference with normal physiological functioning of other organ systems, impairing the health of drug users.

- Many drugs can cross the placental barrier and cause damage to a fetus *in utero*; fetal alcohol syndrome is an example. Other drugs can be transmitted to infants through breast milk.

- **Drug abuse**, also called **substance abuse**, has a tremendous cost to society.

CONNECTIONS TO OTHER CHAPTERS

Chapter 1	'The mind is the same thing as the brain' is an example of a research paradigm.
Chapter 1	Attempts to limit the effects of drug abuse raise numerous ethical issues.
Chapter 8	Drug use can affect the physiological regulation of sex hormones. Drug use can also affect sexual behavior, including sex abuse and unprotected intercourse.
Chapter 9	Certain drugs (such as alcohol) can increase the risk of unwanted pregnancy by impairing judgment; others can do so by interfering with hormonal birth control. Still other drugs can damage the reproductive system and thus impair fertility.
Chapter 10	Drugs interfere with nutrient pathways at many levels.
Chapter 11	Most drugs are plant products or are derived from plant products.
Chapter 12	Drugs such as tobacco and marijuana contain many cancer-causing agents. Alcohol and tobacco act synergistically as cancer causes.
Chapter 13	Drugs may interfere with the normal processes of the brain.
Chapter 15	The brain and the endogenous opiates interact with the immune system.
Chapter 16	Drug use by injection is a major risk factor in the transmission of HIV infection and AIDS.

PRACTICE QUESTIONS

1. If the toxic dose of caffeine is about 16 g for a person weighing 80 kg (about 180 pounds), how many cups of coffee, consumed at one time, would it take to realize this dose?

2. Which of the following drugs are addictive?

   Cocaine   Heroin   Alcohol   Methadone

   Morphine   Codeine   Tobacco   Meperidine

3. How many equivalent 'drinks' are there in a bottle of wine, which is typically 750 ml? How many 'drinks' are there in a quart of liquor? (1 ounce = 28.4 ml; 1 quart = 32 ounces.)

4. Of the many drugs described in this chapter, which causes the largest number of deaths from people under its influence?

5. Name three drugs that can enter the body through the respiratory system.

6. If you take 100 mg of a drug, how much remains metabolically active in your body after one half-life?

7. If the half-life of the drug in question 6 is 1 hour, and you start with 100 mg of drug, how much drug is left after 2 hours?

8. If you take 100 mg of a drug that has a half-life of 1 hour, and after 1 hour you take another 100 mg of the drug, how much of the drug will be in your body after the second dose?

9. Of all addictive drugs, which is the most widely used in the United States?

10. If a drug is described as an opiate agonist, what does that mean?

11. If another drug is described as an estrogen antagonist, what does that mean?

12. If two drugs interact synergistically, what does that mean?

13. If the body develops a tolerance to a painkilling drug, what does that mean?

Mind and Body

Issues

- What are the causes of disease? Can we promote health, or is it simply the absence of disease?
- Why does one person get sick and not another?
- How do our mental and emotional states affect our physical health?
- How do our body systems communicate with our emotions?
- Can stress make you sick?
- How has psychoneuroimmunology reworked our definitions of disease and its prevention?

Biological Concepts

- Dynamic equilibrium (detection of environmental stimuli, homeostasis)
- Organ systems (immune system, lymphatic circulation, autonomic nervous system, endocrine system)
- Health (specific immunity, inflammation, tissue healing, vaccination, stress response, relaxation response, placebo effect)

Chapter Outline

The Mind and the Body Interact

The Immune System Maintains Health

Cells of the immune system and the lymphatic circulation

Innate immunity

Specific immunity

Immunological memory

Passive immunity

Harmful immune responses and immunosuppression

The Neuroendocrine System Consists of Neurons and Endocrine Glands

The autonomic nervous system

The stress and relaxation responses

The Neuroendocrine System Interacts with the Immune System

Evidence for one interconnecting network

The placebo effect

Effects of stress on health

Conditioned learning in the immune system

Voluntary control of the immune system

15 Mind and Body

*W*hat do we mean by health, and how do we achieve it? What makes one person healthy and someone else chronically ill? Certainly there are many answers to these questions, and we have touched on some of them in previous chapters. A person's genetic heritage plays a role in some diseases (Chapter 3), as does his or her diet (Chapter 10). How and where a person lives is important because they influence the person's exposure to infectious microorganisms and to hazardous chemicals. Our immune system helps to remove damaged tissue and repair the body, preventing some diseases before we know we have been exposed and bringing us back to health after we have been sick. Genetics, nutrition, and exposure to chemicals and microorganisms all affect the functioning of the immune system. It is often the case, however, that some people in a particular area will get sick, while other people in the same area with about the same exposure, diet, and genetic background do not. In the previous chapter we saw that many so-called mental illnesses have mechanisms based in brain biochemistry. In this chapter we examine the theory that a person's mental and emotional states are factors in physical health or disease, and that the mind exerts its effect on the body because it interacts with the immune system.

The Mind and the Body Interact

In the preceding chapters of this book we have discussed some biological fields in which theories have been debated, tested, and modified for 150 years—a long time in the science of biology. In this chapter we discuss a very new field of study, **psychoneuroimmunology**. This new subject area is built upon a *central organizing theory*: the premise that the mind, through the action of the nerves, affects the functioning of the immune system (the body's defense apparatus against harmful or malfunctioning molecules) and therefore affects human health. Each part of the name contributes to the overall meaning: *psycho*, from the Greek word *psyche*, meaning "the mind"; *neuro*, referring to the nerves and the brain; and *immunology*, the study of the immune system. Psychoneuroimmunology is the study of how the nervous and immune systems, which were previously assumed to be independent of one another, interact with and influence each other. Psychoneuroimmunology is thus a new paradigm because it embodies both a new theory and a new field of study (see Chapter 1).

Central to the paradigm are large and important issues with which people have struggled for millennia: What do we mean by **health**? Why do we get sick? How we answer these questions will help to define what areas of investigation are valid with regard to the cure and prevention of disease and the promotion of health.

No theories arise spontaneously, and neither did psychoneuroimmunology. Its roots lie in ancient observations that personality, emotional state, and attitudes influence when and if people get sick and how sick they get. Doctors in China, India, and (later) Greece rejected supernatural forces (the gods, evil spirits, or magic) in favor of natural (biological) forces as the explanations for both health and disease. Each of these traditions maintained that there is a life force or life spirit, called *qi* (or *ch'i*) by the Chinese, *prana* by the Indians, and *pneuma* by the Greeks, that is present in humans and other organisms for the duration of their lives. A person is healthy when the life force is balanced, and unhealthy when it is out of balance. A person's mental and emotional states alter the balance or imbalance of the life force.

The Greek physician and teacher Hippocrates taught that diseases had natural causes and that those causes should be discernible and knowable entirely by rational thought. Hippocrates taught that the body contained four fluids, or *humors* (from the same Greek word that gives us *humid)*. Each of the four humors corresponded to a personality type. Each personality type also predisposed people to a further excess of the corresponding humor, creating diseases, also classified into four types according to which humor was present in excess. A person with an excess of black bile, for example, would have a 'hot' personality and would be prone to 'hot' diseases accompanied by fevers. People who had an even balance of all four humors enjoyed good health and were thus said to be in "good humor."

As European science developed in the 1600s, the criteria for 'knowability' changed. In 1616 an English physician, William Harvey, described the circulation of the blood on the basis of dissection and observation, not purely on rational thought. At about the same time, Galileo invented the thermometer, and this was used by another Italian, Santorio, to demonstrate that people said to have an excess of black bile were no more hot than other people. Not only did these two discoveries falsify the notion of the four humors, they ushered in an era in which hypothesis testing was added to the standards of 'knowing' about human health. Although the scientific method has vastly increased our knowledge of health and disease, a negative aspect has been that things that could not be seen or in some way quantified have come to be viewed as irrelevant to the explanation of health and disease.

CONNECTIONS
CHAPTERS 1, 3, 10, 13

Christianity viewed humans as unchanging reflections of God and therefore beyond the scope of study by the scientific method. The French philosopher René Descartes offered a way around this problem by positing that the mind was separate from the body: the mind was the seat of the spiritual essence, and hence belonged to the realm of the church. The body was a purely physical essence, and hence suitable for scientific study. This split of mind from body (or Cartesian dualism as it is sometimes called) had a profound effect on the study of biology and medicine. As medicine strove to become more of a science, the split became wider. It is just this split, however, that psychoneuroimmunology rejects as artificial because it does not fit the evidence.

In seeking to rejoin the mind to the body, scientists working in this field test three major hypotheses: that the immune system maintains health, that the mind and emotions can affect the functioning of the immune system, and that mental states can therefore affect health. There is an underlying assumption here, as there was in Chapter 13, that all of the functions of the 'mind' (thoughts, emotions, hopes, and dreams) can be studied in terms of brain biochemistry. Some take this assumption one step further, postulating that the mind is more than the brain; it is the integrated, inseparable network that includes the nervous system, the endocrine system, and the immune system. Other scientists in this field would be more likely to say that the psyche (or mind or spirit) affects the brain and the body in ways that can be studied by biology. Think about these distinctions as you proceed through this chapter.

In order to understand how the central theory of psychoneuroimmunology can be tested, we need to learn about the biology of the immune system, the nervous system, and the endocrine system. These three systems in the body have communication as their primary function.

THOUGHT QUESTIONS

1 Think about the difference between the following two statements: (a) The mind *is* the neuroendocrine–immune system and (b) The mind can be studied by studying the neuroendocrine–immune system. Are both scientific statements? Would both allow the formulation of falsifiable hypotheses?

2 Can a person accept the data produced by psychoneuroimmunologists without taking a stand on which of the assumptions stated in thought question 1 is correct?

The Immune System Maintains Health

A healthy multicelled organism can be viewed as an ecosystem of cells in which the parts are in dynamic equilibrium, or **homeostasis**. The immune system is the sense organ that detects whether this homeostasis exists and attempts to bring the organism back to this state if it does not exist. We can talk about homeostasis in many physiological contexts: temperature regulation in organisms that maintain a constant body temperature, for example. Immunological homeostasis is a more general state, suggesting that the cells have a way of asking, "Are we all together?" "Are we in harmony?" The immune system can be viewed as a communication network that carries on a 'conversation' throughout the organism, checking to make sure that all the parts are contributing. In this view, the central function of the immune system is the maintenance of 'self,' meaning the aggregate of cells forming a cooperative unit that we recognize as a multicellular organism. As we see in this chapter, the immune network is aided in this task by its ability to exchange chemical messages with the nervous and endocrine systems.

Cells of the immune system and the lymphatic circulation

Cells of the immune system are primarily **white blood cells**. They are called that because under the microscope they look clear or white by comparison with the oxygen-carrying red blood cells. The immune system consists largely of these mobile cells that travel throughout the body and often are not confined to specific locations. The white blood cells comprising the immune system that are discussed in the ensuing sections of this chapter are shown in Figure 15.1.

White blood cells develop in the bone marrow, spleen, and thymus and are then transported throughout the body by the bloodstream, particularly to the areas of the body that contact the external environment: the skin, the nasal passages and lungs, and the intestinal lining. They leave the blood and 'crawl' through the spaces between the cells in tissues; later they are transported via a second circulatory system called the **lymphatic circulation** to the lymph nodes. Other immune cells may differentiate in the lining of the small intestine and in the lower layers of the skin. The skin and the gut lining are thus important organs of the immune system. These structures of the immune system are shown in Figure 15.2.

The lymphatic circulatory system drains liquid from tissues. The spaces between cells in all tissues are filled with a water-based liquid. (Recall that all cells must be constantly in contact with water both inside and outside the cell, as it is the repulsion of lipid molecules by water that keeps cell membranes intact; Chapter 10, pp. 331–332.) Immune cells, but not red blood cells, move in this liquid, cleaning up any dead or damaged cells. This fluid diffuses from the intercellular spaces into lymphatic capillaries. Once there, the fluid is called **lymph**. From the lymphatic capillaries, lymph drains into larger collecting vessels, the lymphatic vessels. Lymph is chiefly returned to the blood via the thoracic duct, which empties into a large vein near the heart (see Figure 15.2). There is no pump to move fluids through the lymphatic circulation in the way the heart moves fluids through the blood. Muscle contractions and movements of the individual provide what little push this system gets.

Cells and molecules that are not contributing to the homeostasis of the cellular ecosystem are called *nonself*. Thus, dead, damaged, or cancerous cells, as well as molecules from outside the organism, are nonself. Immune cells spend much of their time checking tissues for these nonself molecules. The lymphatic circulation not only helps to remove wastes and damaged cells from tissues, but it solves the problem of how the immune system can monitor all of the cells and molecules in all of the tissues of the body. This is accomplished by the lymphatic circulation bringing molecules to centralized locations (lymph nodes, tonsils, and adenoids; see Figure 15.2) for checking by other immune cells. In an

CONNECTIONS
CHAPTER 10

Figure 15.1

The white blood cells that provide innate and specific immunity.

INNATE IMMUNITY		SPECIFIC IMMUNITY

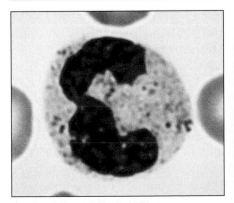

Neutrophil: engulfs and kills microorganisms

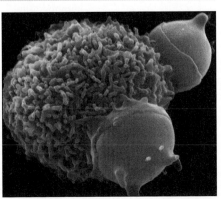

Macrophage: engulfs and disposes of dead cells (shown here) and cell debris. Also involved in inflammation.

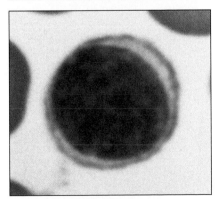

Lymphocyte: secretes antibody (B lymphocytes); kills virally infected cells and cancer cells (T lymphocytes)

Figure 15.2

Macroscopically visible parts of the immune system. The parts of the lymphatic circulation are in blue, and the major lymphatic tissues (thymus, bone marrow, tonsils, etc.) are colored orange.

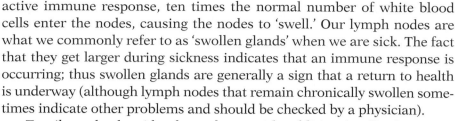

tonsils

lymph nodes

thoracic duct

thymus

heart

bone marrow

adenoids

spleen

small intestine containing immune cells

skin containing immune cells

lymphatic vessels

active immune response, ten times the normal number of white blood cells enter the nodes, causing the nodes to 'swell.' Our lymph nodes are what we commonly refer to as 'swollen glands' when we are sick. The fact that they get larger during sickness indicates that an immune response is occurring; thus swollen glands are generally a sign that a return to health is underway (although lymph nodes that remain chronically swollen sometimes indicate other problems and should be checked by a physician).

Tonsils and adenoids also enlarge with additional cells during an immune response. In the mid-twentieth century, the function of tonsils and adenoids as tissues of the immune system was not known. Tonsils and adenoids were routinely removed from children who had repeated respiratory or middle ear infections because the swelling of these tissues during infections can make children's breathing difficult or block the tube that connects the back of the throat to the middle ear (Chapter 13, p. 480). Fortunately, there are backup systems in the immune tissues so that removal rarely had serious consequences, but today tonsils and adenoids are left in place unless the blockage is extreme.

Innate immunity

One important function of the immune system is to promote growth and repair after injury, whether the injury is due to microorganisms or to physical damage. This process is a capacity we are born with; hence the part of the immune system that carries out this function is called the **innate immune system**. The mobilization of innate immune cells to get rid of microorganisms or damaged cells, and to repair wounds is called **inflammation**. Small molecules called **cytokines** are also involved in inflammation helping with cell-to-cell communication. In the first century A.D., the Roman physician Celsus described the "four cardinal signs of inflammation": *rubor* (redness), *calor* (heat), *dolor* (pain), and *tumor* (swelling). If you have ever had a scraped knee or a splinter in your finger, you no doubt have experienced these cardinal signs.

The injured area becomes red, hot, sore, and slightly swollen; all symptoms are caused by changes in local blood vessels.

The various processes of inflammation that lead to healing are illustrated in Figure 15.3. Inflammation is carried out by white blood cells

Figure 15.3

The stages of inflammation.

(A) A wound, with or without bacteria, starts inflammation.

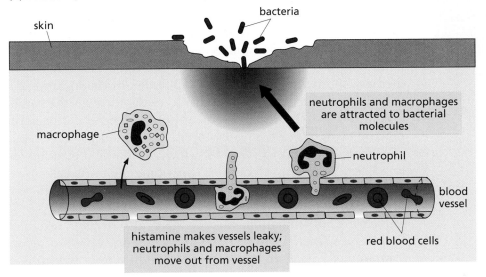

(B) Immune cells heal the wound.

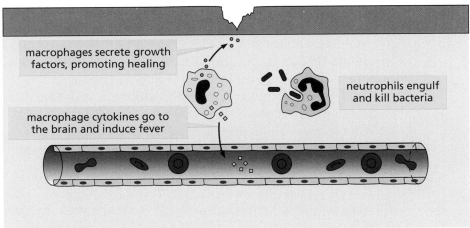

(C) Macrophages initiate specific immunity.

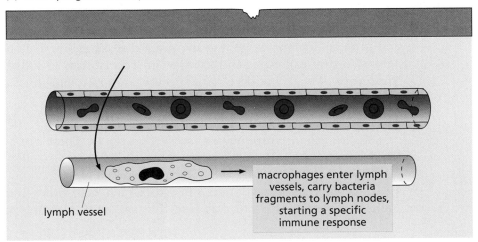

called macrophages and neutrophils that can engulf bacteria and damaged tissue in a process called **phagocytosis**. Cell damage from a wound changes the acidity (pH; Figure 10.9, p. 345) of the local fluids, activating various chemicals. Some of these chemicals attract macrophages and neutrophils to the area (Figure 15.3A). Other chemical factors constrict the blood vessels beyond the site of the wound, causing blood to build up in the capillaries close to the wound. These changes in blood flow result in the redness, heat, and swelling of inflammation. The localized swelling puts pressure on nerve endings causing pain, the fourth characteristic of inflammation. Still other chemical factors, particularly histamine, increase the permeability of the capillaries in the local area. The neutrophils and macrophages are able to crawl through the capillary walls and into the tissue. There the neutrophils remove the bacteria and the macrophages remove the damaged tissue. Macrophages also secrete chemicals called growth factors, which are the first step in wound healing. These stimulate cell division, providing offspring cells to replace the damaged cells, for example the skin cells lost because of the wound (Figure 15.3B).

During inflammation, macrophages also secrete several cytokines that induce fever (Figure 15.3B), the symptom we often recognize as a symptom of having an infection. These induce fever by acting on part of the brain called the hypothalamus (see Figure 13.15, p. 484). The increased body temperature inhibits the growth of bacteria and also enhances the immune response to the bacteria. Human body temperature is normally 98.6°F (37°C). Small increases enhance immunity; however, increases above 105°F (40.6°C) can result in convulsions or death. This is the first of many examples we will see in which products of the immune system act on the cells of the nervous system.

Finally, the macrophages enter the lymphatic vessels and carry bacteria (or other nonself molecules) to the lymph nodes (Figure 15.3C). Here the macrophages show the bacterial molecules to other immune cells, triggering another type of immunity called specific immunity.

Specific immunity

Humans and other vertebrates actually have two immune systems. One is the innate immune system we are born with and which we have just discussed. The second immune system, called the specific immune system, is found only in vertebrates. There are many infectious diseases in which our first exposure to a microorganism results in disease, but also builds up a protective immunity that helps us recover from the disease and protects us against getting the same disease again. This protection against future exposures to the same microorganism is called **specific immunity**. We are not born with specific immunity but must acquire it by exposure to microorganisms during our lifetime.

The removal of antigen. The white blood cells of the specific immune system are called **lymphocytes** (see Figure 15.1). There are two types: the B lymphocytes (**B cells**), which make blood proteins called antibodies (explained below), and the T lymphocytes (**T cells**), some of which kill infected cells directly and some of which help other immune responses. These cells protect us from disease by responding to specific

antigens, an antigen being any molecule that is detected by the immune system. Many antigens recognized by the immune system are parts of whole bacteria, viruses, or cancer cells. The specific immune response removes or blocks these to prevent disease or promote recovery from disease. The several mechanisms by which the activated specific immune system does so are summarized in Figure 15.4.

A type of T cell called cytotoxic T cells can directly kill cancer cells or cells infected with a virus (Figure 15.4A). In contrast, B cells do not kill antigens directly. Instead they make and secrete **antibodies** that are specific for a particular antigen, which then circulate in the body fluids. Antibodies are proteins present in the blood, lymph, and other body fluids that bind to specific antigens. Once antibody is bound to the bacteria or virus, the antibody can then combine with other blood proteins called **complement**. The antibody–complement combination kills bacteria by making holes in their cell membranes and can also inactivate viruses that are not yet inside cells (Figure 15.4B).

In addition, antibodies and complement can coat bacteria, allowing the bacteria to be engulfed and killed by the white blood cells called neutrophils. Unlike lymphocytes, neutrophils cannot bind to most bacteria directly. Instead, they have receptors that bind to one end of antibody molecules. As a result, once a bacterium has been coated with antibodies, the other ends of the antibody molecules can be bound by a neutrophil, which will then take up the bacterium, kill it and digest it (Figure 15.4C).

Figure 15.4

Ways in which the immune system can eliminate nonself antigens.

(A) Antigen-specific cytotoxic T cell killing a cancer cell.

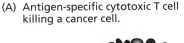
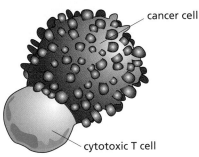

cancer cell

cytotoxic T cell

(B) Antibody and complement forming holes in a bacterial membrane.

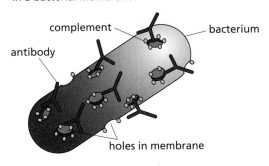

complement

bacterium

antibody

holes in membrane

(C) Neutrophil phagocytosing an antibody-coated bacterium.

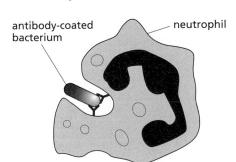

antibody-coated bacterium

neutrophil

(D) Antibody blocking of bacterial adherence to host tissue, allowing the bacterium to be washed away.

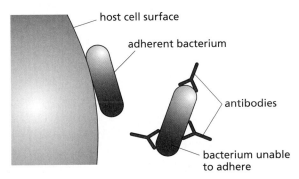

host cell surface

adherent bacterium

antibodies

bacterium unable to adhere

(E) Antibody inhibition of toxin activity.

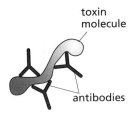

toxin molecule

antibodies

Some antibodies work not by killing microorganisms but by preventing their binding to the host. Most microorganisms cannot initiate disease without attaching to the host; this is especially true of respiratory viruses and oral bacteria. Antibodies bound to these organisms block their adherence, preventing disease. Antibodies present in the mucous linings of the respiratory tract and the throat and mouth are especially important in blocking the adherence of organisms trying to gain entry through those routes, causing them to pass harmlessly through the body and to be excreted as waste (Figure 15.4D).

Other antigens, such as toxins, are soluble molecules secreted by bacteria, not parts of bacterial cells, and it is the toxin, rather than the whole bacteria, that induces the disease. Toxins are inactivated by having specific antibody bind to them (Figure 15.4E). Many of our most successful vaccines actually stimulate the production of antitoxins, that is, antibodies against toxins. For example, the lethal results of diphtheria, a disease that formerly produced epidemics that killed many hundreds of thousands of people, result from the action of a bacterial toxin. Immunity conferred by diphtheria vaccine produces an antitoxin that prevents the disease and has virtually eradicated diphtheria from areas of the world where people are vaccinated against it.

Antigen discrimination. The specific immune system has three unique characteristics that allow the development of protective immunity.

1. It can distinguish one molecule from another.
2. It can further distinguish self molecules from nonself.
3. It shows memory of having seen particular molecules in the past.

Each of these is explained by a singular characteristic of lymphocytes. Each individual lymphocyte can bind to only one specific antigen. The population of lymphocytes as a whole, however, can recognize a vast diversity of antigens. This combination of individual cell specificity and population diversity allows the immune system to distinguish one antigen from another.

Antigen receptor specificity. Each B or T cell has receptors that can bind to only one specific antigen, yet the immune system as a whole is able to recognize over 10^{11} (100,000,000,000) different antigens. If each different receptor resulted from a different gene, as is typical for many proteins, we come up against a large problem: the number of different antigen receptors is more than 1 million times greater than the total number of genes in humans. However, this huge array of antigen receptors is made from just a few genes. Something happens to these antigen receptor genes that is not known to happen in any other genes. During the differentiation of each B and T cell, the DNA in the antigen receptor gene rearranges, and some strings of nucleotides are cut out. This is different from the alternative ways of processing mRNA (Chapter 4, pp. 116–117) that can result in different proteins being synthesized at different times in a cell. Here it is the DNA sequence that is changed, not the mRNA. Thus the rearrangements are permanent in that cell and its offspring cells. Each mature T cell or B cell is therefore able to synthesize a unique protein that functions as a receptor for only one specific antigen.

CONNECTIONS
CHAPTER 4

The DNA is rearranged randomly in each developing T or B cell so the antigen receptor proteins made on one lymphocyte differ from those on another. The whole population of T and B cells thus contains cells that can bind to over 10^{11} different antigens. Not only are the rearrangements random, but they happen *independently* of the person's being exposed to any particular antigen. *Even before* being infected by a particular type of virus, for example, a person has a small number of lymphocytes that can bind to that virus.

Self/nonself discrimination. Immature T cells develop in the bone marrow, then travel through the blood to an organ called the thymus (see Figure 15.2), where they complete their differentiation into T cells. Those T cells capable of reacting against self are eliminated, an important process that usually protects you from reacting against the tissues of your own body. The immune system's capability of discriminating self from nonself thus results from the selection of a population of responding cells from those that arose randomly, and is a characteristic of the *population* of cells but not of any single cell.

Like T cells, the developing B cells undergo selection to eliminate any B cell that could make antibody against self molecules. (For a visual representation of population selection of specific immune cell clones see our Web site, under Resources: Immune selection.)

Immunological memory

A person is not born with immunity to specific antigens, but is born with the capacity to acquire it. This capacity is the result of the two processes just explained: production of the vast numbers of different antigen receptors by DNA rearrangements, followed by selection of a population of lymphocytes that cannot bind to self molecules. Lymphocytes are released into the bloodstream, where some circulate at all times, ready to go into action when needed. These lymphocytes are carried by the bloodstream to the lymph nodes, tonsils, and adenoids (see Figure 15.2), where they encounter nonself antigens brought from the tissues by the lymphatic circulation. Most of the B cells cannot bind to the antigen and so continue on their way. However, a small number of B cells can bind, and they do so, which triggers them to begin to divide, producing a group of identical B cells able to bind to the antigen. This group of cells then differentiates into two types. One type of activated B cell produces the antibodies that help destroy the antigens as we saw previously (see Figure 15.4). The other type are long-lived **memory cells** that help with later attacks by the same antigen. A visual aid to understanding this process can be found on our Web site, under Resources: B cell activation.

T cells do the same. If you are exposed to a particular virus (such as influenza virus), a group of cytotoxic T cells that recognize the virus will develop and kill it off. In addition, a group of memory T cells will be ready to make a fast response on the next exposure to the same virus.

Antigen recognition and cell division take some time, so the first time a person encounters a particular infectious microorganism, there is time for the microorganism to make that person sick before the immune system fights it off (Figure 15.5). In other words, on the first exposure, the immune system may not block the microorganism fast enough to prevent

Figure 15.5

The activity of the immune system changes in response to a second exposure to an infectious microorganism.

the disease, but then, as specific immunity develops, it is able to stop the infection, thereby ending the disease. Some of the specific B and T cells produced in that first encounter remain in the body as memory B and T cells, sometimes for the lifetime of the individual. The second time the person is exposed to that same virus, there are more memory cells than on the first encounter, so the body responds more quickly and more strongly. In most cases this heightened response is enough to prevent the illness from occurring on the second exposure.

B and T cells also encounter self molecules in the lymph nodes, but remember that in a fully functional immune system the B or T cells that could have bound to self molecules have been eliminated before they matured. If self molecules become altered, B and T cells will exist that can bind and thus remove the altered molecules; self cells that have become transformed (precancerous) or damaged are removed in this way.

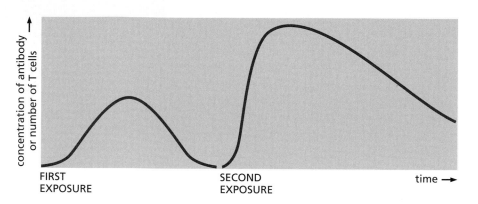

FIRST EXPOSURE

SECOND EXPOSURE

time →

Immunization. Remember that we are not born with specific immunity, but with the potential to develop it. We develop specific immunity only to those things to which we are exposed in our individual lifetimes, and so one person's 'immune repertoire' will not be the same as another's.

This is the basis for **immunization**, also called **vaccination**. A person is given molecules from various disease-causing bacteria or viruses, but in a form that will not cause disease. The person's immune system responds to this artificial challenge and establishes a group of memory cells, which stand ready to protect the person from later exposure to the real bacterium or virus. The vaccine, in other words, substitutes for the first exposure shown in Figure 15.5. If a person is later exposed to live, virulent forms of the microorganisms used in the vaccine, they have the rapid and strong response typical of the second exposure in Figure 15.5, protecting them from disease. Immunization, working as it does *with* the disease-preventive powers of the body, has become one of our most effective ways of preventing many infectious diseases.

Because the activation of each immune response is antigen-specific, several immune responses can be occurring at one time. Because each antigen-specific response occurs essentially independently of any others, responding to one bacterial or viral infection does not prevent responses to others. This also means that it is safe to receive more than one vaccination at a time. (For more on how specific immune responses are independently activated and suppressed, see our Web site, under Resources: Suppression.)

Passive immunity

If we have to *acquire* our own specific immunity, why aren't we killed by our first exposure to bacteria when we are very young? First of all, newborns do temporarily have some specific immunity transferred from their mothers; some antibodies cross the placenta, enter the fetal circulation, and can protect an infant for six months or so after birth. Other antibodies can be passed from mother to child in breast milk, particularly in the first week or two of breast-feeding (Figure 15.6). Remember that these antibodies protect the infant against antigens for which the mother has acquired immunity, which may or may not be the antigens to which the infant is exposed. This is a type of **passive immunity**, meaning antigen-specific immunity transferred from another individual. Passive immunity is only temporary. Antibody proteins, like all proteins in the body, are constantly being degraded, and the recipient of antibody has not acquired the mother's antigen-specific B cells to produce more. Passive immunity can also be transferred from one adult to another by transfusing blood containing specific antibody or by giving an antibody-containing fluid called gamma globulin, which is derived from blood.

In addition to passive immunity transferred from its mother, a baby is born with innate immunity. The parts of the innate immune system operate even on our first exposure to some, but not all, new infections. The blood proteins called complement are able to bind to and kill some types of bacterium or to inactivate some types of virus without the aid of antibodies. Viruses nonspecifically induce lymphocytes to secrete a cytokine (called **interferon**) that prevents the replication of other virus strains as well as the virus strain that induced its secretion. Cells capable of engulfing particles (macrophages and neutrophils) can engulf some types of bacteria and fungi without antibodies and thus seem to be homologous to the nonspecific immune system of invertebrates. Because innate immunity does not involve antigen-specific T cells or B cells, it does not show memory and is not stronger on the second exposure to the same antigen.

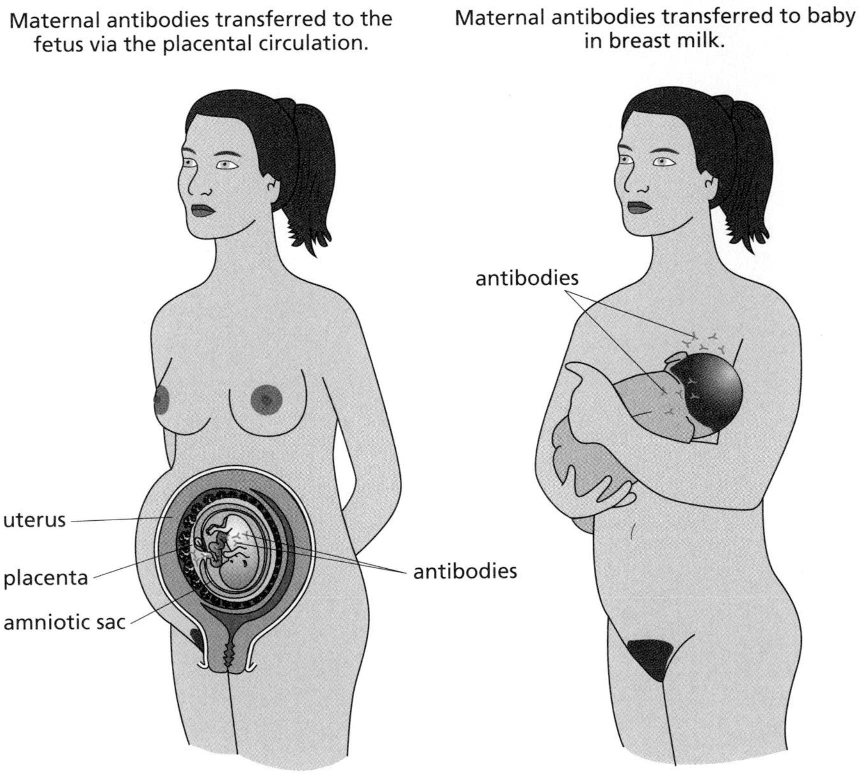

Figure 15.6

Passive immunity transferred from mother to child.

Maternal antibodies transferred to the fetus via the placental circulation.

Maternal antibodies transferred to baby in breast milk.

antibodies

uterus

placenta

amniotic sac

antibodies

Harmful immune responses and immunosuppression

Almost every biological system is changeable to some degree; healthy organisms are constantly adapting and adjusting to their environments.

The immune system is probably among the most changeable of all the body's systems. Some or all parts of the immune system can be either inhibited or strengthened by both the internal and external environments of the organism.

The widespread metaphorical view of the immune system as our defender against disease may lead us to assume that the immune system is always protective. There are several situations when it is not. The abnormal reactions of the immune system are generally against a specific antigen or a small number of antigens, while other antibodies and immune cells function normally. In other cases, the whole immune system can be suppressed.

Autoimmune diseases. **Autoimmune diseases** result when the immune system begins to make an immune response to self, resulting in both antibodies and cytotoxic T cells that react with antigens in the body's own tissues. The immune cells and/or antibodies then try to rid the body of these antigens as though they were nonself, resulting in damage to the body's own tissues. Although the mechanisms that produce tissue damage are known for some autoimmune diseases, the factors that trigger autoimmunity are unknown.

Multiple sclerosis is an autoimmune disease in which some cytotoxic T cells develop that are specific for an antigen on the insulating myelin sheath around nerve cells in the brain (Chapter 13, pp. 466–467). These cytotoxic T cells migrate to the brain, where they kill the cells bearing their antigen. The ensuing damage to the nerve sheaths causes a variety of problems, depending on exactly which nerves have been affected.

In insulin-dependent diabetes mellitus (IDDM), self-reactive T cells develop as well as B cells that produce antibody to self. These destroy the cells in the pancreas that produce the hormone insulin. Because insulin controls the cellular uptake of glucose, its absence produces severe consequences throughout the body.

Allergies. People who suffer from allergies do so because their immune systems react atypically to some antigens from which the host does not need protection (pollen or dust mites, for example). The atypical response produces a special type of antibody called IgE, specific for these antigens (which are called **allergens**). IgE binds to cells of the immune system called **mast cells**. When the person later encounters the same allergen, the allergen binds to the IgE on the mast cells, triggering the explosive release of histamine. Histamine is one of the chemicals that has a positive role in the first stages of inflammation, making blood vessels leaky to allow the entrance of neutrophils and macrophages into the tissue (see Figure 15.4A). In an allergic reaction, however, larger amounts of histamine are suddenly released (Figure 15.7), producing the various symptoms of allergy. Whether an allergic response produces runny eyes, sneezing, or shortness of breath (such as in asthma) depends on the tissue in which the mast cells were triggered (the eyes, nasal lining, or the lungs). Because the symptoms are produced by histamine, antihistamine medications stop the symptoms by blocking the binding of histamine to cells in the blood vessels. Antihistamines do not prevent the immune response or the release of histamine by the mast cells.

Other allergies, such as skin rashes in reaction to poison ivy, latex, or the dyes and other chemicals in cosmetics or clothing, are mediated by T cells, not by IgE antibodies and mast cells. Consequently, antihistamines do not block these reactions, although they may alleviate the associated itching. Severe T-cell-mediated allergic responses are treated with steroid drugs, which we discuss later.

Because each allergy is an antigen-specific immune response, it shows memory and a greater response on the next exposure, which is why people's allergies can worsen over time. Although there are thousands of different substances that produce allergy in some people, each person with allergies is usually bothered by only a few. The severity of allergy to one substance will not predict the severity of allergy to some other substance, because each is a separate, antigen-specific response. We are not yet able to predict who will become allergic or what they will become allergic to, but there does seem to be some inherited component because allergies do run in families.

Transplant rejection. Organ transplantation is a very effective (although very expensive) way of replacing damaged organs. To be successful, the organ donor and recipient must 'match' in a series of cell-surface molecules. These "transplantation antigens" are proteins coded for by about a half-dozen genes, most of which have several hundred known alleles. All of a person's dozen antigens will be expressed on their cells. Because there are so many alleles, the chances of an unrelated donor having the same set of antigens are slim. All of the genes are located close together on a chromosome, so are generally inherited as a unit. Two siblings thus have a 25% chance of having the same set of antigens.

If donor tissue has the same transplantation antigen type as the recipient's own tissue, the transplant is accepted. However, if the donor organ is of a different transplantation antigen type, the recipient's immune system will make an immune response against it. Both T cells and antibody will arise and kill the transplant. This is why organ banks attempt to find the closest possible match when an organ becomes available. A perfect match is very rare, and so an organ recipient must take drugs for the rest of their life, which suppress the reactivity of their immune system.

Figure 15.7

Allergic release of histamine.

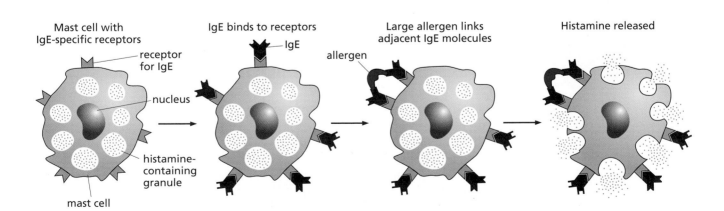

Mast cell with IgE-specific receptors — receptor for IgE — nucleus — histamine-containing granule — mast cell

IgE binds to receptors — IgE

Large allergen links adjacent IgE molecules — allergen

Histamine released

Immunological tolerance. When immune responses to a specific antigen are inhibited, it is called immunological **tolerance**. Such tolerance can be induced. For example, people who suffer from allergies can often be desensitized, that is, made non-responsive to those particular antigens. The desensitization procedure consists of giving the person repeated small doses of the substance he or she is allergic to. The procedure must be carried out very carefully because giving the wrong dose, either too much or too little, will make the allergy worse, not better. Because both allergy and its desensitization are antigen-specific, the procedure must be carried out for each separate allergen. Induced tolerance to one allergen does not change the ability of the individual to react to other allergens or to truly harmful antigens. Once established, tolerance is long lasting and prevents that allergic response for years or decades.

Immunosuppression. Other factors can inhibit the functioning of all or parts of the immune system to all antigens and are thus said to produce **immunosuppression**. Because immunosuppression inhibits the workings of innate immunity or of all B cells or all T cells, rather than just antigen-specific groups, it is generally correlated with an increase in disease.

Many environmental pollutants and other chemicals suppress the whole immune system. Other immunosuppressive factors include alcohol, cocaine, and heroin. Chronic use of these leads to increased incidence of infectious disease and, in the case of alcohol, increased incidence of cancer (Chapter 12). Particular foods have not been found to be immunosuppressive, but too much food (overnutrition) has been. People who are obese have a higher incidence of infection-related sickness and death. Chronic protein undernourishment, even of a moderate nature, impairs the ability of the immune system to fight off infectious diseases. The high rate of mortality in undernourished infants is partly a result of immunosuppression, leading to a high incidence of infections. Deficiencies of many of the micronutrients, including magnesium, selenium, zinc, copper, vitamin A, and vitamin C (Chapter 10, pp. 336–341), impair various aspects of the immune system. Many elderly people become deficient in one or more of these micronutrients and consequently become immunosuppressed. Many of the infectious diseases of the elderly can be minimized in frequency and severity if nutrition is adequate.

In contrast, in autoimmunity, where the immune system has turned against itself, or in organ transplantation, suppression of immune reactivity may be needed to prevent tissue damage. Because immunosuppression is nonspecific it also inhibits protective immunity. Because it thereby also increases susceptibility to infectious disease, the risk–benefit ratio of immunosuppressive drugs must be carefully assessed in each case.

Psychological factors, such as stress, change the internal environment and are among the factors that can induce immunosuppression. Psychological factors can also positively affect the immune system, in which case the effect is called immune potentiation. We examine psychologically produced immunosuppression and immune potentiation in more detail after we have examined the workings of the neuroendocrine system.

THOUGHT QUESTIONS

1 Since the beginnings of immunology at the turn of the century, its language has often been very military. The immune cells are said to "protect us from invaders" or to "kill off foreign antigens." Locate an immunology textbook or an article on immunology from the popular press. Can you find examples of military language in these accounts?

2 Make a list of the thoughts that come to your mind when the word *self* is used in a nonimmunological context. Make a second list for your thoughts about what *self* means in immunology. Is there any overlap between your two lists?

3 We need words to convey what we imagine the immune system to be doing, based on experimentation and hypothesis testing. Do you think that the new imagery of the immune system as a communications system has taken hold at this time because we are in the "information age," or because new discoveries brought about a need for new terminology? To what extent are scientific terms metaphors for reality and to what extent are they models? Can the words we choose cloud our view or prevent us from being open-minded about new hypotheses?

4 When people get a bacterial or viral infection, they often get a fever. Why? Why do people sometimes get a fever after a vaccination?

The Neuroendocrine System Consists of Neurons and Endocrine Glands

The **endocrine glands** are a series of organs that secrete chemical products directly into the bloodstream. (In contrast, glands that release their secretions into the digestive tract or through the skin are called **exocrine glands**.) These secreted chemicals alter the function of target organs and thus can be said to carry messages from one organ to another. The chemicals secreted by endocrine glands are called hormones. Because hormones are distributed throughout the body by the bloodstream, the target or receiving cells can be far removed from the endocrine glands.

It was once common for the endocrine glands to be described as an endocrine system because the actions of these glands were thought to be separate from the other known communication system of the body, the nervous system. In the past few decades, biologists have discovered that several hormones originally thought to be secreted only by cells of the endocrine glands are also secreted by brain cells. Other endocrine secretions are chemically related to the neurotransmitter substances originally thought to be secreted only by the cells of the nervous system. Today, the endocrine and the nervous systems are considered to be so completely intertwined that many scientists now refer to them collectively as the **neuroendocrine system**.

All vertebrate nervous systems have a central nervous system (CNS), consisting of a brain and spinal cord, and a peripheral nervous system (Chapter 13, pp. 465–466). The peripheral nervous system connects the

CONNECTIONS

CHAPTER 13

CNS to the more distant parts of the organism (the sensory receptors, muscles, glands, and organs) and is itself composed of the **somatic nervous system** and the **autonomic nervous system**.

The autonomic nervous system

The autonomic nervous system has been considered, at least by Western scientists, to be largely involuntary, carrying signals to and from the gut, blood vessels, heart, and various glands, and thus regulating the internal environment of the body. *Autonomic* literally means "self-governing" or "self-regulating," reflecting the fact that the autonomic nervous system can work by itself without any input from the centers of conscious awareness in the brain. The autonomic nervous system can thus regulate body functions even while we are distracted or asleep. In contrast, the somatic nervous system, largely under conscious, voluntary control, carries signals to and from the skeletal (voluntary) muscles, skin, and tendons.

The sympathetic and parasympathetic divisions. The autonomic nervous system consists of two functionally separate divisions, called the sympathetic and the parasympathetic nervous systems. As is shown by Figure 15.8, the two divisions of the autonomic nervous system have opposite effects. In general, the neurons of the **sympathetic nervous system**, and its principal neurotransmitter norepinephrine, ready the organism for heightened activity, while neurons of the **parasympathetic nervous system**, and its neurotransmitter acetylcholine, do the opposite. Tissues and organs throughout the body receive nerve endings and neurotransmitters from both of these divisions of the autonomic nervous system.

During moments of physical exertion or emergency, the hypothalamus at the base of the brain signals the sympathetic system to dominate, stimulating dilation of the pupils and bronchi, conversion of the storage molecule glycogen to glucose or conversion to energy, acceleration of the heartbeat, and slowing of the digestive processes. During rest, or when the organism is not receiving much sensory input, the parasympathetic system is predominant. Saliva, bile, and stomach enzyme secretion are stimulated, as is rhythmic muscle contraction (peristalsis) in the stomach and intestine, while the pupils of the eye constrict, the bronchi of the lungs constrict, and the heart rate slows. When the stimuli that produced the sympathetic response are no longer present, the parasympathetic system predominates once again. In extreme cases, this switch can be rapid, producing a rebound effect, as when a person feels woozy or faint after an emergency situation is over.

Fight-or-flight. Imagine that you are crossing a street. You hear a loud noise! You turn your head suddenly, and a large truck is heading right at you! Your heart begins to pound faster, your sweating increases, your voluntary muscles are stimulated (and their threshold for action is lowered), your breathing speeds up, and your digestive organs stop digesting your last meal. (You may even feel nauseous in extreme cases.) All these are the results of stimulation of different organs and tissues by the sympathetic nervous system and its neurotransmitter, norepinephrine (also called noradrenaline). In general terms, the sympathetic nervous system prepares the body for reactions that require large amounts of energy and oxygen for voluntary muscle contraction, including the 'fight-or-flight' response, so called because the individual is primed to fight hard or to get away fast.

Rest and ruminate. Now imagine having finished a sumptuous candle-light dinner with your favorite food, elegant service, and music playing

Figure 15.8

Functions of the sympathetic and parasympathetic nervous systems.

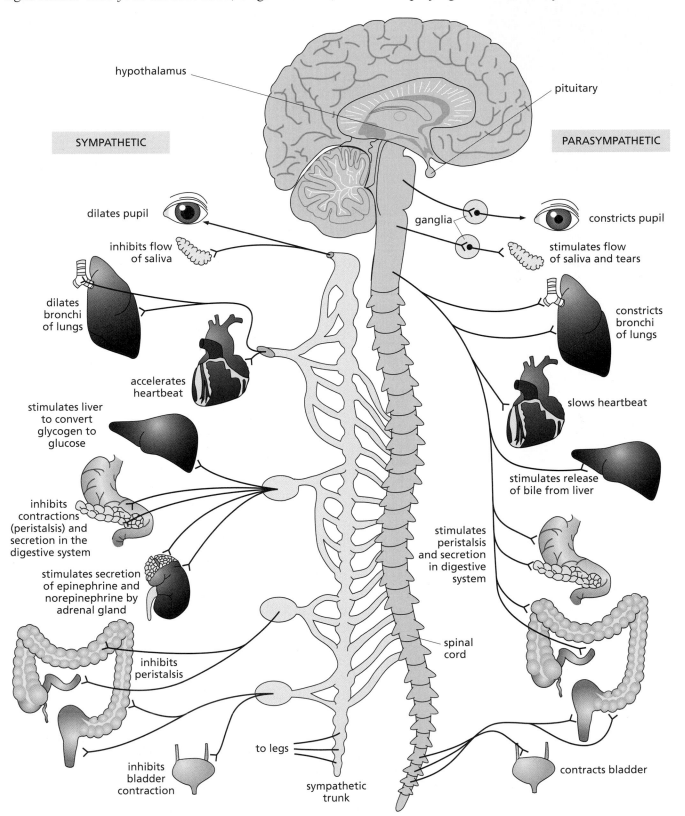

SYMPATHETIC

PARASYMPATHETIC

hypothalamus

pituitary

dilates pupil

ganglia

constricts pupil

inhibits flow of saliva

stimulates flow of saliva and tears

dilates bronchi of lungs

constricts bronchi of lungs

accelerates heartbeat

slows heartbeat

stimulates liver to convert glycogen to glucose

stimulates release of bile from liver

inhibits contractions (peristalsis) and secretion in the digestive system

stimulates peristalsis and secretion in digestive system

stimulates secretion of epinephrine and norepinephrine by adrenal gland

inhibits peristalsis

spinal cord

inhibits bladder contraction

to legs

contracts bladder

sympathetic trunk

softly in the background. You are relaxing in a comfortable chair or sofa with your favorite drink in your hand and wonderful company nearby. Just thinking about this scene (or reading this paragraph to yourself slowly and calmly) can relax you, cause your heartbeat and your breathing to slow down, your sweating to stop, your voluntary muscles to relax (and to raise their threshold for action), and your blood to be diverted to digestive organs, which are now digesting the sumptuous meal. These are all the effects of the parasympathetic nervous system and its neurotransmitter, acetylcholine. The parasympathetic division prepares the body to 'rest and ruminate', activities that use less oxygen while replenishing the body's store of energy supplies.

As you may have been able to demonstrate to yourself as you imagined the scenes described above, the actual frightening or relaxing situation need not be present. The fight-or-flight response can be triggered by just thinking of tense situations or by watching a frightening movie. Similarly, a rest-and-ruminate response can be brought about just by relaxing comfortably and imagining a relaxing, pleasurable situation. The triggers for these responses can thus originate completely within the brain of the person imagining them.

The stress and relaxation responses

Research on the fight-or-flight response by Dr. Hans Selye, a Canadian (Austrian-born) physiologist, showed that it is the first step of a larger series of physiological reactions. These reactions are produced by chemicals secreted by the nervous system and also by the endocrine and immune systems. The process begins when some stimulus or force, a **stressor**, causes the body to deviate at least temporarily from its normal state of balance. The body's response to this deviation from homeostasis is called **stress** or the **stress response**. Stress consists of physiological and immunological changes that allow our bodies to fight off or remove ourselves from stressors and return to homeostasis, and thus stress can be a useful response. However, when stress persists too long, it can become harmful, causing disease or even death.

Alarm. Alarm, the first stage of the stress response, includes the fight-or-flight response. The hypothalamus stimulates the sympathetic neurons to secrete norepinephrine, stimulating an endocrine gland called the adrenal gland to secrete epinephrine (also called adrenaline). Epinephrine and norepinephrine together bring about the physiological changes known as fight-or-flight. In addition, sympathetic neurons release norepinephrine directly into the lymphoid organs (the spleen, thymus, and lymph nodes), stimulating these organs to release their store of lymphocytes into the bloodstream. As the lymphocytes are released, the organs in which they were stored decrease in size. The hypothalamus also secretes a hormone called adrenocorticotropic hormone (ACTH), which stimulates the adrenal gland to produce corticotropin-releasing hormone (CRH), which in turn stimulates steroid hormones, particularly cortisol, to be released from the cells of the outer layer of the adrenal gland. The alarm phase of the stress response is shown in pink in the upper part of Figure 15.9.

Resistance. If the stress continues, the body enters the second stage, resistance, in which resources are mobilized to overcome the stressor and regain homeostasis. Epinephrine acts on the heart to increase the heart rate (the number of contractions per minute) and cardiac output (the strength of the contractions). Norepinephrine increases the flow of blood to the heart and muscles for possible increased activity, while constricting other vessels, diminishing the flow of blood to the gut, skin, and kidneys. If the stressor is a disease or injury, inflammation begins and innate immune cells are chemically attracted to the inflamed area. New stores of steroid hormones are synthesized, keeping blood levels of these hormones elevated. Chronic elevation of cortisol begins to reduce antibody and cytotoxic T cell activity. Cortisol also suppresses inflammation. The resistance phase of the stress response is shown in red in the middle part of Figure 15.9.

Exhaustion. If the stressor is not successfully overcome, the stress response reaches its third phase, exhaustion. The steroids made in the resistance phase are used up and the animal is unable to make more. During the exhaustion phase, the action of the sympathetic nerves tapers off, while the endocrine organs take over. Adrenal hormones stimulate

Figure 15.9

The phases of the stress response. Stressors induce a variety of physiological changes, including immunological changes, which are mediated by the sympathetic nervous system and adrenal hormones. CRH is corticotrophin releasing hormone.

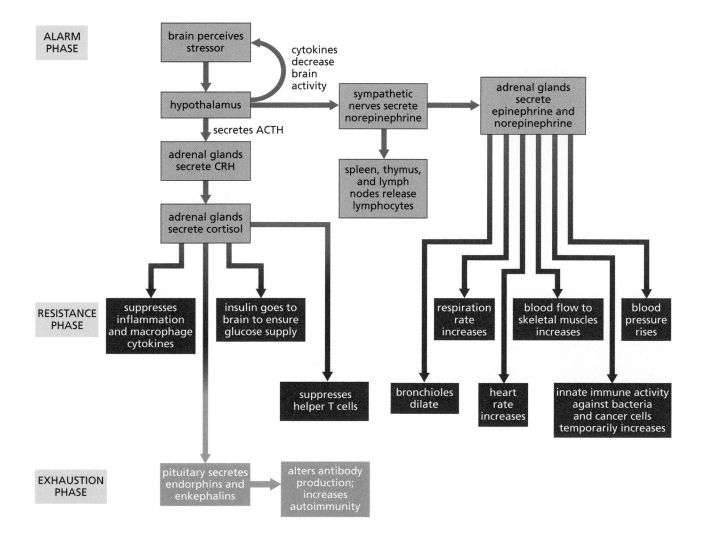

another endocrine gland, the pituitary gland, to secrete chemicals (endorphins and enkephalins) that are structurally related to opioid drugs (Chapter 14) and alter the activity of neurons and of immune cells.

The pituitary was once called "the master gland" because its hormone secretions controlled the activity of many other endocrine glands, but it is now known that the pituitary itself is under the control of the brain via the hypothalamus. Chemicals secreted by the hypothalamus also act on other cells within the brain itself, changing the activity level of neurons, and this change affects behavior, heat production, and many other functions.

Some stress responses are due to actual physical danger; but stress responses can also result from the demands of work or school (e.g., deadlines or examinations), or the actions of the people in our lives. Whether these demands are real or perceived, they can have the same biological consequences: the physiological changes that characterize the stress response.

The relaxation response. When actual physical danger has passed, the stress response will abate. During a stress response, the adrenal and pituitary glands mediate the response of the sympathetic nervous system. This response will gradually be reversed by the actions of the parasympathetic nervous system. This reversal is called the **relaxation response**.

It is sometimes more difficult to turn off our mentally induced stresses. Since the stress response can be mentally induced, it has been hypothesized that the relaxation response should be mentally inducible as well. Some cultures and some religions have been more open to this idea than others. Many practices that are aimed at evoking this relaxation response, including Yoga, transcendental meditation, and others, originated in Asian traditions. Hindu yogis have learned how to consciously control the actions of their autonomic nervous systems and to bring about levels of activity even below the normal levels for the resting state. Measurements made by Western scientists have shown that these yogis are able to lower their blood pressure, breathing, oxygen consumption, heart rate, and metabolic rates.

Other cultures, including many Western cultures, have been less open to the idea that the relaxation response can be controlled. The traditional definition of the autonomic nervous system emphasized that it governed involuntary functions over which we do not have conscious control and this had the unintended effect of discouraging research on any possible interactions of the autonomic nervous system with our emotions and other conscious body states. Recently, however, some of the methods for conscious control of autonomic processes are being borrowed from other cultures. Some athletes have learned meditation, while others have learned to invoke specific mental imagery in order to put one's body as well as mind in a certain state of relaxed determination to succeed in sport. Western medicine has begun to use similar techniques to help cancer patients in fighting their cancers, as described further in a later section on mental imaging.

Less conscious strategies can also produce the relaxation response. Studies using measurements of blood pressure and other physiological indicators have shown that contact with pets can reduce stress and bring about relaxation. Older people who keep pets have been shown to live longer than those who do not, even when comparison is made between people of comparable health status initially. Heart-attack victims are much less likely to have a second heart attack if they care for a pet. Studies such as these show a statistical correlation between two factors. From such data by themselves we cannot assign a causal relation between the two; that is, we can not say that pets caused the increased longevity or improved health.

THOUGHT QUESTIONS

1 Is adrenaline (epinephrine) a hormone or a neurotransmitter?

2 How does thinking about an annoying or threatening event produce stress?

The Neuroendocrine System Interacts with the Immune System

We have discussed how the nervous system and the endocrine system communicate with each other. In recent decades, it has also become apparent that both of these systems also communicate with the immune system. The emerging concept is that the nervous, endocrine, and immune systems do not exist as separate entities but are one interacting communications network. Because, as we have seen, the immune system protects against disease, communication between the brain and the immune system suggests a mechanism by which the mind can affect health.

Psychoneuroimmunology has an appealing central model, suggesting that the mind and body are intertwined. But models or theories that are appealing or that fit with our common sense do not always stand up to scientific scrutiny. Is there any scientific evidence to support the hypothesis that the nervous, endocrine, and immune systems are interconnected and that the mind therefore affects health? Yes. We examine the evidence in this section. Furthermore, we extrapolate findings to discuss the possibility that a person can learn to control their immune system.

Evidence for one interconnecting network

The first line of evidence comes from the discovery that the immune, endocrine and nervous systems use the same cytokines for cellular communication. In addition, there is both structural and functional evidence, gathered from *in vitro* studies ("in glass," that is, studies in test tubes), and *in vivo* studies ("in life," that is, studies in animals and humans).

Shared cytokines. The term 'cytokine' includes molecules that are secreted by cells and function in communication with other cells. The ability of any cell to respond to a particular cytokine is dependent on whether the cell has receptors for that cytokine. Therefore, for the cytokines secreted by the nervous, endocrine, or immune systems to have an effect on the other systems, one must ask if there are receptors for them on cells of the other systems. The answer is yes, there are. American pharmacologists Candace Pert and Solomon Snyder discovered in the early 1980s that there were receptors on immune cells for endorphins, cytokines produced by nerve cells. Since that time, receptors have been found for many other cytokines that interconnect the nervous, immune, and endocrine systems. B lymphocytes have receptors for many cytokines produced by nerve cells (neurotransmitters), including norepinephrine and enkephalins. The presence on immune cells of receptors for these neurotransmitters suggests that immune cells could respond to these transmitters, as does the finding that the number of these neurotransmitter receptors per lymphocyte increases during immune activation. Lymphocytes also possess receptors for several endocrine cytokines induced during stress (ACTH and steroid hormones) or in pain (beta-endorphin). Neurons in the hypothalamus have receptors for the cytokines secreted by immune cells, and it is the effect of these cytokines on the hypothalamus that induces fever. The neuroglial cells that feed the neurons in the brain (Chapter 13, p. 466) also have receptors for immune cytokines, as do some endocrine cells.

Nerve endings in immune organs. Evidence of another sort came from studies done by American scientist David Felten on the nerve supply of the organs of the immune system. These studies used a technique called immunohistochemistry. Immune organs from animals are frozen and sliced very thin *(histology)* and then are stained with antibodies *(immuno-)* bound to enzymes *(chemistry)*. Such techniques borrow the exquisite antigen specificity of the immune system. Antibodies that recognize some molecule you want to detect are produced artificially in cell cultures. The antibody used by Felten's group recognizes and binds to an enzyme used in the synthesis of norepinephrine but does not bind to other chemicals. Thus, the antibodies give specificity to the technique, just as they do during an immune response in the body. The assay enzyme allows detection by producing a color change in the location where the antibody has bound. Some typical results are shown in Figure 15.10.

While most nerve cells terminate in synapses with other nerve cells or on muscle cells (Chapter 13, pp. 485–487), Felten's studies showed nerve cells terminating and releasing norepinephrine in proximity to the cells of the immune system. Neurons of the sympathetic nervous system were found to terminate in the immune organs, such as the spleen, thymus, lymph nodes, bone marrow, and lymphoid tissue in the gut. The sympathetic nerve cells release norepinephrine, and the immune cells in these organs have receptors for norepinephrine.

CONNECTIONS
CHAPTER 13

Analysis of cytokine function in animals. Demonstrating that a receptor is present, or even that both the receptor and its cytokine are present together in a tissue, does not by itself show that any effect follows the binding of the cytokine to its receptor. For that, experiments known as functional assays (techniques used to measure a response) need to be done. Often such studies are carried out using animals as experimental models.

Using functional assays some hormones have been shown to have effects on the immune system. During acute bacterial infection, the secretion of adrenal and pituitary hormones increases. Functional studies showed that the effect is not a direct one. If pituitary cells are stimulated with bacterial molecules *in vitro* they do not secrete these hormones. When immune cells are exposed to these bacterial products, however, they secrete cytokines. If immune-cell cytokines are administered to animals, the level of pituitary hormone in the blood increases.

Neuroendocrine cytokines have also been shown to have effects on the immune system. Among the cytokines secreted by the brain cells are the enkephalins and endorphins. When enkephalins are given to living rats, immune responses are altered, including antibody responses and T cell-mediated responses. Interestingly, low doses of enkephalin increase antibody production, while high doses suppress it. Low doses of enkephalins also increase some destructive aspects of the immune response, including both allergic and autoimmune responses, while high doses of enkephalins depress those responses.

Just as low doses of enkephalins may actually strengthen aspects of the immune system, so too does short-lasting stress. Engulfment of bacteria by white blood cells in mice (see Figure 15.4C) is increased by short-term stress brought on by conflict (see Figure 15.9). The ability of immune cells to kill tumor cells can be increased by restraining rats on a single day so as to increase their stress response. However, several days of restraint-induced stress causes a decrease in tumoricidal activity.

The steroid hormones secreted by the adrenal cortex during the stress response can have inhibitory effects on the immune system. Steroid hormones such as corticosterone given to animals decrease the numbers of cells in lymphoid organs and suppress the secretion of immune cytokines (see Figure 15.9). Daily fluctuations (circadian rhythms, Chapter 13, pp. 492–496) in the plasma levels of corticosterone also correlate with the circadian rhythm of the numbers of B and T cells circulating in the blood. Steroids also increase the susceptibility of the animals to disease and activate latent infections (infectious organisms that have been present but have not brought about disease now do so).

A compound chemically similar to the stress hormones, hydrocortisone, is used medicinally to block inflammation. Remember that normal

Figure 15.10

Immunohistochemical staining showing the presence of the neurotransmitter norepinephrine in the rat spleen. Large black arrows: blood vessels. Small red arrows: nerve fibers synthesizing norepinephrine.

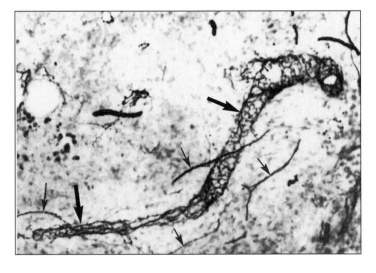

inflammation is the healing phase of the immune response (see Figure 15.3). In insect bites, severe poison ivy, athletic injuries, and rheumatoid arthritis, the swelling and pain are the result of the inflammatory response. In these situations, the annoying or harmful symptoms are actually an indication that the immune system is at work. Thus, a person who chooses to take an anti-inflammatory drug is choosing to suppress the healing processes of the immune system in order to suppress the negative symptoms of inflammation. The symptoms may be so severe that immunosuppression is needed, but long-term immunosuppression by corticosteroids are likely to have adverse consequences on other aspects of health.

The placebo effect

In clinical trials testing new drugs, a common experimental design is for one group of people to receive the test drug and for another group (the **control group**) to receive a **placebo**, a preparation that is similarly colored and flavored but that does not contain the test ingredient. For the drug to be considered effective there must be a statistically significant difference in outcome between the group receiving the experimental drug and the control group receiving the placebo.

Studies like this were initially designed to demonstrate whether particular drugs were effective. What they have also shown, over and over again, is that the people who receive the placebo in such tests have a significant change from the baseline values of whatever parameters are being measured, an effect known as the **placebo effect** (Figure 15.11). They also experience many 'side effects,' although they have not received a drug. In experiments on pain perception, people who are given placebos instead of painkillers very often experience a reduction in pain. People who have purchased street drugs will often feel the reaction they seek even when the drugs are in such low concentration that no real effect could be produced.

Such placebo effects have often been considered an annoyance in research; that is, they make it more difficult to demonstrate the 'real' effects of test compounds. The existence of such effects, however, is additional evidence that the mind can bring about physiological changes in the body, in addition to the effects of the stress response and the relaxation response.

Effects of stress on health

To what extent does the stress response influence human health? The answer is, to a great extent. Stress (i.e., the stress response) is an important risk factor in heart

Figure 15.11

The placebo effect.

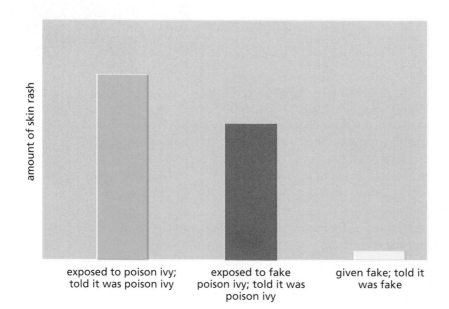

amount of skin rash

exposed to poison ivy; told it was poison ivy

exposed to fake poison ivy; told it was poison ivy

given fake; told it was fake

disease, and there is considerable evidence that people exposed to chronically high levels of stress are statistically more likely to become ill with infectious diseases, to remain ill for longer periods, and to suffer more severe consequences, even death.

The biochemical events of stress can be started by psychological factors, and prolonged stress can suppress the immune system. There have been many demonstrations of immune suppression in people undergoing various types of stress. It follows that long-term psychological stressors might produce conditions in which disease can develop. In one experimental design, blood samples are taken from young, basically healthy students during exams, and various immune parameters are compared with baseline levels measured in blood samples from the same students one month before exams. In this type of experiment, each person serves as his or her own control, minimizing differences due to factors other than the tension of exam situations. Exam periods in these experiments are characterized by an increase in adrenal gland hormones and a decrease in immune cell activity.

There were also decreases in the number of another type of lymphocyte called helper T cells. Helper T cells do not eliminate antigen; instead, they secrete cytokines which 'help' boost the strength of the responses of cytotoxic T cells and B cells to antigens (see Chapter 16, p. 575). Without these helper T cells, the immune response of the cytotoxic T cells and B cells is often not strong enough to prevent disease. Thus when stress decreases helper T cells, this results in an overall decrease in both B and T cell responsiveness. In these studies, such short-term stress was correlated with an increase in disease, primarily upper respiratory tract infections. Many college health centers report increase in student admissions for infectious diseases during exam periods.

CONNECTIONS CHAPTER 16

Another study examined the immune function and health status of men who had separated or divorced within the previous year. The experimental design was different from the one used with the medical students. One group of people, divorced men, was compared with people in a control group of married men. Not only were the divorced men's immune systems found to be impaired, they experienced a greater number of illnesses than did the controls. Comparisons were also made between those men who had not initiated the separation or divorce and those who had. Those who had not initiated the break were significantly more immunosuppressed and had more illnesses than those who did. Studies of this type employ statistical methods for determining whether the differences between groups are greater than could have been predicted by chance only.

Other researchers have found that elderly people who had been caring for a spouse with Alzheimer's disease demonstrated a decrease in three different measures of immune function. The elderly people undergoing prolonged stress got sick more often than those in the control group (people of similar age and health status but who are not caring for a spouse with Alzheimer's disease). The immune systems of these caregivers stayed depressed after the death of the spouse.

Several studies have shown decreased immune function in people with clinical depression. A prospective study showed that people who were depressed had a higher incidence of cancer 17 years later. A

prospective experimental design is one in which a group of people are examined first (using either physical or psychological exams) and then their outcomes are monitored at various later times. One strength of a prospective study is that no one knows ahead of time who will be sick and who will not; baseline data are taken before the outcomes are known. One weakness is that the percentage of people in any particular group who will get a particular disease may be very low, so that the number of people in the study must be very large. Many people will leave the study for unrelated reasons. Many other factors can influence the outcome; to some extent this can be corrected by statistical methods, but only those factors that have been identified can be factored out by statistical methods.

What mechanisms could bring about cancer after clinical depression? Immune activity is compromised, and other functions are also impaired, including levels of an enzyme that repairs damaged DNA. Breaks and misreadings of DNA occur rather frequently, but normally several 'proofreading' mechanisms check the DNA and repair most of the mistakes; mistakes that remain uncorrected are capable of transforming cells. The suppressed immune system then fails to remove these transformed cells before they have become established as cancer (Chapter 12).

Individual variation in the stress response. There are many types of stressors. The effect of stressors on health is highly variable, because the effects are modified by many additional factors. For example, of the people exposed to infectious mononucleosis, prevalent in college-age populations, not everyone becomes sick, and of those who do, some become sicker than others. It is important whether the stress occurs before or after the immune challenge started. The severity and duration of the stress are also important. Genetic factors have some role; in animal studies, different strains of mice (each inbred to minimize genetic variation within the strain) respond differently to stressors. Psychological factors are just as important. If an animal is able to establish coping behaviors, the effect of the stress period on the immune system will be lessened.

Personality profiles and life events have some bearing on disease susceptibility and disease progression in humans as well. Testing methods have been developed for quantifying the psychosocial impacts of life events (one such questionnaire can be found on our Web site, under Resources: Stressful life events). The use of such methods has indicated that certain life events can increase the probability that cancer will develop, although the results have been highly variable from study to study.

We cannot predict how much immunosuppression will be sufficient to result in disease in a given person. Both the degree and duration of suppression that result in disease are likely to be different for different people. Some studies suggest that coping styles can help regulate the degree of impact that stressful events will have on individual health. When psychological tests are given to matched sets of cancer patients (with the same kind of cancer and in comparable stages of the disease), those with more optimistic or aggressive personalities show higher survival rates than those who are more easily resigned to what they perceive to be their fate. Other factors, such as environmental pollutants, drugs, alcohol, and malnutrition, may also weaken the immune system.

If a person's immune system is already weakened by one or more of these factors, the additional immunosuppressive effects of stress are more likely to result in disease.

Conditioned learning in the immune system

If brain cells can interact with immune cells and psychological factors can influence the onset and progression of disease, could a person learn how to control his or her own immune response? As odd as this idea may sound, evidence in support of it is accumulating.

Classical or Pavlovian conditioning is a type of unconscious learning in which an organism learns to associate one stimulus with another. After such conditioning, presentation of the second stimulus will bring on the physical effects of the first stimulus (Chapter 13, pp. 489–490). Robert Ader, a psychiatrist, and Nicholas Cohen, an immunologist, worked together to try to explain why some of Ader's mice had been dying unexpectedly in his studies on a drug called cyclophosphamide. This drug suppresses the immune system and in fact is given to recipients of organ transplants so that their immune systems do not reject their transplants. Ader's mice had been receiving cyclophosphamide along with saccharin in their drinking water. Later the mice received only the saccharin, but their immune systems again became suppressed as they had when on the cyclophosphamide, even though saccharin itself has no effect on the immune system. What Ader and Cohen demonstrated in several controlled studies is that the dying mice had been conditioned. After the mice learned to associate the immunosuppressant chemical, cyclophosphamide, with the saccharin, the immunosuppressant effect could be produced by giving them only the saccharin water without cyclophosphamide.

These experiments demonstrating conditioned immunosuppression have been repeated with several other paired stimuli. Such conditioning has been shown to improve the health of mice with autoimmune disease; preliminary results suggest that conditioned immunosuppression is also useful in the treatment of people with autoimmune disease.

If animals are challenged with antigen paired with another stimulus, classical conditioning that boosts immunity can be demonstrated (Figure 15.12). In a normal immune response, a second exposure to the same antigen would produce an increase in specific immunity to that antigen. In a typical conditioning experiment, the animals are given their first exposure to an antigen paired with a conditioned stimulus such as saccharin. Saccharin itself does not induce antibody for the antigen (Figure 15.12A). When saccharin is given at the same time as the first exposure to antigen, animals make antibody to the antigen (Figure 15.12B). After conditioning, when the animals are exposed to saccharin, they react as though they were being exposed for a second time to the antigen. Saccharin induces an increased secretion of antibody for the antigen, without a second exposure to antigen (Figure 15.12C).

Voluntary control of the immune system

Other work has shown that people can learn to voluntarily regulate many of the physiological processes mediated by the autonomic nervous system. People can learn to regulate the temperature of their hands, their blood pressure, their heart rate, and their galvanic skin resistance (resistance to electrical conductivity, which is a measure of the amount of sweat on the skin). Because the immune system communicates with the autonomic system, these findings raised the hypothesis that parameters of the immune system may also be subject to voluntary control. Several studies have shown that this is possible using different self-regulation procedures, including relaxation, mental imaging, biofeedback, and emotional support. Voluntary potentiation of several immune parameters has been shown, including white blood cell engulfment of bacteria, antibody production, lymphocyte reactivity, and natural killer cell activity. Women with metastatic breast cancer, all of whom were receiving medical treatment of their cancers, survived longer if they were part of support groups than if they were not.

Mental imaging. In one study, subjects were asked to make mental images of their neutrophils becoming more adherent, a cellular process that might make neutrophils more efficient at getting to a disease site. When the adherence of their neutrophils was measured after several sessions of practicing such imaging, it was significantly increased in comparison to the neutrophils from people who had simply relaxed without forming the mental image of their neutrophils. Adherence was also measured in neutrophils from a third group who made mental images but did not have the practice sessions. Neutrophils from this group did not show an increase in adherence, showing that effective imaging takes a period of training.

Biofeedback. In biofeedback, measurement devices are placed on people so that they can monitor the results of their self-regulation. Biofeedback has proved to be effective for some people in the manage-

Figure 15.12

Classical conditioning of the immune response.

(A) Prior to conditioning:
conditioned stimulus (saccharin) given alone

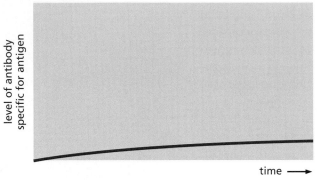

no immune response; antigen-specific antibody does not increase significantly

(B) Conditioning:
conditioned stimulus given together with unconditioned stimulus (antigen 1)

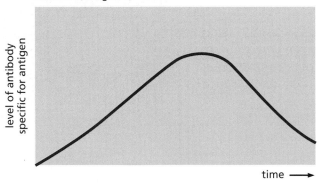

antigen induces an immune response; antigen-specific antibody increases

(C) Postconditioning:
conditioned stimulus again given alone

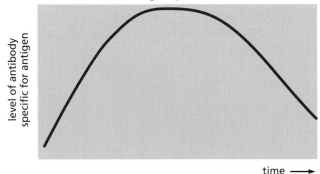

conditioned stimulus induces an immune response to antigen; antigen-specific antibody increases

ment of chronic pain and migraine headaches. Although it is too soon for there to be conclusive data on whether voluntary regulation of immunity will translate into improved health, preliminary studies suggest that it will. People with HIV infection and AIDS have remained healthier when they have used these techniques.

Studies on populations. If the mind and the emotions can influence the immune system, and the immune system helps to fight off many diseases, how far can the disease-fighting process be controlled by the mind? In recent years, statistical evidence has been accumulating from studies in both the United States and China to show that dying patients can exercise control over their disease processes to the extent that they actually influence the time of their death.

Large-scale studies are often done on entire populations by studying death certificates and comparable records. When the date of death is examined for a large number of patients in a population at large, several interesting regularities appear. For instance, the overall mortality rate is lower for a period of several days to either side of each person's birthday, and this is compensated for by an increasing mortality rate about a week or two later. Such data make it appear that dying patients are eager to survive to reach their birthdays and that they can postpone the inevitable by as much as a week or two. Other studies have shown similar statistical effects demonstrating the ability of people to postpone the time of their death until after holidays or family events (e.g., weddings) of special importance to them. In China, this effect even has a name: the Harvest Moon phenomenon. The Harvest Moon Festival is a traditional family celebration in which the oldest and most respected woman in each family is expected to prepare a large feast to celebrate with her entire family. Studies of death certificates show that older Chinese women have a reduced mortality rate around the time of this festival. It is difficult to determine whether this effect is primarily the result of the activities in which the matriarch engages, the increased esteem or importance that she receives, her desire not to disappoint others, or simply the anticipation of the big event. Studies on Jewish populations have shown a similar decline in mortality around the time of Passover.

A complex interaction between expectation and mortality has also surfaced in a recent study in China. Traditional Chinese astrology divides the calendar into 12-year cycles. Each year is represented by a different animal, and people born in that year are said to be under control of that animal's influence, with which certain diseases are associated. In this study, elderly Chinese patients were surveyed to see whether or not their disease matched the predictions of Chinese astrology. Patients with a disease that matched their astrological year were then compared with patients having the same disease but a different astrological year. The patients whose diseases matched their astrological year experienced higher mortality, and this effect was proportional to the patient's belief in traditional Chinese astrology. Presumably, patients who believed that they had the disease that was fated for them in the stars more willingly gave up the struggle and resigned themselves to an earlier death. Although these phenomena are well documented, the mechanism(s) by which they are produced are not known.

THOUGHT QUESTIONS

1 Given that chronic or severe stress generally weakens the immune system, how might such an apparently harmful relationship have evolved?

2 Is stress always harmful?

3 List the types of experiments done in psychoneuroimmunology. Do the results of any of these falsify the hypothesis that the mind and the body interact? Do any of these results prove that the mind and the body interact?

4 Is the Cartesian concept of the mind the same thing as the brain? Is *mind* simply the name we give to the *workings* (or functions) of the brain?

Concluding Remarks

Since the mid-nineteenth century, Western medicine has tended to view disease as having external causes. The ascendancy of this view can be traced back to the work of Louis Pasteur who championed the theory of "specific etiology" as part of the germ theory of disease. In this theory, each disease has one specific and identifiable cause. This theory led to much highly successful research that associated single species of microorganisms (bacteria, viruses, and parasites) with specific diseases. Such research ultimately produced vaccines and antibiotics, which have successfully controlled many infectious diseases. However, many of the diseases that are still without effective cures today are chronic diseases (cancer and heart disease, for example) that do not seem to have simple, single causes. Maybe a new concept of disease and of health is needed to find therapies and preventive measures for these diseases.

Psychoneuroimmunology is redefining our concepts of health and disease. Scientists in this field are using new technologies to reexamine some old concepts of disease causation. Working at the same time as Pasteur, another French scientist, Claude Bernard, questioned what it meant for a microorganism to 'cause' a disease. He observed that there were very great differences in individual response to microorganisms; some people got sick and even died, while other people who were also exposed did not get sick. Bernard, a physiologist, developed an alternative theory, that of the *milieu intérieur* or inner environment, as being an equal determinant in whether or not a person became sick. The past century of research in immunology and more recently in psychoneuroimmunology suggests that even the diseases for which an infectious agent is known are not caused by the microorganism alone, but rather by the outcome of a complex process in which the microorganism disturbs homeostasis while host mechanisms attempt to restore it. Bernard's theory is compatible with a view of the neuroendocrine–immune communication network as a sensory organ by which deviations from homeostasis are detected and corrected.

Psychoneuroimmunology also borrows from Chinese traditional medicine and ayurvedic medicine in India, as well as other Asian traditions that view health as the balance of life forces. Like these, the psychoneuroimmunology paradigm uses a *functional* model of the body, a model that regards the body as an entity that is in a constant state of change. Health is the state in which these forces are in balance, in homeostasis; disease is the state in which they are not. African traditions in which a person's health and well-being are seen to be influenced by the social environment in which the person lives coincide with the psychoneuroimmunology view that mental and emotional factors can affect health and disease. Within the psychoneuroimmunology paradigm, scientists are testing hypotheses suggested by these ancient traditions. Widening one's point of view and being open to new ideas are integral parts of science. New hypotheses are formed by the melding of ideas and must then be followed by the hard, and often slow, work of hypothesis testing.

Chapter Summary

- The immune system is a system that works to detect and correct deviations from **homeostasis** within the organism. It is composed of white blood cells that travel throughout the body in the blood and in the **lymphatic circulation**.

- **Health** is the ability to return to homeostasis.

- **Innate immunity** is present from birth and does not depend on exposure to develop. Innate immunity rids the body of some pathogens, and initiates **inflammation** and wound **healing**.

- **Specific immunity** is acquired by the individual after exposure to specific **antigens** such as bacteria or viruses. By forming specific groups of **B cells** and **T cells**, the immune system retains a memory of the encounter so that it can react faster and more strongly to subsequent exposures to that antigen. This is the basis for **immunization**: an artificial first exposure gives protection against later natural exposure to the same disease.

- Products of the specific immune response (**antibodies** and cytotoxic T cells), rid the body of the specific antigen that induced their production.

- **Passive immunity** is specific immunity developed in one person and transferred to another, for example by transferring antibody to them.

- Specific immunity can be selectively turned off. This is called development of **tolerance**.

- The immune system interacts with the **autonomic nervous system** and the neuroendocrine system in an integrated and multidirectional way. Communication among these systems is mediated by chemicals called **cytokines**.

- Factors that interrupt this communication network will prevent the restoration of homeostasis within the organism, producing disease.

- Mental states can affect the functioning of the immune system and can either increase or decrease its disease-fighting activity, a theory known as **psychoneuroimmunology**.

- The **stress response**, through the action of the sympathetic nervous system and stress hormones, can produce **immunosuppression**, while the **relaxation response**, through the action of the parasympathetic nervous system, can reverse this process.

- A **placebo** is a compound without physiologic effect that is given, for example, to the control group in a clinical test. Mental expectations result in people experiencing physiologic symptoms in response to the placebo, a result known as the **placebo effect**.

CONNECTIONS TO OTHER CHAPTERS

Chapter 1	Psychoneuroimmunology is a good example of a new paradigm.
Chapter 2	The great variety of antigen receptor proteins which can each recognize one of the huge variety of antigens in the world, is based on the rearrangements of just a few genes.
Chapter 5	The immune system has evolved.
Chapter 7	The ability to form an immune response to any particular antigen depends on cell-surface proteins. The allele frequencies of the genes that code for these proteins vary among populations, making some populations more susceptible than others to a particular disease.
Chapter 8	The molecules that block successful organ transplantation between mismatched donor and recipient are the same molecules that allow kin selection in many species.
Chapter 10	Poor nutrition suppresses the immune system.
Chapter 12	Suppressed immune function greatly increases the risks for cancer.
Chapter 13	The brain can affect many immune functions.
Chapter 14	Many drugs suppress the immune system. Many drugs mimic some of the activities of the autonomic nervous system.
Chapter 16	Immunosuppression is characteristic of AIDS.
Chapter 19	Pollution can suppress immune function.

PRACTICE QUESTIONS

1. What type of white blood cell secretes antibodies?

2. What type of white blood cell can kill cancer cells?

3. What type of white blood cell can engulf bacteria and kill them?

4. What type of cell releases histamine?

5. What part of the body is acted on by macrophage cytokines to induce fever?

6. Which part of the autonomic nervous system mediates the flight-or-flight response?

7. Which part of the autonomic nervous system mediates the relaxation response?

8. Which of the following are parts of the immune system: bone marrow, spleen, skin, intestines, bladder, tonsils, eyes?

9. What types of scientific evidence suggest that the brain and nervous system interact with the immune system?

10. How do the terms *stress* and *stressor* differ?

11. What are the stages of the stress response, and what characterizes each stage?

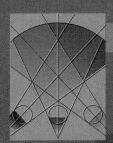

Issues

- How did the discovery of AIDS and HIV develop? What other diseases helped or hindered scientists and doctors in the discovery?

- What do HIV tests tell us?

- Will there be a cure for AIDS? What about this disease makes a cure so difficult?

- Will there be a vaccine to prevent AIDS? If a vaccine is possible, will it solve all the problems associated with AIDS?

- Will studying HIV teach us all we need to know about the AIDS pandemic? What social factors contribute to the global spread of AIDS?

Biological Concepts

- Health and disease (immune system, pathogens and response, receptors, Koch's postulates, routes of transmission, behavior)

- Biodiversity (viruses)

- Scales of size

Chapter Outline

AIDS Is an Immune System Deficiency

AIDS is caused by a virus called HIV

Discovery of the connection between HIV and AIDS

Establishing cause and effect

Viruses and HIV

HIV Infection Progresses in Certain Patterns, Often Leading to AIDS

Events in infected helper T cells

Progression from HIV infection to AIDS

Tests for HIV infection

A vaccine against AIDS?

Drug therapy for people with AIDS

Knowledge of HIV Transmission Can Help You to Avoid AIDS Risks

Risk behaviors

Communicability

Susceptibility versus high risk

Public health and public policy

Worldwide patterns of infection

16 HIV and AIDS

*A**IDS is a disease caused by the virus HIV. HIV undermines the immune system, leaving the infected person vulnerable to other diseases. As we saw in Chapter 15, the immune system has several ways of protecting the body from disease. When people have AIDS, their immune systems no longer function properly, so they are at risk of becoming ill from infections that would barely affect a healthy person. Many people with AIDS, for which there is currently no cure, suffer long and painful deaths. AIDS first received public attention in 1981. It quickly became one of the most feared and widely discussed diseases of our time. As of June 2002, a total of 23 million people have died of AIDS. Worldwide, more than 11 million children have been orphaned because their parents have died of AIDS. HIV is spread from person to person in infected body fluids. Sexual contact is one of the main routes of transmission; thus AIDS is a sexually transmitted disease. The spread of AIDS has been accompanied by the spread of misconceptions concerning the disease. In this chapter we summarize what is known about this dreaded disease and also address certain misconceptions.*

AIDS Is an Immune System Deficiency

The acronym **AIDS** stands for **Acquired *I*mmuno*D*eficiency Syndrome**. *Acquired* means that the illness is not genetically inherited as the result of a defective DNA message. *Immunodeficiency* means that some part of the immune system is not functional, and *syndrome* means that a wide range of symptoms are associated with the disease. People whose immune systems are deficient can become seriously ill with infectious diseases or cancers.

The body has many ways of protecting itself from diseases. The skin protects the body's surface against entry of bacteria, viruses, or other microorganisms. The mouth, vagina, and many other potential entry points are coated with mucous secretions that continually wash away adherent bacteria, inhibit bacterial growth, and promote healing.

A more specific type of protection is afforded by the immune system. An organism's immune system distinguishes between molecules that are part of the organism (self) and ones that are not (non-self; see Chapter 15). Non-self molecules include bacteria, viruses, and molecules made by cancer cells. Many non-self molecules trigger an immune response that inactivates or destroys the non-self molecules (see Chapter 15).

An **immunodeficiency** is an absence of one or more of the normal functions of the immune system. People who are immunodeficient get sick more often than people with healthy immune systems, and their illnesses last longer and are more severe.

How does someone become immunodeficient? Some of the many causes are inherited and some are environmental. One type of inherited immunodeficiency, the severe combined immune deficiency syndrome (SCIDS), caused by a lack of the enzyme adenosine deaminase (ADA), is discussed in Chapter 4 (pp. 100–102). Inherited immunodeficiencies are rare; much more common are those that are acquired as a result of environmental exposures. The functioning of the immune system can be depressed, for example, by alcohol and drugs such as cocaine and marijuana, psychological stress and depression, cigarette smoke and other pollutants, and malnutrition (either total calorie deficit or micronutrient malnutrition; see Chapter 10). Clearly, AIDS is not the only kind of immunodeficiency, but it is among the most severe. Many immunodeficiencies are temporary and reversible: if the causative factor is removed, the immune system recovers. AIDS is long-lasting; the immune system does not recover, and the disease is fatal.

AIDS specifically targets the lymphocytes called **helper (CD4) T cells**. These cells and the cytokine called interleukin-2 that they secrete are necessary for both the B lymphocyte and **cytotoxic (CD8) T cell** responses of the immune system (Figure 16.1). As you probably recall from Chapter 15, B cells make antibodies, our main defense against bacteria and fungi.

Figure 16.1

Interactions between lymphocytes and their cytokines.

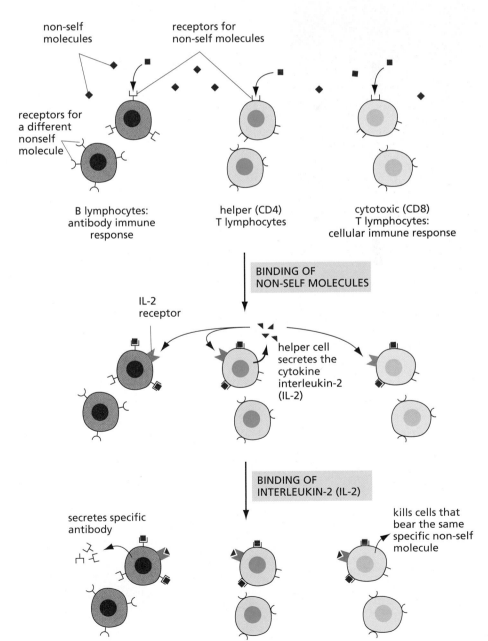

Helper T cells, B cells, and cytotoxic T cells have receptors for non-self molecules. Within each cell population, individual cells have receptors for different non-self molecules.

B lymphocytes and cytotoxic T lymphocytes require two signals to become functional immune cells: (1) binding specific non-self molecule to their receptors, causing receptors for IL-2 to be brought to the cell surface, and (2) binding of IL-2 to IL-2 receptors.

Helper T cells that have bound their specific non-self molecule are the source of IL-2; cells that bind IL-2 are activated. Thus removal of helper T cells shuts down both B-cell and cytotoxic T-cell activity.

Cytotoxic T cells protect against viral infections and against cancer cells. When helper T cells are destroyed in AIDS, both the B cell and cytotoxic T cell arms of the immune system are lost, leaving the individual vulnerable to bacterial, fungal, and viral infections and to cancer. In addition, the neutrophil and macrophage white blood cells of non-specific immunity (see Figure 15.1, p. 543) are also severely weakened. Macrophages, like helper T cells, can be directly targeted, and without antibody, neutrophils cannot engulf bacteria. An infection that would be minor in a person with a healthy immune system can quickly become life-threatening in a person with AIDS.

AIDS is caused by a virus called HIV

We now know that AIDS is an infectious disease caused by a virus known as HIV (Human Immunodeficiency Virus). However, when the syndrome first began to appear in the United States at the very end of 1980, the cause, and even the fact that it was an immunodeficiency, was not known. In this section we trace the steps that led to the identification of this immunodeficiency and its causative agent. How were hypotheses suggested? How were these hypotheses tested? What types of evidence are necessary to call something the 'cause' of a disease? Does such evidence rule out other hypotheses? We then look at the virus itself and how it lives in human cells.

Discovery of the connection between HIV and AIDS

In June of 1981, five cases of pneumonia in San Francisco were reported to be associated with a microorganism called *Pneumocystis carinii*. The report appeared in *Morbidity and Mortality Weekly Review* (*MMWR*), a publication of the Centers for Disease Control and Prevention (CDC), which tallies all cases of sickness (morbidity) and death (mortality) in the United States due to certain kinds of diseases called reportable diseases. They are called 'reportable' diseases because a physician seeing a patient with one of these diseases is legally obliged to report it to the CDC. (The numbers of cases are reported; patients' names are not.) Reportable diseases listed in *MMWR* include most of the serious contagious diseases caused by microorganisms. Each issue of *MMWR* also contains articles written by alert clinicians who have observed patterns of disease that are unusual for a particular geographic area or season, or are too frequent, or are occurring in an age group that does not usually get the disease. Tallies of reportable illnesses are often the first indication of an unusual spread of a known disease or the appearance of a new disease. The statistical study of information about the occurrence and spread of diseases in whole populations is called **epidemiology**.

Pneumonia, a disease characterized by fluid in the lungs, can be caused by many different bacteria and viruses. The five cases of *Pneumocystis* pneumonia reported in *MMWR* were quite unusual because *Pneumocystis* is a parasite that is neither a bacterium nor a virus. All five cases were from a single geographic area and close together in time, and thus represented what epidemiologists call a **case cluster**.

Later in the summer of 1981, a dermatologist in New York City, Dr. Alvin Friedman-Kien, noticed an unusual cancer, called Kaposi's sarcoma, among many of his young homosexual male patients, a finding that he reported in *MMWR*. Kaposi's sarcoma, a cancer of the cells lining the walls of blood vessels, causes red or purple raised patches on the skin. Kaposi's sarcoma was rare in the United States and had previously been found only in elderly men of Italian or Eastern European Jewish descent. The Kaposi's sarcoma seen in the reported cluster was far more aggressive than that seen in elderly men, meaning that it spread much faster and was present in the internal organs, not just on the skin. Aggressive Kaposi's sarcoma is, however, seen in kidney transplant patients, who take medication to suppress their immune systems, which would otherwise reject the transplanted tissue. This fact suggested the hypothesis that the Kaposi's sarcoma becomes aggressive when the immune system is suppressed.

The immunodeficiency hypothesis. Were the Kaposi's cases in any way related to the unusual pneumonia cases? Did immunodeficiency underlie both the unusual pneumonias and the unusual cancers? Both the *Pneumocystis* pneumonia patients and the Kaposi's patients were found to have severely decreased numbers of helper T cells. As we have seen above, helper T cells are central in the immune system; a person lacking helper T cells has a suppressed immune system (see Figure 16.1). The evidence thus fit the hypothesis that the cancer and the pneumonia belong to a syndrome resulting from the same underlying mechanism, namely immunodeficiency. This syndrome was given the name AIDS. Reported cases accumulated quickly: 87 in the first six months of 1981, 365 in the first six months of 1982, and 1215 in the first six months of 1983. AIDS had been given a name, but its cause was still unknown. Funding for research was slow in coming, as was interest on the part of many scientists. Even after the cause was established, there was silence on the part of government agencies. Various projects, such as the AIDS quilt (started in 1987; Figure 16.2), were started by private individuals and nongovernmental organizations to increase public awareness and push for increased research.

Lifestyle hypotheses. Epidemiologists gathered information from AIDS patients, trying to establish any common links between the cases: had they all been exposed to the same chemical agent? Did they all live in the same geographical area or in the same household? Were the patients known to one another? For a while, because the first AIDS patients were homosexual men, the search was for common lifestyle factors, on the assumption that there is such a thing as a 'homosexual lifestyle,' one characterized by some drug or dietary factor that was shared by most or all homosexual men.

Figure 16.2

The AIDS quilt, commemorating those who have died of AIDS. In 1999 the quilt had grown to be 42,960 panels commemorating 83,279 names.

Several researchers hypothesized that the immunodeficiency was due to an overload of the immune system by chronic exposure to non-self molecules via promiscuous sexual activity. Others doubted these hypotheses because the effects seemed to be specifically targeted on one type of cell, the helper T cell; they searched instead for infectious microorganisms that homed in on this type of cell and that might be transmitted by sexual contact.

The viral hypothesis. Was the infectious microorganism a bacterium, a fungus, a protozoan, or a virus? Support for the hypothesis of a viral agent came when some cases of AIDS were reported among hemophiliacs. People with hemophilia lack the genes that code for certain blood proteins necessary for forming blood clots after an injury. Hemophilia can be life-threatening because the person can bleed to death from a minor cut or scrape. As protection, hemophiliacs are given blood-clotting proteins from other people. This works well, but only temporarily; transferred clotting agents, like all proteins in the body, are eventually broken down by protein-degrading enzymes. The clotting factors must therefore be supplied repeatedly to hemophiliacs. These clotting factors are obtained from blood pooled from many donors and filtered to remove bacteria and fungi. Viruses, however, can pass through the filters. Because hemophiliacs receiving a filtered blood product were contracting AIDS, it was reasoned that the infectious agent could be a virus.

One laboratory that was studying viruses at the time was the National Cancer Institute's Laboratory for Tumor Cell Biology, headed by Robert Gallo. The occurrence of AIDS in hemophiliacs convinced Gallo that the infectious agent must be a virus. Gallo's laboratory was studying **retroviruses**, a type of virus whose genetic information is RNA that is copied to make DNA. One retrovirus being studied was the human T-cell leukemia virus, HTLV-I. This retrovirus was known to cause a form of leukemia (a blood cell cancer) associated in some patients with a mild immunodeficiency. It could not be said, however, whether the immunodeficiency seen in the HTLV-I patients was caused by the virus or was the result of the cancer or some other factor. Nevertheless, because it was known that HTLV-I was transmitted from person to person by sexual contact and that it specifically attacked T cells, it fit the pattern seen for AIDS.

In 1983 Luc Montagnier and his co-workers at the Pasteur Institute in France found important new evidence in the tissues of a patient with chronically swollen lymph nodes, a condition common in the early stages of AIDS. (The lymph nodes are the structures that temporarily enlarge when the body is fighting an infection; people often refer to them as 'swollen glands.') The scientists found the enzyme reverse transcriptase in the lymph tissues. Reverse transcriptase is used by retroviruses to produce DNA from RNA. Montagnier's group had not yet found the virus, just one of its enzymes, but the reverse transcriptase was strong evidence of the presence of a retrovirus.

The presence of any retrovirus can be detected by finding reverse transcriptase (as Montagnier's group had done), but the identification of a specific retrovirus requires testing of large quantities of viruses, which are obtained by growing them in laboratory culture. Viruses cannot replicate outside a host cell, but these host cells may be grown in the laboratory rather than in an animal. Gallo's laboratory had developed a method for growing human T cells in the laboratory, and for growing HTLV-I in those cells. Antibodies were made to HTLV-I grown this way. These antibodies were used to show that the new retrovirus from AIDS patients was not HTLV-I.

When scientists tried to grow the new retrovirus from AIDS patients in human T cells in the laboratory, it killed the cells. Mikulas Popovic in Gallo's lab found a type of leukemia T cell in which the new retrovirus could be grown without killing the host cells. Once the method for growing quantities of the new virus had been developed, antibodies were made that were specific for it. These antibodies were then used by Gallo's group to test viruses isolated from three groups of people: a control group that consisted of healthy heterosexuals, a group of AIDS patients, and a group of patients with AIDS-related complex or ARC, a set of symptoms assumed to be an early stage of AIDS. The viruses isolated from AIDS patients and some ARC patients were identified as being the same as the new virus. None of the healthy subjects had this new virus.

Using the specific antibodies, scientists found the virus in 80–100% of AIDS patients, in varying percentages of people in certain defined risk groups, and only rarely in healthy individuals outside the risk groups. These results were strong evidence that this new retrovirus was associated with AIDS. The retrovirus found by Gallo and the retrovirus found earlier by Montagnier were determined to be two strains of the same virus. Each group had given their virus a different name and each group wanted the name they had chosen to become the standard. The International Committee on the Taxonomy of Viruses studied the naming problem and decided in 1986 that neither name should be used. They assigned a new name to the virus, **human immunodeficiency virus** or **HIV**.

Establishing cause and effect

Just because a microorganism is associated with a disease does not mean that it causes the disease. How do we know that HIV is the cause of AIDS? There is a set of rules that have traditionally been used to identify a microorganism as the cause of a particular disease. These rules were formulated in the late 1800s by Robert Koch, a German physician, and have come to be known as **Koch's postulates**:

- First, the microorganism suspected as the causative agent must be present in all (or nearly all) animals or people with the disease.

- Second, the microorganism must not be present in undiseased animals.

- Third, the microorganism must be isolated from a diseased animal and grown in pure culture (that is, a culture containing no other microorganisms).

- Fourth, the isolated microorganism must be injected into a healthy animal, the original disease must be reproduced in that animal, and the microorganism must be found growing in the infected animal and be reisolated from it in pure culture.

Koch used these rules to show that the bacterium *Bacillus anthracis* was the cause of anthrax, a fatal disease in sheep that was decimating European herds in the 1870s. Koch later used his postulates to identify a bacterium now known as *Mycobacterium tuberculosis* as the causative agent for human tuberculosis. Koch received a Nobel Prize for his demonstration of the causes of anthrax and tuberculosis.

Limitations to Koch's postulates. While these rules are straightforward to state, they are difficult to fulfill. Every animal is host to many bacteria, so finding one bacterial species that is present only in diseased animals is not easy. Koch and his contemporaries developed bacterial culture media and techniques that enabled them to grow pure cultures of certain bacteria, but the growing of many other bacterial species (such as those killed by exposure to air) required technology that did not exist in Koch's time. In studying anthrax, a disease in sheep, the fourth postulate—requiring the production of the disease in an experimentally infected animal—was straightforward. In Koch's later studies on tuberculosis, ethical considerations dictated that the fourth postulate could not be fulfilled by infecting a healthy human, so an animal was used instead as a model, or experimental patient.

Despite the difficulties associated with meeting the requirements of Koch's postulates, they have been very useful. About a dozen or so infectious diseases are controllable by vaccination, including rabies, poliomyelitis, whooping cough, tetanus and diphtheria. The infectious microorganisms responsible for these diseases and others were identified on the basis of Koch's postulates. A microorganism that has been shown to cause a disease is called a **pathogen**.

There are other diseases for which these postulates have not been demonstrated. Most bacteria can grow outside cells, so they can be grown in 'pure cultures' if the proper growth conditions can be found. As we have seen with human retroviruses, viruses grow only inside a host cell, so that difficulties in growing the infectious agent in culture became even more acute when viral, rather than bacterial, diseases were studied. Because many bacteria and viruses that infect one species do not infect other species, it is not always possible to find animal models for human diseases, as Koch did for tuberculosis. Further, even the best animal models can never reproduce all of the aspects of a human infection. Some diseases may be caused by the interactions of more than one species of bacterium or virus; in such cases, Koch's last postulate would not work because the organisms isolated in pure cultures would no longer have the same effect unless combined. In many diseases there are healthy carriers (people who are infected and can transmit the infection to someone else, but do not become ill themselves), so the second postulate is not met. Other diseases, such as some cancers, may be multifactorial, meaning that no single factor is causative by itself. Nevertheless, Koch's postulates are the standards by which scientists establish cause and effect for most *infectious* diseases, meaning those diseases that can spread by infection with a microorganism.

HIV and Koch's postulates. Koch's postulates are the accepted standard of proof for asserting cause and effect in infectious disease. To say that HIV *causes* AIDS thus carries with it the implication to many in the scientific community that Koch's postulates have been fulfilled. This implication has been contested by some scientists, most notably by the American virologist Peter Duesburg, who maintains that three out of the four postulates have not been met. The presence of the virus itself was difficult or impossible to demonstrate in some people. The basis for considering someone to be HIV-infected is the presence in their blood of antibodies to the virus, not the virus itself. Duesburg maintains that Koch's first postulate (that the suspected causative agent must be present in all people with the disease), has therefore not been fulfilled. Most scientists no longer find this particular point controversial because newer, more sensitive techniques (such as the polymerase chain reaction described in Chapter 3, p. 81) have been used to detect the virus directly. The virus has now been found in virtually all persons with AIDS. We also now realize that, during the asymptomatic phase of HIV infection when the number of viruses in the blood is low, the virus can be found in high numbers in the cells of the lymph nodes. The third postulate (that the suspected agent must be grown in pure cultures) has also been difficult to satisfy. Although methods were developed for growing HIV in human T cells in culture, Luc Montagnier demonstrated that HIV could only be made to replicate in those cells if there were other infectious agents present. The fourth postulate (that injecting the suspected agent into an animal must produce the disease) has not been fulfilled. Only two nonhuman animals have been found in which HIV will grow (chimpanzees and macaque monkeys), and when these animals are infected with HIV, they do not develop AIDS.

Although many scientists agree that Koch's postulates have not been entirely fulfilled (and perhaps cannot be), the failure to fulfill them does not prove that the suspected agent does *not* cause the disease. Gallo has pointed out that Koch's postulates cannot even be strictly applied to the diseases that Koch himself studied. Many people, for example, can be healthy carriers of the tuberculosis bacterium, and so the second rule is not always met.

Criteria other than Koch's postulates. Because of the limitation of Koch's postulates, other criteria have been suggested for establishing causality, particularly of viral diseases.

The criteria used by Gallo for stating that HIV is the sole cause of AIDS are as follows:

- HIV or antibody to HIV is found in the vast majority of persons with AIDS.

- HIV is found in a high percentage of people with ARC.

- HIV is a new virus and AIDS is a new disease.

- Wherever HIV is found, AIDS develops; where there is no HIV, there is no AIDS.

- People who received transfusions of blood contaminated with HIV developed AIDS.

- HIV infects CD4 helper T lymphocytes, a cell type depleted in AIDS.

- On autopsy, HIV is found in the brains of people who have died of AIDS, and dementia (loss of brain cell function) is a symptom of AIDS.

Necessary causes and sufficient causes. As the above discussion points out, it is not always easy to identify cause and effect. There are some scientists who do not agree that Gallo's criteria are sufficient to establish cause. There are others who say that Koch's rules are the correct criteria, but that, in the case of HIV and AIDS, the criteria have not been met, particularly rule 4. At present, the vast majority of scientists agree that HIV is a *necessary cause* of AIDS; that is, someone who is not infected with HIV will not get AIDS. However, not all scientists agree that HIV is the sole or *sufficient cause* of AIDS (that is, no other factors are required) because there is a great difference in the course of the infection among various persons with AIDS.

There are several lessons to be learned here. Hypotheses about cause and effect of diseases are debated both in written articles and at scientific meetings. Many additional hypotheses are suggested and tested, and the number of scientists involved is often (as in this case) very large. Eventually a consensus develops as to the explanation that best fits the data. Consensus is never 100% agreement, but public policy and public health decisions must be made nevertheless. The accumulation of data may settle a given controversy (e.g., in the case of HIV and AIDS), but at any time new data may arise that require a change in the consensus explanation. Scientific debate and research must always be open to new possibilities because the findings of science are always tentative or provisional.

Saying that a particular virus causes a disease, especially a disease with such a diffuse group of symptoms as AIDS, really tells us little or nothing by itself. All of the 'how' questions remain. How does the virus infect? How do cellular effects progress to clinical symptoms? How can the disease be prevented or stopped? How is the infection transmitted? How contagious is it? How likely is it that HIV infection will become AIDS? How do people cope with such a disease? It is interesting that the articles written by scientists, either for other scientists or for the public, are almost entirely devoted to answering the first three questions, while the literature written by nonscientists is much more concerned with the last four. We examine each of these questions later in this chapter.

Viruses and HIV

Many human diseases are caused by viruses. AIDS is one such disease; measles, mumps, polio, and herpes are also caused by viruses. In addition to this role in disease, viruses are interesting to biologists because they challenge our understanding of what it means for something to be alive (Chapter 1, p. 11). **Viruses** are bits of either DNA or RNA that cannot reproduce by themselves but can replicate inside a cell (called the host cell) by using the biochemical machinery of the host. Biologists define a living organism as one that can reproduce itself, which viruses cannot; yet once inside a host, viruses can cause the host to replicate the virus, something that is not a characteristic of any known nonliving thing. What is the structure of a virus and how do viruses accomplish this?

The viral life cycle. A virus consists of nucleic acid, an outer protein shell and, in some viruses including HIV, a phospholipid bilayer membrane called the viral envelope that also contains some viral proteins. A particular virus has either DNA or RNA, and the nucleic acid is either single-stranded or double-stranded, distinctions that are used in viral classification. Viral genomes vary in size: some have only enough nucleic acid to code for 3–10 proteins, while others code for 100–200 or more proteins. HIV, the virus that causes AIDS, is at the small end of this size range. Viruses are very small, even in comparison with bacteria. A single

CONNECTIONS
CHAPTER 1

human cell is typically 10 µm (10 millionths of a meter) in diameter. Magnifying this human cell 100,000 times would make it a meter wide; at the same magnification a bacterium would be about the size of a football and a virus would be only the size of an M&M candy.

Some viruses can survive outside cells, but no virus can replicate unless it is inside a host cell. Human cells, animal cells, plant cells, and bacteria can all serve as hosts to viruses. For each virus there are only certain species that can serve as its host, and within an individual of the host species only certain types of cells can be host cells. This is because to enter a cell the virus must attach to some molecule on the host cell surface. Each type of virus is able to bind only to specific host cell molecules, which are usually membrane proteins. The species of virus that can adhere to dog or cat cells, for example, usually cannot adhere to human cells, which explains why we usually cannot catch viral diseases from our pets.

After a virus attaches to a host cell, the viral nucleic acid, sometimes with some viral proteins, enters the cell's cytoplasm, usually with the help of energy derived from the host cell. In viruses whose nucleic acid is DNA, many copies of the virus are made, using the host's molecular machinery for DNA replication. Viruses whose nucleic acid is RNA may replicate their RNA genome directly or convert their RNA into DNA by using the host's machinery to make more viral particles. The final stage of the viral life cycle consists of the release of viruses from the cell by the rupturing (also called lysis) of the cell or by the budding out of viruses through the host cell membrane. The new viruses can then infect other cells and repeat their life cycle.

HIV structure and life cycle. HIV is an enveloped virus whose genome consists of two copies of a single strand of RNA. The structure of HIV is shown in Figure 16.3. The virus is surrounded by a phospholipid bilayer envelope. In the viral envelope are proteins that are important for attaching to and entering specific host cells. Inside the viral envelope are two protein layers. Inside the inner protein layer is the viral genome consisting of two identical strands of RNA.

The HIV life cycle is shown in Figure 16.4 and is typical for many retroviruses. Proteins in the viral envelope bind to the CD4 protein found on only a few types of human cells: helper T cells and some macrophages, a type of white blood cell important in inflammation and in engulfing pathogens (see Figure 15.1, p. 543). Helper T cells, macrophages, and cells related to macrophages are thus the only cells that become directly infected by HIV. Other species of

Figure 16.3

The structure of HIV.

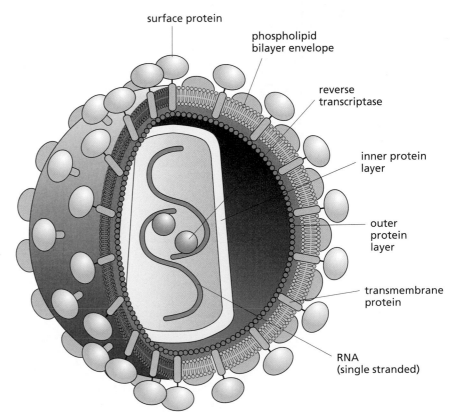

surface protein

phospholipid bilayer envelope

reverse transcriptase

inner protein layer

outer protein layer

transmembrane protein

RNA (single stranded)

mammals also have CD4 and helper T cells but the structure of CD4 in each species is different so that HIV attaches only to human CD4. Once attached to the host cell, HIV enters by fusion of its viral envelope with the host's plasma membrane. This fusion step requires other proteins in the host cell membrane.

As we have already mentioned, HIV is a retrovirus, a type of virus whose genetic information is RNA. For reproduction of the virus, the

Figure 16.4

The retroviral life cycle.

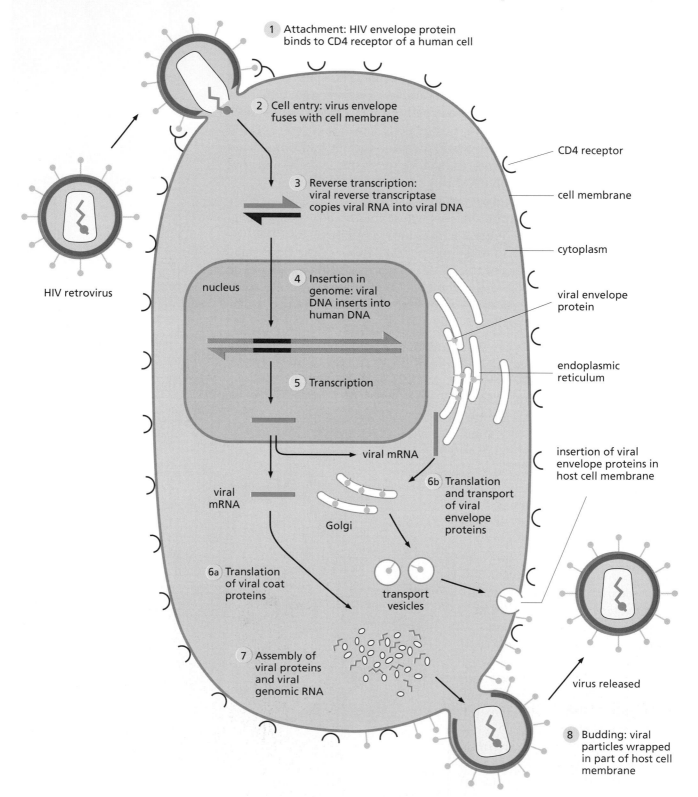

1 Attachment: HIV envelope protein binds to CD4 receptor of a human cell

2 Cell entry: virus envelope fuses with cell membrane

3 Reverse transcription: viral reverse transcriptase copies viral RNA into viral DNA

CD4 receptor

cell membrane

cytoplasm

HIV retrovirus

4 Insertion in genome: viral DNA inserts into human DNA

nucleus

viral envelope protein

endoplasmic reticulum

5 Transcription

viral mRNA

insertion of viral envelope proteins in host cell membrane

viral mRNA

6b Translation and transport of viral envelope proteins

Golgi

6a Translation of viral coat proteins

transport vesicles

7 Assembly of viral proteins and viral genomic RNA

virus released

8 Budding: viral particles wrapped in part of host cell membrane

RNA must first be used to make DNA; because this is the reverse of the usual DNA-to-RNA process of transcription (see Chapter 3, pp. 64–66), the RNA-to-DNA process is called **reverse transcription**. After entry into the cell, a viral protein, the enzyme reverse transcriptase, becomes activated. This enzyme uses the viral RNA as a template and synthesizes complementary DNA. The first DNA strand is, in turn, the template for synthesis of the second strand of DNA. The now double-stranded DNA is incorporated into the host's DNA. The viral RNA is meanwhile broken down or degraded.

Once in the host's DNA, the viral DNA can be transcribed by host enzymes into many copies of viral messenger RNA (mRNA). Some viral mRNA is then translated in the host cell cytoplasm to be the protein coats of new virus particles. Other viral mRNA is translated into viral envelope proteins. Like host cell membrane proteins, viral envelope proteins are made on the host cell's endoplasmic reticulum, transported through the Golgi, then carried via transport vesicles to the plasma membrane. Multiple copies of viral envelope proteins build up on the surface of the host cell (see Figure 16.4). Viral DNA is also transcribed as a whole to make copies of the viral RNA genome. The viral genomic RNA with its protein coat joins a portion of the host cell plasma membrane containing viral envelope proteins. The virus then buds out, carrying along a piece of the host cell membrane, which becomes the viral envelope. A photograph taken at very high magnification through an electron microscope shows HIV budding from a helper T cell (Figure 16.5A). On leaving the host cell, the virus is a mature, cell-free virus, ready to infect a new host cell. Many new viruses bud out of a single helper T cell (Figure 16.5B). HIV can also lyse (rupture) the T cell.

Figure 16.5

HIV budding from a helper T cell. (A) Budding and mature HIV. The mature viruses are free of the cell; the viral protein can be seen inside the viral envelope (a piece of the cell membrane that the viruses have taken with them as they budded out). (B) Many HIV particles bud from the same cell.

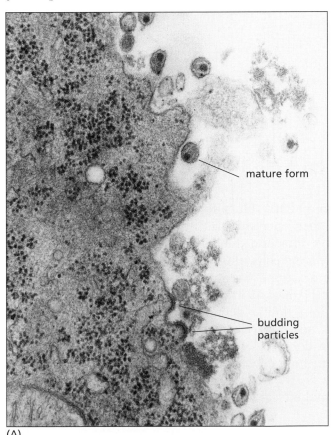

mature form

budding particles

(A)

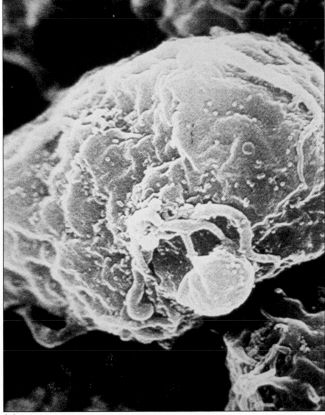

(B)

The HIV envelope and genome contain all of the molecular determinants that make the virus virulent (able to cause disease), infective (able to enter a cell), cell-specific (entering only certain types of cells), and cytopathic (able to kill or inactivate the host cell). The HIV genome also contains some regulatory genes, meaning genes that turn other genes 'off' or 'on.' All of the HIV regulatory genes seem to be essential for the life cycle; thus, they are possible targets for drugs or vaccines because blocking any essential step would inhibit the whole cycle.

Both types of HIVs currently known, HIV-1 and HIV-2, have the same basic structure, life cycle, and routes of transmission, but the rate of transmission of HIV-1 is 5–10 times higher than that for HIV-2. People infected with HIV-1 are 3–8 times more likely to have a decrease in helper T cells, lose immune function, and progress to AIDS than people infected with HIV-2.

How does HIV bring about the disease called AIDS? We examine this question next.

THOUGHT QUESTIONS

1 How might scientists decide whether or not two strains of virus are actually the same?

2 What are some of the reasons why evidence of the types called for by Koch's postulates may be impossible to obtain for some infectious diseases?

3 What are some differences between bacteria and viruses?

4 If an RNA strand from HIV contains the base sequence AAUGCA, what would be the base sequence on the first strand of DNA produced by reverse transcription? What would be the sequence of the second DNA strand transcribed from the first one? (You may need to review material from Chapter 3 to answer this question.)

HIV Infection Progresses in Certain Patterns, Often Leading to AIDS

We have already seen that HIV binds to cells that have a CD4 molecule on their surfaces. It then enters these cells, replicates, and goes on to infect more cells. HIV infection diminishes both the number and the activity of the CD4-bearing cells, thus reducing their ability to perform their disease-fighting functions. Because CD4-bearing helper T cells are central to both arms of the specific immunity (see Figure 16.1), their elimination results in immune deficiency, which in turn results in disease. This process can be described and studied at many levels. At the cellular level, just how does HIV eliminate helper T cells? At the organismal level (the person) how does infection progress to disease? At the population level, how is HIV transmitted? We look in this section at the cellular and organismal effects and in a later section at the population effects.

Events in infected helper T cells

How does HIV eliminate helper T cells? There are several different mechanisms.

1. Direct killing. As we have already seen, HIV can directly kill the cell it has entered by rupturing it. Repeated budding out of replicated viruses also eventually kills the cell (Figure 16.6).

2. Cell suicide (**apoptosis**). HIV may also change a helper T cell so that when it responds to another infection, it commits suicide instead of dividing. Healthy T cells, when activated by a pathogen, begin to synthesize DNA and divide. Under the same conditions, HIV-infected helper T cells undergo a process in which the DNA breaks up into small fragments and the cell dies.

3. Killing by cytotoxic T cells. Once viral proteins that are made in the helper T cell show on its membrane, they mark these helper T cells as targets for killing by cytotoxic T cells (see Chapter 15, p. 547), a process that eliminates many viral infections. With HIV, however, the target of cytotoxic T cells is the infected helper T cell, which adds to the decrease in helper T cells.

4. Cell fusion. HIV carries in its viral envelope a protein that helps it bind to the CD4 protein on the surface of the cell it infects. The infected host cell also expresses some of this viral envelope protein in its plasma membrane before new viruses bud out (see Figure 16.4.) The infected host cell can thus bind to the CD4 protein on the surface of other, uninfected helper T cells. The plasma membranes of the infected and uninfected host cells then fuse, bringing about cell fusion and spreading the virus to a new cell. This fusion can be repeated until a multinucleated 'giant cell' is formed. Although these giant cells are still alive, they can no longer perform the immunological activities of normal helper T cells.

5. Indirect inactivation. Certain strains of HIV cause the production of the wrong cytokines or inhibit the production of the cytokines needed for T cell growth. Without these cytokines, the helper T cells cannot divide to perform their normal disease-fighting processes or to replace cells lost to HIV-induced lysis and apoptosis.

Although these mechanisms can be demonstrated in laboratory experiments, it is not certain that they all occur in infected human hosts. We do not yet know which of these mechanisms causes the greatest loss of T cells or of T cell functions. The result, however, is the same: the loss of healthy, active helper T cells results in immune deficiency. Many of

Figure 16.6

Many HIV particles emerging from a helper T cell. The dark circles are holes in the cell membrane left when the viruses bud out. These holes eventually kill the cell.

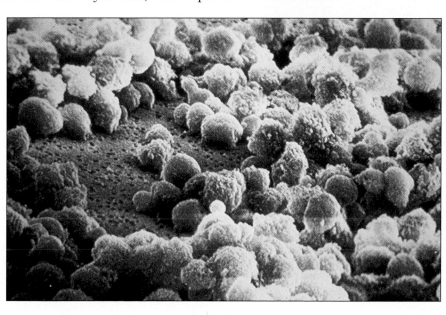

these events also occur in HIV-infected CD4-bearing macrophages, thus impairing the antigen nonspecific, innate portion of the immune system as well.

Progression from HIV infection to AIDS

When a person becomes infected with HIV, the virus keeps spreading to more and more cells. When an HIV infection has progressed to AIDS, 1–10% of the helper T cells have become infected. How does infection of some cells progress to disease in a person?

Three stages of HIV infection. The progression from HIV infection to AIDS follows three stages: the initial infection, an asymptomatic phase, and a third phase called disease progression (Figure 16.7). An infected person can transmit HIV to another person at any of the three stages, but is most likely to do so in the first and third stages, when the numbers of cell-free viruses and infected cells in the body fluids are highest.

In the initial stage, virus levels in the blood are high. As more and more helper T cells are infected and killed, the helper T cell count begins to drop. The initial infection may be accompanied by flu-like symptoms—fever, swollen lymph nodes, and fatigue—which, because they are similar to the symptoms for many other diseases, are often not diagnosed as being an acute HIV infection. The initial infection also stimulates two types of immune response. Antibodies to HIV are produced by B cells and there is an increase in HIV-specific cytotoxic T cells that can kill cells containing HIV. These processes are initially able to contain the HIV. The levels of virus in the blood decrease. New helper T cells develop to replace those that were killed and helper T cell counts return to normal.

With the decrease in virus in the blood and the return of helper T cells, the person enters the second, or asymptomatic, phase, which can last for a few months to many years, with 10 years being typical. The levels of virus in the blood decrease, but the viral population in the lymph nodes continues to grow by viral replication and infection of new helper T cells. By binding to the virus, the antibodies that were synthesized in the initial phase can neutralize the virus, that is, prevent it from infecting

Figure 16.7

The course of HIV infection.

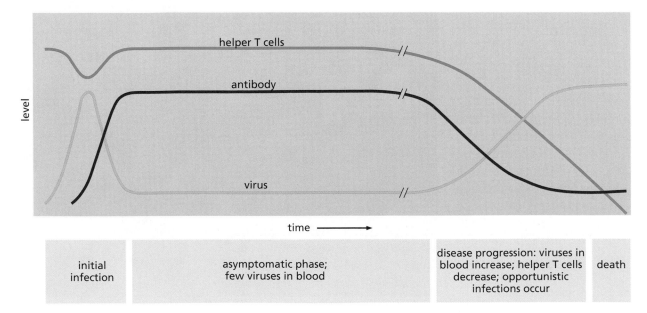

more cells. During the asymptomatic phase, the immune system is still able to keep the infection under control, so the person does not feel ill.

In the third phase, the levels of virus in the blood increase once more, while the numbers of helper T cells in the blood decrease. It is uncertain what triggers the onset of the third phase, although malnutrition, stress or other immunosuppressive factors seem to hasten the onset. Originally it was assumed that HIV infection inhibited the maturation of new helper T cells. It is now known that the maturation rate is actually normal or greater than normal, but the rate of cell death is so great that the overall helper T cell population decreases, particularly as an HIV-infected person progresses to AIDS. The virus mutates to forms that no longer match the antibodies produced in the acute phase, which thus cannot bind to the virus and neutralize it. As we have seen, the host's own cytotoxic T cells turn against the helper T cells. A person is defined as having progressed to AIDS when his or her CD4 helper T cell count (also called the T4 count) falls from a normal value of 1000 cells per microliter of blood to less than 200. Death generally follows when the level of helper T cells declines still further in the third phase.

Other cells bearing CD4 may also be infected and inactivated by HIV. As mentioned previously, macrophages are white blood cells that engulf and remove pathogens and damaged cells or molecules, and secrete cytokines that strengthen immune responses. The elimination of macrophage cells contributes to the risk for infections that may, in turn, lead to the patient's death.

Opportunistic infections and other symptoms. There are microorganisms that are always present in a person or the environment but are kept in check by a healthy immune system, so that they seldom cause illness. Infections caused by these microorganisms in immunodeficient people (called **opportunistic infections**) are one of the primary symptoms of AIDS. These infections can be very severe and even fatal in a person with AIDS. People with intact immune systems need not fear catching these opportunistic infections from a person with AIDS. In the United States, typical opportunistic infections that accompany AIDS are *Pneumocystis* pneumonia, caused by *Pneumocystis carinii*, and fungal infections with *Toxoplasmodium* or *Histoplasmodium*. Recall that the appearance of a cluster of cases of this rare *Pneumocystis* pneumonia was the first hint of AIDS. Another fungus called *Candida* (a yeast) causes mild infections of the mouth, esophagus, or vagina in the absence of AIDS, but *Candida* infections in people with AIDS are much more severe. The same is true of viral diseases including shingles, cytomegalovirus eye infections, and herpes viruses. In Africa, the more common opportunistic infection accompanying AIDS is tuberculosis, a bacterial infection caused by *Mycobacterium tuberculosis*. Tuberculosis is a disease in which there are active periods and periods of remission; HIV infection increases the frequency of reactivation of tuberculosis and also the mortality rate. Worldwide, tuberculosis is the leading cause of death in HIV-infected people.

AIDS patients may also suffer from high fevers, night sweats, general weakness, mental deterioration (dementia), and severe weight loss, although these last two symptoms may not develop for a long time. Dementia may be related to the elimination of macrophagelike cells from the brain. In the gut, there is a type of CD4-bearing cell that has a role in the absorption of nutrients; elimination of these cells may be related to the weight loss.

Variations in disease progression. Many people infected with HIV develop AIDS-related complex (ARC), a set of symptoms milder than AIDS. Originally it was thought that ARC was a pre-AIDS condition and that everyone who had ARC would end up with AIDS. The CDC did not initially require the reporting of ARC, assuming that these cases would later be reported as AIDS cases, which resulted in underestimates of HIV infection rates. As time passed, researchers noticed that several people died while still showing only the symptoms of ARC, not AIDS. Distinctions are no longer made between ARC and other categories of HIV infection; they are all simply called HIV infection. Does everyone infected with HIV get AIDS? Does everyone with AIDS die from the disease? We do not have definitive answers to these questions. The speed with which HIV infection progresses to disease varies greatly. Some people have been infected with HIV for many years without developing AIDS. Several studies have shown that nonprogressive HIV infections are often characterized by a very small amount of the virus, but it is not clear whether this reflects a low infective dose initially or an immune system that has successfully kept the viral population low. A long-term study of HIV-infected homosexual men in San Francisco showed that, after 12 years, 65% had progressed to AIDS, but 35% had not. It may yet turn out that the progression from HIV infection to AIDS is not inevitable. Certainly, the avoidance of other immunosuppressive factors, including drugs, alcohol, and stress, can help to maintain health (see Chapters 14 and 15).

CONNECTIONS CHAPTERS 14, 15

Researchers studied several dozen professional sex workers (prostitutes) in west Africa who were infected with the less virulent strain HIV-2. Significant findings of this study are that HIV-2 infection seemed to offer these women some protection against the more virulent strain HIV-1 and that they were less sick than people infected with HIV-1. They did have high rates of infection for other sexually transmitted diseases, falsifying the hypothesis that the lower HIV-1 rates were simply the result of safer sex practices.

Another group of people was found who remained uninfected with HIV despite numerous exposures. These people turned out to have genetic mutations of molecules (called chemokine receptors) that are co-receptors for cellular infection by HIV. This means that the infection process is greatly enhanced when CD4 molecules and the co-receptors are both present on host cells. People with heterozygous deletions of these co-receptors generally stay asymptomatic for longer, while those with homozygous deletions often remain uninfected with HIV despite repeated exposures. However, it seems probable that deletions do not offer protection against all strains of HIV.

Tests for HIV infection

How can people tell whether they are infected with HIV? The B lymphocytes of a person infected with HIV respond to the virus. This response, which is the basis of most testing for HIV, takes a couple of weeks or months and results in the production of antibodies to HIV in the blood. The development of specific antibodies is called **seroconversion**, and once the antibodies have developed the person is said to be HIV positive (HIV⁺).

There are two common tests for HIV, the ELISA test and the Western blot test, as described on our Web site (under Resources: Tests for HIV infection). Both of these tests detect antibodies to HIV. As mentioned earlier, there are now tests to detect the virus itself, but these tests are very expensive, so most HIV tests are still based on the presence of antibodies to HIV.

For every diagnostic test there exist the possibilities of **false positives**, test results that are positive when the person does not really have the condition, and **false negatives**, test results that are negative when the person really does have the condition. The frequency of false negatives determines the **sensitivity** of the test; the frequency of false positives determines the **specificity** of the test. The reliability of a given test depends on both its sensitivity and its specificity. The more sensitive a test, the less often it will miss a truly positive case; the more specific a test, the fewer will be the cases that are truly negative but that are reported as positive. Every diagnostic test must be thoroughly tried on samples from thousands of individuals whose actual status is known before the test can be sold. These trials must be conducted blind; that is, the person doing the testing cannot know during the trials whether the test samples came from persons infected with HIV. Afterwards, the true infection status (known beforehand but concealed from the researchers) is compared with the status revealed by the test. In this way the frequency of false results can be quantified.

For HIV testing, the ELISA test is done first. The sensitivity of the ELISA test is high—less than 1% false negatives—but it is not very specific: there can be as many as 2–3% false positives. For this reason, when an ELISA result is positive, the result is rechecked with a Western blot test, which rarely gives false positives. Why not use the Western blot as the initial test? The reason is that Western blots are more costly and technically more difficult. Even the ELISA test is too costly for widespread use in many countries.

Both the ELISA and Western blot tests are based on the detection of antibodies specific for HIV. A second generation of tests based on the polymerase chain reaction (PCR; see Chapter 3, p. 81) may replace both these antibody-based tests. The main advantage of PCR tests is that they detect viral RNA rather than antibody, so they can give results soon after infection, rather than weeks or months later.

Should everyone be tested for HIV? One reason against testing everyone is the frequency of false results. The problem of false results is more severe when the true frequency of infection in the test population is lower. A little mathematics will illustrate the point. The frequency of false positives in HIV ELISA tests is between 2% and 3%, while the frequency of false negatives is less than 1% for an overall inaccuracy of about 3%, or 3 false tests out of every 100 tests done. The true frequency of HIV infection in the overall U.S. adult heterosexual population is 15 per 100,000 (Table 16.1). Therefore every 100,000 tests should reveal an average of 15 true cases and 3000 false positives. The false positives translate to a failure rate of 99.5% for the test (3000 false positives out of 3015 positive test results). If, on the other hand, the true rate of infection is 1 in 3, as it is estimated to be among U.S. injection drug users, then every 100,000 tests will produce an average of 33,000 true cases and 3000 false positives, a failure rate of 8% for the test. Calculate the failure rates on your own for some other sets of conditions using Table 16.1. Should everyone be tested when the true frequency of infection in a population is low?

A vaccine against AIDS?

A highly successful strategy for the prevention of many infectious diseases has been vaccination. A vaccination is really a controlled exposure of a person to molecules similar or identical to those carried by the pathogen. The material to which the person is exposed is called the vaccine, so named because the first successful vaccine (which was against smallpox) used the vaccinia virus from the sores of infected cows (Latin *vacca*). Exposure to a vaccine stimulates the immune system to make an immune response to the molecules, and vaccination is therefore also called immunization. The pathogen itself is not used, so the person is not given the disease. The vaccine may be another microorganism, closely related to the pathogen but nonvirulent to humans, as when vaccinia from cows was used to protect against smallpox. (Smallpox vaccination succeeded in eliminating this disease from the globe; therefore smallpox vaccinations are no longer routinely given.) A vaccine may be the pathogen itself but treated so as to make it nonvirulent or kill it. Older vaccines used whole microorganisms, but today molecules vital to the pathogen's life cycle or to its ability to cause disease are more frequently used instead. Several laboratories are attempting to develop vaccines that would prevent HIV infection (pre-exposure immunization) or would prevent the progression of HIV infection to AIDS (post-exposure immunization).

There are many biological barriers to developing vaccines against AIDS. These roadblocks include genetic variation of the virus, a lack of knowledge about which immune responses are protective against HIV, and a lack of animal models in which to test trial vaccines.

HIV nucleic acid sequences change very rapidly. Reverse transcription is error-prone, with 1–5 mutations per round of reverse transcription. In part this is because the RNA is single stranded; there is no complementary strand on which to make corrections. In addition, there are no correcting and editing enzymes like those that keep the mutation rate low in DNA replication. Because there are two single strands of RNA per

Table 16.1

Prevalence of HIV infection. A prevalence rate of 1 in 109, or 917 per 100,000 means that, in a population of 100,000, 917 people (or 1 in every 109) are expected (on average) to have the condition being discussed. Data based on 1998 statistics.

GROUP (ADULTS 15–49)	PREVALENCE	NUMBER PER 100,000	PERCENTAGE
U.S. general heterosexual	1 in 6666	15	0.015
U.S. college students	1 in 500	200	0.2
U.S. prison population	1 in 495	202	0.2
U.S. male homosexuals*	1 in 5	20,000	20
U.S bisexuals, infrequent homosexuals	1 in 20	5000	5
U.S. injection drug users	1 in 3	33,300	33
Worldwide	1 in 109	917	0.9

*Rate among men who have sex with men and who were attending clinics for sexually transmitted diseases.

virus particle, these strands can recombine, further adding to genetic diversity. The virus thus evolves rapidly within a single host. The amount of genetic change within the HIV in a single patient over a 10-year period of disease is estimated to equal millions of years of change in the human species. The enormous resulting genetic variation presents a major problem for the design of vaccines. Is it possible to develop one vaccine that could stimulate a protective immune response in every person vaccinated and that would continue to protect infected people as the viral nucleic acid sequences changed? The answer right now is "maybe": maybe there are some sequences that do not change very much or for which changes have no effect on recognition by the immune system. The latter is possible because the immune system actually recognizes protein *shapes*, not sequences of nucleic acids or amino acids. A change in nucleic acid sequence may cause one amino acid to be substituted for another in the protein during its synthesis, but some substitutions do not alter the shape of the completed and folded protein. If the shape did not change, the immune system would still recognize the altered protein.

Not all immune responses against HIV are protective, as can be seen by the fact that HIV-infected people develop antibody and CD8 cytotoxic T cell immune responses to HIV but still eventually get AIDS. Proteins that function in the viral life cycle are targets for vaccine development, but it is not known whether these will stimulate protective responses. Stimulating an immune response by vaccination may actually trigger progression to AIDS in someone already infected with HIV, as happened in one documented case in which an HIV-positive person rapidly progressed to AIDS after a smallpox vaccination.

The lack of animal models is a significant problem. The effects on each step in an immune response can be studied *in vitro*, but protection from disease can be evaluated only in an animal that gets the disease. Ethical considerations call for extreme caution in the testing of vaccines on human volunteers in a disease known to have a high percentage of fatalities and for which there is no known cure.

In 1994, several vaccines were tested in small-scale trials on humans in Europe and North America. These vaccines did not prove to be entirely protective: a few individuals contracted HIV after vaccination. They did not get HIV from the vaccine; rather, the vaccine failed to protect them from transmission by the routes described in the next section of this chapter. The National Institutes of Health did not allow larger-scale tests to proceed in the United States. The World Health Organization took a different stand and has allowed vaccine tests to be conducted, with Uganda and Thailand chosen as the locations. These tests are continuing (as of the year 2002), and the results are not yet known. The governments in these countries have welcomed these tests because, if successful, the vaccines would confer protection against the HIV strains prevalent in Africa and Asia. So far, 90% of research has focused on the subtype common in Europe and North America; although present elsewhere, this subtype is not the one common in other parts of the world. Thailand's public health officials are additionally interested because drug therapies, which we look at in the next section, are too expensive. Prevention by education and vaccination remains the only affordable option.

Drug therapy for people with AIDS

Very few drugs are helpful against viral diseases. Antibiotics, which are highly effective against bacterial diseases, do not work against viruses. As detailed knowledge becomes available about the few enzymes of its own that HIV has, new drugs may be developed to target these enzymes specifically.

Antiviral drugs. One type of drug consists of **reverse transcriptase inhibitors** that are similar in chemical structure to parts of DNA. These drugs block the reverse transcription of viral RNA to DNA (Figure 16.8), thus preventing the virus from making the DNA it needs to complete its infective cycle (see Figure 16.4). One such drug is zidovudine or ZDV (trade name Retrovir, formerly known as azidothymidine, or AZT). Zidovudine greatly reduces the rate of HIV transmission from pregnant women to their babies before and during birth. Didanosine (Videx, formerly called dideoxyinosine or DDI) is another drug that inhibits reverse transcriptase.

The **protease inhibitors** are a newer type of antiviral drug. Later in the viral life cycle than the reverse transcription step, an enzyme called a protease is required to trim newly translated proteins into their functional form (see Figure 16.4, step 6). Blocking this step stops viral replication and infectivity. HIV protease is very different from human protease enzymes, reducing the effects of the drug on the human host.

A still newer class of anti-HIV drugs are the fusion inhibitors that prevent the virus from entering host cells. After HIV binds to the CD4 molecule on helper T cells or macrophages, it enters the cells by triggering the fusion of its membrane with the host cell membrane. Fusion requires host cell proteins, and it is these proteins that are targeted by the drug, blocking viral entry into the cell. The first of the fusion inhibitor drugs goes by the name T-20. It has worked well in clinical trials, even on patients whose HIV have become resistant to other anti-viral drugs. It is expected that it will be licensed for clinical use soon.

Combination therapy. With its rapidly changing genome, HIV has the potential to evolve resistance to any particular drug quickly. In combination therapy the ideal is to combine two drugs that work by different mechanisms. If resistance to one mechanism of action evolves, the other drug will still be effective. Combination therapy against HIV at

Figure 16.8

How the drug zidovudine (ZDV) interrupts the replication of the HIV virus.

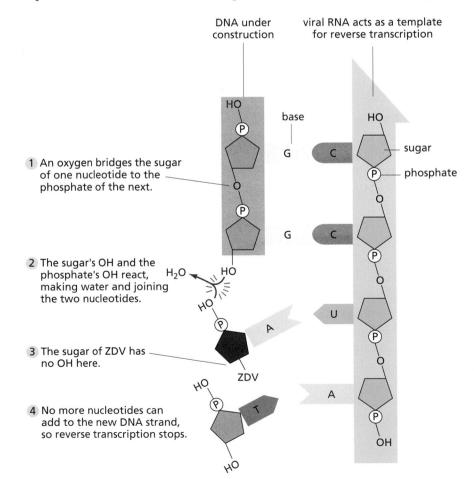

1 An oxygen bridges the sugar of one nucleotide to the phosphate of the next.

2 The sugar's OH and the phosphate's OH react, making water and joining the two nucleotides.

3 The sugar of ZDV has no OH here.

4 No more nucleotides can add to the new DNA strand, so reverse transcription stops.

DNA under construction

viral RNA acts as a template for reverse transcription

base

sugar

phosphate

first used two or more drugs with the same mechanism of action, that is, two or more reverse-transcriptase inhibitors, because they were the only drugs available. Now, more than two drugs are included in what is called the 'drug cocktail,' and the combination includes both reverse-transcription inhibitors and protease inhibitors. These combinations have been very successful, reducing viral loads below detectable levels in a high percentage of HIV-infected people.

Drug combinations reduce the probability of HIV's becoming resistant and rebounding, but these drugs must be taken on schedule. A typical regimen may involve taking 16 pills at 6 different times during the day. Many regimens are even more complex. This complexity, combined with unpleasant side effects of various kinds, and the enormous costs of the drugs ($70,000 to $150,000 per year) makes therapy difficult for many patients. Yet many HIV-infected people—even some who have progressed to AIDS—have been restored to functional lives. It is not yet known how long a person would need to stay on therapy. Reducing virus to 'below detectable limits' does not necessarily mean that the virus is gone; some people's viral loads have returned when they stopped taking the drugs. So, although these drugs have certainly been a source of optimism, we cannot yet call them a 'cure' for HIV or AIDS.

Prevention and treatment of opportunistic infections in persons with AIDS. Nearly all AIDS deaths are caused by opportunistic infections. Attempting to prevent opportunistic infections in people with AIDS is thus very important. During the phase of CD4 helper T cell depletion, people are very susceptible to infectious diseases carried by people who are not infected by HIV. A cold or influenza can have grave consequences in an immunodeficient person. Bacteria picked up from food can be equally hazardous. In the developed world, therapeutic drugs are available for the treatment of many of the opportunistic infections, such as a combination of the drugs trimethoprim and sulfamethoxazole for *Pneumocystis* pneumonia. However, in many parts of the world, such drugs are unavailable because of their cost. The only other way to stop AIDS deaths in populations is by preventing HIV transmission, which is the topic of the next section.

THOUGHT QUESTIONS

1 What is the difference between HIV infection and AIDS?

2 What lifestyle choices can a person make to decrease their chances of becoming immunodeficient? Would those choices also be important for an HIV-positive person?

3 What steps in the ELISA test determine its specificity for HIV? What steps determine its sensitivity?

4 If a vaccine against AIDS were developed, how would you go about testing it? Remember that AIDS develops only in people, so animals cannot reliably be used as subjects. Would your test have a control group? How would you ensure that the conduct of the test was ethical?

Knowledge of HIV Transmission Can Help You to Avoid AIDS Risks

The general term for the transfer of a pathogen from one individual to another is **transmission**. How is HIV transmitted from one person to another? HIV does not have any other animal hosts and does not remain infective in water or in air. It can pass from one person to another only in certain body fluids—blood, semen, and vaginal fluids. To enter another person, these fluids containing HIV or HIV-infected cells must rapidly come in contact with the rectal mucosa or with the bloodstream of the other person via breaks in the mucous membranes or skin. HIV must make rapid contact with cells bearing the CD4 molecule. Cell-free HIV does not remain infectious very long if the fluids are outside a person, for example in a blood spill. Washing with ordinary soap and water kills cell-free HIV when it is outside a person, because soap dissolves the viral lipid envelope. A mother can transmit HIV to her unborn fetus or to her baby during delivery, and a few cases are known in which HIV has been transmited in breast milk. HIV is not found in feces or urine. There are small numbers of HIV particles in the saliva or tears from 1–2% of HIV-infected people, but saliva contains antiviral activity, and HIV has never been known to be transmitted through saliva, including human bites, or via tears. How do the fluids containing HIV get passed? The percentage of AIDS cases in the United States transmitted by various routes is shown in Figure 16.9. This figure represents total numbers of persons with AIDS in 2001.

Risk behaviors

When scientists use the term **risk** of HIV infection, they mean the mathematical probability of transmission of the infection. Various behaviors or activities have been grouped into 'risk categories,' based on what is known about transmission routes. High-risk behaviors are those that give a high probability of transmission from an infected person to an uninfected person. Notice that risk is now categorized by specific **risk behaviors** and not by population groups. The safest way to make choices about these behaviors is to assume that every person whose HIV status is unknown to us may be HIV infected and a potential source of HIV transmission.

Category I: high-risk behaviors. Transmission while engaging in a **high-risk behavior** is very likely. These behaviors, which account for over 97% of all AIDS cases, are:

1. Behaviors in which the passage of blood, vaginal fluids or semen is very likely, such as anal or vaginal intercourse with an infected person without the protection of a condom (unsafe sex).
2. Injection drug use in which needles or syringes are shared with an infected person.
3. An infected mother's going through pregnancy and giving birth without receiving zidovudine treatment.

Anal intercourse is more risky than vaginal intercourse because semen contains an enzyme (collagenase) that breaks down the lining of the rectum and exposes blood vessels, a form of injury to which the vagina is much more resistant. In addition, vaginal intercourse is not as likely to result in HIV infection as anal intercourse because the cells of the rectal mucosa have the CD4 molecule on their surfaces and so can be infected with HIV, whereas the intact epithelial lining of the vagina is a significant barrier. The risk of male-to-female transmission of HIV infection during vaginal intercourse is greater than that for female-to-male transmission.

Category II: likely-risk behaviors. HIV transmission has been documented for routes in this category, but with lower frequency:

1. Anal or vaginal intercourse using a condom (safer sex, not safe sex). Condoms do not make intercourse completely safe. Condoms fail as birth control for about 10% of couples who use them; this means they can also fail to prevent HIV transmission. Although the rate of failure seems to be low (less than 5%) for condoms *used properly*, many users do not exercise proper care in putting condoms on or taking them off, so estimates of failure rates *in general use* can be as high as 20%. Although reliable data of this kind are difficult to obtain, one study found that 17% of women whose husbands were HIV positive became infected despite proper and consistent condom use.

2. Breast-feeding (transmission to a baby from an infected mother).

3. Receiving a blood tranfusion or organ transplant. This risk was high before 1987, but is now very low in the United States because of careful screening of blood products and donated organs. The risk remains higher elsewhere, where blood donors and blood products are not screened as stringently (see below).

4. Artificial insemination. As with blood transfusion, the risks are now low when donated semen has been tested for HIV.

5. Infection of health care professionals by needlestick injuries.

6. Dental care by an infected worker. There is a single cluster of six cases involving only one dentist; no other cases are known in which an infected health care worker has transmitted HIV to a patient.

7. Deep kissing. A single case has been documented and, because both people had severe periodontal disease (bleeding gums), HIV was probably transmitted through blood, not saliva.

Category III: low-risk behaviors. This category includes routes that are biologically plausible, but no cases have been confirmed.

1. Sharing toothbrushes or razors or other implements that may be contaminated with infected blood.

2. Being tattooed or body pierced to produce ornamental scars or for jewelry.

3. Receiving tears or saliva.

Figure 16.9

Routes of AIDS transmission in the United States, based on 2001 cumulative statistics from the Centers for Disease Control and Prevention. The percentages indicate the proportion of cases transmitted by each route.

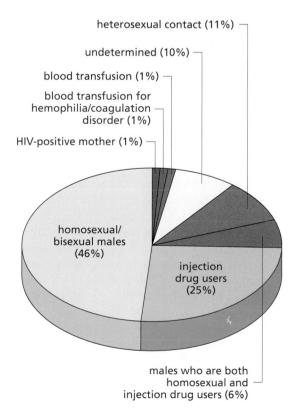

heterosexual contact (11%)

undetermined (10%)

blood transfusion (1%)

blood transfusion for hemophilia/coagulation disorder (1%)

HIV-positive mother (1%)

homosexual/bisexual males (46%)

injection drug users (25%)

males who are both homosexual and injection drug users (6%)

4. Oral sex. Although there is anecdotal evidence of transmission via oral sex, there are no cases in which transmission by this route is documented.

Category IV: no-risk behaviors. Transmission of HIV from engaging in the following behaviors with an infected person is considered not biologically possible:

1. Shaking hands
2. Sharing a toilet
3. Sharing eating utensils
4. Being sneezed on
5. Working in same room
6. Handling the same pets
7. Close-mouthed kissing (kissing with no exchange of saliva or blood).

Also, exposure to mosquitoes or other biting insects is a no-risk behavior, as explained in Box 16.1.

BOX 16.1 Can Mosquitoes Transmit AIDS?

Two frequently asked questions are, "Why don't mosquitoes transmit HIV?" and "How do we know that they don't?" Epidemiological evidence shows that the frequency of AIDS and of the number of unexplained cases is no higher in mosquito-infested areas of the United States than in other areas. In Africa, the people with AIDS are mostly babies and sexually active young adults; mosquitoes do not bite people in these groups more frequently than they bite other people. On all continents, children who are not yet sexually active often get mosquito bites, but they do not get AIDS unless they are infected from their mothers at birth. Laboratory experiments have shown that although HIV and HIV-infected cells may be taken up by mosquitoes who bite infected people, HIV is not transmitted to other people through mosquito bites. Several factors help to explain this: HIV cannot replicate in mosquitoes or survive long in their bodies (because mosquitoes are not a host for HIV), and the amount of blood ingested ($3–4\mu l$, millionths of a liter) is too small to contain enough HIV or HIV-infected cells to infect a person. Saliva (including mosquito saliva) may also have substances that inhibit the virus. Furthermore, because of the many biological factors involved in the transmission of a disease by an insect, it is highly unlikely that a single mutation in either the mosquito or the virus would significantly alter this situation.

A study done at the Institute for Tropical Medicine in North Miami, Florida, proposed in 1986–1987 that the high rate of AIDS in the town of Belle Glade, Florida, could be attributed to the squalor and crowding of its people and to the mosquitoes breeding in nearby swampy lands, "where 100 insect bites a day are not unusual." The U.S. Centers for Disease Control and Prevention (CDC) studied this situation and concluded that the high incidence of AIDS in Belle Glade was attributable to sexual contact and shared needles, not insects.

Transmission of HIV by other blood-sucking animals such as bedbugs (which are insects) and ticks (which are more closely related to spiders) has also been ruled out. A tick that is endemic in the same parts of Africa where AIDS is common carries enough blood and live virus to make transmission theoretically possible. However, the possibility does not fit with the epidemiology: children below the age of sexual activity do get bitten in significant numbers but do not get AIDS.

Communicability

Another question people ask about HIV is, "How contagious is it?" meaning, "If I am exposed, how likely is it that I will become infected?" The term used by the medical community to mean the likelihood of transmission after exposure to HIV is **communicability**. (**Contagious** simply means 'capable of being transmitted;' it does not refer to the probability of transmission.) The concept of communicability, or likelihood of transmission, is directly related to the concept of risk. High-risk behaviors (Category I above) increase the probability of transmission.

There are at least two ways to answer the question of communicability; one takes an epidemiological approach and the other a microbiological approach. The epidemiological approach compares the number of encounters with the number of infections throughout the population or within certain population subgroups. It is difficult to designate the probabilities for HIV transmission because there is a period of weeks or months before antibody develops, and there is often a period of years between infection and disease symptoms, during which people may not know they are HIV infected. The number of encounters with HIV is often not known for an individual, and is even less known for all the people constituting a population.

Probabilities are, however, more accurately known for some routes of transmission than for others. For example, there is a 30–35% chance that an infected mother will transmit the virus to the fetus if she is not treated with the drug zidovudine. The efficiency of transmission of HIV by various routes is shown in Table 16.2. An efficiency of over 90% means that for every 100 exposures more than 90 will result in an HIV infection.

The microbiological approach to determining communicability is to quantify (measure numerically) what is called the infective dose, the number of pathogenic particles that must be transferred to result in an infection of an individual. This value is not known for HIV, although one estimate is that the transfer of 10,000–15,000 HIV particles can establish an infection. Very early in an HIV infection, in the weeks or months before antibody develops, and also very late, when the CD4 helper T cell count is low and the antibody concentration has dropped, the number of HIV particles in blood and genital fluids is much higher than at other stages (see Figure 16.7). The number of HIV particles in general is higher in semen than in vaginal fluids, but each varies at different stages of infection. One study found 4.2 million HIV particles per milliliter of blood on average in people with AIDS, but this can vary tremendously.

Table 16.2

Estimated efficiency of transmission of HIV by various routes.

ROUTE OF TRANSMISSION	EFFICIENCY (%)
Vaginal sexual intercourse (unprotected)	0.1
Receptive anal intercourse (unprotected)	0.5–3
Mother to child during pregnancy or childbirth (without use of zidovudine)	30–40
Transfusion of infected blood	>90
Transfusion of screened blood	0.00015–0.0002
Intravenous injection with infected needle	1–2
Needlestick with infected needle	0.4

From accidents in which health care workers have been exposed to infected blood, it is known that there is a higher probability of infection when a person has been splashed with large quantities of blood onto open sores in the skin and a lower probability when they have been pricked by a needle. For every 250 reported needlesticks, there has been one transmission (0.4% efficiency). The effect of blood volume can also be seen in Table 16.2, where the transfer of greater quantities of blood by intravenous injection or transfusion increases the efficiency of transmission.

Viral load is certainly a factor in determining the efficiency of HIV transmission, but only one of many factors. The precise infective dose is not known and probably varies from one person to another. For example, persons with open genital sores due to other sexually transmitted diseases such as syphilis or herpes are 10–20 times more likely to become infected than other people. Gonorrhea or chlamydia infections increase the probability of HIV transmission threefold or fourfold.

In general, it may be said that HIV is much less communicable than a virus such as hepatitis (another virus spread by contact with contaminated blood), but the communicability varies with the risk behavior. In comparison with the 0.4% efficiency of HIV transmission from a needlestick, the efficiency of transmission of hepatitis B virus by this route is 6–30%. Transmission of opportunistic microorganisms from a person not infected with HIV to an HIV-infected person is much more likely than transmission of HIV from an HIV-infected person to an HIV-uninfected person. Because people with AIDS have severely impaired immune systems, their risk of catching diseases from other people is very high.

Susceptibility versus high risk

What is the difference between the terms susceptibility and high risk? To examine this question, let us go back and look further at how knowledge of AIDS and HIV developed. The fact that early cases were reported in hemophiliacs suggested an infectious cause for AIDS. As more cases were reported, the affected individuals seemed to fall into five groups, which became known among epidemiologists as 'the five H's:' homosexual males, hemophiliacs, heroin addicts, Haitians, and hookers (prostitutes). From an epidemiological perspective, these categories served a useful purpose to describe groups within which cases were showing up. But what was useful scientifically turned out to have negative social consequences. The terms quickly became imprinted in the minds of scientists and the public, allowing complacency on the part of people who were not in these groups and prejudice against those in the groups. To epidemiologists, the five H's were merely a convenient way to designate clusters of reported cases of a mysterious, new syndrome. Such identification did not in itself imply anything about cause and effect or about transmission or about all of the people in the groups, but it was useful in suggesting hypotheses that could be tested.

To epidemiologists, the term **high-risk group**, as applied in connection with a particular disease, simply means that there is a higher frequency of the disease among members of that group (frequency equals the number of people with the infection divided by the number of people in the group). Use of the term implies nothing about the possible reasons for the increased frequency, which may stem from increased exposure or increased susceptibility or both. *Exposure* to a disease means coming into contact with the disease agent. Increased exposure can sometimes be due to shared behaviors, but there are many other possible explanations. A disease may have a higher frequency in a certain group of people if all the people in that group came from the same geographic location so that they were exposed to the same toxic chemical, or if they all ate food from the same source and so were exposed to the same food-borne pathogen. It does not mean that these people are more susceptible; anyone else exposed to the same factors would also have become sick. **Susceptibility** to a disease means the ability to contract that disease *if exposed*. Humans are susceptible to HIV and most other animals are not. Susceptibility can vary from one person to another, and it can be genetically or environmentally influenced (malnutrition, for example, may make a person more susceptible to many infectious diseases).

Several misperceptions about AIDS resulted from the early identification of specific high-risk groups. First, some people not in the identified high-risk group assumed they were not susceptible. Some people assumed that every person within a high-risk group was equally likely to be infected (and that people outside these groups were unlikely to carry the disease). Haitians, in particular, suffered adverse consequences by being classified as 'high risk.' In efforts to screen blood donors before the cause of the disease was known and before appropriate tests were available for screening blood, all Haitians were barred from donating blood in the United States. In the resultant hysteria, some Haitians were evicted from their homes and lost their jobs, Haiti's tourist trade collapsed, and Haitian dictator Jean-Claude Duvalier's state police rounded up and incarcerated homosexuals in Haiti. Haiti's ambassador to Washington wrote a letter published in the *New England Journal of Medicine* deploring the damage done by North American semantic carelessness. As he pointed out, being from a certain country does not contribute to disease in the way that socially acquired behaviors do (for example, having multiple sex partners or using intravenous drugs).

The fifth H, hookers, always seemed problematic because the other risk categories were predominantly or exclusively male. Why did so few women contract the disease at first? If women could contract the disease, why only prostitutes? There was a period when women were thought of as 'carriers' even though they were dying of AIDS themselves. Scientists now think that women are just as susceptible as men to HIV infection, but that the epidemic in the United States began among homosexual men and spread only slowly to women.

Once mechanisms of transmission are known, it becomes more appropriate to focus on high-risk behaviors than on high-risk groups. However, the frequency of infection within a discernible group of people can sometimes have a role in an individual's risk. People within a high-risk group are at risk to the extent that they engage in high-risk behaviors. Their risk may be increased to the extent that their partners in high-risk behaviors are also members of a group in which the frequency of infection is high. A higher population frequency of a disease increases risk by increasing the chance of encountering an infected person, not by altering any individual's susceptibility. (Remember that membership in a

Figure 16.10

HIV/AIDS in different parts of the world. The percentages shown on the map indicate the proportions of the adult population (age 15 and older) currently living with HIV/AIDS. The area of each circle is proportional to the total number of people with HIV/AIDS (estimated at the end of 1999 when the global total was 65.3 million people). In each circle, the red segment represents women with HIV or AIDS, and the blue segment represents men.

group, either a group with a high frequency of infected individuals or one with a low frequency, does not tell you whether a particular individual is or is not infected.)

As we have seen, in the United States infection rates were higher in homosexual men. The lower frequency of HIV infection in females in the United States led many people to assume that women were less susceptible. Therefore, when more women began to fall ill, there was a further misconception that the virus must have mutated to change its infectivity, and if it could mutate once, it could mutate again, and heterosexuals would be susceptible. Women and heterosexual men have always been susceptible to HIV infection, as amply demonstrated by the pattern of the infection in Africa, where the numbers of men and women infected have been about equal (Figure 16.10). The pattern of infection in the United States has changed over time, but it is possible to explain all of the changes on the basis of frequency of HIV in various subpopulations, not on the basis of changes in infectivity of the virus.

Transmission of HIV by vaginal intercourse is not as likely as it is by anal intercourse, but this does not mean that vaginal sex is safe, only that the number of infections per number of encounters is lower. It also does not mean that women are less susceptible than men. It seems that, if there are breaks in the vaginal epithelium (for example, as a result of other sexually transmitted diseases), women are just as likely to be infected as men. So, again, the risk is related to particular practices, not to differences in susceptibility, and these practices carry comparable risks for all groups of people. For example, data collected in both the United States and Africa seem to show that anal sex is just as risky for females as for males.

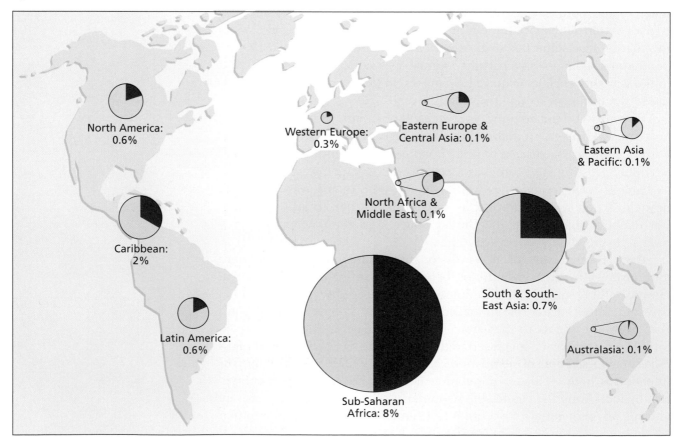

Epidemiology shows that there is a positive correlation in both sexes between the rate of HIV infection and the number of sexual partners; that is, persons with greater numbers of partners have higher rates of HIV infection. There is also a positive correlation between the infection rate and the frequency of previous infections with other kinds of sexually transmitted diseases. People who have contracted any other sexually transmitted disease have already engaged in behavior that puts them at risk for HIV infection. Because properly used condoms can, in general, greatly reduce transmission rates for other sexually transmitted diseases as well as for HIV, someone who has contracted a sexually transmitted disease has probably not used a condom during intercourse. Moreover, they have further increased their risk for HIV infection because the presence of open genital sores greatly increases the probability that contact with HIV will result in HIV infection.

The use of illegal drugs (injectable or not) also increases the rate of infection, particularly in women. In New York City, 32% of female crack cocaine users were HIV positive, compared with 6% of other women. It has not been shown that crack is a cofactor (a factor that increases susceptibility), but the subculture in which crack is used is often one of a high incidence of sexual activity and of sexually transmitted diseases. On college campuses, where the 'drug of choice' is frequently alcohol, the impaired judgment that accompanies alcohol (or other drug use) is a factor working against sexual abstinence or the practice of 'safer sex.'

Another aspect to 'risk' is a person's ability or inability to say "no" to high-risk behaviors. The ability of a person to say no is termed his or her refusal skills. Economic and cultural factors can put severe limitations on a person's refusal skills. In some cultures, for example, women may not be able to insist that their male sexual partners use a condom. Education about the risks of HIV and AIDS must do much more than provide people with information about transmission routes, as we discuss below.

Public health and public policy

Whereas medicine deals with individual cases of disease, public health deals with populations and seeks to minimize the levels of particular diseases in those populations. Many public health efforts require legislation and most require funding, so they are most often conducted by governments or large organizations. Many nongovernmental organizations (NGOs) have been crucial in educating the public about AIDS and in caring for persons with AIDS and their families. Early in the epidemic, they were also crucial in pressuring governments for more funding for research on this disease.

History of public health responses to disease. In each nation in which the AIDS epidemic has spread, the governmental response was molded by the unique history and social customs of that nation. Some nations sought to restrict the immigration of HIV-infected people; others did not. Some jurisdictions segregated certain types of AIDS patients, while others did not. Hospital care and medical insurance for AIDS patients varied greatly from one country to another. Some nations instituted needle-exchange programs for drug addicts; others did not.

A bioethical principle known as the harm principle provides a moral limit on the exercise of freedom of individuals when others may be

injured. By this principle, a person with AIDS could morally be prevented from deliberately spreading the disease but could not be prevented from working or attending school or living in a particular place. The legal specifics of the response to AIDS and of requirements for testing and notification in the United States may be found on our Web site (under Resources: HIV laws).

Guidelines for handling blood. People likely to come in contact with blood—for example, dentists and surgeons as well as their auxiliary workers; sports coaches and trainers; and security personnel—are now required to follow a series of guidelines known in the United States as the *Universal Precautions for Blood-Borne Pathogens*. The term 'universal' refers to the fact that all blood must be handled as though it were infected. The guidelines, which are intended to limit the transmission of any blood-borne pathogen, include wearing gloves when handling blood (Figure 16.11), and further covering when handling large quantities of blood. The guidelines also specify procedures for cleaning blood spills, reporting accidents and injuries in which workers have come into contact with blood, and for educating workers about the risks in handling blood. The U.S. guidelines were developed by the CDC, and the Occupational Safety and Health Administration (OSHA) has been charged by the federal government with monitoring the compliance of employers with these guidelines. Every college and university, for example, must have an infection control plan.

Figure 16.11

Health care workers, researchers, and others who handle human blood must follow the *Universal Precautions for Blood-Borne Pathogens*.

The risk of transmission of HIV through blood transfusion depends a good deal on the methods of blood donor selection. In the United States, blood donors are volunteers. Blood is screened for antibody to HIV, but blood donors are not. The behaviors that transmit HIV are explained to blood donors, and they can anonymously tag their donated blood as having come from a person involved in high-risk behaviors. In the United States, the risk of transmission from a blood transfusion is currently estimated to be 1 in 450,000 to 650,000 (see Table 16.2).

In countries in which donors are paid, the safety of the blood supply is much less assured than in countries in which donations are voluntary. In India, 30–50% of blood donors are professionals who sell their blood an average of 3.5 times per week. In one city in India, 200 professional donors were screened, and 86% were found to be HIV positive.

Access to health care. In the United States, the CDC develops the criteria that define AIDS. The criteria have changed as new information has become known. The wording of the definition is important because people with AIDS are eligible for some types of care from the government that other people are not. AIDS was originally defined as a set of symptoms including particular opportunistic

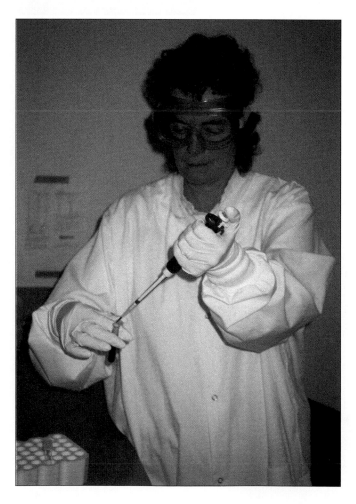

infections. The CDC definition now includes all those persons who are HIV positive and have a CD4 helper T cell count below 200.

HIV-positive women have a poorer prognosis (predicted outcome) than HIV-positive men, both in the United States and elsewhere in the world. This is probably a result of women's generally having poorer access to medical care for the infections that accompany AIDS. The care of people with AIDS has put a strain on public health monies and personnel. In some countries of Africa, more than half of all public health expenditures are for AIDS.

Educational campaigns. In educational campaigns aimed at increasing AIDS awareness, some organizations distribute free condoms and promote their use, while others emphasize abstinence. The former Surgeon General of the United States, C. Everett Koop, stated that the only safe sex is a faithfully monogamous relationship with a faithfully monogamous uninfected partner, and the next best thing is the use of a condom.

Education about HIV and its transmission changed the behavior of homosexual men so that the rates of infection within this group began to decline. This subgroup is generally well educated and has provided many model programs that have been copied in educational efforts to reach other groups. Because the factors guiding people's private behavior differ from one group to another (on the basis of language, income, geography, religion, and cultural background), educational campaigns need to be designed for each different locale and target group.

Information is not the same thing as education. Giving people information about how HIV is spread may not help unless the reasons underlying their high-risk behaviors are addressed. The motivations of people having consentual sex differ from the motivations of commercial sex workers (prostitutes) and street children having sex for survival. Many teenagers and young adults engage in sexual activity (often including high-risk activity), and those who do not are frequently subjected to very strong peer pressure to conform. Education often includes strategies for raising self-esteem and providing support for avoiding high-risk behaviors.

Worldwide patterns of infection

HIV is now a pandemic, a worldwide epidemic. AIDS is now the fourth leading cause of death worldwide and killed 3.1 million people in the year 2002 alone. Forty-two million people are currently infected and 14,000 new infections occur each day. More than three million children are infected, and in many countries AIDS has reversed the hard-won decreases in the infant mortality rate. In addition, 11 million children have been orphaned by the death of parents from AIDS. In the United States, a cumulative total of almost 800,000 people have been diagnosed with AIDS. As many as 1 million people may be HIV infected and 300,000 are currently living with AIDS in the United States. Worldwide, however, 95% of cases are among people in the developing countries.

Obviously, statistics on AIDS vary according to the way in which AIDS is defined. The World Health Organization (WHO) criteria for diagnosing someone with AIDS are very different from the CDC criteria. The WHO criteria, based on symptoms, not on HIV status or T cell counts, are used in countries where monetary or technical considerations make testing for HIV infection impossible. Worldwide surveillance of numbers of AIDS cases is therefore not the same as surveillance of HIV infection, which must be estimated from the numbers of AIDS cases (Figure 16.11). Moreover, many people with AIDS are difficult to distinguish from people who are immunodeficient from other causes, such as undernourishment, and so may or may not be counted as AIDS cases. The numbers may thus underestimate the true AIDS incidence.

AIDS is now the leading cause of death in Africa. Africa also has the highest prevalence of AIDS (the number of infected people divided by the total population), and in most areas the rate is still increasing. In the five years from 1997 to 2002, the number of infected people increased by 30% in Africa, from 23 million to 30 million. Sixteen African countries have a prevalence rate above 10%; in five of these the prevalence is greater than 20%, with Botswana the highest at 42% among sexually active adults. The first rate is still slightly higher than the death rate so that the population of each of these countries is growing, although slowly. Population growth is currently 0.08% in South Africa, for example. Sickness of people living with HIV and deaths from AIDS are having a devastating effect on the economies of these countries, particularly because the prevalence of HIV is highest among working age people. Labor-intensive areas such as food production and mining have been especially hard hit, leading to further hardships as food becomes scarce and mineral exports cannot pay for needed medicine. Education of the next generation is being strained at all levels. Universities have been greatly affected as sickness and death among faculty, staff and students has soared.

HIV infection is not uniformly prevalent throughout Africa. Several countries in West Africa have lower rates owing to early and sustained implementation of prevention programs. Prevalence is also low in North Africa. Recently, however, the adult infection rate in both Nigeria and Ethiopia has exceeded 5%, making public health officials worry that areas of Africa that previously had less AIDS may be catching up.

One African country that has kept its AIDS rate very low is Senegal. The government and local religious leaders in this predominantly Muslim country have cooperated in mounting a strong educational campaign with a consistent message. Also, professional sex workers (prostitutes) are offered free condoms, free monthly checkups, free lessons about AIDS prevention, and payments for child support if they attend monthly clinics. As a result of this campaign, Senegal has one of the lowest HIV infection rates in Africa.

China, Russia and India now each have more than 1 million people with HIV or AIDS. HIV infections are increasing most rapidly in Asia, with Thailand, Myanmar and Cambodia being the most severe. Because the pandemic reached Asia later than it did the other continents, the prevalences are still low (1.4, 1.6 and 2.9 percent respectively).

In Africa the rate of infection is about equal in men and women, while in North America, South America, Europe and Australia it is higher in men. In the United States, a majority of infected women are between the ages of 15 and 25, and 80% of newly infected women are either intravenous drug users or sexual partners of intravenous drug users. From 1996 to 1997, new AIDS cases in the United States declined for the first time since the epidemic began (although the numbers of cases among women increased). The numbers did not decrease further in 1998 and

1999, with the numbers of new cases remaining about equal to those in 1997. In developed countries, the number of AIDS deaths has decreased as new drugs have been developed and made available. Because of the high cost of these drugs, however, most infected people in the developing world have not benefited from them. In 2001, the United Nations began The Global Fund to Fight AIDS, Tuberculosis and Malaria, but pledges so far do not come close to the $7 to $10 billion need annually.

THOUGHT QUESTIONS

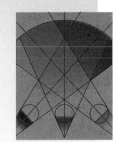

1 What advice would you give to college students about the best ways to avoid getting AIDS? How might you modify the advice for different groups of people?

2 What are the misconceptions surrounding HIV and AIDS? Have they changed over the years?

3 How are the Center for Disease Control and Prevention's criteria for AIDS different from those of the World Health Organization? Why are they different?

4 Is medical research disease-specific? In other words, will the knowledge gained from studying one disease be applicable only to that disease?

5 HIV infection is associated with 'risk behaviors.' To what extent is 'risk' the result of an individual's choice? To what extent do societal factors remove choice from individuals?

Concluding Remarks

What does the future hold? Promising drug therapies now exist and more treatments may be on the way. Prevention, however, remains the best hope for the control of HIV infection. Effective prevention programs will require that research scientists, medical professionals, and educators work together, rather than in isolation. Common language must be found in which these groups can communicate with each other and with the people that they serve. Education cannot be unidirectional: professionals educating the people on the street must also learn from those people.

As Jonathan Mann said in his opening remarks to the VIII International Conference on AIDS in 1992:

> "We have seen important success in basic and applied research, yet that research in isolation from concern about access to its achievements has severely limited its impact on lives of people with HIV... If we believe that the entire problem of AIDS is really only about a virus, then we really only need a virucide or a vaccine. Yet if AIDS is deeply, fundamentally about people and society and if societal inequity and discrimination fuel the spread of the pandemic then, to be effective against AIDS, we would have to address these issues."

(Jonathan Mann headed the WHO AIDS office from 1987 to 1990. He was killed in an airplane crash in 1998.)

Chapter Summary

- The antibodies and **cytotoxic T cells** of the immune system protect the body from disease, but they are produced only in the presence of the cytokine interleukin-2 secreted by the **CD4 helper T cells**.

- When some part of the immune system fails to work, an **immunodeficiency** results. Immunodeficiency leads to an increased probability and severity of sickness.

- **Human immunodeficiency virus** (**HIV**) is the **virus** that causes **acquired immunodeficiency syndrome** (**AIDS**) by destroying CD4 helper T cells and other immune cells displaying CD4 molecules.

- **Epidemiology**, **Koch's postulates** and other types of evidence helped to establish HIV as the cause of AIDS.

- HIV is a retrovirus that uses **reverse transcription** to convert its RNA genome to DNA and then uses the host's cellular machinery to replicate itself. Microorganisms that are normally kept in check by healthy immune systems can cause serious and possibly lethal opportunistic infections when the immune system is compromised by AIDS.

- Most HIV testing detects antibodies to HIV; some tests detect virus itself. All tests have rates of **false negative** results (related to the **sensitivity** of the test) and **false positive** results (related to the **specificity** of the test) that must be considered when interpreting a test result.

- HIV **transmission** is via behaviors in which blood, semen, or vaginal fluids are passed from an infected person to an uninfected person. Essentially everyone is **susceptible** to HIV infection if they receive body fluids from an infected person. **High-risk behaviors** (those with the highest likelihood of transmission) include unprotected sexual intercourse and sharing injection drug needles.

- Since 1981, AIDS has spread to become a worldwide pandemic, burdening individuals and their loved ones and also public health resources.

- There are currently no cures, although new drugs seem promising as therapies. Vaccines are currently being tested.

- Prevention is crucial, but depends on many groups of people listening to each other, learning from each other and working together.

CONNECTIONS TO OTHER CHAPTERS

Chapter 3	HIV carries out reverse transcription, which produces DNA from an RNA template, the opposite of normal transcription.
Chapter 9	Condoms are useful for both contraception and preventing the transmission of sexually transmitted diseases, including AIDS.
Chapter 10	A decline in appetite (and therefore a decline in nutritional status) often occurs in the late stages of AIDS.
Chapter 12	A healthy immune system kills many cancer cells. Kaposi's sarcoma is an otherwise rare cancer that occurs more frequently in AIDS patients.
Chapter 13	Dementia from loss of brain cells is one of the late-developing symptoms of AIDS.
Chapter 14	The abuse of injectable drugs is a major risk factor for AIDS.
Chapter 15	The immune system, which normally protects us from infection, is greatly impaired by HIV infection.

PRACTICE QUESTIONS

1. What nucleic acid is in the genome of HIV?

2. Where does the phospholipid bilayer envelope of HIV originate?

3. What does the enzyme reverse transcriptase do? Is reverse transcriptase an enzyme used by the virus, the host cell or both?

4. Why does HIV infect only cells that carry human CD4 molecules on their surface?

5. What cellular machinery of the host cell does HIV use to replicate itself?

6. What criteria are used for saying that a person is HIV positive?

7. What criteria are used to say that a person has gone from being HIV infected to having AIDS?

8. Define risk.

9. What behaviors produce the greatest risk of HIV transmission? Why?

10. How do reverse transcriptase inhibitor drugs stop HIV replication?

11. How do protease inhibitor drugs stop HIV replication?

12. What species is/are susceptible to HIV infection? What species is/are susceptible to AIDS?

13. Among humans, who is susceptible to HIV infection?

14. When body fluids such as blood are spilled, can cell-free HIV be killed by soap and water? Why or why not?

15. What are the advantages to an ELISA test for HIV infection and what are the disadvantages? What are the advantages and disadvantages of the Western blot test? The PCR test?

Issues

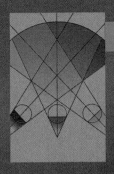

- What factors have led to changes in the global patterns of infectious disease?
- What can be done to reduce the spread of diseases?
- How may microorganisms be used as bioweapons?

Biological Concepts

- Health and disease (pathogens and response, routes of transmission, behavior)
- Evolution (virulence, host–pathogen co-evolution)
- Diversity (viral, bacterial, protist, prion, fungi, parasite)

Chapter Outline

Organisms from Many Kingdoms and Phyla Can Cause Disease

Characteristics of pathogens

Evolution of virulence

Factors governing the spread of pathogens

Intentional transmission turns diseases into bioterrorism

Some Diseases That Spread by Direct Contact Are Increasing in Prevalence

The major sexually transmitted diseases

Factors increasing prevalence

Tuberculosis

Food-borne Disease Patterns Reflect Changes in Food Distribution

One example: variant Creutzfeld–Jakob disease

Social and economic factors contributing to disease outbreaks

Improvements needed

Waterborne Diseases Reflect Changes in Lifestyle and Climate

Cholera

Giardiasis

Legionnaires' disease

Other waterborne diseases

Ecological Factors Especially Affect Patterns of Vector-borne Diseases

West Nile virus

Leishmaniasis

17 New Infectious Threats

In Chapter 16 we saw how a new disease quickly spread around the globe. HIV and AIDS are not the only diseases that have seen major changes in patterns of infection worldwide. These infectious diseases are not just caused by viruses (as we saw for AIDS); some are caused by bacteria. Unlike viral diseases, bacterial diseases can generally be treated by antibiotics. For the last half of the twentieth century people thought that antibiotics would make widespread bacterial disease a thing of the past. Many species of bacteria have, however, become resistant to antibiotics.

In this chapter we mainly focus on human diseases. Keep in mind, however, that all organisms can suffer from disease. Production of disease-resistant strains of plants has been a major goal of both traditional plant breeding and of genetic engineering of plants (see Chapter 11). Diseases of agriculturally and domestically important animal species have long been a focus of research in veterinary medicine. Diseases in wild animal and plant species are a research focus in ecology because of their importance to the health of wild populations and because of their interaction with human disease.

Although a particular organism can be demonstrated to be the causative agent of a disease (Chapter 16), there are many other factors that contribute to the spread of the disease in populations. The development of treatments for specific diseases in individuals depends on knowledge of the pathogen. However, development of public health strategies to prevent the spread of disease depend just as much on knowledge of the routes of transmission of the pathogen. Some diseases, most notably smallpox, have been eliminated entirely. However, according to the CDC, some 30 new infectious diseases, including AIDS and SARS, have emerged in the past few decades as technology and lifestyles have changed. Other infectious diseases have been with us for millennia; they may cycle through long periods of quiescence, only to re-emerge as epidemics once more.

Organisms from Many Kingdoms and Phyla Can Cause Disease

In every phylum, the vast majority of species do not cause any disease, either in humans or in any other organisms. However, many phyla contain some species or strains within species that can cause disease. Some of these species are poisonous; that is, they produce chemicals that have adverse effects. Examples include insect or snake venoms as well as plant poisons. 'Red tide' is another example. These marine algae (unicellular plants) produce toxins. When shellfish such as clams and oysters eat the algae, there is no harm to the shellfish, but the algal toxins become concentrated in the shellfish tissue and can cause acute sickness and even death in a person who eats the shellfish.

As harmful as these organisms can be, they do not cause infections. The term **infection** implies that an organism finds an ecological niche in which to grow within or on another species, causing it harm. A

pathogen is an organism that can cause an infection and produce a **disease**, that is, a condition in which normal activity is prevented, and which, at the extreme, may cause death. Diseases that are present persistently, and usually at a low level, in a local area are called endemic diseases. Measles in the United States is an **endemic** disease. When an infectious disease increases in frequency and spreads to new geographic areas or to new populations, it becomes an **epidemic** disease. The recent outbreak of severe acute respiratory syndrome (SARS) is an epidemic disease. When a disease becomes frequent worldwide, as AIDS has, it is a **pandemic** disease.

Many organisms colonize humans, but the vast majority are not pathogens. Many bacteria, for example, are necessary to our health and well-being, either because they produce some product we need (see Chapter 10) or because they occupy a niche preventing pathogens from infecting us. Examples of the latter include the normal bacteria of the mouth that help prevent the infections that cause tooth decay and gum disease, as well as the intestinal bacteria that help prevent some of the infections described later in this chapter.

Many types of organisms are pathogens. Pathogens can be viruses, bacteria (kingdom Eubacteria), protists (kingdom Protista), and fungi (kingdom Mycota). Examples of pathogens described in other chapters include the virus HIV (Chapter 16) and the protist *Plasmodium*, which causes malaria (Chapter 7). Organisms from phyla within the kingdom Animalia, such as nematode worms (phylum Nematoda) and tapeworms (phylum Platyhelminthes), also cause disease. These are a specific type of pathogen called a **parasite**, an organism that lives within another by using the host as its nutrient source. Pathogens from within these many kingdoms and phyla are illustrated in Figure 17.1.

Figure 17.1

Pathogens are found in many kingdoms and phyla of living things.

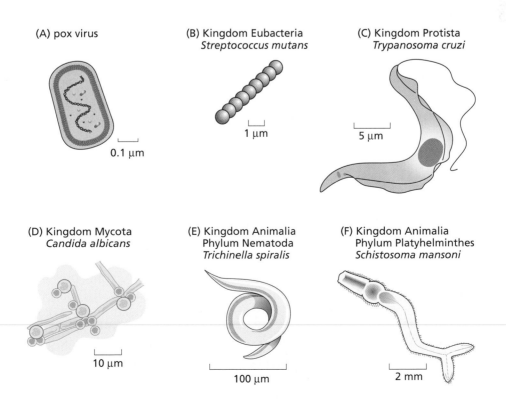

(A) pox virus

0.1 µm

(B) Kingdom Eubacteria
Streptococcus mutans

1 µm

(C) Kingdom Protista
Trypanosoma cruzi

5 µm

(D) Kingdom Mycota
Candida albicans

10 µm

(E) Kingdom Animalia
Phylum Nematoda
Trichinella spiralis

100 µm

(F) Kingdom Animalia
Phylum Platyhelminthes
Schistosoma mansoni

2 mm

CONNECTIONS
CHAPTER 16

Characteristics of pathogens

As mentioned earlier, although pathogenic organisms exist within many phyla, the vast majority of organisms are not pathogenic to humans. Pathogens must have some way of entering the host, such as being breathed in via the respiratory tract, or being taken in with food or water through the digestive system. Many pathogens have specific molecules that can bind to host molecules (as we saw for HIV binding to CD4 in Chapter 16). These attachment mechanisms are very specific, and are one reason why a particular pathogen can only attack a particular host species.

Once inside, pathogens must adapt to life within the environment of the host. Pathogens that cause human disease must be able to live and replicate at human body temperature. Because humans are aerobic (adapted to living in atmospheric oxygen), most organisms that are pathogenic for humans are also aerobic. Pathogens that enter the gut must have a way of surviving in the stomach's low pH and resisting the various digestive enzymes.

Pathogens must also have some mechanism for at least temporarily escaping the immune system. The host immune system produces peroxides to attack bacteria. Most bacteria that can live in humans have an enzyme called catalase that breaks down the peroxides formed during metabolism in the presence of atmospheric oxygen. Bacterial catalase overcomes this host defense. Sometimes pathogens shift to making new shapes of proteins, so antibodies that the host made to the first shape of protein do not work against the new proteins (called **antigenic drift**). Such changes can lead to epidemics, because there will be many newly susceptible hosts for the altered pathogen. Others can change their protein structure even more radically by combining new gene segments. Virus are adept at this process (called **antigenic shift**), particularly when a single host is infected simultaneously with more than one species of virus. These can lead to pandemics, because almost all hosts will be newly susceptible after these large-scale changes in pathogen structure. Some pathogens hide by covering their surface protein antigens in a carbohydrate capsule or by living inside a human cell, where they become inaccessible to the immune system. Still others can make enzymes that destroy parts of the immune system, such as enzymes that degrade antibodies.

Finally, for a disease to spread within a population, there must be a mechanism for the pathogen to exit from one host and be carried to the next. Some pathogens are excreted with feces, some are sneezed or coughed out of the respiratory tract, and others require intimate person-to-person contact.

A successful pathogen will therefore be adapted to gain entry to the host, to adhere and replicate within the host, to avoid or trick the immune response and finally, to exit from the host. Many non-pathogenic organisms, particularly bacteria, live on or in the host and share these adaptive characteristics. In addition, a pathogen also has some characteristic that makes it **virulent**, that is, able to cause disease. Some pathogens directly damage host tissue, such as skin, muscle or bone. Others induce such a high fever that they cause brain damage in the host. The bacterium *Neisseria meningitidis*, that causes the brain infection called meningitis, is an example.

Evolution of virulence

Host species and their pathogens have co-evolved a delicate balance. As we saw in Chapter 7, infectious disease is a significant force for natural selection in human evolution. Adaptation to life within a human is, however, a significant selective pressure for the evolution of the pathogen.

Virulence is the ability of a pathogen to overcome host defenses, thereby causing serious illness or death. From an evolutionary standpoint, virulence poses a severe problem for the pathogen: if it kills its host, it deprives itself of a suitable habitat and food supply. Clearly, a pathogen that causes minimal harm to its host is assured a longer time span for itself and its offspring to continue living in the same place than a pathogen that kills its host. HIV, for example, is related to several viruses that infect higher primates such as monkeys and apes. These viruses are nonvirulent—they spread from one host to another without causing serious illness or death. Thus, they have been around long enough—thousands of years at the very least—for symbiotic relationships with their hosts to have evolved.

Evolutionary biologists who study bacteria and viruses believe that the evolution of virulence is related to the pathogen's fitness, meaning its capacity to leave offspring (see Chapter 5). When a new strain originates, it must compete with the older, nonvirulent strains. A virulent pathogen can proliferate rapidly in a host, but if it spreads from host to host at a slow rate, it might kill off its hosts before being able to colonize new ones; such a virulent strain will be less fit and will soon die out. A virulent strain that spreads rapidly from host to host will soon outcompete its nonvirulent relatives. If the process is rapid enough, an epidemic occurs.

In the long run, natural selection favors the evolution of host defenses against the pathogen, including both physiological and chemical defenses. These changes in the host reduce microbial virulence directly, and the adoption of host behaviors less conducive to the pathogen's spread also slows transmission. When transmission slows, strains that are less virulent once again become more fit than the virulent strains, and the cycle repeats.

You will recall that evolutionary change is dependent on genetic change. Virulent strains will be selected in circumstances in which the rate of transmission is rapid. Public health efforts to change human behaviors and slow transmission will, in contrast, select for less virulent mutations. Thus virulence and pathogenicity are not unchanging traits of an infecting organism; rather, they have to do with the dynamic relationship between the pathogen and its host. The host and pathogen relationship is constantly changing, thereby changing the severity and nature of infectious disease in the host population, leading to the emergence and re-emergence of infectious diseases.

Factors governing the spread of pathogens

There are four types of factors that influence the spread of pathogens. The first, as we have seen, is a combination of susceptible hosts with microorganisms virulent for that host. Host susceptibility and pathogen virulence are each, in part, genetically controlled and host and pathogen exert selective pressures on each other.

Second is a factor called 'herd immunity,' the proportion of nonsusceptible hosts within a population. The lower this proportion, the more likely is the spread of disease. Herd immunity can be increased by vaccination campaigns. Even if every individual is not vaccinated, as long as a high percentage are, the spread of disease will be limited. Measles and polio, for example, are viral diseases for which there are very effective vaccines. Millions of people in the world still die of these diseases each year, and many more are disabled. Polio causes paralysis of the lungs or of the limbs, and young people are most susceptible. If a pregnant woman contracts measles, her fetus may be damaged, sometimes causing deafness. In areas where there is vaccination coverage, in one generation, people have forgotten that these are serious pathogens. Some individuals now choose not to be vaccinated, often for religious reasons or because there is some very low risk of complications from vaccines. In doing so, however, they increase the risk for the whole group. Individual risk of disease is only low when herd immunity is high, so if enough people refuse vaccination, they also increase their individual risk.

A third factor in the spread of pathogens is weather. Disasters such as floods can spread some diseases, if, for example, floods spread sewage into drinking water supplies. More long-term effects are starting to become apparent from climate change as a result of global warming (Chapter 19). This is already having an impact on the ecology of local regions, changing the niches that may serve as reservoirs for pathogens. This can include changes in water temperature, as many pathogens live in water. It can also include changes in the geographical spread and local size of animal populations. Some human diseases are caused by pathogens that can also live in another animal host. Rabies is an example. This viral disease can infect several other animals. Infection in domestic pets can be controlled by vaccination of the pets. But rabies spreads in populations of wild animals, and climate change affects the geographic distribution of these animals. Physiological stress from drought or starvation as a result of changes in their food supply can weaken animals' immunity, increasing the percentage of diseased animals. Other human diseases are spread by **vectors**, organisms that carry the pathogens without getting the disease themselves. As we have already seen for malaria (Chapter 7), insects can be significant vectors for human pathogens. Climate change alters the range and population size of these vector species.

Finally, disease is spread by various means that carry pathogens from an infected person to a new susceptible host. These are referred to as the **routes of transmission**. Some routes of transmission are direct; that is, from one person to another by means of physical contact (handshakes, sneezing, kissing, or sexual contact). Other routes of transmission are indirect, requiring another species or some object for transmission. Indirect routes include insect carriers (vectors), food or water contamination, and needles or syringes,.

Regardless of the route, the size of an epidemic is increased by two factors. One is the increase in the size of human populations. The other is the extent of travel, so that disease is now more often spread from one population to another than it would have been in the past. In 1950 there were a total of 5 million international arrivals at all destinations around

CONNECTIONS
CHAPTERS 7, 19

the world. By the year 2000, that number had grown to 800 million. The numbers of people who have migrated as a result of war or famine or economic dislocation are larger than at any other time in world history.

In terms of disease, as in so many other ways, the world population is now one interconnected population. The health, social and economic impacts of infectious disease in one part of the world quickly have impacts on the health of societies and economies around the world. In 1995 the World Health Organization, in recognition of this, drafted a resolution entitled *Communicable diseases prevention and control: new, emerging and re-emerging infectious diseases* (WHA48.13). This resolution ushered in a new era of public health measures designed to increase disease surveillance (the detection of disease patterns in populations) and improve control programs. Applied research is leading to more effective and low-cost ways to prevent the spread of disease. These efforts have been aided by computer databases and the sharing of such databases by governments and non-governmental organizations around the world. The infrastructure for local health care providers to report disease cases to centralized public health systems has become better. In addition, geographical information system computer software has allowed those data to be instantaneously mapped so disease trends can be spotted quickly.

In the remainder of this chapter, we examine various infectious organisms that can spread through human populations and cause disease. We also examine some of the factors that change the balance between pathogens and their hosts. Before we do so, we discuss the threat from pathogens that can be introduced into the human environment intentionally.

Intentional transmission turns disease into bioterrorism

The year 2001 has, unfortunately, given us an example of what can happen if some person decides to spread a disease intentionally. On a small scale, a deranged individual who knows they have an infectious disease may deliberately expose others in the hope of making them suffer. But the incident that began in October of 2001 was different from this. Someone prepared large quantities of a bacteria called *Bacillus anthracis*, and distributed it widely through the U.S. postal system, giving many people the disease anthrax, and leading to 23 deaths and great economic and social disruption. Intentional transmission to large numbers of people for the purpose of spreading disease or suffering is called **bioterrorism**.

The idea of bioterrorism is not new. During the bubonic plague in thirteenth-century Europe, some armies used catapults to hurl the dead bodies of plague victims over castle walls at their enemies. The Allies were testing anthrax for use as a weapon in World War II. But October 2001 marked the first time that bioterrorism was carried out on a large scale.

Anthrax. Anthrax is ideally suited to bioterrorism. If it is grown in a pure culture under the right conditions it will form spores (Figure 17.2). The DNA becomes condensed, and the cytoplasm becomes dehydrated. Bacterial spores are exquisitely resistant to killing by heat or by freezing or by drying, any of the factors that kill most pathogens in the environment. Anthrax spores can survive in the soil for at least decades, and spores of

Figure 17.2

Spore formation compared with normal bacterial cell division.

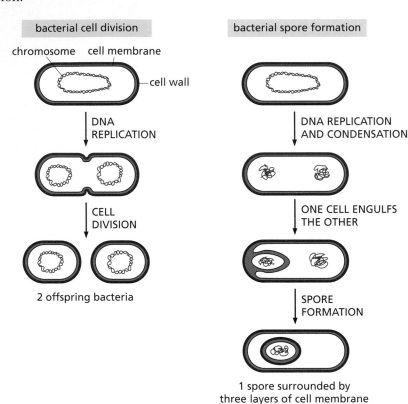

bacterial cell division

chromosome   cell membrane

cell wall

DNA REPLICATION

CELL DIVISION

2 offspring bacteria

bacterial spore formation

DNA REPLICATION AND CONDENSATION

ONE CELL ENGULFS THE OTHER

SPORE FORMATION

1 spore surrounded by three layers of cell membrane and a cell wall

CONNECTIONS ▶ CHAPTER 4

some bacteria can even survive for centuries. They are very tiny, so in their purified form they can form a powder that will easily spread at the slightest movement. Some strains of *Bacillus anthracis* are more deadly than others, and the one that was spread in the postal system in 2001–2002 was the most deadly combination of fine aerosol-forming spores and high infectivity. Different isolates of *Bacillus anthracis* can be identified, using the same techniques outlined in Chapter 4 for identifying individual humans (see Figure 4.6, p. 106). The *Bacillus anthracis* genome contains repeat sequences, and different isolates have variable numbers of these repeats. The bacteria from the October 2001 assault match a strain that originated from a research laboratory in Ames, Iowa, but it has not yet been determined who released it on the world.

When the spores enter a human or animal host, they rehydrate and begin to divide. The bacteria cannot divide in the soil, only in a host animal. The bacteria then secrete a three-protein toxin. One protein binds to host cell membranes and triggers the entry of the other two proteins into the cell cytoplasm. One of these proteins is an enzyme that degrades a host protein necessary for cell signaling. The consequences depend on where in the body the spores entered. In cutaneous anthrax, the spores enter through the skin, usually at the site of an existing break in the skin. This form results in local lesions, and if treated with antibiotics, the person usually recovers. In another form, inhalation anthrax, spores are breathed into the lungs, and the disease results in swelling and hemorrhage in the lungs and a high death rate (including 45% of the bioterrorism exposure cases), even with antibiotic treatment.

Smallpox. Another organism that people fear as a bioterrorism agent is smallpox. Smallpox is caused by a virus and has a high fatality rate. It was already prevalent in India and China 1000 years B.C. By 700 A.D. it was in Europe. The Spanish carried it to Central America in 1520, leading to the death of 3.5 million Aztec Indians within 2 years. Similar epidemics occurred in South America in the 1530s. Throughout the seventeenth and eighteenth centuries, epidemics continued in Europe, killing 400,000 people a year. It is thought that smallpox has killed more people than all other infectious diseases combined.

From ancient times, it was also known that those few people who lived through a smallpox infection were protected from future infections.

Prevention by deliberate exposure to dried crusts from the pox was practiced in China for millennia and in the Middle East and West Africa at least as long ago as the 1600s. The practice was finally adopted in Britain and its North American colonies in the early 1700s. It worked frequently, but not always, and many people contracted the disease rather than being protected. (We now know that this was because the crusts usually contained 'attenuated' virus, virus that was no longer virulent, but sometimes contained live virus.) It was also well known to dairy herdsman and milkmaids by the 1700s that exposure to cowpox, a similar, but far less deadly, disease of cows would protect against smallpox. In 1796, a British physician named Edward Jenner used this knowledge to immunize an 8-year-old boy with material from the cowpox lesion of a milkmaid. The child was later exposed to material from a smallpox patient and proved to be fully protected from disease. Such experiments would never be approved by ethics review boards today, as children are not deemed capable of giving informed consent (see Chapter 1), and because there was no known treatment should exposure lead to disease. But in a way, our ethical stance today is allowed by the luxury of living in a world from which smallpox has been eradicated. Jenner's experiments began the practice that was named 'vaccination,' after the Latin word for 'cow,' *vacca*. Vaccines were gradually developed that offered lifelong protection to virtually all people who were vaccinated. By the 1950s it was known that smallpox is spread only from one person to another, and there are no vectors or other reservoir hosts for the virus. The combination of lifelong immunity and the lack of other reservoirs made it possible to eradicate the disease entirely. In 1958, virologist V.M. Zhadanov of the USSR proposed such an eradication to the World Health Assembly of the United Nations. A stable freeze-dried vaccine was developed in the 1950s that could be carried without refrigeration into remote areas around the world. The eradication campaign began in earnest in 1967. The last cases in Europe occurred in Yugoslavia in 1972, and the last case in the world occurred in 1977 in Somalia. In October, 1979, the World Health Organization declared the world free of the disease.

No smallpox cases have occurred in the United States since 1949, and vaccination of children in the United States was discontinued in 1971, and of hospital employees in 1976. Elsewhere in the world immunization has also been stopped. The small risk from reactions to the vaccine is now higher than the zero risk from the disease. Because we no longer immunize, herd immunity is lost.

Smallpox virus stocks are stored in two places: the Centers for Disease Control and Prevention (CDC) in the United States and the Research Institute for Viral Preparations in Moscow. These were scheduled to be destroyed a few years ago, but by then people had begun to be concerned about the possibility that if anyone else had the virus illegally there would be no stocks from which to reestablish the manufacture of vaccines. Consequently, they have not been destroyed. Smallpox is greatly feared for its potential for use in bioterrorism because its spread is rapid once a few people are infected, and the fatality rate is so high. Because anyone born after 1970 in the United States is a susceptible host, concern is heightened; yet, there remains no credible evidence that anyone has the virus or is thinking of using it. Therefore, the threat, although high in theory, remains a potential threat, not an actual threat.

THOUGHT QUESTIONS

1 What are the differences between the terms 'virulence' and 'pathogen'?

2 Are the same species of organisms pathogens in every person?

3 How should the rights of individuals be balanced against the needs of society as a whole in deciding who should be vaccinated against a particular disease?

4 What methods are in place to control bioterrorism?

5 Should everyone be vaccinated against pathogens that are potential bioterrorism threats? Why, or why not? Can people be vaccinated after an outbreak has started?

Some Diseases That Spread by Direct Contact Are Increasing in Prevalence

Figure 17.3

Worldwide incidence (new cases per year) of some sexually transmitted diseases (STDs). Data are from the World Health Organization Report from 1996, with the exception of HIV/AIDS, which are data from 1998.

Some pathogens are spread directly from one person to another via sexual contact. As we saw in Chapter 16, AIDS is a sexually transmitted disease, but there are many others. In the United States and worldwide, the number of new cases per year (**incidence**) of AIDS is lower than that of many other sexually transmitted diseases (Figure 17.3). Each year in the United States, 3 million people are infected with *Chlamydia*, and there are more than 300,000 cases of gonorrhea and 120,000 new cases of syphilis, with the incidence of all of these rising among teenagers and young adults. Also, about 31 million people in the United States carry type 2 herpes simplex (the most common genital herpes virus), and 500,000 new cases are reported to the CDC annually. Except for AIDS and untreated syphilis, most sexually transmitted diseases are not fatal, but they have other serious consequences including (for different diseases) sterility, paralysis, arthritis, and chronic pain in adults, as well as severe disease in newborns when transmitted from the mother. We will first examine what some of these diseases are and then look at the factors that are leading to their increasing prevalence.

The major sexually transmitted diseases

The sexually transmitted diseases (STDs) shown in Figure 17.3 fall into several categories. Some are bacterial (chlamydia, gonorrhea, and syphilis), some are viral (genital warts, genital herpes, and AIDS), and still others are parasitic protists (trichomoniasis), as is summarized in Table 17.1.

Figure 17.3 bar chart: annual number of new cases worldwide (millions)

- trichomoniasis: 170
- chlamydiasis: 89
- gonorrhea: 62
- genital herpes: 20
- syphilis: 12
- AIDS: 6

Table 17.1

Summary of some sexually transmitted diseases.

DISEASE	TYPE OF PATHOGEN	NAME OF PATHOGEN	U.S. INCIDENCE RATE (CASES PER 100,000 ADULT AND ADOLESCENT POPULATION)
chlamydia	bacterium	*Chlamydia trachomatis*	207
gonorrhea	bacterium	*Neisseria gonorrhoeae*	123
syphilis	bacterium	*Treponema pallidum*	3.2
genital herpes	virus	herpes simplex virus-2	(n.a.)
AIDS	virus	HIV	14.7
trichomoniasis	protist	*Trichomonas vaginalis*	(n.a.)

Compiled from CDC 1997 Surveillance Report and from 2001 data from www.cdc.gov/hiv/stats (n.a. means incidence rate not available).

Chlamydia. A microbiology textbook from 1973 describes this STD as "not common except in individuals who are highly sexually promiscuous" (B.D. Davis, R. Dulbecco, et al., Microbiology. Hagerstown, MD: Harper & Row, Publishers, Inc., 1973). That statement remains true; what has changed is the number of people who would be considered 'promiscuous' by the standards of 1973. Chlamydia is now the most frequently reported infectious disease in the United States, of the diseases for which reporting of cases to the CDC is required. It has increased from 3.2 cases per 100,000 population in 1984 to 207 cases per 100,000 in 1997. Well over 500,000 cases were reported in the United States in 1997 alone. Because 75% of infected women and 50% of infected men show no symptoms, it is assumed that many more infections exist, possibly as many as 3 million per year in the United States. Teenage girls have the highest rates of infection: 15–19-year-old girls account for 46% of the infections, and 20–24-year-old women another 33%.

Chlamydia is an infection of the lower genital tract, rectum, or throat by the bacteria *Chlamydia trachomatis*. The bacterial life cycle alternates between a hardy, extracellular stage and an intracellular stage. The extracellular stage is engulfed by the host cell, forming a vesicle (see Figure 15.4C), but rather than being killed, it lives and replicates inside the vesicle. *Chlamydia* lacks the genes for ATP synthesis and so is an 'energy parasite,' robbing the host cell of the ATP it produces. After several generations, the noninfectious intracellular forms shrink up and are released as the infectious extracellular form by rupture of the host cell (Figure 17.4).

Damage caused by the rupturing of cells results in pelvic inflammatory disease in 40% of women with untreated cases, one-fifth of whom will become permanently infertile. *Chlamydia* infection is one of the major causes of female infertility. Pelvic inflammatory disease from *Chlamydia* infection can also result in life-long severe pelvic pain and in life-threatening tubal pregnancy. In fact, it is the leading cause of pregnancy-related death among U.S. women. Men can develop infections of the urethra or swelling and pain in the testicles. Unprotected sex, which allows transmission of any STD, also allows the transmission of HIV. If a person is infected with *Chlamydia*, they are three to five times more likely to become infected with HIV after exposure.

In addition to the sexually transmitted genital infections, *Chlamydia trachomatis* can also cause trachoma, an eye infection that is the most common cause of blindness worldwide. It is a different strain of the same species of bacteria that is responsible for most eye infections, a strain that can be readily spread from one person to another by touch or by flies. The genital strain can, however, be passed from an infected mother to her baby at birth, causing eye infections and pneumonia in newborns and chronic respiratory infections in children.

Gonorrhea. Gonorrhea is an infection of the urinary and genital tissues by a Gram-negative bacterium called *Neisseria gonorrhoeae*. It works its way past the mucous membranes and into the spaces between epithelial cells. There it induces an acute inflammatory response by the immune system, resulting in a discharge of pus from the urethra or the vagina. Despite the reaction of the immune system, protective immunity does not develop because the bacterium is capable of shifting its antigens, that is, changing the structure of the cell-surface proteins it makes. People can therefore be re-infected multiple times. In untreated males, scars may develop in the prostate and urethra, and, in untreated females, scarring of the fallopian tubes may result in permanent infertility. The rectum and the throat can also be infected, so the bacteria can be spread by anal sex (heterosexual or homosexual) and by oral sex, in addition to vaginal sex. The infection can spread to the blood, heart, and brain, and can also lead to painful arthritis in the joints. Infected mothers can pass the bacteria along to their babies at birth, and *N. gonorrhoeae* infections can result in blindness in the newborn.

Syphilis. This ancient disease is caused by the spiral bacterium *Treponema pallidum*. The infection results in open sores on the genitals, vagina, anus, or rectum and also on the lips and mouth. It is spread by direct

Figure 17.4

Life cycle of the bacterium *Chlamydia trachomatis*. Lysosomes normally kill bacteria that are inside vesicles, but *Chlamydia* has a virulence factor that prevents this.

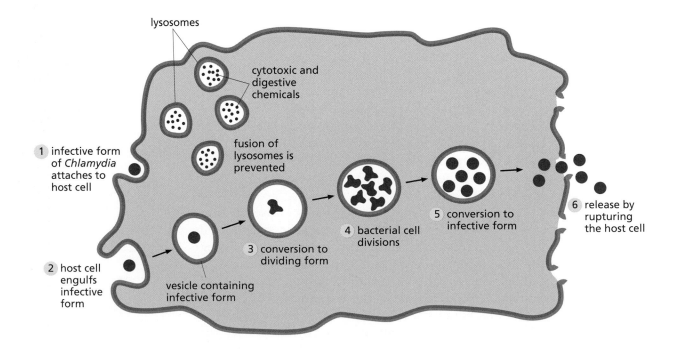

contact with these sores. Therefore, vaginal, anal, or oral sex can spread the bacteria. If untreated, the initial sores progress to the secondary stage consisting of a rash and sometimes a fever, sore throat, patchy hair loss, weight loss, muscle aches, and tiredness. Obviously, many of these symptoms can also be associated with many other diseases besides syphilis, so a person may not be aware of their infection or it may be misdiagnosed. Infected people, whether they are aware of their infection or not, can easily pass the bacteria to their sexual partners. Outward symptoms then disappear, but the infection spreads internally, damaging internal organs including the brain, blood vessels, liver, and bones, and resulting eventually in loss of muscle coordination, numbness, paralysis, dementia, and even death.

Pregnant women infected with syphilis often have stillbirths or babies who die shortly after birth. Babies who live may be infected and, if untreated, the babies will show brain problems, including seizures and developmental delays.

Viral STDs. Genital warts caused by papilloma virus and herpes infections caused by herpes simplex virus (HSV) type 1 or type 2 are the two most common viral STDs. One-fifth of adolescents and adults in the United States are estimated to be infected with HSV type 2, which causes a genital infection. Herpes infections cause blisters but the virus can also be released through skin that looks unbroken. HSV type 1 causes blisters in the mouth and lips, commonly referred to as cold sores, that shed virus and cause a genital infection during oral sex. HSV type1 can also be spread by saliva. While HSV does not produce as severe disease as the bacterial STDs, it can be painful, and if passed from an infected mother at birth it can kill a newborn baby. Herpes does, however, make people more susceptible to HIV infection after exposure. If a person is infected with both HIV and HSV, their HIV becomes more easily transmissible.

Parasitic STDs. The most common of these, and the most common STD of any kind, is trichomoniasis, an infection caused by a microscopic flagellated protozoan called *Trichomonas vaginalis*. It is most common among 16–35-year-old women, especially those who have had multiple sex partners. The infection results in a vaginal discharge or vaginal itching, and may also make intercourse painful. Men can also be infected, although they generally do not show any symptoms.

Factors increasing prevalence

The **prevalence** of a disease is the number of people who have the disease at any given time. The prevalence of a disease is generally higher than the incidence, because prevalence includes both new cases and cases that have not been cured. The global incidences of common STDs are summarized in Figure 17.3. The prevalence of each of these is higher than the number shown for the incidence of each. The prevalences of all STDs are rising, and there are many factors contributing to this rise.

Antibiotics are medicines that kill or stop the growth of bacteria. Different antibiotics work by different mechanisms, but in each case they are targeted at some molecule or pathway found only in bacteria. That way, they exert little direct effect on the human host. Because chlamydia, gonorrhea, and syphilis are bacterial diseases, most cases can be success-fully treated with antibiotics if treatment is started early enough. However, because so many people do not show symptoms from these infections, they often do not seek treatment and remain chronically infected and infectious to others.

For an antibiotic to stop bacteria, it must be taken up by the bacteria. Regardless of whether the antibiotic is taken orally or given intravenously, it is delivered to the site of the infection by the circulatory system. Most bacteria live and grow outside the cells of their host, making access of the antibiotic easier. *Chlamydia* infection is more difficult to treat with antibiotics because of its intracellular growth. Many antibiotics do not reach the insides of host cells, so antibiotics for *Chlamydia* must sometimes be continued for several weeks.

Antibiotics put a strong selective pressure on bacterial populations, selecting for mutated variants that are resistant to the antibiotic. Many strains of bacteria, including those that cause syphilis and gonorrhea, are now resistant to many antibiotics, severely limiting the options for treatment of a disease caused by these strains. For example, 29% of isolates of *N. gonorrhoeae* have become resistant to all of the penicillins or tetracyclines, or to both groups. Many are now also resistant to a newer family of antibiotics, the fluoroquinolones. Penicillin antibiotics block the synthesis of bacterial cell walls, but resistant bacteria counter by being able to break the penicillin molecule, rendering it inactive. Tetracycline antibiotics inhibit bacterial protein synthesis by binding to bacterial ribosomes. Bacteria become resistant by acquiring the gene for a membrane protein that pumps the antibiotic back out of the bacterial cell. The fluoroquinolones, which include such drugs as ciprofloxacin, work by preventing the bacterial chromosome from replicating. A bacterium has a single chromosome in the form of a ring. To replicate, the double helix must unwind, but this presents problems for a circular chromosome. Consequently, bacteria have a special enzyme that breaks the chromosome, allowing its unwinding for replication. Fluoroquinolones, in principle, block this enzyme; however, some strains of *Neisseria* are now resistant. The incidence of gonorrhea in the United States had declined 75% from 1974 to the 1990s, but is now again on the rise, largely due to the increase of antibiotic resistance in *Neisseria gonorrhoeae*. Genes for resistance can be carried on either the bacterial chromosome or the separate small circles of DNA, the plasmids. Because plasmids are easily transferred from bacterium to bacterium and even from one bacterial species to another, horizontal transfer of resistance is especially rapid for plasmid-mediated resistance. These various mechanisms by which antibiotics stop bacteria, and the paths by which bacteria develop antibiotic resistance, are illustrated in Figure 17.5.

Any given mixture of bacteria is a heterogeneous population containing varying degrees of resistance. It is important that antibiotic treatment be continued for the full number of doses. If it is not, the most resistant bacteria are much more likely to remain and grow to cause a relapse of the infection (Figure 17.6).

In contrast to the bacterial STDs, there is no drug to eradicate herpes or papilloma, which are viruses. Antiviral medications can shorten an outbreak, but do not remove the virus.

Only one antibiotic, metronidazole, is active against the protozoan *Trichomonas vaginalis*. *Trichomonas* can become resistant to metronidazole. There is no other drug that will work, leaving a higher dose or a longer time course as the only choices. Metronidazole, especially in high doses, has many adverse side effects. The numbers of white blood cells of the immune system can be temporarily lowered, leaving the person vulnerable to other infections. If alcohol is consumed, a violent reaction ensues. Also, metronidazole kills off so many of the normal protective bacteria that the fungus *Candida* can overgrow and cause additional discomfort.

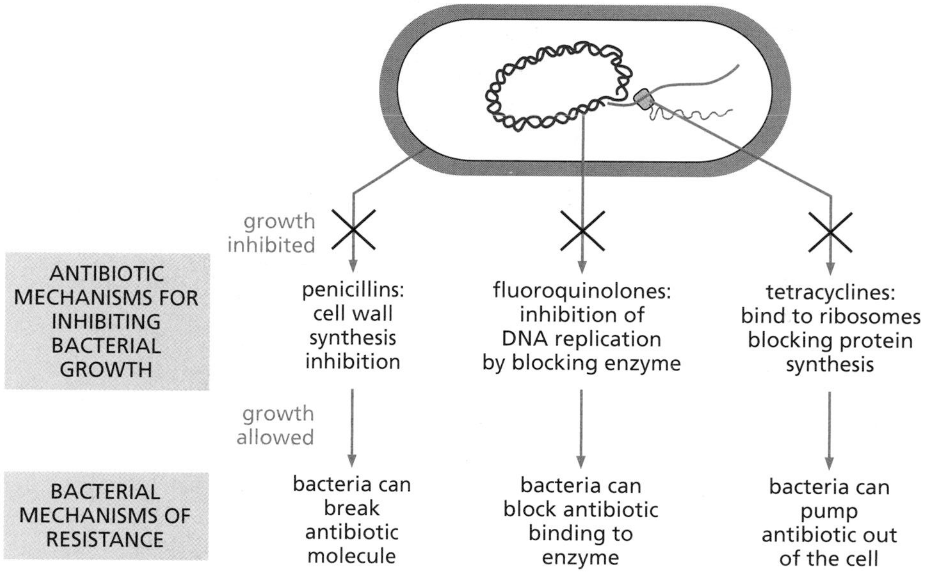

The increases in human population density and the changes in attitudes towards sexual activity that have occurred in the past few decades have increased the transmission of STDs in many parts of the world. One of the largest risk factors for STDs is having multiple sexual partners. People might not think of themselves as having multiple sexual partners if they are monogamous for a time and then move on to a new monogamous relationship, but one's risk increases with each change in partners. The risk of transmission of a STD from a former partner to the next also increases. Many STD pathogens can remain infective for decades and all sexual partners over this long period are therefore at risk.

Figure 17.5

Sites and mechanisms of action of antibiotics and pathways to bacterial resistance to antibiotics.

ANTIBIOTIC MECHANISMS FOR INHIBITING BACTERIAL GROWTH

BACTERIAL MECHANISMS OF RESISTANCE

growth inhibited

growth allowed

penicillins: cell wall synthesis inhibition

fluoroquinolones: inhibition of DNA replication by blocking enzyme

tetracyclines: bind to ribosomes blocking protein synthesis

bacteria can break antibiotic molecule

bacteria can block antibiotic binding to enzyme

bacteria can pump antibiotic out of the cell

Abstinence or lifelong monogamy with an uninfected partner will prevent STDs. Consistent use of condoms is quite effective, although not 100% (see p. 597). Seeking treatment early, when there is any suspicion of having contracted an STD, will help protect oneself and others; however, because of the risk from resistant pathogens, treatment is never as good an option as prevention. For any STD, it does little good to treat only the initial patient. That person will be asked for the names of all of their sexual contacts, so that each of them may also be treated. In the instances where people do not know their sexual contacts, or do not know how to locate them for notification, further infections will continue. When people continue to be sexually active while they are taking antibiotics, the spread of antibiotic-resistant STDs is enhanced, because bacteria will have been transmitted before it is obvious that the bacteria have become resistant.

Tuberculosis

Person-to-person transmission is required for any pathogen to succeed if it cannot live outside a host organism and has no hosts other than humans. Not all of these need to be transmitted by sexual contact, however. One example is tuberculosis caused by the bacterium *Mycobacterium tuberculosis*, which grows intracellularly. Infections with this bacterium can occur in many parts of the body, with different sets of symptoms, but the most common and the most infectious is pulmonary tuberculosis, an infection of the lungs. The infection leads to permanent damage to the lung tissue. It is spread by droplet contamination from coughing. The droplets are so small that they hang in the air for a few hours, and the bacteria remain infectious. Another person can therefore pick up the bacteria by breathing the droplets without being in direct contact with the infected person.

Tuberculosis cases declined greatly in industrialized countries until the 1980s, but there was a large resurgence during the 1980s and 1990s. This is a primary example of a 're-emerging disease.' Worldwide, there are 8 million new cases per year and there were 2 million deaths from tuberculosis in the year 2000 alone, about the same as the number of deaths from AIDS. Part of the increase in deaths from tuberculosis is due to its association with HIV infection in many parts of the world. The disease is much more aggressive in immunocompromised persons. Tuberculosis can be treated with antibiotics, but eradicating the bacterium is difficult because of its intracellular location. During the long treatment time, antibiotic resistance often develops. It is estimated that, even in susceptible strains, one bacterium in a million is resistant. That bacterium can be selected over time, so a person may at first get better and then relapse, requiring treatment with more than one antibiotic working by different mechanisms. Population density and poverty that results in large numbers of people sharing a living space increase transmission of tuberculosis, as does drug resistance that develops if a person does not finish the course of treatment.

Both the health impact and the economic impact of this disease are enormous, a fact that has been recognized by both the World Bank and the World Health Organization. An international standard

Figure 17.6

Evolution of resistance during antibiotic treatment.

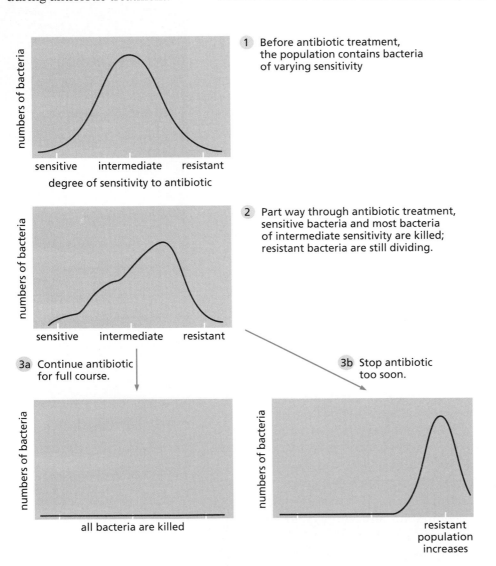

1 Before antibiotic treatment, the population contains bacteria of varying sensitivity

2 Part way through antibiotic treatment, sensitive bacteria and most bacteria of intermediate sensitivity are killed; resistant bacteria are still dividing.

3a Continue antibiotic for full course.

3b Stop antibiotic too soon.

treatment has been developed, called the DOTS (directly observed treatment short-course). Diagnosis is by direct microscopic observation of the bacterium in smears from the lung fluid coughed up by people with the disease. Patients are then given a short course of antibiotics, but are directly observed during treatment to be sure that the drugs are taken. In the United States, individuals are required to receive treatment, and public health officials deliver the drugs to the person to ensure compliance. The World Bank included DOTS as one of the essential clinical services necessary for economic development. The World Health Organization has included tuberculosis (TB) as one of the three major diseases targeted by the Global Fund to Fight AIDS, TB, and Malaria. DOTS is considered one of the most effective public health strategies ever devised. Because the disease is only transmitted from one person to another, the chain of transmission can be broken by decreasing the percentage of people with the disease.

THOUGHT QUESTIONS

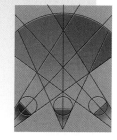

1 Which disease might be more difficult to eradicate from a population, an STD or tuberculosis? Why?

2 What differences might there need to be in public health strategies in dealing with STDs or with tuberculosis?

Food-borne Disease Patterns Reflect Changes in Food Distribution

Many other infectious diseases are more readily spread than STDs because they can be carried in ways other than by direct person-to-person contact. Many pathogens can survive and grow on the surfaces of foods, and eating or handling the foods can spread the bacteria to people. Such pathogens can then be spread from one person to another, but more commonly a disease cluster results from several people eating the same batch of contaminated food.

The major food-borne pathogens are summarized in Table 17.2. The incidence of many of these has increased in the past 25 years. *Campylobacter jejuni*, the leading cause of food-borne bacterial infection in the United States, was not even recognized as a cause of human illness until the late 1970s. Although these infections are common, severe sickness or death is much more likely in people who have weakened immune systems. People with AIDS must be very careful in the preparation of the foods they eat; so must people whose white blood cell count is temporarily low from chemotherapy treatment for cancer. One of the other significant causes of weakened immunity is age, so food-borne illnesses are becoming more prevalent as populations age. Ironically, people are living longer because of the control of many types of infectious disease (smallpox, typhoid,

Table 17.2

The major food-borne pathogens in the United States

BACTERIUM	INCIDENCE	DEATHS PER YEAR	FOOD SOURCES
Campylobacter jejuni	4,000,000	200–1000	Unpasteurized milk, undercooked poultry
Salmonella	2,000,000	500–2000	Raw or undercooked eggs, poultry, other meats
Escherichia coli O157:H7	20,000	100–200	Undercooked ground beef, fresh produce, unpasteurized milk or fruit juice
Vibrio species	10,000	50–100	Raw seafood (mollusks, crustaceans, fish)

Data from W.M. Scheld, W.A. Craig, and J.M. Hughes (eds.), Emerging Infections 2. *Washington, D.C.: American Society for Microbiology Press, 1973.*

diphtheria, and polio to name a few), so they now live long enough to become susceptible to food-borne diseases In 1900 less than 5% of the U.S. population was over the age of 65; in 2000 more than 15% was.

One example: variant Creutzfeld–Jakob disease

Evidence has been accumulating to consider Creutzfeld–Jakob disease (CJD) a food-borne disease. CJD is a very rare brain disease that is invariably fatal. The pathogen responsible does not fall into any of the categories of organisms previously mentioned. The agent appears to be a **prion**, an altered shape of a normal protein that becomes self-replicating (Figure 17.7). Prions are therefore not organisms at all, but can spread from one person to another with the characteristics of an infection, so prion diseases are considered to be infectious diseases. The prion leads to nerve deterioration. An older form of CJD is believed to arise by mutation of the gene for the normal protein. Such mutation is very rare and so the number of cases per year was less than one per million in the United States and elsewhere, and the cases could not be connected to one another. Starting in 1995 in Britain, however, there was a sudden increase in the number of cases. The infection occurred in young, healthy people and progressed much more rapidly than spontaneous CJD; it was given the name 'variant CJD.'

It now appears that humans contract this disease by eating meat from beef cattle that have bovine spongiform encephalopathy (BSE). 'Bovine' refers to cows, an 'encephalopathy' is pathology in the lining of the brain, and 'spongiform' refers to the spongy look of the affected brain tissue. BSE is more widely known as 'mad cow disease.' The BSE prion can be detected in the brain, spinal cord, retina, and bone marrow of infected cows and it causes a brain disease that makes cows act very strangely and leads eventually to their death. The prion probably also arose by spontaneous mutation in cattle, but it is likely to have spread by the practice of using meat and bone scraps to make animal feed that was fed back to more cattle. Large-scale agricultural

Figure 17.7

Prions are altered forms of proteins that self-replicate in huge numbers, forming crystalline protein rods.

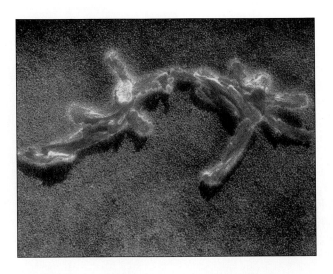

techniques probably contributed to its spread. By 1997, in Britain 170,000 cows were infected and efforts began to stop its spread among cows. In 1997 the U.S. Food and Drug Administration prohibited the use of mammalian protein in the manufacture of food for ruminants (cows, sheep, and goats). In June 2000, the European Union Commission on Food Safety and Animal Welfare strengthened its measures to prohibit the use of meat from the spinal column or brains of cows, sheep, and goats as animal feed. (Sheep also carry the prion, but it does not cause the same disease in them).

Initially, no one knew that the prion could also be spread to humans. The incubation period between infection and symptoms is several years, making discovery of the connection much more difficult. Between 1995 and June of 2002, a total of 124 cases of variant CJD have been confirmed, 98% of them in Britain. There are no cases of variant CJD where there is no BSE. Regulations have strengthened as more evidence has accumulated that the human prion is an infection acquired from eating meat from diseased cows, particularly brains or hamburger that contains nerve tissue. Meat from the spinal column of cows, sheep, and goats is prohibited in human food in member countries of the European Union as of October 2001. All cows over 30 months of age must be tested, and killed if they are found to carry the prion.

CJD is not the only prion disease known. The disease that led to the initial discovery of prions was kuru, a degenerative nervous disorder formerly spread by cannibalism among the Fore people of New Guinea. For his study of this disease, American physician Daniel Gajdusek received the Nobel prize in 1976.

Social and economic factors contributing to disease outbreaks

In many respects, food has become safer as sanitation, refrigeration, and better canning practices have developed. However, in many industrialized countries eating customs have changed drastically so that a much higher percentage of meals are eaten away from home. Indeed, 80% of outbreaks of food-borne illnesses are traced to sources outside the home. Furthermore, food production and distribution systems have changed enormously. People are eating many more fresh fruits and vegetables, and meats such as ground beef, than ever before. The fresh foods are arriving out of season from locations around the globe. A 1990 outbreak of 245 cases of *Salmonella* across 30 states was traced to a single source of cantaloupe imported from Central America. Similarly, a 1995 *Salmonella* outbreak was traced to one source of orange juice from within the United States and resulted in 63 severe infections across 21 states. The short amount of time between harvest and shipment to distant markets can mean that more bacteria are still alive when the food is purchased. Bacteria on the outside of a fruit can be transferred inside when the fruit is cut, and then multiply if the food is held at room temperature.

Centralization of food processing has resulted in outbreaks being larger and more widespread when they occur. An outbreak of *Salmonella* in the United States in 1994 resulted from the transport of ice cream premix in a tanker truck that had not been thoroughly disinfected after carrying raw liquid eggs. After the ice cream was produced and distributed,

224,000 cases of *Salmonella* infection resulted. *Escherichia coli* is a bacterium that is present in the guts of humans and many other animals. Some strains of *E. coli* are very virulent, including strain O157:H7, which was responsible for a 1993 outbreak traced to a fast food restaurant chain in Washington state. Trimmings from many beef cattle were included in one 2000-pound lot of ground beef. This lot was then distributed among hamburger patties delivered to several restaurants in the chain. Five hundred illnesses, 151 hospitalizations and 3 deaths resulted.

A type of *Salmonella* (called *Salmonella* serotype Enteritidis) has been found to infect the ovaries of chickens, from which it can be transmitted into the egg contents within intact eggshells. Farming practices have changed such that a typical hen house may now contain 100,000 hens and many houses may be serviced by the same machinery, making possible widespread contamination that was not possible when a hen house had 500 hens, or when each farm had its own small hen house.

Improvements needed

More thorough testing of foods and food sources would help to prevent and monitor outbreaks. In the 1990s, because of budget cuts within state public health agencies, 12 states in the United States no longer had any personnel monitoring food-borne illnesses. New detection methods using bacterial DNA amplification by PCR are very sensitive, but are time-consuming and expensive. Testing is also complicated by the pooling of ground beef from animals, or the pooling of eggs from many chickens into a single transport truck. The level of bacteria on an infected animal may be low enough not to be detected once it is combined with many others; however, the bacteria can grow and contaminate the whole lot. The surface of animal carcasses can be disinfected by antimicrobial rinses, and if a few bacteria remain on the surface they are readily killed by cooking. When beef is ground into hamburgers, any remaining surface contamination is brought inside the meat where it will not be killed unless the meat is thoroughly cooked all the way through. Time and temperature monitoring of foods is also important, in commercial processing and distribution, and in one's home.

As fruits and vegetables are harvested, they are washed, but often washed with water from farm ponds, and this water is not chlorinated. Chlorination of water used in processing and for making the ice on which foods are transported would stop many cases of contamination. Disinfection of farm machinery and transport vehicles between uses happens frequently, but not always.

As our food supply has become globalized, the World Trade Organization has recognized the sanitary and food safety standards developed by the World Health Organization. The World Trade Organization has no authority over member governments to accept these standards, but those that follow the standards allow the distribution only of products that meet the requirements of the standard.

The last step in the chain of food safety is what you do in your own home. Contamination of food or spreading of pathogens from contaminated food can be prevented by washing your hands and utensils thoroughly with soap and water. Refrigeration slows the multiplication of pathogens in stored food. Thorough cooking kills most bacteria.

THOUGHT QUESTIONS

1 Mayonnaise and chocolate mousse are often made with raw eggs. What special precautions should be taken with these foods if you wanted to serve them in a picnic on a hot day?

2 Some people currently receive injections of botulinum toxin (Botox) to remove wrinkles and sags. Botox is produced by *Clostridium botulinum*, an anaerobic bacterium that causes botulism, a food-borne illness. In botulism, the toxin results in a high rate of fatality. What is the mechanism of action of the toxin? Why can it result in death in one context and smoothing of wrinkles in a different context?

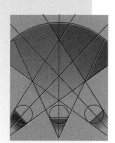

Waterborne Diseases Reflect Changes in Lifestyle and Climate

Contaminated water is one of the major sources of infectious diseases. One-third of the people on Earth do not have access to clean water. Some waterborne diseases are very ancient, but outbreaks have become larger owing to the lack of access to clean water that is associated with poverty and increased population density. Other waterborne diseases are very new, and reflect changes in technology and lifestyle.

Cholera

Cholera has been recognized as a disease since the early 1800s. Indeed, five pandemics had been documented before the discovery of the bacterial cause of the disease. Robert Koch, a German physician, isolated and described *Vibrio cholerae* as the causative agent of cholera in 1886, one of the important milestones in establishing the germ theory of disease during the 1880s.. Prior to that time, people had recognized that the disease was spread by 'bad' water, and this recognition led to the sanitary movement that started municipal water treatment and sewage treatment. Water treatment is still one of the most effective ways of stopping cholera. Numerically, cholera is one of the most important re-emerging diseases, but it is almost unknown in areas of the world with adequate water treatment. Globally, we are now in the seventh pandemic, which started in 1961. In 1970 the disease returned to West Africa, where no cases had been seen for over 100 years. In 1991 the disease also returned to Latin America, where again there had been no cases in a century. One of the reasons for its resurgence is illustrated by the method by which *Vibrio* was introduced into the warm waters of the Gulf of Mexico, off the coast of the United States. A large ship dumped its ballast water in the Gulf and the water was contaminated with bacteria. It was standard practice for ships to take on water as ballast weight when they were not

full of cargo, and then pump it out to reduce the ships' weight when they entered a port to take on cargo. Increases in the sizes of ships and the amount of water they carry as ballast increases the likelihood for enough contamination to be spread to establish a viable bacterial population in new water. *Vibrio cholerae* also lives in warm water and can be carried by warm water crustaceans (shrimp and crayfish), so that another source for transmission is by eating undercooked crustaceans. As global water temperatures rise as a result of global warming (see Chapter 19), the geographic areas are expanding where *Vibrio* can live in rivers and brackish water (water with low salt content, where rivers meet the ocean).

Vibrio cholerae causes an intestinal infection. The bacteria are completely specific for humans, and will not infect the guts of any other animals. This is because infection requires adherence of the bacteria via a pilus: a hair-like extracellular structure whose end protein binds specifically to human gut mucosal cells. Once bound, the bacteria begin to reproduce and then produce a toxin that induces a watery diarrhea. The severe dehydration that results leads to death in 25% of infected people.

The toxin and the protein needed for adhesion are carried close together in the *Vibrio* genome. The expression of these genes is turned on and off together in response to environmental conditions. The virulence genes can be transferred horizontally to nonvirulent strains of *Vibrio* by a bacterial virus (bacteriophage). There is no effective vaccine against cholera because the bacteria are capable of large shifts in their outer molecules. One example is a variant that caused a 1992 outbreak. These bacteria were not recognized by antibodies that would bind other *Vibrio* strains. In this new strain, a large part of the outer membrane had been replaced by a separate, nonmembrane-associated polysaccharide capsule. Such a large deletion/replacement mutation has not been found in any other bacterium. Protective immunity typically develops when antibodies are formed that can bind to some molecule on the outer surface of a bacterium. If the entire outer molecule structure changes, immunity must develop all over again.

The course of the disease is very rapid, often not allowing time for immunity to develop. A toxin that causes diarrhea can quickly spread the infection to new hosts if the host population density is great enough. From the standpoint of the evolution of virulence, such a toxin is an advantage to the pathogen only when host density is high, so that the bacterium can spread to new hosts as it kills its first host. Cholera also places a selective pressure on humans. Those people with blood type O get the most severe diarrhea, and the frequency of blood type O has been found to be very low in areas of the world, such as the Bay of Bengal, where cholera is present at a low endemic rate.

Water treatment is an effective public health strategy to prevent cholera. As with most infectious diseases, prevention of disease by prevention of transmission is more effective, and certainly more cost-effective, than treatment of people after they are sick. We might even say for waterborne diseases that 100 plumbers would have a greater effect on disease rates than 100 doctors. In urban areas where water systems are not adequately maintained, water pressure can be intermittent, and a pressure drop in the water outflow pipes can pull sewage back into the water pipes if sewage pipes are adjacent to them, leading to large outbreaks. Communal bathing and laundry facilities can spread the bacteria, and even clean water can be easily contaminated if water is scooped from a storage container by someone's bare hands.

For cholera, the other effective public health strategy is oral rehydration therapy. This is very inexpensive, and replaces the electrolytes and fluids lost during diarrhea. With rehydration, the mortality rate drops from 25% to less than 1%. However, oral rehydration therapy requires sterile fluids, which are not always available in poor countries.

Giardiasis

Although not as widespread as cholera and not as deadly, other waterborne infections have emerged. One such disease is giardiasis, and its emergence is related in part to its relative resistance to standard water-treatment procedures.

Giardiasis is a severe diarrhea caused by an intestinal parasite called *Giardia intestinalis*. *Giardia* is an unusual organism, and is interesting from the standpoint of evolutionary biology. It is single-celled, yet has its DNA enclosed in a nucleus. In fact it is unique in having two nuclei, each containing a full genome. It also has a cytoskeleton, and the presence of the nucleus and cytoskeleton make it eucaryotic. However, it does not have any of the other membrane-bound organelles usually found in eucaryotic cells: no mitochondria, chloroplasts, endoplasmic reticulum or Golgi. It may therefore represent a very early stage in the evolution of eucaryotic organisms (Figure 17.8).

Humans infected with this parasite are unlikely to be fascinated by its evolutionary oddity. The diarrhea it causes is severe, and untreated will last for several weeks, leading to significant weight loss and dehydration. It lives and reproduces in the intestines of humans and other ani-

Figure 17.8

Giardia intestinalis, a single-celled parasite with two nuclei and a cytoskeleton, yet no other subcellular organelles.

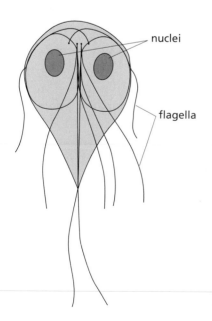

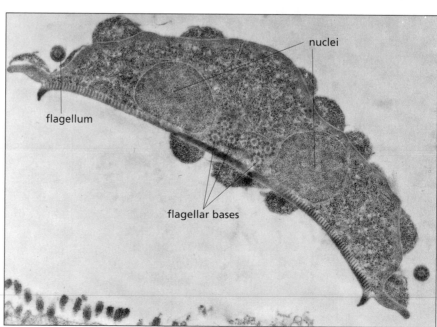

2 µm

mals, and is spread to water by fecal contamination. The parasite has an outer covering that enables it to stay alive for long periods in water, and hikers and others pick up the parasite by drinking water out of streams and rivers, throughout the United States and worldwide. It is, in fact, one of the most common waterborne diseases in the United States. Hikers filtering water will remove this parasite, but water treatment with iodine does not always kill it. *Giardia* also survives in swimming pools, hot tubs, and Jacuzzis as it is relatively resistant to disinfection by chlorination. Hot tubs are a particular problem because the water temperature causes rapid evaporation of the chlorine. Now that this has been recognized, separate standards have been set for chlorination of hot tubs to maintain chlorine at levels effective against most bacteria. But even at those levels *Giardia* may survive. The parasite can also be spread directly by feces. It can be spread by fecal contamination of food, either direct contamination, or contamination by washing food with contaminated water, which emphasizes the connection between waterborne and food-borne routes of disease transmission. Infected people should help to protect others by not swimming for several weeks after the symptoms have ended. Hand washing is very effective at preventing its spread.

Legionnaires' disease

In 1976, several members of the American Legion became seriously ill at a convention at a hotel in Philadelphia. Several people developed severe pneumonia and some of them died. Pneumonia is the accumulation of fluid in the lungs; it can be caused by many different bacterial species or viruses. It was quickly determined, however, that the Legionnaires were not ill with any known microbial source of pneumonia. The hotel closed and the search began for an explanation. The cause was found very rapidly, in one of the great success stories of public health epidemiology. The cause turned out to be a previously unknown species of bacteria that was given the name *Legionella pneumophila*. It was so different from any known bacteria that it was placed in its own family, the Legionellaceae.

We assume that the bacterium has been with us for a while, but probably not in high enough density to be a problem. A condition of modern life caused the emergence; namely that *Legionella* thrives in the water-cooling circulation of air-conditioning systems. It is not passed from one person to another, but is only transmitted by contaminated water. Showerheads and fine mist from whirlpool baths are additional sources. Most people who are exposed do not become sick, but those with a weakened immune system due to age, smoking, and use of alcohol are more susceptible. (Many of the Legionnaires at this convention were older individuals, and quite a few of them smoked or used alcohol, or both.) The pneumonia is fatal in about 10% of cases. Initial flu-like symptoms are followed by difficulty in breathing and also by mental changes including memory loss, disorientation or even hallucination. Since 1976 several outbreaks have occurred, including two in Britain in 2001 and 2002. Some outbreaks have occurred on cruise ships, underscoring the need for the disinfection of air-conditioning systems and for the location of ventilation-system air intakes at a distance from the cooling towers of the air-conditioning system.

Other waterborne diseases

The diseases discussed above are certainly not an exhaustive list of waterborne diseases. Hepatitis A, a viral disease of the liver, is a serious problem. Viruses can live in water, but the disease is also highly contagious and spreads from person to person. It is most common where crowding and poor sanitation facilitate transmission. As we saw, giardiasis is a waterborne parasitic disease, one of the few likely to be encountered by people in the United States and Europe. Cryptosporidiosis has also emerged in recent years. In fact it is one of the most common causes of waterborne disease in the United States; it is also found throughout the rest of the world. It is caused by a parasite that spreads when a water source is contaminated, usually with the feces of infected animals or humans. In 1993 in Milwaukee, Wisconsin, there was an incident in which 400,000 people were affected. In many parts of the world, other waterborne parasites remain the most serious infectious threats. These include *Schistosoma* flatworms (phylum Platyhelminthes), also called bloodflukes (see Figure 17.1). Infection with these flukes causes the disease schistosomiasis (also called bilharziasis), leading to chronic disability and liver damage. The disease has been recognized since the time of the pharaohs, but the worm that causes it was identified in 1851 by a German doctor named Theodore Bilharz. Two hundred million people are infected, making it the second most important tropical disease after malaria (see Chapter 7). This disease has been successfully controlled in Asia, the Americas, North Africa, and the Middle East, only to emerge in new areas. Areas that did not previously harbor the bloodflukes do now, largely as a result of large-scale water projects. Diama Dam on the Senegal River in East Africa led to endemic infections in Mauritania and Senegal, two countries in Africa where the disease was previously rare. Similarly, Lake Volta in Ghana, the largest human-built lake in Africa, has proved a fertile breeding ground for the worms, and villages along its shore have infection rates as high as 90%. Another important factor has been mass movements of refugees, which have recently spread the worms and the disease into Somalia and Djibouti.

CONNECTIONS CHAPTER 7

THOUGHT QUESTIONS

1 In what ways does disease transmission by food differ from transmission by water?

2 In what ways do food and water routes of transmission interact?

Ecological Factors Especially Affect Patterns of Vector-borne Diseases

Vector-borne diseases are those in which the pathogen is carried from one host to another by some other animal that is not itself infected. Often the vector is an insect (phylum Arthropoda, class Insecta), but other sorts of animals, such as snails and ticks, can also serve as vectors. The pathogen can be a parasite as we saw for malaria (see Figure 7.8, p. 228), but it can also be a bacterium or a virus.

An ancient example of a vector-borne bacterial disease is bubonic plague, the 'black death' that killed up to one-third of all people in certain European populations during the twelfth to fourteenth centuries. The disease is now known to be caused by the bacteria *Yersinia pestis*, which is transmitted primarily by fleas. The bacteria can be directly transmitted from one person to another, but only if the material from the open sores were to get inside a cut in the skin. The bacteria, like most bacteria, do not have the ability to penetrate the natural barrier of the skin. It is the flea bite that provides the mechanism for transmitting the bacteria past this barrier. Vector-borne diseases often have a secondary host, another species that can be infected by the pathogen and thus acts as a reservoir of the pathogens in nature. For plague, the secondary hosts are rats. When there is a secondary host, disease eradication is difficult because not only must the vector be controlled, but so also must the secondary host reservoir. Plague continues to exist and small outbreaks occur from time to time, but a combination of rat control, flea control, and treatment of the human patients with antibiotics ensures that outbreaks do not become epidemics and pandemics as they have in the past.

West Nile virus

A vector-borne disease currently receiving much publicity in the United States and Canada is West Nile virus. As the name implies, the pathogen is a virus and it causes a brain inflammation that can be fatal. During 2002 there were over 100 deaths in the United States from a virus that was unknown in North America before 1999. It was first identified in New York City and in 3 years' time has spread from the east to the west coast; cases have been confirmed in 32 states. The first human cases in Canada were in 2001. Thus, West Nile virus is considered an emerging disease in the United States and Canada, although it has existed in Africa for a long time. It is not known how it was carried to the United States, but it is believed to have been discovered early after its arrival by epidemiologists noting the unusual death of birds. Because it was detected early on, its spread has been well documented, and the speed of its spread across the country has surprised public health epidemiologists. It is spread by mosquitoes of the genus *Culex*, different from the *Anopheles* mosquito that spreads malaria (Chapter 7). West Nile virus cannot spread from person to person, either directly or by way of mosquitoes. The virus does not replicate to high numbers in people, so people are considered '**dead-end hosts**,' that is, they get sick but they do not contribute to disease transmission. Even in areas where the virus is present, most mosquitoes do not carry enough virus to make a person sick, and most people do not become ill even if mosquitoes do inoculate them with virus during a mosquito bite. Of those who do become ill, over 95% recover. Horses are another 'dead-end host' but appear to be much more susceptible, and 40% of infected horses die. Birds are the **primary host**, that is, the host in which viruses replicate and from which they are spread by mosquitoes to humans and horses. Over 110 species of birds are known to be hosts for the virus. The virus kills jays and crows, which was noted in New York City at the start of the outbreak, but most birds most show no sign of infection. As we saw in Chapter 16, viruses have a life cycle at the cellular level (see Figure 16.4, p. 584), but they may also have a much larger-scale life cycle in the environment (Figure 17.9). Controlling transmission thus depends on mosquito control. Individuals can protect themselves from infection by using insect repellent or by staying indoors at hours when mosquitoes are active (peak biting times are from dawn to dusk). Such measures will not prevent the spread of the virus in nature because humans are not primary hosts.

As this example illustrates, control of the transmission of vector-borne diseases depends largely on a knowledge of the ecology of the vectors and primary hosts. Factors that increase the population size of non-human host species will probably increase spread. Factors that separate humans from the primary host species or from the vector will control a disease outbreak, but will not eliminate the pathogen from the environment.

Figure 17.9

Life cycle of West Nile virus.

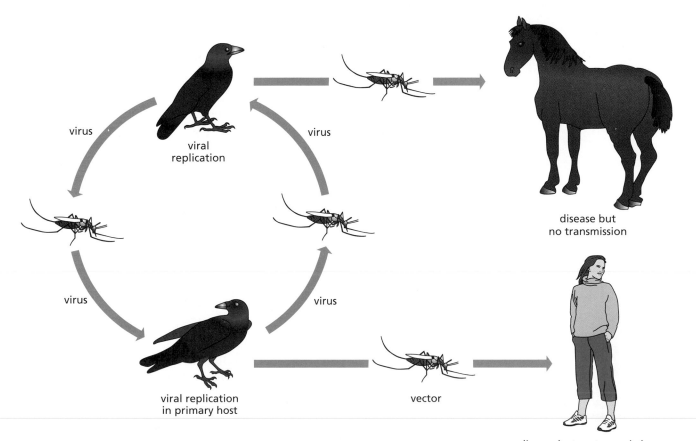

Leishmaniasis

Far more common and far more deadly worldwide is leishmaniasis, a disease caused by the protozoan parasite *Leishmania*. This parasite is spread by over 30 different species of phlebotomine sandfly. There are many hosts, in addition to humans, including rodents and dogs. The sandfly, by carrying blood from one person to another or from an animal to a person, can transmit the parasite. The disease itself takes different forms, one involving fevers, weight loss and anemia, the other resulting in disabling skin lesions.

There has been a significant expansion since 1993 of the regions in which the *Leishmania* parasites are endemic. This has led to a parallel increase in the number of human cases, with 1.5 million to 2 million new cases per year. One factor in this expansion has been large projects like dams and irrigation systems that have led previously unexposed people into new geographic areas to farm. The parasite was already endemic in these areas, and was thereby provided with a new pool of susceptible human hosts. This was followed by massive human migrations from rural to urban areas, spreading the parasite into new locales. Factors contributing to these migrations include deforestation and war. There is, for example, currently an epidemic of 200,000 cases in Kabul in Afghanistan. Leishmaniasis is now endemic in 88 countries in Africa, Asia, Europe, and North and South America, wherever the climate is warm enough to support the life cycle of the sandfly vector.

Another factor has been the incidence of co-infection with HIV. Most people who are bitten by sandflies carrying *Leishmania* do not acquire the parasite; however, when a person is also HIV-infected they are 100–1000 times more likely to pick up the parasite and to develop leishmaniasis. Likewise, if they are already infected with the parasite, the onset of AIDS after an HIV infection is much more rapid. *Leishmania* and HIV co-infection is the greatest problem in southwestern Europe (Spain, Portugal, Italy, and France), where over 70% of the co-infected people are intravenous drug users, so the parasite and HIV are being spread by blood in contaminated needles, rather than by the sandfly.

This vector-borne disease is thus a good illustration of the fact that biological factors, social factors and economic factors all interact in the spread of disease. All will need to be taken into account in trying to control disease.

THOUGHT QUESTIONS

1 What is the difference between the terms 'host' and 'vector'?

2 What factors influence the geographical distribution of vectors? Are these the same factors that influence the geographical distribution of hosts?

Concluding Remarks

Although this chapter has examined new infectious threats, we wish to emphasize that the epidemics and pandemics of today, although not to be ignored, are nothing like those of the past. Twenty million people died in an influenza epidemic in the early 1900s. Millions used to die every year of diseases we no longer even hear of. Highly effective vaccines exist for the prevention of many viral and bacterial diseases. Understanding of the routes of pathogen transmission has led to more effective measures for hygiene control. Worldwide surveillance and reporting of disease incidence has made it possible to control most outbreaks before they become epidemics. For most existing diseases, we know what to do to stop their spread. What remains is to extend access to clean food and water and to adequate nutrition to the areas of the world where these do not exist, the same areas where many diseases remain endemic. Control of pollution will also be important, because pollution weakens immunity, thereby increasing host susceptibility. We examine the topic of pollution further in Chapter 19.

Chapter Summary

- **Pathogens** are organisms that cause **disease**.

- To cause a disease, a pathogen must first cause an infection in a host; that is, it must adhere, colonize, and reproduce within the host.

- Organisms from many phyla can be pathogens. These include viruses, bacteria, protists, fungi, and animals, as well as self-replicating proteins (**prions**).

- The capability of a pathogen to cause disease is called its **virulence**.

- The **incidence** of a disease is the number of new cases in a given period.

- The **prevalence** of a disease is the total number of cases in a population.

- Pathogens spread through a population by various means, collectively called the **routes of transmission**. Transmission may be direct from person to person, or indirect via water, food, objects, or **vectors** (another species that can carry and transmit the infective organism).

- An **endemic** disease is one that persists locally, often at a low prevalence. An **epidemic** disease is of higher prevalence and in a larger area, while a **pandemic** is a global epidemic.

- Environmental, social, and economic factors that change the population size and overall health of the host population, the availability of vectors, and the frequency of transmission will cause the emergence and re-emergence of infectious diseases.

CONNECTIONS TO OTHER CHAPTERS

Chapter 3	Susceptibility of individual hosts varies genetically, as does virulence of pathogens.
Chapter 5	Virulence follows a long-term pattern in its evolution.
Chapter 7	Malaria has been a force of natural selection in human evolution.
Chapter 10	Poor nutrition increases susceptibility to infectious disease.
Chapter 15	Immunocompromised people are more susceptible to infectious disease.
Chapter 16	People with AIDS become susceptible to many other infections.
Chapter 19	Pollution and global warming can both contribute to the spread of disease.

PRACTICE QUESTIONS

1. What is the difference between the terms prevalence and incidence? Under what conditions will prevalence increase if incidence remains the same?

2. What is the difference between the incidence and the incidence rate of a disease? What additional information do you need to know to calculate an incidence rate that you do not need to know to make a table of disease incidence?

3. Why does the risk of antibiotic resistance rise if people do not finish taking all of their antibiotic pills?

4. What do the bacteria that cause chlamydia, tuberculosis and Legionnaires' disease have in common?

5. Could bubonic plague be used effectively in bioterrorism? Why, or why not?

Issues

- What is biodiversity? How is it measured?

- What do we lose if we lose biodiversity? Why should humans be concerned?

- How do humans contribute to loss of biodiversity or to habitat destruction?

- How do economic disparities among people influence habitat destruction?

- How do economic disparities among nations influence habitat destruction?

- How can societies limit the threats to habitats and to biodiversity?

Biological Concepts

- Biodiversity (species diversity, species richness)

- Scales of time (geological time, human experience)

- Extinction

- Specialization and adaptation

- Community structure (mutualisms and other interactions between species)

- Biogeography (biomes)

- Ecosystems (habitats, habitat alteration, human influences, biosphere)

- Conservation biology (renewable and nonrenewable resources, sustainable and nonsustainable uses)

18 Biodiversity and Threatened Habitats

*In the shadow of trees over 60 m tall (more than 200 feet, or as high as a 17-story building), workers use bulldozers and other heavy equipment to clear a 100-m-wide (312-foot-wide) path through a tropical forest. They are building a new road that will bring commerce and communications to the people of the region and will enable them to send their agricultural products, crafts, and minerals to markets in faraway countries. For each kilometer (0.62 mile) of roadway, they are destroying 10 hectares (about 24.7 acres) of tropical rainforest. The building of the road brings many high-paying jobs to the workers who build it, and the road will also open up new land for agriculture and human settlement, a process that will destroy even more forest. The trees are important in themselves, and also because they provide **habitat** (a set of environmental conditions that make up a place to live) to thousands of species.*

*The number and variety of species in a place are referred to as biological diversity or, simply, **biodiversity**. Biodiversity is measured most easily by the number of distinct species present. More broadly, biodiversity also includes genetic diversity within species and also ecological diversity within habitats or ecosystems. In this chapter we consider the importance of biodiversity, the conditions that support biodiversity, some of the threats to biodiversity, and some ways in which humans can reduce these threats.*

Recall from Chapter 5 that species are reproductively isolated groups of interbreeding natural populations. Most new species originate by a process of geographical speciation, in which reproductive isolation evolves during a period of geographic separation. The processes that give rise to new species increase biodiversity, while the processes that result in the extinction of species decrease biodiversity.

Biodiversity Results from Ecological and Evolutionary Processes

There are nearly 1,500,000 species of organisms currently known to science. More than half of these (53.1%) are insects, and another 17.6% (approximately 250,000 species) are vascular plants, so that over 70% of all known species are either vascular plants or insects (Figure 18.1). Animals other than insects make up 281,000 species, or about 19.9% of the total, and the remaining 9.4% are fungi, algae, protozoans, and various procaryotes (most of them microscopic).

Our knowledge is far from complete. Various estimates put the total number of species—known and not yet identified—at between 5 and 30 million. Such estimates are extrapolations from the few studies in which an effort has been made to identify every species in a given area. In one such study, between 4000 and 5000 species were found in a single gram of sand. In another study, 5000 species were found in a gram of forest soil, and these 5000 were almost completely different from the species found in the gram of sand. The more thoroughly we look, the more species we find. Some recent studies suggest that bacteria and other procaryotic organisms are particularly underrepresented and that many hundreds of thousands of undescribed procaryotic species await discovery.

Factors influencing the distribution of biodiversity

Despite our incomplete knowledge of the extent of biodiversity, we have begun to use what we do know to test hypotheses about the factors that contribute to its richness. Present levels of biodiversity are the result of many processes. Above all, the process of speciation increases biodiversity, and extinction decreases biodiversity. Because these are both evolutionary processes, the study of biodiversity depends on an understanding of evolution. To understand biodiversity, we also need to become familiar with the ecological concepts of communities, ecosystems, climate, and

Figure 18.1

Numbers of species currently known in the major groups of organisms.

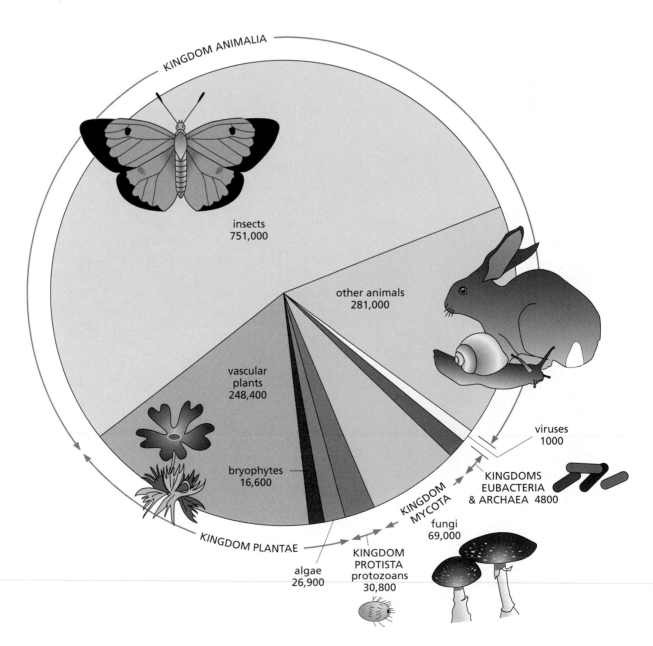

energy flow. These concepts will help us to explain patterns in the distribution of biodiversity across the Earth's surface and also how our own species interacts with the myriad other species that inhabit this planet. By looking at present-day biodiversity in different places, we can test certain hypotheses that give a better understanding of evolutionary and ecological processes that influence biodiversity.

Communities and ecosystems. Biological diversity implies more than just the number of species; also important are the numerous habitats and the diverse ways of life within each. In any given place, many species live together and interact with one another in an organized **community**. Each species also has its own **niche**, meaning it has a unique way of life and role in the life of the community. Each species (or its products) provides part of the niche of many other species in its community. An increase or decrease in the population size of any one of the species is therefore likely to have consequences for all the others. A community plus the physical environment surrounding it and interacting with it is called an **ecosystem**. In addition to living species, ecosystems include the soil, water, rocks, and atmosphere in which those species live. The largest ecosystem of all, that of the entire planet, is called the biosphere (see Chapter 19).

When a new species originates (see Chapter 5), it must find its niche and fit into the ecosystem in which it lives. Otherwise it fails to flourish in that ecosystem and quickly becomes extinct. If, however, a new species finds a niche that integrates it into the ecosystem, its population increases and the ecosystem becomes more complex owing to its presence. Biodiversity is thus a measure of ecosystem complexity.

Energy and biodiversity. One of the major determinants of biodiversity is latitude. Tropical ecosystems contain a much richer diversity of species and genera than temperate-zone ecosystems. The richest diversity on land is in tropical rainforests, while the richest marine ecosystems are those of warm-water coral reefs (Figure 18.2). Some large taxonomic groups, including many entire families and several orders, are confined to tropical ecosystems, and nearly every major group reaches its maximum diversity in the tropics. In marked contrast are the Arctic and Antarctic ecosystems, which are relatively sparse in biodiversity and in ecological complexity.

Why should this be true? One of the great theoretical problems in evolutionary biology is why species diversity is greater in the tropics. One possible explanation is provided by the 'energy–stability–area theory' of bio-

CONNECTIONS
CHAPTERS 5, 19

Figure 18.2

Coral reef diversity in the Red Sea, Egypt.

diversity, which begins with the observation that the species-rich tropics receive the greatest amounts of solar energy and also a more continuous level of solar energy. Each biological population requires a certain minimum amount of energy to maintain a population size capable of reproducing itself. Photosynthesizing plants capture the energy that they need directly from sunlight; most other species obtain their energy from food (see Chapter 11). For an equal amount of nutrients (an important proviso), greater quantities of biomass (mass of living things) grow in areas that are the hottest (receive the most solar energy) and the most humid (have a constant supply of water for photosynthesis). In addition, in tropical regions the climate varies little throughout the year and is also more stable over the centuries or across geological time. Within each unit of area there are many more niches than would be present in an equal-sized portion of the temperate zone, each niche differing slightly from the next in the amount and type of energy that is available. Tropical species can specialize to fill these different stable niches in many ways, living at different heights in a vertically stratified forest, occupying different kinds of microhabitats, exploiting different food resources, or attracting different species of pollinators.

Genetic diversity is another important aspect of biodiversity. If a species is to be resilient in its response to shifts in habitat, it must have the genetic resources that might allow it to display new phenotypes. In many species, however, shrinking populations have reduced the amount of genetic diversity. Humans have also reduced the genetic diversity of many domestic species of animals and plants; one instance of this is discussed.

Interdependence of humans and biodiversity

The study of biological diversity is as old as our need for food, clothing, shelter, and medicines, because all of these things, as well as tools and weapons, are made from the millions of other species that inhabit our planet. At the same time, humans have tilled farms, grown crops, and built cities and roads. All of these activities have altered many ecosystems, some of them profoundly, and have thus influenced biodiversity. Human activities affect biodiversity, and biodiversity affects our lives in return.

The biological value of preserving species. The preservation of biological diversity is important for many reasons, of which three broad types can be distinguished. First, our ignorance as to which species might be beneficial is a reason for simply preserving all species. We know that many plants have yielded important drugs; other plants have yielded important foods, dyestuffs, paper, and rubber. Among the species now living but poorly known, some probably possess a wealth of new possibilities for such uses; therefore we must preserve them all for the sake of those that may someday prove useful to humans. For example, cures for AIDS or cancer may lie hidden in the depths of the rainforest.

A second reason for preserving biodiversity pertains to the wild relatives of our domesticated species. The store of genetic variation, and therefore the possible number of genetic traits from which to choose, is greatly reduced in each of our domestic species, and is much greater in their wild relatives. It is therefore in our long-range best interests to preserve the wild relatives of all domestic species and varieties, so that newly discovered desirable properties (or properties that become desirable) can be bred into domestic stocks from these relatives.

For example, corn (*Zea mays*, also called maize) is one of the world's most valuable domesticated species of plants, but the domesticated variety is an annual plant that must be replanted each year at considerable labor and expense. In the 1970s, however, a wild relative named *Zea diploperennis* was discovered growing in the Mexican state of Jalisco, confined to a small mountain tract. The discovery was made just days before the land was scheduled to be cleared, which would have wiped the species out. *Z. diploperennis* was found to be resistant to a number of diseases that afflict domestic varieties. Best of all, unlike all other species and varieties of corn in the world, the newly discovered species grows as a perennial, meaning that an individual plant produces corn year after year without replanting. If some of these genetic traits could be introduced into domestic corn, either by breeding or by genetic engineering, the new strains could represent billions of dollars' worth of savings for the farmers of all corn-producing regions. Had the Jalisco corn not been discovered in time, an important genetic reserve for this important domestic species would have been lost forever. This is just one instance; similar arguments can be given for the preservation of the genetic resources of other species in zoos, botanical reserves, and gene banks, but the most cost-effective way to preserve these genetic resources is to promote the survival of the wild species or varieties in their natural habitats.

Preserving ecosystem stability. A third reason for preserving biological diversity is that species affect one another. *There are no ecosystems that are made up of only one or a few species.* Recall that a community is a group of species whose needs are interdependent. Stable communities are stable in part because materials are recycled: many producer, consumer, and decomposer species (see Chapter 11, pp. 368–369) are present. A small group of species is much less likely to form a complete and stable community than a larger one. Multiple species of each kind make the stability less likely to be disrupted. For example, multiple prey species provide a more stable food supply for predators because the predators can survive a scarcity of one prey species by switching to other species for their food.

Many communities are unstable in the sense that the removal of just one 'keystone' species can cause the balance among dozens of other species to collapse, so that the disappearance of one species causes other extinctions and leads to other drastic changes. Some of these changes may even affect the physical environment, as when the removal of beavers causes dams not to be built and allows water to flow more freely. Other communities, such as tropical rainforests, are thought to be more stable than this, a consequence of the large number and variety of species. In a typical rainforest there are hundreds of species of trees, with no single species constituting more than 5% or so of the total. There are also hundreds of bird species, thousands of insect species, and a large diversity of other animals and plants, some of them illustrated later in this chapter.

The health of animals, including humans, is promoted by the variety of plants available for them to eat or to climb or to nest in. Likewise, the health and well-being of many of the plants depends on the variety of animals that can pollinate them, disperse their seeds, or fertilize the ground near their roots with their feces and other remains. While it is obvious that the survival of any of these species depends on the survival of the ecosystem as a whole, it is equally true that the stability of the community (and often of the entire ecosystem) depends on the survival of certain key species or groups of species.

Many evolutionary and ecological processes increase biodiversity. Life as we know it depends on the rich biodiversity of the world's major ecosystems, yet we are currently in an era when biodiversity is rapidly declining. What forces are producing this decline? Will the current decline in biodiversity match those of the great mass extinctions of the past? These are some of the issues addressed in the next section.

THOUGHT QUESTIONS

1 Why would the perennial growth habit in the corn from Jalisco be considered a valuable trait? Under what agricultural conditions (and in what nations) would this trait be especially valuable?

2 How easily could the genes for perennial growth be identified? How easily could these genes be introduced from the Jalisco corn into the domestic varieties? (You will probably need to review parts of Chapters 2, 4, and 11 to answer this question.)

3 What would be the effects of clearing 100 square kilometers (100 km^2) from a small rainforest 5000 km^2 in size (about 2000 square miles, approximately the size of Delaware)? What would be the effect of clearing 1000 km^2 of this same forest?

Extinction Reduces Biodiversity

Evolutionary change often produces new species and thus increases biodiversity. But evolution can also lead to the disappearance of species, a phenomenon called **extinction**. The species that we see around us today are in fact only a small proportion of all those that have ever lived.

Species that have no living members are said to be extinct. Extinct species are known through the fossil record. As an example, consider the Age of Reptiles, or Mesozoic era, a time interval from approximately 200 to 65 million years ago (see Figure 5.8, p. 137). Of all the species that lived during that time, none are still alive today—they are all considered extinct. These species did not die out all at once but rather a few at a time, although many perished in a mass extinction at the end of the Cretaceous period. The reasons behind these extinctions differed from case

CONNECTIONS
CHAPTER 5

to case and are imperfectly known for most species. In some cases the fossil record shows that a competing group of species appeared on the scene shortly before the extinction occurred. In other cases no specific cause can be identified.

Types of extinction

If all Mesozoic species are now extinct, how did life manage to persist? To answer this question, we must distinguish between two major types of extinction. First we must recognize the concept of a lineage, which is an unbroken series of species arranged in ancestor-to-descendant sequence, with each later species having evolved from the one that immediately preceded it. If we had a complete record of the history of life on this planet, every lineage would extend back in time to the common origin of all earthly life. Working forward from earlier times, each lineage extends either to a species alive today or to one that has become extinct.

When an entire lineage has died out without issue, no living descendant species exist. We call this true extinction. Many groups of organisms have, according to current theories, undergone true extinction, meaning that no living species are descended from them—their lineages have ended. Among these groups are the trilobites and conodonts shown in Figure 18.3.

When a species no longer exists, its lineage may continue in the form of descendant species. The ancestral horse, *Hyracotherium* (formerly called *Eohippus*), is extinct in one sense: there are none alive today, but they have living descendants, the modern horses. This type of change, called pseudoextinction or phyletic transformation, occurs when a species evolves into something recognizable as a different species. Many of the traits of *Hyracotherium*, and the genes that contributed to these traits, persist among modern horses. In some cases we do not know whether an extinct group has undergone true extinction or only pseudoextinction—the evidence does not permit us to make a clear choice. For example, the dinosaurs are no longer alive, but if living birds are their descendants, as many scientists believe, then the dinosaurs are only pseudoextinct.

Analyzing patterns of extinction

Has extinction occurred randomly? Does the probability of extinction remain constant over time and from place to place? Some late twentieth-century biologists have hypothesized that extinction occurs at random over vast time periods. This hypothesis has been tested in a mathematical model that compares actual extinctions with theoretical predictions based on a model of random extinctions. Comparison of actual data with a theoretical model is a good research strategy (see Chapter 1) because it allows us to identify both circumstances for which the model holds and circumstances for which it does not (and for which additional explanations are therefore needed).

Several studies have compared extinction in the fossil record of particular animal groups with the random model. Most of these comparisons counted all extinctions together, without distinguishing true extinction from pseudoextinction. The fossil record of mammals from the Tertiary period is one of several such groups that conforms to the random extinction model; that is, the actual extinctions match the rates predicted by the model. Many instances of nonrandom extinction have also been discovered. For example, many early invertebrate groups suffered most of their extinction early in their history rather than at a constant rate through time.

Comparison with mathematical models has revealed two major types of departure from randomness: situations in which fewer species become extinct than random models predict, and situations in which more species become extinct.

Living fossils. We first examine situations in which the frequency of extinctions is reduced. 'Living fossils' are species or genera that have survived for many millions of years without true extinction and with only minimal pseudoextinction, meaning that very little morphological change separates the living species from their fossil relatives. Several of these living fossils are described below (the time periods mentioned are shown in Figure 5.8, p. 137):

Figure 18.3

Extinct species of the Paleozoic era. The fossils shown here represent groups that were abundant during Paleozoic times. All these groups suffered considerable extinction at the end of the Permian period. The trilobites (belonging to the phylum Arthropoda) and conodonts (jaw structures belonging to the phylum Chordata) are truly extinct groups with no living descendants. The other groups have some living species, but the species alive today are not the same as the Paleozoic species.

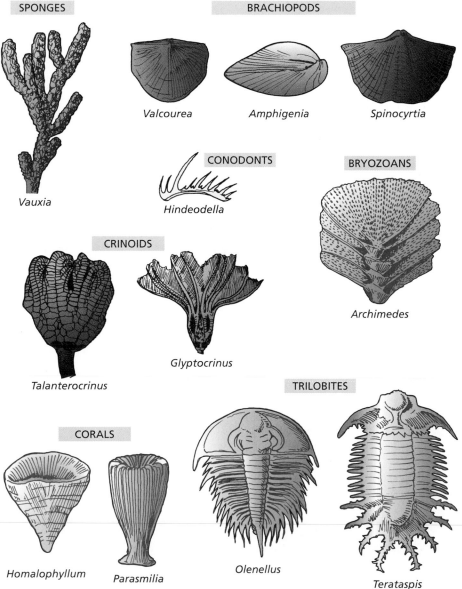

SPONGES

BRACHIOPODS

Valcourea   *Amphigenia*   *Spinocyrtia*

Vauxia

CONODONTS

Hindeodella

BRYOZOANS

CRINOIDS

Archimedes

Talanterocrinus   *Glyptocrinus*

TRILOBITES

CORALS

Homalophyllum   *Parasmilia*   *Olenellus*   *Terataspis*

Psilophyton, a primitive vascular plant (kingdom Plantae, phylum Psilophyta) closely resembling the earliest land plants of the Silurian period.

Ginkgo, a tree (kingdom Plantae, phylum Ginkgophyta) native to China, which closely resembles its Mesozoic ancestors and is planted in many urban areas around the world because it tolerates urban pollution (Figure 18.4).

Lingula, a type of brachiopod (kingdom Animalia, phylum Brachiopoda) with a wormlike body enclosed in a two-valved shell that has no hinge and with a feeding structure that strains suspended particles from the water (Figure 18.4).

Neopilina, a deep-water mollusk (kingdom Animalia, phylum Mollusca) with a low-domed conical shell resembling that of the extinct genus *Pilina*, one of the most primitive mollusks.

Limulus, the horseshoe crab (kingdom Animalia, phylum Arthropoda), which closely resembles its Paleozoic ancestors (Figure 18.4).

Latimeria, a large, rare Indian Ocean fish (kingdom Animalia, phylum Chordata, class Osteichthyes, order Crossopterygii) belonging to a group (the coelacanths) whose other members became extinct during the Mesozoic era (Figure 18.4).

What might make a taxon of organisms less likely to suffer extinction? These living fossils share several characteristics: they all have locally large populations (with a sufficient gene pool to maintain a large amount of genetic variation); they are all adapted to dependably persistent habitats (such as deep ocean waters); those that are animals do not depend on a narrow range of food species (some of them will eat anything within a certain size range); and they all have reproductive stages (pollen, spores, or larvae) that are dispersed mechanically by wind or ocean currents rather than by other species. If there is any secret to long-term survival, these species have surely stumbled upon it.

Mass extinctions. Departing in the other direction from the random extinction model are examples in which many more species became extinct within a geologically short interval of time than would

Figure 18.4

Living fossils that have avoided extinction over long periods of time.

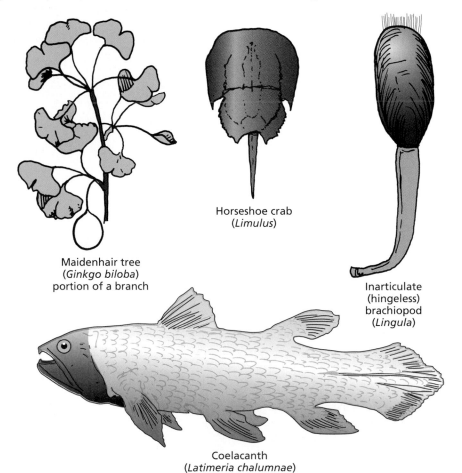

Maidenhair tree
(*Ginkgo biloba*)
portion of a branch

Horseshoe crab
(*Limulus*)

Inarticulate
(hingeless)
brachiopod
(*Lingula*)

Coelacanth
(*Latimeria chalumnae*)

have been predicted by the model. We call these mass extinctions. There was one such event at the end of the Cretaceous period, which was also the end of the Mesozoic era. There was another, even larger, mass extinction at the end of the Permian period (see Figure 18.6).

Mass extinctions devastate biodiversity. Over half of the families and 85% of the genera became extinct in the last 5 million years of the Permian period, and much higher percentages were lost in some classes and phyla. Although many more species and genera perished in this mass extinction than in any other, the Permian event has attracted much less attention than other mass extinctions because nearly all the species were unfamiliar types of organisms (for example, crinoids and brachiopods; see Figure 18.3) that lived in underwater habitats such as the shallow inland seas that were abundant at that time.

The mass extinction at the end of the Cretaceous period has attracted the most attention because many well-known animals became extinct at this time: dinosaurs (the reptilian orders Saurischia and Ornithischia), flying reptiles (order Pterosauria), several types of marine reptiles (orders Sauropterygia, Ichthyopterygia, and others), ammonoids (phylum Mollusca, class Cephalopoda, mostly with large, coiled shells), and several groups of fish, plants, and other organisms (Figure 18.5).

Possible causes of mass extinctions. The fossil record shows at least five mass extinctions in which many families of marine organisms died

Figure 18.5

Extinct species belonging to groups that died out completely at the end of the Cretaceous period.

ammonoids

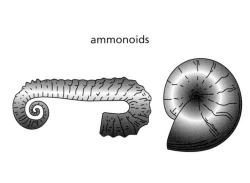

giant ammonoid

artist's view of various Mesozoic reptiles

out (Figure 18.6). The rates of extinction happening today are as great as the rates during these mass extinctions. Many scientists have therefore concluded that a sixth great mass extinction is currently in progress.

What could cause such high rates of extinction? There are several hypotheses including: warming or cooling of the Earth, changes in seasonal fluctuations or ocean currents, and changing positions of the continents (plate tectonics). Biological hypotheses include ecological changes brought about by the evolution of cooperation between insects and flowering plants or of bottom-feeding predators in the oceans. Some of the proposed mechanisms require a very brief period during which all extinctions suddenly took place; other mechanisms would be more likely to have taken place more gradually, over an extended period, or at different times on different continents. Some hypotheses fail to account for simultaneous extinctions on land and in the seas. Each mass extinction may have had a different cause. Evidence points to hunting by humans and habitat destruction as the likely causes for the current mass extinction.

American paleontologists David Raup and John Sepkoski, who have studied extinction rates in a number of fossil groups, suggest that episodes of increased extinction have recurred periodically, approximately every 26 million years since the mid-Cretaceous period. The late Cretaceous extinction of the dinosaurs and ammonoids was just one of the more drastic in a whole series of such recurrent extinction episodes. The possibility that mass extinctions may recur periodically has given rise to such hypotheses as that of a companion star with a long-period orbit deflecting other bodies from their normal orbits, causing some of them to fall to Earth as meteors, wreaking widespread devastation upon impact.

The asteroid impact hypothesis. Of the various hypotheses attempting to account for the late Cretaceous extinctions, the one that has attracted

Figure 18.6

Changes through time in the number of families of marine organisms. Mass extinctions are indicated by numbered triangles.

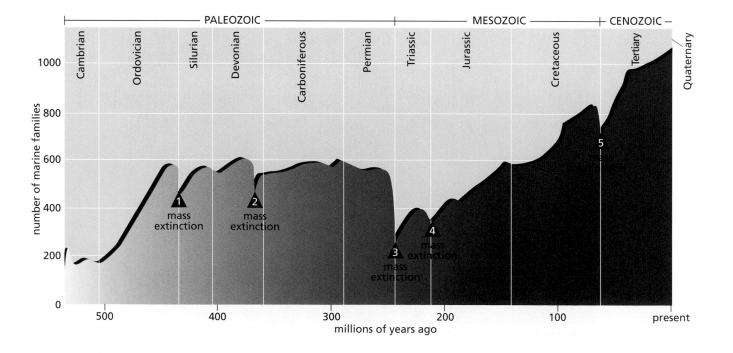

the most attention in recent years is the asteroid impact hypothesis first suggested by Luis and Walter Alvarez. According to this hypothesis, the Earth collided with an asteroid with an estimated diameter of 10 km, or with several asteroids, the combined mass of which was comparable. The force of collision spewed large amounts of debris into the atmosphere, darkening the skies for several years before the finer particles settled. The reduced level of photosynthesis led to a massive decline in plant life of all kinds, and this caused massive starvation first of herbivores and subsequently of carnivores. The mass extinction would have occurred very suddenly under this hypothesis.

One interesting test of the Alvarez hypothesis is based on the presence of the rare-earth element iridium (Ir). The Earth's crust contains very little of this element, but most asteroids contain a lot more. Debris thrown into the atmosphere by an asteroid collision would presumably contain large amounts of iridium, and atmospheric currents would carry this material all over the globe. A search of sedimentary deposits that span the boundary between the Cretaceous and Tertiary periods shows that there is a dramatic increase in the abundance of iridium briefly and precisely at this boundary (visible in Figure 1.4, bottom row, middle panel, p. 17). This iridium anomaly offers strong support for the Alvarez hypothesis. However, no asteroid itself has ever been recovered.

CONNECTIONS CHAPTER 1

An asteroid of this size would be expected to leave an immense crater, even if the asteroid itself was disintegrated by the impact. The intense heat of the impact would produce heat-shocked quartz in many types of rock. Also, large blocks thrown aside by the impact would form secondary craters surrounding the main crater. To date, several such secondary craters have been found along Mexico's Yucatan peninsula, and heat-shocked quartz has been found both in Mexico and in Haiti. A location called Chicxulub, along the Yucatan coast, has been suggested as the primary impact site.

Quaternary extinctions. There were many extinctions during the past 2 million years, an interval that includes the Pleistocene epoch and its several glacial episodes, plus the Recent epoch (beginning with the end of the last ice age, about 10,000 years ago). A number of species of large mammals, reptiles, and flightless birds became extinct during these 2 million years, and several hypotheses have been advanced as explanations. The most obvious hypothesis attributes the extinctions to changes in climate and the advance and retreat of Pleistocene glaciers. But paleontologists who have carefully examined the fossil record point out that this hypothesis fails to explain the timing of the extinctions, most of which did not coincide with the extremes of temperature.

A second hypothesis is that newly introduced species brought about extinction through increased competition. Widespread glaciation caused a decline in sea levels, which resulted in the emergence of land bridges, including those across Panama, the Bering Strait, and the English Channel. Many species were thus introduced from one landmass to another, and the newly introduced species competed with other species already present and established new predator–prey interactions. Many species could not adjust to the new conditions and became extinct as a result. This hypothesis has been used to explain many of the animal extinctions that followed the emergence of the land bridge across Panama, connecting South America (previously an island continent) with North America. Most of this group of extinctions, however, happened early in the Pleistocene, leaving another large group of later extinctions still to be explained. We examine these most recent extinctions next.

The human role in extinctions. From the comparison of species that became extinct in the past 50,000 years with others that did not become extinct, an interesting pattern emerges: only large, conspicuous species of mammals, reptiles, and flightless birds became extinct, while smaller animals (including rodents, bats, and small birds) or marine animals suffered very little extinction or else none at all. This pattern suggests yet another hypothesis: that the activities of humans, including both hunting and alterations of habitat, had a large role in the extinctions of the past 50,000 years.

A great deal of circumstantial evidence favors the hypothesis of extinction by human agency: in those places where the time of first human arrival can be dated (e.g., Madagascar, New Zealand, and certain Pacific islands), dense piles of animal bones accumulated beginning at the times of human arrival, and most of the extinctions took place soon afterwards, within several hundred years of the arrival of humans at each place (Figure 18.7). Moreover, the species that became extinct were those that humans would be apt to hunt, mostly large herbivores, whereas most species that would have been difficult to hunt, or too small to be worth hunting, survived. On the island continent of Australia, for example, the arrival of humans some 50,000 to 60,000 years ago was soon followed by the extinction of 20 species of giant kangaroos, along with a marsupial lion, a marsupial wolf or tiger (*Thylacinus*, surviving into the 1800s on Tasmania), and the giant, cow-sized herbivore *Diprotodon*. On New Zealand and the Hawaiian Islands, the extinction of flightless birds began with human arrival and was nearly complete by the time of European discovery.

Several vertebrate species have become extinct within historic time: the dodo (a large, flightless bird) and the passenger pigeon are two famous examples. Many other species of birds and also many plants and insects also became extinct. Hawaii was home to some 50 species of land birds at the time of its European discovery in 1778. Humans and the animals that they have introduced since that time have caused the extinction of one-third of these species, and archaeologists have shown that the indigenous Hawaiians had hunted an additional 35–55 species to extinction by before the arrival of Captain Cook. On New Zealand, the giant moa (an ostrich-like flightless bird) was hunted to extinction by the Maoris, the indigenous people of that island nation. Much the same thing happened on Madagascar, where an even larger flightless bird, the elephant bird, had become extinct long before European colonists arrived.

Most recently, and on a worldwide scale, it is now estimated that more than 100,000 plant and animal species became extinct during the decade of the 1980s. Nearly all of these species were on land and were scattered among many families, so they do not show on the graph of Figure 18.6, which counts only marine families. In the United States, many species of migratory songbirds have been greatly reduced in numbers and several have become extinct. Bachman's warbler, a bird species that was once common throughout the southeastern United States, was last seen in the 1950s. The Ohio River and Lake Erie once had dense populations of 78 species of freshwater mollusks, of which 19 are now extinct and another 29 are rare. Biologists now believe that a single dam, Wilson Dam, on the Tennessee River, caused the extinction of 44 species of freshwater mollusks when it was built. In Africa's Lake Victoria, over 100 species of fishes became extinct after the introduction of the Nile perch, a large, predatory sport fish introduced to the lake in 1959.

Species threatened with extinction today

A species threatened with extinction is called an **endangered species**. An example is the northern spotted owl (*Strix occidentalis*) of California, Oregon, and Washington. These owls nest only in old pine trees in 'old growth' forests, meaning forests with many species of trees of mixed ages and sizes, including very old trees (up to 200 years old or more) and also dead wood, both standing and fallen. A large area of 'old growth' forest is needed to support enough rodents, small birds, and other wildlife for each individual spotted owl to find enough food. The species is currently endangered because of logging of the Pacific coastal forests for timber.

Various governments and international organizations maintain official lists of endangered species. The United States, for example, maintains a list of endangered species in the United States as part of the Endangered Species Act. International lists are maintained by such organizations as the International Union for Conservation of Nature and Natural Resources. These lists differ from one another because they are based on different criteria.

Predictors of extinction. How do we know whether a species is endangered? One indication that a species may be endangered is a reduction in its numbers. The extinction of a species is nearly always preceded by its becoming rare, and rarity may thus be the prelude to extinction. However, the only type of rarity qualifying as the harbinger of extinction, and therefore qualifying a species as endangered, is the kind in which the entire species is represented by only

Figure 18.7

The extinction of many large mammals and flightless birds followed soon after human arrival in many parts of the world. In Africa, however, extinctions took place more gradually because humans had been present for a much longer period.

gradual declines with humans present for whole period

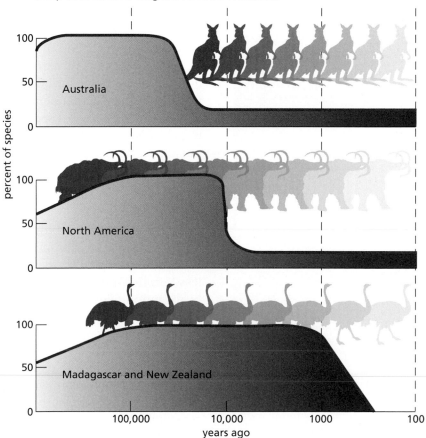

sharp declines coinciding with arrivals of humans

one or a few populations, all of which are small. Once the population size falls below a certain minimum, several factors can increase the risk of extinction. One of these is genetic drift (see Chapter 7, pp. 220–221), a pattern in which allele frequencies in small populations change erratically and not necessarily adaptively, which may hasten extinction. A second factor is that in very small populations there are more matings between related individuals than in large populations, and this inbreeding makes homozygous recessive traits more frequent in these populations than in non-inbred populations. Because many homozygous recessive traits are harmful, they reduce survival and thus cause the population to decline further and possibly to die out. A third factor is that environmental fluctuations (resulting from changing seasons, weather phenomena, and so forth) may favor different genotypes at different times. Large populations that contain many different genotypes are thus more likely to survive, but populations with less genetic diversity are more susceptible to extinction.

Extinction of a niche. Other indications that a species is threatened with extinction are the disappearance (or impending disappearance) of its habitat, or the disappearance of another species on which its niche depends. A remarkable example of the latter is the tambalacoque tree (*Sideroxylon* or *Calvaria grandiflorum*) of the island of Mauritius in the Indian Ocean. None of the seeds of these long-lived trees were ever observed to germinate, even when planted. A botanist who studied these seeds noticed that they have a very hard outer husk that mechanically prevents the seed within from breaking through. The seeds could be made to germinate by grinding away some of the outer husk before planting the seeds. The same effect was obtained by feeding these seeds to turkeys, a bird about the size of the dodo. The dodo was a type of bird that lived on the island of Mauritius before Europeans and their domestic animals caused its extinction around 1681. We now believe that the dodos fed on the seeds of the tambalacoque tree, and the hard outer husk was an adaptation that permitted the seeds to survive the digestive action of the dodo's stomach. (Seed-eating birds often intentionally swallow and retain pebbles; these pebbles work like pulverizing machines in their muscular stomachs to abrade such things as tough seeds.) Once the dodos became extinct, the tambalacoque trees were unable to reproduce. Young tambalacoque trees are now growing with the aid of humans, saving the species from the threat of certain extinction.

Species currently in danger. The list of endangered species is long and growing. Among mammals, the list includes giant pandas, gorillas, orangutans, elephants, manatees, caribou, timber wolves, and dozens of less familiar species. There are also a large number of endangered fishes, birds, amphibians, reptiles, insects, and plants. The International Council for Bird Preservation estimates that close to 2000 species of birds have already become extinct in the past 2000 years, mostly driven by human agency, and that 11% of the living species (1029 of 9040) are endangered.

A few species that were once listed as endangered have, for the moment, been saved from the brink of extinction, but only when a concerted effort has been made to conserve the species and its habitat. For example, bald eagles (*Haliaeetus leucocephalus*) have been protected and in some cases released into the wild from captivity. Because their numbers are once again sufficient to maintain stable populations, they are no longer considered an endangered species.

Some species, of course, are endangered because of indiscriminate hunting, fishing, or poaching. These are mostly large and conspicuous organisms. However, a much larger threat to biodiversity lies in the destruction of natural habitats, a process that threatens thousands of species at once, including those that are inconspicuous and poorly known. Among these small and inconspicuous organisms, the number of endangered species is certainly many more than those on any official list, and many of these inconspicuous organisms will undoubtedly become extinct before they have even been discovered and named.

The whole strategy of saving individual species is being rethought. In the decades since the first legislation on endangered species, scientists have come to realize that individual species cannot be saved without saving their habitats. The focus is therefore shifting to habitat preservation.

THOUGHT QUESTIONS

1 Experts still disagree on whether the great extinctions of the past occurred suddenly or gradually. If a new article claimed new evidence for either of these hypotheses, how would you go about evaluating this evidence?

2 The presence or absence of dinosaur fossils is used to determine whether a certain bed of rock is Cretaceous or Tertiary. How would this practice influence research on the question of whether dinosaur extinction came about gradually (at different times in different places) or suddenly (simultaneously everywhere)?

3 In 1995, the U.S. Department of the Interior downgraded the legal status of bald eagles (*Haliaeetus leucocephalus*) from 'endangered' to 'threatened,' and in 1999 they were removed from the endangered species list altogether. Why is the U.S. Government involved in this matter? Bald eagles also live in Canada and part of northern Mexico. Does the U.S. Government have any authority to declare the bald eagle 'endangered' or 'threatened' in those places?

Some Entire Habitats Are Threatened

Among the most serious threats that any species faces is the destruction of its habitat. Of all the endangered species in the world, an estimated 73% are endangered because of the destruction of their habitats. Habitat destruction threatens many species at once, often thousands at a time. Norman Myers, a British ecologist, has identified 18 areas of the world as 'hot spots' (Figure 18.8) where threatened habitats put many thousands of species at the risk of extinction simultaneously. One reason that so many species are simultaneously threatened is that these places contain many species that are found nowhere else.

Habitats do not exist independently but are associated into ecosystems. Groups of ecosystems that are similar in type are called **biomes**.

Each biome contains ecosystems in different parts of the world that are similar in climate and contain similar habitats and species assortments. Notice that in Figure 18.9 most biomes are restricted to certain bands of latitude, sometimes in both the Northern and Southern Hemispheres. Individual species differ from place to place, but a particular biome has certain proportions of ecological types, such as large herbivores and tall grasses on tropical grassland (savanna), or large stands of pine and other evergreen conifers on taiga. Temperate shrubland (also called chaparral or Mediterranean) has mild, rainy winters and supports many shrubs and small trees, such as olives. Some other examples of biomes are desert, rainforest, and tundra (a cold, treeless land with low-growing vegetation only). Marine biomes also exist, but they are less clearly defined.

What impact does habitat destruction have on human lives? How should humans value habitat? In this section we examine two specific types of habitat destruction: the destruction of tropical rainforests and the expansion of deserts.

The tropical rainforest biome

Tropical rainforests are a biome that encompasses habitats of low latitude (from about 15° S to 25° N), continually warm temperatures, and year-round high precipitation. Average temperatures are usually about 25 °C, typically fluctuating only from about 22 to 27 °C with seasonal extremes no lower than 20 °C. Precipitation is high throughout the year, with annual totals of 1800 mm (about 71 inches) or more. Although there is seasonal variation, no month averages less than 60 mm (2.36 inches) of rainfall. Under these conditions, humidity stays moderate to high at all times.

Figure 18.8

'Hot spots': habitats containing many species found nowhere else and threatened with extinction from human activity.

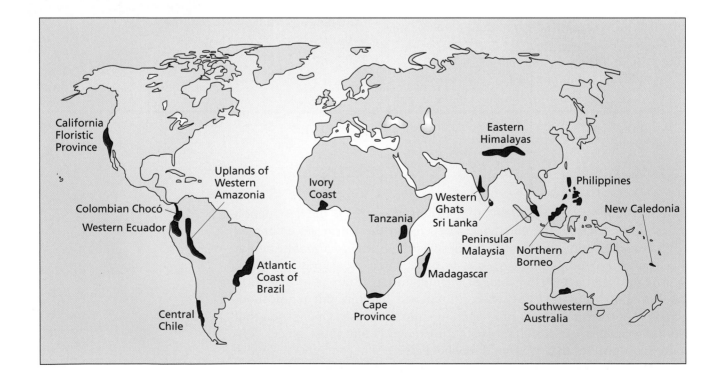

The most conspicuous rainforest vegetation consists of tall trees, 30–60 m (about 100–200 feet) in height, up to the height of a 17-story building. As Figure 18.10 shows, their leafy tops make a continuous canopy through which tree-living animals can roam widely without ever descending to the ground. The tallest of the trees may protrude above this canopy level. The taller trees are often buttressed at their bases with a variety of woody supports that give their base a fluted rather than cylindrical shape (see Figure 18.10A and C). Much of the rainforest receives little direct sunlight below the canopy, and most of the plants here are shade tolerant. Many plants adapted to the understory have huge, dark green leaves that capture a maximum amount of light, and some are now used as indoor office and lobby decorations precisely because they don't need much sunlight.

Tropical rainforests are vital to the biosphere because almost half of all the photosynthesis that plants perform on our planet takes place in tropical rainforests. This photosynthesis is the principal safeguard against a global increase in carbon dioxide (as well as the principal source of atmospheric oxygen). Unfortunately, human activities are already producing carbon dioxide at a rate so fast that global photosynthesis rates cannot keep up. Scientists believe that the increasing levels of atmospheric carbon dioxide are leading to global warming (see Chapter 19). Tropical

Figure 18.9

The world's terrestrial biomes.

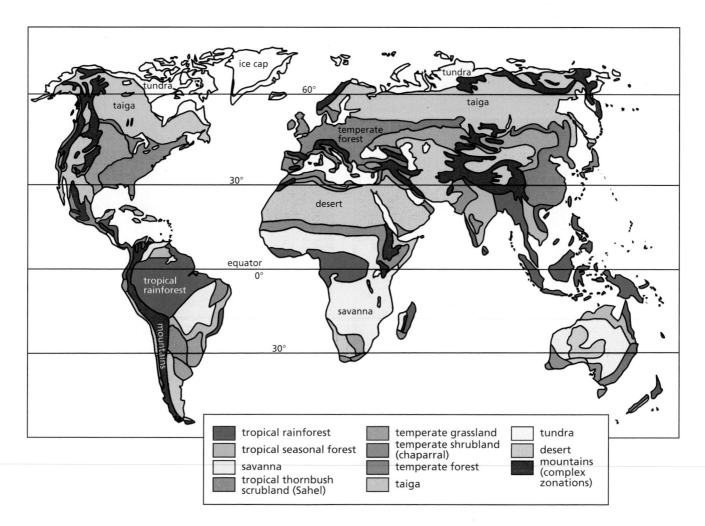

■ tropical rainforest	■ temperate grassland	□ tundra
■ tropical seasonal forest	■ temperate shrubland (chaparral)	■ desert
□ savanna	■ temperate forest	■ mountains (complex zonations)
■ tropical thornbush scrubland (Sahel)	■ taiga	

Figure 18.10

Rainforests.

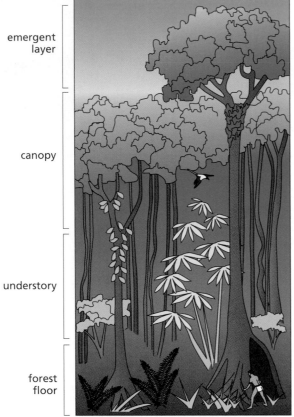

emergent
layer

canopy

understory

forest
floor

(A) Profile view of a rainforest, showing
vertical stratification.

(C) Trees in tropical rainforests often
have buttress supports at the base
of their trunks.

(B) The interior of a Costa Rican rainforest. Notice the marked
differences in light levels from the canopy above to the
understory below.

rainforests are thus an important global resource whose continued health benefits the entire planet, not just the countries in which the rainforests are found.

Rainforest biodiversity and ecological diversity. As we have seen, the tropics contain many more species than do temperate and polar regions. Habitat destruction in the tropics therefore affects many more species than it does at higher latitudes. Tropical rainforests are particularly vulnerable because they contain so many different habitats, and they are being destroyed at an accelerating rate. American ecologist E.O. Wilson estimates that one Peruvian farmer clearing some rainforest land to grow food for his family will cut down more kinds of trees than are native to all of Europe.

Rainforests have a high diversity of both plant and animal species, with no single species forming more than a small fraction of this rich diversity. In addition to the many hundreds of species of trees and other plants, rainforests are home to several hundred kinds of birds, numerous reptiles, amphibians, fishes, and mammals, and thousands of insect species (Figure 18.11). Some biologists have estimated that more than half of the world's species live in the rainforests and nowhere else. For example, Costa Rica, a nation about the size of West Virginia, has 820 species of birds, more than the whole of the United States and Canada combined!

The variety of rainforest trees makes for a variety of habitats for other plants, such as orchids, that use these trees for support; plants that use other plants for support are called **epiphytes** (Figure 18.12). There are also many different habitats for other plant and animal species that have adapted to life on the forest floor, in the canopy, in the understory, or in the well-lit clearings created by river banks, landslides, or fallen trees. Numerous decomposer species, including ants, termites, fungi, and bacteria, favor the fast decomposition and recycling of nutrients. The great diversity of small-scale habitats within a rainforest contributes to the maintenance of biodiversity among species.

Complex interactions among species. One result of this great diversity in tropical rainforests is the large number of ecological interactions among the many species present, including predation, competition, and **mutualism** (see Chapter 10, p. 347). Most tropical plants have elaborate mechanisms to ensure seed dispersal by animals (dispersal by wind is much less effective amidst tall trees). For example, shrubs of the genus *Piper* (see Figure 18.11) have fruits that are eaten by bats, which disperse the seeds in their feces. More elaborate are the many mutualisms between ants (kingdom Animalia, phylum Arthropoda, class Insecta, order Hymenoptera) and the plants that they inhabit. Many plants provide their resident ant populations with food and with places to live. In the case of the bull's horn acacia (*Acacia cornigera*, kingdom Plantae, phylum Anthophyta, class Dicotyledonae), ants of the genus *Pseudomyrmex* live in the base of swollen thorns, and the plants provide them with sugary food from nectaries and protein-rich globules at the tips of the leaflets (Figure 18.13). In return, the ants vigorously defend the plants against other insects, and swarm to bite any large herbivore that would feed on the plant more destructively. Acacia trees of this species

CONNECTIONS
CHAPTER 10

Figure 18.11

Some of the biodiversity of tropical rainforests.

Guzmania nicaraguensis, showing bright yellow flowers surrounded by modified leaves called bracts, whose bright red color attracts the hummingbirds that pollinate this species. Bright red colors are common in bird-pollinated plants.

Spathophyllum, showing leaves damaged by the feeding activities of herbivores.

Piper, a tree whose inflorescences reach upward.

A tarantula from Costa Rica.

A walking stick insect, camouflaged to resemble the lichens that grow on many tree trunks.

A butterfly feeding on a colorful flower.

Bronze beetle.

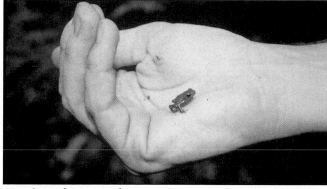
Strawberry frog, one of many poison arrow frogs (*Dendrobates*) whose skin secretes distasteful chemicals that deter predators. Many of these skin secretions are used as dart poisons or arrow poisons by Native Americans because they contain curare, a poison that can paralyze muscles by inhibiting acetylcholinesterase.

Emerald boa, one of many tropical snakes.

White-throated toucan, a large bird of the parrot family.

A tree sloth and her young, who spend most of each day hanging as you see here.

Jaguar.

whose resident ant populations have been experimentally removed are soon eaten by goats, deer, or other mammalian herbivores.

As an example of an even more complex interaction, consider the figs of the genus *Ficus* (kingdom Plantae, phylum Anthophyta, class Dicotyledonae), an extremely successful group of tropical shrubs and trees that often grow to great heights. The edible part of a fig is an aggregate fruit called a receptacle; this fruit is unusual in that several dozen flowers are contained within it (Figure 18.14). Many of these flowers are home to the tiny fig wasps of the genus *Blastophaga* (phylum Arthropoda, class Insecta, order Hymenoptera) that pollinate the plants.

Figs produced in the cooler months bear winter receptacles containing mostly sterile flowers but a few fertile male flowers. A female wasp lays her fertilized eggs in the winter receptacle before she dies. The wasps develop within the sterile flowers inside the fig throughout the winter months and in the spring emerge from their pupae (cocoons). The male wasps emerge first, wingless and nearly blind. They move around inside the fig looking for female wasps, which are still in their pupae. A male then chews his way into a female pupa and inseminates the female

Figure 18.12

Life on rainforest tree trunks.

Orchids and other epiphytes growing on a tree trunk in Central America.

Lianas (woody vines) growing in Costa Rica along with the aerial roots of epiphytic plants growing high above.

before she emerges; he then dies without ever leaving the fig. The female wasp emerges later; as she leaves the fig, she picks up pollen from the male flowers located near the exit.

The newly emerged female wasp flies around in search of fresh spring-season figs, which have a second type of receptacle containing both fertile female flowers and sterile flowers. The female wasp enters a fig and roams around inside as she lays her eggs by the hundreds and meanwhile pollinates the female flowers with pollen she has picked up from the male flowers in the winter receptacle. The tiny new wasps that emerge then repeat the process and produce a second generation of wasps.

Female wasps emerging late in the year lay their eggs in a third type of receptacle that contains only sterile flowers. The wasps emerging as this third generation lay their eggs in the winter receptacles, completing the yearly cycle. Three types of receptacles are thus home to three generations of fig wasps each year, with male flowers appearing only in the first (winter) type of receptacle and female flowers (and thus seeds) only in the second. All three types of receptacles contain sterile flowers, which alone support the development of new fig wasps. The wasps develop only within these sterile flowers, which they also use as food. The figs are pollinated only by the wasps, each species of figs generally supporting its own species of wasps.

This story of complex interactions does not end there, for the seeds of the figs will not grow if they fall beneath the tree that bore them. The established trees have such an overwhelming competitive advantage that the offspring have little or no chance of competing successfully for

Figure 18.13

Mutualism between the bull's horn acacia (*Acacia cornigera*) and ants of the genus *Pseudomyrmex*.

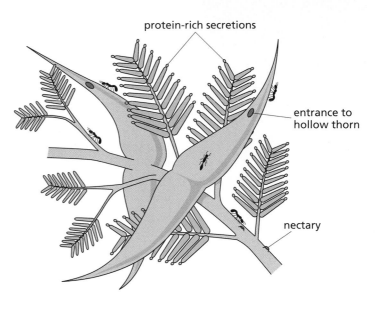

protein-rich secretions

entrance to hollow thorn

nectary

Figure 18.14

Cutaway view of a fig (*Ficus*), showing a female fig wasp (*Blastophaga*) ready to enter. An enlarged view of the female wasp is shown; the male has a similar body but is wingless. One of the many flowers inside the fig is shown in red.

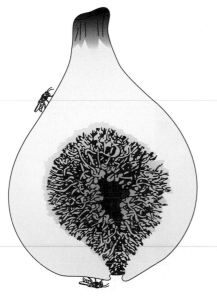

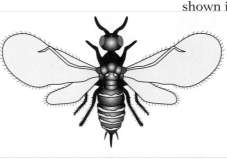

moisture and nutrients if they fall and germinate near their parents. The seeds that succeed are therefore the ones that have dispersed. The seeds of some fig species germinate and grow first as epiphytes upon the branches of other trees and later grow roots reaching down into the ground; to germinate properly, the seeds of these species must find their way to above-ground perches.

The service of dispersing the fig seeds is performed by animals, with different species of animals scattering the seeds of different figs in different areas. In parts of Indonesia and Malaysia, the best dispersers of fig seeds are orangutans, *Pongo pygmaeus* (phylum Chordata, class Mammalia, order Primates, family Pongidae). These large apes practice quadrumanual clambering, a form of locomotion through the trees in which the orangutan's weight hangs from a branch, often supported by the feet as well as the hands (Figure 18.15). Quadrumanual clambering requires a lot of energy, especially for a large animal. An orangutan, if it is to avoid starving, must therefore eat enough in any one place to sustain it on its high-energy journey to the next place, which may be miles—and days—away. Wild orangutans are nearly always hungry and are continually wandering in search of energy-rich foods, the most preferred of which are usually figs.

The wide-roaming habits of orangutans are ideal for the figs, for the orangutans consume hundreds of figs, containing tens of thousands of seeds, then wander for miles through the forest. When an orangutan defecates amidst the branches, it leaves behind hundreds of fig seeds, together with a supply of moist, nutrient-rich fertilizer that helps the seeds to sprout and establish themselves and eventually replenish the supply of fig trees in a forest.

The orangutans must cope with the fact that the fig trees flower and bear fruit at different times of the year. Because there are many species of figs, each bearing fruits in a different seasonal pattern, a resourceful orangutan can usually manage to find some trees bearing fruit in almost every month of the year. Among orangutans, there is thus a selective advantage in having a good spatial memory—a mental map of the forest covering many square miles. The orangutan with the best chance of survival is the one whose intelligence allows it to remember where to find the most fig trees, when each was last visited, how far advanced its figs had grown, and when the time would be optimal to visit each particular tree again.

The interrelatedness of the lives of figs, wasps, and orangutans (and goats, birds, monkeys, humans, and other seed-dispersing species in the

Figure 18.15

The orangutan, *Pongo pygmaeus*, showing quadrumanual clambering.

places where the orangutans do not live) shows how complex life may be among the species of the rainforest. Destroy a few fig trees, and many orangutans may starve. The removal of a few orangutans may decrease the ability of fig trees to disperse their seeds, which would also diminish their ability to provide homes for fig wasps and food for the insect-eating species that feed on the wasps. The destruction of a portion of the rainforest thus has consequences far beyond the portion actually cleared, for it diminishes the health of the whole ecosystem for miles around.

Slowness of ecological succession. When a large tree falls in the rainforest, it creates a small clearing. Human activity may create larger clearings, some of which are later abandoned. Certain pioneer species of shade-intolerant plants are adapted to take advantage of such clearings and create a new ground cover. These species take advantage of the open sunlight while they can, only to be replaced at a later stage by the more shade-tolerant species. Only after many years can any of these grow back to the height of the original tree that had fallen. In this way, entire communities are replaced by a **succession** of other communities that take over, one after another. Although a few pioneer species will colonize an area in a few months or years, tropical botanists have found that it may take centuries for the rainforest to regain its former canopy height and species density. Because the small-scale conditions may be different from those in previous successions, tropical rainforests, even if they regrow, may not become the same community as the forest that was there before. Small clearings in the rainforest are much more likely to grow back than large ones. Many large clearings may not grow back at all.

Deforestation. Around the world, rainforests are being destroyed at an alarming rate. One expert on tropical rainforests estimates that they are disappearing at the rate of 150,000 km^2/year (410 km^2/day), an area equivalent to the size of Manhattan island every 3.5 hours, and that at least 40% of the world's rainforests have already been destroyed!

Tropical deforestation has many causes. Among the causes that have been identified are the growth of human populations (see Chapter 9), the spread of agriculture, the desire of farmers to earn a better living, the attitude that humans are entitled to dominion over nature, and the quest for corporate profits from timber, minerals, or agricultural activities. In the twentieth century, the United Fruit Company (Chiquita Banana) cleared large rainforest areas in Central America and replaced them with banana plantations, while the Goodyear Rubber Company cleared large parts of west African rainforests to establish rubber plantations.

CONNECTIONS
CHAPTER 9

Among the most destructive uses of the rainforest are the exploitation of nonrenewable resources on a one-time basis. Some rainforests are cleared because of the mineral wealth that lies beneath them. In other cases, it is the trees themselves that are being harvested for timber (Figure 18.16). Trees are renewable given a sufficient length of time, but regrowth of tropical rainforests is usually very slow. Many logging operations destroy the trees much faster than they can grow back, so that the forest cannot long sustain continued logging on such a scale.

The largest amount of rainforest destruction is carried out for agricultural reasons. The land cleared from the rainforests is put to agricultural use as grazing land for cattle or as farmland for crops, either for local human consumption or for commerce. In many places, new fields are cleared by cutting down and burning forest (see Figure 18.16), a practice called slash-and-burn agriculture. In many cases, fields created in this way are fertile for only a decade or two, so new fields are then cleared to replace them.

It may seem odd that rainforests that appear so lush will not grow back quickly and will not support agriculture for long. This is because the soil beneath many rainforests is very poor in nutrients, for several reasons. One reason is that tropical ecosystems recycle most nutrients in the forest litter before they reach the soil. Another reason is that the continuous and heavy rainfall washes soluble and partly soluble minerals from the upper layers of the soil, a process called leaching, which leaves few mineral nutrients behind. What remains behind after extensive leaching is in many cases a dark or reddish, very hard, mineral-poor soil (called laterite). Most attempts to grow crops on lateritic soils have been very disappointing because of their low nutrient content and frequent mudslides.

Figure 18.16

Tropical deforestation.

Trees cleared to make way for agriculture.

Timber ready for transport. These operations destroy many hectares of forest at a time; the results include a loss of topsoil and extensive erosion, often leaving the land unsuited for agriculture or human habitation.

Slash-and-burn agriculture quickly depletes the soil of its nutrients, requiring the fields to be abandoned and new fields to be cleared by burning more forest.

These slopes in Costa Rica were once forested, but the harvesting of trees has increased the frequency of landslides and the rate of erosion.

In rainforest regions with lateritic soils, agricultural use of the land quickly exhausts the low nutrient content of the soil. The amount of rainfall is often too high for most crops, and problems with drainage and erosion also arise in many places where the soil holds water poorly. In Madagascar, Haiti, and many other places, rainforests cleared for agricultural use have given way to widespread erosion in which thousands of tons of red, lateritic soil wash annually into the sea. Madagascar also has one of the world's fastest growing populations; its 1999 population of 14 million is expected to double in 22 years, putting even more demands on precious tropical habitat.

Rainforest destruction is often permanent. Because rainforest vegetation contributes (through leaf transpiration) to the rain clouds that maintain the rainy climate, any large-scale clearing of the rainforest is bound to alter the delicately balanced water cycle. Large areas cleared from rainforests tend to suffer from greatly reduced precipitation, and the land soon becomes unsuitable for crops or even for cattle grazing.

Deforestation is a problem at all latitudes, not just in the tropics. Norway, Russia, Canada, and the United States all have vast forests that consume carbon dioxide and contribute oxygen and water to the atmosphere. These forests are also being harvested for their wood (including both timber and pulpwood for making paper) and cleared for agricultural and other uses. Northern forest timber resources can be renewable if properly managed, especially if only some of the trees are cut in any one location. In contrast, removal of all the trees (clear-cutting) from large tracts of land is damaging to forest ecosystems and to the global atmosphere that all forests support. Even if the forest is replanted, it is often with a single species of tree. This is a form of monoculture (see Chapter 11, pp. 388–389); it may maintain a supply of one tree species, but it does not restore biodiversity or habitat variety. Species displaced from their habitat may not come back because the new habitat is very different from the old habitat. Some species, such as the spotted owl mentioned earlier, are particularly sensitive to these changes and can thus be used as sentinel species that allow us to monitor the health of whole ecosystems, as explained later. Of course, species driven to extinction are lost forever.

CONNECTIONS
CHAPTER 11

Desertification

Land that supports the richness of life of a rainforest can be transformed into a desert capable of supporting very little life. Destruction of rainforest can begin this process, which is called desertification. To understand it, we need to learn more about zones of climate and vegetation in tropical regions. In this section we first use Africa as an example and take a close look at desertification there. We then extend our view worldwide and consider whether humans can reverse desertification.

Climatic zones of Africa. The global atmospheric patterns shown in Figure 18.17 conspire to rob the regions around 30° N and 30° S latitude of all their moisture on a continuing basis: prevailing high air pressure creates winds that evaporate moisture from the land and transport it away from these regions. The northern half of Africa lies in the belt at 30° N latitude and is occupied by the world's largest desert, the Sahara.

Figure 18.17

Global patterns of prevailing winds. Notice that winds blow away in both directions from latitudes 30° S and 30° N, carrying moisture away from these latitudes and creating desert regions.

To the south of the Sahara, much of Africa is characterized by a series of parallel zones differing in moisture and thus differing in vegetation (Figure 18.18): there is tropical rainforest along the coast from Sierra Leone to Gabon (and east across the Congo Republic); then, heading north, patches of forests interrupted by more open land; then a more open woodland with scattered trees and shrubs only, giving way to a tropical grassland (savanna) further inland, then a type of dry pastureland, the Sahel; and finally the desert. Each band has more rainfall than the one to its north, and less than the band to its south. The overall pattern has local exceptions where the land is mountainous, but in general it prevails across most of Africa north of the Equator. Each band supports a distinctive kind of vegetation and a distinctive human culture. Each of these bands is also a biome, similar to other ecosystems in other parts of the world (see Figure 18.9).

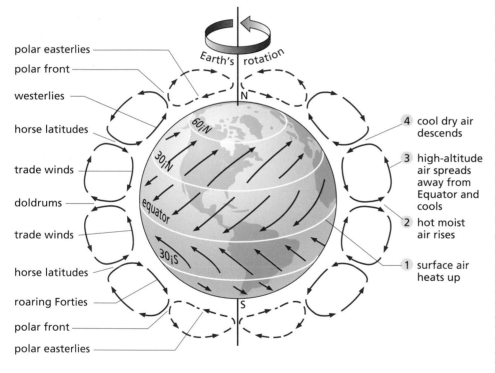

polar easterlies
polar front
westerlies
horse latitudes
trade winds
doldrums
trade winds
horse latitudes
roaring Forties
polar front
polar easterlies

Earth's rotation

60° N
30° N
equator
30° S

N
S

4 cool dry air descends
3 high-altitude air spreads away from Equator and cools
2 hot moist air rises
1 surface air heats up

These zones of vegetation and rainfall are not static, but are slowly changing. The Sahara is very slowly advancing southward by the process of desertification, and the other vegetation zones are moving southward with it. Desertification has also taken place in other directions where the Sahara reaches westward to the shores of the Atlantic and eastward to the shores of the Red Sea and beyond into the Arabian Peninsula. Archaeological excavations confirm that these lands were all much wetter and the vegetation was lush only a few thousand years ago, as the bones (and cave paintings) of hippopotami and crocodiles attest.

How does desertification take place? At least within the Sahel, an important factor promoting desertification is the overgrazing of pasture lands by flocks of domestic animals—goats, cattle, camels, sheep, and other species—that removes the land's vegetative cover. Without the many plant roots that held the soil and its moisture, the precious topsoil blows away. The land, which can no longer support plant life, becomes a desert.

Another important process takes place farther south, where rainforests and tropical woodlands are cleared for agricultural use, often by the slash-and-burn agriculture described earlier. After a few decades of agricultural use, the land is abandoned and new land is cleared. There are several unfortunate consequences of this method of agriculture: the abandoned land is never totally reclaimed by forest ecosystems; the agricultural land has much less ability to retain moisture than the forests that it replaces; and the dry pasture of the Sahel replaces much of the abandoned fields.

Desertification around the world. The problem of desertification is not limited to Africa, although the advance of the Sahara claims more new land each year than all the other deserts of the world combined. The Mojave Desert in Southern California and the Great Indian Desert (along the India–Pakistan border) are two other deserts that are advancing on adjacent agricultural land. The situation in India may be broadly similar to that in much of Africa. The situation in the western United States is somewhat different because desertification is only in its early stages in most places and because most of the problems seem to be associated with the use of underground water reserves for irrigation and for domestic use in cities like Los Angeles. Farmers and ranchers throughout the western United States use aquifers (underground water deposits) for irrigation. In many cases, these aquifers are either shrinking or becoming saltier as ocean waters (e.g., from the Gulf of California) encroach farther inland. If water use exceeds the natural capacity of aquifers to refill, desertification will result.

The immediate effects of desertification are the loss of cropland and rangeland, an effect that is felt keenly but locally. In a few cases, there is also increasing conflict over water rights—for example, between California and Arizona over the use of the Colorado River and between Turkey and Syria over the use of the Euphrates River. Much more serious, however, are the long-term effects of desertification. With reduced vegetation cover, the ground retains less moisture. This means that the air above can become drier, and rain clouds are far less likely to form. The absence of rain clouds results in reduced rainfall, which in turn accelerates the process of desertification.

Prospects for reversing desertification. Can desertification be arrested or turned back? Yes, but only very slowly, very expensively, and with a concerted effort over many years. Israel has had great success in 'making the desert bloom,' turning desert and scrubland (like the Sahel) into agricultural land. One key to this process is irrigation, using water from rivers or lakes or desalinated sea water. Irrigation is always an expensive undertaking—particularly so in a dry climate—and natural water supplies set limits to what can be sustainably farmed with the help of irrigation. Israel is a relatively prosperous nation that has the

Figure 18.18

African vegetation zones.

The Sahel, with a dust storm in the distance.

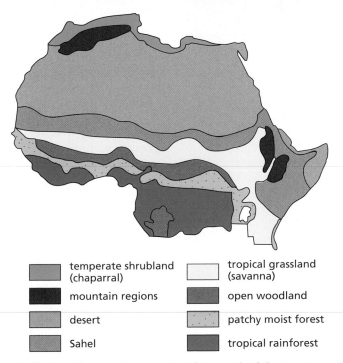

temperate shrubland (chaparral)	tropical grassland (savanna)
mountain regions	open woodland
desert	patchy moist forest
Sahel	tropical rainforest

Distribution of vegetation zones in Africa north of the Equator.

economic resources to reclaim desert lands on a scale appropriate to its small size. Many of the world's desert regions are in poor nations that do not have the financial resources to repeat Israel's successful experiment in reclaiming desert lands for agricultural use.

Valuing habitat

Although we have examined only two of the world's many biomes, some of the conclusions that we have drawn pertain to other types of ecosystem as well. In particular, habitat destruction threatens many ecosystems, whether they are coastal wetlands, pine forests, or the African Sahel. Why does this habitat destruction continue?

Humans often make decisions by first assigning a value, consciously or unconsciously, to such things as happiness, land, money, and even life itself. Decisions are then made by choosing the alternative that has maximum value. Under this system, bad decisions often result from attributing too much or too little value to something, and conflicts may arise if different people assign very different values to the same thing. For example, different people attribute different values to rainforest habitats and to the need to sustain the habitability of our planet.

Ways of assessing value. Philosophers often distinguish between intrinsic value, the value that something has as an end in itself, and instrumental value, the value that something has as a means to some other end. Dollar bills, for example, have no intrinsic value; they are valued only because of what we can buy with them, and they would be useless in a society that did not accept them in trade. Later in this section we examine the instrumental value of various habitats as places where valuable resources can be obtained. Before we do so, let us also point out that many people also value other living species, and entire ecosystems, as having a high intrinsic value. The habitat that sustains living ecosystems is likewise valued intrinsically by many people. Also, in the view of many people, no species has a right to destroy another species or to deprive it of its habitat or its means for continued existence.

Value does not only include the value of something to our species alone. Another type of value may be biological value: the interdependence of species, genetic diversity, ecological diversity, and the resulting dynamic stability of the biological community.

Habitat destruction and ethics. Habitat destruction can take many forms, including the clear-cutting of forests and the draining of swamps. In some cases, the destruction takes place to permit the building of housing tracts or shopping centers. In other cases, land is cleared for agricultural use. In still others, extractive industries such as mining or logging simply exploit the land on a one-time basis for its mineral wealth or its standing crop of trees. Many of the social, political, and economic forces that impinge on tropical rainforests often threaten the destruction of other habitats as well.

When we pause to consider what the many cases of habitat destruction have in common, we soon realize that the same ethical issues recur in case after case. How important are natural communities? How important are their habitats? Is it more important to leave nature undisturbed or to feed an expanding human population? Is it more important to preserve natural habitat or to satisfy people's demands for agricultural land, timber, or housing? To what extent do the answers to the previous questions depend on the quality of the soil, the economics of the country in question, or other factors? To what extent do the answers depend on how much we value other species in addition to our own? Do other species have value apart from their relationships to humans?

All of these are basically ethical questions (see Chapter 1), or parts of a larger, all-embracing ethical question: is it better to preserve a particular ecosystem in its 'natural' state, or is it better to convert the area into agricultural or similar use? Viewed one nation at a time, the forces that push toward one alternative or the other weigh heavily against many natural ecosystems. The pressure of an increasing human population, the need for land and food, the need for income, and the need for economic development are all obvious to the people living near the habitats that come under pressure. Measured against all these forces is the value of undisturbed wilderness, a value that is not always obvious or locally appreciated. Of course, things are never that simple: many Brazilians want to preserve their rainforest habitats, even if this means that economic development cannot proceed quite as fast as other Brazilians would like.

When we consider the worldwide ecosystem of the Earth as a whole, however, the balance seems to shift in the other direction, in favor of preserving the natural environment. The advance of the Sahara, or the destruction of rainforests, threatens the planet with consequences far greater than the continuation of poverty and underdevelopment in any one country. The case for Brazil can easily be argued in these terms: the preservation of the Amazon rainforest is best for the planet as a whole, and the economic best interests of Brazil would be viewed as secondary if the good of the planet were given priority. Perhaps this makes sense to North American environmentalists and philosophers, but it is certain to be a very unpopular attitude in Brazil! It is the Brazilians who are largely in control of their rainforest, and they are likely to resent any suggestion that they sacrifice the well-being of their nation's economy for the 'greater good' of a global environment that the wealthier nations of the north have already started to destroy. A similar argument can be directed against the industrial nations of North America and Europe: a reduction of resource consumption by these nations would reduce pollution, reduce the trend toward global warming, and benefit the planet as a whole. Large tracts of land in Australia, Argentina, and the United States are devoted to cattle ranching, an activity that produces far less food per unit area than if that same land were used for growing crops. More of the world's hungry could be fed if lands now used for cattle ranching were instead used to raise wheat or corn, but ranchers (whose interests are often supported by their governments) are not likely to give up their way of life for that reason alone. They argue that much of the land now used for ranching is so used precisely because it is unsuitable for growing crops economically.

It is easy, on utilitarian principles (see Chapter 1), to argue that the good of the planet should take precedence over the economic well-being of any single nation or occupational class. However, on just about any principle of fairness, it is just as easy for Brazilians to argue that they should not bear the entire burden for a sacrifice that benefits the whole world. If the world benefits from the Brazilian rainforest, then the world

CONNECTIONS   CHAPTER 1

should somehow pay to maintain it in its natural state. If rainforests offer such good protection against global warming, then all nations should contribute to rainforest conservation, perhaps in proportion to the amount of carbon dioxide that they generate. Currently, the United States produces the largest amount of carbon dioxide in relation to its population of any nation on Earth. Also, Brazilians can point to the logging operations that destroy forests at an alarming rate in the United States and other northern countries. Why, they ask, should one nation be told to cease cutting its forests when other nations continue to cut theirs?

Habitat destruction versus sustainable use. It does not take long to realize that many forms of habitat destruction are driven by very short-sighted goals. As an example, the harvesting of slow-growing trees brings only short-term gains, and only to a small number of people (those in a single industry, sometimes only a single company), but the damage that it causes both to the local economy and to the biosphere as a whole may be irreversible. The same can be said of cutting down the rainforest for the planting of those crops that grow poorly on lateritic soils.

Easter Island in the southeast Pacific Ocean shows us a particularly gruesome lesson in the consequences of habitat destruction. When humans first arrived, the island supported a rainforest that was home to many edible plant species as well as numerous species of birds and other animals. The Easter Islanders, a Polynesian people, prospered for several hundred years, building the large stone statues for which the islands are now famous. But instead of conserving the rainforest and living off its rich resources, the Easter Islanders cut much of it down, until too little was left to sustain the edible bird and plant species, which slowly disappeared along with the forest itself. Like other rainforests, the one on Easter Island had created its own rain clouds, producing the conditions for high rainfall, but when it was destroyed, less rain fell on agricultural crops. With vanishing timber supplies, the Easter Islanders could scarcely find enough wood to build the boats needed to sustain their fishing activities. As all food supplies dwindled, the Easter Islanders began to starve and the last survivors resorted to cannibalism before they were rescued by the arrival of Europeans. Nor is this case unique. The Maori, a Polynesian people of New Zealand, hunted many of the native species of their islands to extinction or nearly so, and were beginning to show signs of starvation and decline when Europeans arrived. Other islands in the Pacific were subject to similar exploitation, although nowhere else did the process go as far as it did on Easter Island.

What we need instead of destructive uses are **sustainable** uses of forests, uses that allow people to derive profit from maintaining the rainforest instead of from destroying it. Most nations that contain rainforests are nonindustrialized, and many of them are also poor. A sustainable economic use of the rainforest would be an economic incentive to maintain the forest rather than to destroy it. It is therefore in the long-term best interests of all nations and all people to help tropical nations to develop such sustainable uses.

One example of sustainable use is the gathering of small amounts of high-income rainforest products such as pharmaceutical plants. For example, the anticancer drugs vinblastine and vincristine, derived from a rainforest plant (the Madagascar rose periwinkle), account for sales of $180 million a year. In 1991 the pharmaceutical firm of Merck and

Company entered into a million-dollar agreement with Costa Rica's National Institute of Biodiversity. Scientists working for the Institute were to identify as many rainforest plants as they could (the total number is estimated to be 12,000), extract samples from them, and send the more promising ones to Merck for tests of their medicinal value. Merck's $1 million investment should be compared with their annual sales close to $15 billion in 1994. In 1990, Merck sold $735 million worth of just one drug, Mevacor (lovastatin), a cholesterol-lowering drug derived from a soil fungus. If even one new drug discovered by scientists under this agreement brings in a small fraction of this amount, Merck will recoup its original investment many times over.

Certain kinds of rainforest agriculture can be sustainable, but others are not, and much remains to be learned from experimentation and frequent reevaluation. Most promising are the attempts at mixed uses of rainforest habitats, allowing tall trees, shorter trees and shrubs, and smaller plants to persist side-by-side. Coffee, vanilla, cocoa, cashews, bananas, and certain spices are potential candidates for such experimental attempts, and tropical botanists can help to identify others. Many uses of rainforest plants are traditional, but they could benefit economically from improvements in harvesting, transport, and marketing. Agricultural scientists and business interests could help to develop new markets, which would bring much-needed income and provide local people with an economic incentive to maintain the forest ecosystem. Tropical plants that could easily be marketed more widely include amaranth (a nutritious and drought-resistant grain), fruits such as durians and mangosteens, and the winged bean (*Psophocarpus tetragonolobus*) of New Guinea (kingdom Plantae, phylum Anthophyta, class Dicotyledonae). This last species is a fast-growing plant that produces spinachlike leaves, young seed pods that resemble green beans, and mature seeds similar to soybeans, all without the use of fertilizers. Hundreds of other fruits are grown and eaten in the tropics but are only rarely exported.

An important goal of such efforts would be to identify which plants might profitably be grown or harvested in a given region without harm to the environment. New ways must be found to exploit the rainforests without destroying them—to develop rainforest ecosystems into sustainable resources for both local and worldwide benefit.

Sustainable use and habitat preservation make sense economically in nearly all ecosystems, not just in rainforests. The city of New York was planning to spend between $6 billion and $8 billion to build a new water filtration and treatment plant, until officials discovered that they could accomplish the same goals by spending only $1.5 billion to help preserve the natural watersheds of the Catskills and the Delaware Basin, two sources of naturally filtered water supplies. In Hawaii, the Maui Pineapple Company runs the Pu'u Kukui Watershed Preserve, in which they protect native plants that maintain both soil and groundwater that sustains nearby pineapple plantations. "Without it," says a company official, "rain would run off into the ocean."

Ecotourism. If Brazilians and other tropical nations are to preserve the rainforest instead of destroying it, they must have economic incentives to do so. One type of economic incentive is the small but growing market for ecologically based tourism, also called green tourism or ecotourism. Ideally, this type of tourism seeks to make as little impact on the natural environment as possible. Ecotourism is now Costa Rica's second largest source of foreign income, behind coffee and ahead of bananas. Ecotourism is also a major source of revenue in Kenya. By comparison, the rainforests of Brazil afford a largely untapped tourist resource. Of course, some land would have to be set aside for airports, roads, and hotels. Beyond this initial investment, however, ecotourism would provide economic incentives for leaving the rest of the rainforest untouched. Ecotourism not only provides a country with income, it also gives that country an economic incentive to preserve its own natural heritage for the benefit of all.

As we search for ways to stop the destruction of ecosystems, we must realize that no solution will work if the rich and poor nations continue at odds with one another. If battles continue over short-range economic interests, the planetary ecosystem will be threatened, along with human survival.

THOUGHT QUESTIONS

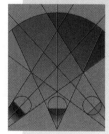

1 Is a biodegradable insecticide completely harmless? If it were used over a wide area of a forest, might it cause the extinction of an insect species confined to that area? What effect would such a loss have on other species? Do you think the use of such insecticides in research carries a certain amount of risk? If so, how can this risk be minimized?

2 Do you think undisturbed habitats have intrinsic value, or only instrumental value? In other words, is habitat valuable as an end in itself, or only because of the uses to which it might be put? Think of other things that you value intrinsically, such as close family members. Does habitat have the same kind of intrinsic value? In what ways are the values similar? In what ways are they different?

3 Are there things of intrinsic value that you personally would be willing to 'do without' if it meant preserving more habitat for other species? Will it be possible to preserve habitat for other species without humans changing their use of habitat?

4 How much habitat destruction do you think takes place at the hands of wealthy people, and how much at the hands of poor people? (The answer may differ from country to country.) What would it take to secure the cooperation of both rich and poor people in an effort to halt desertification or rainforest destruction? Could you easily appeal to both rich and poor together, or would it be easier to appeal to the two groups separately?

5 The ecologist E.O. Wilson has estimated that one Peruvian farmer clearing land to grow food cuts down more species of trees than are native to all of Europe. Do you think there may have been greater numbers of trees in Europe before human populations grew to their present density? More species of trees?

6 Are habitats being destroyed near where you live or go to college? What are they? What factors contribute to the destruction?

7 Does ecotourism sometimes do harm? How? Can the harm be minimized? Can it be eliminated entirely? Think of some examples of ecotourism and other forms of rural tourism. On the whole, do you think that the benefits of ecotourism outweigh the harm?

Concluding Remarks

Human activity is contributing to a rapid decline in biodiversity comparable to the extinction rates of the mass extinctions in the geological past. The high rate of extinction of species is made even higher by the destruction of entire habitats, such as in rainforests. The quest for corporate profits motivates some of this destruction, but so does the need for food and living space for human populations. The reduction of biodiversity makes ecosystems less stable and makes our planet less habitable for humans and for many other species. Can we value the continued existence of other species, not just humans? Can we learn that the continued existence of humans depends on our living in natural balance with other species? The survival of our planet and its ecosystems depends on the choices that we make. To many people, 'long-range planning' means thinking only one year into the future, but the choices that we make today often have consequences that last for decades, centuries, or even longer. Technology may soon link humans everywhere into a single global community, but we should also remember that ecosystems link our species to the rest of life.

Chapter Summary

- **Biodiversity** is measured by the number and variety of species, of which 70% are either insects or vascular plants.

- Each species occupies a **niche** within a **community**. A community and its physical environment interact as an **ecosystem**. One type of community may replace another in an ecological process called **succession**.

- Speciation increases biodiversity, while **extinction** decreases it.

- **Endangered species** are those threatened with extinction, often because their populations are too small and genetically too homogeneous to adapt to change.

- **Habitats** support the interdependent lives of biological communities of species.

- Groups of ecosystems that are similar but in different geographic areas are called **biomes**.

- Destruction of a habitat threatens the survival of all the species that live in it. Overuse, rather than **sustainable** use, is endangering many habitats worldwide.

- Biodiversity is greatest in tropical rainforests and in coral reefs, largely because of the great amount of solar energy and climate stability near the Equator. The expansion of human populations and agricultural lands threatens many rainforests. Because a majority of the world's photosynthesis and oxygen production occurs in rainforests, their destruction is a worldwide threat to the atmosphere and to the entire global ecosystem or biosphere.

CONNECTIONS TO OTHER CHAPTERS

Chapter 1	Pollution and habitat destruction violate several ethical injunctions including the principle to do no harm to others.
Chapter 4	Genomes are an important biodiversity resource.
Chapter 5	Species arise through evolution; biodiversity increases as a result.
Chapter 6	Classifications must be modified as newly discovered species are inserted.
Chapter 7	Allele frequencies change in populations.
Chapter 9	The expansion of human populations put increased pressure on ecosystems all over the world.
Chapter 11	Plants are energy producers in every ecosystem and thus support many other species.
Chapter 19	Pollution threatens the habitats of many species.

PRACTICE QUESTIONS

1. Which has greater biodiversity, a habitat in which there are 1500 resident species, most of which are procaryotes, or a habitat in which there are 1500 species, most of which are plants?

2. Rainforest is being destroyed at the rate of 410 km^2/day. How many square miles per day is that? How many acres? How big is your campus, town or city?

3. What percentage of animal species are birds? How many times more species of insects are there than species of birds?

4. What biome is your college located in? Where on the globe are comparable biomes located?

5. How do wind patterns contribute to deserts being located at 30° N and 30° S latitudes and rainforests being located along the Equator?

6. What percentage of the world's bird species live in Costa Rica? What percentage live in Hawaii?

7. If 500,000 genera were living in the Permian period, and if 85% of them became extinct in the last 5 million years of that period, then how many genera became extinct, on average, during each million-year interval? If there were five species per genus, on average, then how many species per million years became extinct? How does this rate compare with the estimated 100,000 species extinctions for the decade of the 1980s?

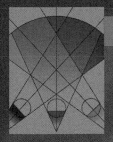

Issues

- Is it possible to live in the industrial world without polluting?
- How can societies limit the threats to ecosystems?
- How do economics influence pollution?
- Who should pay for cleanup and remediation?
- How can societies limit the threats to the atmosphere?
- How can people in one place help to control pollution that originates elsewhere?

Biological Concepts

- Ecosystems and biosphere
- Environmental factors (biotic, abiotic)
- Evolution (origin of life)
- Human influences on the biosphere
- Atmosphere
- Matter (water cycle, carbon cycle)
- Chemical and physical basis of biology (molecular structure, oxidation and reduction reactions)
- Energy and metabolism (autotrophs, heterotrophs)
- Valuing habitat
- Conservation biology (renewable and nonrenewable resources, sustainable and nonsustainable uses)

Chapter Outline

The Biosphere: Land, Water, Atmosphere, and Life

The development of the atmosphere and of life

Evidence of early life on Earth

The water cycle

Pollution Threatens Much of Life on Earth

Sources and indicators of pollution

Toxic effects

Pollution prevention

Human Activities Are Affecting the Biosphere

Aquatic pollution and its biological effects

Bioremediation

Air pollution

Acid rain

Atmospheric ozone

CO_2 and global warming

19 Protecting the Biosphere

Human populations have important impacts on ecosystems, both locally and globally. Just to meet basic human needs, our quest for food, drinking water, and places to live creates various disturbances in local ecosystems. Human agriculture and housing for people are two important ways in which we will always alter our local ecosystems. In addition, people's demands for industrial products and for energy create further impacts. In all these things, the growth of human populations (see Chapter 9) puts ever-increasing stress on world ecosystems. Some of these impacts were described in Chapter 18; others, especially those related to pollution, are described in this chapter.

*The Earth's land, water, atmosphere and its living things form a global ecosystem called the **biosphere**. As part of this ecosystem, the atmosphere supports life in the sense that all animals and many other organisms would soon die without oxygen. The atmosphere also supports life in that it maintains the Earth's surface temperature within a certain range. In another sense, however, it is life on planet Earth that supports Earth's atmosphere, because all the important atmospheric gases exist in equilibrium with the activities of living organisms: plants and other photosynthesizing organisms produce the oxygen, while a series of bacteria regulate the nitrogen.*

Human activities can have a great and varying influence on a local scale. People can cut down trees, clear land, plant fields, dig mines, or erect buildings of many kinds. Human activities can also leave behind many waste products, whether intentionally or unintentionally. On a global scale, human activities can also change the Earth's atmosphere and thus threaten the stability of many ecosystems and myriad species of life on Earth. One such change involves the destruction of atmospheric ozone (O_3), a process that increases the ultraviolet radiation reaching the surface, causing higher rates of mutation and increased cancer rates. A different change comes from the buildup of carbon dioxide, a change that may raise global temperatures to a point that would cause widespread extinctions, devastating crop losses, and the flooding of most coastal cities. Photosynthesis by plants, especially forest plants, may help to limit the buildup of carbon dioxide that contributes to global warming. The health of our atmosphere thus depends on the continued health of major ecosystems in the tropics and elsewhere. These are among the issues that we examine in this chapter.

The Biosphere: Land, Water, Atmosphere, and Life

The Earth and all its living species form a giant ecosystem known as the **biosphere**. As an ecosystem, the biosphere includes all animals, plants, and other organisms, and also all of the Earth's land features, its oceans and fresh waters, and the atmosphere. All of these interact with one another and form an interconnected whole. Each mountain, pond, forest, and biological species helps to reshape all the rest and is in turn reshaped by them. No full understanding of any one of them can be gained without considering all the others and their past histories.

The Earth's landforms have been shaped through time by purely physical processes such as earthquakes, and also by the activities of organisms. Lichens and fungi secrete chemicals that slowly dissolve certain rocks, and plant roots wedge themselves into cracks and help to break up larger rocks into smaller ones. Burrowing animals help to reshape the soil, while plant roots hold the soil and protect it from washing away by erosion. The activities of aquatic organisms may either loosen sediments or build up new structures such as coral reefs. The biological activities of these organisms may alter the chemistry of the water itself. Organisms also interact with the atmosphere and use its gases; slowly they may modify the composition of the atmosphere over long spans of time.

The development of the atmosphere and of life

The Earth's atmosphere supports all its ecosystems because most forms of life require oxygen. Species that do not need oxygen interact with those that do, and they require other atmospheric gases besides. The atmosphere also maintains the Earth's surface temperature within a certain range, cooler than our planet would be with a much denser atmosphere and much warmer than our planet would be with no atmosphere at all. We have only to look to our nearest neighboring planets to view some of the alternative possibilities: Venus, with a much denser atmosphere, is unbelievably hot, while Mars, with almost no atmosphere, is inhospitably cold. In another sense, however, it is life on planet Earth that supports Earth's atmosphere, because all the important atmospheric gases exist in equilibrium with the activities of living organisms: plants and other photosynthesizing organisms produce the oxygen, while a series of bacteria regulate the nitrogen (see Chapter 11).

The present atmosphere of planet Earth is about 78% nitrogen (N_2), 21% oxygen (O_2), less than 1% argon (Ar), and much smaller amounts of water vapor (H_2O), carbon dioxide (CO_2), and other gases (Figure 19.1). While the element nitrogen (N) is essential to all living things, nitrogen gas is in a form that few organisms can use directly (see Chapter 11). Argon gas is chemically unreactive and has no known influence on the activities of organisms. On

Figure 19.1

Chemical composition of the atmosphere.

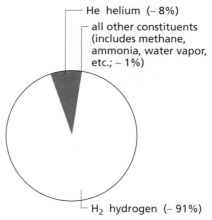

Probable composition of original reducing atmosphere (value approximate, similar to composition of Jupiter and Saturn)

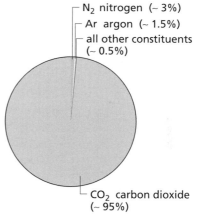

Probable composition after evolution of heterotrophic organisms but before the evolution of photosynthesis (approximate values)

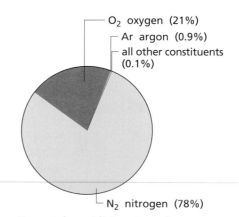

Present-day oxidizing atmosphere

the other hand, oxygen is absolutely essential to the continued existence of animals and many other forms of life present today. Oxygen is also important for many of the chemical reactions that happen naturally at the Earth's surface, including fire and other types of combustion, as well as rust and other forms of slower oxidation. The normal form of oxygen consists of molecules of O_2, oxygen atoms bonded together in groups of two. There is also a more reactive form of oxygen called **ozone**, a form of oxygen abundant at high altitudes. Ozone molecules are O_3, oxygen atoms bonded together in groups of three.

Life has changed the atmosphere. A good deal of evidence suggests that the atmosphere has not always had its present composition. Modern thoughts on the evolution of Earth's atmosphere began with the biological theories of the early twentieth-century Russian biochemist Aleksandr Oparin. Most scientists at the time accepted the French microbiologist Louis Pasteur's conclusions that living organisms always came from pre-existing organisms. But where had the first living organisms come from? Oparin, a Marxist, rejected the possibility of a divine or other miraculous creation. Had life always existed? The evidence from astronomy was that the Earth originated under conditions that could not have supported life. Oparin also knew that the conditions in interplanetary space, such as extreme cold (around –270 °C), utter dryness, and constant bombardment by high levels of ultraviolet radiation, were incompatible with all forms of life. These facts convinced Oparin that life could not have come through space from anywhere else. Life must have originated on planet Earth.

Pasteur had said, "there is now no circumstance known in which it can be affirmed that microscopic beings came into the world without...parents similar to themselves." From this, Oparin reasoned that, if present conditions do not permit organisms to originate from nonliving matter, then life must have originated on Earth under conditions very different from those that prevail today, at a time when the atmosphere had a very different composition from what it has now. From studying what was known about the chemical composition of the solar system, Oparin postulated an early atmosphere rich in hydrogen, the most abundant material in the Solar System and in the Universe as a whole. Chemists refer to such hydrogen-rich conditions as **reducing** conditions, in contrast to the **oxidizing** (high-oxygen) conditions of our present atmosphere. Under the early reducing atmosphere postulated by Oparin, no free oxygen was present. Instead, each element existed largely in its most reduced form, in combination with hydrogen (H). Thus, most carbon was combined with hydrogen into methane (CH_4), most nitrogen was combined into ammonia (NH_3), and most oxygen was combined with hydrogen to form water vapor (H_2O).

Working with the assumption of an early atmosphere consisting of the gases hydrogen (H_2), ammonia (NH_3), methane (CH_4), and water vapor (H_2O), Oparin figured out some of the chemical reactions that he thought might have produced simple biological molecules such as sugars and organic acids. He theorized that these molecules would have built up in the primitive oceans once the Earth's temperature permitted water to become liquid. By the slow accumulation of these molecules, the world's oceans would have become like a "hot, dilute soup." Oparin published his detailed hypothesis in book form in 1935, and the book was soon translated into English under the title *The Origin of Life*.

The origin of life: testing Oparin's ideas. In 1952, an American biochemist named Stanley Miller decided to test Oparin's model experimentally. He built several types of apparatus, such as the one shown in Figure 19.2. The apparatus was filled with hydrogen, ammonia, methane, and water vapor, the four gases that had been postulated by Oparin. The gases reacted in a chamber in which electric sparks simulated atmospheric lightning. Reaction products were cooled so that water became liquid, simulating rain. Compounds that formed in the reaction chamber dissolved in the water, which collected in the lower part of the apparatus, simulating the ponds and oceans of the primitive Earth. A heat supply vaporized some of the liquid and returned it to the reaction chamber. Miller circulated his reaction mixture for several days and withdrew samples to analyze the results. Among the compounds that had been formed he found amino acids (building blocks for proteins), simple sugars, and most of the building blocks for DNA and RNA. The building blocks for all major biological molecules had thus been formed, without the aid of any organisms, under conditions simulating those of the primitive Earth.

Miller then proceeded to repeat his experiment under somewhat altered conditions, and several other scientists have also done so. They found that differences in the starting materials did not greatly affect the experimental results, as long as sources of hydrogen, oxygen, nitrogen, and carbon were all present. For example, carbon dioxide or another carbon compound could substitute for methane, and nitrogen gas or nitrogen oxides could substitute for ammonia. An energy source was also needed, but the particular kind of energy was not important. An ultraviolet light source could successfully substitute for the electric sparks; so could natural sunlight, even with no other heat source.

To show that the reaction products were produced by the experimental reactions, and not by any biological contaminants, Miller also did a control experiment: he repeated all experimental conditions except for the electric sparks. Without the energy source provided by these electric sparks, no reaction products were detected.

Experiments simulating primitive Earth conditions have succeeded in producing so many biologically important products that the following conclusion is now inescapable: all of the molecules important to life *could have been produced* in a lifeless environment on the primitive Earth, starting with nothing more than a few basic gases mixed together as a simulated reducing atmosphere. However, while these experiments show us what *could* have happened, they cannot show us what actually *did* happen.

Figure 19.2

Stanley Miller's experiment, in which amino acids and other molecules used by living organisms were produced. Heating the flask at the lower left boils the water and keeps the mixture circulating in the direction shown by the arrows. Reactions take place in the spark chamber and reaction products are condensed and recirculated. Valve A is used to sterilize the apparatus and to introduce the starting materials; valve B is used to withdraw samples of the reaction products.

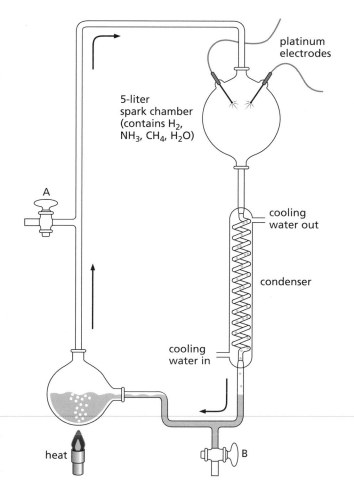

platinum electrodes

5-liter spark chamber (contains H$_2$, NH$_3$, CH$_4$, H$_2$O)

A

cooling water out

condenser

cooling water in

cooling water in

heat

B

Evidence of early life on Earth

Is there any evidence that the events recreated in Miller's flask actually did happen on the primitive Earth? There is some evidence, but it is incomplete and indirect. Meteorites falling to Earth from elsewhere in the solar system sometimes contain organic (carbon-containing) compounds indicative of a Miller-style synthesis in other parts of the universe, including compounds that organisms now on Earth produce only rarely or not at all. Ancient rocks on Earth over a billion (10^9) years old contain various kinds of 'chemical fossils,' compounds that give us clues to the conditions that prevailed at the time when these compounds were formed. Our present atmosphere differs greatly from the primordial atmosphere postulated by Oparin because free oxygen (O_2) is now abundant. The oldest rocks on Earth contain compounds that would not have persisted under oxygen-rich conditions, and thus seem to have been deposited at a time when the atmosphere contained little or no oxygen.

One important finding is that chemicals formed from the breakdown of chlorophyll molecules are not present in the oldest rocks. Chlorophyll is a key molecule in photosynthesis, a reaction by which plants and blue-green bacteria (Cyanobacteria) produce O_2 (see Chapter 11). The breakdown products of chlorophyll first appeared in the geological record at about the time that the reducing atmosphere began to change slowly, over a period of about a billion years, to the oxidizing conditions that now prevail.

If the atmosphere now differs so drastically from the primordial atmosphere postulated by Oparin, how and when did the change occur? In particular, how did the present oxidizing conditions replace the earlier reducing conditions? Most scientists who have investigated this question have concluded that life itself is primarily responsible.

The first forms of life. The first forms of life had to live under reducing conditions in which no oxygen was present, conditions that are called **anaerobic**. A variety of bacteria are capable of living under anaerobic conditions. The first organisms, simpler than any modern bacteria, would have been enclosed by membranes and would have contained nucleic acids capable of passing on genetic information, but their mechanisms of obtaining and using energy would have been very different from the methods used by most organisms alive today. Most of the early bacteria would have been heterotrophs, meaning that they had to derive all their energy from the high-energy organic molecules that they found in their environs. Nowadays, such molecules are in most cases produced by other organisms, but the first organisms would have had to rely on the organic molecules that had formed without life (abiotically), under conditions like those simulated by experiments similar to Miller's. For many thousands or maybe millions of years, the supply of these molecules may have been adequate for heterotrophic life to expand and perhaps to flourish. (We can't tell for sure, because such organisms leave very few fossil traces of their existence.) Eventually, however, the organisms expanded to the point that the limited amount of energy-rich chemicals in the environment were just not enough. This was possibly the first global environmental crisis in the history of life on Earth.

We can imagine several possible responses to this crisis. Some organisms may have discovered a way to attack and devour other organisms, getting their nutrients from their prey. Organisms of this type continued to eat one another, but the total quantity of living organisms (the total biomass) that the planet could support remained limited. Those organisms more efficient at eating one another, or at making do with what little food they could find, did better than their less efficient competitors. Quite possibly, none of the organisms were able to adapt to this way of life, and all eventually perished. In that case, a second abiotic synthesis would have taken place all over again, and perhaps a third, until finally some group of organisms succeeded in evolving the means to feed on other organisms.

Even for organisms that ate other organisms, however, the possibilities would have been strictly limited. Other organisms would be encountered only so often, and only some of their constituents would be usable. Also, after a period of evolution had elapsed, more and more organisms would have evolved defenses against predators, and the amount of biomass would still have been limited. The problem may even have become worse, because some of the waste products of metabolism included gases such as CO_2, which simply escaped into the atmosphere, taking carbon out of reach of most organisms.

Evolution of photosynthesis and its atmospheric effects. A more permanent solution to the limited supply of energy-rich chemicals occurred much later, when some organisms produced the first chlorophyll-like molecules, about 4 billion years ago. Such a molecule would enable life forms to use solar energy in the form of light, which would get around the problem of using the limited number of other organisms. All forms of photosynthesis require the use of some chemical to supply hydrogen atoms and act as an electron acceptor (see Chapter 11). Most bacterial forms of photosynthesis use hydrogen sulfide (H_2S) for this purpose, or iron compounds, or organic molecules such as nicotinamide adenine dinucleotide (NAD); NAD was already present in organisms as derivatives of the nucleic acids. The primitive forms of photosynthesis did result in some minor changes in the atmosphere, perhaps including the use of some atmospheric CO_2 as a raw material.

The greatest change to the Earth's atmosphere resulted from the evolution of a new and more efficient kind of photosynthesis, using a new and different hydrogen donor (see Chapter 11). The new source of hydrogen was water, the most abundant hydrogen source on Earth. The splitting of water in photosynthesis generated a new atmospheric gas: oxygen (O_2). The first organisms to evolve this kind of photosynthesis were blue-green bacteria (Cyanobacteria). During the next 2 billion years or so, these blue-green bacteria became the dominant form of life on Earth, decreasing the abundance of atmospheric CO_2 to a small fraction of its former level, and slowly generating more and more oxygen. Calculations of the photosynthetic capabilities of these blue-green bacteria show that they were well capable of generating all the oxygen in the Earth's atmosphere within half a billion years or so.

As the Earth's atmosphere became more and more oxygen-rich, additional changes began to occur. Capture of certain wavelengths of ultraviolet light by oxygen breaks up the O_2 molecules into a pair of highly reactive oxygen atoms (O). These then react with the nearest other molecule to produce new compounds. If oxygen atoms react with water, they produce hydrogen peroxide (H_2O_2), which acts as a bacterial poison. If oxygen atoms react with oxygen molecules (O_2) in the presence of ultraviolet light, they produce ozone (O_3). Eventually, the production of ozone in this manner would slowly give rise to an ozone layer in the stratosphere and screen out additional ultraviolet light from reaching the Earth's surface. However, before such a layer existed, much more ultraviolet light would have penetrated the atmosphere and reached the Earth's surface, splitting more oxygen molecules and making more hydrogen peroxide. Hydrogen peroxide is toxic to most bacteria. Some bacteria are capable of breaking down small amounts of hydrogen peroxide with the aid of enzymes that are chemically similar to chlorophyll and other pigments.

Scientists believe that the present composition of the Earth's atmosphere reflects the activities of living organisms (see Figure 19.1). The earliest forms of life on Earth used up much of the methane and ammonia in the synthesis of organic compounds. Their waste products contributed to a buildup of atmospheric CO_2. The evolution of photosynthetic organisms gradually depleted the atmosphere of its carbon dioxide, replacing it with oxygen and giving our atmosphere its modern composition. Differences in density have caused the gases in the atmosphere to form several major layers that we describe later (see Figure 19.7).

The atmosphere sustains life and is sustained by life. The atmosphere of the Earth and the living things on the Earth have always influenced each other to such an extent that life and the Earth's atmosphere should be considered parts of an integrated whole. The very air we breathe is a product of biological activity, and the stability of the planet's overall temperature is in part the result of atmospheric gases that are maintained in equilibrium with biological systems. This general idea has been promoted by James Lovelock and Lynn Margulis. Lovelock called it the **Gaia hypothesis**, after Gaia (or Gaea), the Greek goddess of the Earth after whom the science of geology is named. Life on Earth has drastically changed our atmosphere from a reducing one (hydrogen-rich) to an oxidizing one (oxygen-rich). Many of the chemicals that existed in reduced forms in primordial times are now more abundant in their oxidized forms: carbon as carbon dioxide rather than methane, and oxygen as molecular oxygen and ozone in addition to water. These changes, produced by living organisms, have made the planet even more habitable for the growth of other living organisms. Ozone, for example, captures ultraviolet light up in the stratosphere, greatly reducing the formation of peroxides at the surface. Several atmospheric gases, such as the CO_2 produced by many organisms, trap solar energy within the troposphere and reflect it back to Earth, producing the so-called "greenhouse effect" described later in this chapter. This greenhouse effect makes the Earth warmer than it would otherwise be, and more hospitable to life as a result.

The water cycle

Water is essential to human life and to the lives of most other creatures on Earth. Irrigation is essential to agriculture in many places, and many thousands of animals and plants die each year under drought conditions.

Water evaporates from the surface of the oceans and from all bodies of fresh water. Much water is also transpired through the leaves of plants

(Chapter 11) or released from the respiratory activities of nearly all species of organisms. The resulting water vapor mixes into the atmosphere. At altitudes where the temperature is cold enough, much of this water vapor condenses to form ice crystals. In other places, water droplets form where the air is saturated with moisture. Ice crystals and water droplets form clouds as they accumulate. As air masses move around, some of these clouds cool down a bit and are no longer able to hold as much moisture as before. The excess moisture then falls as precipitation in the form of rain or snow. Much of this precipitation falls back into bodies of water, but some of it also falls upon the land, where it runs downhill in streams and rivers, accumulating in ponds, lakes, and oceans. Water thus recycles throughout the world's ecosystems (Figure 19.3).

Figure 19.3

The water cycle.

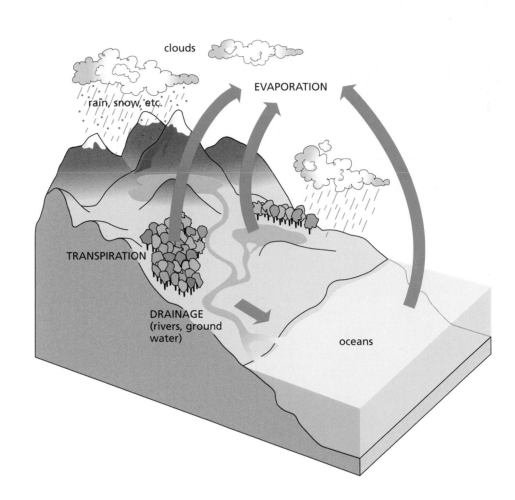

THOUGHT QUESTIONS

1 In what way is the atmosphere part of the biosphere?

2 What would Earth's atmosphere be like if life had not evolved? What would it be like if heterotrophic life had evolved, but not photosynthesis?

3 Which step in the origin of life do you think was the most important? How much do we know about each step in the process of life's origin? What means of investigation do you think will bring us additional knowledge?

4 How do you think the properties of life listed in Chapter 1 originated?

5 Are there limitations on the types of organisms that could evolve on Earth? How would we investigate such limitations? What limitations, if any, would apply to organisms on other planets or in other solar systems? How similar to organisms on Earth would such extraterrestrial organisms necessarily be?

Pollution Threatens Much of Life on Earth

Toxic dumps are places where waste disposal creates environmentally hazardous conditions. Millions of tons of waste materials are intentionally discarded every year. Other environmental problems arise from accidents. In 1989, the tanker *Exxon Valdez* ran aground in Alaska and spilled thousands of tons of oil. Many other tanker accidents and smaller oil spills have occurred as well. Toxic dumps and oil spills are two of the many kinds of pollution that threaten our environment. Other types of pollution include groundwater contamination, fertilizer buildup in soils, and the release of harmful gases into the atmosphere from vehicles, smokestacks, and other sources.

The original meaning of the verb *pollute* was to contaminate or make dirty. Today, **pollution** can be defined as anything that is present in the wrong quantities or concentrations, in the wrong place, or at the wrong time. Although there is room for people to disagree about acceptable quantities, usually there is general agreement that pollution exists when it affects the health of humans or other organisms. Oil spilled from tankers or drilling operations can kill thousands of aquatic birds, mammals, fish, and other organisms. Lead, cadmium, and other heavy metals in drinking water, food, or house paints can cause brain damage and other neurological defects. Toxic dumps can poison people and raise cancer rates. In some of the worst cases, numerous unidentified chemicals were dumped together into the same toxic waste site, forming a 'witch's brew' that underwent further and often unpredictable chemical reactions to produce additional hazards. In many countries, it is now illegal to dispose of many chemicals except by government-approved methods.

Sources and indicators of pollution

When we flush the toilet or send our garbage to a landfill, we are contributing to the accumulation of solid and liquid wastes. When we drive our cars, we are contributing to the pollution of the air that we all must breathe. All of us contribute to pollution in many different ways. Even our breathing releases carbon dioxide into the atmosphere.

Does this mean that every act mentioned in the previous paragraph is an immoral act? Certainly not. In order to live, we must breathe, eat, urinate, and defecate. When we eat, we throw away inedible parts (skin, bones, pits, rinds, shells), packaging materials, and unfinished remains. To go on living, we must continue to pollute in certain ways. So why is there such a fuss?

Pollution: a problem of quantities. Pollution in most cases is a problem of quantities, and sometimes also of location and rates of accumulation. Clearly, you don't want your garbage to accumulate in your living room. Suppose, for the moment, that garbage disposal services were not available, and you had to dispose of your household garbage yourself, as many people in rural places still do. You could perhaps bury it in the backyard, or just toss it away. If you only tossed away bones, rinds, and other biodegradable materials (things that can be broken down by bacteria, fungi, and other decomposer organisms), then you might be able to dispose of your own garbage in this way. Up to a point, that is. Your backyard might have enough decomposer organisms to break down and recycle your own personal wastes, or perhaps your family's wastes. Of course, this depends in part on the size of your family, and also on the size of your backyard. Clearly, your backyard would not be able to handle the garbage produced by an entire town or city.

Just about every known pollutant is harmless in some sufficiently small quantity. Pollutants become bothersome or toxic as quantities increase. In most cases, measuring pollution means measuring quantities, for pollution is a matter of quantities.

How do we know whether a habitat is polluted? If land, air, or water is polluted, how can we tell to what extent? If pollution cleanup is attempted in a particular place, how do we measure the success of the cleanup effort? Most of the specific answers depend upon the measuring of particular chemical substances. First, a particular chemical pollutant or breakdown product must be identified as an indicator of the pollution in question. Second, the concentration of this particular chemical must be measured repeatedly at various places and times.

Sentinel species. More general indicators of pollution are also needed, especially pollution from hazards that have not yet been clearly identified. A number of environmentally sensitive species have been suggested as possible general indicators of pollution. These **sentinel species** serve the same role as the canaries that coal miners often took with them into the mines. Because the canaries were extremely sensitive to methane and other dangerous gases present in coal mines, the health of the canary reassured the miners, and the sickness or death of a canary was always viewed by the miners as a danger signal. Frogs and other amphibians are sentinel species that can warn us of the deterioration of freshwater habitats, just as the spotted owl can be considered a sentinel species whose numbers are indicative of the health of old-growth Pacific forests. Dolphins have occasionally been suggested as sentinel species for marine habitats because environmental pollution can stress the immune systems of these marine mammals and raise their rate of infectious diseases. The grounding of marine mammals on beaches may also be a stress-related phenomenon that reflects marine pollution.

Toxic effects

The field of toxicology deals with the damage done to human or animal health by various quantities of poisons, including environmental pollutants. Toxic damage can affect any of the body's systems, but the nervous and reproductive systems are especially vulnerable in many species. Chemical tests can often detect the quantities of environmental pollutants, but biological tests (bioassays) are in many cases even more sensitive. Bioassays can even be used to monitor the effects of hazards that have not yet been chemically identified, or hazards that may contain a complex mix of several chemical substances.

As an example of an environmental pollution hazard, consider the chemicals known as dioxins. Dioxins are chlorine-containing organic compounds that arise as inadvertent by-products from the use of chlorine as a bleach in the paper-making process. There are several different dioxins, of which the most toxic is 2,3,7,8-tetrachloro-dibenzo-*p*-dioxin. Dioxins are often found in freshwater ecosystems downstream from industrial sources, and in these places they can poison and sometimes kill fish and other aquatic animals. Dioxins are chemically similar to 'agent orange,' a toxic substance used as a defoliant herbicide during the Vietnam War (1960–1975). Dioxins are among the many toxic substances that have estrogenic effects, meaning that they can affect the reproductive systems of many species, including humans. Knowing of these toxic effects, many companies have been testing other types of bleaching processes for paper, and some paper is no longer bleached at all.

Other types of toxic chemicals are intentionally used, for example, as pesticides, and they then become hazardous to human health when they accumulate in unwanted places. Some long-lasting toxic pollutants accumulate in biological tissues, and are subject to biomagnification (see Chapter 11), a process that increases their concentration as they pass through the food chain. Pollutants subject to biomagnification include heavy metals such as mercury and many insecticides such as DDT. Most insecticides and many heavy metals are neurological poisons. Many long-lasting pesticides, including DDT, are now banned in the United States and in many other countries. Pollution from insecticides can be greatly reduced by avoiding the long-lasting chemicals that accumulate in biological tissues and are biomagnified, and by switching instead to biodegradable pesticides that break down in biological systems.

Pollution prevention

Pollution awareness is a crucial component in the prevention of pollution. Unless we realize how we are polluting, it is unlikely that we will take any corrective action to pollute less. We all should inform ourselves about the disposal of our garbage and our industrial waste, about the emissions from our cars and from nearby (and distant) smokestacks, and about the cleanliness of our beaches, playgrounds, drinking supplies, and foods.

Because pollution is a consequence of the ways in which we live and work, there are certain things that we can each do to reduce the amount of pollution that we cause. Most of these measures also have further benefits, such as saving money or contributing to human health. For example, car-pooling saves money, while bicycling to work saves even more money and contributes to health and fitness as well. Recycling saves money, too, and we can all recycle such items as paper, bottles, and cans. Making an aluminum soda can out of recycled materials requires only a small fraction of the electricity needed to make the same can from aluminum ore. Many industries are now responding to market forces by using recycled materials in their products and by advertising their use of biodegradable or other 'Earth-friendly' materials.

Costs and benefits. Ethical decisions about pollution are decisions about quantities. The determination of both costs and benefits associated with pollution thus depends upon the measurement of quantities. Whether costs are measured in dollars, in lives lost, or in reduced human health, what counts is that they are measured. The same is true of the benefits of preventing or cleaning up pollution, which can be measured in dollars, in lives saved, or in increased health or enjoyment.

One general problem is that costs are often easier to identify and measure than benefits, and are subject to much less uncertainty. The costs of curtailing or modifying a manufacturing process can easily be measured in terms of costs of new equipment, costs to operate the equipment, and either increased wages (if additional employees are needed) or reduced employment (if an activity is curtailed). These costs are fairly certain, and are easily measured. The benefits to the ecosystem are less certain and are harder to measure: if a particular set of changes is implemented, will the ecosystem recover? What levels will the pollutants reach at some distance from the source? How many lives will be saved or how much disease prevented as a result? What dollar value should be placed on the prevention of disease, on the bird or fish population, or on the health of the environment?

Translating 'quality' of health or enjoyment into something that can be quantified is a relatively new and important specialty within economics. One way of measuring these intangibles (certainly not the only way, but definitely one of the easiest ways) is by measuring the average dollar amount that people are willing to pay to obtain them. To take one small example, the value of a better neighborhood can be measured by the additional amount that people are willing to pay to live in it, compared with some other neighborhood. The value of a clean environment can be measured by the amounts of money that people are willing to sacrifice to live in it, work in it, or visit it.

Attempts to measure environmental quality can sometimes be misunderstood. When we try to measure environmental quality, we need to consider the value of the entire ecosystem, not just the sentinel species used as a measuring stick to monitor quality. Salmon have a certain value as a commercial food species, but they have a far greater importance as a general indicator of pollution or of the health of a river ecosystem. If salmon populations are used to measure pollution or pollution abatement in freshwater ecosystems, it is the value of the entire ecosystem that should be counted as a benefit, not just the commercial value of the salmon fishery. Likewise, bald eagles and spotted owls (neither of which have commercial value) can be used to indicate the general health of ecosystems, and it is the health of these entire ecosystems that should be counted as a benefit, not just the value that we place on the eagles or owls. The death of the canary in the mine means much more to the miners than the price of a replacement bird.

THOUGHT QUESTIONS

1 How does your town or city dispose of trash? Is there a recycling program? What gets recycled? What is not currently recycled in your community, but could be?

2 Is it possible to live in the industrial world without polluting?

3 Think of at least two important sources of pollution in the region where you live or go to school. Is there a way to reduce this pollution? What practices would have to change?

4 Materials that cannot easily be recycled are generally taken to places specifically set aside as dumps. If such facilities are needed, where should they be located?

Because dumps are generally unsightly, unclean (sometimes toxic), and subject to noisy traffic, people generally don't want to live near them. The motivating desire to have these facilities somewhere else is often symbolized by the phrase 'not in my back yard' (NIMBY). One result is that dumps are often located wherever people have the least political clout, sometimes giving rise to charges of 'ecological racism.' What further social problems arise from the widespread application of the NIMBY principle? Is there any socially responsible way to locate an undesirable facility? What is the best way to reduce the need for such facilities in the first place?

Human Activities Are Affecting the Biosphere

Human activities can change the biosphere in many ways. Humans clear many habitats for housing, for industrial development, and for the planting of crops. Many other human activities affect the quality and availability of water. Some human activities pollute the land, water, and atmosphere. Atmospheric pollution can impair respiration and can affect the acid content of rain and snow. Some atmospheric pollution can also result in the destruction of atmospheric ozone, a process that can raise the rates of mutation and cancer through increases in the ultraviolet radiation reaching the Earth's surface. Increased release of carbon dioxide and certain other gases may threaten to raise global temperatures. The health of the biosphere is thus at stake in what we do.

Aquatic pollution and its biological effects

Materials produced by human activities can enter the water directly or can be washed into the water from soil or from disposal sites. Dissolved and suspended materials in the water can be taken up by organisms and can interact with biological systems. Some materials are beneficial, but many others can cause harm. Aquatic pollution derived from human activity includes agricultural runoff, industrial wastes, human sewage, and accidental spills. In Chapter 11 we discussed how agricultural fertilizers can run off and enter aquatic ecosystems, causing algal blooms that may sometimes kill fish and other animals. Toxic chemicals that can enter aquatic ecosystems include pesticides from agricultural fields, dioxins from paper factories, polychlorinated biphenyls (PCBs) from electrical insulation, heavy metals from various industrial processes and from mining activities, and a variety of other chemicals including solvents. Certain conditions also allow pathogens to enter water and to proliferate and cause disease, as we saw in Chapter 17.

Pollution of water supplies can limit the availability of water for irrigation or for human consumption. Water pollution may be a source of political conflict between downstream states or countries and those located more upstream. In many cases, the downstream users seek protection from upstream uses that might affect water flow or water quality. Such 'water wars' are an important source of friction between California and Arizona, between Israel and its neighbors, and between Turkey and Iraq.

Any material can be accidentally spilled, especially during its transport, but spills of oily substances are usually of greatest concern because they are transported in very large quantities and because they spread out very rapidly in water. When oil is spilled at sea, little oil dissolves in the water. Oil is a mixture of many different chemical compounds, the exact mixture depending on where the oil came from and whether or not it has been refined. Most of the compounds are nonpolar, and so are not very soluble in water, which is very polar (see Chapter 10, p. 328). Because crude oil is not very soluble in water and because it is less dense than water, it floats on top, gradually spreading across the surface to form a 'slick.' Although slicks may be only a few molecules thick, they interfere

CONNECTIONS   CHAPTERS 10, 11, 17

with organisms that need to absorb oxygen at the water surface. Evaporation removes the smaller, lighter-weight components of the slick; the components left behind may then be more dense than water and sink, eventually blocking bottom-dwelling organisms' access to oxygen. Oil pollution can clog the gills and feeding surfaces of many fishes and other aquatic organisms; it can also harm the insulating fur or feathers of marine mammals and birds, making them less able to escape, to keep warm, or to resist diseases. Evaporation, sinking, or dispersal of oil may make it disappear from view, but the oil molecules are unchanged and can still cause harm until they are cleaned up.

Bioremediation

Some types of pollution can be reduced, and habitats restored, with the help of living organisms. **Bioremediation** is an approach in which the decomposition activities of living organisms are put to work in cleaning up contaminated soil and water. **Biodegradation** refers to the natural decomposition processes that go on without human intervention; bioremediation, in contrast, implies the manipulation of biodegradative processes by humans. Bioremediation has been used in the cleanup of oil and chemical spills, and in wastewater treatment. The aim of bioremediation is to change the molecules into something harmless by using the chemical reactions carried out by decomposer organisms, some of which are very effective at this task.

Bioremediation of oil spills. Bioremediation is often applied to problems of oil contamination on land, but in this section we discuss the challenge of cleaning up aquatic oil spills. As we saw in Chapter 11 (pp. 368–369), certain organisms derive their energy by breaking down complex molecules. These are the decomposers, mostly bacteria and fungi, which keep both energy and matter cycling through the biosphere.

Oil-degrading bacteria and fungi are found in all types of aquatic habitats, both freshwater and marine, and include representatives of over 70 different genera. No one species can degrade all the molecular compounds in oil. Even when bacteria capable of producing oil-degrading enzymes are present where oil has spilled, the rate and extent of biodegradation depends on many environmental factors, including temperature, the amount of oxygen, and the availability of other nutrients. Because many of these factors are unpredictable, the success of biodegradation at any given site is also unpredictable.

Probably the most critical factor for biodegradation is the availability of nutrients such as nitrogen, phosphorus, and iron. These elements are necessary for microbial synthesis of proteins and nucleic acids, and as enzyme cofactors, much as they are in other living organisms (see Chapters 10 and 11).

Most types of bioremediation attempt to enhance the natural processes of biodegradation. Three basic strategies are employed at a spill site: (1) enrichment of rate-limiting nutrients, (2) introduction of bacteria, and (3) introduction of genetically engineered bacteria.

In concept, nutrient enrichment is much like the fertilization of soil (see Chapter 11, p. 383). It assumes that oil-degrading microorganisms adapted to the local conditions are already present. Supplying nutrients

CONNECTIONS CHAPTERS 10, 11

supports the faster reproduction of degradative bacteria, and the hope is that the bacterial population will grow quickly and to a great enough density to overcome the large quantity of oil. After the *Exxon Valdez* spill in 1989, the concept was extensively tested by the Environmental Protection Agency (EPA), in conjunction with Exxon and the state of Alaska. In these tests, the rate of biodegradation on a 110-mile stretch of nutrient-enriched beach was accelerated twofold to fourfold for a period of at least 30 days. A second application of nutrients after 3 to 5 weeks accelerated the rate even more. Similar measures have been used to clean up other oil spills, such as after the Persian Gulf war of 1991.

The bacterial species indigenous to a spill area may not have the right enzymes to degrade the compounds in the type of oil in that spill. Each decomposer species is able to make specific enzymes that can degrade some specific shapes of molecules but not others. Even if the right species are present, their numbers may not be sufficient to degrade the oil at an appreciable rate. Hence, it may be advantageous to introduce into a spill site a mixed bacterial population containing many species able to digest more of the oil components than would any single species. Some people advocate using genetic engineering to produce bacteria that can make a greater range of enzymes. The first patented bacterial species was one that had been genetically engineered to degrade oil. This approach may also have potential for remediation of biohazardous wastes other than oil.

Although the concept of introducing bacteria into oil spills seems plausible, we have not yet perfected the process. Two species of commercially available bacteria were introduced on beaches polluted by the *Exxon Valdez* spill, but (in contrast to the boosting of locally occurring bacteria by fertilizing) there was no significant enhancement of biodegradation compared with that on untreated beaches. It is possible that introduced bacteria might increase to numbers great enough to upset the ecological balance of other organisms, but it is more likely that the introduced population would die off as soon as the oil was used up. Initial experience suggests that a greater problem than overgrowth may be getting an introduced population to survive long enough to degrade all the oil.

Bioremediation of wastewater. Wastewater includes any water that has been used, for whatever purpose. All wastewater treatment depends on the biological activity of microorganisms. Sewage is water that contains human fecal wastes; if not treated, sewage is a major route for the spread of infectious diseases (Chapter 17). Ingestion of fecally contaminated water spreads bacterial diseases such as cholera, shigellosis, typhoid fever, and viral diseases such as hepatitis A. Swimming or wading in contaminated water puts people at risk of such parasitic diseases as schistosomiasis. Fecally contaminated water is concentrated by filter-feeding animals, so that mollusks collected from polluted waters may be highly infectious to people who eat them.

CONNECTIONS   CHAPTER 17

Gray water is water that has been used for bathing, washing clothes, or other uses in which there is no contact with human wastes. In virtually all homes and other buildings, gray water joins sewage in common drains; in some cities, rainwater from storm drains and industrial effluents also joins sewage drains. Thus, chemical pollutants and fecal pollutants are mixed together and all water must be treated as wastewater, although the particulate, or solid, content of the wastewater may be very low (about 0.03%).

Streams and soil often contain both minerals that trap some pollutants and bacteria capable of biodegrading waste materials. For very low human population densities of times past, natural filtering and biodegradation by streams and soil was adequate sewage treatment. Even today, in many areas where soil is suitable and population density is not too great, the wastes from homes can be collected and treated by septic tanks and leach beds (Figure 19.4). The treated water goes back into the ground, not back into the home. This type of system can only be effective when the wastewater input does not exceed the biodegradation capacity of the soil bacteria. Putting things down the drain that kill the soil bacteria make the system nonfunctional.

A slightly larger-scale sewage system can separate fecal material from wastewater in a shallow lagoon in which the wastewater can be contained for about 30 days. This is sufficient time for solids to settle and for sunlight, air, and microorganisms to kill the bacteria and viruses from human and animal sources, including many microorganisms that cause human disease. Wastewater lagoons are actually complex ecosystems. Algae grow in a wastewater lagoon, using carbon dioxide to conduct photosynthesis and producing oxygen in the process. The oxygen produced by the algae keeps the lagoon aerated, allowing the growth of aerobic bacteria that digest organic matter and kill fecal bacteria. Simple sewage systems work well if they are not overloaded. They can be overloaded by excess rainwater, an increase in users from a growing population, or an increase in wastewater production per person.

Where simple systems prove inadequate, centralized water-treatment plants are needed. Municipal water-treatment plants are further examples of bacterial biodegradation, and they treat water in two or three stages.

Primary wastewater treatment is essentially the same as the settling process in a wastewater lagoon. The part that does not settle quickly, which is called the effluent, goes on to secondary treatment. Here, in large aeration tanks, air or oxygen is bubbled through the effluent, allowing aerobic bacteria to remove up to 90% of the organic wastes. The sludge that settles undergoes further digestion by anaerobic bacteria (bacteria that do not require oxygen). Some of these bacteria degrade the organic matter to organic acids and

Figure 19.4

Septic tanks and leach beds.

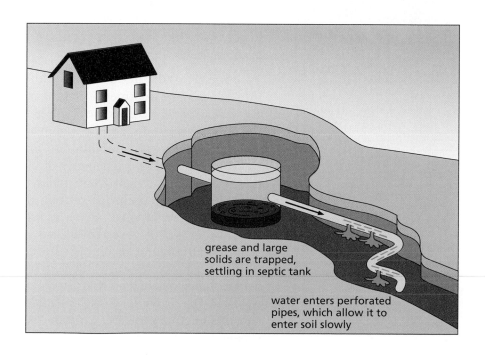

grease and large solids are trapped, settling in septic tank

water enters perforated pipes, which allow it to enter soil slowly

carbon dioxide. Methane-producing anaerobic bacteria digest the organic acids and release methane gas (CH_4), which is used as fuel for running the treatment plant. The sludge that remains is dried and incinerated or put in landfills. It contains valuable nutrients, about one-fifth the amount in an equal volume of commercial lawn fertilizer, and soil conditioners, much the same as compost, and is sometimes used as such.

All municipalities in the United States must have at least primary and secondary wastewater treatment. Where secondary treatment is the last step, the secondary treatment effluent is disinfected with chlorine and discharged into some body of water or sprayed onto fields designated for that purpose.

In some cities, the secondary effluent goes on to tertiary treatment, where nitrogen can be removed by denitrifying bacteria (see Chapter 11, pp. 374–375). These bacteria convert nitrogen compounds to gaseous ammonia, which evaporates into the air. The remaining organic material is removed by nonbiological means such as fine filtration or the electrical precipitation of dissolved ions. Water is then chlorinated before being discharged. Tertiary treatment, although very costly, results in water that is once again fit for drinking.

Some towns are experimenting with using marshlands to treat wastewater. Because many natural wetlands have been destroyed by pollution or development, towns have created new wetlands. Wastewater goes through primary treatment and then is pumped into these marshes. Marsh plants remove nitrogen and phosphorus, break down sewage, and even filter out toxic chemicals. Preliminary treatment may be done in greenhouses. In tanks containing such plants as water hyacinths and cattails, algae and microorganisms decompose the organic materials, which are then used as nutrients by the plants. The water then passes to other tanks containing snails and zooplankton that consume the algae and microorganisms; the zooplankton are then eaten by fish, and the water is further purified by marshlands. Such systems work well, but only if the quantities of wastewater do not exceed the capacity of the treatment ecosystem. The demonstrated success of such small-scale treatment facilities emphasizes that there are many different, ecologically sound ways to treat wastewater.

Treatment of drinking water. Although wastewater that has passed through tertiary treatment is theoretically fit to drink, municipal drinking water usually comes from other sources and is treated in separate treatment plants. Incoming water, which usually comes from natural streams and lakes, is stored in reservoirs to allow the particulates to settle. Particles of dirt, particularly of clay, as well as bacteria and viruses that are too fine to settle, are removed by flocculation with aluminum potassium sulfate (alum), a process that was known to reduce the incidence of cholera long before it was known that cholera is caused by bacteria. Water is then passed over beds of sand or diatomaceous earth that adsorb microorganisms on their surfaces (Figure 19.5). Flocculation and filtration remove several pathogens (such as the protozoan *Giardia*) that are not killed by chlorination.

Drinking water is usually treated with chlorine (Cl_2) to kill disease-causing microorganisms. Chlorine is very reactive and toxic to other organisms, including humans, if its concentrations are too high. The risk of these toxic effects is currently considered acceptable because the benefits of disinfection are so much greater. However, chlorine levels in the water must be carefully monitored to be sure that they are high enough to effectively disinfect the water, yet not so high as to be toxic. Some water-treatment facilities have begun to use ozone (O_3) as a disinfectant, producing the ozone electrically on site. This is highly effective, although very expensive.

Many treatment facilities also add fluoride (Fl^-) to drinking water (see Chapter 10, p. 341). The fluoride ion can substitute for phosphate in the hydroxyapatite crystal that makes up tooth enamel. The resultant enamel is less soluble than ordinary enamel in the acidic pH produced by dental cavity-causing bacteria. Fluoride has the greatest effect in children as their adult teeth are forming through their early years, but it has decay-preventing benefits in adults as well, because enamel is constantly remodeled even in adults.

Air pollution

The air we breathe is necessary to human life, but some human activities have begun to change the atmosphere. Natural ecosystems can generally handle the gases that we exhale. The other products of biological activities, including agriculture, are usually recycled naturally and safely as long as these products do not accumulate too much in any one place.

Air pollution affects the air that we breathe, both indoors and outdoors, and causes many unwanted changes in the atmosphere. Some of these changes are local only, but others have worldwide consequences.

Automobile exhausts and industrial plants are major sources of outdoor air pollution. Large forest fires can also pollute the atmosphere sporadically in the regions where they occur. Many outdoor air pollutants are oxides of carbon (carbon monoxide and carbon dioxide), oxides of nitrogen, and oxides of sulfur released as combustion products. Ozone is released to the air by many processes, but automobile exhausts release more ozone than any other source. Certain other gases, such as chlorine, benzene, and hydrogen sulfide, are also released by some industrial processes. When the concentrations of these pollutants get high, many people begin to suffer from respiratory ailments. Smog, for example, is an ozone-laden mixture of smoke and fog; it can be particularly harmful to lungs because it contains dangerous amounts of ozone. (In contrast, ozone in the stratosphere protects us from dangerously high levels of cancer-causing ultraviolet radiation.) In addition to the gases that we have mentioned, outdoor air pollution can also include industrial soot,

Figure 19.5

Diatoms: single-celled, ocean-dwelling organisms (kingdom Plantae, division Chrysophyta) with elaborate silicon-impregnated cell walls. Diatoms accumulate in huge numbers and can be mined as diatomaceous earth from land that was formerly on the ocean floor. The enormous surface area of the many separate microscopic pieces gives diatomaceous earth a great capacity for adsorption; hence its use in water purification in water-treatment plants and in swimming pools.

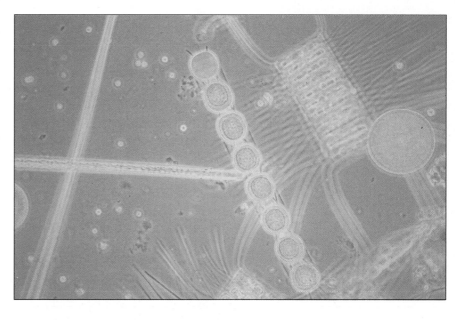

bacterial pathogens, mold spores, and allergens such as pollen. The U.S. Environmental Protection Agency estimates that 6.6 metric tons of particulate air pollution are released each year in the United States alone.

Air pollutants can become trapped in the ventilation systems of buildings and cause indoor air pollution. Indoor air pollution can include many of the same components as outdoor air pollution (including pollen), plus additional bacteria, plus pollutants such as asbestos, benzene, or formaldehyde formed by the decomposition of materials used in building construction. Buildings with poorly designed ventilation systems can impair the health of the people working in them, a phenomenon sometimes called 'sick building syndrome.' Indoor air pollution may be one of the causes of a 70% increase in the incidence of allergies in the United States over the past two decades. Studies in England show that people working in older buildings with open windows as a source of ventilation suffer fewer respiratory illnesses than people working in newer buildings with recirculated air. Second-hand cigarette smoke is another form of indoor air pollution. A large-scale English study showed that cancer rates were much higher among nonsmoking married women whose husbands smoked than among a matched group of women married to nonsmokers, presumably because the women in the first group were breathing the polluted indoor air containing the second-hand smoke (see Chapter 12). Campaigns to banish smoking from restaurants and other public places are motivated by such health considerations.

Acid rain

One of the most widespread and best understood pollution problems is that of acid rain. (We should really speak of 'acid deposition,' because much of the problem comes from acid snow, acid fog, and dry deposition of acid dust or condensation.) The acidity of a substance is measured on a standard pH scale (see Figure 10.9, p. 345), where the lower the value, the higher the acidity (the more hydrogen ions). Rainwater of pH 5 or below is considered acid rain, and values as low as 2.7 have been measured on occasion. Recall that the pH scale is a logarithmic scale, so that a lowering of pH by one unit corresponds to a tenfold increase in the concentration of acid, and a pH of 3.0 is therefore 100 times as acidic as a pH of 5. Because most enzymes work optimally only within a very narrow pH range, and because the structure of biological molecules may change at different pH values, a small change in pH can have an enormous effect on living organisms (Figure 19.6).

Chemical tests of acid rain show that the source of most of the acidity is either sulfuric acid (H_2SO_4) or nitric acid (HNO_3). Acid rain across the eastern United States and the Scandinavian countries is primarily from sulfuric acid, while the acid content of acid rain in western United States in locations such as Denver, Colorado, is more than 50% nitric acid, derived primarily from the nitric oxides produced by automobile exhausts.

Sulfuric acid pollution begins when materials containing sulfur are burned in air, forming sulfur dioxide (SO_2). The sulfur dioxide combines with additional oxygen to make sulfur trioxide, which then combines with water to make sulfuric acid. Sulfur is a common impurity in many coal deposits, and is also present in many of the ores from which lead, zinc, nickel, and certain other minerals are commonly obtained. In many parts of the United States, Canada, England, and Germany, the mining of metals from sulfur ores generates sulfur dioxide that ends up as acid rain in such downwind locations as the northeastern United States and Sweden. The burning of high-sulfur coal in electrical power plants is an even larger source of sulfur dioxide pollution that eventually falls as acid rain. This type of acid rain is a political as well as an environmental problem because the governments of New York and of Sweden are relatively powerless to control pollution that originates in Illinois, Indiana, or Germany. The United States and Canada have accused one another of being major sources of cross-border acid rain pollution.

Acid rain is a problem wherever there are large numbers of factories or automobiles. In many parts of Asia and Latin America, countries burn

Figure 19.6

Some of the effects of acid rain.

trout taken from a lake in Ontario, Canada, in 1979, at pH 5.4

trout taken from the same lake in 1982, at pH 5.1

trees killed by acid rain in North Carolina

stone sculpture dissolving in acid rain

more coal as they industrialize, and more gasoline as their populations become more prosperous. Acid rain is therefore becoming more of a problem in these countries, most acutely in China.

Acid rain erodes and slowly dissolves marble and limestone statues and buildings (see Figure 19.6), including those of the famous Parthenon in Athens, Greece. In localities where the rock formations are predominantly limestone and other carbonate rocks, acid rain is neutralized as it runs through these rocks or as it percolates through the soils derived from the weathering of these rocks. However, granitic rocks, such as those that predominate in New England, northern New York State, Scandinavia, and parts of Ontario, have little or no capacity to neutralize acid rain. As a result, acid deposition in those areas accumulates in ponds and lakes to levels that kill fish (see Figure 19.6). The Adirondacks of New York State contain hundreds of lakes and ponds that once teemed with fish, but whose fish populations have completely died out because of the acidity of the water.

Atmospheric ozone

The Earth's current atmosphere is organized into several major layers (Figure 19.7). The lowest layer, of most immediate impact on human

Figure 19.7

Structure of the Earth's atmosphere.

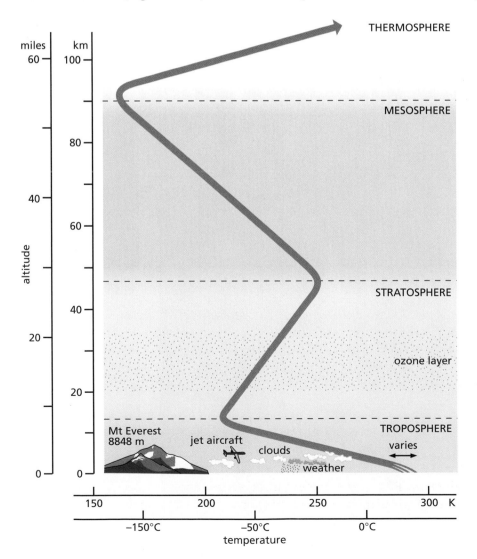

activity, is the troposphere, containing the majority of atmospheric gases. Weather phenomena and all routine types of aircraft flight are confined to the troposphere. Above the troposphere lies the stratosphere, the zone in which ozone accumulates. Still higher are the mesosphere and the thermosphere.

The two most abundant gases in the atmosphere are nitrogen and oxygen. Most oxygen consists of molecules of O_2, oxygen atoms bonded together in groups of two, but ozone (O_3) is a more reactive form in which oxygen atoms are bonded together in groups of three. If you have ever smelled a sharp, pungent, 'electrical' smell around a hot iron or toaster, that is the smell of the small concentration of ozone produced when electrical activity dissociates oxygen molecules:

$$O_2 \rightarrow O + O; \text{ then } O + O_2 \rightarrow O_3$$

Ozone is the most important and most dangerous constituent of urban smog, produced when thousands of automobile exhausts are trapped beneath a layer of warm air (a temperature inversion). When enough nitrogen oxides accumulate from these automobile exhausts, a photochemical (light-induced) reaction produces ozone as follows:

$$NO_2 + \text{light} \rightarrow NO + O$$
$$O + O_2 \rightarrow O_3$$

Ozone at ground level is hazardous to human health because it causes irritation of the eyes and respiratory passages, and oxidative damage to many mucous membranes. Death may occur by respiratory failure, especially in individuals with preexisting respiratory problems. Despite the harm that ozone produces at ground level, it is beneficial—even essential—at high altitude. In the Earth's stratosphere, oxygen molecules bombarded by the sun's ultraviolet radiation slowly rearrange to form ozone. The ozone becomes concentrated into a distinctive layer within the stratosphere (see Figure 19.7). The ozone layer actually contains only a small fraction of ozone, but this fraction is very distinctive because ozone is quite rare in the other layers. By absorbing ultraviolet light with wavelengths of about 40–320 nm, the ozone layer protects the Earth's surface and all living organisms against the harmful effects of ultraviolet light in these wavelengths (see Chapters 7 and 12). The existence of an ozone layer has been known for decades and is not disputed.

CONNECTIONS
CHAPTERS 7, 12

Certain chemicals have the potential to do great harm to the ozone layer. This fact was first realized by some theoretical chemists who tried to determine the fate of certain industrially produced molecules. The molecules in question are called **chlorofluorocarbons** or CFCs, small molecules composed of carbon and hydrogen (hydrocarbons) to which chlorine and fluorine atoms are attached. These molecules had been used since the late 1930s as nontoxic refrigerants, fluids pumped through refrigerators and air conditioners in their cooling cycles. Early refrigerators had used either ammonia or sulfur dioxide for this purpose, but both were toxic and chemically reactive, so CFC refrigerants (commonly known as 'freons') were hailed as safe substitutes when they first came into use in the 1930s. In the 1950s, other uses were developed for CFCs, especially as propellants for insecticides, hair sprays, deodorants, and other aerosol products. CFCs were preferred in these applications because they were chemically unreactive. In spray cans, the CFC propellants could be used with a wide variety of products and not react with them.

What, however, becomes of the CFCs when they are released from aerosol spray cans or from discarded refrigerators and air conditioners? Remember that CFCs were used because they were chemically unreactive. When released into the atmosphere, the CFCs would not react with other atmospheric gases—they would simply dissipate into the atmosphere and mix by wind action with the other gases until they were spread across the entire planet. Moreover, they would stay in the atmosphere and not 'rain out' because they were not soluble in rain water. After a period of time measured in decades, these molecules would eventually be carried aloft by upward air currents until they reached the stratosphere. Once the CFCs rose above the ozone layer (but not sooner), they would be broken down by ultraviolet light, releasing many chemically reactive breakdown products, including such free radicals as chlorine monoxide (ClO), monatomic chlorine (Cl), and monatomic oxygen (O). Two American chemists, F.S. Rowland and Mario Molina, theorized in 1973 that these free radicals would react repeatedly with ozone, destroying one ozone molecule after another.

The predicted destruction of ozone could not be immediately confirmed when these results were first published. However, the most alarming prediction was that it would take from 40 to 150 years for the CFCs *already released* into the lower atmosphere to migrate into the stratosphere and destroy much of the ozone layer, *even if no additional CFCs were ever released*. In other words, if these predictions were correct, the harm had already been done, although it would take decades to become apparent. The predicted destruction of the ozone layer would allow a large increase in the amount of ultraviolet light reaching the Earth's surface, causing severe itching of the eyes and a large increase in the incidence of skin cancer, particularly among light-skinned people (see Chapters 7 and 12). Mutation rates would also increase, and the ultraviolet light would also cause more ozone production in the atmosphere at ground level, resulting in increased respiratory problems.

When the Rowland–Molina model of ozone destruction by CFCs was first proposed, industry spokespeople were quick to point out that the model was untested and that there were few data to support it. After all, no damage to the Earth's ozone layer had yet been measured. The world's largest manufacturer of CFCs, the Dupont Corporation, in a display of corporate responsibility, pledged in 1975 that it would stop making CFCs *if* it could be shown that these chemicals did in fact threaten the Earth's ozone layer.

From the time when Rowland and Molina made their predictions until the destruction of stratospheric ozone was detected over Antarctica in 1984, various environmental activists raised cries of alarm and proposed the banning of CFCs. Industry voices countered with calls for more data and further study. Both sides examined the Rowland–Molina model, and many new scientists became increasingly interested in this area of research. As additional chemical reactions were considered, the models became more numerous and more complex. The various models predicted different *rates* of ozone depletion, but all predicted that some damage to the ozone layer would sooner or later take place. The news media began reporting 'debates' over CFCs, sometimes exaggerating the extent of disagreement among scientists, but they also stirred up public interest in the ozone layer. Ordinary citizens with no interest in the details of atmospheric chemistry began to hear that some scientists thought that CFCs were damaging the ozone layer. Although industry began by dragging its feet or in some cases by actively opposing any ban, many Earth-conscious consumers began to shun spray-can products. A few small communities banned their sale. A few companies began to advertise that their products did *not* use CFCs, the first major example of what came to be known as 'green marketing.' In 1975, the Johnson Wax Company announced that they would no longer use CFCs in their spray cans. Once consumers started buying sprays that did not contain CFCs, other spray-can manufacturers followed suit. After a few years, very few spray cans remained that used CFC propellants, *even though no widespread ban had been implemented*. In this case, market forces had caused an industry to discontinue use of an environmentally destructive product even before there was a legislative ban, and (more remarkably) even before the scientific community had reached any general consensus on the issue! Congress finally did pass legislation against the use of CFC spray cans, but the measure was by then largely symbolic because such sprays had already all but disappeared from the market.

The case of CFCs illustrates the more general problem of making public policy decisions in the face of scientific uncertainty (see Chapter 1). In this case, there was a claim that a product was a potential long-range threat to the Earth's ozone layer, but the claim could not yet be verified. As we discussed in Chapter 1, controversies of this sort are now commonly divided into three parts: scientific issues, science policy issues, and policy issues. The important scientific issue in this case was the correctness of the Rowland–Molina model, for which critical data were largely unavailable because the predicted damage to the ozone layer lay many years into the future. In terms of science policy issues, the economic costs of banning CFCs were easy to measure, but the environmental costs of continued use were not. In the face of this uncertainty, most people examined the issue at the level of the science policy questions: what harm would be done if we banned CFCs and the ban later turned out to be unnecessary? On the other hand, what harm would occur if the predictions were correct and we waited several years for more data before implementing a ban? In this case, many people decided that the costs of a CFC ban were bearable, while the possible costs to society of a large but uncertain amount of future damage to the ozone layer were great. In 1978, Congress banned the use of CFC propellants in spray cans. Most unusual and most significant was the fact that this decision was made even before convincing evidence was available of any damage to the environment.

Evidence supporting the Rowland–Molina theory. The hypothesis that CFCs would damage the ozone layer was largely untested, but it did not remain that way for long. Balloons were sent aloft to measure CFC concentrations, and satellites began gathering data on ozone concentrations, but nobody noticed any changes immediately.

Late in 1984, scientists working for the British Antarctic Expedition in Halley Bay, Antarctica, noticed that their annual measurements of atmospheric constituents had been undergoing an alarming change: there was a definite trend toward a decline in stratospheric ozone concentration, and about 40% of the ozone over Antarctica has already been

CONNECTIONS   CHAPTER 1

Figure 19.8

Changes in the levels of atmospheric ozone, as measured at Halley Bay, Antarctica, in October of each year. One Dobson unit corresponds to a layer of pure ozone 0.01 mm thick, corrected to 0 degrees Celsius and sea level atmospheric pressure.

destroyed (Figure 19.8). For the next two years, scientists debated the significance of this 'ozone hole' that appeared each year during the Antarctic spring in October. Other ground-based observations confirmed the existence and seasonal growth of this ozone hole, and satellite-based observations soon confirmed the results as well. In 1987, airborne recording devices confirmed the presence over Antarctica of chlorine monoxide (ClO), one of the free radicals that Rowland and Molina had predicted. Depletion of the ozone layer was already occurring, and human-made (anthropogenic) chlorine was the major culprit.

The annual appearance of the 'ozone hole' is no longer disputed, nor is the role of anthropogenic chlorine compounds in destroying the ozone that produces this hole. For their theoretical model of atmospheric ozone depletion, Roland and Molina were awarded a Nobel Prize in 1995.

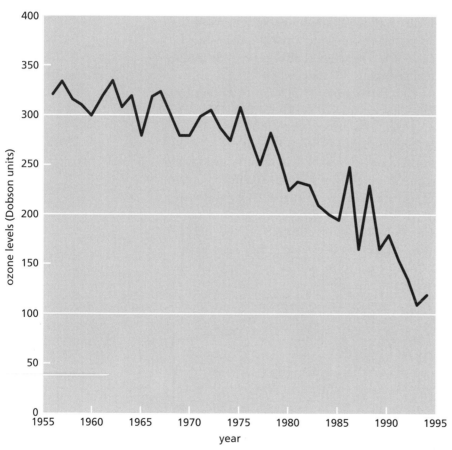

In September of 1987, representatives from 43 nations met in Montreal, Canada, and agreed to curtail most CFC use. The total elimination of all CFCs was stated as a long-range goal. Because of the Montreal Protocol, spray cans no longer use CFCs. In the United States and most of Europe, anyone who dismantles or recharges a refrigerator or air conditioner containing CFCs is required to capture the CFCs into waste containers and not release them into the atmosphere. However, older models sold to the third world are subject to no such restrictions and are probably destined some-day to release their refrigerants to the atmosphere. Ozone-friendly refrigerants and propellants have been developed as alternatives to CFCs. The computer and electronics industries, which formerly used CFCs as solvents to clean their circuits, have switched to other solvents, as has the dry-cleaning industry.

Recently, some scientists have reported that the increase in atmospheric CFCs is beginning to slow down, and the Antarctic ozone hole may have begun to decrease.

CO_2 and global warming

In Chapter 11 we saw that carbon dioxide gas (CO_2) is used in photosynthesis. The carbon is used to synthesize carbohydrates and subsequently other biological molecules. When these molecules break down during respiration, CO_2 is released back to the atmosphere, completing the carbon cycle. CO_2 is also released when any carbon-containing compound is burned or when it decays and is oxidized more slowly. Most of the carbon compounds used as fuel are fossil fuels—the dead remains of plants and other forms of life fossilized many millions of years ago.

The development of life on Earth initially caused an increase in CO_2 and certain other gases because CO_2 was a product of many biological processes. The evolution of photosynthetic organisms caused a decrease in CO_2 and an increase in oxygen. The burning of fossil fuels adds greatly to atmospheric CO_2, and global CO_2 levels are currently on the rise and have been blamed for causing a **global warming** trend. The U.S. Environmental Protection Agency estimates that about 6.3 billion metric tons of CO_2 are released into the atmosphere worldwide by human activities each year.

When sunlight hits the surface of any planet, much of the energy is absorbed and warms the planet, only to be re-radiated at a later time as heat. Without a planetary atmosphere, the heat (or infrared radiation) would simply escape into space. For a planet of a given size, composition, and distance from the sun, scientists can easily calculate or measure the amount of incident solar radiation and the amount of heat that would be given off in the form of infrared radiation; these calculations (called a heat budget) can be used to predict a planet's temperature.

The Earth's atmosphere modifies this heat budget in several ways. The most important modification is that certain atmospheric gases absorb the infrared radiation given off by the planet's surface, and much of this absorbed heat is re-radiated back down to the planet. A certain amount of heat (or infrared radiation) is therefore trapped by the atmosphere and stays with the planet instead of being lost into space. The effect is similar to that of a greenhouse, and is usually called the **greenhouse effect** (Figure 19.9). Venus, with a much denser atmosphere

Figure 19.9

How a greenhouse captures heat. Sunlight penetrates the atmosphere or the walls of a greenhouse. When the sunlight strikes opaque objects, much of the energy is converted into infrared (heat) radiation, which becomes trapped inside the greenhouse or inside the atmosphere. Not to scale.

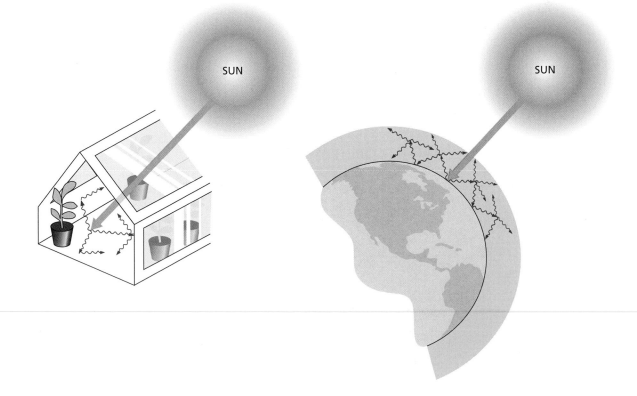

(mostly CO_2) has an even greater greenhouse effect and consequently a temperature much higher than would otherwise be predicted for a planet of its size, composition, and distance from the sun.

The existence of an atmospheric greenhouse effect was first proposed as a hypothesis by the French mathematician and physicist Jean B. Fourier, who also compared this effect to the operation of a greenhouse. In a lengthier explanation given around 1860, the Irish physicist John Tyndall explained that certain gases (among which he named CO_2, water vapor, and ozone) trap the sun's energy by absorbing the heat radiating upwards from the Earth's surface and radiating it back to Earth, thus warming the planet instead of letting the heat escape.

Increasing atmospheric concentrations of CO_2. In 1896, Swedish scientist Svante Arrhenius (who later won the Nobel prize for his work on chemical equilibria) explained that, by burning wood and coal, we were "vaporizing our coal mines into the air," thus increasing the CO_2 content of the atmosphere. The burning (rapid oxidation) of any organic (carbon-containing) material, including wood, coal, or oil, causes oxygen to combine with carbon and form atmospheric CO_2. Arrhenius also concluded that this change in atmospheric CO_2 was capable of raising the Earth's temperature. The effects of such a temperature increase were not yet apparent, but became gradually known with advances in the atmospheric sciences since the 1950s. Among other effects, global warming was predicted to cause longer and more devastating droughts in semi-arid grassland areas, an expansion of the world's deserts by an accelerated rate of 'desertification' (see Chapter 18), an increase in the number and severity of forest fires, and a partial melting of the polar ice caps. The melting of glaciers and polar ice would also cause a significant rise in sea levels, inundating many of the world's large coastal cities. Pathogens that thrive in warmer climates would become more widespread, and the epidemic diseases that they cause might become more frequent and more devastating (Chapter 17). As climates shifted, many existing species would become extinct, and biodiversity (Chapter 18) would suffer a decline of uncertain scope.

While nobody doubted that industry and home heating were releasing CO_2 into the atmosphere, some scientists were hopeful that the world's plants would be able to use up the extra CO_2 by stepping up photosynthesis. Other scientists argued that CO_2 would become trapped as insoluble carbonate salts (principally calcium carbonate, $CaCO_3$) in the shells of marine invertebrate animals and in underwater carbonate deposits. In 1957, oceanographers Roger Revelle and Hans Suess explained that the ocean's capacity to absorb additional CO_2 was limited, and that at least half of the increase in CO_2 emissions would show up as increases in atmospheric concentrations. Still, many scientists were uncertain as to how much of the global CO_2 output would be naturally recycled or absorbed and how much would accumulate and contribute to global warming. In fact, the debate continues over the role that the oceans may have in slowing down (but not stopping) the global increase of atmospheric CO_2.

From the 1950s onward, C. David Keeling of the Scripps Oceanographic Institution measured atmospheric concentrations of CO_2 at various locations, at various times of day and in various seasons. Levels rise everywhere during the night and drop during the day because of the daily cycle of photosynthetic activity in plants (see Chapter 11), but the daily afternoon lows and seasonal spring lows were close to 315 parts per million (ppm) all over the world, and the levels recorded at mid-ocean stations (such as Mauna Loa in Hawaii) closely approximate the global averages. As Keeling continued to keep CO_2 records, he noticed a continued upward movement of the seasonal cycles: the spring lows were at 315 ppm in 1957, but crept upwards to 350 ppm by 1987 and continued on an upward trend (Figure 19.10).

In Greenland, scientists drilling core samples through glacial ice meanwhile developed techniques for analyzing the composition of tiny air bubbles trapped beneath the surface, bubbles that contained minute atmospheric samples from past centuries. First in Greenland, and later in Antarctica, scientists were able to confirm that global CO_2 concentrations were on the rise. (Readings taken in remote areas, far from any industrial sources, are much more likely to reflect global averages.) Before the industrial revolution, CO_2 concentrations were only 280 ppm, meaning that human industry had caused approximately a 25% increase in CO_2 concentrations since that time. What is more, the record extended back for some 160,000 years, and the rise and fall of CO_2 concentrations matched the known increases and decreases in the Earth's temperature during that time. Although we are not sure of the causes of these earlier CO_2 fluctuations, a correlation between atmospheric CO_2 concentrations and global temperatures had finally been demonstrated: they both rise and fall together, in accordance with the hypothesis of a greenhouse effect. Subsequent observations continue to confirm such a link.

Remediation of excess CO_2 by plants. Green plants can reduce the CO_2 concentration in the atmosphere through photosynthesis, a process that consumes CO_2 and releases oxygen (Chapter 11). Animals and other heterotrophs use O_2 and release CO_2 during the respiration that produces the ATP they need (Chapter 10). Globally, photosynthesis and respiration are balanced, although the balance shifts back and forth. In higher latitudes, photosynthesis decreases in the winter; this is why Keeling's curve rises and falls annually. The largest amount of photosynthesis takes place in tropical latitudes, especially in the rainforests. About half of all photosynthesis

Figure 19.10

Annual fluctuations and persistent long-term increases in CO_2 concentrations, as measured at the Mauna Loa Observatory in Hawaii.

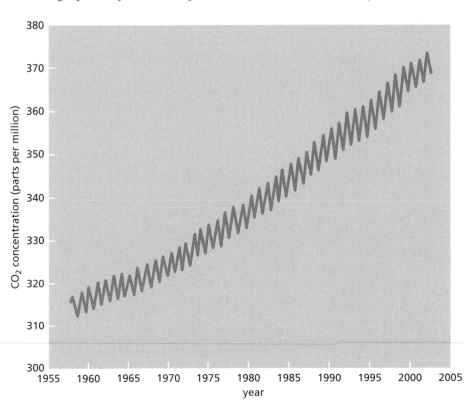

on Earth occurs in these tropical rainforests, and much of the remainder occurs in other forest regions. The gas exchange of CO_2 for O_2 during photosynthesis is comparable to the gas exchange of O_2 for CO_2 that occurs in our lungs. Thus the rainforests are sometimes refered to metaphorically as the 'lungs' of our planet.

CONNECTIONS   CHAPTERS 9, 18

Unfortunately, rainforests are being destroyed at the astounding rate of about three football fields a *second* (see Chapter 18). Much of this destruction is taking place in Brazil, a developing nation in need of agricultural land. The destruction often takes the form of burning, using fires that are so large that they can be photographed from space. Burning the rainforests not only adds further CO_2 to the atmosphere, but it also removes the very plants that could have remediated much of the increase.

Although rainforests occur in many other nations, international attention has focused on Brazil because Brazil contains, in its Amazon region, approximately half of the world's rainforests. Ecologists from all over the world (the developed nations primarily) have urged Brazil to stop destroying its rainforests. Brazil, however, is a rapidly growing third-world nation trying to modernize. Its population is growing at a rapid rate (see Chapter 9), and two of its cities (São Paolo and Rio de Janeiro) are among the world's largest. Mining companies and timber companies stand to make millions or even billions of dollars from the continued destruction of the rainforests. Landowners (including the Brazilian government) stand to make even more money from the development of agricultural land, and their impatience has often led to the burning of the forests, a process much more rapid than the harvesting of timber. Many Brazilians argue that they are only striving to achieve the material standard of living that the industrialized nations have enjoyed for decades. They point to America's feeble efforts to curb automobile pollution and urban smog as evidence of a lack of commitment to solving ecological problems that conflict with economic self-interest. Auto exhausts produce carbon monoxide gas and nitrogen oxides, all of which contribute to global warming. Nations like Brazil, India, and Indonesia have asked why they should stifle their development to forestall a problem created largely by the industrial nations. This argument raises again some general ethical issues of policy choice: how much suffering should one nation endure for the good of the planet? Why should any poor nation make sacrifices to its economic future, while the wealthy nations seem unwilling to make comparable sacrifices to their own economic well-being? In one example of a cooperative partnership, Papua New Guinea has agreed to a moratorium on cutting down its rainforests as a source of mahogany, and some of its former customers, including New Zealand, are subsidizing this effort.

Measurements continue to be made of CO_2 concentrations in the atmosphere, dissolved carbonates in the oceans, carbonate sediments, the biomass of the world's organisms, and the rates of exchange by which carbon travels among these and several other sources. About one-third of anthropogenic CO_2 (produced by the burning of fossil fuels and by tropical deforestation) is absorbed by the oceans, but atmospheric CO_2 continues to increase.

The atmosphere as part of the global ecosystem. In addition to CO_2, there are other 'greenhouse gases' that can contribute to global warming. Of the gases that occur naturally in our atmosphere, the principal contributor to this greenhouse effect is CO_2. The contributions of several other gases to the greenhouse effect are shown in Table 19.1. Although these other gases are much less abundant than CO_2, their ability to absorb infrared heat waves (and thus to contribute to the greenhouse effect) is much greater than that of CO_2, often many hundreds or thousands of times greater (see Table 19.1). Some of these gases result from the metabolic activity of living organisms, but a great amount is also anthropogenic in origin. For example, the same CFCs that destroy ozone also contribute to global warming and are thus doubly dangerous.

Human activities continue to add both CO_2 and other greenhouse gases to our atmosphere in record amounts; these other gases include methane, CFCs, and the oxides of nitrogen. Methane is added by industry and also by domestic cattle and other large animals raised for food. It is also produced in swamps, marshes, rice paddies, and other wetlands. Eastern Europe has been mentioned as a significant methane source because of natural gas leaking from its pipelines and storage tanks. CFCs are still used in some air conditioners and are released whenever older-model refrigerators or air conditioners are dismantled. Automobile exhausts still produce several kinds of nitrogen oxides. Pollution control laws in the United States, the world's largest user of automobiles, have started to address the question of automobile exhaust gases, but the laws are still very weak. For one thing, the laws pertain only to *new* cars, while older, more heavily polluting vehicles continue to be sold second-hand to many non-industrialized countries.

The existence of a greenhouse effect is no longer disputed among scientists, nor is it disputed any longer that we are compounding the effect by releasing CO_2 and other greenhouse gases into the atmosphere. Even Dixie Lee Ray, a steadfast critic of the environmentalist movement, writes as follows: "We need to remember that our Earth, together with its enveloping atmosphere, does indeed constitute a 'Greenhouse.'...The Greenhouse theory holds that an increase in the concentration of any of the greenhouse gases will lead to increased warming. No one disputes this, but the question is how much will it warm and are there naturally occurring corrective phenomena?" (Ray & Guzzo, 1993, p. 17). In other words, there is general agreement that a release of CO_2 and other greenhouse gases will intensify the greenhouse effect and will increase global temperatures. One question that continues to be disputed among scientists is how severe the increase in global warming will be and how rapidly it will occur.

Meanwhile, our production of greenhouse gases continues largely unabated, and rainforest destruction adds to the problem by robbing the Earth of its ability to cope with the CO_2 buildup, making a natural balance ever harder to achieve. Not surprisingly, Keeling's curve snakes steadily upward to new heights each year. With the increase in CO_2 and other greenhouse gases, the global warming trend continues, and no immediate end for it is in sight. As human activity changes the atmosphere, so the changed atmosphere will also affect human activity.

Table 19.1

Ability of gases to absorb infrared heat waves and add to greenhouse effect (relative to CO_2).

O_2 oxygen	Negligible
N_2 nitrogen	Negligible
Ar argon	0
CO_2 carbon dioxide	1
CH_4 methane	20
N_2O nitrous oxide	250
Chlorofluorocarbons	20,000

THOUGHT QUESTIONS

1 How can one persuade legislators in one jurisdiction to spend money on anti-pollution measures if most of the damage occurs in other jurisdictions? For example, if acid rain in Norway and Sweden originates mostly in Germany, how can the people in Norway and Sweden influence legislators to change German laws or industrial practices?

2 What would be the composition of our atmosphere if plants had not evolved?

3 How long has the Earth had a greenhouse effect? In what ways is the greenhouse effect helpful? In what ways is it harmful?

4 What would have a greater impact on atmospheric levels of CO_2 and other greenhouse gases: slowing down production of these gases, or slowing down the destruction of trees that reduce some of these gases? How would you go about measuring the rates of the relevant processes?

5 Are there any ways that human societies can meet their needs for fuel without producing CO_2? Do modern societies need to consume fossil fuels to maintain their standard of living? Are there any societies with a high standard of living but low fuel consumption?

6 If half of global photosynthesis comes from sources other than tropical rainforests, why has so much attention been focused on these tropical rainforests? Are other scientific principles involved? Are nonscientific considerations involved? What kinds of information would you need to make these comparisons?

7 When oil from the tanker *Aragon* washed ashore on beaches in Spain, nutrients were sprayed on the beaches as part of an attempt at bioremediation. The results were not as good as in the *Exxon Valdez* oil spill. What factors might explain why nutrient enrichment might work for one oil spill and not for another?

8 In what ways is wastewater treatment efficient? In what ways is it inefficient? Would it increase efficiency to separate gray water from wastewater? Could this easily be done?

Concluding Remarks

The causes of pollution are many: industrial activity, agriculture, and also our own everyday activities such as driving our cars and heating our homes. As pollutants build up, more evidence continues to accumulate that we are poisoning much of the Earth's water supply and the very air we breathe, and each increase in human populations only compounds these problems. Fortunately, many forms of ecological damage can be remediated once the damage has stopped, and biologists are devising new ways to use living organisms to help in remediation.

Ozone in the stratosphere screens out the more dangerous wavelengths of ultraviolet light. Chlorofluorocarbons cause depletion of this ozone layer and will continue to do so for many decades. Conditions over Antarctica have already resulted in the appearance of an 'ozone hole' over that continent every spring. Ozone is a natural product of atmospheric oxygen. Neither ozone nor oxygen existed in the atmosphere originally, but were produced as a result of the activities of living organisms, especially blue-green bacteria and photosynthetic plants. These same plants keep carbon dioxide levels from rising too rapidly. Human activities, however, including industrial pollution and the burning of millions of acres of rainforest, have greatly added to the carbon dioxide levels and have simultaneously reduced the availability of plant life to cope with this increase. The continued buildup of carbon dioxide and the consequent global warming is altering the entire biosphere. The trend toward global warming has closely followed changes in carbon dioxide concentrations for the past 160,000 years. This global warming trend is continuing, and no immediate end for it is in sight.

The atmosphere and the living organisms of the Earth both belong to the same vast ecosystem. Changes in either usually result in changes in the other because the interrelations are so numerous. For example, the world's forests, especially the rainforests, contain plants capable of reducing global carbon dioxide levels, given enough time. To give plants a chance to halt the process of global warming, we must give these natural ecosystems a chance to work.

Can we learn to live in ways that pollute less? The following sentiments, apparently written in 1972, are often incorrectly attributed to Seattle, chief of the Suquamish people of Puget Sound:

> The earth is our mother. What befalls the earth befalls all the sons of the earth.... Man did not weave the web of life, he is merely a strand in it. Whatever he does to the web, he does to himself.

Chapter Summary

- All living species belong to a worldwide ecosystem called the **biosphere,** which also includes the land, water, and atmosphere.

- Scientists think that Earth's atmosphere was originally hydrogen-rich (reducing). Our present oxygen-rich (oxidizing) atmosphere is in large measure the result of photosynthesis.

- **Pollution** is a matter of quantities. Ethical decisions about pollution are also decisions about quantities because neither costs nor benefits can be measured without measuring all the quantities involved. Local sources of pollution can have widespread effects on the biosphere. Sustainable practices can reduce or prevent pollution. **Sentinel species** can be used to monitor for pollution.

- Living organisms can restore polluted areas by **biodegradation** and **bioremediation**, but prevention of pollution is far less costly than restoration afterwards.

- **Ozone** in the upper atmosphere protects us from dangerous levels of ultraviolet radiation. Chlorofluorocarbons can damage this ozone layer.

- Because of the **greenhouse effect**, increases in carbon dioxide emissions will cause a **global warming** trend.

CONNECTIONS TO OTHER CHAPTERS

Chapter 1	Ozone depletion and global warming are examples of scientific theories that are developed and tested.
Chapter 1	Pollution and habitat destruction violate several ethical injunctions including the principle to do no harm to others.
Chapter 1	Public policy decisions must sometimes be made before all the risks and benefits are certain.
Chapter 3	Certain pollutants increase mutation rates.
Chapter 3	The Earth's ozone layer protects us against dangerously high levels of ultraviolet radiation.
Chapter 5	The Earth's organisms have co-evolved with its atmosphere.
Chapter 7	Ultraviolet light is more dangerous for light-skinned people.
Chapter 9	Population growth increases fuel consumption, CO_2 build-up, and global warming, and puts increased pressure on ecosystems all over the world.
Chapter 11	Plant photosynthesis is a global source of oxygen and a process that reduces atmospheric CO_2.
Chapter 11	Bacteria and other organisms contribute to the nitrogen cycle and other cycles that maintain our present atmosphere.
Chapter 12	Chemical pollutants and ultraviolet radiation can increase cancer rates. The ozone layer protects us from skin cancer.
Chapter 15	Chemical pollutants and ultraviolet radiation can suppress our immune systems.
Chapter 17	New infectious threats will emerge at a much higher rate if global climate change continues.
Chapter 18	Photosynthesis in tropical ecosystems can help reduce CO_2 levels and limit global warming to some extent, but only if these ecosystems remain healthy. The destruction of forest plants greatly accelerates global CO_2 build-up.

PRACTICE QUESTIONS

1. How many times more H^+ ions are there in acid rain at pH 5 than in water at neutral pH? How many times more at pH 5 than in ordinary water at pH 6?

2. Which is more acidic, pH 4 acid rain composed mostly of sulfuric acid, or pH 4 acid rain composed mostly of nitric acid?

3. What are the similarities between bioremediation and biodegradation? What are the differences?

4. How does an increase in ultraviolet light increase cancer rates? For which cancers? (You might wish to review material from chapters 7 and 15 in answering this question.)

5. In what ways could temperature affect the efficiency of biodegradation of an oil spill?

Glossary

Abortion: Expulsion or removal of a fetus from the womb prematurely.

Acquired characteristics: Physiological or other changes developed during the lifetime of an individual.

Actin: A contractile protein found in many eucaryotic cells, especially muscle cells, forming part of the cytoskeleton.

Action potential (spike): A large reversal of polarization in a nerve cell membrane, resulting in a nerve impulse.

Active transport: Use of energy to transport a substance, often from an area where it is in lower concentration to an area where it is in higher concentration. Active transport is performed by membrane proteins called transporters.

Acute effect: An effect that ceases soon after its cause is removed.

Adaptation: (1) Any trait that increases fitness or increases the ability of a population to persist in a particular environment. (2) A physiological change in response to a stimulus that prepares the body to better withstand or react more vigorously to similar stimuli.

Addiction: A strong psychological and physiological dependence.

Additive effect: A physiological response produced by two drugs given together that is the same as the sum of the effects of each drug given separately.

Age pyramid: A diagram that represents the age distribution of a population by a stack of rectangles, each proportional in size to the percentage of individuals in a particular age group.

Age structure: The distribution of members of a population into different age groups.

Agonist: A drug that stimulates a particular receptor or that has a stated effect.

AIDS (Acquired ImmunoDeficiency Syndrome): Impairment of most parts of the immune system resulting from infection with human immunodeficiency virus, accompanied by opportunistic infections or rare cancers and leading, in most cases, to death.

Adult stem cell: A cell in an adult organism that still retains the capacity to differentiate and form cells of several different kinds.

Algae: Photosynthetic plants not differentiated into tissues and having reproductive cells (gametes) not surrounded by protective nonreproductive cells.

Alkaptonuria: A genetic condition (inborn error of metabolism) in which urine turns dark upon exposure to air because of the body's inability to break down a compound called homogentisic acid.

Allele: One of the alternative DNA sequences of a gene.

Allele frequency: The frequency of an allele in a population, or the fraction of gametes that carry a particular allele.

Allen's rule: In any warm-blooded species, populations living in warmer climates tend to have longer and thinner protruding parts (legs, ears, tails, etc.), while the same parts tend to be shorter and thicker in colder climates.

Allergen: Anything provoking an allergic response by the immune system.

Alloparental behavior: Caring for young individuals to which one is not genetically related.

Altruism: Any act that increases another individual's fitness but lowers or endangers one's own fitness.

Alveoli: Small pouches or cavities, especially the air-filled pouches in which gas exchange occurs in the lungs.

Anaerobic: Conditions when oxygen is not present, or organisms that can live under such conditions.

Analogy: Resemblance resulting from similar evolutionary adaptation, as in wings of similar shape made of different materials.

Anaphase: The stage of cell division in which paired chromosomes or chromatids separate and begin to travel in opposite directions.

Angiogenic growth factors: Cytokines that induce the growth of blood vessels nearby.

Anisogamy: A condition in which the two types of gametes (eggs and sperm) differ in size and other characteristics.

Anorexia nervosa: A psychological eating disorder characterized by self-imposed starvation.

Antagonist: A drug that inhibits another or that inhibits a particular receptor.

Antagonistic interaction: A combined effect in which two drugs together produce less of a physiological response than either drug given separately.

Anthropogenic: Caused by human activity.

Antibodies: Proteins, secreted by lymphocytes during an immune response, that bind specifically to the type of molecule that induced their secretion, thus helping to protect the body from disease.

Anticodon: A three-nucleotide sequence in a transfer RNA molecule that pairs with a messenger RNA codon.

Antigen: Any molecule or part of a cell that is detected by the immune system.

Antigenic drift: Spontaneous single nucleotide mutations that change the shape of antigenic molecules within a species, often reducing the effectiveness of previously acquired immunity within a host.

Antigenic shift: Genetic reassortment of large sequences of nucleotides that result in major changes in the antigen shapes on the cells of the organism, making the organism 'invisible' to previously immune hosts.

Antioxidant: A substance that prevents oxidation of a molecule by an oxidizing agent.

Anus: The terminal outlet of the digestive system.

Apoptosis: A process of cellular self-destruction brought about when DNA breaks into numerous fragments.

Arteries: Blood vessels that carry blood away from the heart.

Artificial selection: Consistent differences in the contributions of different genotypes to future generations, brought about by intentional human activity.

Asexual reproduction: Reproduction (i.e., increase in the number of individuals) without the recombination of genes.

Atherosclerosis: Deposits of fat and cellular debris, which may become calcified, on the interior walls of arteries.

Autoimmune diseases: Any disease in which the immune system attacks 'self' cells not normally recognized by healthy immune systems.

Autonomic nervous system: Part of the peripheral nervous system that regulates 'involuntary' physiological processes of the body; consists of the parasympathetic and the sympathetic divisions.

Autosomal chromosomes: Chromosomes other than those involved in sex determination.

Autotroph: An organism capable of making its own energy-rich organic compounds from inorganic compounds.

Axon: An extension of a nerve cell that carries an impulse away from the nerve cell body.

B lymphocyte (B cell): A type of lymphocyte that makes antibodies.

Balanced polymorphism: A situation in which different alleles of a gene persist in a population because of the superior fitness of the heterozygous condition.

Barrier methods: Birth control methods in which a barrier is inserted across the path of sperm to prevent the sperm from reaching the egg.

Basement membrane: A membrane to which cells are attached at their base, especially in epithelial (sheetlike) tissues.

Basal metabolic rate: The rate at which the body uses energy when awake but lying completely at rest.

Benign: See *Benign tumor*.

Benign tumor: A tumor that has not broken through its basement membrane.

Bergmann's rule: In any warm-blooded species, populations living in colder climates tend to have larger body sizes, compared with smaller body sizes in warmer climates.

Beta carotene: A yellow–orange plant pigment that is a vitamin A precursor.

Bilateral symmetry: Symmetry in which the right and left halves of the body are mirror images.

Biodegradation: Breaking down a chemical by biological action.

Biodiversity: The number and variety of biological species, their alleles, and their communities.

Biofeedback: Monitoring one's own physiological activity as an means of learning how to modify this activity, e.g., to reduce stress.

Bioinformatics: The study of molecular sequences (such as those in nucleic acids) and the devising of methodologies (often using computers) to search and compare these sequences.

Biological determinism: The belief that one's physical characteristics and behavior are unalterably determined by one's genetic makeup.

Biological value: The value of a species in terms of its role in its ecosystem.

Biology: The scientific study of living systems.

Biomagnification: The increasing concentration of pollutants as one proceeds up the food energy pyramid from one trophic level to the next.

Biome: A group of similar ecosystems in various locations around the world.

Bioremediation: Human manipulation of naturally occurring biodegradation processes, as in the cleanup of environmental pollution.

Biosphere: The ecosystem that includes the whole Earth and its atmosphere.

Bioterrorism: The use of biological agents to spread harm and also fear of greater harm in human populations.

Birth control: Any measure intended to prevent unwanted births or to reduce the birth rate.

Birth rate (B): The number of births in a given time period divided by the number of individuals in a population at the beginning of that period.

Blastula: An embryonic stage consisting of a hollow ball of cells.

Bottleneck effect: A type of genetic drift that occurs when a population is temporarily small.

Branching descent: Descent in which certain species share common ancestors, creating a treelike pattern overall.

Bulimia: A psychological eating disorder characterized by an overeating binge, followed by self-induced vomiting or laxative abuse.

Cancer: A group of diseases characterized by DNA mutations in growth control genes and in which some cells divide without regard to the growth control signals of other cells.

Canopy: See *Continuous canopy*.

Capillaries: The smallest blood vessels, generally with walls only one cell thick.

Carbohydrates: Polar molecules used by organisms as energy sources and consisting of carbon, hydrogen, and oxygen, with hydrogen and oxygen atoms in a 2:1 ratio.

Carcinogen: A physical, chemical, or viral agent that induces cancer; its action is called carcinogenesis.

Cardiovascular disease: Any degenerative (age-related) disease of the heart or blood vessels, including atherosclerosis, arteriosclerosis, heart attack, and stroke.

Carrier protein(s): A membrane protein that carries a substance across a membrane by facilitated diffusion.

Carrying capacity (K): The maximum population size that can persist in a given environment.

Cartesian dualism: Descartes' theory of the mind and body as totally separate.

Case cluster: A group of patients with the same disease and seemingly something else in common that may provide clues as to the disease's causation.

Categorical imperative: Kant's ethical criterion, that an act is good or bad depending on whether you want everyone to copy it.

Cecum: A blind-ended pouch at the start of the large intestine.

Cell: The smallest unit of living systems that shows the characteristics of life; can either be free-living or part of a multicelled organism.

Cell cycle: The process by which a cell divides into two cells.

Census: Any enumeration (counting) of the members of a population.

Central nervous system: The brain and spinal cord.

Cerebellum: Part of the hindbrain that controls muscular coordination and balance.

Cerebral cortex: The outer layer of the cerebrum.

Cerebral hemispheres: The two halves of the cerebrum.

Cerebrospinal fluid: A fluid contained within the cavities of the brain and spinal cord, from which nutrients and oxygen diffuse to the neurons.

Cerebrum: The part of the forebrain controlling conscious activity and thought; it is the major part of the brain in humans.

Channels: Porelike openings in a membrane, through which ions or other particles can move by passive diffusion.

Chemical digestion: The use of enzymes and chemical reactions to break down food into molecules absorbable by cells.

Chlorofluorocarbons: Organic compounds derived from hydrocarbons by replacing some hydrogen atoms with chlorine and others with fluorine.

Chlorophyll: A green pigment molecule that traps light in the light reactions of photosynthesis.

Chloroplasts: Photosynthetic organelles containing chlorophyll.

Cholesterol: A lipid with a multiringed structure, found in the cell membranes of most animal cells.

Chromosomal aberrations: Sudden changes in chromosome sequences large enough to contain many genes. Examples include chromosomal inversions, translocations, duplications, and deletions.

Chromosomes: Elongated structures that contain DNA; in eucaryotic cells, the chromosomes are located in the nucleus and contain protein as well as DNA.

Chronic effects: Lasting or life-long effects.

Cilia: Hairlike processes on the surface of some eucaryotic cells, each typically containing two protein strands in the center surrounded by nine doublet strands.

Circadian rhythm: A biological change whose pattern repeats approximately every 24 hours.

Circulatory system: A fluid (such as blood) and a series of vessels or other means by which the fluid can transport materials around the body.

Clade: A branch of a family tree formed by a single species and all its descendants.

Cladistics: A method of classification in which taxa are made to correspond to clades.

Class: A subdivision of a phylum, typically containing several orders.

Classical conditioning: A form of learning in which one stimulus (the conditioned stimulus) that repeatedly precedes or accompanies another (the unconditioned stimulus) becomes capable of evoking the response originally elicited only by the unconditioned stimulus.

Classification: An arrangement of larger groups of species that are subdivided into smaller groups on the basis of some organizing principle or theory.

Cline: A gradual geographic variation of a trait within a species.

Clone: The genetically identical cells or organisms derived from a single cell or individual by cell division or asexual reproduction.

Cloning: Production of a new individual having the complete genome of another individual.

Codominant: Alleles capable of producing different phenotypic effects simultaneously.

Codon: A coding unit of three successive nucleotides in a messenger RNA molecule that together determine an amino acid.

Coenzyme: A nonprotein substance needed for an enzyme to function.

Colon: The large intestine, excluding the cecum.

Communicability: The probability that a disease-causing microorganism will be transferred from one individual to another, either directly or indirectly.

Community: A group of species that interact in such a way that a change in the population of one species has consequences for the other species in the community.

Complement: Blood proteins that, usually in combination with antibody, can destroy some bacteria and viruses.

Complete protein: A protein that contains all of the amino acids considered essential for human nutrition.

Composting: A managed process that uses microorganisms to help break down wastes (garbage, leaf litter, lawn clippings) into material that can be spread on fields as a fertilizer.

Concentration gradient: A situation in which the concentration of a substance is different in different locations or on opposite sides of a membrane.

Conditioned withdrawal symptom: Symptoms of drug withdrawal brought about by psychological stimuli or memories of drug-related behavior.

Condom: A latex or other barrier to the passage of sperm, worn as a covering over the penis.

Consumers: Organisms that obtain their energy from other organisms and use much of this energy in biosynthesis; heterotrophs other than decomposers.

Contact inhibition: The inability of a normal cell to divide if it is touching other cells.

Contagious: Capable of spreading from one infected individual to another, either directly or indirectly through another species.

Continuous variation: Variation in which in-between values are always possible, such as a length of 23.15 cm between the values 23.1 and 23.2.

Contraceptive: Any method that prevents conception (fertilization).

Control group: In an experiment, a group used for comparison. For example, if animals are experimentally exposed to a drug, then a control group might consist of similar animals not exposed to the drug but treated the same in every other way.

Convergence: Independent evolution of similar adaptations in unrelated lineages.

Corpus luteum: Progesterone-secreting scar tissue, formed within the ovary by a follicle after its egg has been released.

Correlation by fossils: Judging geological formations to be of the same age if they contain fossil assemblages with many of the same or similar species.

Crossing-over: The rearrangement of linked genes when homologous chromosomes break and recombine.

Cytokines: Chemicals that carry information from one cell to another but have no nutritional value or enzymatic activity of their own.

Cytoplasm: The portion of a cell outside the nucleus but within the plasma membrane.

Cytoskeleton: A protein network in the cytoplasm of eucaryotic cells that provides support for the cell and allows directed movement of chromosomes, organelles, and the cell itself.

Cytotoxic (CD8) T cells: Lymphocyte cells of the immune system that react specifically to a nonself molecule, becoming activated to kill cells bearing that molecule.

Dark reactions: Those reactions of photosynthesis that do not require light, including the reactions in which carbon dioxide is incorporated and carbohydrates are produced.

Data: Information gathered so as to permit the testing of hypotheses.

DDT: Dichloro-diphenyl-trichloroethane, a long-lasting pesticide that is not biodegradable and therefore accumulates in biological systems.

Dead-end host: A host in which a pathogen reaches the end of its life cycle and is not transmitted to subsequent hosts.

Death control: Any measure that reduces the death rate.

Death rate (*D*): The number of deaths in a given time period divided by the number of individuals in the population at the beginning of that period.

Declarative learning: Conscious remembrance of persons, places, things, and concepts, requiring the actions of the hippocampus and the temporal regions of the brain.

Decomposers: Organisms that break down organic matter into simpler molecules.

Deduction: Logically valid reasoning that guarantees a true conclusion whenever the premises are true.

Demographic momentum: A temporary population increase that can be predicted in a population that has more prereproductive members and fewer postreproductive members than a population with a stable age distribution would have.

Demographic transition: An orderly series of changes in population structure in which the death rate decreases before a similar decrease occurs in the birth rate, resulting in a population increase during the transition period.

Demography: The mathematical study of populations.

Dendrites: Nerve cell processes that receive signals and respond by conducting impulses toward the nerve cell body.

Deontological: A type of ethics in which the rightness or wrongness of an act is judged without reference to its consequences.

Deoxyribonucleic acid (DNA): A nucleic acid containing deoxyribose sugar and usually occurring as two complementary strands arranged in a double helix.

Dependence: Inability to carry out normal physiological functions without a particular drug.

Depolarization: The disappearance of a separation of unequal electrical charges.

Depression: A mental disorder characterized by low levels of serotonin and other neurotransmitters, by lack of motivation, and, in severe cases, by suicidal thoughts and actions.

Desertification: The processes whereby habitats are replaced by an advancing desert.

Determined: A state of development in which the future identity of a cell's progeny is predictable.

Determinism (genetic determinism): The belief that an individual's characteristics are wholly determined by its genes.

Deuterostome: Having an embryo in which the mouth forms at the opposite end from the entrance to the gastrula's interior cavity.

Differentiation: The process of becoming different; a restriction on the set of future possibilities for a cell's progeny.

Diffusion: A process in which molecules move randomly from an area of high concentration to an area of low concentration until they are equally distributed.

Digestive system: A system of organs responsible for breaking down food into simpler materials and absorbing those materials.

Diploid: Possessing chromosomes and genes in pairs, as in all somatic cells.

Disease: Any condition of an organism in which normal biological function is lessened or impaired.

Discontinuous variation: 'Either/or' variation in which intermediate conditions usually do not exist, as in the presence or absence of a disease.

DNA marker: Any part of DNA whose chromosomal location is known, permitting it to be used to help locate genes.

DNA probe: Any DNA fragment that has both (1) a sequence complementary to that of a gene or marker, and (2) some ability to be made visible (e.g., by radioactive labeling). Probes are used to determine the locations of the sequences to which they are complementary.

DNA polymerase: An enzyme that builds a new stand of DNA complementary to a preexisting strand used as a template.

Dominant: A trait that is expressed in the phenotype of heterozygotes; an allele that expresses its phenotype even when only one copy of the allele is present.

Dopamine: A neurotransmitter synthesized from the amino acid tyrosine and terminating in an amino group.

Dose: The amount of a drug given at one time.

Doubling time: The time required for the number of individual units in a population to double.

Down syndrome: A complex syndrome that includes varying degrees of mental retardation, plus eyes with epicanthic folds and in some cases heart malformations, arising usually from trisomy of chromosome number 21, or less often from other chromosome abnormalities.

Drug: Any chemical substance that alters the function of a living organism other than by supplying energy or needed nutrients.

Drug abuse (substance abuse): Excessive use of a drug, or use which causes harm to the individual or society.

Economic impact level (EIL): The smallest population level of a pest species that reduces crop yields by an unacceptable amount.

Ecosystem: A biological community interacting with its physical environment.

Echolocation: A navigation system used by bats and certain whales in which sounds are emitted and their echoes are used to locate food, obstacles, and other objects.

Ectoderm: The outer layer of cells in an embryo, forming such structures as the nervous system and the outer layers of the skin.

Egg: The female gamete, nonmotile and larger than the sperm because it contains more cytoplasm.

Electrical potential: A form of stored (potential) energy consisting of a separation of electrical charges.

Electrolyte: Charged mineral ions in solution.

Embryonic stem cell: An embryonic cell having the capacity to differentiate into cells of many different kinds.

Endangered species: A species threatened with extinction.

Endemic: Persistently found in a specified location, e.g., a disease that maintains a low to moderate prevalence over a long time.

Endocrine glands: Glands that secrete their products, called hormones, into the blood stream rather than into a duct.

Endocytosis: Bringing a particle into a cell by surrounding it with cell membrane.

Endoderm: The innermost layer of cells of an embryo, forming the interior lining of the gut.

Endogenous opiates: Pain-killing neurotransmitters that act on neurons and on cells of the immune system.

Endorphin: A type of endogenous opiate secreted by the pituitary gland.

Endosymbiosis: A theory that explains the origin of eucaryotic cells from large procaryotic cells that engulfed and maintained smaller procaryotic cells inside the larger cells.

Enkephalins: Small, peptide molecules secreted by the brain as neurotransmitters; a type of endogenous opiate.

Enzyme: A chemical substance (nearly always a protein) that speeds up a chemical reaction without getting used up in the reaction; a biological catalyst.

Epidemiology: The study of the frequency and patterns of disease in populations.

Epidemic: An outbreak of a disease at much greater prevalence than usual.

Epilepsy: A brain disorder characterized by uncontrollable muscle seizures.

Epiphyte: A plant that lives upon and derives support, but not nutrition, from another plant.

Erythrocytes: Red blood cells containing hemoglobin and capable of transporting oxygen.

Essential amino acids: Amino acids essential in nutrition, from which the body can make all other (nonessential) amino acids.

Estrogen: A hormone that stimulates development of female sex organs prior to reproductive age and the growth of an ovarian follicle each month during the reproductive years.

Eucaryotic: Cells with the following properties: they contain various organelles bounded by membranes, including nuclei surrounded by a nuclear envelope; their chromosomes are usually multiple and contain protein as well as nucleic acids; they have a cytoskeleton that is composed of structural and/or contractile fibers of protein.

Eugenics: An attempt to change allele frequencies through selection or changes in fitness. Raising the fitness of desired genotypes is called 'positive' eugenics; lowering the fitness of undesired genotypes is called 'negative' eugenics.

Euphenics: Measures designed to alter phenotypes (producing phenocopies) without changing genotypes.

Euphoria: A feeling of elation and well-being, especially one unrelated to the true state of affairs.

Eupsychics: Social and educational measures that accommodate people with differences.

Eusocial: A form of social organization characterized by overlapping generations (parents coexist with offspring), strictly delimited subgroups (castes), and cooperative care of eggs and young larvae.

Euthenics: Measures designed to assist people to overcome some of the consequences of their phenotypes. Wheelchairs and eyeglasses are examples.

Eutrophication: An ecological succession in which a lake becomes filled with vegetation and eventually disappears.

Evolution: The process of permanent change in living systems, especially in genes or in the phenotypes that result from them.

Excretion: The production of waste products, especially by the kidney, and their subsequent removal from the body.

Exocrine glands: Glands that secrete their products into a duct.

Exogenous: Originating outside the body.

Exon: the functional part of a messenger RNA sequence that remains after certain internal fragments (introns) are deleted.

Exoskeleton: A hard supporting skeleton on the outside the body, as in insects and other arthropods.

Experiment: An artificially contrived situation in which hypotheses are tested by comparison with some known condition called the control condition.

Experimental sciences: Sciences that rely primarily on hypothesis testing by means of experiments.

Exponential growth: A form of geometric growth without any limit, according to the equation $dN/dT = rN$.

Extinction: Termination of a lineage without any descendents.

Extracellular matrix: Material produced by cells but located outside any cell. Connective tissues have large amounts of extracellular matrix.

Fairness: The principle that all individuals in similar circumstances should receive similar treatment.

False negative: A negative test result in a sample that actually has the condition being tested for; indicates a lack of sensitivity of the test.

False positive: A positive test result in a sample that does not actually have the condition being tested for; indicates a lack of specificity of the test.

Falsifiable: Capable of being proved false by experience.

Family: A taxonomic subdivision of an order, containing one or more genera.

Fats: Lipids that are generally solid at room temperatures.

Fatty acids: Long-chain, nonpolar organic acids, released by the digestion of fats or phospholipids.

Feedback mechanism: See *Feedback system*.

Feedback system: Any process in which a later step modifies or regulates an earlier step in the process.

Female athlete triad: The combination of severe weight loss (often anorexia), causing estrogen depletion which results in cessation of menstruation and loss of bone density, occurring in female athletes who overtrain.

Females: Individuals who produce large gametes (eggs).

Fertilization: The combining of a sperm with an egg to form a zygote.

Fertilizer: Any substance artificially furnished to promote the growth of crops.

Fetal alcohol syndrome: Permanent brain damage and mental retardation, accompanied by abnormal facial features, caused by fetal exposure to alcohol while in the uterus.

Fetus: An embryo after all of its organs have formed.

Fitness: The ability of a particular individual or genotype to contribute genes to future generations, as measured by the relative number of viable offspring of that genotype in the next generation.

Flagellum (flagella): (1) On certain eucaryotic cells, a locomotor structure with two protein strands in the center surrounded by nine doublet strands and operating by means of a whiplike action. (2) On some procaryotic cells, a much simpler structure capable of rotary motion.

Forebrain: The front portion of the brain, containing the paired olfactory bulbs, olfactory lobes, cerebral hemispheres, and several unpaired portions including the hypothalamus.

Fossils: The remains or other evidence of life forms of past geological ages.

Founder effect: A type of genetic drift in which the allele frequencies of a population result from the restricted variation present among a small number of founders of that population.

Frameshift mutation: A mutation caused by the addition or deletion of one or a few base pairs in the DNA, causing many mRNA codons to be changed.

Free radicals: Very reactive chemicals containing unpaired electrons.

G6PD deficiency (favism): A deficiency of the enzyme glucose-6-phosphate dehydrogenase (G6PD), producing a blood-cell-rupturing (hemolytic) anemia when certain beans, especially fava beans, are eaten.

Gaia hypothesis: The hypothesis that life on Earth has helped to make conditions on Earth more suitable for life than it had been originally.

Gall bladder: A pouch in which bile accumulates until it is needed.

Gametes: Reproductive cells (eggs or sperm), containing one copy of each chromosome.

Gastrula: An embryonic stage in all animals except sponges, consisting of an outer layer of cells (ectoderm), an inner layer of cells (endoderm) which line a cavity open to the outside, and in many cases also a third or middle layer of cells (mesoderm) between them.

Gene: A portion of DNA that determines a single protein or polypeptide. In earlier use, a hereditary particle.

Gene expression: Transcription and translation of a gene to its protein product.

Gene family: A group of genes, similar in sequence, believed to have originated by successive mutations from multiple copies of an ancestral gene.

Gene pool: The sum total of all alleles contained in a population.

Gene therapy: Introduction of genetically engineered material into an individual for the purpose of curing a disease or a genetic defect.

General adaptation syndrome: A series of physiological reactions produced by stressors, including the three stages of alarm, resistance, and exhaustion.

Genetic drift: Changes in allele frequencies in populations of small to moderate size as the result of random processes.

Genetic engineering: Direct and purposeful alteration of a genotype.

Genetics: The study of heredity, including genes and hereditary traits.

Genome: The total genetic makeup of an individual, including its entire DNA sequence.

Genomics: The study of entire genome sequences

Genotype: The hereditary makeup of an organism as revealed by studying its offspring.

Genus (genera): Taxonomic subdivisions of a family, each containing one or more species.

Geographic isolation: Geographic separation of populations by an extrinsic barrier such as a mountain range or an uninhabitable region.

Global warming: An increase in the overall or average temperature of a planet, such as is caused by the buildup of carbon dioxide and other gases.

Gloger's rule: In any warm-blooded species, populations living in warm, moist climates tend to be darkly colored or black; populations living in warm, arid climates tend to have red, yellow, brown, or tan colors; and populations living in cold, moist climates tend to be pale or white in color.

Glycogen: A carbohydrate consisting of many glucose units linked together, used as a storage molecule in animals and certain microorganisms.

Glycolysis: The breakdown of carbohydrates to pyruvate.

Greenhouse effect: The retention of heat by the atmosphere of a planet, much of the heat being reflect by the atmosphere back to the planetary surface.

Green manure: A natural compost made from plant wastes.

Group selection: Selection that operates by differences in fitness between social groups.

Growth factors: Chemical messengers which signal cells to divide.

Growth rate (r): The population increase during a specified time interval (usually a year) divided by the population size at the beginning of that time interval.

Guard cells: Enlarged cells on the underside of many leaves, whose swelling opens the breathing pores (stomates).

Habitat: The place and environmental conditions in which an organism lives.

Habituation: A form of learning in which an organism learns not to react to a stimulus that is repeated without consequence.

Half-life: For a drug, the time that it takes for the level of the drug in the body to be reduced by half.

Haplodiploidy: A form of sex determination characteristic of the insect order Hymenoptera, in which males have one copy of each chromosome (haploidy) while females have a pair of each type of chromosome (diploidy).

Haploid: Containing only unpaired chromosomes, as in gametes or procaryotic organisms.

Hardy–Weinberg equilibrium: A genetic equilibrium formed in large, randomly mating populations in which selection, migration, and mutation do not occur or are balanced.

Hardy–Weinberg principle: In a large, random-mating population in which selection, migration, and unbalanced mutation do not occur, allele frequencies tend to remain stable from each generation to the next.

Harm reduction: An approach to drug use in society that seeks to minimize the harm done rather than to punish the drug user.

HDLs: See *High-density lipoproteins*.

Health: The ability of an organism to maintain homeostasis or to return to homeostasis after disease or injury.

Helper (CD4) T cells: Lymphocyte cells of the immune system that react specifically to a nonself molecule by secreting interleukin-2, a cytokine needed for full activity of either B cells or CD8 T cells.

Hemoglobin: The oxygen-carrying protein in red blood cells.

Herd immunity: Resistance of a population to a disease by the presence of many individuals who do not transmit it.

Hereditarian: The belief that an individual's characteristics are wholly determined by its heredity.

Heterotroph: An organism not capable of manufacturing its own energy-rich organic compounds and therefore dependent on eating other organisms or their parts to obtain those compounds.

Heterozygous: Possessing two different alleles of the same gene in a genotype.

High-density lipoproteins (HDLs): Proteins that carry lipids away from tissues via the bloodstream; often referred to as 'good cholesterol.'

Higher taxon: Any taxon at a rank higher than that of species, and thus capable of including more than one species.

High-risk behaviors: Behaviors or actions that increase the probability of undesirable outcomes, such as the transmission of a disease.

High-risk groups: A subpopulation of people who share some behavioral, geographic, nutritional, or other characteristic and who have a higher frequency of a particular disease than the general population.

Hindbrain: The rear portion of the brain, containing the cerebellum and medulla.

Hippocampus: A folded structure deep in the interior of the forebrain, important in certain types of memory.

HIV: See *Human immunodeficiency virus*.

Homeostasis: The ability of a complex system (such as a living organism) to maintain conditions within narrow limits. Also, the resulting state of dynamic equilibrium, in which changes in one direction are offset by other changes that bring the system back to its original state.

Homologous: Similar by virtue of common ancestry, as in similar body structures, shared DNA sequences, or chromosomes that carry similar sets of genes and therefore pair with one another during meiosis.

Homology: Shared similarity of structure resulting from common ancestry.

Homozygous: Possessing two like alleles of the same gene in a genotype.

Hormone: A chemical messenger, transported through the blood, that affects the activity of the cells in a target tissue.

Human immunodeficiency virus (HIV): The virus that causes AIDS by infecting and inactivating cells of the immune system that bear a molecule called CD4.

Humus: A dark-colored, nutrient-rich layer of soil in which organic matter is abundant; also called topsoil.

Hydroponics: The practice of growing plants without soil.

Hypertension: High blood pressure.

Hyphae: Tiny, threadlike filaments constituting the vegetative or feeding stage of a fungus.

Hypothalamus: A structure at the base of the forebrain that regulates body temperature and controls the release of various pituitary hormones.

Hypothesis: A suggested explanation that can be tested.

Ileum: The final third of the small intestine, in which most absorption of digestion products takes place.

Immortal: A property of transformed cells that relieves them from having a limit on the number of times they can divide.

Immunity: See *Innate immunity* and *Specific immunity*.

Immunization (vaccination): Artificial exposure to an antigen that evokes a protective immune response against a potential disease-causing antigen similar in structure to the antigen in the vaccine.

Immunodeficiency: A decreased activity of some part of the immune system as the result of genetic, infectious, or environmental factors.

Immunosuppression: Decreasing the strength of future immune functions in any manner that is not antigen-specific.

Implantation: The attachment of an early embryo to the wall of the uterus, where it later forms a placenta.

Incidence: The number or frequency of new cases of a disease or other condition. Compare *Prevalence*.

Inclusive fitness: The total fitness of all individuals sharing one's genotype, including fractional amounts of the fitness of individuals sharing fractions of one's genotype.

Independent assortment, law of: Genes carried on different chromosomes segregate independently of one another; the separation of alleles for one trait has no influence on the separation of alleles for traits carried on other chromosomes. Also called 'Mendel's second law.'

Induction: Reasoning from specific instances to general principles, which can sometimes be unreliable, as in 'these five animals have hearts, so all animals must have hearts.'

Industrial melanism: The evolution of protective dark coloration in soot-polluted habitats.

Infectious: Capable of being transmitted from one individual to others in the same population

Infanticide: The killing of an infant shortly after birth.

Infection: The colonization and growth of a pathogen within a host species.

Inflammation: A physiological response to cellular injury that includes capillary dilation, redness, heat, and immunological activity that stimulates healing and repair.

Informed consent: A voluntary agreement to submit to certain risks by a person who knows and understands those risks.

Innate: Inborn; present from birth.

Innate immune system: See *Innate immunity*.

Innate immunity: Host defenses that exist prior to that individual's exposure to an antigen and are not antigen-specific.

Instinct: Complex behavior that is innate and need not be learned.

Instrumental value: The value that something has only as a means to attain something else of value.

Insulin: A hormone, produced by the pancreas, important in the metabolism and cellular uptake of carbohydrates.

Integrated pest management (IPM): An approach to the management of pest populations that emphasizes biological controls and frequent monitoring of pest populations.

Interbreeding: The mating of unrelated individuals or the exchange of genetic information between populations.

Interferon: A cytokine secreted by lymphocytes that prevents viral replication.

Interphase: The long interval between one mitosis and the next, during which cell metabolism is active but chromosomes are not visible.

Intrauterine device (IUD): Anything inserted into the uterus to prevent implantation or pregnancy.

Intrinsic value: The value that something has on its own, other than as a means to something else.

Intron: A portion of a messenger RNA sequence that is deleted before translation to a protein.

Invertebrate: Any animal not possessing a backbone.

***In vitro* fertilization:** Fertilization that takes place outside the body in laboratory glassware.

Isogamy: A condition in which gametes are all similar in size.

IUD: See *Intrauterine device*.

Karyotype: The chromosomal makeup of an individual.

Kilocalorie (kcal): The amount of energy required to raise the temperature of 1000 grams of water through 1 degree Celsius; equal to 1000 calories.

Kin selection: Selection that favors characteristics that decrease individual fitness but that are nevertheless favored because they increase inclusive fitness.

Kingdoms: The largest taxonomic groups, such as the animal or plant kingdoms, each containing many phyla.

Klinefelter syndrome: A condition arising from the chromosomal arrangement XXY, resulting in a sterile male, often thin, with underdeveloped genitalia and with varying degrees of mental retardation and breast development.

Koch's postulates: A set of test results that must be obtained to demonstrate that a particular species of microorganism is the cause of a particular infectious disease.

Krebs cycle: A series of biochemical reactions that break apart pyruvate and use the chemical bond energy to make some ATP and NADH from ADP and NAD.

K-selection: Natural selection that characterizes populations living at or near the carrying capacity (K) of their environments by favoring adaptations for parental care and efficient exploitation of resources.

Kwashiorkor: A type of protein deficiency in which total calorie consumption is adequate but protein intake is not.

Lactose intolerance: Inability to digest the sugar lactose.

Lateral line system: A sense organ in most fishes, consisting of a series of canals and nerve cells sensitive to the movements of water caused by other fish or by obstacles.

Laterite: A nutrient-poor soil type, usually red, found in many tropical areas with high rainfall. The low nutrient content is usually attributed to the constant leaching of minerals by rainwater.

Law of unintended consequences: A principle of human ecology, that any desired change results in many other changes, some of which may be undesirable; also expressed as "you can't change just one thing."

Learning: The modification of behavior or of memory on the basis of experience.

Left atrium: A heart chamber that receives oxygen-rich blood from the lungs.

Left ventricle: A heart chamber that pumps oxygen-rich blood to the body's organs.

Life expectancy: The average duration of life in a population.

Light reactions: Those reactions of photosynthesis that require light, especially those involved in the splitting of water molecules and the release of oxygen.

Limiting amino acid: An amino acid present in small amounts that, when used up, prevents the further synthesis of proteins requiring that amino acid.

Limiting nutrient: Any nutrient whose amounts constrain the growth of an organism or population; supplying greater amounts of this nutrient therefore allows a population of organisms to increase or grow more vigorously.

Lineage: A succession of species in an ancestor-to-descendent sequence.

Linkage: An exception to the law of independent assortment in which genes carried on the same pair of chromosomes tend to assort together, with the parental combinations of genes predominating.

Lipids: Nonpolar molecules formed primarily of carbon and hydrogen, occurring in cell membranes and also used as energy sources.

Logistic growth: Growth that begins exponentially but then levels off to a stable population size (K), according to the equation $dN/dT = rN (K–N)/K$.

Low-density lipoproteins (LDLs): Proteins that carry lipids to tissues via the bloodstream; often referred to as 'bad cholesterol.'

Lymph: A fluid containing white blood cells, but no red blood cells.

Lymphatic circulation: An open circulatory system in vertebrate animals that gathers intracellular fluid and returns it to the blood along with cells of the immune system.

Lymphocytes: White blood cells that have specific receptors for antigens and are therefore capable of forming an antigen-specific immune response.

Macronutrients: Carbohydrates, lipids, and proteins, collectively.

Macrophage: A large, amoebalike wandering cell that surrounds and destroys cellular debris and bacteria.

Malaria: A parasitic infection, transmitted by mosquitoes, in which a protozoan of the genus *Plasmodium* infects blood cells.

Males: Individuals who produce small gametes (sperm).

Malignant: A tumor that has grown through the extracellular matrix.

Malnutrition: Poor or inadequate nutrition.

Marasmus: A type of malnutrition in which calorie intake and protein intake are both inadequate.

Margin of safety: The ratio between the toxic dose of a drug and its effective dose.

Mast cells: Cells whose release of histamine causes inflammation.

Maternal effect gene: A gene that is transcribed in an egg prior to fertilization.

Mating system: A description of mating behavior in terms of such features as the number of mates chosen (monogamous, polygamous, promiscuous) and the seasonality and permanence of mating units.

Mean value: The arithmetical sum of many values divided by the number of values.

Mechanical digestion: Breaking food into smaller particles by physical means such as chewing and churning, exposing new surfaces for chemical digestion.

Medulla oblongata: See *Medulla*.

Medulla: The innermost part of any organ, such as the medulla oblongata, a portion of the hindbrain that controls breathing and other involuntary activities that continue even during sleep.

Medusa: A freely floating body form in certain Cnidaria in which the mouth is directed downwards; commonly called a 'jellyfish.'

Meiosis: A form of cell division in which the chromosome number is reduced from the diploid to the haploid number. Compare *Mitosis*.

Melatonin: A hormone produced by the pineal body during darkness; its changing levels of concentration entrains the body to follow circadian rhythms.

Memory: The ability to recall past learning.

Memory cells: Cells of the immune system that retain the ability to respond rapidly to an antigen that the body has encountered before.

Mesoderm: The middle layer of tissue in an embryo, giving rise to most of the muscles, skeleton, circulatory system, reproductive system, and significant portions of many other organs.

Messenger RNA (mRNA): A strand of RNA that leaves the nucleus after transcription and passes into the cytoplasm, where it functions in protein synthesis.

Metaphase: The phase of cell division in which all chromosomes line up on a disk-shaped area (the metaphase plate) before separating.

Metastasis: The ability of transformed cells to leave the original tumor, travel through the body, and adhere and form new tumors in other locations.

Microarray: Short pieces of complementary DNA attached to a solid such as a glass slide or a piece of nylon membrane, for the purpose of probing the presence or absence of particular genes in a DNA sample. Microarrays typically contain 6400 probes; a high-denisty version, called a DNA chip, can contain up to 1 million nucleotides per square centimeter.

Micronutrients: Collectively, vitamins and minerals, nutrients needed in much smaller quantities than macronutrients.

Midbrain: The middle portion of the brain, containing most of the reticular formation.

Mimicry: A situation in which one species of organisms derives benefit from its deceptive resemblance to another species.

Minerals: Inorganic (non-carbon-containing) atoms or molecules needed to regulate chemical reactions in the body.

Mitochondria: Organelles in eucaryotic cells that produce most of the energy-rich ATP that the cells use.

Mitosis: The usual form of cell division, in which the number of chromosomes does not change. Compare *Meiosis*.

Model: A mathematical, pictorial, or physical representation of how something is presumed to work.

Monoculture: Growth of only one species in a particular place, as in a field planted with a single crop.

Monogamy: A mating system in which each adult forms a mating pair with only one member of the opposite sex.

Morals: Rules governing human conduct.

Morphological (typological) race concept: A definition of each race by its physical characteristics, based on the assumption that each characteristic is unvarying and reflects an ideal type or *form* shared by all members of the group.

Morphological species concept: A now-discarded concept that defined each species according to its morphological (physical) features.

Mosaicism: The existence of cells or patches of cells that differ genetically from one another within an organism because of changes that took place during that organism's development.

Motor neuron: A neuron that conducts impulses away from the central nervous system.

Mouth: The entrance to the digestive tract or gut.

mRNA: See *Messenger RNA*.

Multiregional model: A model that views the human species as divided into various regional populations that exchange genes with one another frequently enough so that they all evolve together.

Mutagen: An agent that causes mutation in DNA.

Mutation: A heritable change in a DNA sequence or gene.

Mutual aid (mutualism): An interaction between species in which both species benefit from the interaction

Myelin sheath: A lipid-rich covering that surrounds and insulates many neurons.

Myosin: A contractile protein found principally in muscle cells.

Narcotic: Any drug capable of inducing sleep or loss of consciousness.

Natural selection: A naturally occurring process by which different genotypes consistently differ in fitness, i.e., in the number of copies of themselves that they pass to future generations.

Naturalistic sciences: Sciences in which hypotheses are tested by the observation of naturally occurring events under conditions in which nature is manipulated as little as possible.

Negative eugenics: Attempting to change the gene pool of a population by discouraging or preventing certain genotypes from surviving or reproducing.

Negative reinforcement: The removal of an unpleasant stimulus, which may result in learning whatever behavior preceded the removal.

Nerve impulse: An electrical excitation that travels along a nerve cell without decreasing in strength.

Nerve: A bundle of axons outside the central nervous system.

Neuroendocrine system: The nervous system and the endocrine system considered as an interactive whole.

Neuroglia: Cells of the nervous system other than neurons.

Neurons: Specialized cells that conduct nerve impulses along their surface.

Neurotransmitter: Any chemical that transmits a nerve impulse from one cell to another.

Niche: The way of life of a species, or its role in a community.

Nitrogen cycle: A cyclical series of chemical reactions occurring in nature in which nitrogen compounds are built up, broken down, and changed from one form into another with the help of living organisms.

Nitrogen fixation: A process in which atmospheric nitrogen is incorporated into other molecules.

Nonpolar: Having a molecular structure in which electric charges are evenly distributed (or nearly so) across chemical bonds; nonpolar substances are not stable in water because water is a polar solvent.

Normal distribution: A mathematical description of random variation about a mean value.

Normal science: Science that proceeds step-by-step within a paradigm.

Notochord: A stiff but flexible rod of connective tissue that defines the body axis in animals of the phylum Chordata.

Nuclei: Plural of *nucleus*.

Nucleic acids: DNA and RNA; compounds containing phosphate groups, five-carbon sugars, and nitrogen-containing bases.

Nucleotide: Part of a nucleic acid molecule consisting of a phosphate group linked to a five-carbon sugar and then to a nitrogen-containing base.

Nucleus (nuclei): (1) The central part of an animal cell, plant cell, or other eucaryotic cell, containing the chromosomes. (2) Also, a clump of nerve cell bodies in the central nervous system.

Nutrient cycle: Any cyclical series of chemical reactions occurring in nature in which compounds of a specified element are built up, broken down, and changed from one form into another.

Obesity: A condition in which ideal body weight is exceeded by at least 20%.

Oddity problem: A task in which the subject is expected to pick out the one item of a set that is dissimilar to the rest.

Oils: Lipids that are liquid at room temperature.

Oncogene: A mutated growth control gene that leads to the transformation of a cell, which may then lead to cancer.

Operant conditioning: A form of conditioning in which a pleasurable stimulus reinforces any behavior (including spontaneous behavior) occurring just before the stimulus, making that behavior more likely to be repeated.

Opiates: Drugs derived from the opium poppy (*Papaver somniferum*), or similar chemicals produced by the brain itself.

Opportunistic infection: Infection in a host with suppressed immunity, resulting from microorganisms that are normally present in the host's environment but that do not cause disease in a host with a normal immune system.

Order: A taxonomic subdivision of a class, containing a family or a group of related families.

Organ: A group of tissues working together structurally and functionally.

Organelles: Cellular parts consisting of or bounded by membranes.

Organizer: An embryonic tissue whose chemical secretions induce the differentiation of other cells.

Osmosis: Diffusion of water molecules across a semipermeable membrane in response to a concentration gradient of some other molecule or ion.

Osteoporosis: Weakness and brittleness of the bones due to mineral loss.

Ovary: An organ that produces eggs in females.

Over the counter (OTC): Any drug that may legally be sold without a prescription.

Ovulation: The release of an egg from the ovary.

Oxidation: Removal of electrons from an atom or a molecule.

Ozone: A form of oxygen, O_3, containing three atoms at a time; ozone is highly reactive at the Earth's surface, but is more prevalent at higher layers in the atmosphere.

Pancreas: A glandular organ that secretes digestive enzymes and also insulin.

Pandemic: A worldwide epidemic.

Paradigm: A coherent set of theories, beliefs, values, and vocabulary terms used to organize scientific research.

Paradigm shift: The replacement of one paradigm with another.

Parasite: A species that lives in or on another species (the host), to which it causes harm.

Parasympathetic nervous system: A division of the autonomic nervous system that brings about the relaxation response and secretes acetylcholine as its final neurotransmitter.

Parental investment: The energy or resources that a parent invests in the production of offspring and the raising of offspring.

Passive immunity: Antigen-specific immunity acquired by one organism and then transferred to another organism in the form of antibodies or specific immune cells.

Pathogen: A disease-causing organism.

PCR: See *Polymerase chain reaction*.

Pecking order: A linear dominance hierarchy in which one individual is dominant to all others, a second individual to all others except the first, and so on.

Pedigree: A chart showing inheritance of genetic traits within a family.

Peptides: Short strings of amino acids, shorter than proteins.

Perception: The interpretation that the brain gives to a particular stimulus.

Peripheral nerves: Nerves outside the brain and spinal cord.

Peripheral nervous system: The nervous system except for the brain and spinal cord.

Pesticide: A chemical used to kill undesired (pest) organisms.

Phagocytosis: A process in which one cell surrounds, engulfs, and kills another.

Pharmacology: The study of drugs and their effects.

Phenotype: The visible or biochemical characteristics or traits of an organism.

Pheromones: Chemical signals by which organisms communicate with other members of their species.

Phloem: A vascular plant tissue in which photosynthetic products are transported throughout the plant, generally from the leaves downward.

Phospholipids: Molecules containing long, nonpolar hydrocarbon chains attached at one end to a polar phosphate group.

Photosynthesis: A process by which plants and certain other organisms use energy captured from sunlight to build energy-rich organic compounds, especially carbohydrates.

Phylogeny: A family tree or history of a group of organisms, forming a branching pattern of descent in most cases.

Phylum (phyla): A major subdivision of a kingdom, containing a group of related classes.

Pineal body: A structure on the roof of the diencephalon of vertebrate brains that maintains circadian rhythms.

Placebo: A drug formulation lacking the active ingredient being tested.

Placebo effect: Physiological response to a placebo that does not result from the chemistry of the placebo but that often produces the response expected by the subject.

Placenta: A structure appearing in the development of most mammals, composed of tissue derived from both the embryo and the mother's uterine lining, by means of which the embryo is nourished during its development in the uterus.

Plasma: The fluid portion of blood.

Plasmid: A bacterial DNA fragment that can separate from the main chromosome and later reattach at the point of separation.

Platelets: Cellular fragments capable of releasing blood clotting factors in response to injury.

Point mutation: A mutation resulting from a change in a single nucleotide.

Polar: Having a molecular structure in which most bonds have electrons shared unevenly, producing one part of the bond that has more negative charge than another part; water is polar, therefore other polar molecules do not spontaneously separate out from water.

Polarized: Having opposite ends or surfaces differing from one another in electrical charge.

Policy decisions: Decisions that must be made in terms of human preference, especially ethical preference, for one set of consequences over another.

Pollution: Contamination of an environment by substances present in undesirable quantities or locations.

Polyandry: An uncommon mating system in which each mating unit consists of one female and many males.

Polygyny: A mating system in which each mating unit consists of one male and many females.

Polymerase: An enzyme that joins smaller units into larger molecules (polymers).

Polymerase chain reaction (PCR): An artificial replication process in which many copies are made of specific DNA regions.

Polymorphism: The persistence of several alleles in a population at levels too high to be explained by mutation alone.

Polyp: A body form in certain Cnidaria in which the mouth is directed upwards; polyps often grow attached to a surface.

Population: A group of organisms capable of interbreeding among themselves and often sharing common descent as well; a group of individuals within a species living at a particular time and place.

Population control: All measures that limit or reduce the rate of population growth.

Population ecology: The study of populations and the forces that control them.

Population genetics: The study of genes and allele frequencies in populations.

Positive checks: Involuntary measures that limit or reduce population growth, such as famine, war, and epidemic diseases.

Positive eugenics: Attempting to change the gene pool of a population by encouraging desired genotypes to reproduce in greater numbers.

Positive reinforcement: A pleasant or pleasurable stimulus that results in learning.

Postsynaptic cell: A cell that receives a signal across a synapse.

Potentiality: The range of possible futures for a cell's progeny.

Preventive checks: Voluntary measures that reduce population growth, including voluntary abstinence from sexual activity.

Presynaptic cell: A cell that transmits a signal across a synapse.

Prevalence: The number or frequency of existing cases of a disease or other condition at any particular time. Compare *Incidence.*

Primary host: the host in which a pathogen or parasite spends the majority of its life cycle, usually including the reproductive stages.

Prion: A protein capable of producing an infection. Creutzfeld–Jacob disease in humans and BSE (bovine spongiform encephalopathy or 'mad cow disease') are examples of diseases caused by prions.

Procaryotic: Cells containing no cytoskeleton and no internal membrane-bounded organelles, but having a simple nuclear region that is never surrounded by a nuclear envelope, and a single chromosome (usually circular) containing nucleic acid only and no protein.

Procedural learning: Learning how to do things, a process that does not require the hippocampus and is not necessarily conscious.

Producers: Species that use the sun's energy to produce energy-rich organic materials.

Progesterone: A hormone that maintains the uterine lining in its enlarged, blood-rich condition, ready for implantation of a zygote.

Promiscuity: A mating system in which no permanent mating units are formed and in which each adult of either sex mates with many individuals of the opposite sex.

Promoter: A DNA sequence where RNA polymerase binds and where transcription of a gene therefore begins.

Prophase: The first and longest stage of mitosis, in which chromosomes condense and then rearrange before lining up during metaphase.

Prospective experimental design, prospective study: An experimental design in which subjects are chosen beforehand and data are subsequently gathered on events as they happen.

Protease inhibitor: An antiviral drug that interferes with viral replication by inhibiting the processing of its proteins.

Proteins: Molecules built of amino acids linked together in straight chains, which then fold up on themselves to produce complex shapes, functioning most often as enzymes or as structural materials in or around cells or their membranes.

Proteomics: The study of protein sequences

Proton gradient: A form of potential (stored) energy created by a separation or unequal distribution of protons (hydrogen ions).

Proto-oncogene: A normal gene from which an oncogene is derived; it encodes a product that regulates cell division.

Protostome: A type of animal embryo in which the mouth forms from the entrance to the cavity of the embryonic gastrula.

Pseudoextinction: Extinction of a taxon by its evolution into something else, thus continuing to have descendents.

Pseudopod: A temporary extension of cytoplasm that allows a creeping form of locomotion in amoeba-like cells.

Psychoactive drug: Any chemical substance that alters consciousness, mood, or perception.

Psychoneuroimmunology: A theory that postulates that the mind and the body are a single entity interconnected through interactions of the nervous, endocrine, and immune systems.

Puberty: A series of hormonal changes and their consequences associated with the onset of sexual maturity.

Punctuated equilibrium: A theory that describes species as remaining the same over long periods of time and then changing suddenly and giving rise to new species.

Quadrumanual clambering: A form of locomotion in which the body is suspended from three or four limbs that are pulled under tension.

Race: A geographic subdivision of a species distinguished from other subdivisions by the frequencies of a number of genes; a genetically distinct group of populations possessing less genetic variability than the species as a whole. This concept is called the *population genetics race concept* and is distinguished from other, older race concepts by defining race as a characteristic that can only apply to populations, not to individuals. Important older meanings include the following. *Socially constructed race concept*: A definition of an oppressed group and the individuals in that group by their oppressors, using whatever cultural or biological distinctions the oppressors wish to use. *Morphological (typological) race concept*: A definition of each race by its physical characteristics, based on the assumption that each characteristic is unvarying and reflects an ideal type or *form* shared by all members of the group.

Racism: A belief that one race is superior to others.

Rate of concordance: In studies of twins or other matched individuals, the fraction of individuals with a certain trait whose twin (or matched individual) also has the trait.

Reception: Receipt of a stimulus by a sensory cell capable of producing a nerve impulse as a result.

Receptor: A protein or other molecule that binds with a specific drug or other chemical substance and responds to the binding by initiating some cellular activity.

Recessive: A trait that is not expressed in heterozygotes; an allele that expresses its phenotype only when no dominant allele of the same gene is present.

Recommended dietary allowance (RDA): The amount of a vitamin or mineral recommended to be consumed daily to maintain good health in healthy adult humans.

Rectum: The straight part of the large intestine following the colon.

Reducing conditions: Hydrogen-rich conditions in which no free oxygen is present.

Relaxation response: A voluntary, self-induced stimulation of the parasympathetic nervous system in which the stress response is ended, blood pressure and breathing are reduced, the threshold of excitation of nerve cells becomes higher, and digestive activity is stimulated.

Replication: A process in which DNA is used as a template to make more DNA.

Reproductive cloning: Cloning whose purpose is the production of new individuals.

Reproductive isolating mechanism: Any biological mechanism that hinders the interbreeding of populations belonging to different species.

Reproductive isolation: The existence of biological barriers to interbreeding.

Reproductive strategy: A pattern of behavior and physiology related to reproduction.

Resting potential: The difference in electric charge maintained by a nerve cell membrane in the absence of a nerve impulse.

Restriction enzymes: Enzymes that selectively split DNA only at those locations where a particular sequence of bases occurs.

Restriction fragment length polymorphisms (RFLPs): Variations in the lengths of the DNA fragments created using a particular restriction enzyme.

Reticular activating system: A system of neurons, radiating upward from the midbrain, maintaining the body alert and attentive to stimuli.

Reticular formation: A center in the midbrain and part of the hindbrain from which the reticular activating system radiates upward.

Retina: The light-sensitive membrane coating the rear of the eye.

Retrospective study: An experimental design in which data are collected about events that have already happened.

Retroviruses: RNA viruses that begin their reproduction by synthesizing DNA from their RNA.

Reuptake: Absorption of a neurotransmitter by the cell that secreted it.

Reverse transcriptase inhibitors: Anti-AIDS drugs that interfere with the viral enzyme that helps to transcribe viral RNA to DNA.

Reverse transcription: Transcription of complementary DNA from a template of RNA.

Rhythm method: A birth-control method of timed abstinence in which the couple avoids having sex during the time when ovulation is most likely.

Ribonucleic acid (RNA): A nucleic acid containing nucleotides with ribose sugar and usually existing in single-stranded form.

Ribosome: An intracellular particle containing RNA and protein, and serving as the site of protein synthesis during translation.

Right atrium: A heart chamber that receives oxygen-poor blood from the body.

Right ventricle: A heart chamber that pumps oxygen-poor blood to the lungs.

Rights: Any privilege to which individuals automatically have a just claim or to which they are entitled out of respect for their dignity and autonomy as individuals.

Risk behaviors: Behaviors classified according to the likelihood of disease transmission. See *High-risk behaviors*.

Risk: The probability of occurrence of a specified event or outcome.

RNA polymerase: An enzyme that builds a strand of messenger RNA during transcription complementary to a DNA strand by using that DNA strand as a template.

Root nodules: Enlargements on the roots of certain plants that create anaerobic conditions that attract and maintain nitrogen-fixing microorganisms.

Route of transmission: The means by which an infectious disease spreads from one host individual to another.

r-**selection:** Natural selection that characterizes populations living far below the carrying capacity of their environments and favors high rates of reproduction (high *r*) and maximum dispersal ability.

Rule utilitarianism: An ethical system that establishes rules of right and wrong behavior based on their consequences and judges individual acts only according to their conformity to these rules.

Salivary amylase: A starch-digesting enzyme contained in saliva.

Saturated fats: Lipids with no double bonds between their carbon atoms.

Schizophrenia: A disorder that results in an inability to distinguish real from imaginary situations or stimuli, and characterized by frequent auditory or other hallucinations.

Science: An endeavor in which falsifiable hypotheses are systematically tested.

Scientific method: a method of investigation in which hypotheses are subjected to testing by comparison with empirical data.

Scientific revolution: The establishment of a new scientific paradigm, including the replacement of earlier paradigms.

Second messenger: Molecules within the cytoplasm of a cell that carry information from membrane receptors to other locations in the cell.

Secondary sexual characteristics: Features characteristic of one sex but not essential in reproduction; examples include female breasts and male beards in humans, and antlers in male deer.

Segregation, law of: When a heterozygous individual produces gametes, the different alleles separate so that some gametes receive one allele and some receive the other, but no gamete receives both.

Sensation: Perception of a stimulus.

Sensitivity: The smallest amount of some substance that can be detected by a clinical or other test.

Sensitization: A form of learning in which an intense and often aversive stimulus increases subsequent responses to other stimuli.

Sentinel species: A species that can be monitored as an indicator of the health of an ecosystem.

Seroconversion: Development in a person or other host of an antibody specific for some microorganism to which they have been exposed, either through infection or vaccination.

Severe combined immune deficiency syndrome (SCIDS): A genetic condition in which the lymphocytes responsible for antigen-specific immunity fail to develop, leaving an individual extremely susceptible to being killed by infections that would be of little or no consequence in most other individuals. One form of this condition is caused by a lack of the enzyme adenosine deaminase.

Sex chromosome: One of the chromosomes that differ between the sexes, usually distinguished as X and Y.

Sex-linked: Carried on the X chromosome.

Sexual reproduction: Reproduction in which recombination of alleles occurs.

Sexual selection: A process by which different genotypes leave unequal numbers of progeny to future generations on the basis of their success in attracting a mate and in reproducing.

Sickle-cell anemia: A genetic disorder in which hemoglobin A is replaced by hemoglobin S, with resulting sickle-shaped red blood cells having a reduced oxygen-carrying capacity but also a resistance to malaria.

Side effect: A drug effect other than the one for which the drug was intended.

Slash-and-burn agriculture: A form of agriculture in which forests are first cleared by burning.

Social behavior: Any behavior that influences the behavior of other individuals of the same species.

Social organization: A set of behaviors that define a social group and the role of individuals within that group.

Social policy: A formal or informal set of rules under which people make decisions in individual cases.

Sociobiology: The biological study of social groups and social behavior and their evolution.

Sodium–potassium pump: A group of membrane proteins that can actively transport sodium ions from the inside to the outside of a cell, such as a nerve cell, while actively transporting potassium ions in the opposite direction.

Somatic cell: any cell other than an egg or sperm; a diploid body cell.

Somatic nervous system: That part of the nervous system that is under conscious control.

Somite: A block of mesoderm in a developing embryo.

Speciation: The process by which a new species comes into being, especially by a single species splitting into two new species.

Species: Reproductively isolated groups of interbreeding natural populations.

Specific (acquired) immunity: An acquired antigen-specific ability to react to a previously encountered antigen.

Specificity: The degree to which a test detects only the molecule it is meant to detect and not detect other molecules.

Sperm: The male gamete, smaller and more motile than the egg in most species.

Spores: The tiny reproductive stages (usually cold-resistant and dryness-resistant) of fungi, certain plants, and certain bacteria.

Stem cell: An undifferentiated cell that retains the ability to divide and differentiate.

Stereotyped: See *Stereotyped behavior*.

Stereotyped behavior: Behavior that is always performed the same from one occasion to the next and from one individual to the next.

Steroid hormone: Any hormone chemically related to (and synthesized from) cholesterol.

Stomates: Pores on the underside of leaves in most vascular plants, through which gases are exchanged.

Stratigraphy: The study of layered sedimentary rocks.

Stress, stress response: A physiological response or state of heightened activity brought about by the sympathetic nervous system and maintained for a longer time by the immune and endocrine systems.

Stressor: Any stimulus or condition that brings on a stress response.

Subspecies: A geographical subdivision of a species, characterized by less genetic variation within the subspecies than in the species as a whole.

Substance abuse: Use so excessive as to cause harm to the user or to others.

Succession: In ecology, a process by which one community replaces another.

Susceptibility: (1) The likelihood that a person who is exposed to a microorganism will become infected with that organism. (2) The probability that a person will get a particular disease.

Sustainable: Any practice that could continue indefinitely without depleting any material whose supply is limited.

Symbiosis: Any type of interaction between two species living together; mutualism and parasitism are two types of symbiosis.

Symbiotic: Living together as two interacting species.

Sympathetic nervous system: A division of the autonomic nervous system that brings about the fight-or-flight response and secretes epinephrine as its final neurotransmitter.

Synapse: A meeting of cells in which a nerve cell stimulates another cell by secreting a neurotransmitter; the postsynaptic cell must have a receptor to which the neurotransmitter binds.

Synergistic effect: A physiological response to two drugs given simultaneously that is greater than the sum of the effects of the same two drugs given separately.

Synergistic interaction: A combination of two causes that lead to an effect greater than the sum of the effects that would have been produced by the two causes independently; for example, a combined effect in which two drugs together produce a greater physiological response than the sum of the effects of each drug given separately.

T lymphocyte (T cell): A type of white blood cell that helps bring about an antigen-specific immune response without releasing antibodies. See also *Cytotoxic (CD8) T cells* and *Helper (CD4) T cells*.

Taxon: A species or any other collective group of organisms.

Taxonomy: The study of how taxa are recognized and how classifications are made.

Tela choroidea: A thin-walled roof of one the brain cavities, across which nutrients and wastes are exchanged between the blood and the cerebrospinal fluid.

Telomere: A structure at the end of a chromosome whose gradual loss during cell division limits the cell's capacity to go on dividing forever.

Telophase: The last phase of cell division, during which chromosomes gather at opposite ends of the cell just before cytoplasmic division (cytokinesis) begins.

Testis (testes): The male gonads, or sperm-producing organs.

Testosterone: A steroid hormone that produces male primary and secondary sex characteristics.

Thalassemia: A form of anemia, common in many Mediterranean countries, resulting from shortened forms of the beta chain in hemoglobin molecules, and protecting the bearers from malaria.

Theory: A coherent set of well-tested hypotheses that guide scientific research.

Therapeutic cloning: Cloning whose purpose is the treatment of illnesses or other medical conditions.

Threshold: (1) A minimum level of a drug below which no physiological response can be detected. (2) The minimum level of a stimulus that is capable of producing an action potential.

Thymus: A mass of lymphoid tissue where T lymphocytes develop; it develops in the throat but usually migrates to the vicinity of the heart before adulthood.

Tissue: A group of similar cells and their extracellular products that are built together (structurally integrated) and that function together (functionally integrated).

Tissue culture: A growth of cells and tissues in a laboratory, artificially maintained outside any organism.

Tolerance: (1) A condition in which a greater amount of a drug is required to produce the same physiological effect that a smaller amount produced originally. (2) In immunology, acquired unreactivity to a specific antigen after repeated contact with that antigen.

Trace minerals: Minerals needed only in minute amounts.

Tract: A bundle of axons in the central nervous system.

Transcription: A process in which DNA is used as a template to guide the synthesis of RNA.

Transformation: (1) The multistage process that a cell undergoes in changing from a normal cell to an unregulated, less-differentiated, immortal cell lacking contact inhibition and anchorage dependance. (2) In bacteria, a hereditary change caused by the incorporation of DNA fragments from outside the cell.

Transgenic: Containing genes from another species.

Translation: A process in which amino acids are assembled into a polypeptide chain (part or all of a protein) in a sequence determined by codons in a messenger RNA molecule.

Transmission: The transfer of microorganisms from one individual to another; does not imply any particular route by which the transfer may occur.

Transpiration: Evaporation of water from the leaves of plants.

Transpiration-pull theory: The theory that sap ascends trees and other tall plants by being pulled from above by the reduced fluid pressure resulting from evaporative transpiration.

Transporter proteins: Membrane proteins that use energy to transport substances across the membrane from a region of low concentration to a region of higher concentration.

Triglycerides: Lipids formed from glycerol and three fatty acid units at a time.

Trisomy: A chromosomal condition in which three copies of a particular chromosome are present.

Trophic pyramid: A diagram summarizing the amount of food energy present at the level of producer organisms, primary consumers that feed on the producers, secondary consumers that feed on the primary consumers, and so on.

True extinction: See *Extinction*.

Tumor: A solid mass of transformed cells that may also contain induced normal cells such as blood vessels.

Tumor initiator: Agents that begin the process of transformation by causing permanent changes in the DNA; mutagens and radiation are tumor initiators.

Tumor promoter: An agent that completes the process of cell transformation after the process is started by a tumor initiator; tumor promoters are not mutagenic by themselves but cause partly transformed cells to go into cell division.

Tumor suppressor genes: Genes which normally suppress the growth of tumors.

Turgor: Fluid pressure that causes swelling and stiffness in plant cells and some other types of cells.

Turner syndrome: A condition arising from a single unpaired X chromosome (XO), resulting (if untreated) in a sterile female with immature genitals, widely spaced breasts that do not develop fully, webbing of skin at the neck, and varying degrees of mental retardation.

Unsaturated fats: Lipids with one or more double bonds between their carbon atoms.

Utilitarian: A system of ethics in which the rightness or wrongness of an act is judged according to its consequences.

Vaccination: See *Immunization*.

Vascular plants: Plants containing tissues that efficiently conduct fluids from one part of the plant to another.

Vector: (1) A virus used to transfer DNA. (2) An insect or other intermediary that transmits a disease organism.

Veins: Blood vessels that carry blood toward the heart.

Ventral tegmental area (VTA): An area in the midbrain that functions as a positive reward center by generating pleasurable sensations in response to certain stimuli.

Vestigial structures: Organs reduced in size and nonfunctional, but often showing resemblance to functional organs in related species.

Villi: Fingerlike processes, such as those lining the inside of the small intestine.

Virulence: The ability of a microorganism to cause a disease.

Virulent: Capable of causing an infectious disease.

Virus: A particle of nucleic acid (RNA or DNA) enclosed in a protein coat that cannot replicate itself but can cause a cell to replicate it.

Vitamin deficiency disease: A disease caused by insufficient amounts of a vitamin, and cured in most cases by adding the vitamin to the diet.

Vitamins: Carbon-containing molecules needed in small amounts to facilitate certain chemical reactions in the body.

White blood cells (leucocytes): The several types of blood cells that perform various protective (immune) functions but do not possess hemoglobin and do not carry oxygen.

Withdrawal: Physiological changes or unpleasant symptoms associated with the cessation of drug taking.

Xylem: A vascular tissue that gives wood its strength and that conducts watery fluids with dissolved minerals from the roots through the stem to the upper parts of the plant.

Zero population growth: A condition in which a population no longer changes size because its birth rate and death rate are equal.

Zygote: The cell that results when a sperm fuses with an egg, doubling the number of chromosomes.

Credits

Chapter 1

Figure 1.1 ROSE IS ROSE © United Feature Syndicate. Reprinted with permission.

Figure 1.4 Charles Darwin from the Grace K. Babson Collection of the Works of Sir Isaac Newton on permanent deposit at the Dibner Institute and Burndy Library, Cambridge, MA; Gregor Mendel courtesy of the Mendelianum, Museum Moraviae, Brno, Czech Republic; Ernest Everett Just courtesy of the Marine Biological Laboratory, Woods Hole Oceanographic Institution Library, Woods Hole, MA; Barbara McClintock courtesy of Majorie Bhavnani; Luis and Walter Alvarez courtesy of the Lawrence Berkeley National Laboratory, Berkely, CA; Sarah B. Hrdy courtesy of Anula Jayasuriya.

Figure 1.6 Courtesy of the National Archives and Records Administration.

Figure 1.7 Norbert Schäfer/CORBIS.

Figure 1.8 Courtesy of BUAV.

Chapter 2

Chapter opener: Everard Williams Jr/Sharpe and Associates.

Figures 2.1 and 2.2 Portions modified from Postlewait, Hopson and Veres, *Biology! Bringing Science to Life* (McGraw-Hill, 1991), p. 137, Figure 8.3.

Figure 2.4 Modified from Sinnott, Dunn and Dobshansky, *Principles of Genetics*, 5th ed. (McGraw-Hill, 1958), p. 72, Figure 6.1.

Figure 2.5 Courtesy of Kim Findlay.

Figure 2.8 Images by Dr Conly L. Rieder, Division of Molecular Medicine, Wadsworth Center, Albany, New York 12201-0509.

Figure 2.11 Modified from Sinnott, Dunn and Dobshansky, *Principles of Genetics*, 5th ed. (McGraw-Hill, 1958), p. 163, Figure 13.2.

Figures 2.12, 2.15 and 2.17 (left) Photographs courtesy of Laurent J. Beauregard, Genetics Laboratory, Eastern Maine Medical Centre, Bangor, ME.

Figure 2.17 (right) Courtesy of the March of Dimes National Foundation.

Figures 2.20 and 2.21 Modified from Pelczar, Chan and Krieg, *Microbiology Concepts and Applications* (McGraw-Hill, 1993), pp. 44–45, Figures 1.22, 1.23 and 1.24.

Chapter 3

Chapter opener: Ronnie Kaufman/CORBIS.

Figure 3.1 Modified from Postlewait and Hopson, *The Nature of Life*, 3rd ed. (McGraw-Hill, 1992), p. 245, Figure 10.7B.

Figure 3.3 Modified from Postlewait and Hopson, *The Nature of Life*, 3rd ed. (McGraw-Hill, 1992), p. 246, Figure 10.8.

Figure 3.4 Modified from Postlewait and Hopson, *The Nature of Life*, 3rd ed. (McGraw-Hill, 1992), p. 249, Figure 10.11B.

Figure 3.5 E. Kiselva and D. Fawcett/Visuals Unlimited.

Chapter 4

Chapter opener: Reprinted with permission from Mark D. Adams *et al.*, 'Sequence of the Human Genome', *Science*, 291: 1304–1351 © 2001 American Association for the Advancement of Science.

Chapter 5

Chapter opener: Lester Lefkowitz/CORBIS.

Figure 5.3 Photographs courtesy of Gary Feldman.

Figure 5.4 Modified from Postlethwait and Hopson, *The Nature of Life*, 3rd ed. (McGraw-Hill, 1995), p. 860, Figure 39.11.

Figure 5.5 Photographs from the estate of E. B. Ford, courtesy of J. S. Haywood.

Figure 5.6 From Postlethwait and Hopson, *The Nature of Life*, 3rd ed. (McGraw-Hill, 1995), p. 371, Figure 15.7.

Figure 5.7 and 5.9 Modified from Moore, Lalicker and Fisher, *Invertebrate Fossils* (McGraw-Hill, 1957), pp. 336, 338, 334–345, 364, 394, Figures 9-2, 9-3, 9-4, 9-5, 9-23, 9-47.

Figure 5.10 Modified from Colbert and Morales, *Evolution of the Vertebrates*, 4th ed. (Wiley-Liss, 1991), p. 186, Figure 14-2, as redrawn from G. Heilmann, *Origins of Birds* (D. Appleton and Co., 1927).

Figure 5.12 and 5.13 Modified in part from Hopson and Wessells, *Essentials of Biology* (McGraw-Hill, 1990), pp. 736, 734, Figures 40-3, 40-1.

Figure 5.14 From Postlethwait and Hopson, *The Nature of Life*, 3rd ed. (McGraw-Hill, 1995), p. 404, Figure 16.18.

Chapter 6

Chapter opener: Courtesy of the Bruce Coleman Collection.

Figure 6.3 Courtesy of Norman Pace.

Figure 6.6 *Palmaria* courtesy of Robert Thomas; *Peridinium* courtesy of Susan Carty; *Ascophyllum* courtesy of Barry Logan; *Volvox* by Richard Gross/Biological Photography.

Figure 6.7 *Conocephalum* and *Marchantia* courtesy of Robert Thomas; *Polytrichum* by Richard Gross/Biological Photography.

Figure 6.8 *Nephrolepis* by Richard Gross/Biological Photography; *Equisetum*, *Daisy* and *Trillium* courtesy of Robert Thomas; *Pinus* courtesy of Barry Logan; *Rosa* courtesy of

Roberts Botany Slide Collection, Bowdoin College, Brunswick, ME.

Figure 6.12 *Morchella, Rhizopus* and *Amanita* courtesy of Robert Thomas.

Figure 6.13 Sponge courtesy of Frederick Atwood; *Hydra* and jellyfish courtesy of Richard Gross/Biological Photography; anemone by Erwin and Peggy Bauer/Wildstock.

Figure 6.15 Giant flatworm and rotifers courtesy of Frederick Atwood; roundworm by Richard Gross/Biological Photography.

Figure 6.16 Tree snail, shrimp and tropical earthworm courtesy of Frederick Atwood; copepod courtesy of Wilhelm Hagen; water bear courtesy of R.O. Schuster, Bohart Museum of Entomology, University of California, Davis.

Figure 6.17 Crinoid and brittle stars courtesy of Dieter Piepenburg.

Figure 6.18 Kodiak bear, tree frog and coral snake by Richard Gross/Biological Photography; angle fish courtesy of Will Ambrose; penguins courtesy of Cecilie Quillfeldt.

Figure 6.19 Ring-tailed lemur, mandrill and chimpanzee by Erwin and Peggy Bauer/Wildstock; slow loris by David Agee/Anthrophoto; squirrel monkey by Irven DeVore/Anthrophoto.

Figure 6.20 *Australopithecus* (front view) by K. Cannon-Bonventre/Anthrophoto; *Australopithecus* (side view) by Steve Ward/Anthrophoto.

Figure 6.22 *Homo erectus* by D. Cooper/Anthrophoto; *Homo sapiens* by B. Vandermeersch/Anthrophoto.

Box 6.1 Modified from *Functional Anatomy of the Vertebrates, An Evolutionary Perspective*, 3rd ed. by Liem/Bemis/Walker /Grande © 2001. Reprinted with permission of Brooks/Cole, a division of Thomson Learning: www.thomson-rights.com. Fax 800 730-2215.

Box 6.2 Portions modified from Postlethwait and Hopson, *The Nature of Life*, 3rd ed. (McGraw-Hill, 1995), pp. 71, 75, Table 3.1, Figure 3.12 and from Krieg, *Microbiology Concepts and Applications* (McGraw-Hill, 1993), p. 58, Figure 2.2.

Box 6.3 Modified from Hopson and Wessells, *Essentials of Biology* (McGraw-Hill, 1990), p. 337, Figure 19-10.

Chapter 7

Chapter opener Rob Lewine/CORBIS.

Figure 7.3 Modified from Buettner-Janusch, *Origins of Man* (John-Wiley, 1966), pp. 499, 500, 501.

Figure 7.4 Portions modified from Hopson and Wessells, *Essentials of Biology* (McGraw-Hill, 1990), p. 191, Figure 11-16.

Figure 7.5 Modified from Carola, Harley and Noback, *Human Anatomy and Physiology*, 2nd ed. (McGraw-Hill, 1992), p. 571, Figure 18.11.

Figure 7.7 Modified from Stein and Rowe, *Physical Anthropology*, 5th ed. (McGraw-Hill, 1993), p. 190, Figure 8.9.

Figure 7.9 Courtesy of Patricia Farnsworth.

Figure 7.11 Modified from Stein and Rowe, *Physical Anthropology*, 5th ed. (McGraw-Hill, 1993), pp. 126–127, Figure 6.6.

Figure 7.12 Modified from Stein and Rowe, *Physical Anthropology*, 5th ed. (McGraw-Hill, 1993), p. 177, Figure 8.2.

Figure 7.13 Courtesy of Glenbow Archives, Calgary, Canada (ND-7-714).

Chapter 8

Chapter opener: Richard Gross/Biological Photography

Figure 8.1 Modified from Postlewait, Hopson and Veres, *Biology! Bringing Science to Life* (McGraw-Hill, 1991), p. 614, Figures 1, 2 and 3.

Figures 8.2 and 8.3 Erwin and Peggy Bauer/Wildstock.

Figure 8.4 Pelicans by Richard Gross/Biological Photography; wildebeest courtesy of Frederick Atwood; minnows by John D. Cunningham/Visuals Unlimited; gannets courtesy of David Baker.

Figure 8.5 Modified from Hopson and Wessells, *Essentials of Biology* (McGraw-Hill, 1990), p. 840, Figure 45-9.

Figure 8.6 Modified from Postlewait and Hopson, *The Nature of Life*, 3rd ed. (McGraw-Hill, 1990), p. 929, Figure 42.20; photograph courtesy of Paul W. Sherman.

Figure 8.8 Ants courtesy of Frederick Atwood; honeybees by David S. Addison/Visuals Unlimited.

Figure 8.9 Canada geese by Irene Vandermolen/Visuals Unlimited; fur seals by Leonard Lee Rue III/Visuals Unlimited; baboons by James R. McCullagh/Visuals Unlimited.

Figure 8.10 Courtesy of Harlow Primate Laboratory, University of Wisconsin.

Figure 8.11 Modified from Hinde, *Biological Basis of Human Social Behaviour* (McGraw-Hill, 1974, pp. 294–295), Figures 18.1 and 18.2.

Figure 8.12 Baboons grooming by Irven DeVore/Anthrophoto; threat display by Joseph Popp/Anthrophoto; rhesus grooming by Jane Teas/Anthrophoto.

Chapter 9

Chapter opener: Lester Lefkowitz/CORBIS.

Figure 9.1 Photograph by Jeff Greenberg/Visuals Unlimited; graph modified in part from Postlewait and Hopson, *The Nature of Life*, 3rd ed. (McGraw-Hill, 1995), p. 843, Figure 38.15A.

Figure 9.7 Courtesy of David Epel.

Figure 9.8 Modified from Postlewait and Hopson, *The Nature of Life*, 3rd ed. (McGraw-Hill, 1995), p. 335, Figure 14.3A.

Figure 9.9 Modified from Postlewait and Hopson, *The Nature of Life*, 3rd ed. (McGraw-Hill, 1995), p. 338, Figures 14.5A and B.

Figure 9.10 Modified from Postlewait and Hopson, *The Nature of Life*, 3rd ed. (McGraw-Hill, 1995), p. 340, Figure 14.6.

Figure 9.11 Modified from Audeskirk and Audeskirk, *Biology: Life on Earth*, 4th ed. (Prentice Hall, 1996), p. 778, Figure 35-18B.

Figure 9.14 From J. Trussel *et al.*, 'The economic value of contraception: a comparison of 15 methods,' *American Journal of Public Health*, 85: 494–503, 1995, Figure 2.

Figure 9.15 Courtesy of Dennis Graflin.

Figure 9.16 Photograph courtesy of Frederick Atwood.

Chapter 10

Chapter opener: BSIP, Chassenet/Science Photo Library.

Figure 10.1 Portions modified from Postlewait, Hopson and Veres, *Biology! Bringing Science to Life* (McGraw-Hill, 1991), p. 37, Figure 2.20.

Figure 10.3 Portions modified from Postlewait, Hopson and Veres, *Biology! Bringing Science to Life* (McGraw-Hill, 1991), p. 38, Figure 2.21.

Figure 10.4 Portions modified from Carola, Harley and Noback, *Human Anatomy and Physiology*, 2nd ed. (McGraw-Hill, 1992), p. 58, Figure 3.2, and from Postlewait and Hopson, *The Nature of Life*, 3rd ed. (McGraw-Hill, 1995), p. 76, Figure 3.13.

Figure 10.7 From Alberts *el al.*, *Essentials of Cell Biology* (Garland Publishing, 1998), p. 169, Figure 5-27; photograph courtesy of Richard J. Feldman.

Figure 10.8 Portions modified from Vander, Sherman and Luciano, *Human Physiology*, 6th ed. (McGraw-Hill, 1994), p. 562, Figure 8.1 and Carola, Harley and Noback, *Human Anatomy and Physiology*, 2nd ed. (McGraw-Hill, 1992), p. 778, Figure 8.1.

Figure 10.9 Modified from Starr and Taggart, *Biology: The Unity and Diversity of Life*, 8th ed. (Wadsworth Publishing, 1998), p. 32, Figure 2.19.

Figure 10.11 Portions modified from Van Wynsberghe, Noback and Carola, *Human Anatomy and Physiology*, 3rd ed. (McGraw-Hill, 1995), p. 63, Figure 3.4.

Figures 10.12 and 10.13 Modified from Postlewait, Hopson and Veres, *Biology! Bringing Science to Life* (McGraw-Hill, 1991), p. 100, Figure 5.16.

Figure 10.14 Modified from Postlewait and Hopson, *The Nature of Life*, 3rd ed. (McGraw-Hill, 1995), p. 588, Figure 25.10 A.

Figure 10.15 Modified from Hopson and Wessells, *Essentials of Biology* (McGraw-Hill, 1990), p. 519, Figure 29-5.

Chapter 11

Chapter opener: © Holt Studios International.

Figure 11.2 Modified from Hopson and Wessells, *Essentials of Biology* (McGraw-Hill, 1990), p. 139, Figure 8-6.

Figure 11.3 Modified from Hopson and Wessells, *Essentials of Biology* (McGraw-Hill, 1990), p. 138, Figure 8-4.

Figure 11.4 Modified from Hopson and Wessells, *Essentials of Biology* (McGraw-Hill, 1990), p. 143, Figure 8-9.

Figure 11.5 Modified from Starr and Taggart, *Biology: The Unity and Diversity of Life*, 8th ed. (Wadsworth Publishing, 1998), p. 115.

Figure 11.6 Modified from Chrispeels and Sadava, *Plants, Genes and Agriculture* (Jones and Bartlett, 1994), p. 214, Figure 7.16.

Figure 11.7 C.P. Vance/Visuals Unlimited.

Figure 11.8 Pitcher plant by Richard Gross/Biological Photography; drawings modified from Slack, *Carnivorous Plants* (MIT Press, 1980), p. 126; sundew plant (left) by Claude Nuridsany/Science Photo Library; sundew plant (right) by Marie Perenou/Science Photo Library.

Figure 11.9 Modified from Hopson and Wessells, *Essentials of Biology* (McGraw-Hill, 1990), p. 485, Figure 27-1.

Figure 11.10 Modified from Hopson and Wessells, *Essentials of Biology* (McGraw-Hill, 1990), p. 486, Figure 27-2.

Figure 11.11 From Hopson and Wessells, *Essentials of Biology* (McGraw-Hill, 1990), p. 489, Figure 27-5.

Figure 11.12 Sensitive mimosa by Richard Gross/Biological Photography; venus fly trap courtesy of Robert Thomas.

Figure 11.13 Courtesy of David W. Schindler.

Figure 11.15 Richard Gross/Biological Photography.

Figure 11.16 Wheat field courtesy of the Nebraska Wheat Board; lettuce field courtesy of David Handley.

Figure 11.19 A, B and C courtesy of David Handley; D by Richard Gross/Biological Photography.

Figure 11.20 Data from *Agronomy Journal*, vol. 44 (1952), p. 61, with permission from the American Association of Agronomy.

Figure 11.22 Modified from Chrispeels and Sadava, *Plants, Genes and Agriculture* (Jones and Bartlett, 1994), p. 404, Figure 15.2.

Chapter 12

Figure 12.4 Photographs courtesy of Steve Martin.

Figure 12.6 Modified from Postlewait and Hopson, *The Nature of Life*, 2nd ed. (McGraw-Hill, 1992), p. 220, Figure 10.14A.

Figure 12.7 Modified from *New England Journal of Medicine* 332: 986, 1995, Figure 1.

Figure 12.8 Part A modified from Hopson and Wessells, *Essentials of Biology* (McGraw-Hill, 1990), p. 280, Figure 16-12b; part B modified from Hopson and Wessells, *Essentials of Biology* (McGraw-Hill, 1990), p. 289, Figure 17-2.

Figure 12.9 Modified from Gurdon, *Gene Expression During Cell Differentiation* (Oxford University Press, 1973).

Figure 12.11 Modified from Hopson and Wessells, *Essentials of Biology* (McGraw-Hill, 1990), p. 291, Figure 17.4; photograph courtesy of Jonathan Slack.

Figure 12.12 Courtesy of Jean-Paul Revel.

Figure 12.16 Modified from Alberts *et al.*, *Essential Cell Biology* (Garland Publishing, 1998), p. 201, Figure 6-25.

Figure 12.17 Cancer at the cellular level from Alberts *et al.*, *Molecular Biology of the Cell*, 3rd ed. (Garland Publishing, 1995), p. 1262, Figure 24-10.

Figure 12.18 Death rate data from the National Center for Health Statistics; cigarette consumption data from the Centers for Disease Control and Prevention.

Figure 12.19 Data from *Scientific American*, September 1996: 90–1.

Figure 12.20 Modified from R. Doll and A. B. Hill, *British Medical Journal*, 1:1399–1410, 1964.

Figure 12.21 Data from W. J. Blot *et al.*, *Cancer Research*, 48: 3282–3287, 1998.

Table 12.1 Modified from Hopson and Wessells, *Essentials of Biology* (McGraw-Hill, 1990), p. 292, Table 17-2.

Chapter 13

Chapter opener Howard Sochurek/CORBIS.

Figure 13.2 A and B Modified from Vander, Sherman, and Luciano, *Human Physiology*, 6th ed. (McGraw-Hill, 1994), p. 182, Figure 8-3.

Figure 13.3A Modified from Hopson and Wessells, *Essentials of Biology* (McGraw-Hill, 1990), p. 616, Figure 34-3(a).

Figure 13.5 Modified from Postlethwait and Hopson, *The Nature of Life*, 3rd ed. (McGraw-Hill, 1995), p. 699, Figure 31.5.

Figure 13.9B Modified from Campbell *et al.*, *Biology: Concepts & Connections*, 2nd ed. (Benjamin/Cummings, 1997), p. 581, Figure B.

Figure 13.12 Part A modified from Vander, Sherman, and Luciano, *Human Physiology*, 6th ed. (McGraw-Hill, 1994), p. 260, Figure 9-36; parts B and C modified from Vander *et al.*, *Human Physiology* 6th ed. (McGraw-Hill, 1994), p. 262, Figure 9-40; part D photographs courtesy of C. G. Wright, from Rowland *et al.* (eds.), *Hearing Loss* (Thieme Medical Publishing, 1997), p. 208, Figures 7.8 a and c.

Figure 13.14 Modified from Tullar, *The Human Species* (McGraw-Hill, 1977), p. 18, Figure 1-8.

Figure 13.15 Modified from Noback and Demarest, *The Human Nervous System*, 3rd ed. (McGraw-Hill, 1981), p. 6, Figure 1-6.

Figure 13.20 Part A courtesy of Adrian J. Williams, The Lane Fox Unit, St. Thomas' Hospital; part C modified from Vander, Sherman, and Luciano, *Human Physiology*, 6th ed. (McGraw-Hill, 1994), p. 372, Figure 13-6.

Chapter 14

Chapter opener: Matthias Kulka/CORBIS.

Figure 14.1 Portions modified from Vander, Sherman, and Luciano, *Human Physiology* 6th ed. (McGraw-Hill, 1994), pp. 475, 477, Figures 15-1, 15-3(B).

Figure 14.2 Top two figures modified from Hopson and Wessells, *Essentials of Biology* (McGraw-Hill, 1990), p. 599, Figure 33-5(a, b); bottom figure modified from Postlethwait and Hopson, *The Nature of Life*, 3rd ed. (McGraw-Hill, 1995), p. 666, Figure 29.8(B).

Figure 14.9 Modified from Carlson, *Patten's Foundations of Embryology*, 5th ed. (McGraw-Hill, 1988), pp. 279, 281, Figures 7-21(C), 7-22.

Table 14.3 From Segal, *Drugs and Behaviour* (Gardner Press, 1988), pp. 258–259, 369, Tables 11-2 and 15-1.

Chapter 15

Chapter opener: Cristina Pedrazzini/Science Photo Library.

Figure 15.1 Neutrophil by Dr Willis/Visuals Unlimited; macrophage courtesy of Jean-Paul Revel; lympocyte courtesy of Daniel S. Friend.

Figure 15.2 Modified from Van Wynsberghe, Noback and Carola, *Human Anatomy and Physiology*, 3rd ed. (McGraw-Hill, 1995), p. 711, Figure 22.1 A.

Figure 15.4 Portion redrawn from Van Wynsberghe, Noback and Carola, *Human Anatomy and Physiology*, 3rd ed. (McGraw-Hill, 1995), pp. 66–7, Figure 3.5.

Figure 15.7 Modified from Hopson and Wessells, *Essentials of Biology* (McGraw-Hill, 1990), p. 548, Figure 30-14.

Figure 15.8 Modified from Hopson and Wessells, *Essentials of Biology* (McGraw-Hill, 1990), p. 626, Figure 34-14 and Hole, *Human Anatomy and Physiology*, 6th ed. (WCB/McGraw-Hill, 1993), p. 404, Figure 11.39.

Figure 15.9 Modified from Carola, Harley and Noback, *Human Anatomy and Physiology*, 2nd ed. (McGraw-Hill, 1992), p. 534, Figure 17.13.

Figure 15.10 Courtesy of David and Suzanne Felten.

Chapter 16

Chapter opener: UN/DPI Photo.

Figure 16.2 Courtesy of The Names Project Foundation (photograph by Mark Theissen).

Figure 16.3 From Van Wynsberghe, Noback and Carola, *Human Anatomy and Physiology*, 3rd ed. (McGraw-Hill, 1995), p. 752, unnumbered figure.

Figure 16.4 Modified from Postlewait and Hopson, *The Nature of Life*, 3rd ed. (McGraw-Hill, 1995), p. 614, Figure 26.14.

Figure 16.5 Photographs courtesy of Cynthia Goldsmith, Erskine Palmer and Paul Feorino, Centers for Disease Control and Prevention (CDC).

Figure 16.6 Courtesy of Alyne Harrison, Erskine Palmer and Paul Feorino, Centers for Disease Control and Prevention (CDC).

Figure 16.8 Modified from Postlewait and Hopson, *The Nature of Life*, 3rd ed. (McGraw-Hill, 1995), p. 232, Box 9.1, Figure 1.

Figure 16.9 Data from the Centers for Disease Control and Prevention (CDC).

Figure 16.10 Data from Stine, *AIDS update 1999* (Prentice Hall, 1999), p. 238, Figure 8-15.

Chapter 17

Chapter opener: Cristina Pedrazzini/Science Photo Library.

Figure 17.7 Eye of Science/Science Photo Library.

Figure 17.8 G.D. Schmidt and L.S. Roberts, *Foundations of Parasitology*, 4th ed. (Times Mirror/Mosby, 1989); photograph courtesy of Dennis Feely from *Journal of Protozoology* 35: 151–158, 1988.

Chapter 18

Chapter opener: Richard Gross/Biological Photography.

Figure 18.1 Modified from Wilson, *The Diversity of Life* (The Belknap Press of Harvard University Press), p. 134.

Figure 18.2 Hal Beral/Visuals Unlimited.

Figure 18.3 Modified from Moore, Lalicker and Fisher, *Invertebrate Fossils* (McGraw-Hill, 1957), portions of Figures 3-4, 4-17, 4-29, 5-9, 6-20, 6-24, 6-36, 13-7, 13-20, 18-20, 18-29, 23-1.

Figure 18.4 *Ginkgo* and *Limulus* modified from Palmer and Fowler, *Fieldbook of Natural History*, 2nd ed. (McGraw-Hill, 1974), pp. 112, 433; *Lingula* modified from Hyman, *The Invertebrates*, vol. 5 (McGraw-Hill, 1959), p. 519, Figure 183C; *Latimeria* modified from Weichert, *Anatomy of the Chordates* (McGraw-Hill, 1970), p. 27, Figure 2.18.

Figure 18.5 Ammonoids modified from Moore, Lalicker and Fisher, *Invertebrate Fossils* (McGraw-Hill, 1957), Figures 9-40, 9-41; photograph courtesy of Don Prothero; painting by Rudolph Zallinger © the Peabody Museum of Natural History, Yale University.

Figure 18.7 Modified from Wilson, *The Diversity of Life* (The Belknap Press of Harvard University Press), p. 252.

Figure 18.9 From Hopson and Wessells, *Essentials of Biology* (McGraw-Hill, 1990), p.756, Figure 41.7.

Figure 18.10 Part A modified from Postlethwait and Hopson, *The Nature of Life*, 3rd ed. (McGraw-Hill, 1995), p. 896, Figure 41.7A; part B courtesy of Jane Mackarell; part C by Richard Thom/Visuals Unlimited.

Figure 18.11 *Guzmania*, *Spathophyllum*, *Piper* and strawberry frog courtesy of Sharon Kinsman; tarantula courtesy of Jane Mackarell; stick insect, butterfly, bronze beetle, emerald boa, white-throated toucan, and tree sloths courtesy of Frederick Atwood; jaguar by Erwin and Peggy Bauer/Wildstock.

Figure 18.12 Orchids by Max and Bea Hunn/Visuals Unlimited; lianas courtesy of Jane Mackarell.

Figure 18.15 Kjell B. Sandved/Visuals Unlimited.

Figure 18.16 Left-hand photographs by G. Prance/Visuals Unlimited; right-hand photographs courtesy of Sharon Kinsman.

Figure 18.17 Modified from Hopson and Wessells, *Essentials of Biology* (McGraw-Hill, 1990), p. 755, Figure 41-6.

Figure 18.18 Photograph by Leonard Lee Rue III/Visuals Unlimited.

Chapter 19

Chapter opener: Science Photo Library.

Figure 19.5 Courtesy of Elin Haugen.

Figure 19.6 Trees killed by acid rain by Richard Gross/Biological Photography; pH-affected trout courtesy of David Schindler from Schindler *et al.*, *Science*, 228: 1395–1401 © 1985 American Association for the Advancement of Science; stone sculpture by John D. Cuningham/Visuals Unlimited.

Index